B.B.A. LIBRARY – VOLUME 3

FORM AND FUNCTION OF PHOSPHOLIPIDS

B.B.A. LIBRARY

Volume 1
BIOCHEMICAL PROBLEMS OF LIPIDS
edited by A.C. FRAZER

Volume 2
PORPHYRINS AND METALLOPORPHYRINS
Their General, Physical and Coordination Chemistry, and Laboratory Methods
by J.E. FALK

Volume 4
NEW PERSPECTIVES IN BIOLOGY
edited by M. SELA

Volume 5
GLYCOPROTEINS
Their Composition, Structure and Function
edited by A. GOTTSCHALK

Volume 6
MUCOPOLYSACCHARIDES
Chemical Structure, Distribution, and Isolation
by J.S. BRIMACOMBE and J.M. WEBBER

Volume 7
REGULATION OF METABOLIC PROCESSES IN MITOCHONDRIA
edited by J.M. TAGER *et al.*

Volume 8
FLAVINS AND FLAVOPROTEINS
edited by E.C. SLATER

Volume 9
MOLECULAR BIOLOGY OF MUSCULAR CONTRACTION
edited by S. EBASHI *et al.*

Volume 10
REGULATION OF NUCLEIC ACID AND PROTEIN BIOSYNTHESIS
edited by V.V. KONINGSBERGER and L. BOSCH

Volume 11
REGULATORY FUNCTIONS OF BIOLOGICAL MEMBRANES
edited by J. JÄRNEFELT

B.B.A. LIBRARY

VOLUME 3

Second, completely revised and enlarged, edition

FORM AND FUNCTION OF PHOSPHOLIPIDS

edited by

G.B. ANSELL

Department of Pharmacology (Preclinical), The Medical School, University of Birmingham (Great Britain)

J.N. HAWTHORNE

Department of Biochemistry, The Medical School, University of Nottingham (Great Britain)

and

R.M.C. DAWSON

Department of Biochemistry, Agricultural Research Council Institute of Animal Physiology, Babraham, Cambridge (Great Britain)

ELSEVIER SCIENTIFIC PUBLISHING COMPANY

AMSTERDAM – LONDON – NEW YORK

1973

ELSEVIER SCIENTIFIC PUBLISHING COMPANY
335 JAN VAN GALENSTRAAT
P.O. BOX 1270, AMSTERDAM, THE NETHERLANDS

AMERICAN ELSEVIER PUBLISHING COMPANY, INC.
52 VANDERBILT AVENUE
NEW YORK, NEW YORK 10017

LIBRARY OF CONGRESS CARD NUMBER: 72-87950

ISBN 0-444-41055-4

WITH 69 ILLUSTRATIONS AND 58 TABLES

PRINTED IN THE NETHERLANDS

To the memory of Georg Hübscher (1926–1970), a scientist and leader of distinction and a fondly remembered colleague

CONTRIBUTORS TO THIS VOLUME

R.T. AMBRON Department of Biochemistry, Temple University School of Medicine, Philadelphia, Pa.; present address: Department of Neurobiology and Behavior, Public Health Research Institute, New York, N.Y. 10016 (U.S.A.)

G.B. ANSELL Department of Pharmacology (Preclinical), The Medical School, University of Birmingham, Birmingham B15, 2TJ (Great Britain)

D. CHAPMAN Department of Chemistry, Sheffield University, Sheffield (Great Britain)

R. COLEMAN Department of Biochemistry, University of Birmingham, Birmingham B15, 2TT (Great Britain)

R.M.C. DAWSON Department of Biochemistry, Agricultural Research Council Institute of Animal Physiology, Babraham, Cambridge (Great Britain)

J.B. FINEAN Department of Biochemistry, University of Birmingham, Birmingham B15, 2TT (Great Britain)

T. GAILLARD Agricultural Research Council Food Research Institute, Colney Lane, Norwich, NOR 70F (Great Britain)

J.N. HAWTHORNE Department of Chemistry, University of Calgary, Alberta (Canada); present address: Department of Biochemistry, The Medical School, University of Nottingham, Nottingham NG7 2RD (Great Britain)

W.C. MCMURRAY Department of Biochemistry, Health Sciences Centre, University of Western Ontario, London 72, Ontario (Canada)

D. PAPAHADJOPOULOS Department of Experimental Pathology, Roswell Park Memorial Institute, Buffalo, N.Y. (U.S.A.)

R.A. PIERINGER Department of Biochemistry, Temple University School of Medicine, Philadelphia, Pa. (U.S.A.)

H. ROSENBERG Department of Biochemistry, John Curtin School of Medical Research, Institute of Advanced Studies, Australian National University, Canberra (Australia)

SHEILA SPANNER Department of Pharmacology (Preclinical), The Medical School, University of Birmingham, Birmingham B15, 2TJ (Great Britain)

K.P. STRICKLAND Department of Biochemistry, Health Sciences Centre, University of Western Ontario, London 72, Ontario (Canada)

G.A. THOMPSON JR. Department of Botany, University of Texas, Austin, Tex. 78712 (U.S.A.)

D.A. WHITE Department of Biochemistry, The Medical School, University of Nottingham, Nottingham NG7, 2RD (Great Britain)

PREFACE

When *Phospholipids: Chemistry, Metabolism and Function* was published in 1964 the authors commented that there had been no single volume devoted to the subject since the admirable monograph by Witcoff in 1951. Since 1964 the number of papers published on phospholipids has been enormous and yet no single work has attempted to summarise and assess the progress in the field. This we have now attempted by adopting a multi-author format and asking acknowledged experts in particular aspects of phospholipid research to contribute. The original authors have also persuaded Rex Dawson to make an editorial trio.

Since the original edition of the book, whole new subjects have arisen which barely existed in 1964. The phosphonolipids, for example, are now very obviously more than biochemical curiosities, while research in the plant kingdom is revealing many new types of phospholipid to the diligent investigator. New techniques have played a great part in recent developments. Thin-layer chromatography had barely come of age in 1964 and separations of molecular species were still unknown, while the last decade has seen the development of more sophisticated physical methods which have contributed to the tremendous interest in phospholipids contained in biological membranes. Studies of the orientation of lipids and proteins in membranes are likely to provide one of the most exciting developments in biochemistry during the next decade.

The Editors remember the days when one called a preparation 'lecithin' and hoped for the best. Now 1-palmitoyl-2-oleoyl-*sn*-glycero-3-phosphorylcholine can be obtained from a friendly neighbourhood supplier. In short, a lot of the donkey-work has gone out of the subject and progress will undoubtedly be faster in the future.

The Editors offer this volume in the hope that it will be a help to the established investigator and encourage others to enter the field. Although multi-author volumes are notorious for uneven coverage, a special effort has been made to eliminate it in this book and particular attention has been paid to cross-referencing. Special thanks are due to Roger Whiting and John Bleasdale (University of Nottingham) for producing the index. The final responsibility for clarity and comprehensiveness rests, of course, not with the authors but with the editors. The trio regret any discords, defects and omissions and will be glad to have them pointed out.

G.B. ANSELL (Birmingham)
R.M.C. DAWSON (Cambridge)
J.N. HAWTHORNE (Nottingham)

ABBREVIATIONS

For convenience the following abbreviations are used in this book.

CL	Diphosphatidylglycerol
DPI	Phosphatidylinositol monophosphate; diphosphoinositide
EP	Ethanolamine plasmalogen
HDL	High density lipoprotein
LCAT	Lecithin:cholesterol acyl transferase
LDL	Low density lipoprotein
PA	Phosphatidic acid
PC	Phosphatidylcholine (lecithin)
Lyso PC	Lysophosphatidylcholine (lysolecithin)
PE	Phosphatidylethanolamine
Lyso PE	Lysophosphatidylethanolamine
PG	Phosphatidylglycerol
PI	Phosphatidylinositol
PS	Phosphatidylserine
TPI	Phosphatidylinositol diphosphate; triphosphoinositide
TSH	Thyrotrophin
SM	Sphingomyelin
VHDL	Very high density lipoprotein
VLDL	Very low density lipoprotein

For the full chemical names of the phospholipids the reader is referred to Chapter 2.

CONTENTS

List of Contributors to this Volume vi
Preface vii
Abbreviations viii
Contents ix

Chapter 1. Historical Introduction –
by G.B. ANSELL 1

References 7

Chapter 2. The Chemistry of Phospholipids –
by K.P. STRICKLAND 9

1. Classification 10
2. Nomenclature 12
3. Structures, isolation and chemical syntheses of natural phospholipids 17
A. Phosphoglycerides, 17 – (1) Phosphatidic acid and its analogues, 17 – (2) Cytidylic phosphoglyceride, 19 – (3) Choline phosphoglycerides, 19 – (4) Ethanolamine phosphoglycerides, 21 – (5,6) *N*-methyl- and *N,N*-dimethylethanolamine phosphoglycerides, 24 – (7) *N*-acyl ethanolamine phosphoglyceride, 24 – (8) Serine phosphoglycerides, 24 – (9) *N*-2-(hydroxyethyl) alanine phosphoglyceride, 25 – (10) Glycerol phosphoglycerides, 25 – (11) Glycerophosphate phosphoglycerides, 26 – (12) Phosphatidylglycerol phosphoglyceride, 27 – (13) Mono- and diacylglycerol phosphoglyceride, 27 – (14) Glucosaminylglycerol phosphoglyceride, 28 – (15) *O*-Amino acid esters of glycerol phosphoglycerides, 28 – (16) Inositol phosphoglyceride, 29 – (17,18) Inositol mono- and diphosphate phosphoglycerides, 30 – (19) Monomannosyl-hexamannosyl inositol phosphoglycerides, 31 – (20) Glucose phosphoglyceride, 31 – (21) *O*-diglucosylglycerol phosphoglyceride, 32 – B. Phosphoglycolipids, 32 – C. Phosphodiol lipids, 32 – (1) Acyldihydroxyacetone phosphate, 32 – (2) Alkyldihydroxyacetone phosphate, 32 – D. Phosphosphingolipids, 32 – (1) Sphingomyelin, 34 – (2) Ceramide phosphorylethanolamine, 35 – (3,4) Ceramide phosphorylglycerol and phosphorylglycerophosphate, 35 – (5) Ceramide phosphorylinositol-containing lipids, 36 –
4. Effects of hydrolytic and other reagents 37
Alkaline alcoholysis, 37 – Acid hydrolysis, 37 – Oxidation reactions, 37 – Lithium aluminium hydride reduction, 38 – Hydriodic acid, 38 – Acetolysis, 38 –

References 39

Chapter 3. Separation and Analysis of Phospholipids –
by S. SPANNER 43

1. Extraction of phospholipids and removal of non-lipid contaminants 43
Extraction, 43 – Removal of non-lipid contaminants, 44 –
2. Purification of the phospholipid fraction 45
3. Fractionation and analysis of intact phospholipids 46
Aluminium oxide, 46 – Silicic acid, 46 – DEAE cellulose, 52 – Hydroxylapatite, 54 – Extraction and separation of the higher phosphoinositides, 54 –

4. Analysis of intact lipid fractions 56
Fatty acids and aldehydes, 56 – Molecular species, 58 – Bases, 59 –
5. Analysis of phospholipids by selective hydrolysis 61
References 62

Chapter 4. Phospholipid Metabolism in Animal Tissues –
by G.A. THOMPSON JR. 67

1. Introduction 67
2. Biosynthesis 68
Phosphatidic acid, 68 – Phosphatidylcholine, 69 – Phosphatidylethanolamine, 71 – Phosphoinositides, 72 – Phosphatidylserine, 73 – Plasmalogens and glycerol ether phospholipids, 73 – Phosphatidylglycerol, 75 – Diphosphatidylglycerol, 76 – Metabolic interconversions of phospholipids, 77 – Sphingomyelin, 78 – Summary and conclusions, 81 –
3. Metabolic exchange of phospholipids and their components 81
Introduction, 81 – Turnover of portions of phospholipid molecules, 82 – Exchange of entire molecules, 84 – Rates of phospholipid turnover *in vivo*, 84 – Summary and conclusions, 85 –
4. Degradation 86
Introduction, 86 – Hydrolysis of fatty acid esters, 86 – Scission of ether bonds, 89 – Cleavage of the phosphate-glycerol linkages, 91 – Phosphatidate phosphohydrolase, 91 – Cleavage of the phosphate–base linkages, 92 – Dephosphorylation of phosphoinositides, 92 – Degradation of sphingosine bases, 93 – Summary and conclusions, 93 –
References 94

Chapter 5. Specificity of Enzymes Involved in the Metabolism of Phospholipids –
by R.M.C. DAWSON 97

1. Two types of specificity: chemical and physical 97
2. Physical factors affecting enzymatic reactions involving phospholipids 98
Physical form of lipid participating in reactions, 98 – Ratio between substrate mass and area of interface available for enzyme attack, 99 – Influence of molecular interfacial packing on the rate of enzymatic reactions, 101 – Lipid solvent addition, 101 – The effect of the electrostatic field around the phospholipid–water interface, 102 –
3. Measurement of specificity 103
Recorded specificities of enzymes forming and breaking down phospholipids, 105 –
References 114

Chapter 6. Physical Chemistry of Phospholipids –
by D. CHAPMAN 117

1. Solid state behaviour 117
2. Thermotropic mesomorphism 120
Lecithins and phosphatidylethanolamines, 120 –
3. Phospholipid–water systems and lyotropic mesomorphism 123
The structure of the lyotropic phases, 124 – (1) Nerve lipid extracts, 129 – (2) Egg-yolk phosphatidylethanolamine, 130 – (3) Sphingomyelin, 130 – (4) Brain lipid extract, 130 – (5) Mitochondrial lipid extract, 131 – Interactions with metal ions, 132 – Monolayer studies, 134 – Phospholipid–cholesterol interactions, 136 – 'Solutions' of phospholipids in water, 139 – Non-aqueous solvent systems, 140 –
References 141

Chapter 7. Phosphorlipids as Model Membranes: Monolayers, Bilayers and Vesicles – by D. PAPAHADJOPOULOS 143

1. Monomolecular films (methodology and intrinsic properties) 144
2. Bilayer films (methodology and intrinsic properties) 148
3. Phospholipid vesicles (liposomes) 150
4. Permeability properties and interactions with ions 155
5. Ionophores and excitability 157
6. Influence of acyl chain length, unsaturation and cholesterol on permeability 159
7. Interactions of phospholipid membranes with proteins 161
8. Interactions with pharmacologically active compounds 164
9. Concluding remarks 165

References 165

Chapter 8. Phospholipids in Biological Membranes and the Study of Phospholipid–Protein Interactions – by J.B. FINEAN 171

1. General characteristics of membrane lipid 171
2. Membrane extraction and solubilisation 174
3. Studies of intact membrane systems 176
   X-ray diffraction, 177 – Spectroscopic and optical studies, 181 – Electron microscopy, 186 –
4. Lipid-dependent enzyme activities 188
5. Soluble lipoproteins 190
6. Lipid–protein interactions *'in vivo'* 191
   Re-aggregation of dispersed membrane components, 194 –
7. Conclusions 195

References 198

Chapter 9. Phospholipids in Subcellular Organelles and Membranes – by W.C. McMURRAY 205

1. Introduction 205
2. Phospholipid composition of membranes 206
   Mitochondria, 209 – Endoplasmic reticulum, 214 – Golgi apparatus, 216 – Plasma membranes, 216 – Nuclear membranes, 219 – General comments, 222 –
3. Functions of membrane phospholipids 223
   Mitochondrial membranes, 223 – Microsomal and other membranes, 229 –
4. Intracellular synthesis and turnover 231
5. Exchange of phospholipids between membranes 236
6. Membrane biogenesis 240

References 246

Chapter 10. Phospholipid Metabolism in Photosynthetic Plants – by T. GALLIARD 253

1. Composition of plant phospholipids 253
   Fatty acid distribution in plant phospholipids, 260 –
2. Some practical considerations in biochemical studies with plant lipids 267
   Extraction of lipids, 267 – Preparation of cell-free extracts, 267 –
3. Biosynthesis of phospholipids

Acyl-CoA:*sn*-glycero-3-phosphoric acid *O*-acyltransferase, 268 – Diglyceride phosphokinase, 268 – CTP:phosphatidic acid cytidyltransferase, 269 – Phosphatidylglycerol and diphosphatidylglycerol, 269 – Phosphatidylinositol, 270 – Phosphatidylserine, 270 – Phosphatidylcholine and phosphatidylethanolamine, 270 – ATP:choline phosphotransferase, 271 – CTP:cholinephosphate cytidyltransferase, 271 – CDP-choline:1,2-diacyl-*sn*-glycerol cholinephosphotransferase, 271 – Interconversions of nitrogenous phospholipids, 271 – Transphosphatidylation reactions, 272 – Fatty acyl transfer reactions, 273 –

4. Phospholipid catabolism 274
Enzymes attacking the polar group of phospholipids, 274 – Enzymes attacking fatty acyl ester bonds of phospholipids, 276 –
5. Subcellular localization of phospholipid metabolism 278
6. Function of phospholipids 280
Photosynthesis, 280 – Effect of temperature on membrane structure and function, 281 – Role of phospholipids in fatty acid metabolism, 282 – Role of phospholipids in other cellular processes, 283 –
7. Some physiological aspects of lipid metabolism 283
8. Relevance of phospholipid metabolism in edible plants 284
References 284

Chapter 11. Phospholipids in Micro-organisms –
by R.T. AMBRON and R.A. PIERINGER 289

1. Phosphatidic acid 289
2. CDPdiglyceride 292
3. Phosphatidylglycerophosphate, phosphatidylglycerol, and diphosphatidylglycerol . . . 293
4. Phosphatidylethanolamine 293
5. Phosphatidylcholine 295
6. Amino-acyl phosphatidylglycerols 302
7. Glucosaminyl phosphatidylglycerol 305
8. Diglucosyl derivatives of phosphatidylglycerol 306
9. Diether derivatives of phosphatidylglycerol 306
10. Phosphatidylinositol mannosides 307
Chemical structures, 307 – Biosynthesis, 309 –
11. Acylated derivatives of glycerylphosphoryl diglucosyl diglyceride 312
12. Phosphosphingolipids and unusual lipids of microorganisms 314
13. Some factors influencing phospholipid metabolism *in vivo* 316
A. Temperature shifts, 316 – B. Induction of electron transport systems, 318 –
14. Functional aspects of phospholipids in microbial membranes 323
References 327

Chapter 12. Phosphonolipids –
by H. ROSENBERG 333

1. Phosphonolipid classes: discovery and chemistry 333
2. Distribution of phosphonolipids in nature 335
3. Biosynthesis 337
4. Degradation 340
5. Function 341
References 342

Chapter 13. Phospholipids and the Hepato-Portal System –
by R. COLEMAN 345

1. Fat absorption 345
Intraluminal triacylglycerol hydrolysis, 346 – Intraluminal phospholipid hydrolysis, 346 – Physical state of the intraluminal phase, 347 – Uptake of lipids from the intraluminal phase, 347 – Intracellular metabolism of acylglycerols, 348 – Intracellular metabolism of phospholipids, 349 –
2. Chylomicron metabolism 350
Chylomicron composition and structure, 351 – Chylomicron formation, 352 – Chylomicrons in the circulation, 352 – Chylomicron clearance, 353 – Role of the liver in chylomicron metabolism, 353 –
3. Uptake of lipids by the liver 354
Fatty acids, 354 – Phospholipids, 356 –
4. The liver and plasma lipoproteins 356
Physical and chemical characteristics, 356 – Structure, 357 – Metabolism of VLDL, 358 – Site of biosynthesis of VLDL, 359 – Turnover of lipoprotein components, 360 – Functions of plasma lipoproteins, 360 –
5. Fatty liver 361
Occurrence of fatty liver, 361 – Causes of fatty liver, 361 – Interference with lipoprotein production, 362 –
6. Bile formation 363
Composition of bile, 363 – Origin of biliary phospholipid, 364 –
7. Phospholipid in relation to other activities of the liver 365
Drug metabolism, 366 – Starvation, 366 – Regeneration, 367 –
8. Pathways of phospholipid metabolism in liver 367
Phospholipid exchange, 367 – Phospholipid pools, 368 –
9. Overall phospholipid metabolism in liver 369
References 370

Chapter 14. Phospholipids and the Nervous System –
by G.B. ANSELL 377

1. The phospholipid composition of the nervous system 377
Neurones and glia, 378 – Subcellular fractions, 378 – Myelin, 380 – Nuclei, 383 – Mitochondria, 383 – Microsomal fraction, 384 – Synaptosomes, 384 –
2. Biosynthesis of phospholipids in the nervous system 384
Phosphatidic acid, 385 – Phosphatidylcholine, 385 – Ethanolamine phosphoglycerides, 387 – Sphingomyelin, 390 – Phosphatidylserine, 391 –
3. Catabolism 391
4. Phospholipid metabolism *in vivo* 393
The origin and transfer of brain phospholipids, 395 – Phospholipid metabolism in cell culture, 396 –
5. Phospholipids and the developing nervous system 397
Changes in the levels and composition of phospholipids in the developing brain, 397 – Myelin composition during development, 399 – Phospholipid metabolism in the developing brain, 401 – Dysmyelination, 402 – Other inborn errors characterized by dysmyelination, 403 – The effect of some exogenous and endogenous factors on phospholipids in development, 405 –
6. Demyelination 408
Primary demyelination, 408 – Secondary demyelination, 409 –
7. Some aspects of the role of phospholipids in the function of the nervous system 410
Conclusion, 414 –
References 415

Chapter 15. Phospholipid Metabolism and Transport of Materials Across the Cell Membrane –
by J.N. HAWTHORNE 423

1. Transmission processes in nerve tissue 423
(a) Phospholipid metabolism at the nerve ending, 423 – (b) Polyphosphoinositide metabolism and function, 429 – (c) Phospholipids and the 'sodium pump' ATPase, 431 –
2. The secretion of catecholamines by the adrenal medulla 432
3. Zymogen secretion by the pancreas 433
4. Thyrotrophin and the thyroid gland 435
5. Activation of lymphocytes by phytohaemagglutinin 435
6. Phagocytosis and phospholipid metabolism 436
7. Phospholipid changes in denervated muscle 436
8. General comments on the metabolism of phospholipids during transport processes 436
9. Dipalmitoyl lecithin in lung alveoli 437
References 438

Chapter 16. The Phospholipid Composition of Mammalian Tissues –
by D.A. WHITE 441

Notes on the Tables 443
Table 1 – The phospholipid content of tissues 444
Table 2 – Phospholipid composition of brain 446
Table 3 – Phospholipid composition of heart 448
Table 4 – Phospholipid composition of liver 449
Table 5 – Phospholipid composition of kidney 450
Table 6 – Phospholipid composition of spleen 451
Table 7 – Phospholipid composition of lung 452
Table 8 – Phospholipid composition of skeletal muscle 453
Table 9 – Phospholipid composition of erythrocytes and human platelets 454
Table 10 – Phospholipid composition of peripheral nerve 455
Table 11 – Phospholipid composition of pancreas, thyroid, thymus, pituitary and adrenals . . 456
Table 12 – Phospholipid composition of intestinal mucosa, corpus luteum, mammary gland, milk, skin and aorta 457
Table 13 – Phospholipid composition of other tissues 458
Table 14 – Phospholipids of subcellular fractions: nuclei 459
Table 15 – Phospholipids of subcellular fractions: mitochondria 460
Table 16 – Phospholipids of subcellular fractions: microsomal fractions 462
Table 17 – Phospholipid composition of subcellular fractions: miscellaneous 464
Table 18 – Phospholipid composition of myelin from central nervous system 466
Table 19 – Phospholipid composition of peripheral nervous system myelin 467
Table 20 – Fatty acids of phosphatidylcholine of whole tissue and subcellular fractions . . . 468
Table 21 – Fatty acids of diphosphatidylglycerol of whole tissue and subcellular fractions . . 471
Table 22 – Fatty acid composition of phosphatidic acid, phosphatidylglycerol and lyso-compounds of whole tissue and subcellular fractions 472
Table 23 – Fatty acids of phosphatidylethanolamine of whole tissue and subcellular fractions . 473
Table 24 – Fatty acids of phosphatidylinositol of whole tissue and subcellular fractions . . . 475
Table 25 – Fatty acids of polyphosphoinositides of bovine brain 476
Table 26 – Fatty acids of phosphatidylserine of whole tissue and subcellular fractions . . . 477
Table 27 – Fatty acid composition of sphingomyelin of whole tissue and subcellular fractions . 478
References 479
Subject Index 483

Chapter 1

HISTORICAL INTRODUCTION

G. B. ANSELL

Department of Pharmacology (Preclinical), Medical School, Birmingham, B15 2TJ (Great Britain)

In living systems the structures of the cells and tissues are based on large molecules – the proteins, polysaccharides and complex lipids – whilst the organisation of the systems appears to be the function of the nucleic acid complexes. Over the past few decades, detailed studies on proteins, polysaccharides and nucleic acids have demonstrated the patterns of repeating units within an infinite variety of molecular structures. Although most lipid molecules are not as complex or as large as these, a large variety of compounds exists, many of them differing only in the composition of the long-chain fatty acid or aldehyde moieties.

The word lipid (Gr. *lipos*, fat) covers an astonishing variety of compounds, the classification of which has been the subject of some controversy. One of the most widely used classifications is that of Bloor[1] (1925–1926), adapted by Deuel[2] in 1951 in his definitive treatise. In this classification lipids are subdivided as follows.

1. Simple lipids

(a) Neutral fats (glycerol esters of largely long-chain fatty acids).
(b) Waxes (solid esters of long-chain monohydric alcohols).

2. Compound lipids (esters of fatty acids with alcohols, which contain also an additional group)

(a) Phospholipids (lipids containing a phosphate residue).
(b) Cerebrosides and gangliosides (lipids containing a carbohydrate residue).
(c) Sulphatides (lipids containing a sulphate residue).

3. Derived lipids

This group contains products derived from 1 and 2 which still possess lipid-like characteristics. They include long-chain fatty acids, aldehydes and alcohols, sterols, hydrocarbons (carotenoids, Vitamins D, E, and K) and sphingosine.

In broad terms this classification still survives but, as chemical structures have been more precisely defined, so a more rigorous nomenclature has become necessary. So far the recommendations proposed by the Joint Commission of the International Union of Pure and Applied Chemistry and the International Union of Biochemistry[4, 5] remain a 'document for discussion'. A great deal of thought has gone into these proposals and in spite of the criticisms that have been made, they have been widely adopted by journals and are therefore used throughout this book. Undoubtedly, the method devised to describe the stereochemistry of lipids, not least the phospholipids, caused some heart-searching and requires some mental effort to master; it is described in some detail in Chapter 2, pp. 12–17).

In the following historical sketch, as in similar accounts of the historical development of many biochemical topics, the purely chemical investigations predominate. More elaborate accounts are given by Levene and Rolf[6], Maclean and Maclean[7], Thierfelder and Klenk[8], and Wittcoff[9]. The bibliography in Wittcoff's book is especially noteworthy. Some interesting incidental historical information about early investigations on brain phospholipids is given in a review by Tower[10], whilst a dogmatic and polemical, but highly detailed, commentary on early and mid-nineteenth century investigation is given by Thudichum[11] in his original report to the Privy Council in 1874.

Since phospholipid function in nervous tissue is more obscure than extracerebral function it is ironical that the first observation that phosphorus was bound to fatty substances was made on material extracted from brain. This discovery was made by L.N. Vauquelin[12] in 1811 (Fig. 1) who used ethanol as an extractant; ether was not a common solvent at that time and chloroform had not been synthesised. Vauquelin (1763–1829) was a pupil of A.F. de Fourcroy (1755–1809), a well-known French chemist; further details of his activities can be found in Tower's review[10].

In 1834, Couerbe[13] isolated similar substances from brain and his contemporary Frémy[14] (1814–1894) named one of them 'oleophosphoric acid' (1841). Vauquelin, Couerbe and Frémy all obtained brain material which was soluble in hot alcohol and which was deposited on cooling ('matière blanche', 'cérébrote', 'acide cérébrique'); this was the source of Liebreich's 'protagon' (see p. 4). Gobley (1811–1876) isolated from egg-yolk and brain a phosphorus-containing lipid[15,16] in 1846–1847 which he later (1850) called 'lecithin'[17] (Gr. *lekithos*, egg-yolk). In his publication of 1850, Gobley also showed that glycerophosphoric acid could be obtained from lecithin whilst Diaconow[18] who worked in Hoppe-Seyler's laboratory, and Strecker[19] demonstrated the presence of choline in 1868. Strecker[20] had originally isolated this base from hog bile in 1862 (Gr. *cholĕ*, bile). From these observations Diaconow and Strecker were able to deduce the structure of lecithin, although Strecker alone suggested that choline was attached to the phosphate by ester linkage.

It is, however, to J.L.W. Thudichum (1828–1901) that phospholipid chemistry owes some of its most considerable advances and, incidentally, some classical terminology. While studying at the University of Giessen (from which he graduated in medicine in 1851) he came under the influence of the chemist Liebig. On emigrating to London in 1853 he practised medicine privately before taking teaching appointments in London hospitals, but from 1871 he worked largely in his private laboratory. Here he carried out the fundamental work on the chemical constitution of the brain[11,21] which represents, however, only a portion of his researches. His original publications are difficult to obtain, though his book on the chemical constitution of the brain has been reproduced in facsimile[22] and a very readable biography is available which contains a complete bibliography[23]. A shorter but objective account of Thudichum's contribution to biochemistry has been written by McIlwain[24].

50 ANNALES

4°. 100 parties de la même matière jetée successivement dans 200 parties de nitrate de potasse fondu, se sont enflammées avec une grande facilité; il ne s'est presque point produit de fumée, tout a été détruit, et il n'est pas resté la plus légère trace de matière charbonneuse.

Le résidu de cette opération, traité de la même manière que celui de la précédente, a donné la même quantité de phosphate de chaux.

Que conclure de ces expériences, si ce n'est qu'il y a du phosphore combiné avec la matière grasse du cerveau, lequel s'est dissous en même tems que cette dernière dans l'alcool?

En effet, on ne trouve dans le résidu de la combustion de cette substance, ni phosphate de chaux ni de magnésie, et les phosphates alcalins auraient trouvé assez d'eau dans la matière cérébrale pour rester en dissolution dans l'alcool, et pour ne pas s'en précipiter par le refroidissement; aussi retrouve-t-on du phosphate de potasse et des phosphates acides de chaux et de magnésie dans le résidu de l'alcool évaporé qui a servi à traiter le cerveau.

Il faut donc nécessairement admettre dans

Fig. 1. A page from one of Vauquelin's papers in which he described the presence of phosphorus bound to fat in the brain[3]. (Reproduced by kind permission of Masson et Cie., Paris.) This paper is apparently the same as the original publication which was not available to the author.

References p. 7

In the course of his investigations on the brain, Thudichum isolated many phospholipid fractions which he attempted to classify by careful analysis and, among other methods, their nitrogen–phosphorus ratio. Many of his substances including, for example, 'cephalin' which he separated from brain lipids (Gr. *kephalě*, head) and distinguished from lecithin by its relative insolubility in warm ethanol, are now known to be mixtures. In point of fact, although he isolated ethanolamine as a hydrolysis product he did not consider it as a normal constituent of the cephalin but as a decomposition product of choline. It is typical of the difficulties associated with phospholipid separations that it was not until 1913 that Renall[25] and Baumann[26] simultaneously showed that one of the bases in the cephalin fraction was ethanolamine.

Thudichum also isolated from brain tissue a phospholipid which he designated sphingomyelin 'in commemoration of the many enigmas which it presented to the enquirer' (Gr. *sphingein*, to bind tight; *myelos*, marrow). He recognised it as a compound with an N:P ratio of 2:1 'which is a phosphatide, and contains an alcohol, not being glycerol, an acid, and two nitrogenised radicles as proximate constituents' (Ref. 22, p. 118). The occurrence of the alcohol ('sphingol') proved to be incorrect but after alkaline hydrolysis of the sphingomyelin he successfully obtained its two constituent bases,sphingosine and choline ('neurine'), in addition to phosphoric acid and a fatty acid. It is noteworthy that the realised that 'protagon' (Gr. *protos*, first: *agonistes*, a combatant), thought by Liebreich[27] (Fig. 2) to be the mother substance of all phospholipids, was not a distinctive substance, in disagreement with the theories of many, but not all, German chemists. For over twenty years (1874–1896) Thudichum produced his evidence and bitterly stated his case but his assertions were not vindicated until Rosenheim and Tebb showed that protagon was a mixture of sphingomyelin and cerebroside[28]. The problem of protagon, although now only of academic interest, had considerable significance more than sixty years ago and its ramifications were well documented by Maclean and Maclean[7] in 1927.

It is perhaps not unfair to say that from about 1905 until the late 1930s, knowledge of the chemical nature of phospholipids did not greatly advance. Two investigators in this period should, however, be mentioned. Levene carried out extensive investigations on the separation and purification of sphingomyelin and other phospholipids and identified some of the fatty acids present[6]. He was the first to obtain a relatively pure sphingomyelin and showed that its hydrolysis products were exclusively choline, sphingosine, phosphoric acid and fatty acids, one of which was lignoceric acid[29, 30]. Ernst Klenk, who died while this volume was being prepared, first made important contributions to phospholipid and sphingolipid chemistry nearly half a century ago; he was the last link with the pioneering period 1920–1940.

Because of the lack of adequate structural information and reference compounds, investigations of metabolism were sporadic. Advances made in the same period in, for example, the structure of carbohydrates and their metabolic intermediates is in striking contrast with progress in phospholipid biochemistry; yet many of the intermediates in carbohydrate metabolism are distinctly elusive entities, often present in small amounts. Lipids, even minor ones, and their intermediates are equally widely distributed and often present in higher concentrations.

Ueber die chemische Beschaffenheit der Gehirnsubstanz;

von *Oscar Liebreich*,

Assistenten am Schlofslaboratorium zu Tübingen.

So ungemein grofs die Schwierigkeiten sind, die sich bei der chemischen Untersuchung der Nervensubstanz darbieten — wie es alle Forscher einstimmig zugestehen —, schien es doch eines neuen Versuches werth, eine Trennung wenigstens der einigermafsen näher beschriebenen Bestandtheile vorzunehmen, um Genaues über ihr quantitatives Vorkommen zu erfahren.

Wenn ich auch diese Absicht nicht erreicht habe, weil — wie ich später zeigen werde — sie zu erreichen unmöglich ist, so bin ich doch insofern zu einem Resultate gelangt, als ich nachzuweisen vermochte, dafs alle diejenigen Körper, die man als Cerebrin, Cerebrinsäure, Lecithin u. s. w. und als phosphorhaltige Fette bezeichnete, primär nicht im Gehirn existiren. Vielmehr läfst sich ein Theil dieser Körper als reine Zersetzungsproducte eines im todten Gehirn primär vorkommenden Stoffes nachweisen, und ein anderer Theil als nicht hinreichend chemisch characterisirt aus der Reihe wirklich chemischer Individuen zurückweisen.

Es kann hier weder meine Absicht sein, eine historische Darstellung der bisherigen Untersuchungen zu geben, noch die Fehler der Methoden aufzuzählen und zu kritisiren, welche mir bei dem Bestreben, die von früheren Autoren als dem Hirn eigenthümlich beschriebenen Stoffe zu isoliren, aufgestofsen sind. Ich werde mich im Folgenden darauf beschränken, den einen von mir gefundenen Körper und seine nächsten Zersetzungsproducte zu beschreiben und zu erwähnen, in

Fig. 2. The first page of Liebreich's paper[27] in which he put forward the 'protagon' concept. (Reproduced with kind permission of Verlag Chemie, Weinheim/Bergstr.)

References p. 7

The difficulties lay largely in the absence of adequate methods of fractionation and the hazards involved in obtaining components which would crystallise. In 1916 Levene and West[32] stated '. . . the experience on sphingomyelin has shown that often thirty recrystallisations are required before the substance is obtained in a sufficient degree of purity'. The phospholipids do not differ greatly from one another in many physical properties and the fatty acid or aldehyde components confer the properties of intractable oils on many of the impure specimens.

Most of the early methods of fractionation of phospholipid mixtures depended upon the differential solubility of their components. It was observed for example, that sphingomyelin is virtually insoluble in ether and that lecithin is more soluble in warm ethanol than the components of 'cephalin'. Folch[33] brilliantly exploited relative solubilities in 1942 when he successfully separated brain cephalin into three fractions containing respectively ethanolamine, serine and inositol. It is a further comment on the earlier difficulties that in 1914 MacArthur[34] had demonstrated the presence of α-aminocarboxylic acid nitrogen in cephalin but another thirty years were to elapse before the source of this nitrogen was identified as serine.

Greater success was obtained in the nineteen thirties in ascertaining the distribution of lipid components in a sample or in tissues by the estimation of hydrolysis products which were water-soluble and/or capable of chemical determination. Thus the determination of choline as its reineckate became important and this method is, in fact, still used since others are quite complicated (Chapter 3, p. 59). It was more difficult to determine glycerol, ethanolamine and serine and suitable methods did not become available until the fifties. Schmidt and his co-workers[35] devised a method in 1946 whereby the lability of certain phospholipids to mild alkali treatment was used to differentiate them from the alkali-stable sphingomyelin.

It is obvious that the advances made in our knowledge of the chemistry and metabolism of the phospholipids that have taken place in recent years have depended upon the sophisticated methods of separation and analysis that have been developed and which are discussed in Chapter 3. The first use of a chromatographic method in studies on phospholipids was that of Thannhauser and Setz[36] in 1936 who utilised a column of aluminium oxide to remove cerebrosides from a solution of sphingomyelin in light petroleum which was contaminated with these lipids. There was then a considerable lapse of time before chromatography on aluminium oxide and silicic acid was successfully accomplished by a number of workers[37–39]. The paper chromatography of intact phospholipids proved more difficult; it was first carried out on silica-impregnated paper by Lea et al., in England[38] and Marinetti's group in America[40].

In 1950 Baer and Kates[41] observed that brief alkaline hydrolysis of phosphatidylcholine released the fatty acids very much faster than free choline. This finding was exploited by Dawson[42] in 1954 who showed that very brief alkaline hydrolysis of all diacyl phospholipids yielded characteristic water-soluble phosphodiesters which could be separated by paper chromatography. A large number of publications on the distribution and metabolism of phospholipids has stemmed from this method as will be evident from Chapter 3. As far as the author is aware, the first paper on the application of gas-liquid chromatography to the fatty acid composition of total phospholipids appeared in 1958 (ref. 43) and of phospholipid fractions from a chromatography column in 1959 (ref. 44). Soon after the first applications of thin-layer chromatography to phospholipid separations appeared[45–47], the two methods were combined to study the distribution and metabolism of fatty acids and aldehydes in individual phospholipids in small samples of tissue. The application of these techniques has led to the discovery of hitherto unknown phospholipids and thrown new light on established ones. Without them this volume would have been very much shorter.

Over the last thirty years, too, organic chemistry has provided the biochemist with a significant range of synthetic phospholipids for use as reference compounds. Groups led by the late Thomas Malkin (1897–1961) in England, by Baer in Canada and by Van Deenen and De Haas in The Netherlands have made substantial contributions in this area. The work of Van Deenen's group is especially notable for a neat combination of synthetic organic chemistry and enzymology, since phospholipases are used to modify the compounds synthesized.

At this point in time it is perhaps difficult, especially for younger workers in the field, to realise that the biochemistry of the phospholipids was until recently a neglected area, apart from the work of a few dogged enthusiasts. At one time it might fairly be said that biochemists only wanted to remove phospholipids from preparations of other substances. Undoubtedly, improved methodology has been partly responsible for the growth of this subject but considerable stimulation has come from another quarter, namely the desire to understand the structure and function of membranes. Indeed, paralleling the growth in lipid methodology has been the development of methods for the isolation of biological membranes from a whole variety of cells. A great deal is now known about the phospholipid composition of these structures, as will be apparent from several chapters in this book. We are, however, far from clear about the role of phospholipids in lipoprotein structure, a situation which offers the prospect of gainful employment for a lot of future investigators. The roles of membrane phospholipids in electron transport and in the transport of ions and molecules across membranes, not to mention the significance of the range of fatty acids and aldehydes and different base moieties, are all far from clear. There is still a good deal to be ascertained about the form and function of phospholipids. The following chapters not only record progress in the field so far but also define our ignorance more precisely.

REFERENCES

1 W.R. Bloor, *Chem. Rev.*, 2 (1925–1926) 243.
2 H.J. Deuel, *The Lipids*, Vol. 1, Interscience, New York, 1951, p. 3.
3 L.N. Vauquelin, *Ann. Chim. (Paris)*, 81 (1812) 37.
4 IUPAC-IUB Commission on Biochemical Nomenclature. The Nomenclature of Lipids, *Biochim. Biophys. Acta*, 152 (1968) 1.
5 IUPAC-IUB Commission on Biochemical Nomenclature. Amendments (1968) to the Lipid Proposals, *J. Lipid Res.*, 11 (1970) 173.
6 P.A. Levene and I.P. Rolf, *Physiol. Rev.*, 1 (1921) 327.

7 H. MacLean and I.S. MacLean, *Lecithin and Allied Substances*, Longmans, Green, London, 1927.
8 H. Thierfelder and E. Klenk, *Die Chemie der Cerebroside und Phosphatide*, Springer, Berlin, 1930.
9 H. Wittcoff, *The Phosphatides*, Reinhold, New York, 1951.
10 D.B. Tower, *Neurology*, 8, suppl. 1 (1958) 3.
11 J.L.W. Thudichum, *Researches in the Chemical Constitution of the Brain*, Rept. Med. Off. Priv. Council, New Series No.3, App. 5, Eyre and Spottiswoode, London, 1874, p. 113.
12 L.N. Vauquelin, *Ann. Mus. Hist. Nat.*, 18 (1811) 212.
13 J.P. Couerbe, *Ann. Chim. Phys.*, 56 (ser. 2) (1834) 160.
14 E. Frémy, *Ann. Chim. Phys.*, 2 (ser. 3) (1841) 463.
15 M. Gobley, *J. Pharm. Chim. (Paris)*, 9 (ser. 3) (1846) 1, 81, 161.
16 M. Gobley, *J. Pharm. Chim. (Paris)*, 11 (ser. 3) (1847) 409; 12 (ser. 3) (1847) 1.
17 M. Gobley, *J. Pharm. Chim. (Paris)*, 17 (ser. 3) (1850) 401.
18 C. Diaconow, *Zentr. Med. Wiss.*, 6 (1868) 2, 97.
19 A. Strecker, *Ann. Chem. Pharm.*, 148 (1868) 77.
20 A. Strecker, *Ann. Chem. Pharm.*, 123 (1862) 353.
21 J.L.W. Thudichum, *A Treatise on the Chemical Constitution of the Brain,* Ballière, Tindall and Cox, London, 1884; *Die chemische Konstitution des Gehirns des Menschen und der Tiere*, Pietzcker, Tübingen, 1901.
22 J.L.W. Thudichum, *A Treatise on the Chemical Constitution of the Brain* (with a new historical introduction by D.L. Drabkin), Archon, Hamden, Conn., 1962.
23 D.L. Drabkin, *Thudichum: Chemist of the Brain*, University of Pennsylvania Press, Philadelphia, 1958.
24 H. McIlwain, *Proc. Roy. Soc. Med.*, 51 (1958) 127.
25 M.H. Renall, *Biochem. Z.*, 55 (1913) 296.
26 A. Baumann, *Biochem.. Z.*, 54 (1913) 30.
27 O. Liebreich, *Ann. Chem. Pharm.*, 134 (1865) 29.
28 O. Rosenheim and M.C. Tebb, *Biochem. Z.*, 25 (1910) 151.
29 P.A. Levene, *J. Biol. Chem.*, 18 (1914) 453.
30 P.A. Levene, *J. Biol. Chem.*, 24 (1916) 69.
31 E. Klenk, *Z. Physiol. Chem.*, 185 (1929) 169.
32 P.A. Levene and C.J. West, *J. Biol. Chem.*, 24 (1916) 41.
33 J. Folch, *J. Biol. Chem.*, 146 (1942) 35.
34 C.G. MacArthur, *J. Am. Chem. Soc.*, 36 (1914) 2397.
35 G. Schmidt, J. Benotti, B. Hershman and S.J. Thannhauser, *J. Biol. Chem.*, 166 (1946) 505.
36 S.J. Thannhauser and P. Setz, *J. Biol. Chem.*, 116 (1936) 527.
37 D.J. Hanahan, M.B. Turner and M.E. Jayko, *J. Biol. Chem.*, 192 (1951) 623.
38 C.H. Lea, D.N. Rhodes and R.D. Stoll, *Biochem. J.*, 60 (1955) 353.
39 J.M. McKibbin and W.E. Taylor, *J. Biol. Chem.*, 196 (1952) 427.
40 G.V. Marinetti and E. Stotz, *Biochim. Biophys. Acta*, 21 (1956) 168.
41 E. Baer and M. Kates, *J. Biol. Chem.*, 185 (1950) 615.
42 R.M.C. Dawson, *Biochim. Biophys. Acta*, 14 (1954) 374.
43 C.J.F. Böttcher, J.G. Keppler, C.C. ter Haar Romeny-Wachter, E. Boelsma-van Houte and C.M. van Gent, *Lancet*, 2 (1958) 1207.
44 C.E. Rowe, *Biochem. J.*, 73 (1959) 438.
45 H. Weicker, *Klin. Wschr.*, 37 (1959) 763.
46 H. Jatzkewitz and E. Mehl, *Z. Physiol. Chem.*, 320 (1960) 251
47 H. Wagner, L. Hörhammer and P. Wolff, *Biochem. Z.*, 334 (1961) 175.

Chapter 2

THE CHEMISTRY OF PHOSPHOLIPIDS

K.P. STRICKLAND*

Department of Biochemistry, University of Western Ontario, London (Canada)

As outlined in the previous chapter, when the monograph by MacLean and MacLean[1] appeared in 1927, several fairly well-defined phospholipids had been described. These included lecithin, cephalin and sphingomyelin. Subsequent to 1927 and up to the time of the appearance of two well-documented treatises by Deuel[2] and Wittcoff[3] in 1951, several new phospholipids were discovered. Thus, to the above list were added: (*a*) phosphatidic acid which was isolated as its calcium salt from cabbage leaves in 1927 (ref. 4), (*b*) acetal phosphatide, an aldehyde-containing phospholipid isolated from beef heart in 1939 (ref. 5), (*c*) phosphatidylethanolamine, phosphatidylserine and an inositol-containing phospholipid found as components of brain 'cephalin' in 1942 (ref. 6), (*d*) cardiolipin, isolated as its cadmium salt in 1942 and barium salt in 1944 (ref. 7), and (*e*) 'diphosphoinositide' which was isolated as a $Ca^{2\oplus}$–$Mg^{2\oplus}$ complex from brain in 1949 (ref. 8).

Almost without exception the discovery and isolation of the above phospholipids depended on differences in solubility exhibited by the natural phospholipid, or by specific salt forms, in various organic solvents. In the early 1950's a number of developments occurred which greatly facilitated the isolation and analysis of phospholipids. These developments included the application of silicic acid column chromatography[9, 10] and silica-impregnated paper[10, 11] to separations of intact lipids and the development by Dawson[12] of a mild alkaline hydrolysis procedure for the release of water-soluble phosphodiesters (from phospholipids) which could be separated by two-dimensional paper chromatography. The availability of the above techniques and their refinement to a very high level (see for example Kates[13], Rouser et al.[14] and Dawson[15]) and the more recent application to polar lipids of thin-layer chromatography (TLC)[14, 16–18] has permitted the separation and characterization of many new phospholipids over the last fifteen years.

This chapter describes the known naturally occurring phospholipids. Initially some discussion is devoted to the classification and nomenclature of these phospholipids. The latter task has been made easier by the availability in published form of a tentative set of rules for the nomenclature of lipids prepared by the IUPAC–IUB** Commission on Biochemical Nomenclature[19]. The structures of the natural phospholipids, their isolation and chemical syntheses are then briefly described. In the final section of the chapter a

*This chapter was written while a Visiting Professor in the Department of Biochemistry, The John Curtin School of Medical Research, Australian National University, Canberra, Australia.
**International Union of Pure and Applied Chemistry–International Union of Biochemistry.

brief outline is given of the effects of hydrolytic and other degradative reagents on the phospholipids. Because of the large number (now about 40) of phospholipids to be discussed in the limited space available, it has not been possible to cover each one exhaustively. A number of books or general reviews may be consulted for further detail[20–31].

1. CLASSIFICATION

No general agreement exists on the best way to classify the lipids, or any special sub-group such as the phospholipids. The latter possess a number of biological, chemical and physical properties (e.g. solubility in organic solvents, surface activity, component of membranes etc.) in common which are dependent on the presence of hydrophobic (due to glyceride or *N*-acylsphingosine) and hydrophilic (due to phosphomonoester moiety) groups. However, no easy classification emerges based on differences in these properties. The only reasonable approach is one based on the common constituents present. Thus, most classifications contain a broad category for the glycerol-containing phospholipids (phosphoglyceride, glycerophosphatide etc.) and one for the sphingolipids of which the phosphorus-containing sphingolipids form one part.

Table 1 depicts a proposed classification which can be amended to include new phospholipids as they are discovered and their structures are determined. In conformity with the recommendation of the Commission on Biochemical Nomenclature[19] the generic terms, 'phosphoglyceride' and 'phosphosphingolipid' are used to cover the categories described in the preceding paragraph. The term 'phosphoglyceride' signifies any derivative of *sn*-glycero-3-phosphoric acid that contains at least one *O*-acyl, or *O*-alkyl, or *O*-alk-1′-enyl residue attached to the glycerol moiety[19]. The term 'phosphosphingolipid' refers to any lipid containing phosphorus and a long-chain base (LCB). Prefixes have been added to these generic terms to provide categories for those lipids which cannot normally be classified under the above two terms. This classification attempts to take into account all of the known natural phospholipids that have been isolated and characterized. Other designations and trivial names are given in the subsequent discussion on each of the phospholipids.

In recent years a considerable number of lipids have been isolated and characterized which contain phosphorus in the phosphonate, or phosphonic acid form

$$\text{R–CH}_2\text{–}\overset{\displaystyle\text{OH}}{\underset{\displaystyle\text{OR}}{\text{P}}}\text{=O}$$

rather than in the *ortho*-phosphate, or phosphoric acid form

TABLE 1

A CLASSIFICATION OF KNOWN PHOSPHOLIPIDS[a]

Classes[b]	*Analogues*[c]				
A. Phosphoglycerides	(a)	(b)	(c)	(d)	(e)
(*1*) Phosphatidic acids	+	+	+	+	
(*2*) Cytidylic phosphoglycerides (CDPdiglyceride)		+			
(*3*) Choline phosphoglycerides	+	+	+	+	+
(*4*) Ethanolamine phosphoglycerides	+	+	+	+	+
(*5*) *N*-Methylethanolamine phosphoglycerides		+	+		
(*6*) *N,N*-dimethylethanolamine phosphoglycerides		+			
(*7*) *N*-acyl ethanolamine phosphoglyceride		+			
(*8*) Serine phosphoglycerides		+	+		
(*9*) *N*-2-(Hydroxyethyl)alanine phosphoglyceride		+			
(*10*) Glycerol phosphoglycerides		+			+
(*11*) Glycerophosphate phosphoglycerides		+			+
(*12*) Phosphatidylglycerol phosphoglyceride (Diphosphatidylglycerol)		+			
(*13*) Mono- and diacylglycerol phosphoglycerides (lyso bis phosphatidic acids)	+	+			
(*14*) Glucosaminylglycerol phosphoglyceride					
(*15*) *O*-amino acid esters of glycerol phosphoglycerides		+			
(*16*) Inositol phosphoglyceride		+			
(*17*) Inositol monophosphate phosphoglyceride		+			
(*18*) Inositol diphosphate phosphoglyceride		+			
(*19*) Monomannosyl–hexamannosyl inositol phosphoglycerides		+			
(*20*) Glucose phosphoglyceride		+			
(*21*) *O*-diglucosylglycerol phosphoglyceride		+			
B. Phosphoglycolipids					
(*1*) Diacyl (glycerylphosphoryldiglucosyl)glycerol					
C. Phosphodiol lipids					
(*1*) Acyl dihydroxyacetone phosphate					
(*2*) Alkyl dihydroxyacetone phosphate					
D. Phosphosphingolipids					
(*1*) Sphingomyelin (Ceramide phosphorylcholine)					
(*2*) Ceramide phosphorylethanolamine					
(*3*) Ceramide phosphorylglycerol					
(*4*) Ceramide phosphorylglycerophosphate					
(*5*) Ceramide phosphorylinositol-containing lipids					
E. Phosphonolipids					
(See Chapter 12 for classification)					

[a]An attempt has been made to conform to the recommendations of the IUPAC–IUB Commission of Nomenclature[19].
[b]The more general generic terms ('Phosphoglycerides', 'Phosphoglycolipids', 'Phosphodiol lipids' and 'Phosphosphingolipids') may be easily fitted into any general classification of lipids.
[c](a), 1-acyl-, or 2-acyl- (i.e. lyso); (b), 1,2-diacyl- (i.e. phosphatidyl); (c), 1-alk-1'-enyl-, 2-acyl- (i.e. plasmalogen); (d), 1-alkyl-, 2-acyl- and (e), 2,3-dialkyl.

$$\begin{array}{c} \quad\quad\quad\quad\quad\quad OH \\ \quad\quad\quad\quad\quad\quad | \\ R{-}CH_2{-}O{-}P{=}O \\ \quad\quad\quad\quad\quad\quad | \\ \quad\quad\quad\quad\quad\quad OR \end{array}$$

which exists in the above phospholipids. This new group of lipids, referred to as the 'phosphonolipids' is described fully by Rosenberg in Chapter 12 (Table 1).

2. NOMENCLATURE

Until the recent publication on lipid nomenclature[19] considerable confusion has existed in the literature in relation to the correct and precise manner of naming lipids. As described by the Commission[19] and commented upon in the predecessor to this volume[21], the greatest problem is the manner in which stereoisomers should be distinguished, particularly in the case of glycerol.

Naming of lipids containing glycerol

For glycerol at least four different systems of designation have been proposed (see Table 2). From the proposal of Baer and Fischer[32] emerged the system which places the glycerol derivative 'in the same category with that of glyceraldehyde into which it could be transformed by oxidation without any alteration or removal of substituents'. Since by this designation the glycerophosphate, derived from most of the then known natural phospholipids, had an L-α-configuration, these phospholipids were given the L-α-designation[33]. Quite naturally confusion developed over whether α referred to the 1- or 3-position of glycerol and as a result a more conventional nomenclature was devised which still used *D*/*L* prefixes for enantiomorphic forms, but numerals as locators[34]. Although two numeral designations were possible, and even used, there emerged the practice of numbering the carbon, C-1, whose primary hydroxyl group was first substituted. Unfortunately the presence of two systems both using the *D*/*L* terminology left open the possibility of confusion and ambiguity unless care was taken to show clearly the system in use (cf. systems 1 and 2, Table 2).

Ambiguity of the sort described can be avoided with the *R*/*S* system[35, 36] (system 3, Table 2). However, as with the two *D*/*L* systems, when applied to glycerol derivatives, one is confronted with the need to make several changes of prefixes for transformations involving the formation and cleavage of ester and ether bonds at C-1 (or C-3) in chemical and biochemical reactions. This point can be illustrated by examining the names that emerge for the glycerol derivatives shown in Table 2. Thus in the interconversions:

diacylglycerol	$\rightleftharpoons$	phosphatidic acid	$\rightleftharpoons$	glycerophosphate
(II, Table 2)		(III, Table 2)		(I, Table 2)

a number of prefix changes involving inversions are necessary. In system 1 there is a D-α,β-

TABLE 2

NOMENCLATURE OF GLYCEROL-CONTAINING PHOSPHOLIPIDS

Structure	Names given to various glycerol derivatives according to:							
	System (1)– α, β-designations D/L *prefixes (refs. 32, 33)*		*System (2)–numeral locators 1, 2 and 3; D/L-prefixes (ref. 34)*		*System (3)-numeral locators 1, 2 and 3; R/S system*[b] *(refs. 35, 36)*		*System (4)*[c]*–**stereo**-specific numbering (sn) (refs. 19, 37)*	
	C designation	*Names*	*C designation*	*Names*	*C designation*	*Names*	*C designation*	*Names*
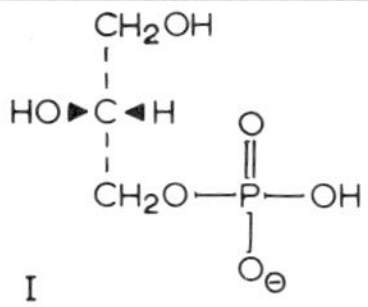 I	α		1		1		1	
	β	L-α-glycerophosphate	2	*D*-glycerol 1-phosphate	2	(*R*)-glycerol-phosphate	2	*sn*-glycerol 3-phosphate
	α		3	(*L*-glycerol 3-phosphate)	3		3	
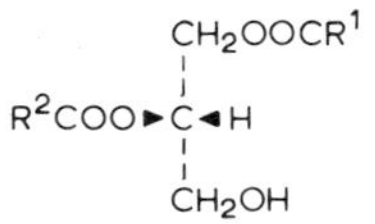 II	α	D-α, β-diacylglycerol	1	*L*-1,2-diglyceride; *L*-1,2-diacyl glycerol; (*D*-2,3-diglyceride; *D*-2,3-diacylglycerol)	1	(*S*)-1,2-diglyceride (*S*)-1,2-diacylglycerol;	1	1,2-diacyl-*sn*-glycerol
	β	(D-α, β-diglyceride)	2		2		2	
	α		3		3		3	
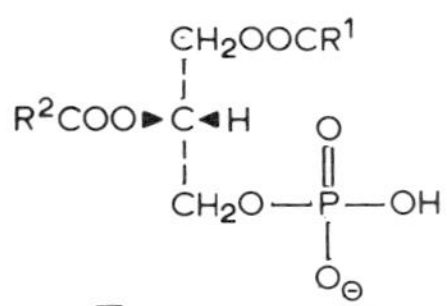 III	α		1	*L*-3-phosphatidic acid; *L*-1,2-diacyl glycerol 3-phosphate; (*D*-1-phosphatidic acid; *D*-2,3-diacyl glycerol 1-phosphate)	1	(*R*)-3-phosphatidic acid; (*R*)-1,2-diacylglycerol 3-phosphate;	1	3-*sn*-phosphatidic acid; 1,2-diacyl-*sn*-glycerol 3-phosphate;
	β	L-α-phosphatidic acid	2		2		2	
	α		3		3		3	

References p. 39

TABLE 2 (*Continued*)

NOMENCLATURE OF GLYCEROL-CONTAINING PHOSPHOLIPIDS

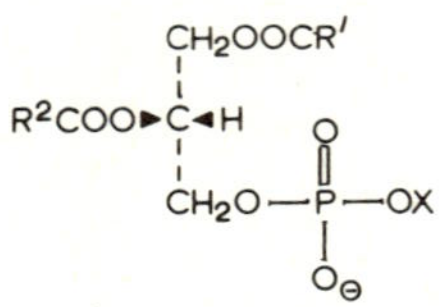	α	L-α-phosphatidyl choline;	1	*L*-3-phosphatidyl choline;	1	(*R*)-3-phosphat-idyl choline;	1	3-*sn*-phosphat-idyl choline;
	β	(L-α-lecithin)	2	(*D*-1-phosphatidyl	2		2	1,2-diacyl-*sn*-
	α		3	choline)	3		3	glycero-3-phosphoryl-choline
IV								
X, $-CH_2CH_2N^{\oplus}(CH_3)_3$	α							

[a]Triangles indicate bonds projecting towards reader; broken lines indicate bonds projecting away from reader.
[b]*R*, rectus or right; *S*, sinister or left.
[c]System (4) is the system recommended for use by the IUPAC–IUB Commission on Biochemical Nomenclature[19].

to L-α- change; in system 2, an *L*-1,2- to *D*-2,3- change, and in system 3 an '*S*' to '*R*' change. As noted by the Commission[19], both the stereospecific action of phospholipases (particularly A) and the designation of a non-random distribution of fatty acids in triacylglycerols requires a system that permits the easy differentiation of ester linkages.

Most problems of designation for the glycerol derivatives are overcome if the fourth system proposed by Hirschmann[37] is used. This system now referred to as *stereospecific numbering*[19] designated by the symbol *sn*, recognises the fact that in most biochemical reactions the two primary carbinol groups are not identical in their reactions. The system designates each carbinol group separately and requires that this designation be maintained regardless of the substitution. The method used to decide which carbinol group should receive the lowest number is based on the priorities assigned in the *R*/*S* system[35, 36]. For glycerol (V), if the secondary hydroxyl group is projected to the left

Carbon number

1 CH_2OH

|

2 HO►C◄H (V)

|

3 CH_2OH

of C-2 in a Fischer projection, then the carbon atom above C-2 is designated C-1 and the one below C-3. When this type of stereospecific numbering is assigned to glycerol, then the prefix *sn* is used before the stem name of the compound. Thus, for example, if the hydroxyl on C-3 is substituted with a phosphate group, the product (I, Table 2) becomes *sn*-glycerol 3-phosphate. If both optical antipodes exist, the prefix *rac* (for *racemo*) is used. Where the configuration is unknown, this should be designated by the prefix '*X*'.

The *stereospecific numbering* system is strongly recommended as the system of choice by the IUPAC–IUB Commission[19]. Comparison of this system with the other three systems in Table 2 confirms the fact that the *sn* system provides a much more precise designation of the carbons of glycerol derivatives. In designations of esters, ethers and other *O*-derivatives of glycerol, the Commission states that Rules 10 and 11 of the rules of Carbohydrate Nomenclature[38] are to be followed. The naming of fatty acids and their radicals, long chain alcohols, amino acids and sugars should adhere to internationally adopted conventions[19]. The use of certain trivial names is permitted, but it is generally recommended that trivial names (and other names) should be defined whenever they are used.

Naming of sphingolipids

Since the discovery in recent years of many compounds structurally related to sphingosine, it was deemed advisable by the Commission[19] to develop a semi-systematic

TABLE 3

NOMENCLATURE OF LONG-CHAIN BASES

	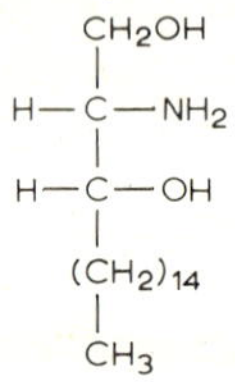	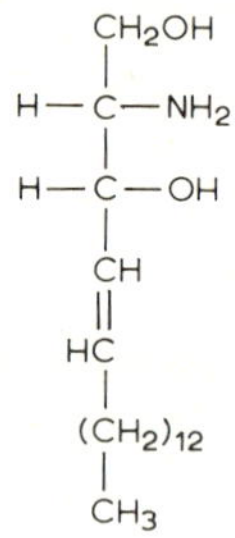	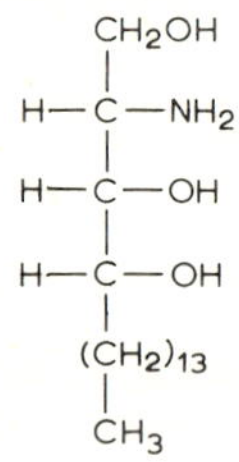
	(VI)	(VII)	(VIII)
	sphinganine	4-sphingenine	4*D*-hydroxysphinganine
Other names	dihydrosphingosine; 2*D*-aminoöctadecane-1,3*D*-diol	sphingosine; *trans*-2*D*-amino-4-octadecene-1,3*D*-diol	phytosphingosine; 2*D*-aminoöctadecane-1,3*D*, 4*D*-triol
	D-*erythro*-2-aminoöctadecane-1,3-diol	*trans*-D-*erythro*-2-amino-4-octadecene-1,3-diol	D-*ribo*-2-amino-octadecane-1,3,4-triol
	(2*S*, 3*R*)-2-amino-octadecane-1,3-diol	*trans* (2*S*, 3*R*)-2-amino-4-octadecene-1,3-diol	(2*S*, 3*R*, 4*R*)-2-amino-octadecane-1,3,4-triol

nomenclature.* They have suggested that the compound known as dihydrosphingosine (VI, Table 3) (also designated 2*D*-aminoöctadecane-1,3*D*-diol; D-*erythro*-2-octadecane-1,3-diol or (2*S*, 3*R*)-2-aminoöctadecane-1,3-diol) should be called 'sphinganine'. The Commission states: 'This name may be modified by prefixes to indicate additional substituents or higher or lower homologues'. For the latter designation the prefix is obtained by deletion of the terminal 'ne' from the systematic name of the hydrocarbons having the same number of carbons as the long-chain bases. Configuration of additional substituents should be specified by the prefixes '*D*' or '*L*' following the number that indicates the position of the carbon substituted. Any deviation in configuration of C-2 and C-3 in sphinganine should be specified (this applies to the *D*/*L* prefix; if the *R*/*S* system is used then the full systematic name with the specific configurations should be given; if the configuration is unknown '*X*' should be used; for *racemic* mixtures *rac* should be used). Where unsaturation is present the names for partly unsaturated compounds are derived from the names of the corresponding saturated compounds by terminations denoting unsaturation, namely 'ene', 'diene', 'yne', etc. Double bonds are presumed to be of *trans* orientation unless designated '*cis*' or '*X*' (for unknown orientation). As an example, the new nomenclature would give sphingosine (the LCB first discovered; VII, Table 3) the new name '4-sphingenine'. This means that, compared to sphinganine (dihydrosphingosine),

*This nomenclature has been criticized (R.M. Burton, *Lipids*, 6 (1971) 862). For a further comment on nomenclature of sphingosine derivatives see K.-A. Karlsson (*Lipids*, 5 (1970) 878).

sphingosine has an identical structure except for a *trans* double bond at C-4. Similarly the third LCB commonly encountered in nature becomes 4*D*-hydroxysphinganine (VIII, Table 3).

Additional rules relevant to particular substituents, such as neuraminic acid, are given in the original publication[19]. In the naming of individual compounds the only trivial name still allowed is sphingosine. If other trivial names are used, they should be defined.

3. STRUCTURES, ISOLATION AND CHEMICAL SYNTHESES OF NATURAL PHOSPHOLIPIDS

The individual phospholipids are considered in the order presented in Table 1. In general, only brief reference is made to the relevant work concerned in the isolation and structural determination of each phospholipid. Except for recent discoveries of new phospholipids more refined procedures involving a combination of separations on silicic acid columns or DEAE cellulose columns and TLC (described in Chapter 3) are available and should be used since by this means purer preparations, often in higher yield, are obtained. The brief comment offered on the chemical syntheses of the natural phospholipids is derived mainly from a number of earlier[20, 22, 23, 39, 40] and very recent[30, 41, 42] reviews.

A. Phosphoglycerides

(*1*) *Phosphatidic acid and its analogues*

Phosphatidic acid and its analogues include the simplest of the phosphoglycerides in which phosphate is in a monoester linkage with the 3-position of (a) 1,2-diacylglycerol to give phosphatidic acid (PA; 1,2-diacyl-*sn*-glycerol-3-phosphate; III, Table 2); (b) 1-, or 2-acylglycerol to give lysophosphatidic acid (lyso PA; 1- or 2-acyl-*sn*-glycerol-3-phosphate; IX, Table 4) and (c) 1-alk-1′- enyl- (or 1-alkyl), 2-acyl-glycerol to give 1-alk-1′- enyl- (or 1-alkyl-), 2-acyl-*sn*-glycerol-3-phosphate (X, XI, Table 4). Chemically this group may be considered to contain the basic phosphoglyceride structures of all the remaining phosphoglycerides. This relationship also seems to hold biologically since PA and its analogues, in general, act as metabolic precursors for the natural phosphoglycerides.

The earlier possibility that PA might be a characteristic plant lipid ended with the demonstration that the PA of plants[4] is largely an artefact formed by phospholipase D action[20, 26]. For some time the existence of PA in fresh mammalian tissues was questioned even though *in vitro* evidence supported its accumulation[12] and synthesis[43]. With the emergence of a key role for PA in the biosynthesis of lipids, an active and successful search was made to show its presence in tissues[44–46]. Hübscher and colleagues[45, 46] successfully isolated PA from liver by silicic acid column chromatography. This preparation contained relatively large quantities of unsaturated fatty acids (mainly linoleic). Quantitative analyses with improved methods now show that most natural sources (mammalian tissues[47–51], insects[29], plants[26], fungi[52] and bacteria[25, 53]) contain small amounts of PA (1–5% of total phospholipids). Apart from chemical syntheses the best way to obtain significant quantities of PA is through the action of phospholipase D on pure natural

Table 4
Structures of the phosphoglycerides

(a) *1-acyl, or 2-acyl (lyso)*

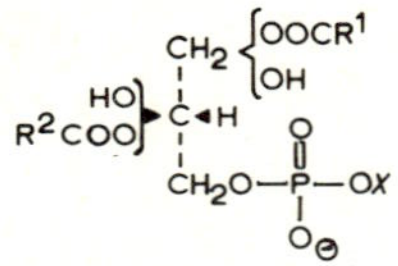

In structure:

IX X, $-H$

XIII X, $-CH_2CH_2N^{\oplus}(CH_3)_3$

XVI X, $-CH_2CH_2N^{\oplus}H_3$

XXXIV X,

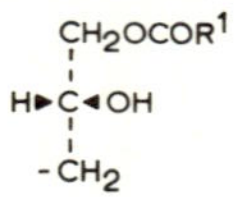

(b) *1,2-diacyl (or phosphatidyl)*

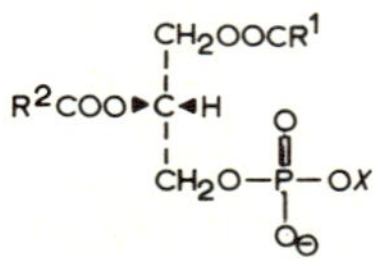

In structure:

III X, $-H$

XII X,

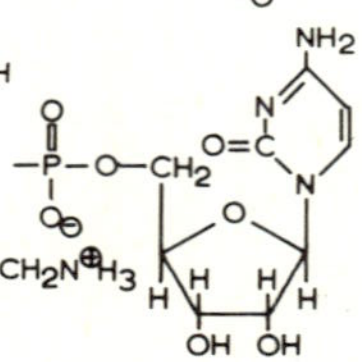

XVII X, $-CH_2CH_2N^{\oplus}H_3$

XX X, $-CH_2CH_2N^{\oplus}H_2CH_3$

XXII X, $-CH_2CH_2N^{\oplus}H(CH_3)_2$

XXIII X, $-CH_2CH_2NHCOR^5$

XXIV X, $-CH_2CH(N^{\oplus}H_3)COO^{\ominus}$

XXVII X, $-CH_2CH_2N^{\oplus}H-CH(CH_3)COO^{\ominus}$

XXVIII
Y, $-H$
XXX
Y, $-O-P(=O)(OH)-O^{\ominus}$
} X,

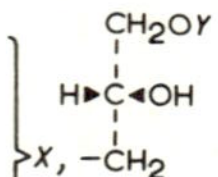

XXXII X,

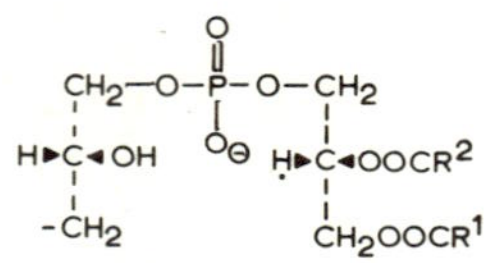

XXXIII X,

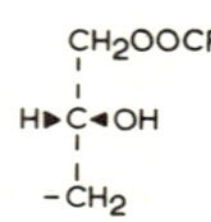

XXXV Y, $-H$
Z, β-D-glucosamine
XXXVI Z, $-H$
Y, β-D-glucosamine
XXXVII Y, $-COCHNH_2R$
Z, $-H$
} X,

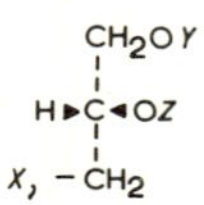

XXXVIII
Y, $-H$; Z, $-H$
XXXIX Y, $-P(=O)(OH)-O^{\ominus}$; Z, $-H$
XL Y and Z, $-O-P(=O)(OH)-O^{\ominus}$
} X,

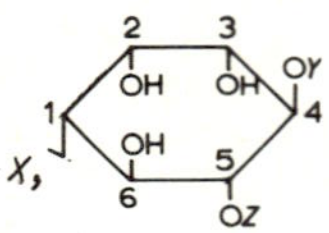

XLI X,

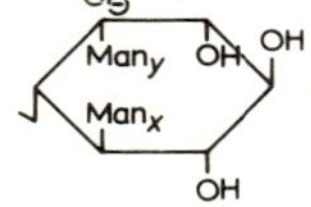

y, 1 or 2
x, 1 – 4

(c) *1-alk-1-enyl, 2-acyl-(or plasmalogen)*

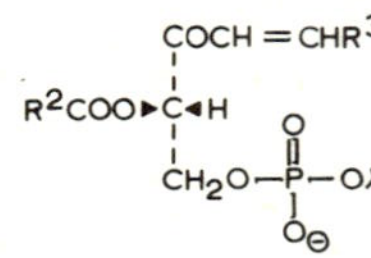

In structure:

X X, $-H$

XIV X, $-CH_2CH_2N^{\oplus}(CH_3)_3$

XVIII X, $-CH_2CH_2N^{\oplus}H_3$

XXI X, $-CH_2CH_2N^{\oplus}H_2CH_3$

XXV X, $-CH_2CH(N^{\oplus}H_3)COO^{\ominus}$

(d) *1-alkyl-, 2-acyl-*

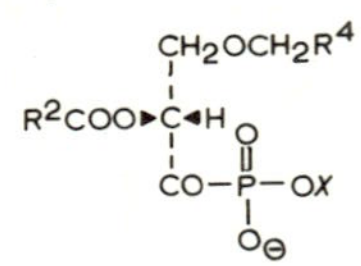

In structure:

XI X, $-H$

XV X, $-CH_2CH_2N^{\oplus}(CH_3)_3$

XIX X, $-CH_2CH_2N^{\oplus}H_3$

XXVI X, $-CH_2CH(N^{\oplus}H_3)COO^{\ominus}$

(e) *2,3-dialkyl-*

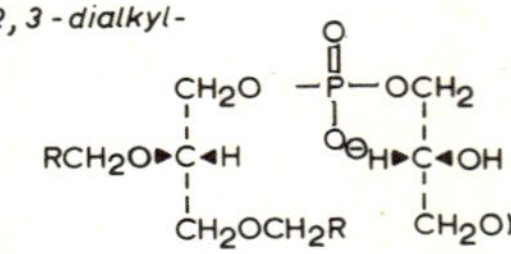

In structure:

XXIX Y, $-H$

XXXI Y, $-O-P(=O)(OH)-O^{\ominus}$

In above RCH_2O, phytanyl group
(i.e. $CH_3CH(CH_3)CH_2(CH_2CH_2CH(CH_3)CH_2)_3CH_2O$)

R^1COO-,	$C_{14}-C_{22}$ acyl group (mainly saturated, some may be odd-numbered or branched)
R^2COO-,	$C_{14}-C_{22}$ acyl group (unsaturated predominate)
$R^3CH=CHO-$,	1-alk-1-enyl (C_{16} or C_{18})
R^4CH_2O,	long-chain alkyl (C_{16}, C_{18})
R^5CO,	acyl group

lecithin, followed by isolation of the PA by extraction and precipitation[54, 55] or by silicic acid and DEAE-cellulose column chromatography[21, 30].

The first reliable chemical synthesis of PA with the correct configuration (i.e. 3-*sn*) was achieved by Baer[40]. Both enantiomeric forms and racemic mixtures of PA were synthesized by phosphorylation of the appropriate diacylglycerol with diphenylphosphoryl chloride and removal of the phenyl groups by catalytic reduction with hydrogen. PA's with unsaturated fatty acids have been prepared by the phosphorylation of diacylglycerol with phosphorus oxychloride. This method often produces undesirable side products which require rigorous purification to remove. An alternative, and often preferable, method which can be used for all acid combinations involves a reaction between 1,2-diacyl-*sn*-glycerol-3-iodohydrin and silver salts of di-*p*-bromobenzylphosphate or di-*t*-butylphosphate[30].

Although two alternative pathways for the *in vitro* formation of lyso-PA (acylation of *sn*-glycerol-3-phosphate[28, 56]; reduction of acyl dihydroxyacetone phosphate[57]) have been described in a variety of membrane or particulate systems[56], it is questionable if this lipid exists in significant amounts in intact cells.

The alk-1′-enyl ether grouping has been detected in the PA fraction of liver[45]. Recently considerable interest has been aroused in this phosphoglyceride because of its potential role as a precursor in the biosynthesis of plasmalogens. To date, only the formation (in liver and brain[57] and Ehrlich ascites[58]) of 1-*O*-alkyl-*sn*-glycerol-3-phosphate and its 2-acyl derivative has been demonstrated.

(2) Cytidylic phosphoglyceride (CDPdiglyceride)

Cytidylic phosphoglyceride (CDPdiglyceride), as structure XII (Table 4) shows, is a condensation product of cytidylic acid (CMP) and PA. This 'liponucleotide', because of its rapid metabolic utilization does not normally accumulate in sufficient amount to permit isolation. Enzyme systems capable of CDPdiglyceride formation (which presumably provide some index of its occurrence) have been described for bacterial, plant and animal preparations[56]. Different CDPdiglycerides varying in their fatty acid composition have been chemically synthesized from CMPmorpholidate and *rac*-PA[59].

(3) Choline phosphoglycerides

At least four types of choline phosphoglycerides occur naturally. These include (a) a *monoacyl derivative* (1- or 2-acyl-*sn*-glycero-3-phosphorylcholine; lysophosphatidylcholine or lysolecithin; lyso-PC; XIII, Table 4); (b) a *diacyl derivative* (1,2-diacyl-*sn*-glycero-3-phosphorylcholine; phosphatidylcholine or lecithin; PC; IV, Table 2); (c) an *alk-1-enyl-acyl derivative* (1-alk-1′-enyl-, 2-acyl-*sn*-glycero-3-phosphorylcholine or choline plasmalogen, CP; XIV, Table 4) and (d) an *alkyl, acyl derivative* (1-alkyl-, -2-acyl-*sn*-glycero-3-phosphorylcholine; XV, Table 4). Of these four, PC occurs more widely and in greatest abundance. The nitrogen-base choline has a pK near 13 and in its ester linkage with the phosphoric acid radical it imparts a strongly zwitterion character to these phosphoglycerides over the entire pH range. The acyl, alk-1′-enyl and alkyl residues tend to range from C_{14} to C_{22} in chain lengths. For PC from many sources saturated fatty acids (16:0 and 18:0) predominate at the 1-position and unsaturated fatty acids (monoenoic 16:1, 18:1, dienoic 18:2 and polyenoic 20:4 etc.) predominate at the 2-position.

(a) *Monoacyl derivative (lyso-PC)* – is a minor component (1 to 5%) of the phospholipids of many tissues[21,47–51,60]. One of the highest concentrations reported (17%) is that found in the chromaffin granules of bovine adrenal gland[28]. There is evidence that 1-acyl, or 2-acyl (or both) derivatives may arise because of the action of specific phospholipases[24,61]. Characterization of lyso-PC has been carried out with physico-chemical methods, but the most sensitive and best methods employ the use of phospholipases A and/or C[30].

Most preparations of lyso-PC from natural sources depend on the conversion of PC to the lyso compound by the action of snake venom phospholipase A and subsequent purification by column and/or preparative TLC. Although a number of chemical syntheses of lyso-PC have been carried out using glycerol derivatives blocked in the 1- or 2-position, the most versatile approach involves the prior synthesis of specific PC's (see next section) and removal of the fatty acid from the 1- or 2-position by action of pancreatic lipase or snake venom phospholipase A[30].

(b) *Diacyl derivative (PC)* – From the identification of hydrolysis products a provisional structure was postulated as early as 1868 (ref. 3). For some time uncertainty existed as to whether there were two forms of lecithin ('α' where phosphorylcholine is at the 3-position and 'β' where it is at the 2-position). However, as stated by Hanahan[20] 'the existence of β-lecithins was based on a misinterpretation of the results of the hydrolytic procedure used in structure proof'. By use of the chemical and biochemical approaches outlined by Hanahan[20,23], natural PC has been shown to have the *sn*-3-configuration. The first confirmation of this structure came through chemical synthesis of the correct enantiomeric form by Baer and Kates[40]. A similar proof has been produced for the deacylation product, *sn*-glycerol 3-phosphorylcholine (GPC).

The first relatively pure preparations of PC were obtained by precipitation from alcoholic solutions of the $CdCl_2$ complex. More recently PC has been isolated from sources such as egg yolk or ox brain (low in alk-1-enyl derivative) by passage through alumina columns to remove the cephalin components followed by separation on silicic acid to remove lyso-PC and sphingomyelin[20,21]. Alternatively the lipid classes may be separated on DEAE cellulose prior to final separation of the PC on silicic acid or by preparative TLC[14]. PC is one of the principal phospholipids of mammalian tissues[47–51,60] and plants[26]. It is generally absent from heterotrophic bacteria and hence, fails to qualify as a universal membrane constituent[28,53]. Studies with phospholipase A show a preponderance of unsaturated fatty acids at the 2-position.

Chemical synthesis of PC has been achieved by the reacylation (with specific acyl chlorides) of GPC (as its $CdCl_2$ complex) derived from the large scale deacylation of pure PC[30]. Other syntheses follow routes similar to those for PA up to the formation of the phosphorylated derivative. In the synthesis of PC the phosphate is blocked except for one reactive chloro- or hydroxyl group which can then be reacted with appropriate reagents to give the choline substituent[22,30]. Mixed acid variations are best achieved through removal of the 1-acyl or 2-acyl groups by pancreatic lipase or phospholipase A and subsequent reacylation of the lyso PC.

(c) *Alk-1′-enyl-, acyl (CP)* – Detailed information on the plasmalogen (including CP) may be obtained from a number of reviews[62–65]. Discussion of their characterization appears in the section on ethanolamine plasmalogen (EP). CP is present in appreciable amount in heart muscle (up to 40% of choline phosphoglycerides)[62,66] and ram seminal fluid[67], and in smaller amounts in the nervous system[68], plasma and red blood cells[59]. The first isolation of pure CP was achieved by Gottfried and Rapport[69] who carried out selective hydrolysis of the PC component of a beef heart lecithin preparation with *Crotalus atrox* venom (which does not react with CP) and then separated the lyso-PC formed from CP on silicic acid. Further discussion on the use of selective hydrolysis for the separation of plasmalogens is given for ethanolamine plasmalogen (EP).

The chemical synthesis of plasmalogens has been difficult to achieve[30,42]. Neutral plasmalogens (1-alk-1′-enyl-, 2,3,-diacylglycerol) have been synthesized and then deacylated at the 3-position by pancreatic lipase to yield 1-alk-1′-enyl-, 2-acyl-glycerol. The latter has been phosphorylated with 2-bromoethylphosphoric acid dichloride. Subsequent quaternization with trimethylamine has yielded the choline derivative.

(d) *1-Alkyl, 2-acyl* – There is good evidence that small amounts of this derivative accompany PC[65,70]. Appreciable quantities (20–50%) of this phosphoglyceride occur amongst the choline phosphoglycerides of Tetrahymena, various species of molluscs, crustacean nerve, bovine erythrocytes and bone marrow. Lesser quantities occur in most animal tissues. The fatty acids are of the unsaturated type in animal tissues and of the saturated type in molluscs. The available evidence indicates that the glycerol ether moiety is of the 1- (or α-) alkyl type since chimyl and batyl alcohols can be derived from the phosphoglyceride. Viswanathan et al.[31,31a] have indicated that characterization of the PC analogues is possible by conversion, after phospholipase C treatment, of these analogues into their diglyceride or 'diglyceride-like' acetates and subsequent separation and characterization of these acetates by two-dimensional reaction TLC and examination of their IR spectra. In two-dimensional reaction TLC[31a] the acetates after application to the thin layer plate are exposed to HCl fumes to cleave the alkenyl or 'aldehydogenic' group prior to running in the first dimension. Then the fatty acid esters of the acetates are inter-esterified on the plate by spraying with sodium methoxide prior to running in the second dimension. The latter workers[31a] have tentatively identified a dialkyl derivative in beef heart lipids.

(4) Ethanolamine phosphoglycerides

Analogous structures and names to those already described for the choline phosphoglycerides also exist for the ethanolamine phosphoglycerides. The only difference in the structures (XVI, XVII, XVIII, and XIX, Table 4) and names is the replacement of choline by ethanolamine. These phosphoglycerides do not exhibit zwitterion properties (as shown) over all pH ranges but may exist in ionic forms. The pattern of fatty acid distribution generally shows a higher degree of unsaturation than that of the corresponding choline phosphoglycerides.

References p. 39

(a) *Monoacyl derivative (lysophosphatidylethanolamine; lyso-PE; XVI)* – Small amounts of lyso-PE have been reported to occur in certain mammalian tissues[49–51] and fungi[52]. Presumably the origin of lyso-PE is similar to that described for lyso-PC.

(b) *Diacyl derivative (phosphatidylethanolamine; PE; 1,2-diacyl-sn-glycero-3-phosphorylethanolamine; XVII)* – This phosphoglyceride was originally named 'cephalin' (a term applied to the alcohol-insoluble phospholipids of tissues). However, with the demonstration of the multiple nature of this fraction[6], this name (although it has persisted) should be discontinued and replaced by the correct name for each component. PE occurs more widely in nature than its choline counterpart. In mammalian and plant tissues PE usually occurs in lesser amounts than PC[26, 47–52, 60, 68]. In bacteria this phosphoglyceride is often the principal phospholipid present in contrast to PC which is rarely present[53]. Thus PE, or a closely related structure seems to be universally present in membranes.

As commented on by several authors[20, 21, 23, 24, 30] PE is not easy to purify from natural sources. Separations from choline phosphoglycerides are relatively easily accomplished on alumina columns, but separation from phosphatidylserine (PS) and EP is difficult. Rouser et al.[71] have described a column method for the separation of PS and PE. Earlier reasonably pure preparations had been obtained from brain, liver and yeast[20]. The latter preparations were sufficiently pure to give optical rotations (+ 6.1°–6.3°) comparable to those for PE synthesized with the 3-*sn*- (or L-α-) configuration. Ansell and Spanner[72a] have recently described a method for the isolation of PE from brain tissue. Studies with phospholipase A also show a preponderance of unsaturated fatty acids at the 2-position.

Several methods have been used for the chemical synthesis of PE[30]. Only recently has reasonable success been achieved starting with GPE. The amino nitrogen is blocked with phthaloyl or trityl groups before acylation. This method yields PE with two identical fatty acids. Mixed acids can be obtained by use of phospholipase A and reacylation. Three other approaches have been utilized to synthesize PE's: (a) the phosphorylation of diacylglycerols with appropriate *N*-substituted derivatives of phosphorylethanolamine or of isopropylidene glycerols with the *N*-substituted derivative followed by removal of the isopropylidene group and acylation, (b) the condensation of PA with appropriate *N*-substituted ethanolamines, and (c) the coupling of 1,2-diacyl-*sn*-iodohydrin with silver salts of suitably protected phosphorylethanolamine derivatives. The latter approach is quite attractive for mixed acid PE's.

(c) *Alk-1′-enyl, acyl derivative (ethanolamine plasmalogen; EP; 1-alk-1′-enyl, 2-acyl-sn-glycero-3-phosphorylethanolamine; XVIII)* – By the mid 1950's a number of studies indicated that the acetal phosphatide isolated after alkaline treatment from beef muscle[5] and beef brain[72] did not possess the properties of native plasmalogen. Klenk and Debuch[62] and Rapport et al.[63] independently showed that plasmalogen contained one fatty acid. Lyso-PC and lyso-PE were obtained (after treatment with acetic acid at 37° to remove the aldehyde group) from beef heart and beef brain preparations, respectively. The aldehydes released were mainly C_{16} and C_{18} with smaller amounts of C_{14}.

Information on the position of the residues, and hence establishment of a 3-*sn*-structure came from several independent studies. Thannhauser et al.[72] isolated GPE from their crystalline preparation of acetal phosphatide (now believed to be lyso-PE) and measured its optical rotation. The latter proved to agree well with the value for chemically synthesized 3-*sn*-GPE[20]. Evidence that an alk-1′-enyl ('vinyl' ether, or α,β-unsaturated ether) chain gave rise to the aldehyde came from studies in which it was shown that the aldehydogenic residue (a) produced an alkyl ether on hydrogenation (b) produced C_{15} and C_{17} fatty acids when attacked by ozone (fatty acids were considered to derive from C_{16} and C_{18} alk-1′-enyl ether linkages) and (c) responded in an iodination reaction which differentiates alk-1-enyl unsaturation from olefinic bonds in a carbon chain. The assignment of the position of the alk-1′-enyl chain proved to be a difficult problem. Periodate oxidation of the saturated glycerol ether (derived after hydrogenation and deacylation of the parent plasmalogen) showed it to possess a linkage at the 1- (or α-) position. Once the specificity of the snake venom phospholipases was firmly established to be at the 2-position, the observations that these enzymes cleave fatty acids from plasmalogens provided confirmatory evidence for a 2-acyl-group. The alk-1′-enyl lipids (including the phosphoglycerides) have been shown to possess the *cis*-configuration[63]. Craig et al.[73] have confirmed with optical rotatory dispersion studies that the absolute configuration for natural plasmalogens is that depicted by structures X, XIV and XVIII.

Plasmalogens occur quite widely in nature[62–65]. EP is present in relatively high concentrations in the white matter (particularly myelin) of the nervous system and in lesser amounts in heart muscle and other tissues of mammals. Many marine animals contain this plasmalogen. EP is difficult to separate from its other analogues. Renkonen[74] indicates that the most promising approaches to the isolation of pure EP are those dependent on selective chemical hydrolysis (e.g. mild alkaline hydrolysis where the order of cleavage of ester bonds is diacyl > alk-1′-enyl, acyl > alkyl, acyl) or enzymatic hydrolysis (certain phospholipase A, C and D preparations attack the diacyl compound more readily than the plasmalogen or alkyl, acyl). Conventional column and thin-layer chromatography is then used to separate the deacylated products from the intact plasmalogens or alkyl, acyl phosphoglycerides.

(d) *Alkyl-, acyl-derivative (1-alkyl-, 2-acyl-sn-glycero-3-phosphorylethanolamine*[21, 23, 65]*; XIX)* – The first direct evidence for the occurrence of an alkyl ether phosphoglyceride came in 1958 when a phosphorylethanolamine derivative of batyl alcohol (glyceryl-1-octadecyl ether) was isolated from egg yolk[74a]. Shortly afterwards evidence was obtained for the presence of a similar lipid in brain and heart. Later Hanahan's group isolated an alkyl, acyl phosphoglyceride from bovine erythrocytes in sufficient quantity to do a fairly extensive study of its properties and structure. The 1-glyceryl ether structure was confirmed and reasonable evidence was presented for a 2-acyl group. Thus, the alkyl ether lipid appeared to have a 1-alkyl-, 2-acyl-*sn*-glycero-3-phosphorylethanolamine structure (XIX, Table 4). In addition to the above sources this phosphoglyceride is reasonably abundant in bone marrow, certain protozoa and molluscs. In brain both an alkyl, acyl and a dialkyl derivative are present[75]. Isolation procedures depend on a preliminary selective hydrolysis by alkali and acid of the diacyl and alk-1′- enyl-, acyl analogues. Both the ethanolamine and choline alkyl, acyl analogues have been synthesized chemically[76]. In the syntheses by Chacko and Hanahan[76] 2-stearoyl selachyl alcohol was converted to the phosphoryl dichloride before reaction with *N*-phthaloylethanolamine or choline iodide.

References p. 39

(5,6) N-methyl- and N,N-dimethylethanolamine phosphoglycerides (N-methyl and N,N-dimethyl phosphatidylethanolamines)

The above classes have either *N*-methylethanolamine or *N,N*-dimethylethanolamine substituted for ethanolamine in their structures (XX, XXI, XXII, Table 4). The *N*-methyl derivative occurs in both the diacyl and alk-1-enyl, acyl forms. In 1959 attention was focused on the possible importance of the above phosphoglycerides in interconversions from PE to PC with the isolation of mono- and dimethylethanolamine from rat liver phospholipids[77]. More recent assessment by Lester and White[78] has shown the above phosphoglycerides to be present in trace quantities in rat tissues and *Hemophilus influenzae* and in significant quantity in yeast. A number of bacteria contain one or both of these lipids[29].

The purest and best characterized preparations are those isolated from choline-requiring mutants of *Neurospora crassa*[79] and from *Clostridium butyricum*[80], yeast[81] and *Brucellus abortus* Bang[82]. In all instances repeated or combination runs involving silicic acid, alumina, DEAE cellulose or TLC were required. Metaperiodate oxidation studies on the deacylation products and hydrolyses of these phosphoglycerides by snake venom phospholipase A strongly suggested a 1,2-diacyl-*sn*-glycero-3-phosphoryl configuration. It is of interest to note that *C. butyricum* accumulates only the *N*-methyl derivative with 78% of it in the plasmalogen form. A number of chemical syntheses have been described for both the *N*-methyl- and *N,N*-dimethyl derivatives[30] which parallel those for PE with the exception that reagents with *N*-methyl or *N,N*-dimethyl substitution (or capable of these substitutions) must be used for the addition of the ethanolamine or phosphoryl-ethanolamine moiety.

(7) N-acyl ethanolamine phosphoglyceride (N-acyl phosphatidylethanolamine; N-acyl PE; XXIII)

In 1965 Bomstein[83] isolated by successive chromatography (over silicic acid, alumina, silica gel G) two unknown phosphoglycerides from wheat flour, one of which he identified as *N*-acyl lyso-PE and the other probably as *N*-acyl PE. Since this report Dawson et al.[84] and Aneja et al.[85] have described the isolation and characterization of *N*-acyl PE from pea seeds and soya beans, respectively. The *N*-acyl PE from peas was characterized by examination of the products released by stepwise degradation with acid and alkali and comparison of its properties with those of chemically synthesized *N*-stearoyl PE. Chromatographic evidence was obtained for the existence of *N*-acyl PE in oats, soya beans and spring (tick) beans. Aneja et al.[85] isolated their *N*-acyl PE from a soya bean 'lecithin' preparation by TLC using silica gel G impregnated with triethylamine. They compared their product with a synthetic *N*-stearoyl-*O*- (1,2-dimyristoyl-*sn*-glycero-3-phosphoryl) ethanolamine. Both p.m.r. spectra and optical rotation values supported the contention that this phosphoglyceride has the structure (XXIII): *N*-acyl-*O*- (1,2-diacyl-*sn*-glycero-3-phosphoryl) ethanolamine.

(8) Serine phosphoglycerides[20–24]

Of the amino acid-containing phosphoglycerides those with serine (particularly the diacyl compound) occur most widely in nature. Evidence exists for at least three different natural serine phosphoglycerides: (a) the diacyl derivative, (i.e. phosphatidylserine; PS; 1,2-diacyl-*sn*-glycero-3-phosphorylserine; XXIV, Table 4), (b) the alk-1-enyl, acyl derivative (serine plasmalogen; SP; XXV, Table 4) and (c) the corresponding alkyl, acyl derivative (XXVI). It is probable that under certain metabolic conditions lysophosphatidylserine may exist.

In 1948 Folch[86] succeeded in isolating from brain 'cephalin' by means of solvent fractionation a preparation of PS that was about 92% pure. The isolated PS was obtained as a $Na^{\oplus}$– $K^{\oplus}$ salt (mainly $K^{\oplus}$). From chemical studies Folch concluded that PS contained two fatty acids (mainly stearic and oleic), glycerophosphate and L-serine bound through its OH group. These data supported a diacylglycerylphosphorylserine structure. The exact structure was established by Baer and Maurukas[87] who hydrogenated natural PS (from brain) and showed that its optical rotation was sufficiently similar to synthetic 3-*sn*-PS to assign the 3-*sn*-configuration. Approaches similar to those available for the chemical synthesis of PE are also available for PS. The main difference is the use of appropriately blocked reagents that give rise to serine or phosphorylserine[30].

PS occurs quite widely in nature but usually in concentrations less than 10% of the total phospholipids. White matter (or myelin) of the nervous system has higher amounts. PS occurs in low amounts in plants[26] and in trace quantities in bacteria[25, 53]. The plasmalogen form is present in very low concentrations in brain[21, 68] and possibly red blood cells[60, 87a]. Isolation of pure natural PS free from PE has only been achieved in recent years through a combination of consecutive runs on columns and alterations in the cationic environment[24, 71].

(9) N-2-(Hydroxyethyl) alanine phosphoglyceride (phosphatidyl-N (2-hydroxyethyl)-alanine)

This new phospholipid was isolated by Kemp and Dawson[88] from a mixed rumen protozoal fraction and also from cultures of *Entodinium caudatum*. The phospholipid (as its calcium salt) separated with PS and phosphatidylinositol (PI) and a pigmented material on silicic acid. On conversion of the lipids to their sodium salts and a repeat separation on silicic acid, a pure phospholipid was obtained. Fatty acid (mostly palmitic, stearic and *trans*-octadec-11-enoic) to phosphorus ratios and the area per molecule of the intact phospholipid in unimolecular films agreed with a diacyl phosphoglyceride structure. The deacylation product (from alkaline hydrolysis) did not correspond to any known phospholipid. It had one net negative charge at pH 3.6 and was cleaved by 1 *M* HCl into α-glycerophosphate and a ninhydrin reacting substance. Subsequent isolation and characterization of this unknown substance by NMR spectra and chemical synthesis established it as L-*N*-(2-hydroxyethyl)alanine. Therefore, structure XXVII (Table 4) is the most probable structure for this new phospholipid.

(10) Glycerol phosphoglycerides (phosphatidylglycerol and dialkyl analogue)[89, 90]

These phosphoglycerides have a second glycerol attached to the phosphate through the 1′-position in phosphatidylglycerol (PG; 1,2-diacyl-*sn*-glycero-3-phosphoryl-1′-*sn*-

References p. 39

glycerol; XXVIII, Table 4) and the 3′-position in the unique dialkyl analogue (2,3-dialkyl-*sn*-glycero-1-phosphoryl-3′-*sn*-glycerol, XXIX, Table 4).

PG, the simplest of the polyglycerophospholipids, was first isolated by Benson and Maruo[91] from Scenedesmus. From examination of the deacylation products of PG and acetonated PG, they concluded that this lipid possessed a phosphatidyl structure. This lipid occurs widely, but great differences exist in the amounts present[24–26, 53, 89–92] in animal (very low amounts and mainly in mitochondria), plant (20–30% of phospholipids, 40–60% in the chloroplast) and bacterial (trace amounts to 70% of total phospholipids) membranes. Although PG was initially isolated by two-dimensional paper chromatography, it is generally better to use repeated chromatography on silicic acid or some combination of DEAE cellulose, silicic acid and TLC.

Biosynthetic studies with mammalian and bacterial systems indicated that PG should have the 3-*sn*, 1′-*sn* structure for its two glycerols[24, 27]. Benson's group[92] considered that plant PG also had this structure, but this was only later proven by Haverkate and Van Deenen[93] for PG from spinach leaves. By examination of the products of chemical degradation and the action of phospholipases C and D, they confirmed the above structure and showed that the main molecular species was 1-linolenyl,2-Δ^3-*trans*-hexadecenoyl-*sn*-glycero-3-phosphoryl-1′-*sn*-glycerol. Several chemical syntheses of PG have been described[30, 40]. Where the stereochemical configuration of natural PG is desired, the best method involves reaction of the silver salt of fully protected *sn*-glycerol 1-phosphate with 1,2-diacyl-*sn*-glycerol 3-iodohydrin followed by removal of the protecting groups.

The dialkyl analogue of PG forms a small (5%) proportion of the phospholipids from extremely halophilic bacteria[13, 90, 94]. The pure lipid has been isolated (by silica-impregnated paper or preparative TLC) and its structure characterized and confirmed by chemical synthesis. Two unique features were observed: (a) the alkyl halides released by HI treatment were shown to derive from identical phytanyl (3′,7′,11′,15′-tetramethylhexadecyl) groups and (b) the stereochemical configuration proved to be opposite to that usually encountered.

(11) Glycerophosphate phosphoglycerides (phosphatidylglycerophosphate and dialkyl analogue)

The structures of these phosphoglycerides are comparable to those of PG with the exception that phosphate is substituted at the 3′- or 1′- positions (XXX, XXXI, Table 4). The main evidence for the existence of phosphatidylglycerophosphate (PGP) has come from studies on its biosynthesis in mammalian and bacterial systems[24, 27–29]. Requirements for these systems support a 1,2-diacyl-*sn*-glycero-3-phosphoryl-1′-*sn*-glycerol 3′-phosphate structure (XXX). Small amounts of PGP have been observed in brain. PGP has been successfully synthesized by reaction of 1,3-diiodo-2-*O*-*t*-butyl glycerol with the silver salt of 1,2-diacyl-*sn*-glycerol-3-(monobenzyl)phosphate and subsequent addition of phosphate at the 3′-position via the silver salt of di-*t*-butylphosphate[30].

The dialkyl analogue of PGP forms 70–80% of the phospholipids of extremely halophilic bacteria (less halophilic have the normal diacyl component). As for dialkyl PG, Kates et al.[13, 90, 94] have isolated this phospholipid in a pure state and characterized its structure as 2,3-diphytanyl-*sn*-glycero-1-phosphoryl-3′-*sn*-glycerol 1′-phosphate. This structure has been confirmed by chemical synthesis.

(12) Phosphatidylglycerol phosphoglyceride (diphosphatidylglycerol)[24, 30, 89, 95]

This phosphoglyceride has phosphatidylglycerol linked to the basic phosphoglyceride (or phosphatidyl) unit thus making it diphosphatidylglycerol (cardiolipin, CL; XXXII, Table 4). Its more specific designation is bis-(1,2-diacyl-*sn*-glycero-3-phosphoryl)-1′,3′-*sn*-glycerol. Each of the phosphates has one dissociable group which accounts for the lipid's acidic properties and ability to bind cations.

Ever since Pangborn's original isolation[7] of cardiolipin (CL) a variety of isolation methods have been[89] and continue to be described[30, 95] for this lipid. The more recent methods use repeated column chromatography or recrystallizations of the barium salt followed by chromatography on ammonia- or bicarbonate-treated silicic acid. Eichberg and Burnham[95] have obtained CL in good yield (160 mg/100 g fresh tissue) from a heart proteolipid preparation by salt denaturation and gel filtration on Sephadex LH-20. Apart from heart where CL reaches 10% of the total phospholipids, animal tissues contain only small amounts (about 5% of total phospholipids) of this lipid, mostly confined to the mitochondria. Plants[26] and bacteria[25, 53] contain anything from trace amounts up to quite significant amounts. Often an inverse relationship between PG and CL concentrations may be seen in bacteria which is dependent on the conditions and phase of growth.

Structural studies have paralleled the improvements in isolation[30, 89]. Early investigations (ref. 89) could not agree on the number of glycerols and phosphates linked together in the backbone. Good provisional evidence for the diphosphatidylglycerol structure came from analytical data and periodate oxidation studies. Furthermore, examination of the products of hydrolysis by 90% acetic acid showed that they were diglycerides, glyceroldiphosphate, but not free fatty acid. From a comparison of the deacylation product with a synthetic preparation, Lecocq and Ballou[95a] concluded that the stereospecificity of the polyglycerolphosphate core corresponded to bis-(*sn*-glycero-3-phosphoryl)-1′, 3′-*sn*-glycerol. The CL from beef heart has been shown by De Haas and Van Deenen[96] to possess the diphosphatidylglycerol structure. They compared the optical rotation, IR spectra and chromatographic behaviour of the natural CL with that of chemically synthesized CL. The most conclusive proof came from studies with phospholipase C which yielded 1,2-diacyl-*sn*-glycerol and *sn*-glycero-1,3-diphosphate from both preparations. Chemical synthesis of the correct configuration was achieved by reaction of one mole of 1,3-diiodo-2-*t*-butyl-isopropanol with two moles of the silver salt of 1-stearoyl-2-oleoyl-*sn*-glycerol 3-(monobenzyl) phosphate. The blocking groups were released by anionic debenzylation and treatment with HCl[30]. The diphosphatidylglycerol structure has been confirmed for preparations from plants and yeast[30].

*(13) Mono- and diacylglycerol phosphoglyceride (semi-lyso and lyso bis-phosphatidic acids)**

In recent years evidence has been obtained which indicates the presence of small

*Semi-lyso bis-PA has been tentatively identified in Mycoplasma (Plackett et al., *J. Bacteriol.*, 104 (1970) 798) and has been isolated from *Salmonella typhimurium* and its structure established as *sn*-3-phosphatidyl-1′-(3′-acyl)*sn*-glycerol (Olson and Ballou, *J. Biol. Chem.*, 246 (1971) 3305).

quantities of semi-lyso (proposed structure XXXIII, Table 4) and lyso bis-phosphatidic acids (proposed structure XXXIV, Table 4) in certain animal tissues[97, 98] and spinach leaves[99]. Body and Gray[97], by separation of dog lung lipids on silicic acid and silica gel H, obtained two phosphoglycerides which they identified, on the basis of alkaline hydrolysis, periodate oxidation and acetolysis, to be semi-lyso and lyso-bis-PA's. Rouser et al.[98], by a combination of Sephadex and DEAE cellulose column chromatography and TLC, observed the presence of a new lipid in the livers of humans suffering from Niemann–Pick disease which they identified by alkaline hydrolysis and acetolysis to be lyso-bis-PA. Similarly, Debuch and Rose[99] obtained a polyglycerolphospholipid from spinach leaves which resembled a semi-lyso-bis-PA with the exception that the second glycerol is possibly linked at its 2-position.

(14) Glucosaminylglycerol phosphoglyceride (glucosaminyl phosphatidylglycerol)

Phospholipids containing glucosamine have been isolated in a pure state from chloroform–methanol extracts of *Bacillus megaterium* on silicic acid columns or silica-impregnated paper and fully characterized[100, 101]. The preparation isolated by Van Deenen's group was cleaved by nitrous acid to give PG and 2,5-anhydromannose, the deamination product of D-glucosamine. The PG was investigated with phospholipases A and C and glycerol 3-phosphate dehydrogenase and shown to be 1,2-diacyl-*sn*-glycero-3-phosphoryl-1′-*sn*-glycerol. Partial acid hydrolysis cleaved off diglyceride and the remaining product was characterized as 2′-*O*-(2-amino-2-deoxy-β-glucopyranosyl)-glycerol. Final proof was provided by chemical synthesis of the correct stereoisomers of glucosaminylglycerol and glucosaminylglycerophosphate. In a separate study McDougall and Phizackerley[101] have isolated two glucosamine phospholipids and established their structures. One is similar to that described above and has the structure (XXXV, 1′-(1,2-diacyl-*sn*-glycero-3-phosphoryl)-2′-*O*-β-(2-amino-2-deoxy-D-glucopyranosyl)-*sn*-glycerol and the second has the structure (XXXVI), 1′-(1,2-diacyl-*sn*-glycero-3-phosphoryl)-3′-*O*-β-(2-amino-2-deoxy-D-glucopyranosyl)-*sn*-glycerol. Structure XXXV is reported to comprise 15% of the total phospholipids under acid culture conditions (∼ pH 5) (ref. 100).

(15) O-Amino acid esters of glycerol phosphoglycerides

In 1962 MacFarlane[102] succeeded in isolating from the lipids of *Clostridium perfringens* by successive runs on silicic acid and magnesium silicate columns ninhydrin-positive phosphoglycerides with properties similar to PG. Very brief treatment with weak alkali released a mixture of amino acids (mainly alanine) and an ether-soluble phosphorus component that was oxidizable with periodic acid. Saponification released glycerylphosphorylglycerol in addition to the amino acid and acid hydrolysis released glycerophosphate. These chemical studies led to the conclusion that the new phosphoglycerides were *O*-amino acid esters of PG. The amino acid was arbitrarily assigned to the 3′-position (XXXVII, Table 4).

Subsequently Van Deenen's group[24, 30, 90, 103] demonstrated that under certain growth conditions (< pH 5), *Staphylococcus aureus* accumulated the lysine ester of PG. In a single run on silicic acid the lysine ester was separated in a pure state. Very mild alkaline treatment released lysine which was characterized as L-lysine by its response to L-lysine carboxylase. The PG released was characterized as described in the previous section. These studies proved that the lipid accumulating was an L-lysine ester of 1,2-diacyl-*sn*-glycero-3-phosphoryl-1′-*sn*-glycerol. A reasonably precise assignment of the *O*-amino ester on the second glycerol has been achieved by comparison of the natural L-alanyl and L-lysyl esters with chemically synthesized phosphoglycerides of the same stereochemical configuration containing these amino acids in the 2′- and 3′-positions[30, 104]. Molotkovsky and Bergelson[104] concluded that alanine and lysine are located at the 3-position. Chemical synthesis of a lipid with the correct configuration was achieved by reaction of the protected silver salt of PA with the appropriate protected amino acid ester of 3-benzyl-*sn*-glycerol-1-iodohydrin and subsequent removal of the protecting groups.

The occurrence of these phosphoglycerides is primarily confined to the Gram-positive bacteria. A variety of amino acid esters have been described but the L-alanyl, L-lysyl and L-ornithyl esters of PG are best characterized. Some evidence has been obtained for a D-alanyl ester[30].

Inositol-containing phospholipids – The early work on the inositol-containing phospholipids is described in detail in other reviews[20, 105]. Prior to the mid 1950's the available evidence indicated that these phospholipids occurred in sources such as tubercle bacillus, soya beans and brain. Due to the inadequacy of techniques to provide pure preparations, little of a definitive nature was achieved in establishing the true structures of these phospholipids until the period 1954–1961. Several reviews describe the interesting and challenging developments associated with the isolation and establishment of the structures of these phosphoglycerides[20, 24, 27, 105–109]. A recent proposal has been made on the nomenclature of the inositol-containing phospholipids which suggests that stereospecific numbering with an *sn*-designation be used for the derivatives of *myo*-inositol[110]. Although not used here, such a proposal has some merit and may prove to be as useful as the *sn*-designation for glycerol. These phosphoglycerides are discussed in the order appearing in the classification in Table 1.

(16) Inositol phosphoglyceride (phosphatidylinositol)

This phosphoglyceride has inositol linked through its 1-hydroxyl to a phosphatidyl group to give the specific structure (XXXVIII, Table 4), 1,2-diacyl-*sn*-glycero-3-phosphoryl-1-*myo*-inositol (phosphatidylinositol, PI). Because of its acidic nature, PI usually occurs as a salt in association with monovalent ($Na^{\oplus}$, $K^{\oplus}$) or divalent cations ($Ca^{2\oplus}$, $Mg^{2\oplus}$).

In attempts to obtain relatively pure PI a variety of solvent and column chromatographic procedures have been applied to the mixed phospholipids of different sources (brain, heart, yeast, wheat germ, peas etc.). The best separations have been achieved by repeated runs on silicic acid columns or a preliminary run on alumina followed by silicic acid[20, 106]. Examination of these preparations with hydrolytic procedures revealed the presence of fatty acids, glycerol, *myo*-inositol and phosphate in a ratio of 2:1:1:1. PI from animal tissues is rich in stearic and polyunsaturated acids. Mild alkaline hydrolysis yielded glycerylphosphorylinositol which was shown to be optically active and to have glycerol in 3- (or α-) linkage to phosphate. Acid hydrolysis under appropriate conditions released inositol phosphate and diglyceride. Work from the laboratories of Ballou and of Brown (ref. 106) established that the inositol phosphate from soya bean, liver and brain preparations contained the D-*myo*-inositol-1-phosphate structure. It is notable that Brown and Clark[111] have isolated pure PI from brain lipids after dinitrophenylation of the nitrogen-containing phospholipids and subsequent separation on silicic acid. These workers established the structure to be that assigned for structure XXXVIII. Chemical synthesis of PI with the correct configuration and attachment of inositol has been difficult. Recently the synthesis of PI containing a phosphatidyl group linked to the 1-position of inositol has been reported[41].

PI occurs widely in nature in amounts usually corresponding to 5–10% of the total phospholipids (for animal tissues see refs. 52–56, 59, 105, 106, for plants see ref. 26, and for bacteria see refs. 25, 53).

(17, 18) Inositol mono- and diphosphate phosphoglycerides (phosphatidylinositol mono- and diphosphates)

These phosphoglycerides have the same structure as PI with the exception that the inositol has additional phosphate groups at the 4-position (inositol-4-phosphate, XXXIX, Table 4) and at the 4,5-positions (inositol-4,5-diphosphate, XL, Table 4). In 1961 attention was focussed on the true nature of the 'diphosphoinositide' fraction of Folch[8,106,112,113]. Ballou and colleagues[106,112] examined the nature of the products released by 2 *N* KOH under reflux. Through the use of acid migration studies, treatment with phosphomonoesterase and periodate oxidation conclusive evidence was obtained for the existence of inositol-1,4-diphosphate and inositol-1,4,5-triphosphate as components of the Folch phosphoinositide complex. Examination of the products released by mild alkali (as done for PI) confirmed a D-*myo*-inositol-1-phosphate structure for the phospholipids that gave rise to the above inositol di- and tri-phosphates. Thus, it was concluded that the phosphoinositide complex of beef brain (i.e. Folch's fraction) contained phosphatidylinositol, phosphatidylinositol monophosphate (1,2-diacyl-*sn*-glycero-3-phosphoryl-1-*myo*-inositol-4-monophosphate; DPI; XXXIX) and phosphatidylinositol diphosphate (1,2-diacyl-*sn*-glycero-3-phosphoryl-1-*myo*-inositol-4,5-diphosphate, TPI, XL). Independently, Dawson and Dittmer[113] showed that the 'Folch fraction' was a complex of the above phosphoinositides and succeeded in isolating pure 'triphosphoinositide'. The latter was characterized to the point where they were able to assign a diacylglycerylphosphorylinositol diphosphate structure. Brown and Stewart[114] have fully characterized the TPI prepared by the method of Dawson and Dittmer and showed it to have the structure given above. Russian workers have synthesized DPI by condensing 1,2:4,5-dicyclohexylidene-*myo*-inositol with phosphorylchloride and 1,2-distearoyl-glycerol[41].

A number of recent improvements in the isolation and purification of the phosphoinositide complex have been described. These include a salt gradient separation on DEAE cellulose[115] and repeated extractions with chloroform–methanol of lyophilized brain preparations and subsequent separation on formalin-treated filter paper[116]. DPI and TPI occur in greatest quantity in the nervous system (primarily confined to white matter or myelin)[106]. There are indications that small amounts are present in other tissues[117] (heart, lung, kidney, liver and red blood cells) and yeast[118]. Recent assessment of the molecular species of PI, DPI and TPI from beef brain shows a similar pattern for each with the 1-stearoyl-2-arachidonoyl-species representing 40% of the total[119].

(19) Monomannosyl–hexamannosyl inositol phosphoglycerides

As noted in the reviews of Ballou and Lee[108] and Lederer[109], by 1960 several laboratories had shown that Mycobacteriae (*M. tuberculosis* and *M. phlei*) contain lipids of the phosphatidylinositol type, in which one to five mannose units are linked to the inositol. These lipids, in a manner similar to the work done on DPI and TPI, were shown to have glycerol linked in the *D*-1-position of inositol. Three of the deacylation products examined contained mannose (mono-, di- and pentamannosides). In an elegant structural determination study Lee and Ballou[108, 120] have established the complete structures of what may be regarded as a family of *myo*-inositol phospholipids with increasing numbers of mannose units. Their characterization was done on the deacylated water-soluble products obtained by chromatography on paper or DEAE Sephadex. The position of substitution on the *myo*-inositol ring was established through periodate oxidation and methylation studies. Structure XLI (Table 4), shows a general structure for the family of mono- to hexamannosides. Mono- has mannose in 1 → 6 linkage and di- has an additional mannose in 1 → 2 linkage with inositol. Tri- to penta- have additional mannoses in 1 → 6, 1 → 6 and 1 → 2 linkages with the mannose at position-6 (i.e. $x = 1$ to 4). The hexamannoside which is not well characterized appears to have an additional mannose in 1 → 2 linkage with the mannose at position-2 ($y = 1$ or 2). P.m.r. spectra have confirmed that all of the glycosidic linkages are in the α-configuration. The complete name for the pentamannoside is 1-[1,2-diacyl-*sn*-glycero-3-phosphoryl]-2-[*O*-α-D-mannopyranosyl]-6-[*O*-α-D-mannopyranosyl-(1 → 2)-*O*-α-D-mannopyranosyl-(1 → 6)-*O*-α-D-mannopyranosyl-(1 → 6)-*O*-α-D-mannopyranosyl]-D-*myo*-inositol.

Pangborn and McKinney[121] have isolated two phosphatidyl dimannosides and two phosphatidyl pentamannosides from *M. tuberculosis*. In each lipid palmitic and tuberculostearic acids accounted for more than 90% of the fatty acids. Mole ratios of fatty acids to P of 3 or 4 for each of the above lipids indicated that acylation must occur beyond the phosphatidyl group. This observation has been confirmed in *M. phlei* by Brennan and Ballou[122].

(20) Glucose phosphoglyceride ('phosphatidylglucose')

Several reports have appeared in the literature on the occurrence of glucose-containing phosphoglycerides in mycoplasma and streptococci (ref. 123). Relatively pure preparations have been obtained by repeated silicic acid and DEAE cellulose column chromatography. Complete characterization has not been achieved on any of these preparations. The preparation tentatively identified as phosphatidylglucose has recently been shown to possess a different structure[123] as discussed in Chapter 11.

(21) O-diglucosylglycerol phosphoglyceride

A diglucosyl phospholipid has been purified out of chloroform–methanol extracts of *Streptococcus hemolyticus* and partially characterized[124]. The deacylation product has been identified as *O*-α-D-glucopyranosyl-(1 → 2)-*O*-α-D-glucopyranosyl-(1 → 1)-glycerylphosphoryl-1′-glycerol. Two fatty acids are present in the structure but their positions are unassigned.

B. Phosphoglycolipids

This class is intended to include the simpler glycolipids [e.g. diacyl(mono- or diglucosyl)glycerol] which may have a phosphate attached to the carbohydrate moiety. Only one preparation appears to qualify in this group. Shaw et al.[123] have recently re-examined their 'phosphatidylglucose' preparation by a variety of degradative procedures and now conclude that a more plausible type of structure is glycerylphosphoryldiglucosyl diglyceride (see also Chapter 11).

C. Phosphodiol lipids

(1) Acyldihydroxyacetone phosphate and

(2) Alkyldihydroxyacetone phosphate

The existence of the above phosphodiol lipids depends primarily on biosynthetic studies carried out in isolated systems. Hajra and Agranoff[125], in studying the nature of a lipid that is labelled from [γ-^{32}P]ATP by guinea-pig mitochondria succeeded in isolating (by silica gel and formaldehyde-treated paper) a new lipid which they characterized as acyldihydroxyacetone phosphate (XLII):

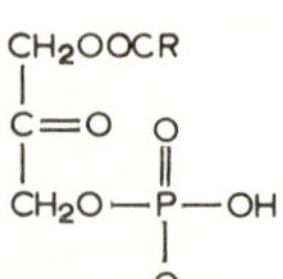

(XLII)

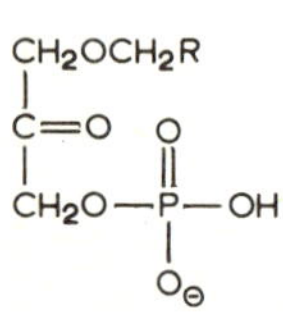

(XLIII)

In studies on alkyl ether biosynthesis two laboratories have described systems in liver and brain[126] and Ehrlich ascites cells[58] which form alkyldihydroxyacetone phosphate (XLIII). Characterization of the product was carried out using reduction with $LiAlH_4$ (to form alkylglycerylether), reduction with NADPH and resistance to acid and alkaline hydrolysis.

D. Phosphosphingolipids

These lipids contain long-chain bases (LCB's) such as sphinganine, 4-sphingenine or 4*D*-hydroxysphinganine linked to certain monophosphate esters (choline, ethanolamine, glycerol and glycerophosphate) through the primary hydroxyl and having a long-chain fatty acid linked to the amino group (i.e. *N*-acyl or amide). Such lipids qualify as candidates under both phospholipids and sphingolipids and appropriately may be classified as phosphosphingolipids. The other sphingolipids include the cerebrosides, sulphatides and gangliosides which fall outside the coverage of this volume. However, many of the problems of isolation and purification of the LCB's and much of their chemistry applies equally to each of the above classes of sphingolipids. A number of earlier reviews (refs. 2, 3; Zabin in refs. 20, 21, 23), and two recent publications (one by

TABLE 5

STRUCTURES OF SOME PHOSPHOSPHINGOLIPIDS

A. General structure

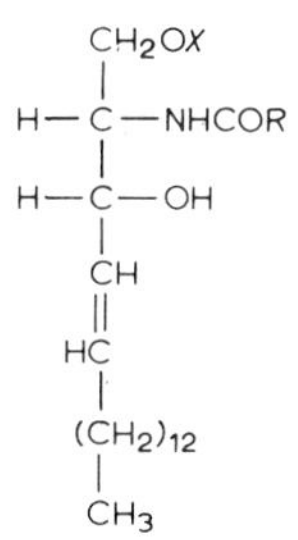

Where RCO = C_{12}—C_{24} acyl group (saturated and monounsaturated in animals; odd numbers and branched in certain bacteria) and

X,

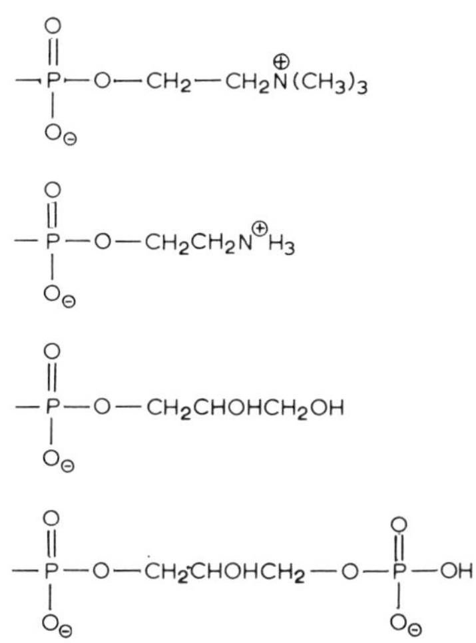

XLIV

XLV

XLVI

XLVII

B. Tetrasaccharide of phytoglycolipid XLVIII

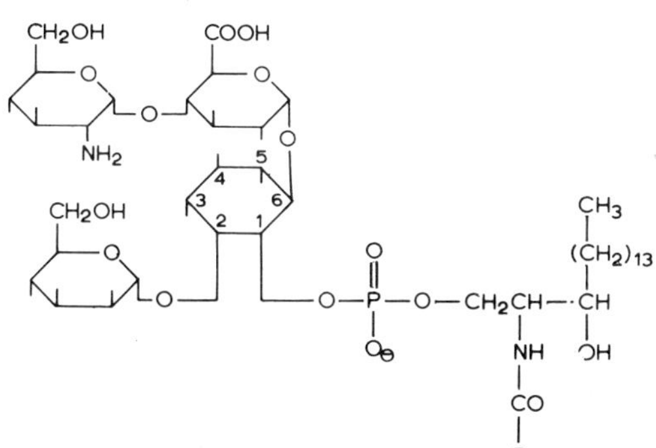

Shapiro[127] and the second, a special issue of 'Chemistry and Physics of Lipids' (Vol. 5, No. 1, 1970) containing 'a collection of papers dedicated to H.E. Carter') have provided much of the information for the presentation on phosphosphingolipids.

(a) Long-chain bases (LCB's)

Sphingosine was isolated from an alkaline hydrolysate of the cerebroside, phrenosin, by Thudichum in 1882 (ref. 127). The complete elucidation of the structure, stereochemistry and chemical synthesis of this LCB was not achieved until early in 1950 (refs. 127, 128). A number of improvements over the earlier isolation procedures (which depended on repeated recrystallizations from solvents such as diethylether) are now generally employed[129]. These include cleavage of the phosphosphingolipid with phospholipase followed by alkaline hydrolysis of the ceramide and subsequent separation of the LCB's by silicic acid column chromatography or TLC before or after conversion to their dinitrophenyl derivatives. The purified bases are further analyzed by GLC of the trimethylsilyl (TMS) derivatives of the aldehydes formed by periodate oxidation. Recently mass spectrometry has greatly facilitated structural studies on natural LCB's.

The number of LCB's characterized has increased greatly with the implementation of the above techniques[129]. The three main structures are those shown in Table 3, although the chain lengths may vary. 4-Sphingenine (i.e. $C_{18:1}$ sphingosine) predominates amongst animal phosphosphingolipids. In addition, small quantities of $C_{18:0}$ sphinganine and the diene, 4,*cis*-14-sphingadienine have been observed. Recently appreciable amounts of 4*D*-hydroxysphinganine (phytosphingosine) have been found in bovine kidney. Traces of bases with *iso*-, *anteiso* chains, odd numbers of carbons, C_{16} and C_{20}–C_{22} chains have been observed. The main LCB's found in plants are 4*D*-hydroxysphinganine and the corresponding monounsaturated base (8-*trans*). The C_{19} analogue and sphinganine have also been observed in plants.

(b) Individual phosphosphingolipids

(1) Sphingomyelin (ceramide phosphorylcholine)[20, 21, 23, 127]

Sphingomyelin (XLIV, Table 5, ceramide-1-phosphorylcholine) primarily occurs in animal tissues. Its relatively high concentration in brain (particularly white matter or myelin) permitted Thudichum to discover this phospholipid in 1884 and establish that its hydrolysis products were sphingosine, fatty acid, phosphoric acid and choline (Chapter 1). Classically sphingomyelin was extracted by hot pyridine from tissue residues left after acetone and ether extraction. Sphingomyelins contaminated with saturated lecithins crystallized out on cooling. More recently sphingomyelin has been isolated (with or without preliminary alkaline treatment to deacylate lecithins) by silicic acid column chromatography, silica-impregnated paper or glass paper, or by TLC.

With the establishment of the nature and structure of the long chain base, the main structural problem that remained for sphingomyelin was the location of the attachment of phosphorylcholine. The latter was proven to be attached at position-1 by oxidative degradation studies with permanganate, periodate and performic acid. With this information and the earlier knowledge that the fatty acid is in amide linkage the structure of sphingomyelin was firmly established to be *N*-acyl-*erythro*-sphingosine-1-phosphorylcholine (ceramide-1-phosphorylcholine). The phosphorylcholine group, as for lecithin, causes sphingomyelin to be an essentially neutral compound which exists in solution in a zwitterionic form. Confirmation of the structure has been achieved through chemical syntheses. Details of the syntheses and some of the limitations are outlined by Shapiro[127]. Specific *threo* and *erythro* (natural) forms were obtained by phosphorylation with β-chloroethyl phosphoryldichloride of optically resolved tartrates of *N*-benzoyl sphingosine sulphates and subsequent conversion of the product to the choline derivative by reaction with triethylamine.

Sphingomyelin occurs in appreciable quantities in myelin, blood and kidney[47, 60, 50] and in lesser quantities (4–10% of total phospholipids) in most other tissues[48, 51, 66]. About two-thirds of the fatty acids of grey matter sphingomyelin consist of stearic acid, whereas in white matter C_{22}–C_{26} fatty acids (mainly 24:0, lignoceric and 24:1, nervonic) form two-thirds of the fatty acids[130]. Sphingomyelins from other tissues show a more uniform distribution with palmitic and nervonic acids in about equal amounts contributing two-thirds of the total[130]. Small amounts of incompletely characterized sphingomyelin have been reported to occur in the pleuropneumonia-like organism *Mycoplasma gallisepticum*[131].

(2) Ceramide phosphorylethanolamine[131–133]

Only quite recently has it clearly been shown that ceramide-1-phosphorylethanolamine (XLV, Table 5, sometimes imprecisely designated sphingoethanolamine) exists in a number of rather diverse species. These include certain snail shellfish (Mesogastropoda)[132], certain insects (the blowfly and possibly the housefly)[133], rumen protozoa[134] and in certain anaerobic bacteria (*Bacteroides melaninogenicus* and *ruminicola*)[131]. In all of the above studies the mixed lipids were subjected to preliminary separations on silica gel columns or plates and the peak containing the 'alkali-stable' lipids was re-run after a mild alkaline hydrolysis to destroy any contaminating monoacyl or diacyl lipids.

Both water-soluble and ether- or chloroform-soluble substances, obtained after strong acid or alkaline hydrolysis, were examined. Ethanolamine and phosphorylethanolamine were shown to be present in the water-soluble fraction. Fatty acids in the organic layer proved to be mainly $C_{20:0}$ (51%) and $C_{22:0}$ (35%) for the lipid from blowfly and C_{12} to C_{21} with a large proportion branched and of an odd number of carbon atoms for the lipid from *B. melaninogenicus*. The LCB corresponded most closely to sphingosine in both the snails[132] and blowfly[133]. In the above anaerobic bacteria the LCB's are C_{17}–C_{21} in length and many have a methyl branch. Periodate oxidation studies indicated that the phosphorylethanolamine was linked to the 1-position of the LCB.

(3,4) Ceramide phosphorylglycerol and phosphorylglycerophosphate*

In a manner essentially similar to that described for ceramide phosphorylethanol-

*Ceramide phosphorylglycerol containing an *N*-hydroxyacyl group and an *O*-acyl group (position not fully identified) has been isolated from a strain of Mycoplasma and its structure tentatively identified by Plackett et al. (*J. Bacteriol.*, 104 (1970) 798).

amine, White's group[131, 135] also succeeded in isolating and characterizing two new glycerol-containing phosphosphingolipids. Periodate oxidation studies before and after strong acid or alkaline hydrolysis and the examination of the products of mild acid-hydrolysis (glycero-3-phosphate and glycerol-1,3-diphosphate) provided evidence that these new lipids were ceramide-1-phosphoryl-1′-*sn*-glycerol (XLVI, Table 5) and ceramide-1-phosphoryl-1′-*sn*-glycero-3-phosphate (XLVII, Table 5).

(5) Ceramide phosphorylinositol-containing lipids (phytoglycolipid and related lipids)

For a number of years Carter and associates[136, 137] have studied the ceramide and inositol-containing phospholipids of plants. Through counter-current distribution (CCD) they showed that many plants contained two main fractions: phosphatidylinositol and phytoglycolipid. The latter contained *N*-acyl phytosphingosine phosphorylinositol linked to a number of carbohydrates. The inositol lipids of flax were shown to contain a third fraction (ceramide–phosphate–polysaccharide). In initial studies a preliminary alkaline hydrolysis was necessary, but more recently it has been possible to carry out the separation without the hydrolysis step by first doing a CCD separation of the $Ca^{2\oplus}$–$Mg^{2\oplus}$ salts and then a subsequent CCD separation of the $Na^{\oplus}$ salts[137]. This technique revealed that the ceramide–phosphate–polysaccharide fraction was also present in a number of plants seeds (corn, soybean, flax, safflower).

By means of a variety of studies involving periodate oxidation and sodium borohydride reduction, permethylation, peracetylation and the use of p.m.r. spectra Carter's group[136] were able to determine the complete structure of a tetrasaccharide phytoglycolipid (XLVIII, Table 5). It has a basic ceramide-1-phosphoryl-1-*myo*-inositol structure. Three additional carbohydrate residues are linked to give the following tetrasaccharide: 2-mannosido-6[*D*-glucosamido-(1 → 4)*D*-glucuronido]-inositol.

Recently an incompletely characterized inositol-containing phosphosphingolipid has been isolated from baker's yeast[138]. The available evidence suggests that the lipid was a mannosyl-diinositol, diphosphoryl ceramide (designated mannose-(inositol phosphate)$_2$-ceramide).

4. EFFECTS OF HYDROLYTIC AND OTHER REAGENTS

In the analyses and structural investigations performed on phospholipids a variety of chemical, as well as enzymatic, degradative procedures have been utilized. A brief resumé is given of the effects of some of the more common chemical reagents used (see Snyder[139] and Viswanathan[31, 64] for recent summaries).

Alkaline alcoholysis

Mild alkaline alcoholysis at 37° (0.025–0.1 *M* NaOH in methanolic or ethanolic solutions) is sufficient to split ester linkages of fatty acids[15, 21]. At room temperature some selectivity is seen in the susceptibility of phosphoglycerides (the order is diacyl > alk-1'- enyl, acyl > alkyl, acyl)[74]. Under the above conditions most of the water-soluble phosphodiesters released (GPC, GPE, etc.) are relatively intact (some cyclic glycerophosphate may derive from GPC and GPI). With more vigorous alkaline hydrolysis the phosphodiesters are cleaved further via cyclic intermediates. With those phosphodiesters containing a nitrogen base, glycerophosphate (2-, 3-mixture) and the free base are formed. With the remaining phosphodiesters, where the second substituent has an hydroxyl function, then phosphomonoesters of this substituent may preferentially form. Proposed mechanisms for base hydrolysis as applied to the phosphoglycerides are discussed in the following reviews[20, 21, 105, 106, 140]. Phosphosphingolipids are not affected unless subjected to strong alkaline conditions. A recommended procedure for LCB release employs a preliminary degradation by phospholipase to yield ceramide which is then hydrolyzed by alkali[129].

Acid hydrolysis[20, 21, 31, 139]

Mild acid hydrolysis (trichloroacetic acid, acetic acid, HCl and a little $HgCl_2$) results in the complete cleavage of alk-1-enyl bonds producing long-chain aldehydes. With increasing strength of acid and heating (e.g. 2 *N* HCl or glacial acetic acid at 100°) diglyceride and inositol phosphate are formed from PI and diglyceride and glyceroldiphosphate are formed from diphosphatidylglycerol. Acid hydrolysis of phosphodiesters such as GPC also occurs with cyclic intermediate formation[20]. Strong acid hydrolysis (6 *N* in H_2O, 2 *N* in methanol) at various temperatures and times depending on the lipid leads to thc release of bases (including LCB) and fatty acids from all phospholipids. Glycerophosphate and inositol phosphates, despite the acid migration of the phosphate group, are resistant to acid hydrolysis[106, 141] but decompose on prolonged hydrolysis at high temperatures[21].

Oxidation reactions[31, 139]

Permanganate–periodate oxidation, chromic acid oxidation and reductive ozonolysis may all be used for the location of double bonds. Periodate oxidation may be carried out on any substances containing two (or more) adjacent hydroxyl groups or adjacent amino and hydroxyl groups (e.g. glycerol[21], glycerophosphate[21], inositol and some of its phosphates[106, 142], LCB's[143]). As pointed out earlier in a number of sections, this technique has proven to be particularly useful in the elucidation of the structures of a number of phospholipids.

References p. 39

Both glycerol and *sn*-glycerol-3-(or -1-) phosphate (once released by appropriate hydrolytic or enzymatic procedures) readily undergo oxidation by periodate at room temperature[21,144]. Under these conditions glycerol yields two molecules of formaldehyde and one molecule of formic acid whereas the above glycerolphosphates yield one molecule of formaldehyde and formic acid (the latter in proportion to the glycolaldehyde phosphate that undergoes hydrolysis). Inositol and those phosphates which react with periodate require longer times at elevated temperatures (e.g. for pH 4.5, 2 h at 65°) to obtain complete enough oxidation for reliable work[142,144]. Provided there are free vicinal hydroxyls on a substituted inositol, it should undergo ring cleavage with periodate. Inositol and its monophosphates should undergo complete cleavage to formic acid (one mechanism gives formic acid mainly plus some carbon dioxide[142]). Inositol di- and tri-phosphates, if oxidized with excess periodate and the products formed are reduced with borohydride and dephosphorylated with phosphomonoesterase, yield a variety of polyols whose structure depends on the position and number of the original phosphate groups[106]. Thus, among the diphosphates *ortho*-substitution would yield a tetritol, *meta*-substitution a pentitol and *para*-substitution only formic acid and inorganic phosphate. Similarly the triphosphates would yield pentitols, hexitols or unaltered inositol. LCB's such as sphingosine and dihydrosphingosine are readily oxidized by periodate (e.g. 60 min at room temperature)[143]. The products formed are one molecule of formaldehyde, one molecule of formic acid, one molecule of NH_3 and the appropriate long-chain aldehyde (*trans*-2-hexadecenal or hexadecanal). It is of interest to note that for nearly all of the above examples appropriate assay methods have been developed which are dependent on the actions of periodate just described[144].

Lithium aluminium hydride reduction ($LiAlH_4$)[139] (also Chapter 3)

$LiAlH_4$ removes acyl and phosphoryl-base groups from lipids. The acyl groups are reduced to their corresponding alcohols. $LiAlH_4$ leaves intact alkyl ether and alk-1-enyl ether (including the unsaturation) linkages. Recently Snyder et al.[145] have used a new reducing agent termed 'Nitride' [$NaAlH_2(OCH_2CH_2OCH_3)_2$] and shown it to be considerably better than $LiAlH_4$ for the reduction of phospholipids. 'Nitride' gives a much improved yield of *O*-alk-1-enyl glycerol.

Hydriodic acid (HI)[139]

This reagent cleaves alkyl ether linkages yielding alkyl halides which may be subsequently analyzed by gas-liquid chromatography.

Acetolysis[141]

Treatment of phosphoglycerides with acetic acid–acetic anhydride (or its equivalent) under reflux conditions for several hours results in the formation of the appropriate diglyceride acetate by acetolysis. A number of modifications to give quantitative yields have been made. These include the replacement of acetic acid by trifluoroacetic acid and the insertion of a saponification step (after the acetolysis) followed by an acetylation step.

REFERENCES

1 H. Maclean and I.S. MacLean, *Lecithin and Allied Substances*, Longmans, Green, London, 1927.
2 H.J. Deuel Jr., *The Lipids*, Vol. 1, Interscience, New York, 1951, pp. 4 and 405–473.
3 H. Wittcoff, *The Phosphatides*, Reinhold, New York, 1951.
4 A.C. Chibnall and H.J. Channon, *Biochem. J.*, 21 (1927) 233.
5 R. Feulgen and T. Bersin, *Z. Physiol. Chem.*, 260 (1939) 217.
6 J. Folch, *J. Biol. Chem.*, 177 (1949) 497, 505.
7 M.C. Pangborn, *J. Biol. Chem.*, 153 (1944) 343.
8 *J. Folch,* J. Biol. Chem. *177* (1949) 497, 505.
9 D.J. Hanahan, M.B. Turner and M.E. Jayko, *J. Biol. Chem.*, 192 (1951) 623.
10 C.H. Lea, D.N. Rhodes and R.D. Stoll, *Biochem. J.*, 60 (1955) 353.
11 G.V. Marinetti and E. Stotz, *Biochim. Biophys. Acta*, 21 (1956) 168.
12 R.M.C. Dawson, *Biochim. Biophys. Acta*, 14 (1954) 374.
13 M. Kates, in G.V. Marinetti (Ed.), *Lipid Chromatographic Analysis*, Vol. 1, Dekker, New York, 1967, p. 1.
14 G. Rouser, G. Kritchevsky and A. Yamamoto, in G.V. Marinetti (Ed.), *Lipid Chromatographic Analysis*, Vol. 1, Dekker, New York, 1967, p. 99.
15 R.M.C. Dawson, in G.V. Marinetti (Ed.), *Lipid Chromatographic Analysis*, Vol. 1, Dekker, New York, 1967, p. 163.
16 W. Schlemmer, *Bull. Soc. Ital. Biol. Sper.*, 37 (1961) 134.
17 H. Wagner, L. Hörhammer and P. Wolff, *Biochem. Z.*, 334 (1961) 175.
18 O. Renkonen and P. Varo, in G.V. Marinetti (Ed.), *Lipid Chromatographic Analysis*, Vol. 1, Dekker, New York, 1967, p. 41.
19 IUPAC–IUB Tentative Rules for *The Nomenclature of Lipids*, *Biochim. Biophys. Acta*, 152 (1968) 1. (See *J. Lipid Res.*, 11 (1970) 173 for Amendments and lists of additional journals.)
20 D.J. Hanahan, *Lipid Chemistry*, Wiley, New York, 1960.
21 G.B. Ansell and J.N. Hawthorne, *Phospholipids–Chemistry, Metabolism and Function*, Elsevier, Amsterdam, 1964.
22 L.L.M. van Deenen and G.H. de Haas, *Advan. Lipid Res.*, 2 (1964) 167.
23 D.J. Hanahan and H. Brockerhoff, in M. Florkin and E. Stotz (Eds.), *Comprehensive Biochemistry* (Vol. 6), *Lipids and Amino Acids and Related Compounds*, Elsevier, Amsterdam, 1965, p. 83.
24 L.L.M. van Deenen and H.G. de Haas, *Ann. Rev. Biochem.*, 35 (1966) 157.
25 M. Kates, *Advan. Lipid Res.*, 2 (1964) 17.
26 M. Kates, *Advan. Lipid Res.*, 8 (1970) 225.
27 K.P. Strickland, in P. Bernfeld (Ed.), *Biogenesis of Natural Compounds*, 2nd Revised ed., Pergamon, Oxford 1968, p. 103.
28 H. Goldfine, *Ann. Rev. Biochem.*, 37 (1968) 303.
29 M. Kates and M.K. Wassef, *Ann. Rev. Biochem.*, 39 (1970) 323.
30 A.J. Slotboom and P.P.M. Bonsen, *Chem. Phys. Lipids*, 5 (1970) 301.
31 C.V. Viswanathan, *Chromatog. Rev.*, in the press.
31a C.V. Viswanathan, F. Phillips and O. Lundberg, *J. Chromatog.*, 38 (1968) 267.
32 E. Baer and H.O.L. Fischer, *J. Biol. Chem.*, 128 (1939) 475.
33 E. Baer and D. Buchnea, *J. Am. Chem. Soc.*, 81 (1959) 1758.
34 M.L. Karnovsky, G. Hauser and D. Elwyn, *J. Biol. Chem.*, 226 (1957) 881.
35 R.S. Cahn, C.K. Ingold and V. Prelog, *Experientia*, 12 (1956) 81.
36 R.S. Cahn, C.K. Ingold and V. Prelog, *Angew. Chem. Intern. Ed. Engl.*, 5 (1966) 385.
37 H. Hirschmann, *J. Biol. Chem.*, 235 (1960) 2762.
38 *Rules of Carbohydrate Nomenclature*, *J. Org. Chem.*, 28 (1963) 281.
39 T. Malkin and T.H. Bevan, in R.T. Holman, W.O. Lundberg and T. Malkin (Eds.), *Progress in the Chemistry of Fats and other Lipids*, Vol. 4, Pergamon, London, 1957, p. 97.
40 E. Baer, in R.T. Holman, W.O. Lundberg and T. Malkin (Eds.), *Progress in the Chemistry of Fats and other Lipids*, Vol. 6, Pergamon, London, 1963, p. 33.

41 B.A. Klyashchitskii, S.D. Sokolov and V.I. Shvets, *Russ. Chem. Rev. (Engl. Transl.)*, 38 (1969) 345.
42 C. Piantadosi and F. Snyder, *J. Pharm. Sci.*, 59 (1970) 283.
43 A. Kornberg and W.E. Pricer Jr., *J. Biol. Chem.*, 204 (1953) 345.
44 L.E. Hokin and M.R. Hokin, *J. Biol. Chem.*, 233 (1958) 800.
45 G. Hübscher and B. Clark, *Biochim. Biophys. Acta*, 41 (1960) 45.
46 G. Hübscher, *Biochim. Biophys. Acta*, 52 (1961) 571.
47 G. Rouser and A. Yamamoto, *Lipids*, 3 (1968) 284.
48 G. Rouser and R.D. Solomon, *Lipids*, 4 (1969) 232.
49 C.F. Baxter, G. Rouser and G. Simon, *Lipids*, 4 (1969) 243.
50 G. Rouser, G. Simon and G. Kritchevsky, *Lipids*, 4 (1969) 599.
51 G. Simon and G. Rouser, *Lipids*, 4 (1969) 607.
52 R.D. Bowman and R.O. Mumma, *Biochim. Biophys. Acta*, 144 (1967) 501.
53 M. Ikawa, *Bacteriol. Rev.*, 31 (1967) 54.
54 M.J. Coulon-Morelec and M. Faure, *Bull. Soc. Chim. Biol.*, 49 (1967) 273.
55 M. Faure and J. Troestler, *Bull. Soc. Chim. Biol.*, 49 (1967) 279.
56 W.J. Lennarz, *Ann. Rev. Biochem.*, 39 (1970) 359.
57 A.K. Hajra, *Biochem. Biophys. Res. Commun.*, 39 (1970) 1037.
58 R.L. Wykle and F. Snyder, *J. Biol. Chem.*, 245 (1970) 3047.
59 H.H. Bishop and K.P. Strickland, *Can. J. Biochem.*, 48 (1970) 269.
60 J.H. Williams, M. Kuchmak and R.F. Witter, *Lipids*, 1 (1966) 89, 391.
61 R.J. Rossiter and K.P. Strickland, in A. Lajtha (Ed.), *Handbook of Neurochemistry*, Vol. III, Plenum, New York, 1970, p. 467.
62 E. Klenk and H. Debuch, in R.T. Holman, W.O. Lundberg and T. Malkin (Eds.), *Progress in the Chemistry of Fats and other Lipids*, Vol. 6, Pergamon, Oxford, 1963, p. 1.
63 M.M. Rapport and W.T. Norton, *Ann. Rev. Biochem.*, 31 (1962) 103.
64 C.V. Viswanathan, *Chromatog. Rev.*, 10 (1968) 18.
65 F. Snyder, in R.T. Holman (Ed.), *Progress in the Chemistry of Fats and other Lipids*, Vol. X, pt. 3, Pergamon, Oxford, 1969, p. 287.
66 T.W. Scott, B.P. Setchell and J.M. Bassett, *Biochem. J.*, 104 (1967) 1040.
67 E.F. Hartree and T. Mann, *Biochem. J.*, 80 (1961) 464.
68 G.Y. Sun and L.A. Horrocks, *Lipids*, 3 (1968) 79.
69 E.L. Gottfried and M. Rapport, *J. Biol. Chem.*, 237 (1962) 329.
70 O. Renkonen, *Biochim. Biophys. Acta*, 125 (1966) 288.
71 G. Rouser, J. O'Brien and D. Heller, *J. Am. Oil Chem. Soc.*, 38 (1961) 14.
72 S.J. Thannhauser, N.F. Boncoddo and G. Schmidt, *J. Biol. Chem.*, 188 (1951) 417, 423, 427.
72a G.B. Ansell and S. Spanner, in R. Fried (Ed.), *Methods in Neurochemistry*, Vol. 1, Dekker, New York, 1971, p. 31.
73 J. Cymerman-Craig, D.P.G. Hamon, K.K. Purushothaman, S.K. Roy and W.E.M. Lands, *Tetrahedron*, 22 (1966) 175.
74 O. Renkonen, *Advan. Lipid Res.*, 5 (1967) 329.
74a H.E. Carter, D.B. Smith and D.N. Jones, *J. Biol. Chem.*, 232 (1958) 681.
75 L.A. Horrocks and G.B. Ansell, *Biochim. Biophys. Acta*, 137 (1967) 90.
76 G.K. Chacko and D.J. Hanahan, *Biochim. Biophys. Acta*, 164 (1968) 252.
77 J. Bremer and D.M. Greenberg, *Biochim. Biophys. Acta*, 35 (1959) 287.
78 R.L. Lester and D.C. White, *J. Lipid Res.*, 8 (1967) 565.
79 M.O. Hall and J.F. Nyc, *J. Lipid Res.*, 2 (1961) 321.
80 N.A. Baumann, P.O. Hagen and H. Goldfine, *J. Biol. Chem.*, 240 (1965) 1559.
81 R. Letters, *Biochim. Biophys. Acta*, 116 (1966) 489.
82 O.W. Thiele, D. Busse and K. Hoffmann, *European J. Biochem.*, 5 (1968) 513.
83 R.A. Bomstein, *Biochem. Biophys. Res. Commun.*, 21 (1965) 49.
84 R.M.C. Dawson, N. Clarke and R.H. Quarles, *Biochem. J.*, 114 (1969) 265.
85 R. Aneja, J.S. Chadha and J.A. Knaggs, *Biochem. Biophys. Res. Commun.*, 36 (1969) 401.
86 J. Folch, *J. Biol. Chem.*, 174 (1948) 439.
87 E. Baer and J. Maurukas, *J. Biol. Chem.*, 212 (1955) 25, 39.

87a D.K. Belsare and D.R. Chowdhuri, *Lipids*, 3 (1968) 21.
88 P. Kemp and R.M.C. Dawson, *Biochem. J.*, 113 (1969) 555.
89 M.G. Macfarlane, *Advan. Lipid Res.*, 2 (1964) 91.
90 W.J. Lennarz, *Advan. Lipid Res.*, 4 (1966) 175.
91 A.A. Benson and B. Maruo, *Biochim. Biophys. Acta*, 27 (1958) 189.
92 A.A. Benson and M. Miyano, *Biochem. J.*, 81 (1961) 31P.
93 F. Haverkate and L.L.M. van Deenen, *Biochim. Biophys. Acta*, 106 (1965) 78.
94 C.N. Joo and M. Kates, *Biochim. Biophys. Acta*, 176 (1969) 278.
95 J. Eichberg and J.D. Burnham, *J. Lipid Res.*, 11 (1970) 386.
95a J. Lecocq and C.E. Ballou, *Biochemistry*, 3 (1964) 976.
96 G.H. de Haas, P.P.M. Bonsen and L.L.M. van Deenen, *Biochim. Biophys. Acta*, 116 (1966) 114.
97 D.R. Body and G.M. Gray, *Chem. Phys. Lipids*, 1 (1967) 254.
98 G. Rouser, G. Kritchevsky, A. Yamamoto, A.G. Knudson Jr., and G. Simon, *Lipids*, 3 (1968) 287.
99 H. Debuch and E. Rotsch, *Z. Physiol. Chem.*, 347 (1966) 79.
100 J.A.F. op den Kamp, P.P.M. Bonsen and L.L.M. van Deenen, *Biochim. Biophys. Acta*, 176 (1969) 298.
101 J.C. MacDougall and P.J.R. Phizackerley, *Biochem. J.*, 114 (1969) 361.
102 M. Macfarlane, *Nature*, 196 (1962) 136.
103 U.M.T. Houtsmuller and L.L.M. van Deenen, *Biochim. Biophys. Acta*, 106 (1965) 564.
104 J.G. Molotkovsky and L.D. Bergelson, *Chem. Phys. Lipids*, 2 (1968) 1.
105 J.N. Hawthorne, *J. Lipid Res.*, 1 (1960) 255.
106 J.N. Hawthorne and P. Kemp, *Advan. Lipid Res.*, 2 (1964) 127.
107 R.M.C. Dawson, in H. Kindl (Ed.), *Cyclitols and Phosphoinositides*, Vol. 2, Pergamon, Oxford, 1966, p. 57.
108 C.E. Ballou and Y.C. Lee, in H. Kindl (Ed.), *Cyclitols and Phosphoinositides*, Vol. 2, Pergamon, Oxford, 1966, p. 41.
109 E. Lederer, *Chem. Phys. Lipids*, 1 (1967) 294.
110 B.A. Klyashchitskii, V.I. Shvets and N.A. Preobrazhenskii, *Chem. Phys. Lipids*, 3 (1969) 393.
111 D.M. Brown and B.F.C. Clark, *Nature*, 194 (1962) 1081.
112 H. Brockerhoff and C.E. Ballou, *J. Biol. Chem.*, 236 (1961) 1907.
113 R.M.C. Dawson and J.C. Dittmer, *Biochem. J.*, 81 (1961) 540.
114 D.M. Brown and J.C. Stewart, *Biochim. Biophys. Acta*, 125 (1966) 413.
115 H.S. Hendrickson and C.E. Ballou, *J. Biol. Chem.*, 239 (1964) 1369.
116 V.Ya. Dvorkin and G.V. Kiselev, *Biokhimiya*, 34 (1969) 1245.
117 J.N. Hawthorne and R.H. Michell, in H. Kindl (Ed.), *Cyclitols and Phosphoinositides*, Vol. 2, Pergamon, Oxford, 1966, p. 49.
118 R.L. Lester and M.R. Steiner, *J. Biol. Chem.*, 243 (1968) 4889.
119 B.J. Holub, A. Kuksis and W. Thompson, *J. Lipid Res.*, 11 (1970) 558.
120 Y.C. Lee and C.E. Ballou, *Biochemistry*, 4 (1965) 1395.
121 M.C. Pangborn and J.A. McKinney, *J. Lipid Res.*, 7 (1966) 627.
122 P. Brennan and C.E. Ballou, *J. Biol. Chem.*, 242 (1967) 3046.
123 N. Shaw, P.F. Smith and H.M. Verheij, *Biochem. J.*, 120 (1970) 439.
124 I. Ishizuka and T. Yamakawa, *J. Biochem. (Tokyo)*, 64 (1968) 13.
125 A.K. Hajra and B.W. Agranoff, *J. Biol. Chem.*, 243 (1968) 1617.
126 A.K. Hajra, *Biochem. Biophys. Res. Commun.*, 37 (1969) 486.
127 D. Shapiro, *Chemistry of Sphingolipids*, Hermann, Paris, 1969.
128 H.E. Carter, D.S. Galanos and Y. Fujino, *Can. J. Biochem. Physiol.*, 34 (1956) 320.
129 K.A. Karlsson, *Chem. Phys. Lipids*, 5 (1970) 6.
130 E. Svennerholm, S. Stälberg-Stenhagen and L. Svennerholm, *Biochim. Biophys. Acta*, 125 (1966) 60.
131 J.P. La Bach and D.C. White, *J. Lipid Res.*, 10 (1969) 528.
132 T. Hori, M. Sugita and I. Arakawa, *Biochim. Biophys. Acta*, 152 (1968) 211.
133 R.M.C. Dawson and P. Kemp, *Biochem. J.*, 106 (1968) 319.
134 R.M.C. Dawson and P. Kemp, *Biochem. J.*, 105 (1967) 837.

135 D.C. White and A.N. Tucker, *Lipids*, 5 (1970) 56.
136 H.E. Carter, D.R. Strobach and J.N. Hawthorne, *Biochemistry*, 8 (1969) 383.
137 H.E. Carter and A. Kisic, *J. Lipid Res.*, 10 (1969) 356.
138 S. Steiner, S. Smith, C.J. Waechter and R.L. Lester, *Proc. Natl. Acad. Sci. (U.S.)*, 64 (1969) 1042.
139 F. Snyder, *Methods Cancer Res.*, 6 (1971) 399.
140 D.M. Brown, *Ann. N.Y. Acad. Sci.*, 165 (1969) 687.
141 R.N. Roberts, in G.V. Marinetti (Ed.), *Lipid Chromatographic Analysis*, Vol. 1, Dekker, New York, 1967, p. 447.
142 T. Posternak, *The Cyclitols*, Hermann, Paris, and Holden-Day, San Francisco, 1965.
143 C.C. Sweeley and E.A. Moscatelli, *J. Lipid Res.*, 1 (1959) 40.
144 J.C. Dittmer and M.A. Wells, in J.M. Lowenstein (Ed.), *Methods in Enzymology*, Vol. XIV, *Lipids*, Academic Press, New York, 1969, p. 482.
145 F. Snyder, M.L. Blank and R.L. Wykle, *J. Biol. Chem.*, 246 (1971) 3639.

Chapter 3

SEPARATION AND ANALYSIS OF PHOSPHOLIPIDS

SHEILA SPANNER

Department of Pharmacology (Preclinical), The Medical School, Birmingham B15 2TJ (Great Britain)

The rapid expansion of chromatography in the last ten years, particularly thin-layer and gas-liquid chromatography, has made it possible to analyse phospholipid mixtures and to isolate individual molecular species. Analysis has also been aided by our improved knowledge of the specificity of a number of the phospholipases.

To examine and determine the phospholipids in detail it is essential to extract them from the parent tissue and to free them from lipid and non-lipid contaminants. They may then be separated into groups homogeneous with respect to their base moiety. It is, however, extremely difficult to sub-fractionate these groups and keep each lipid intact. For complete analysis a certain degree of degradation is, at present, obligatory; this degradation may be either enzymatic or by acid or alkaline hydrolysis.

For information additional to that covered in this chapter, the reader is referred to books by the following authors; Ansell and Hawthorne[1], James and Morris[2], Marinetti[3], Fried[4] and to various volumes of *Advances in Lipid Research* and the *Progress in the Chemistry of Fats and other Lipids.*

1. EXTRACTION OF PHOSPHOLIPIDS AND REMOVAL OF NONLIPID CONTAMINANTS

Extraction

There is, at the moment, no known method for the selective extraction of the phospholipids from tissues. The total lipid components of tissue, many of which occur as protein-lipid complexes, can, however, be extracted, and from these mixtures the phospholipids can be isolated. Many solvent systems have been tried and for the early methods the reader is referred to Ansell and Hawthorne[1]. The most successful and the most extensively used is the chloroform-methanol mixture devised by Folch, Lees and Sloane Stanley[5]. This solvent will extract all the known phospholipids except the polyphosphoinositides. For these the addition of acid to the chloroform-methanol is obligatory[6, 7].

In the extraction method devised by Folch et al.[5], the tissue is homogenised in a Waring type blendor with 20 vol. (w/v) of chloroform:methanol (2:1, v/v) to give a single phase system. The homogenate is filtered and the polyphosphoinositides are extracted from the residue at 37° with chloroform:methanol (2:1, v/v) containing 0.25% (v/v) hydrochloric acid[7]. The technique of Folch et al.[5] was developed for brain tissue but it has proved suitable for a wide range of both plant and animal tissues. With some of the more intractable tissues, it is necessary to use a more vigorous homogenisation than can be obtained with the Waring Blendor. The Vertis 45 (The Vertis Co. Inc., Gardiner, N.Y., U.S.A.) has been used for skin and bone. With heart tissue, a preliminary mincing in a domestic mincer helps the subsequent homogenisation and extraction.

There have been numerous modifications to the procedure of Folch et al.[5] to make it more suitable for a specific tissue. In the extraction of erythrocyte ghosts, Dawson, Hemington and Lindsay[8] found that the standard chloroform:methanol (2:1, v/v) extraction produced a 'putty-like' mass. However, if the ghosts were blended with 7 vol. of methanol before the addition of 14 vol. of chloroform, the extraction was complete. Wren and Mitchell[9] powdered whole flies in dry ice in a Servall stainless steel homogeniser before making repeated extractions of the powder with chloroform:methanol (2:1, v/v). Bligh and Dyer[10] developed a two-phase system of chloroform, methanol and water. The final concentration of these, including the water content of the tissue was 20:20:18, by vol. respectively. This two-phase system has been modified and is a useful method of extracting lipids after studies *in vitro* where the water content may be high. 4 vol. of chloroform:methanol (2:1, v/v) added to the incubation system followed by warming and shaking will extract most of the phospholipids into the lower chloroform phase. The two-phase system has, however, been criticised by Palmer[11] as failing to extract all the acidic lipids into the chloroform phase.

Plant tissue poses another extraction problem. Some seeds can be extracted by the method of Folch et al.[5] but certain seeds and many plant tissues contain very active phospholipases which are not inactivated by chloroform or methanol and which readily break down the phospholipids during this extraction procedure. This can be overcome by homogenising the tissue with iso-propanol, filtering and re-extracting the residue with chloroform:iso-propanol (2:1, v/v) and pooling the two extracts. Iso-propanol deactivates the phospholipases[12].

Removal of non-lipid contaminants

The protein-lipid complexes which are extracted from the tissues in all of the above procedures, are decomposed by taking the extract down to dryness under reduced pressure and repeating this several times after the addition of water-wet chloroform. The protein is insoluble in chloroform:methanol and can be filtered off. Other non-lipid contaminants can be washed out of the de-proteinised extract with a salt solution[5].

Another method for the removal of non-lipid contaminants is that of Wells and Dittmer[13]. The de-proteinised lipid extract in chloroform:methanol:water (60:30:4.5, by vol.) is passed through a column of Sephadex G25-fine. The effluent together with a small volume of chloroform:methanol (2:1, v/v) eluate, contains the lipids free from non-lipid contaminants. Turner and Rouser[14, 15] used a similar method in their studies on the lipids of plasma and erythrocytes. Sephadex G25 was also used by Williams and Merrilees[16] to remove water and non-lipid contaminants from their lipid extracts of leaves. Sephadex was added to the filtered chloroform:methanol (2:1, v/v) extract and the mixture taken to dryness. Chloroform was added and the suspension again taken to dryness. Finally the Sephadex was suspended in chloroform and poured into a glass column. The lipid was eluted with chloroform and the non-lipid contaminants were eluted with methanol and finally with water.

2. PURIFICATION OF THE PHOSPHOLIPID FRACTION

The lipid extract, free of protein and other non-lipid contaminants, contains only a part of its weight as phospholipid. The remainder is made up of free fatty acids, cholesterol and its esters, other steroids, carotenoids and cerebrosides. These non-phosphorus containing lipids can be removed before the lipid is analysed or they may be removed at a later stage in the analysis.

One method which has been used extensively in studies on heart muscle, is to treat the lipid extract in chloroform, with 10 vol. of acetone. The precipitate which forms is re-dissolved in chloroform and a further 10 vol. of acetone are added. This second precipitate contains all the phosphorus-containing lipids, freed from most of the non-phosphorus-containing lipids[17].

In 1952, Borgström[18] showed that the neutral lipids could be separated from the phospholipids on a column of silicic acid. In this method, the neutral lipids were eluted with chloroform and the phospholipids with methanol. Other solvents have been used. In their studies with Gram-positive bacteria, Vorbeck and Marinetti[19] passed the lipid, suspended in chloroform, through a column of silicic acid which had been prepared as a slurry, in chloroform. The neutral lipids were not adsorbed and the glycolipids were eluted with chloroform:acetone (1:1, v/v) and acetone alone. The phospholipids were then eluted with chloroform containing increasing percentages of methanol. In this method, there may be a loss of diphosphatidylglycerol which is eluted, in part, with acetone. A variation of this method was described by Rouser, Kritchevsky and Yamamoto[20].

Fillerup and Mead[21], working with plasma, also used a column of silicic acid to remove lipid contaminants of the phospholipid fraction. They eluted the contaminants with light petroleum containing increasing percentages of diethyl ether up to 50%. The phospholipids were eluted with 25% methanol in diethyl ether. Robertson and Lands[22], working with liver, removed the neutral lipids with ether alone.

The total lipid or phospholipid extract may now be analysed in detail. There are several approaches to this but they can be summarised in two main categories. The lipid extract can be subjected to chromatography and the intact lipid, thus obtained, analysed, for phosphorus, nitrogen, free amino groups, glycerol, acyl ester linkages, alkenyl ether linkages, fatty acids, fatty aldehydes and inositol (for details see below). The other approach is that of the selective hydrolysis of the lipid extract[23]. This method yields information about the hydrophilic part of the molecule alone. Without doubt the first approach gives the most information but it can be tedious and the chances of loss, particularly of lipids in low concentrations, are great.

References p. 62

3. FRACTIONATION AND ANALYSIS OF INTACT PHOSPHOLIPIDS

There are numerous column chromatography methods for the separation of individual phospholipids. In the complete analysis of tissues, a combination of more than one method gives the best result. Alumina, silicic acid, magnesium silicate (Florisil), diethylaminoethyl (DEAE) cellulose, triethylaminoethyl (TEAE) cellulose and hydroxylapatite have all been used in the fractionation of phospholipids. Different tissues and availability of materials are but two of the criteria that influence the choice of chromatography to be used. Some of the characteristics of the stationary phases and the type of separations possible with them are given below.

Aluminium oxide

The use of alumina was pioneered by Long, for the separation of brain phospholipids and the method was first described in detail by Davison and Wajda[24]. Provided that the column is run quickly, there is no degradation of the phospholipids. Phosphatidylserine is, however, very firmly bound to the column and is almost impossible to elute quantitatively. More recently Luthra and Sheltawy[136] have described a method for separating phospholipids, including phosphatidylserine, on aluminium oxide by eluting with solvents containing ammonium salts. Letters[25] has separated ethanolamine, monomethylethanolamine, dimethylethanolamine and choline-containing phospholipids of yeast on alumina and Ansell and Chojnacki[26] have used a similar method for the separation of the same lipids from liver. The main factor in the separation is the base moiety of the lipid so that, with chloroform:methanol (1:1, v/v), for example, the four choline-containing lipids – phosphatidylcholine, choline plasmalogen, lysophosphatidylcholine and sphingomyelin – are eluted together. It is, however, possible to separate some of these components with careful fraction collecting and monitoring on thin layer plates (see below). The fractions must be collected through filter paper to remove traces of alumina, particularly if the further analysis of the fractions involves the evaporation of the fraction to dryness. Under these conditions the presence of alumina does cause a degradation of the lipid.

If, after chloroform, chloroform:methanol (4:1, v/v) is used instead of 1:1, v/v, the phosphatidylcholine is eluted first, followed by the sphingomyelin and the two lipids can be separated (unpublished findings). Chloroform:methanol (4:1, v/v) followed by chloroform:methanol (1:1, v/v) then 6% (v/v) water in chloroform:methanol (1:1, v/v) followed by 10% water (v/v) in chloroform:methanol (1:1, v/v) results in a better separation of the phospholipids. The phosphatidylethanolamine (10% water in chloroform:methanol (1:1, v/v)) recovery is better and the lipid appears to be uncontaminated by other lipids (unpublished results).

The use of aluminium oxide for the separation of phospholipids on TLC has been described by Rao, Ramachandran and Cornwell[27].

Silicic acid

Column chromatography

It was established by Borgström[18] in 1952, by Fillerup and Mead[21] and by Lea, Rhodes and Stoll[28], that a column of silicic acid could be used to separate the phospholipids from the 'neutral' lipids, i.e. cholesterol and fatty acids, of a tissue extract. Lea et al.[28] also demonstrated that it was possible to separate the phosphatidylethanolamine from the phosphatidylcholine of egg yolk on a column of silicic acid by eluting the two lipids with chloroform:methanol (4:1, v/v).

However, when this work was extended to include more complex mixtures of lipids, an adequate separation of the components was difficult to achieve. Hanahan, Dittmer and Warashina[29] found that, when using a stepwise gradient of increasing concentrations of methanol in chloroform, there was a considerable tailing and overlapping of the fractions from liver lipid. Wren and Mitchell[9] in their studies with Drosophila had the same problem. Bader and Morgan[30] had more success in fractionating heart phospholipids on silicic acid by eluting the fractions with chloroform-methanol as a continuous gradient. They demonstrated five fractions: at 4% methanol in chloroform, phosphatidic acid, phosphatidylglycerol and phosphatidylinositol, at both 10% and 14% phosphatidylethanolamine and phosphatidylserine, at 35% phosphatidylcholine and at 50% sphingomyelin. A similar method was used by Nielson[31] when preparing cardiolipin from heart lipid extracts; it could be eluted from a column of silicic acid with chloroform:methanol (19:1, v/v) and (9:1,v/v). Letters[25] also eluted the cardiolipin of yeast from silicic acid with 6% (v/v) methanol in chloroform. Columns of silicic acid have been little used for the fractionation of total phospholipid extracts of brain. Svennerholm[32], however, carried out a preliminary fractionation of brain lipids on silicic acid again with methanol in chloroform as the eluting solvent. Cholesterol was eluted with 2% methanol, ethanolamine, serine and inositol containing phospholipids with 20% methanol and choline-containing phospholipids with 80% methanol in chloroform (cf. Bader and Morgan[30]).

Other solvent systems have been tried in an attempt to elute fractions of greater homogeneity. Rouser, O'Brien and Heller[33] found that the fraction containing both phosphatidylserine and phosphatidylethanolamine could be re-chromatographed on a column of silicic acid. The fraction was applied to the column in chloroform:methanol (4:1, v/v) containing 1% (w/v) ammonium hydroxide. The phosphatidylethanolamine was then eluted with chloroform:methanol (4:1, v/v) and the phosphatidylserine with methanol. Gray[34] applied a total lipid extract of lung tissue to a column of silicic acid in chloroform. The neutral lipids were eluted with chloroform:methanol (49:1, v/v) and cholesterol with tetrahydrofuran:benzene (1:4, v/v). After the addition of tetrahydrofuran:methylal (dimethoxymethane):methanol:water (10:6:4:1, by vol.) to the column, four peaks were eluted; (*1*) the acidic phospholipids (including cardiolipin and a small amount of the phospholipids of fraction 2), (*2*) phosphatidylglycerol + phosphatidylinositol, (*3*) phosphatidylinositol and (*4*) phosphatidylethanolamine. Phosphatidylcholine was eluted together with sphingomyelin by chloroform:methanol:water (4:16:1, by vol.).

Some workers have added filter aids like 'Celite' or 'Hyflo super cel' to the silicic acid. Ways and Hanahan[35] isolated the phospholipid fraction from the total lipid extract of red blood cells by chromatography on silicic acid. They then applied this fraction to a column containing silicic acid:Hyflo super cel (2:1, w/w) prepared with

References p. 62

chloroform. The column was washed through with chloroform followed by chloroform: methanol (9:1, v/v). The ethanolamine phospholipids were eluted with chloroform: methanol (6:1, v/v). This was followed by ethyl acetate:methanol (3:2, v/v) which eluted two separate peaks; the first containing phosphatidylserine and phosphatidylinositol and the second the choline-containing phospholipids. Rathbone[36] also used a silicic acid–Hyflo super cel mixture in his detailed studies on the fractionation of the various salt forms of phosphatidylserine.

Hall and Nyc[37] separated the phospholipids of *Neurospora crassa* as homogenous fractions on a column of silicic acid. The major phospholipids present are phosphatidylmonomethylethanolamine, phosphatidyldimethylethanolamine and phosphatidylcholine and these were eluted using a stepwise gradient of methanol in chloroform.

Columns of silicic acid have been used extensively to purify fractions previously isolated by other chromatographic procedures. It was mentioned above that Rouser et al.[33] separated the ethanolamine- from the serine-containing phospholipids, which had been eluted from one column of silicic acid, on a second column, with a slight change of solvent composition. Both Rouser, Kritchevsky, Heller and Lieber[38] and O'Brien, Fillerup and Mead[39] purified the choline-containing fraction, obtained on silicic acid, to give homogenous fractions of phosphatidylcholine and sphingomyelin free from each other and from the contaminating neutral lipid. Silicic acid has also proved to be a useful tool in the separation of lysophosphatidylcholine and lysophosphatidylethanolamine from the diacyl or 1-alkenyl-2-acyl analogue[40–42]

Thin-layer chromatography

Thin-layer chromatography (TLC) on silicic acid has proved to be an efficient monitoring system for phospholipids. The number of solvent systems tried and used successfully is so vast that it is impossible to cover all the variables. Some well-tried systems are given in Table 1. For an excellent review of the methods for TLC up to 1961, the reader is referred to the paper by Mangold[54]. Silicic acid as a thin layer on a glass plate is a much more efficient tool for the separation of phospholipids than it is in a column as will be obvious from a study of the R_F-values given in Table 1. Another advantage is the 'second dimension' so that lipids, not separated by one solvent can almost certainly be separated by a second one (Table 2). It may be necessary to degrade the lipid before the second run (see Section 4). After drying the plates, the lipids are located, scraped off the plate with the silicic acid and the phosphorus content is determined. An extremely sensitive method using Rhodamine 6 G has been developed by Schiefer and Neuhoff[59] for the determination of phospholipids previously separated by TLC on microplates (32 × 24 mm). TLC is used extensively to examine the components of the eluates from columns.

An interesting use of TLC was described by Cherayil and Scaria[53]. These workers applied frozen brain sections directly to the silica gel plate and ran the plate in various solvent systems. The phospholipids were extracted and separated as the solvent moved up the plate.

TABLE 1

SINGLE DIMENSIONAL THIN LAYER CHROMATOGRAPHY OF PHOSPHOLIPIDS

Figures in parentheses refer to the solvent systems listed below.

	R_F-values																
	(1)	*(2)*	*(3)*	*(4)*	*(5)*	*(6)*	*(7)*	*(8)*	*(9)*	*(10)*	*(11)*	*(12)*	*(13)*	*(14)*	*(15)*	*(16)*	*(17)*
Phosphatidylcholine	0.24	0.38	0.32	0.30	0.34	0.33	0.16	0.31	0.33	0.26	0.17	0.92	0.31	0.51	0.22	0.69	0.27
Phosphatidylethanolamine	0.46	0.79	0.87	0.42	0.45	0.40	0.45	0.54	0.58	0.52	0.42	0.64	0.96	0.89	0.32	0.43	0.29
Lysophosphatidylcholine	0.03	–	0.10	–	0.11	0.15	0.04	–	0.15	–	0.01	–	0.08	0.17	–	0.43	–
Sphingomyelin	–	0.22	0.19	–	0.18	0.15	0.09	0.11	0.20	–	–	–	0.17	0.31	0.14	0.53	0.15
Phosphatidylserine	–	0.59	0.59	–	0.06	0.07	0.25	0.06	0.17	–	–	0.28	0.46	0.31	0.06	0	0.37
Phosphatidylinositol	0.08	–	0.48	0.10	0.11	0.15	0.34	0.12	0.20	0.13	0.12	0.02	0.46	–	0.11	0	0.32
Phosphatidic acid	0.71	–	–	0.02	0	0	0.82	–	0	0.87	0.68	0.12	0.93	0.07	–	0	–
Cardiolipin	–	–	–	–	0.55	0.50	0.72	0.80	0.84	0.72	–	–	0.93	0.07	0.45	0.24	–
Phosphatidylglycerol	–	–	–	–	–	–	0.61	–	–	0.39	–	–	–	–	–	–	–

Solvent systems:

(1) $CHCl_3$:MeOH:acetic acid:water, 65:15:10:4, by vol.[12].
(2) $CHCl_3$:MeOH:acetic acid:water, 25:15:4:2, by vol.[43].
(3) $CHCl_3$:MeOH:acetic acid:water, 50:25:7:3, by vol.[44].
(4) $CHCl_3$:MeOH:10.5 *M* NH_4OH, 60:25:4, by vol.[12].
(5) $CHCl_3$:MeOH:10.5 *M* NH_4OH, 17:7:1, by vol.[45].
(6) $CHCl_3$:MeOH:6.5 *M* NH_4OH, 17:7:1, by vol.[45].
(7) $CHCl_3$:MeOH:4 *M* NH_4OH, 75:37:7, by vol., after 9 min pre-treatment in tetrahydrofuran:MeOH, 3:1, v/v.[46].
(8) $CHCl_3$:MeOH:water, 65:25:4, by vol.[47].
(9) $CHCl_3$:MeOH:water, 14:6:1, by vol.[48].
(10) $CHCl_3$:acetone:MeOH:acetic acid:water, 5:2:1:1:0.5, by vol.[49].
(11) Diisobutylketone:acetic acid:water, 80:50:7, by vol.[12].
(12) Phenol in water, 436:100, w/v:10.5 *M* NH_4OH, 99:1, v/v.[50].
(13) $CHCl_3$:MeOH:acetic acid:saline, 50:25:8:4, by vol.[51].
(14) $CHCl_3$:MeOH:10.5 *M* NH_4OH, 60:35:8, by vol.[51].
(15) $CHCl_3$:MeOH:9.7 *M* NH_4OH:water, 140:50:7:3, by vol., followed in the same direction after drying with $CHCl_3$:MeOH:acetic acid:water, 160:20:4:1.2, by vol[52].
(16) $CHCl_3$:MeOH:water:pyridine:10.5 *M* NH_4OH, 65:27.5:4:2:2, by vol.[27].
(17) $CHCl_3$:MeOH:water, 65:25:4, by vol.[43, 53].

Stationary phases

(1) to *(12)* incl., silica gel G.
(13) to *(14)*, silica gel HR.
(15) silica gel H with Florisil.
(16) aluminium oxide: calcium sulphate, 9:1 w/w.
(17) hydroxylapatite.

References p. 62

TABLE 2

TWO DIMENSIONAL THIN LAYER CHROMATOGRAPHY OF PHOSPHOLIPIDS

Figures in parentheses refer to solvent systems listed below.

	R_F-values												
	(1)		*(2)*			*(3)*		*(4)*		*(5)*		*(6)*	
	(i)	*(ii)*	*(i)*	*(iia)*	*(iib)*	*(i)*	*(ii)*	*(i)*	*(ii)*	*(i)*	*(ii)*	*(i)*	*(ii)*
Phosphatidylcholine	0.39	0.60	0.26	–	0	0.29	0.49	0.29	0.39	0.67	0.61	0.24	0.53
Phosphatidylethanolamine	0.59	0.62	0.61	0.08	0.12	0.49	0.43	0.51	0.56	0.43	0.09	0.35	0.60
Lysophosphatidylcholine	0.24	0.34	–	–	–	0.09	0.35	–	–	0.43	0.38	–	0.27
Sphingomyelin	–	–	0.14	–	0	0.21	0.41	0.20	0.39	0.54	0.54	0.15	0.38
Phosphatidylserine	0.33	0.49	0.44	0.02	0	0.52	0.32	0.18	0.51	0	0	0.05	0.15
Phosphatidylinositol	0.33	0.17	0.31	0.13	0.20	0.40	0.56	0.27	0.49	0	0	0.08	0.21
Phosphatidic acid	0.59	0.49	–	–	–	0.94	0.78	–	–	0	0	–	–
Cardiolipin	0.69	0.45	0.94	0.93	0.83	0.86	0.90	0.84	0.64	0.25	0	–	–
Phosphatidylglycerol	0.47	0.42	0.48	0.40	0.29	–	–	–	–	–	–	–	–
Lysophosphatidylethanolamine	0.45	0.32	–	–	–	–	–	–	–	–	–	–	0.28
Phosphatidyldimethylaminoethanol	0.59	0.62	–	–	–	–	–	–	–	–	–	–	–

(1) $CHCl_3$:MeOH:acetic acid:water, 50:20:7:3, by vol. *(i)*, then $CHCl_3$:MeOH:40% aqueous methylamine:water, 13:7:1:1, by vol. *(ii)*, silica gel H or HR[55].

(2) $CHCl_3$:MeOH:water, 65:25:4, by vol. *(i)*, then tetrahydrofuran:methylal:MeOH:water, 10:6:4:1, by vol. *(iia)* or tetrahydrofuran:methylal:MeOH:2M-NH_4OH, 10:5:5:1, by vol. *(iib)*[34], silica gel H.

(3) $CHCl_3$:MeOH:acetic acid:water, 250:74:19:3, by vol. *(i)*, then $CHCl_3$:MeOH:7 *M* NH_4OH, 230:90:15, by vol. *(ii)*[56], silica gel G in 0.01 *M* Na_2CO_3.

(4) $CHCl_3$:MeOH:water, 65:25:4, by vol. *(i)*, then *n*-butanol:acetic acid:water, 60:20:20, by vol. *(ii)*, silica gel G:magnesium sulphate 9:1, w/w[57].

(5) $CHCl_3$:MeOH:water:pyridine:10.5 *M* NH_4OH, 65:27.5:4:2:2, by vol. *(i)*, then $CHCl_3$:MeOH:water, 60:40:6, by vol. *(ii)*, aluminium oxide:calcium sulphate (9:1, w/w)[27].

(6) $CHCl_3$:MeOH:10.5 *M* NH_4OH, 65:25:4, by vol. *(i)*, treatment with HCl then $CHCl_3$:MeOH:10.5 *M* NH_4OH, 100:50:12, by vol. *(ii)*, silica gel G[58].

Silica-impregnated paper

Paper, impregnated with silicic acid (Whatman Chromedia S.G. 81),is also a useful method for the separation of intact phospholipids. The method was first described by Lea et al.[28] and has been developed extensively by Marinetti and his colleagues. Much of the work has been reviewed[60]. Marinetti and his colleagues used di-isobutylketone:acetic acid:water in preference to the chloroform-methanol solvent of Lea et al.[28]. The method takes longer than the TLC method but there is a good separation of phospholipids (Table 3).

TABLE 3

CHROMATOGRAPHY OF PHOSPHOLIPIDS ON SILICIC ACID-IMPREGNATED PAPER

Letters in parentheses refer to the solvents listed below.

	R_F-values			
	(*a*)	(*b*)	(*c*)	(*d*)
Phosphatidylcholine	0.37	0.42	0.44	0.38
Phosphatidylethanolamine	0.50	0.70	0.72	0.51
Lysophosphatidylcholine	0.18	–	0.13	0.15
Sphingomyelin	0.31	0.28	0.29	0.26
Phosphatidylserine	0.45	0.60	0.29	0.46
Phosphatidylinositol	0.21	0.09	0.22	0.26
Phosphatidic acid	0.78	–	–	–
Cardiolipin	0.57	–	0.87	0.59
Phosphatidylmonomethylethanolamine	0.50	–	–	–
Phosphatidyldimethylethanolamine	0.48	–	–	–
Lysophosphatidylethanolamine	0.30	–	–	–

(*a*) diisobutyl ketone:acetic acid:water, 40:20:3, by vol.[60].
(*b*) diisobutyl ketone:acetic acid:water:benzene, 160:50:7.5:7, by vol.[61].
(*c*) $CHCl_3$:MeOH:water, 65:25:5, by vol.[62].
(*d*) diisobutyl ketone:acetic acid:water, 40:25:5, by vol.[62].

Detection and qualitative analysis of lipids from thin layer and paper chromatography

There are many methods available for locating the lipids on the chromatograms. Both charring with sulphuric acid and iodine vapour are good general, albeit unspecific, indicators. Rhodamine 6 G in ethanol and 0.2% (w/v) alcoholic 2′7′-dichlorofluorescein[63] have also been widely used and these will react with all lipids to give a complex which fluoresces under UV light. Rouser et al.[57] spray their chromatograms with H_2SO_4-$K_2Cr_2O_7$, char under controlled conditions, then pass the plate through a densitometer to give a quantitative result. Another general spray is the ammonium bisulphate spray described by Adams and Sallee[46]. More specific are the molybdate sprays for phosphorus described by Dittmer and Lester[64], Vaskovsky and Kostetsky[65], Goswami and Frey[66] and Vaskovsky and Svetashev[137], ninhydrin for the free amino groups and Dragendorff's reagent for choline-containing phospholipids[67].

References p. 62

Of all these methods, iodine is the only one which usually does not modify the lipid molecule. It can therefore be used to locate the lipid; the spots are then marked and the iodine is allowed to fade before the areas required are scraped off and the lipid analysed. It is possible, but not entirely satisfactory, to oxidise the silicic acid-bound lipid with perchloric acid to release the phosphorus. Two rather safer methods have been developed. Rosenthal and Ching-Hsien Han[68] release the inorganic phosphorus using 1 ml of a mixture containing 10 *N* H_2SO_4:0.1 *M* periodic acid : 95% v/v ethanol (3:1:6, by vol.) in a boiling water bath for 45 min. This is suitable for all the phospholipids except cardiolipin and sphingomyelin. Parker and Peterson[69] heat the silicic acid in a tube for 3 h at 250° in a heating block with 0.5 ml conc. H_2SO_4. After the addition of 2 to 3 drops of hydrogen peroxide, the tubes are returned to the heating block, now at 160°, for 1 h. The lipids can be eluted from the silicic acid in a small column as described by Cuzner and Davison[45] or by the one-solvent system of Abramson and Belcher[56]. These last workers elute all phospholipids with 4 × 2 ml of chloroform:methanol:water:formic acid (97:97:4:2, by vol.). By eluting the lipid, a more detailed analysis is possible (see Section 4).

DEAE cellulose

The use of DEAE cellulose chromatography for the separation of lipid classes was pioneered by Rouser, O'Brien, Heller and Lieber[70]. As with other media, there is a certain degree of grouping within the fractions (Table 4) so subsequent chromatography on

TABLE 4

CHROMATOGRAPHY OF PHOSPHOLIPIDS ON DEAE CELLULOSE[57] (I)

15 g DEAE cellulose acetate in $CHCl_3$ in a column 2.5 × 20 cm, to which was added 150 mg beef brain total lipid.

Solvent	*Volume (ml)*	*Fractions*
$CHCl_3$:MeOH, 9:1, v/v	440	cholesterol, cerebrosides, PC, SM
$CHCl_3$:MeOH, 7:3, v/v	550	PE
MeOH	350	non-lipid
$CHCl_3$:acetic acid, 1:6, v/v	500	PS
Acetic acid	500	gangliosides
MeOH	400	non-lipid
$CHCl_3$:MeOH, 4:1 v/v, containing 20 ml of 28% aqueous ammonium hydroxide per litre and 0.01 *M* NH_4 acetate	750	sulphatides, PI, PA, TPI, CL

other columns (see above) or on TLC plates is necessary in order to achieve homogeneity. The method has been used extensively for the separation of phospholipids from many animal tissues with very little change from the original solvent pattern.

Gluck, Kulovich and Brody[71] have, however, developed a very different system for a micro method for the measurement of lung tissue phospholipids. After preparing the DEAE celluose acetate in the usual way[38], a column, 0.7 × 9.5 cm of DEAE cellulose, suspended in benzene:acetone (9:1, v/v) was prepared. The lipids were eluted as indicated in Table 5, and the fractions subdivided by TLC. A disadvantage of the method is the toxicity of the benzene vapour.

TABLE 5

CHROMATOGRAPHY OF PHOSPHOLIPIDS ON DEAE CELLULOSE[71] (II)

250 mg of DEAE cellulose acetate in benzene:acetone (9:1, v/v) in a column 0.7 × 9.5 cm to which was added 5–10 mg total lipid.

Solvent	*Volume*[a]	*Fractions*
Benzene:acetone, 9:1, v/v	10	free fatty acids, cholesterol, monoglycerides
Benzene:acetone, 8:2, v/v	10	di- and tri-glycerides
$CHCl_3$:MeOH, 7:3, v/v	10	PC, PE, lysoPC, SM
Ethylacetate:diethylether, 1:1, v/v	10	non-lipid
Ethylacetate:diethylether, 1:1, v/v with 0.05 *N* NH_4OH	10	PS, PI, CL

[a]Column volumes.

A rather specialised method using DEAE cellulose has been developed by Hendrickson and Ballou[72], for the separation of the phosphoinositides with a continuous, linear gradient of increasing concentrations of ammonium acetate. This method was developed primarily as a preparative procedure.

More recently Turner and Rouser[14, 15] described a lipid separation on TEAE cellulose for the analysis of the lipids of erythrocytes[14] and blood plasma[15]. It appears to give a better separation of the lipids than can be obtained with DEAE cellulose. For example, the choline lipids are separated from the neutral lipid contaminants (Table 6).

TABLE 6

CHROMATOGRAPHY OF PHOSPHOLIPIDS ON TEAE CELLULOSE[14]

Solvent[a]	*Fraction*
$CHCl_3$	neutral lipids, cholesterol, cerebrosides, triglyceride
$CHCl_3$:MeOH, 9:1, v/v	choline lipids
$CHCl_3$:MeOH, 2:1, v/v	ceramides
$CHCl_3$:MeOH, 2:1, v/v with 1% acetic acid	PE
Acetic acid	PS
MeOH	non-lipid
$CHCl_3$:MeOH, 4:1, v/v with 20 ml 28% ammonium hydroxide per litre and 0.1 *M* potassium acetate	PA, PI

[a]8 column volumes of each solvent.

References p. 62

Hydroxylapatite

Column chromatography

Another recent development in column chromatography is that described by Slomiany and Horowitz[47]. They apply an extract of serum phospholipids to a column of hydroxylapatite $[Ca_{10}(OH)_2\ (PO_4)_6]$ in chloroform and elute the column with chloroform followed by increasing concentrations of methanol in acetone, methanol alone and finally with alkaline ether-ethanol. Apart from the co-chromatography of phosphatidylserine and phosphatidylinositol in the final fraction, the phospholipid classes appear to be fairly well separated from each other. It has the added advantage of separating phosphatidylcholine from sphingomyelin.

TABLE 7

CHROMATOGRAPHY OF PHOSPHOLIPIDS ON HYDROXYLAPATITE[47]

20 g of Bio gel HTP in $CHCl_3$ in a column 1.2 × 35 cm to which was added 250 mg total lipid.

Solvent		*Volume (ml)*	*Fraction*
$CHCl_3$		250	neutral lipid
Acetone:EtOH,	9:1, v/v	250	CL
	7:3, v/v	300	PC
	5:5, v/v	250	SM
	3:7, v/v	250	cerebrosides
MeOH		300	cerebrosides and PE
Diethyl ether:EtOH:aqueous 0.04 *N* KOH, 10:7:5, by vol.		650	PS, PI

Thin-layer chromatography

Hydroxylapatite was used for TLC before it was used for column chromatography. Hofmann[73] separated liver phospholipids using a mixture of hydroxylapatite:calcium sulphate (100:8, w/w) in 1,2-dichloroethane:ethanol:water (262:170:24, by vol.).

Paper chromatography

Kennedy and Collier[74] described a method for separating intact phospholipids by ascending chromatography on Whatman No. 1 paper in organic solvent systems. The spots were located with solochrome dyes.

Extraction and separation of the higher phosphoinositides

It has been stated that the procedure of Folch et al.[5] for the extraction of phospholipids fails to extract the higher phosphoinositides. These phospholipids are readily hydrolysed *post mortem* so the following procedure for their extraction and separation has been developed.

The tissue is rapidly frozen *in situ*, then removed from the body, and homogenised in chloroform:methanol (1:1, v/v) followed by a re-extraction of the residue with chloroform:methanol (2:1, v/v). These two extractions remove most of the phospholipid. The higher phosphoinositides are now extracted with chloroform:methanol (2:1, v/v) containing 0.25% (v/v) conc. hydrochloric acid. This extraction is carried out at 37° for 20 min, and is repeated a further twice with fresh solvent. The pooled extracts are taken to dryness *in vacuo* and the residue dissolved in chloroform:methanol (2:1, v/v). This is washed twice with 10 mM calcium chloride[6,7,75].

The washed extract can now be separated into its components by chromatography on either formaldehyde-treated paper[76] or by TLC[77]. With formaldehyde-treated paper and the ether-equilibrated upper phase of a mixture of *n*-butanol:acetic acid:water (4:1:5, by vol.), the R_F-values are as follows: phosphatidylinositol (MPI), 0.47; phosphatidylinositol monophosphate (DPI), 0.41; phosphatidylinositol diphosphate (TPI), 0.36. The inositol-containing spots can be visualised with a 0.002% (w/v) solution of Nile Blue in 0.1 N sulphuric acid. There are two suitable solvent systems for the separation of the inositol-containing phospholipids on TLC. The silica gel H is prepared with 1% (w/v) potassium oxalate. The R_F-values in the two solvents are as follows:

	MPI	DPI	TPI
$CHCl_3$:MeOH:4 N NH_4OH (9:7:2, by vol.)	0.78	0.36	0.14
n-propanol:4 N NH_4OH (2:1, v/v)	0.55	0.43	0.29

Cicero and Sherman[78] have devised a GLC method for the separation of the phosphoinositides. The washed extract is lyophilised and left overnight at room temperature in *N,O*-bis(trimethylsilyl)trifluoroacetamide containing 1% (v/v) trimethylchlorosilane: dry pyridine (2:1, v/v). The three phosphatidylinositides can then be separated by GLC.

Determination of inositol

Free *myo*-inositol can be formed from the intact lipid by acid hydrolysis, and its concentration measured by bioassay. The most convenient methods are those of Norris and Darbre[79], using *Schizosaccharomyces pombe*, and Campling and Nixon[79a] using *Kloeckera brevis*. Several colorimetric methods for the determination of inositol have been developed (e.g. ref. 80) but in all of them, sugars interfere.

In 1965 four groups of workers reported a technique for the GLC determination of inositols as their trimethylsilyl derivatives[81–84]. There are minor differences between the four methods but the essential details are the same. The intact lipid is hydrolysed to release the inositol which is converted to the trimethylsilyl derivative in the presence of a mixture of hexamethyldisililazone–trimethylchlorosilane in dry pyridine. The derivatives are then separated by GLC and the concentration of the inositols determined by comparison with standard samples.

References p. 62

In a recent paper[84a], a method for the analysis of the molecular species of the mono-, di- and triphosphoinositides from bovine brain was described. The lipids were isolated by the method described above, and converted to their diglycerides by the action of a phosphodiesterase. This gave a 60–70% hydrolysis. The diglycerides were then acetylated and separated by argentation TLC. The solvent system contained 0.7 or 0.8% methanol in chloroform. The bands were visualised, scraped from the plates and suspended in diethyl ether:methanol:glacial acetic acid (60:40:1, by vol.), 5 ml to 0.5 g of silicic acid. The mixture was diluted to 25% water and extracted 3 times with an equal volume of petroleum ether. This extract was then analysed by GLC. Two further methods for the fractionation of phosphatidylinositol into molecular species by TLC on silver-impregnated silicic acid, have been described by Luthra and Sheltawy[138]. In one method the phosphorus as well as the fatty acids remain intact and in the other the phosphorus and inositol ring are preserved.

4. ANALYSIS OF INTACT LIPID FRACTIONS

It has been shown above that no one chromatographic method has as yet been devised which will separate a total phospholipid extract into single, homogenous fractions. Two-dimensional TLC comes nearest to being the ideal method. However, none of the chromatographic methods which have been discussed, separate the 1-alkenyl-2-acyl phospholipid from the diacyl phospholipid. If these are the only two lipids present in a sample, a determination of the alk-1′-enyl ether groups[85] will give the concentration of the 1-alkenyl-2-acyl phospholipid, and the 1,2-diacyl phospholipid can be determined by difference.

The lipids can also be separated by modified TLC[54, 86, 87]. The lipids are separated on a plate of silica gel G in chloroform:methanol:15 *N* ammonium hydroxide (65:25:4 by vol.), dried, and the plate is exposed to hydrochloric acid fumes. This breaks the alk-1-enyl ether linkage of the plasmalogen converting it to the lyso-phospholipid. The 1,2-diacyl phospholipid is unaffected by the acid. The plate is then developed in the second dimension in chloroform:methanol:15 *N*-ammonium hydroxide (100:50:12, by vol.). This solvent system clearly separates the intact diacyl lipid from the lyso-lipid (Table 2). They are located with iodine vapour, the spots scraped from the plate and the concentration of each lipid is determined by the phosphorus content[58]. It does not appear to be possible to separate the two lipids without partially degrading one of them, though Hoevet, Viswanathan and Lundberg[17] have achieved a partial separation of the choline analogues by argentation TLC.

Fatty acids and aldehydes

Any phospholipid, purified by chromatography, will normally still be a heterogenous mixture containing a wide variety of fatty acids and, in the case of the plasmalogens, fatty aldehydes. These may be removed from the parent molecule by hydrolysis and converted to their methyl esters or dimethyl acetals, respectively. These are separated, identified and the percentage of each measured by GLC.

Fatty acids

The fatty acids are analysed as their methyl esters by GLC. The method of Ways, Reed and Hanahan[88] is simple and reproducible. The phospholipid is refluxed for 4 h at 60–70° in benzene:methanol:conc. sulphuric acid (0.5:2.0:0.04, by vol.). After cooling, the methyl esters are extracted into light petroleum (b.p. 40–60°) and the extract washed with water. The washed extract is dried with anhydrous sodium sulphate and a sample injected on to a column of 10% SE 30 on 100/120 mesh diatomite C and run at 230°. The method recommended by the American Oil Chemists Society for the preparation of methyl esters is that using boron trifluoride in methanol[89].

Fatty aldehydes

In the presence of acid-methanol, plasmalogens give methyl esters of the fatty acids and dimethyl acetals of the fatty aldehydes[59,90]. These are then analysed by GLC.

It was shown, however, by Stein and Slawson[91] and by Mahadevan, Viswanathan and Phillips[92] that, during GLC, the dimethyl acetals tended to yield methyl-1-alkenyl ethers or the parent fatty aldehyde. Stein and Slawson[91] found that this change was dependent to a great extent upon the type of column, and in particular the nature of the stationary phase used for the separation.

In view of the apparent instability of the dimethyl acetals, Ferrell, Radloff and Radloff[93] have developed a method in which the intact plasmalogen is degraded with hydrochloric acid gas and the products, free fatty aldehydes and lysophospholipid, are separated on a column of silicic acid. The free aldehydes are analysed by GLC on 16% Apiezon M on 60–80 mesh Gas Chrom Q at 199°. The methyl esters of the fatty acids obtained from the lysophospholipid are analysed in the usual way. Wood and Harlow[94] also give a system for the separation of free aldehydes on GLC. They use a column of 15% ethylene glycol succinate–methyl silicone polymer (EGSS–X) on 100–120 mesh Gas Chrom P, at 130–180°.

Other GLC methods include that of Rao et al.[27] in which the fatty aldehydes are converted to cyclic acetals before GLC and that of Johnson, Pearson and Dugan[95] in which the conversion is to the dimethylhydrazones of the fatty aldehydes.

Some tissues, particularly those of the invertebrates, may contain a high concentration of ethanolamine phospholipid in the 1-alkyl-2-acyl form. The reader is referred to a recent review by Snyder[96] for the determination of this lipid.

Snyder, Blank and Wykle have recently described[97] a method for separating the alkyl-acyl glycerophosphorylethanolamines from the alkenyl-acyl glycerophosphoryl-ethanolamines. The mixture is reacted with 70% Vitride [$NaAlH_2(OCH_2CH_2OCH_3)_2$] in benzene. This removes the fatty acid from the 2-position and the phosphorylethanol-amine from the 3-position. The two glycerols (alkyl- and alkenyl-) may be separated by TLC on silica gel G in diethyl ether:acetic acid (100:1, v/v). The 1-alkenyl-glycerol has an R_F of 0.45 and the 1-alkyl-glycerol of 0.35. This method appears to be more quantitative than the much used earlier one of Thompson and Lee[98] who degraded the lipids with $LiAlH_4$ to give glyceryl cyclic acetal from plasmalogens and glyceryl ethers from the saturated ether lipids. The products were separated by TLC on silica gel in light petroleum (b.p. 30–60°):diethyl ether:acetic acid (30:70:1, by vol.), with R_F values of 0.50 for the glyceryl cyclic acetal and 0.38 for the glyceryl ether.

References p. 62

Molecular species

With GLC, it is necessary to degrade the intact phospholipid to determine the fatty acids and aldehydes but with the development of argentation TLC on silica gel[99], it became possible to separate intact phospholipids into their molecular species by TLC[100, 101].

The plates are prepared either by treating them with a solution of silver nitrate or by suspending the silicic acid in a solution of silver nitrate before spreading it on the plate. (If the spreader is made of aluminium, the second procedure is inadvisable.) The plates are then activated in the normal way. With this medium it is possible to separate phosphatidylcholine, phosphatidylethanolamine and phosphatidylinositol into sub-fractions dependent upon the degree of unsaturation of their fatty acids. The length of time the plates are activated appears to affect the separation[101]. Early work on argentation TLC has been reviewed by Morris[99].

Several different solvent systems have been used. Van Golde and Van Deenen[102] sub-fractionated liver phosphatidylcholine in chloroform:ethanol:water (65:35:4, by vol.) and phosphatidylethanolamine from *E. coli* in chloroform:ethanol (98:2, v/v)[103]. Arvidson[101] used chloroform:methanol:water (55:35:7, by vol.) in separating liver phosphatidylethanolamine and Hopkins, Sheehan and Lymen[104], chloroform:methanol:acetic acid:water (25:25:4:2, by vol.).

Various additive procedures followed by TLC have been tried such as ozonolysis[105, 106] and mercuric acetate[107, 108] but argentation TLC has tended to supersede these methods.

All these methods have the disadvantage of failing to distinguish between the fatty acids on the 1- and 2-positions of the diacyl phospholipids. The use of phospholipases together with one or more of the above methods, has made it possible to distinguish between the two fatty acids, and this has led to a much greater knowledge of the fatty acid composition of some of the phospholipids. Phospholipase A_2, C and D have all been used to give 1-acyl-glycero-3-phosphorylcholine (ethanolamine), 1,2-diacyl-*sn*-glycerol and 1,2-diacyl-*sn*-glycerophosphate respectively. These compounds can then be analysed and separated using GLC and TLC. Several workers have described the structure of the phosphatidylcholines using such a method[102, 103, 109–111]. A summary of the method is as follows:

Phospholipid extract

↓ TLC

phosphatidylcholine extracted, decomposed
with phospholipase C

↓

1,2-diglyceride

↓ argentation TLC

molecular sub-fraction acted upon with
pancreatic lipase$_2$ –GLC.

A similar system may be envisaged for other diacylphospholipids and also for plasmalogens.

A review of the work on the analysis of the molecular species up to 1967 is given by Renkonen[112]. In this review[112], Renkonen mentions the work of Collins who achieved a partial separation of the phosphatidylcholines using a 120-tube counter-current apparatus. The solvent used was chloroform:methanol:water (124:70:7, by vol.) containing the antioxidant *n*-propyl-gallate. Collins[113] has also used the counter-current apparatus to separate the acidic phospholipids cardiolipin, phosphatidic acid and phosphatidylserine after converting them to their dinitrophenyl derivatives. A sample of total lipid (less than 1 g) is dissolved in benzene and left at room temperature with 1-fluoro-2:4-dinitrobenzene and triethylamine for 2 h.

Bases

Ethanolamine and serine

Ethanolamine and serine can be determined after their release from the lipid by acid hydrolysis. In the method of Axelrod, Reichenthal and Brodie[114], the bases are converted to their dinitrofluorobenzene derivatives. The ethanolamine derivative is extracted into chloroform while the serine derivative may be extracted from the remaining aqueous phase by acidifying and shaking with methyl isobutyl ketone. After washing the extract, the colour of the complex is read in aqueous acid in a spectrophotometer at 420 nm. The method is suitable for 5–50 μg of base. In most of the chromatographic methods described for the separation of intact lipids, the ethanolamine-containing phospholipids are separated from those containing serine. It is therefore possible to use the method of Siakotos[115]. The intact lipid (0.05–0.40 μmoles) is treated with neutralised hyamine hydroxide, the trinitrobenzene sulphonic acid derivative is made and this is measured spectrophotometrically.

Some of the simpler plants like the yeasts contain a high percentage of phosphatidylmonomethylethanolamine and phosphatidyldimethylethanolamine. Letters[25] demonstrated that these lipids could be separated from each other and from phosphatidylcholine and phosphatidylethanolamine on alumina (see Section 3). Lester and White[116] have demonstrated that after hydrolysing a total phospholipid extract with aqueous 6 *N* hydrochloric acid, it is possible to extract the ethanolamine, monomethylethanolamine and dimethylethanolamine from the neutralised hydrolysate with tertiary butanol and to separate these compounds by GLC on Chromasorb P (80–100 mesh) coated with 25% Dowfax 9N9 and 2.5% sodium hydroxide.

Choline

It is difficult to find a method for the determination of choline which is specific and sensitive. Methods have been developed which necessitate the isolation of the base from the lipid and the subsequent formation of the reineckate or periodide[1, 117]. A recent enzymic radioassay method using choline kinase and [^{32}P]orthophosphate has been developed by Reid, Hanbrick and Krishna[118]. Probably the most accurate method, however, for phosphatidylcholine is to treat the lipid with alcoholic alkali which readily releases the water-soluble glycerophosphorylcholine. This can be isolated by paper chromatography (Table 8) and the phosphorus content determined[121].

References p. 62

TABLE 8

PAPER CHROMATOGRAPHY OF THE WATER-SOLUBLE ALKALINE BREAKDOWN PRODUCTS OF PHOSPHOLIPIDS

Letters in parentheses refer to the solvents listed below.

	R_F-values					
	(*a*)	(*b*)	(*c*)	(*d*)	(*e*)	(*f*)
Glycerophosphorylcholine	0.45	0.93	0.77	0.67	–	–
Glycerophosphorylethanolamine	0.45	0.61	0.63	0.47	0.55	0.53
Glycerophosphoryldimethylethanolamine	0.56	0.88	–	–	–	–
Glycerophosphorylmonomethylethanolamine	0.68	0.84	–	–	–	–
Glycerophosphorylserine	0.15	0.27	0.30	0.50	0.17	0.22
Glycerophosphorylinositol	0.22	0.15	0.20	0.45	–	0.11
bis-Glycerophosphorylglycerol	0.22	0.28	0.28	0.60	0.10	–
Glycerophosphorylglycerol	0.54	0.45	0.26	0.60	0.35	0.39

(*a*) iso-propanol:water:ammonium hydroxide, 7:2:1, by vol.[25].
(*b*) phenol:acetic acid:EtOH:water, 80:10:12:20, by vol.[25].
(*c*) phenol(water-saturated):acetic acid:EtOH, 100:10:12, by vol.[23].
(*d*) MeOH:formic acid:water, 80:13:7, by vol.[23].
(*e*) phenol(water saturated):0.88 ammonium hydroxide, 99.88:0.12, v/v[119].
(*f*) phenol:water, 100:38, w/v[120].

Sphingosine

Sphingosine can be released from sphingomyelin and other sphingosine-containing phospholipids, with 6 *N* hydrochloric acid[122]. It may then be determined by one of a number of rather unspecific colorimetric techniques. Not one of these techniques distinguishes between the various sphingosines e.g. dihydrosphingosine, phytosphingosine.

However, Sweeley and Moscatelli[123] devised a method for separating a mixture of long chain bases by GLC and determining their concentration quantitatively. Since then several methods have been developed for the determination of the sphingosines.

Sphingosine, released from the sphingomyelin with hydrochloric acid, is converted to the corresponding aldehyde with sodium periodate. The aldehydes are separated by GLC on 8% diethylene glycol succinate polyester on 115–150 mesh silanized Gas Chrom P at 160° or 130° (ref. 124) or on 15% diethylene glycol adipate polyester on 60–80 mesh Chromasorb W[123]. Greene, Kaneshiro and Law[125] converted the sphingosines to fatty acids with periodate–permanganate reagent and subsequently to the methyl esters with diazomethane before separating them by GLC.

Gaver and Sweeley[126] found that the presence of water during the hydrolysis of the sphingomyelin, prevented the production of methylsphingosine. Sphingomyelin was hydrolysed in a reagent containing 8.6 ml of 11 *N* hydrochloric acid with 9.4 ml of water made up to 100 ml with methanol. For every 2 mg of sphingomyelin, 1 ml of this reagent was required, and the reaction was allowed to proceed at 70° for 18 h. The sphingosines were purified on a column of Amberlite CG 45 [OH^-] resin then converted to their trimethylsilyl derivatives. These can be separated by GLC on a column of 2.5% SE-30 on 100–120 mesh silanized Gas Chrom S at 180°.

The sphingosines can be separated by TLC on silica gel containing 10% calcium sulphate[126a]. The plate is developed in chloroform:methanol:2 *N* ammonium hydroxide (40:15:1.5, by vol.). This will separate the sphingosine, methylsphingosine and dihydrosphingosine. The spots can be visualised with iodine vapour or one of the fluorescein dyes then eluted with chloroform:methanol (1:1, v/v) and a colorimetric determination like that of Lauter and Trams[127] applied.

5. ANALYSIS OF PHOSPHOLIPIDS BY SELECTIVE HYDROLYSIS

It has been shown above, that with a combination of column and thin-layer chromatography, it is possible to separate and to isolate intact phospholipids. The concentration of the lipid can then be determined by one of the methods outlined.

There is, however, another approach whereby the total lipid extract is hydrolysed by selective hydrolysis to give the water-soluble glycerophosphate esters of the parent lipids and these may be separated by paper chromatography and determined by their phosphorus content. This approach has been developed by Dawson and his colleagues[8, 128]. The method takes no note, of course, of the highly complex pattern of fatty acids and fatty aldehydes. An outline of the method is given below:

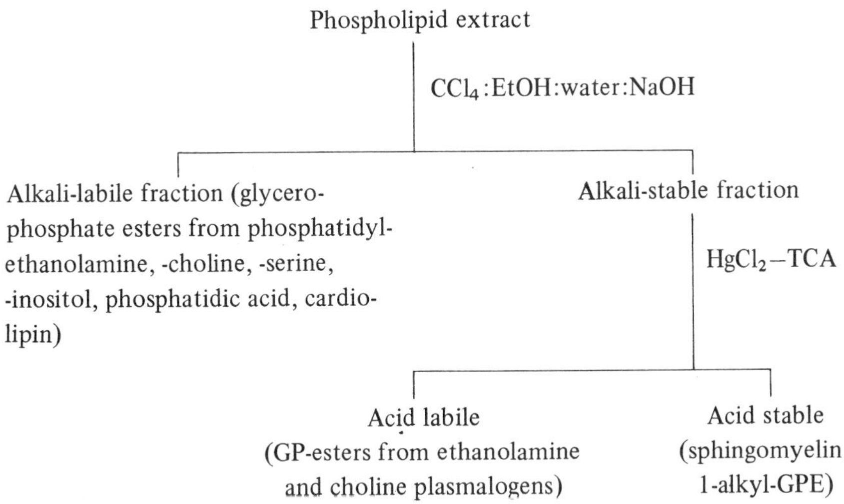

Very many modifications of the method have been described. Dawson and Clarke[129] hydrolyse the total phospholipid extract for 1 h (ref. 130) and separate the alkali-labile fraction by two-dimensional paper chromatography (Table 8). The alkali-stable fraction, however, is subjected to TLC on silica gel F 254 plates in chloroform:methanol:diethylamine:water (110:50:8:5, by vol.). This separates the sphingomyelin (R_f 0.22), the deacylated 1-alkyl- and 1-alkenyl-3-glycerophosphorylethanolamine (R_F 0.55) and the 1-alkyl- and 1-alkenyl-3-glycerophosphorylcholine (R_F 0.11).

Another modification is that of Pries, Aumont and Böttcher[131] who did a preliminary column separation of their phospholipids on silicic acid before hydrolysis. This method could be adapted for the determination of the phospholipids after any of the column procedures described in Section 3.

Wells and Dittmer[132] have described a microanalytical method for determining the brain phospholipids, employing selective mild alkaline and acid hydrolysis followed by the chromatography of the water-soluble esters on anion exchange resins. The phospholipids which are stable to this mild treatment are separated on a column of silicic acid:Hyflo super cel (2:1, w/w). After washing through with chloroform:methanol: water (11:1.0:0.12, by vol.), the 1-alkenyl-3-glycerophosphorylethanolamine and the 1-alkyl-3-glycerophosphorylethanolamine are eluted sequentially with chloroform: methanol:water (6:1:0.07, by vol.).

The combination of GLC and mass spectrometry is now being developed for the detailed analysis of phospholipids. The phospholipids are extracted and deacylated to give the glycerophosphate esters of the bases[133]. These are isolated by passing them through Dowex 50 [H^+] and are then converted to their trimethylsilyl derivatives. The individual components are separated by GLC and the individual peaks analysed by mass spectrometry. This method has been applied to extracts of mixed lipids[133], the phosphoinositides[134] and cardiolipin[135].

REFERENCES

1 G.B. Ansell and J.N. Hawthorne, *Phospholipids – Chemistry, Metabolism and Function*, Elsevier, Amsterdam, 1964.
2 A.T. James and L.J. Morris, *New Biochemical Separations*, Nostrand, London, 1964.
3 G. Marinetti, *Lipid Chromatographic Analysis*, Vol. 1, Arnold, London, 1967.
4 R. Fried, *Methods of Neurochemistry*, Vol. 1., Dekker, New York, 1971.
5 J. Folch, M. Lees and G.H. Sloane Stanley, *J. Biol. Chem.*, 226 (1957) 497.
6 J.C. Dittmer and R.M.C. Dawson, *Biochem. J.*, 81 (1961) 535.
7 M. Kai and J.N. Hawthorne, *Biochem. J.*, 98 (1966) 62.
8 R.M.C. Dawson, N. Hemington and D.B. Lindsay, *Biochem. J.*, 77 (1960) 226.
9 J.J. Wren and H.K. Mitchell, *J. Biol. Chem.*, 234 (1959) 2823.
10 E.G. Bligh and W.J. Dyer, *Canad. J. Biochem. Physiol.*, 37 (1959) 911.
11 F.B. St. C. Palmer, *Biochim. Biophys. Acta*, 231 (1971) 134.
12 B.W. Nichols, in A.T. James and L.J. Morris (Eds.), *New Biochemical Separations*, Nostrand, London, 1964, p. 321.
13 M.A. Wells and J.C. Dittmer, *Biochemistry*, 2 (1963) 1259.
14 J.D. Turner and G. Rouser, *Analyt. Biochem.*, 38 (1970) 423.
15 J.D. Turner and G. Rouser, *Analyt. Biochem.*, 38 (1970) 437.
16 J.P. Williams and P.A. Merrilees, *Lipids*, 5 (1970) 367.
17 S.P. Hoevet, C.V. Viswanathan and W.O. Lundberg, *J. Chromatog.*, 34 (1968) 195.
18 B. Borgström, *Acta Physiol. Scand.*, 25 (1952) 101.
19 M.L. Vorbeck and G.V. Marinetti, *J. Lipid Res.*, 6 (1965) 3.
20 G. Rouser, G. Kritchevsky and A. Yamamoto, in G.V. Marinetti (Ed.), *Lipid Chromatographic Analysis*, Vol. 1, Arnold, London, 1967, p. 99.
21 D.L. Fillerup and J.F. Mead, *Proc. Soc. Exptl. Biol. Med.*, 83 (1953) 574.

22 A.F. Robertson and W.E.M. Lands, *Biochemistry*, 1 (1962) 804.
23 R.M.C. Dawson, N. Hemington and J.B. Davenport, *Biochem. J.*, 84 (1962) 497.
24 A.N. Davison and M. Wajda, *J. Neurochem.*, 4 (1959) 353.
25 R. Letters, *Biochim. Biophys. Acta*, 116 (1966) 489.
26 G.B. Ansell and T. Chojnacki, *Biochem. J.*, 98 (1966) 303.
27 P.V. Rao, S. Ramachandran and D.G. Cornwell, *Lipids*, 3 (1968) 187.
28 C.H. Lea, D.N. Rhodes and R.D. Stoll, *Biochem. J.*, 60 (1955) 353.
29 D.J. Hanahan, J.C. Dittmer and E. Warashina, *J. Biol. Chem.*, 228 (1957) 685.
30 H. Bader and H.E. Morgan, *Biochim. Biophys. Acta*, 57 (1962) 562.
31 H. Nielson, *Biochim. Biophys. Acta*, 231 (1971) 370.
32 L. Svennerholm, *J. Neurochem.*, 11 (1964) 839.
33 G. Rouser, J. O'Brien and D. Heller, *J. Am. Oil Chem. Soc.*, 38 (1961) 14.
34 G.M. Gray, *Biochim. Biophys. Acta*, 144 (1967) 511.
35 P. Ways and D.J. Hanahan, *J. Lipid Res.*, 5 (1964) 318.
36 L. Rathbone, *Biochem. J.*, 85 (1962) 461.
37 M.O. Hall and J.F. Nyc, *J. Lipid Res.*, 2 (1961) 321.
38 G. Rouser, G. Kritchevsky, D. Heller and E. Lieber, *J. Am. Oil Chem. Soc.*, 40 (1963) 425.
39 J.S. O'Brien, D.L. Fillerup and J.F. Mead, *J. Lipid Res.*, 5 (1964) 329.
40 O. Renkonen, *Acta Chem. Scand.*, 17 (1963) 275.
41 O. Renkonen, *Acta Chem. Scand.*, 17 (1963) 634.
42 G.B. Ansell and S. Spanner, in R. Fried, (Ed.), *Methods of Neurochemistry*, Vol. 1, Dekker, New York, 1971, p. 31.
43 Y. Tamai, S. Matsukawa and M. Satake, *Brain Res.*, 26 (1971) 149.
44 V.P. Skipski, R.F. Peterson, J. Sanders and M. Barclay, *J. Lipid Res.*, 4 (1963) 227.
45 M.L. Cuzner and A.N. Davison, *J. Chromatog.*, 27 (1967) 388.
46 G.M. Adams and T.L. Sallee, *J. Chromatog.*, 49 (1970) 552.
47 B.L. Slomiany and M.I. Horowitz, *J. Chromatog.*, 49 (1970) 455.
48 R.C. Roozemond, *J. Histochem. Cytochem.*, 15 (1967) 526.
49 M. Lepage, *Lipids*, 2 (1967) 244.
50 C.M. Redman and R.W. Keenan, *J. Chromatog.*, 15 (1964) 180.
51 F. Snyder, in H. Busch (Ed.), *Methods in Cancer Research*, Vol. VI, Academic Press, London, 1971, p. 399.
52 N.M. Neskovic and D.M. Kostic, *J. Chromatog.*, 35 (1968) 297.
53 G.D. Cherayil and K.S. Scaria, *J. Lipid Res.*, 11 (1970) 378.
54 H.K. Mangold, *J. Am. Oil Chem. Soc.*, 38 (1961) 708.
55 G.S. Getz, S. Jakovcic, J. Heywood, J. Frank and M. Rabinowitz, *Biochim. Biophys. Acta*, 218 (1970) 441.
56 D. Abramson and M. Blecher, *J. Lipid Res.*, 5 (1964) 628.
57 G. Rouser, C. Galli, E. Lieber, M.L. Blank and O.S. Privett, *J. Am. Oil Chem. Soc.*, 41 (1964) 836.
58 L.A. Horrocks, *J. Lipid Res.*, 9 (1968) 469.
59 H.-G. Schiefer and V. Neuhoff, *Z. Physiol. Chem.*, 352 (1971) 913.
60 G.V. Marinetti, *J. Lipid Res.*, 3 (1962) 1.
61 S.-S. Tsao and W.E. Cornatzer, *Lipids*, 2 (1967) 41.
62 G.V. Marinetti, *J. Lipid Res.*, 6 (1965) 315.
63 D.C. Malins and H.K. Mangold, *J. Am. Oil Chem. Soc.*, 37 (1960) 576.
64 J.C. Dittmer and R.L. Lester, *J. Lipid Res.*, 5 (1964) 126.
65 V.E. Vaskovsky and E.Y. Kostetsky, *J. Lipid Res.*, 9 (1968) 396.
66 S.K. Goswami and C.F. Frey, *J. Lipid Res.*, 12 (1971) 509.
67 W.D. Skidmore and C. Entenman, *J. Lipid Res.*, 3 (1962) 471.
68 A.F. Rosenthal and S. Ching-Hsien Han, *J. Lipid Res.*, 10 (1969) 243.
69 F. Parker and N.F. Peterson, *J. Lipid Res.*, 6 (1965) 455.
70 G. Rouser, A.J. Bauman, G. Kritchevsky, D. Heller and J.S. O'Brien, *J. Am. Oil. Chem. Soc.*, 38 (1961) 544.
71 L. Gluck, M.V. Kulovich and S.J. Brody, *J. Lipid Res.*, 7 (1966) 570.
72 H.S. Hendrickson and C.E. Ballou, *J. Biol. Chem.*, 239 (1964) 1369.
73 A.F. Hofmann in A.T. James and L.J. Morris (Eds.), *New Biochemical Separations*, Nostrand, London, 1964, p. 262.

74 G.Y. Kennedy and R. Collier, *Comp. Biochem. Physiol.*, 7 (1962) 179.
75 R.M.C. Dawson and J. Eichberg, *Biochem. J.*, 96 (1965) 634.
76 H. Wagner, A. Lissau, J. Hölzl and L. Horhammer, *J. Lipid Res.*, 3 (1962) 177.
77 F. Gonzalez-Sastre and J. Folch-Pi, *J. Lipid Res.*, 9 (1968) 532.
78 T.J. Cicero and W.R. Sherman, *Biochem. Biophys. Res. Commun.*, 42 (1971) 428.
79 F.W. Norris and A. Darbre, *Analyst*, 81 (1956) 394.
79a J.D. Campling and D.A. Nixon, *J. Physiol.*, 126 (1954) 71.
80 M.K. Gaitonde and M. Griffiths, *Analyt. Biochem.*, 15 (1966) 532.
81 D.R. Flint, T.-C. Lee and C.G. Huggins, *Federation Proc.*, 24 (1965) 662.
82 Y.C. Lee and C.E. Ballou, *J. Chromatog.*, 18 (1965) 147.
83 W.W. Wells, T.A. Pittman and H.J. Wells, *Analyt. Biochem.*, 10 (1965) 450.
84 R.N. Roberts, J.A. Johnston and B.W. Fuhr, *Analyt. Biochem.*, 10 (1965) 282.
84a B.J. Holub, A. Kuksis and W. Thompson, *J. Lipid Res.*, 11 (1970) 558.
85 E.L. Gottfried and M.M. Rapport, *J. Biol. Chem.*, 237 (1962) 329.
86 K. Owens, *Biochem. J.*, 100 (1966) 354.
87 C.V. Viswanathan, F. Phillips and W.O. Lundberg, *J. Chromatog.*, 35 (1968) 66.
88 P. Ways, C.F. Reed and D.J. Hanahan, *J. Clin. Invest.*, 42 (1963) 1248.
89 W.R. Morrison and L.M. Smith, *J. Lipid Res.*, 5 (1964) 600.
90 J.W. Farquhar, *J. Lipid Res.*, 3 (1962) 21.
91 R.A. Stein and V. Slawson, *J. Chromatog.*, 25 (1966) 204.
92 V. Mahadevan, C.V. Viswanathan and F. Phillips, *J. Lipid Res.*, 8 (1967) 2.
93 W.J. Ferrell, D.M. Radloff and J.F. Radloff, *Analyt. Biochem.*, 37 (1970) 227.
94 R. Wood and R.D. Harlow, *J. Lipid Res.*, 10 (1969) 463.
95 C.B. Johnson, A.M. Pearson and L.R. Dugan Jr., *Lipids*, 5 (1970) 958.
96 F. Snyder, in G. Pataki and A. Niederwieser, (Eds.), *Progress in Thin Layer Chromatography and Related Methods, Ann Arbor Science, 1969.*
97 F. Snyder, M.L. Blank and R.L. Wykle, *J. Biol. Chem.*, 246 (1971) 3639.
98 G.A. Thompson Jr. and P. Lee, *Biochim. Biophys. Acta*, 98 (1965) 151.
99 L.J. Morris, *J. Lipid Res.*, 7 (1966) 717.
100 G.A.E. Arvidson, *J. Lipid Res.*, 6 (1965) 574.
101 G.A.E. Arvidson, *Europ. J. Biochem.*, 4 (1968) 478.
102 L.M.G. van Golde and L.L.M. van Deenen, *Biochim. Biophys. Acta*, 125 (1966) 496.
103 L.M.G. van Golde and L.L.M. van Deenen, *Chem. Phys. Lipids*, 1 (1967) 157.
104 S.M. Hopkins, G. Sheehan and R.L. Lyman, *Biochim. Biophys. Acta*, 164 (1968) 272.
105 O.S. Privett and M.L. Blank, *J. Am. Oil Chem. Soc.*, 40 (1963) 70.
106 C.F. Wurster Jr. and J.H. Copenhaver Jr.,*Biochim. Biophys. Acta*, 98 (1965) 351.
107 R.L. King and J.A. Clements, *J. Lipid Res.*, 11 (1970) 381.
108 D. Malins, in R.T. Holman (Ed.), *Progress in the Chemistry of Fats and Other Lipids*, vol. 8, Pergamon, Oxford, 1966, p. 307.
109 L.M.G. van Golde, W.A. Pieterson and L.L.M. van Deenen, *Biochim. Biophys. Acta*, 152 (1968) 84.
110 A. Montfoort, L.M.G. van Golde and L.L.M. van Deenen, *Biochim. Biophys. Acta*, 231 (1971) 335.
111 K.A. Devor and J.B. Mudd, *J. Lipid Res.*, 12 (1971) 396.
112 O. Renkonen, *Advan. Lipid Res.*, 5 (1967) 329.
113 F.D. Collins, in A.T. James and L.J. Morris (Eds.), *New Biological Separations*, Nostrand, London, 1964, p. 380.
114 J. Axelrod, J. Reichenthal and B.B. Brodie, *J. Biol. Chem.*, 204 (1953) 903.
115 A.N. Siakotos, *Lipids*, 2 (1967) 87.
116 R.L. Lester and D.C. White, *J. Lipid Res.*, 8 (1967) 565.
117 C. Long, R. Odavic and E.J. Sargent, *Biochem. J.*, 102 (1967) 216.
118 W.D. Reid, D.R. Haubrich and G. Krishna, *Analyt. Biochem.*, 42 (1971) 390.
119 A.R. Tarlov and E.P. Kennedy, *J. Biol. Chem.*, 240 (1965) 49.
120 T.G. Tornabene and J.E. Ogg, *Biochim. Biophys. Acta*, 239 (1971) 133.
121 L. Ernster, R. Zetterström and O. Lindberg, *Acta Chem. Scand.*, 4 (1950) 942.
122 E. Robins, O.H. Lowry, K.M. Eydt and R.E. McCaman, *J. Biol. Chem.*, 220 (1956) 661.

123 C.C. Sweeley and E.A. Moscatelli, *J. Lipid Res.*, 1 (1959) 40.
124 E.A. Moscatelli and J.R. Mayes, *Biochemistry*, 4 (1965) 1386.
125 M.L. Greene, T. Kaneshiro and J.H. Law, *Biochim. Biophys. Acta*, 98 (1965) 582.
126 R.C. Gaver and C.C. Sweeley, *J. Am. Oil Chem. Soc.*, 42 (1965) 294.
126a K. Sambasivarao and R.H. McCluer, *J. Lipid Res.*, 4 (1963) 106.
127 C.J. Lauter and E.G. Trams, *J. Lipid Res.*, 3 (1962) 136.
128 R.M.C. Dawson, *Biochem. J.*, 75 (1960) 45.
129 R.M.C. Dawson and N. Clarke, *J. Neurochem.*, 18 (1971) 1313.
130 G.B. Ansell and S. Spanner, *J. Neurochem.*, 10 (1963) 941.
131 C. Pries, A. Aumont and C.J.F. Böttcher, *Biochim. Biophys. Acta*, 125 (1966) 277.
132 M.A. Wells and J.C. Dittmer, *Biochemistry*, 5 (1966) 3405.
133 J.H. Duncan, W.J. Lennarz and C.C. Fenselau, *Biochemistry*, 10 (1971) 927.
134 T.J. Cicero and W.R. Sherman, *Biochem. Biophys. Res. Commun.*, 42 (1971) 428.
135 T.J. Cicero and W.R. Sherman, *Biochem. Biophys. Res. Commun.*, 43 (1971) 451.
136 M.G. Luthra and A. Sheltawy, *Biochem.J.*, 126 (1972) 251.
137 V.E. Vaskovsky and V.I. Svetashev, *J. Chromatog.*, 65 (1972) 451.
138 M.G. Luthra and A. Sheltawy, *Biochem. J.*, 126 (1972) 1231.

Chapter 4

PHOSPHOLIPID METABOLISM IN ANIMAL TISSUES

GÜY A. THOMPSON JR.*

Department of Botany, University of Texas, Austin, Tex. 78712 (U.S.A.)

1. INTRODUCTION

A full understanding of phospholipid function requires considerably more insight than can be gained through simply compiling an inventory of the types present in a tissue. The function of phospholipids in a cellular membrane is dependent upon the dynamic state of their metabolism. It has been firmly established that phospholipids in membranes and in the lipoproteins of tissue fluids are in a constant state of metabolic flux, with the rates of synthesis and degradation often varying with the physiological activity of the organism. In the face of these fluctuations of biosynthesis and catabolism, the cell is somehow capable of maintaining a finely regulated balance among its various structural lipids, so that each cell type preserves its characteristic lipid distribution. We are only beginning to investigate the factors responsible for maintaining this delicate equilibrium.

The format for this chapter is in general based on the plan used in a recent contribution to the series *Comprehensive Biochemistry*[1]. It describes the avenues used by cells in synthesizing and degrading phospholipids. The basic reactions involved have been fairly well elucidated, but considerable attention still needs to be directed towards gaining a more detailed insight into the processes. For example, a great deal of experimental effort is being aimed at learning how cells adjust the phosphoglyceride-bound fatty acids in such a way as to achieve a characteristic (and different) assortment of fatty acyl groups in each phospholipid class. Locating a degree of specificity for individual fatty acids in one or more steps of biosynthesis or catabolism in undoubtedly the key to understanding this phenomenon.

It seems certain that the orientation of research in this area will increasingly shift from the almost completed task of characterizing pathways to the more complex problem of how these reactions operate in concert to regulate membrane properties of the cell. At thc moment, however, meaningful information bearing on physiological control of the reactions is scarce indeed.

*Supported in part by NSF grants GB-16363 and GU-1598 and Robert A. Welch Foundation grant F-350.

2. BIOSYNTHESIS

Phosphatidic acid

Phosphatidic acid plays a key role as an intermediate in the synthesis of triglycerides and most phospholipids. As such it does not ordinarily accumulate in significant amounts but is maintained at a constant low level in tissues by carefully regulated processes of formation and utilization.

Four pathways have been described for the synthesis of phosphatidic acid. The first was discovered by Kornberg and Pricer[2] in 1953. A system from liver was shown to catalyze the rapid esterification of *sn*-glycero-3-phosphate utilizing the coenzyme A thioesters of long-chain fatty acids and producing phosphatidic acid.

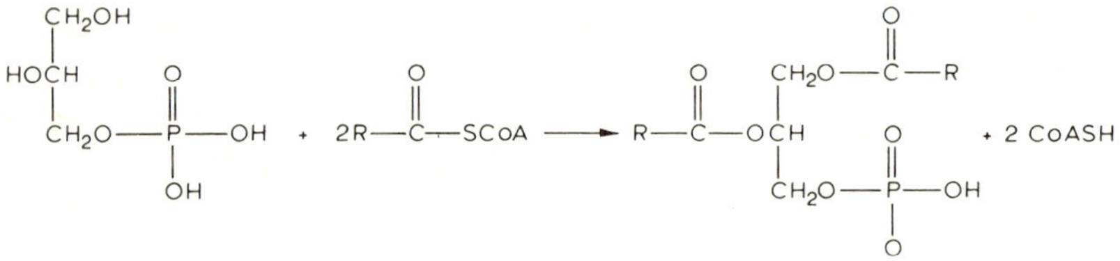

The coenzyme A derivatives of fatty acids arise either by *de novo* synthesis from acetate within the cell or by the action of microsomal and mitochondrial enzymes which couple free fatty acids with coenzyme A in an ATP-requiring reaction[3].

In bacterial systems the acyl carrier protein (ACP)-bound fatty acid is more active than fatty acyl-CoA in the acylation of glycerophosphate. When utilizing acyl-ACP, there is a striking preference for the enzyme to esterify the 2 position of 1-acyl-*sn*-glycero-3-phosphate with unsaturated fatty acids[4]. This provides an explanation for the observed preferential positioning of unsaturated fatty acids at the 2-position and saturated fatty acids at the 1 position of glycerol in natural phospholipids.

Fatty acyl-acyl carrier proteins have not yet been shown to function in this way in animal cells. There are conflicting reports concerning a similar specificity in fatty acid positioning during phosphatidic acid biosynthesis from fatty acyl-CoA. Some experiments have suggested little specificity during *de novo* synthesis[5,6]. However, it now appears that in rat liver, saturated fatty acids and unsaturated fatty acids can be directed principally to the 1 and 2 positions, respectively, of phosphatidic acid[7]. Whether both acylations are catalyzed by the same enzyme, acyl-CoA:*sn*-glycero-3-phosphate *O*-acyltransferase (EC 2.3.1.15) is not firmly established, and comparisons are difficult due to differing substrate solubilities. Whatever the exact mechanism, this pathway is probably a major determinant, along with certain transacylases to be discussed later, of the characteristic phospholipid distribution pattern of saturated and unsaturated fatty acids.

In addition to the direct acylation of *sn*-glycero-3-phosphate, the two related pathways shown below for phosphatidic acid formation have been demonstrated.

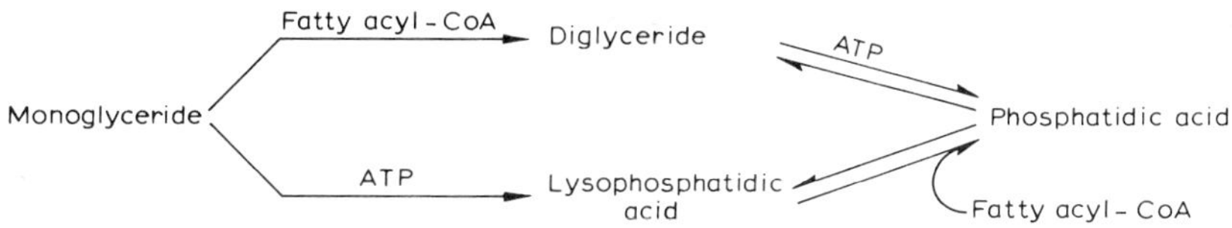

The physiological significance of the reaction shown on the lower line has not been studied extensively, and the observed acylation of lysophosphatidic acid may actually represent an intermediate step in the acylation of *sn*-glycero-3-phosphate. On the other hand, many tissues are known to possess the enzyme diglyceride kinase, which phosphorylates diglycerides with the participation of ATP. The increased phosphatidic acid turnover in cells stimulated to carry out certain membrane-transport processes (Chapter 15), such as salt or protein secretion, has been reported to occur through increased activity of diglyceride kinase and an associated enzyme, phosphatidic acid phosphatase[8]. However, other investigators find no significant enhancement of diglyceride kinase activity during zymogen secretion[9]. Thus further work would seem necessary in order to clarify the physiological role of this phosphatidic acid 'cycle' in membrane transport phenomena.

One further source of phosphatidic acid involves fatty acyl-CoA and dihydroxyacetone phosphate. The scheme illustrated below was initially demonstrated by Hajra and Agranoff[10] in liver mitochondria from the guinea pig.

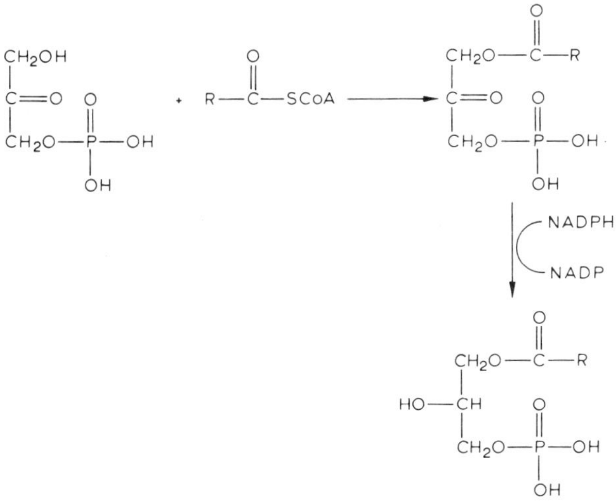

It has also been reported to occur with liver microsomes, where either NAPH or NADPH can provide the needed reducing potential[11]. Although efforts have been made to assess the quantitative importance of this pathway[12, 13, 13a], the results are still inconclusive.

Phosphatidylcholine

Phosphatidylcholine (lecithin) is the principal phosphorus-containing lipid of most animal tissues. As such, it was the first phospholipid whose biosynthesis was studied extensively. Synthesis proceeds along the pathway outlined below.

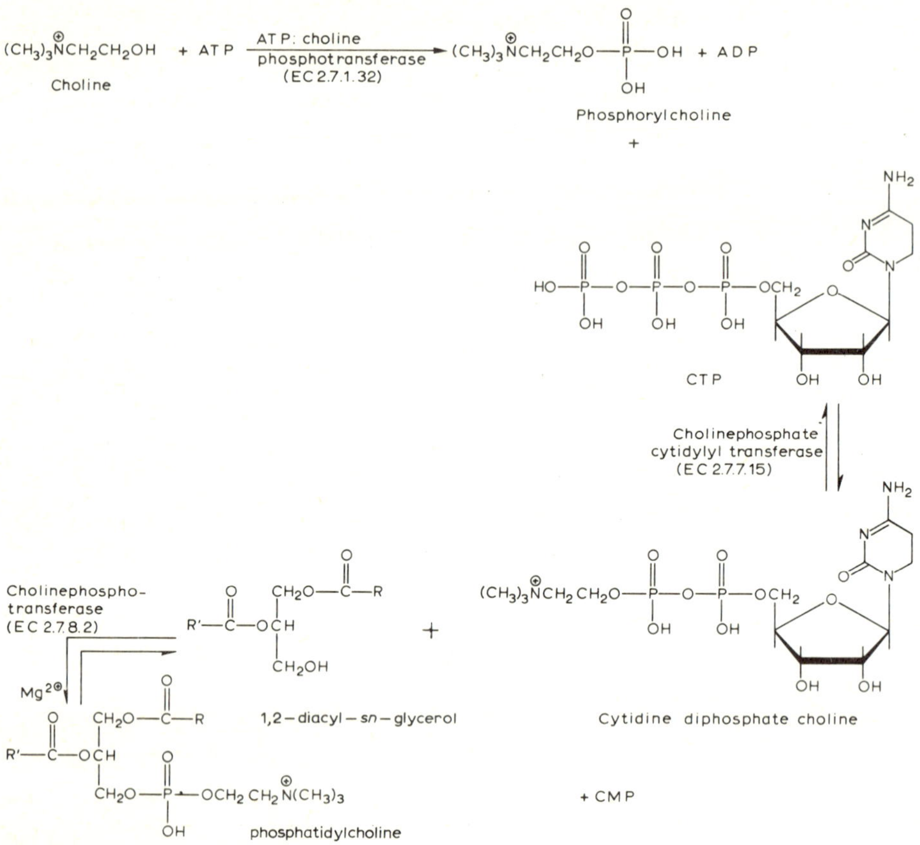

The formation of phosphorylcholine was described in 1953 by Wittenberg and Kornberg[14]. The enzyme responsible for its synthesis requires $Mg^{2\oplus}$ and can also transfer ethanolamine (2-aminoethanol) or dimethylethanolamine, although choline is the preferred substrate *in vitro*. Using phosphorylcholine labelled with both ^{14}C and ^{32}P, Kornberg and Pricer[15] demonstrated in 1952 the capacity of a liver enzyme to form lipid retaining the same ^{14}C : ^{32}P ratio as in the starting material. The lipid was subsequently characterized by Kennedy and Weiss[16] as phosphatidylcholine. Characterization of CDPcholine as the active intermediate in lecithin biosynthesis was followed by the use of synthetic CDP-choline in proving its role as the choline phosphate donor.

The terminal step in phosphatidylcholine formation involves the interaction of CDP-choline with 1,2 diacyl-*sn*-glycerol (diglyceride), which is derived principally from the dephosphorylation of phosphatidic acid. The experiments of Bjørnstad and Bremer[17] established that the reaction is freely reversible *in vivo*. This exchange reaction has been examined as a possible regulation point for determining the types of bound fatty acyl groups utilized for lecithin synthesis. Diglycerides for which the enzyme had a low affinity might be channeled into other reactions, such as the synthesis of triglycerides or phosphatidylethanolamine. However, the reactivity of the enzyme towards several diglycerides has been tested *in vitro*[18], and little specificity is apparent.

Lecithin is also formed by the methylation of phosphatidylethanolamine, *S*-adenosylmethionine being the methyl donor. This pathway is discussed later in the chapter.

Phosphatidylethanolamine

The tables of natural lipid patterns in Chapter 16 demonstrate that phosphatidylethanolamine is among the most abundant phospholipids. More commonly than with choline phosphoglycerides, the ethanolamine phosphoglycerides in a given tissue are a mixture of diacyl species and monoacyl monoether species (including both alkyl and alkenyl ethers), each type making up a considerable percentage of the total. The presence in the precursor diglyceride of an ether linkage (whose formation will be discussed later) does not appear to influence greatly the sequence of reactions illustrated below.

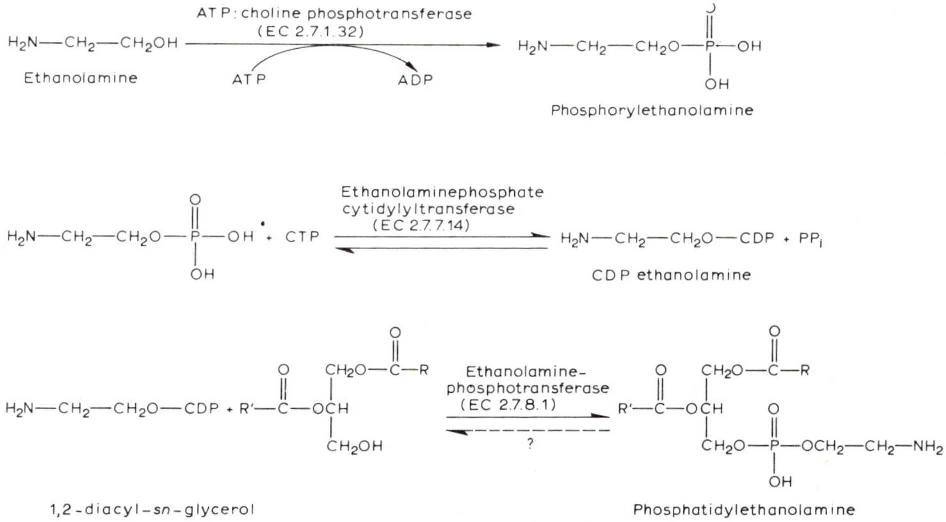

The pathway for the biosynthesis of phosphatidylethanolamine *de novo* is similar to that already described for phosphatidylcholine. The phosphorylation of ethanolamine is probably carried out by the same enzyme responsible for phosphorylcholine synthesis, though phosphorylethanolamine can also be produced enzymatically from sphingosine[56,119]. CTP is required in the activation of phosphorylethanolamine before ethanolamine phosphotransferase can couple it to diglyceride[16]. According to the work of Bjørnstad and Bremer[17], the latter reaction is much less reversible than the analogous step in phosphatidylcholine

References p. 94

biosynthesis. This might explain the relatively slow rate observed *in vitro* for radioactive diglyceride incorporation into phosphatidylethanolamine[18].

In addition to the pathway just considered, phosphatidylethanolamine can arise through the decarboxylation of phosphatidylserine. The significance of this process will be considered in the discussion of lipid conversions.

Phosphoinositides

The biosynthesis of inositol phospholipids also involves cytidine nucleotides but, surprisingly, the mechanism of their involvement is quite unlike that found to operate in the formation of ethanolamine and choline phosphatides. During phosphatidylinositol synthesis, a microsomal enzyme requiring divalent cations promotes the reaction of CTP with phosphatidic acid, yielding CDPdiglyceride and inorganic pyrophosphate[19, 20].

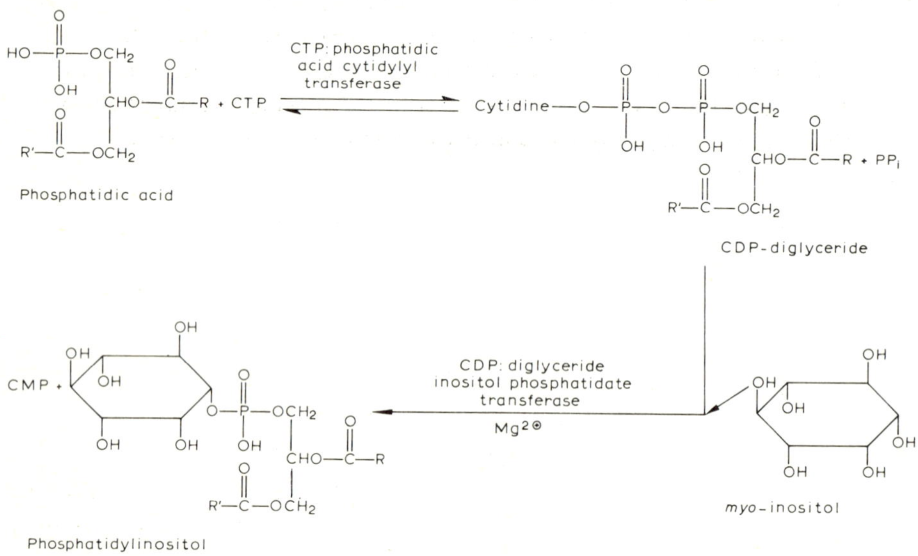

A direct reaction then occurs between CDP-diglyceride and inositol, giving rise to phosphatidylinositol and CMP[19, 21]. The enzyme responsible is also localized in microsomes and appears to prefer CDP-diglycerides having oleic acid at the 2-position rather than palmitic or stearic acid[22].

The biosynthesis of the polyphosphoinositides (di- and tri-phosphoinositide) proceeds by a sequential phosphorylation of phosphatidylinositol, first in the 4- and then in the 5-position.

Phosphatidylinositol + ATP → diphosphoinositide + ADP
Diphosphoinositide + ATP → triphosphoinositide + ADP

The enzyme catalyzing the initial phosphorylation, phosphatidylinositol kinase (ATP: phosphatidylinositol-4-phosphotransferase), requires $Mg^{2\oplus}$ for activity. Unlike most lipid-synthesizing enzymes, it seems to be localized in the plasma membrane of mammalian cells[23]. The formation of triphosphoinositide is achieved by the enzyme diphosphoinositide kinase (ATP: diphosphoinositide 5-phosphotransferase), which may also be localized in the plasma membrane, at least in kidney[24].

Labeling experiments with $^{32}P_i$ have shown that the phosphate groups bound at the 4-and 5-positions of brain triphosphoinositide turn over at a much faster rate than any other brain lipid components. The abundance of triphosphoinositide in the myelin membranes of nervous tissue and the rapid rate of its formation during the early stages of nerve myelination imply that this lipid is primarily a component of the myelin sheath[25].

Phosphatidylserine

Unlike the phospholipids considered so far, phosphatidylserine is not synthesized in animal systems by the interaction of phosphatidic acid or diglyceride with cytidine nucleotide derivatives. Such a sequence is present in bacteria, where phosphatidylserine arises through the reaction of L-serine with CDPdiglyceride (see Chapter 11). Phosphatidylserine is formed in animal cells by an exchange reaction in which pre-existing phosphoglycerides, principally phosphatidylethanolamine, exchange their nitrogenous base with free L-serine as follows[26]:

$$\text{phosphatidylethanolamine} + \text{L-serine} \rightleftharpoons \text{phosphatidylserine} + \text{ethanolamine}$$

In rat liver the exchange enzyme is localized in the microsomal fraction[27]. Further characteristics of the base exchange process will be considered later.

Plasmalogens and glycerol ether phospholipids

Only recently have we come to understand the fundamental biosynthetic steps involved in forming lipids containing α-glycerol ethers, i.e., derivatives having a 1-*O*-alkyl or a 1-*O*-alkenyl grouping. Although there are still a number of unsolved problems with regard to the biosynthesis of the alkenyl ethers (plasmalogens), the essential reactions of ether bond formation have largely been elucidated, principally through the efforts of Snyder and his associates. A monograph reviewing ether-lipid biochemistry has recently been edited by Snyder[28].

It has been known for many years that the unique ether-linked side chain of glycerol ether-containing phospholipids can arise from precursors similar to those involved in diacyl phospholipid formation. In 1969 Snyder et al.[29] demonstrated synthesis of the ether linkage by a cell-free system utilizing glyceraldehyde-3-phosphate and long chain fatty alcohols. Further studies with this same system from mouse preputial gland tumors[30] and with cell-free preparations from rat liver[31] revealed dihydroxyacetone phosphate (DHAP) to be an even more direct precursor of the glyceryl moiety than glyceraldehyde phosphate.

Our current understanding of the reactions is as follows. DHAP combines first with a fatty acyl-CoA molecule, producing acyl-DHAP. The acyl group is then exchanged for a fatty alcohol[32]. *O*-alkyl-DHAP is subsequently reduced to *O*-alkyl-glycerol phosphate, with

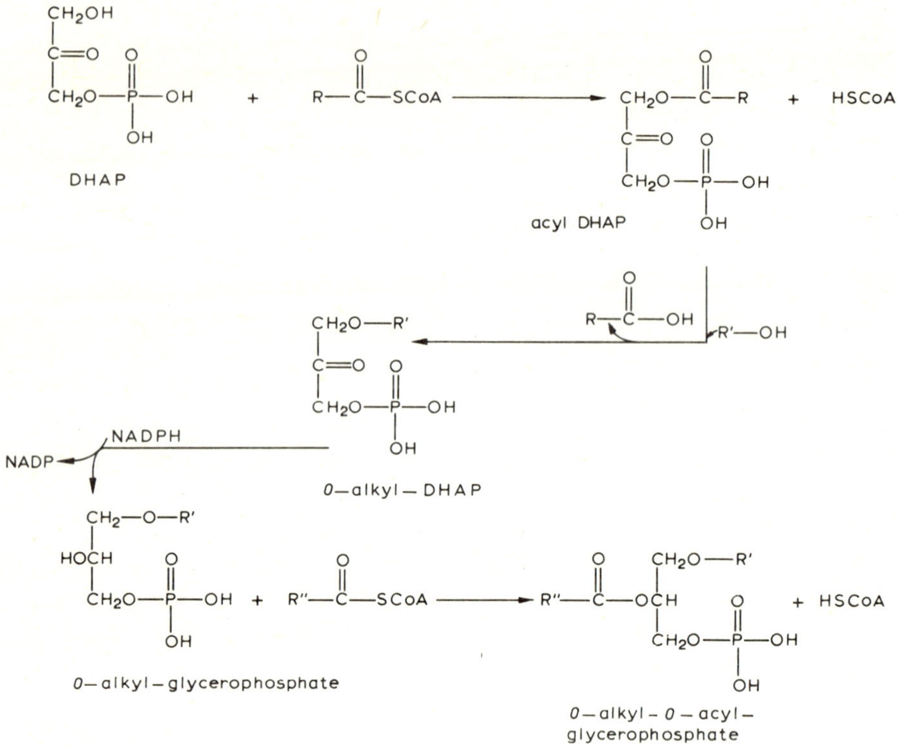

NADPH acting as H-donor. After acylation at position 2, the resulting analog of phosphatidic acid enters the familiar pathways for conversion to one of several possible phospholipids, principally those containing choline, ethanolamine or serine.

The enzymes catalyzing these reactions are microsomal and have been shown to occur in a variety of animal tissues[28]. Although the mechanism of ether bond formation is still incompletely understood, studies with ^{18}O have revealed that the ether oxygen atom originates from the fatty alcohol[33]. Such a hydrogen–oxygen scission of an alcohol is similar to that found during the esterification of carboxylic acids with alcohols. Further insight into the mechanism of ether bond formation may emerge from the observed stereospecific loss of one hydrogen atom attached to the dihydroxyacetone phosphate carbon atom which is directly involved in forming the ether linkage[33a].

Most of the findings reported from studies with mammalian systems can be explained by involving the scheme outlined above. On the other hand, there are data suggesting that a different pathway is involved in the synthesis of bacterial ether-lipids[34, 35].

One major unresolved question of ether-lipid metabolism concerns the manner in which the α, β-unsaturated ether side chain of plasmalogens is made. For a reason yet unknown, plasmalogens sustain a much slower incorporation of precursors into their carbon skeleton than do diacyl lipids or mono-alkyl mono-acyl lipids in all animal systems so far examined.

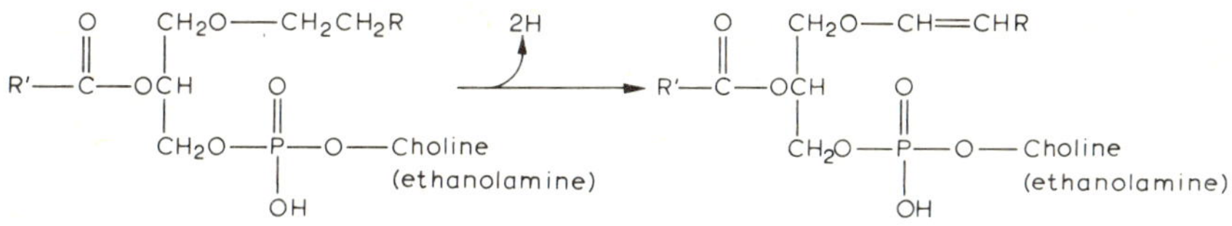

Many investigators believe that plasmalogens are produced by a direct desaturation of the analogous alkyl glycerol ether lipid. The first rigorous proof that this conversion can occur was described by Thompson[36]. He observed the incorporation of fed 1-*O*-[1-^{14}C]-hexadecyl-2-[^{3}H] glycerol (chimyl alcohol) into phospholipids of the terrestrial slug *Arion ater* and the subsequent appearance of both isotopes in the glyceryl alk-1-enyl moiety of plasmalogens. Because the ^{14}C/^{3}H ratio of the plasmalogen derivative was similar to that of the initial substrate, it was concluded that the alkyl ether is converted to the alkenyl ether without cleavage of the ether bond.

More recent studies *in vivo* with tissues as diverse as ascites tumor cells[37], rat brain[38], and the slime mold *Physarum polycephalum*[39] are best interpreted in the way outlined above. A particularly enlightening study by Stoffel and Lekim[38] made use of appropriately labeled radioactive precursors to demonstrate the stereospecificity of the dehydrogenation reaction. The α, β-unsaturated linkage of the alkenyl side-chain is formed by *cis* elimination of hydrogen atoms having a (1S, 2S) configuration, producing a *cis* double bond as illustrated below.

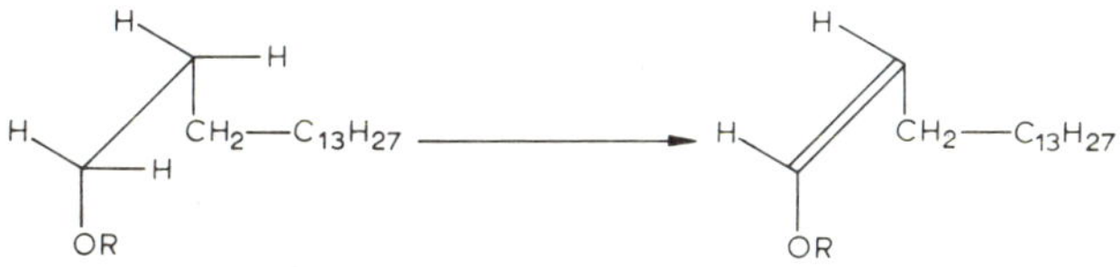

R, glycerol derivative.

Although support for this scheme of plasmalogen formation is increasing, there are still some reports, as discussed by Snyder[28], which are in disagreement.

An important step towards resolving the controversy has been made by the development of a cell free system from Ehrlich ascites cells[40]. Hopefully, details of the biosynthetic pathway will soon be forthcoming.

Phosphatidylglycerol

A number of animal tissues, including rat and chicken liver[41], guinea pig liver[42], rat heart[43], and sheep brain[44], have been shown capable of the synthesis of phosphatidylglycerol *in vitro*. In all instances the synthetic enzymes are located predominantly in mitochondria. The steps involved are as follows:

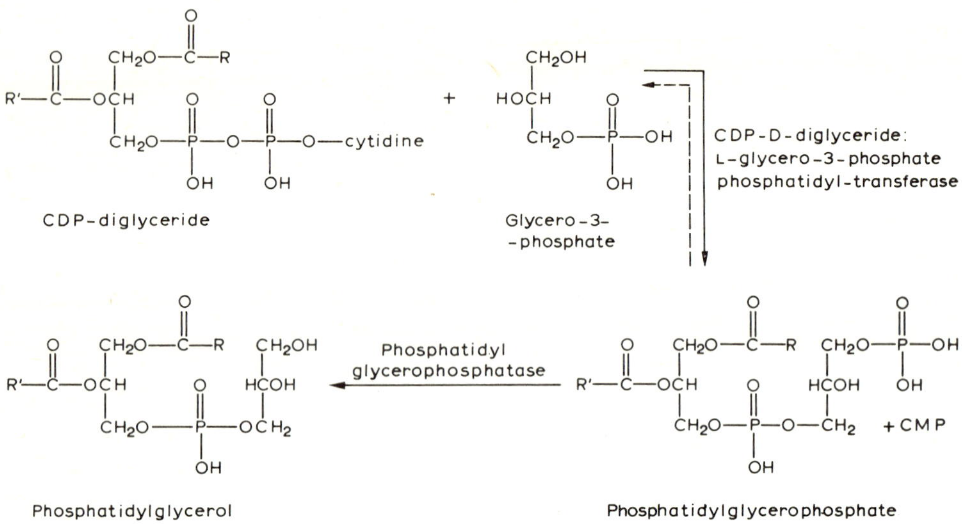

The first reaction has been shown to be reversible in *E. coli* preparations, but whether this is so in animals is not known. Phosphatidylglycerophosphate formed by the mitochondrial system shows the same fatty acid distribution as the added CDPdiglyceride[44]. The addition of sulfhydryl poisons preferentially inhibits the phosphatase reaction, and the resulting accumulation of phosphatidylglycerophosphate provides further evidence for its role as an intermediate.

Diphosphatidylglycerol

In animal cells diphosphatidylglycerol (cardiolipin) has the unique property of being localized almost exclusively in mitochondria. Its synthesis *in situ* within mitochondria has been widely postulated, but it has only been recently, that Davidson and Stanacev[42] have shown that diphosphatidylglycerol is formed in isolated guinea pig liver mitochondria by the sequence of reactions found previously to function in *E. coli*[45].

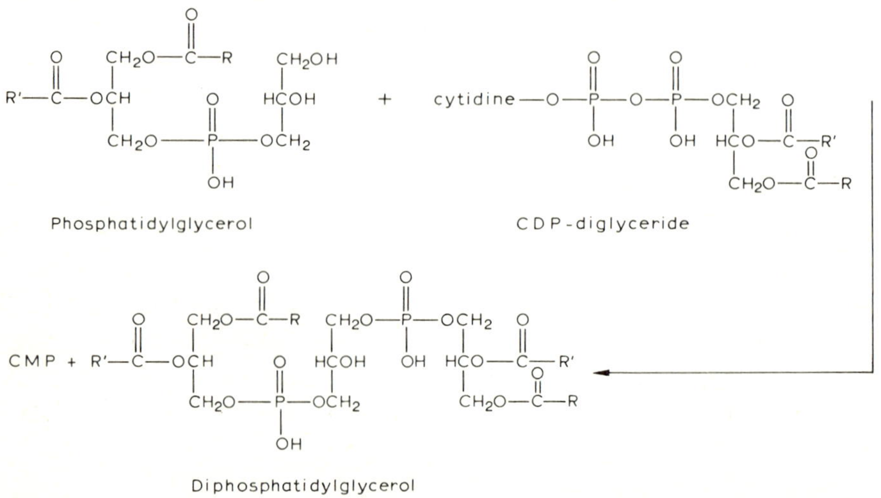

Previous failures to effect diphosphatidylglycerol synthesis *in vitro* were apparently caused by the inability of added CDPdiglyceride to penetrate to the site of enzymatic activity. If CDPdiglyceride is generated within the mitochondria by supplementing the incubation mixture with CTP, significant amounts of diphosphatidylglycerol are recovered[42]. Unexpectedly, recent experimentation has shown the true mechanism of diphosphatidylglycerol formation in *E. coli* to be

$$2 \text{ phosphatidylglycerol} \rightarrow \text{diphosphatidylglycerol} + \text{glycerol}$$

However, the pathway originally postulated for *E. coli*, involving CDPdiglyceride, has been confirmed in liver mitochondria[45a].

Metabolic interconversions of phospholipids

Thus far we have concerned ourselves primarily with the *de novo* synthesis of phospholipids from precursors of lower molecular weight. In addition, animal cells are capable of converting certain intact phospholipids into others by reactions involving the nitrogen-base moieties. While these reactions have the potential for altering the cell's lipid distribution, there is no net synthesis of phospholipid.

The most widely studied of these biological transformations involves phosphatidylcholine, phosphatidylethanolamine, and phosphatidylserine. It has been shown[46,47] that liver enzymes can catalyze the decarboxylation of serine to ethanolamine and the subsequent stepwise methylation of ethanolamine to form choline while the nitrogenous bases are still covalently bound in phospholipids.

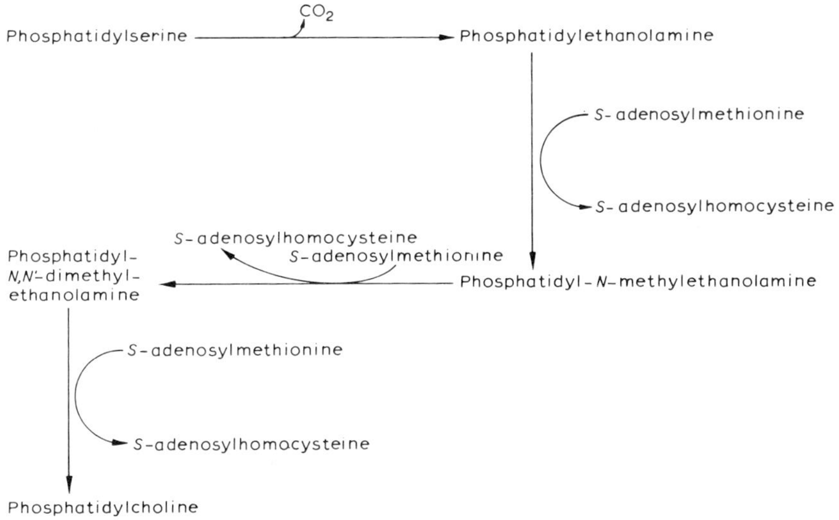

Whereas the decarboxylation of phosphatidylserine is catalyzed by a mitochondrial enzyme[27], the stepwise methylation of phosphatidylethanolamine appears to be mediated

References p. 94

by two microsomal enzymes, one catalyzing the first methyl addition and a second carrying out the final two methylations[48]. Although convincing support for this sequence was derived from work with mutant strains of Neurospora[49] (see also Chapter 11), the inability to isolate an enzyme system active in performing the initial methylation step *in vitro* still prevents confirmation of the pathway in animals.

Surprisingly, among tissues of higher animals only liver derives a major fraction of its lecithin via the methylation of phosphatidylethanolamine[17]. Considering the role of liver as a major producer of phosphatidylcholine for transport to plasma and bile, it is tempting to speculate that the two biosynthetic pathways might serve different cellular needs. However, information derived from study of [1,2-^{14}C] choline, [1,2-^{14}C] ethanolamine, and [CH_3-^{14}C] methionine incorporation patterns *in vivo* indicates that lecithin arising from either pathway enters a common pool which is in rapid equilibrium with plasma[17]. This explanation may be oversimplified, for indications of separate liver lecithin pools, perhaps characterized by different fatty acid composition, have been reported[50].

There is evidence that the methylation pathway becomes more active during choline deficiency[51, 52]. This potential for augmenting the organism's choline supply during times of need may represent the principal value of the methylation pathway.

In addition to the pathway just discussed, interconversions of phospholipids can occur by direct base exchange. Thus free serine, choline, and ethanolamine can exchange with the existing base of a phosphatide to yield the corresponding phospholipid[53]. The serine–ethanolamine exchange and possibly the others as well are catalyzed by a microsomal enzyme[27]. With the possible exception of ethanolamine, there is no ATP requirement for the exchange. The reactions are strongly cation-dependent, all but inositol requiring $Ca^{2\oplus}$ for maximal activity. While it was at first thought that a separate enzyme is involved for the exchange of each base, several procedural refinements, including the replacement of *tris* (hydroxymethyl) aminomethane, which can itself participate in the exchange reaction, by another buffering compound, have increased the likelihood that all reactions are catalyzed by a single enzyme[54]. Some information is available regarding specificity of the bases for different types of phospholipid, but conclusive evidence is lacking[54a].

Sphingomyelin

Details of recent work on the metabolism of sphingosine and its analog dihydrosphingosine have been adequately reviewed elsewhere[54b, 54c], and only the bare essentials of the reactions will be discussed.

In general, data gained with the two principal systems, the yeast *Hansenula ciferri*[55] and various tissues of the rat[56], are in close agreement. Carbon atoms 1 and 2 of sphingosine bases are derived from carbon atoms 2 and 3 of serine, and carbon atoms 3–18 arise from palmitic acid. For some years it was thought that palmitaldehyde functions as an intermediate in the biosynthetic sequence, but recent studies have established the following reactions catalyzed by microsomal enzymes[55, 56].

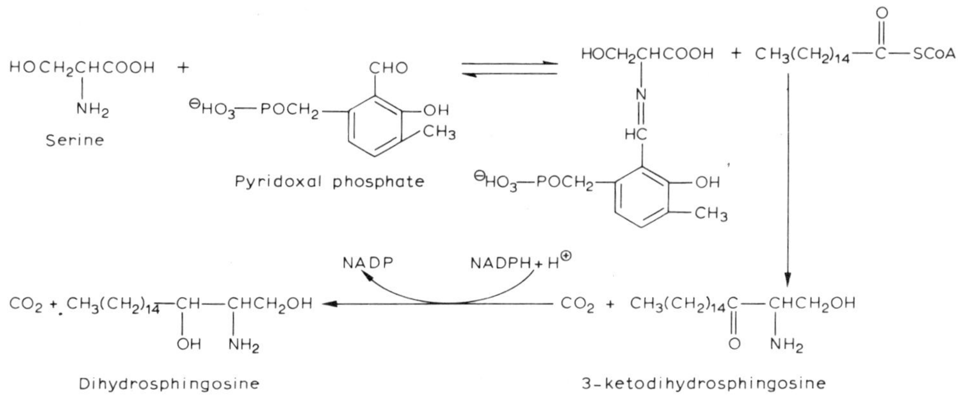

Pyridoxal phosphate-bound serine reacts with palmitoyl-CoA, producing 3-ketodihydrosphingosine, which is reduced in the presence of NADPH to yield dihydrosphingosine. The step at which the 4,5 double bond of the more common base sphingosine is inserted has not been precisely defined, but current evidence suggests that desaturation occurs at the fatty acyl-CoA level, before condensation with serine takes place[55].

Sphingosine-type bases of varying chain lengths and degrees of unsaturation have been isolated from animal lipids[57]. It is likely that these are formed by the above pathway, utilizing other fatty acyl-CoA precursors.

The next step in sphingomyelin biosynthesis involves the coupling of a fatty acid residue to the amino group of sphingosine to form ceramide. It appears that this microsomal reaction can be catalyzed by either of two apparently different enzymes. One of these, studied in preparations from liver and brain[58], carries out the reaction using as substrates sphingosine and fatty acyl-CoA. Free fatty acid cannot be utilized. Both

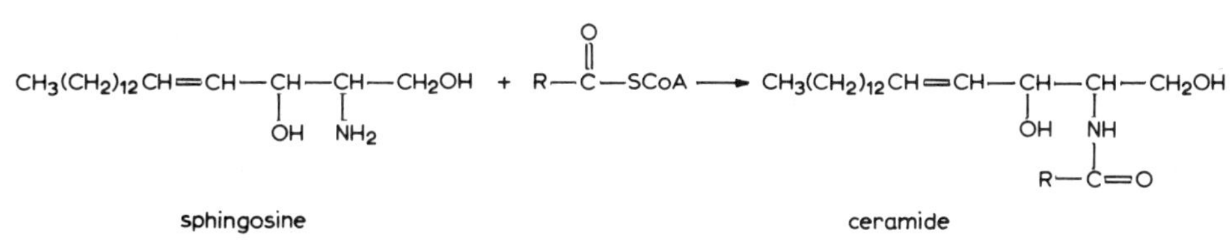

erythro-sphingosine, the natural isomer, and *threo*-sphingosine can be acylated.

Gatt[59] has characterized a different enzyme purified from rat brain. This enzyme will utilize free fatty acid, and no requirement for CoA or ATP could be demonstrated. The reaction is freely reversible and probably takes part in the normal process of sphingomyelin catabolism.

Once formed, the *N*-acyl derivative of the long chain base (ceramide) can react with CDPcholine under the influence of the enzyme CDPcholine:ceramide cholinephosphotransferase (EC 2.7.8.3). Sribney and Kennedy[60] found the enzyme to be of widespread occurrence in animal tissues. After some technical difficulties in determining

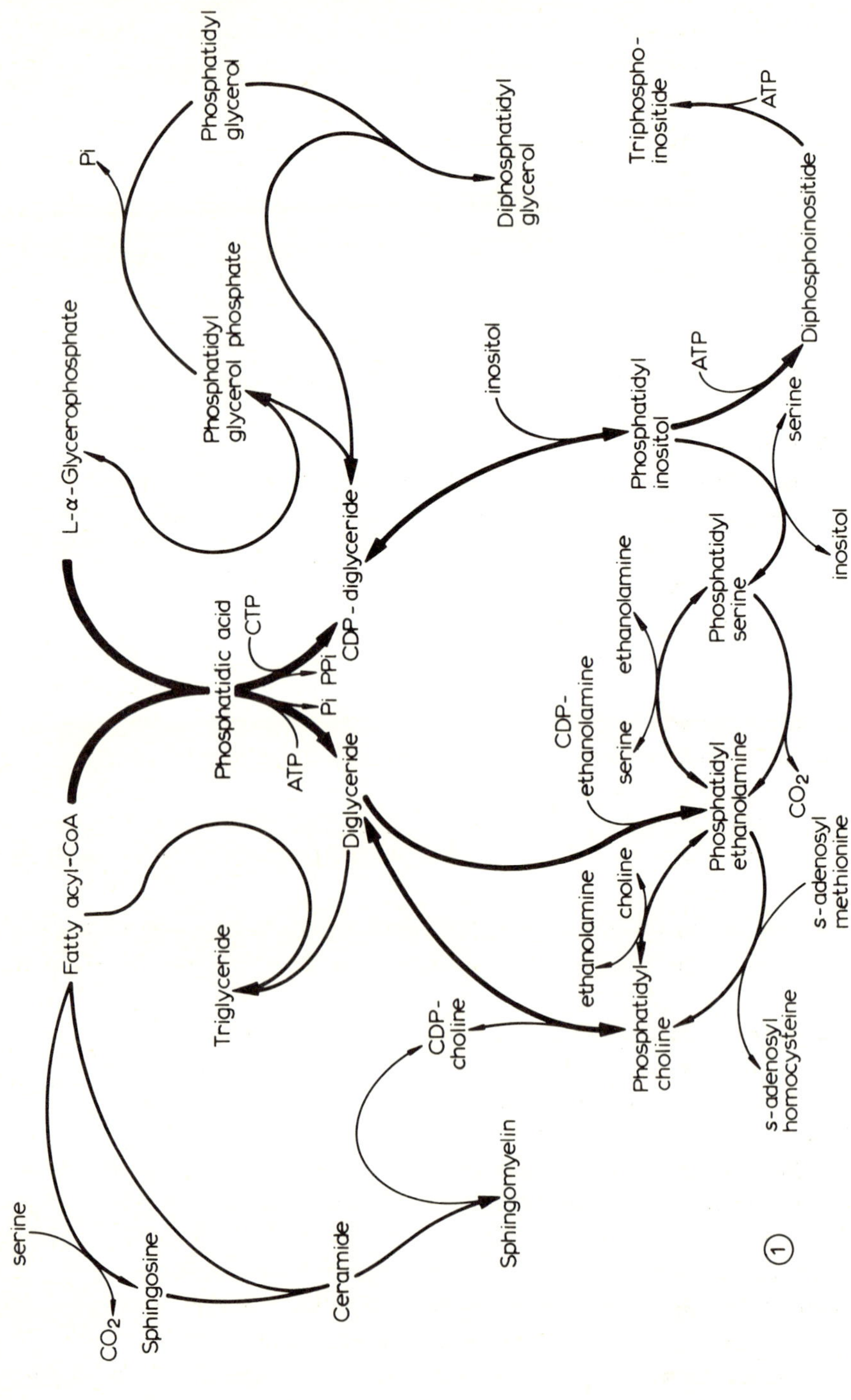
Fatty acyl-CoA
L-α-Glycerophosphate
Phosphatidic acid
ATP
CTP
Pi
PPi
Diglyceride
CDP-diglyceride
Phosphatidyl glycerol phosphate
Pi
Phosphatidyl glycerol
Diphosphatidyl glycerol
Triglyceride
inositol
Phosphatidyl inositol
ATP
Diphosphoinositide
ATP
Triphospho-inositide
serine
inositol
Phosphatidyl serine
ethanolamine
CDP-ethanolamine
serine
Phosphatidyl ethanolamine
CO_2
choline
ethanolamine
s-adenosyl methionine
Phosphatidyl choline
s-adenosyl homocysteine
CDP-choline
serine
CO_2
Sphingosine
Ceramide
Sphingomyelin

1

substrate specificity, it has been established that the natural ceramides, containing *erythro*-sphingosine, can take part in the reaction[61].

A possible alternative pathway for sphingomyelin synthesis has been observed in homogenates of young rat brain[62]. Sphingosylphosphorylcholine is coupled with fatty acyl-CoA to yield sphingomyelin. There is no indication as yet that this pathway makes a significant contribution to sphingomyelin production in animal tissues.

Diringer et al.[62a] have raised the possibility of still another mechanism for sphingomyelin biosynthesis. In cultured virus-transformed mouse cells, sphingomyelin seems to arise through the transfer of phosphorylcholine from lecithin to sphingosine or *N*-acylsphingosine.

Several laboratories have reported relatively small amounts of a sphingomyelin analog in which choline is replaced by ethanolamine. This compound is presumably formed by a pathway similar to that operative for sphingomyelin.

Summary and conclusions

It has been convenient for us to consider separately the biosynthetic pathway for each major phospholipid. In the cell, of course, all pathways operate simultaneously and in close proximity to one another. Precursors and products alike are shared in producing a balanced output for membrane turnover and expansion. Fig. 1 (as modified from a previous publication[1]) is intended to illustrate some of the principal biosynthetic interrelationships.

Once the finished phospholipid molecules are released from their sites of synthesis and inserted into a lipoprotein membrane, they might be expected to remain unchanged until they are replaced by the normal turnover processes. However, there are numerous indications that further modification of intact phospholipids can occur at any time during their continued existence. In some cases the changes may consist of alterations in the bound fatty acids, ranging from peroxidation of double bonds[63] to alkylenation[64] or desaturation[65] of the acyl groups. In other instances, glyceryl alkyl ether phospholipids are converted to plasmalogens after a lengthy residence in some cellular membrane. It is not known whether such alterations occur *in situ* or after the molecules have been removed from a membrane to allow enzymatic attack.

These reactions are a part of phospholipid metabolism, albeit a little studied part. We may expect that increased attention will be devoted to these potentially important modifying reactions in future years.

3. METABOLIC EXCHANGE OF PHOSPHOLIPIDS AND THEIR COMPONENTS

Introduction

The concept of metabolic turnover is universally accepted by the scientific community[66]. With the exception of DNA, all molecular components are periodically replenished, both in growing and in non-growing cells. The rates of molecular replacement

References p. 94

vary, depending upon the nature of the molecule involved. The life of phospholipids in the cell is known to be shorter than that of most of the cell's other structural elements.

Metabolic turnover of proteins is an all-or-nothing process, in the sense that degradation of a protein molecule, once begun, continues to completion. Replacement requires the formation of an entirely new molecule. Lipids, on the other hand, can often be renewed by a mechanism which is metabolically more economical. There can be an enzymatic replacement of only a certain molecular component, while the greatest part of the complex lipid molecule is reused.

At this stage of our knowledge, it is impossible to gauge the relative importance of this selective turnover of components as compared with the replacement of whole lipid molecules. However, it is possible to outline in some detail the several mechanisms by which turnover can proceed. A discussion of the cellular consequences of such reactions will be found in Chapter 9.

Turnover of portions of phospholipid molecules

Acyl groups

The exchange of fatty acyl groups of phospholipids is ordinarily accomplished by the consecutive action of two enzymes. First an intact phospholipid is deacylated in either the 1- or the 2-position by phospholipase A_1 or A_2. This is followed by the reacylation of the lysophosphatide by an acyl-CoA:phospholipid acyltransferase. The acyltransferase reaction has been studied extensively by Lands and his associates. The reaction may be expressed as follows:

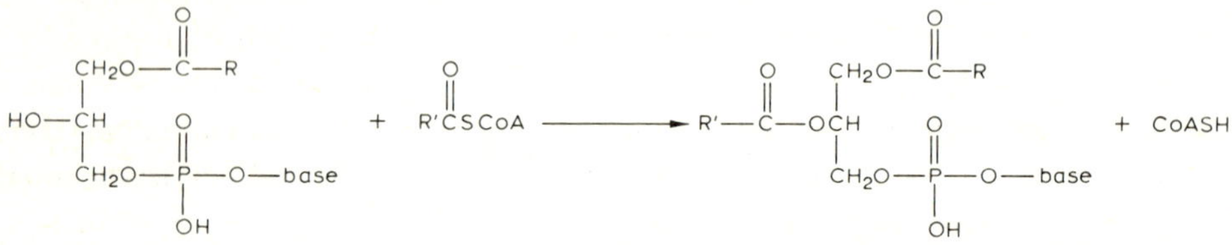

The substrates studied most frequently have been 1- or 2-lysolecithin or lysophosphatidylethanolamine. There have been indications that acyltransferase activity actually represents a family of enzymes, each specific for a single substrate[67]. However, more evidence is needed to substantiate this view. It does seem clear that the enzyme(s) is different from those acylating glycero-3-phosphate[68].

There is some uncertainty as to the principal subcellular site of acyltransferase action. Eibl et al.[69] reported the enzyme to be microsomal while Waite et al.[70] found activity in both microsomes and the outer mitochondrial membranes.

The microsomal activity is the most carefully studied. There is a marked specificity for esterifying saturated fatty acids at the 1-hydroxyl position and unsaturated fatty acids at the 2-hydroxyl position[71]. The selectivity of the enzyme depends strictly upon the position of the hydroxyl group to be acylated and is not influenced by the nature of the acid attached at the other position of the lysophosphatide. The specificity of the liver microsomal enzyme(s) in transferring test acyl-CoA derivatives to the 1-acyl-*sn*-glycero-3-phosphorylcholine establishes the following preferences in transfer: oleate > linoleate > laurate > palmitate ⩾ stearate[72]. Acting on 2-acyl-*sn*-glycero-3-phosphorylcholine, the same enzyme preparation showed this order of preference: stearate > palmitate > laurate ⩾ oleate > linoleate. The physiological significance of these *in vitro* findings is not yet clear because the acyltransferase specificity is variable depending upon the incubation conditions employed[72a].

Significant differences have been detected in the specificity for acylating lysophosphatidylcholine and lysophosphatidylethanolamine[73]. For example, when the two lysophospholipids are incubated together with rat liver microsomes, linoleoyl-CoA contributes non-selectively to both lipids while α-linolenoyl-CoA and arachidonoyl-CoA are predominantly utilized by the ethanolamine lipid. In addition to its reactivity towards the substrates described above, the acyltransferase from erythrocytes is also active with lyso-cholineplasmalogen but, surprisingly, the liver enzyme does not acylate the latter derivative[74].

One of the most clearly demonstrated instances where acyltransferases are of quantitative significance is that of the erythrocyte membrane. Here drastic changes in diet alter the fatty acid composition of phospholipids more rapidly than can be accounted for by net phospholipid synthesis. The reaction rate and specificity of the membrane-bound acyltransferases were shown to be adequate for effecting the observed changes[75]. Thus the activity of these acyltransferases can account for the modification of membrane phospholipids in a way that might change the physical characteristics and therefore the physiological properties of the membranes.

The system of fatty acid exchange in erythrocytes constitutes a special case, since there is little phospholipase A activity available in the cell for the generation of lysophosphatides[76]. It is probable that red cell lysolecithin is derived from plasma through the action of the enzyme lecithin: cholesterol acyltransferase, which also serves as the major source of plasma cholesterol esters[77].

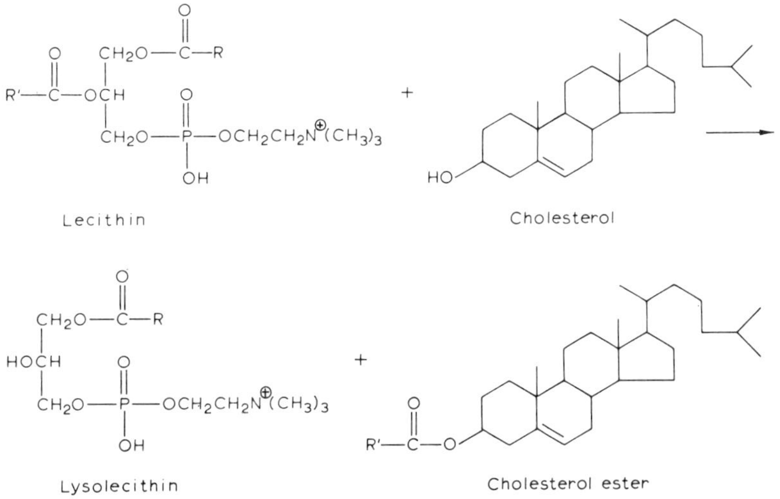

References p. 94

In addition to the formation of diacyl phospholipids from lysophospholipids and acyl-CoA, a second pathway for lysophospholipid acylation has been reported. The post-microsomal supernatants of rat liver[78, 79] and yeast[79] catalyze the exchange of a fatty acid from one lysolecithin molecule to another, yielding lecithin and glycero-3-phosphorylcholine. The quantitative importance of this reaction in intact animal tissues is probably not as great as that of the exchange mediated by acyl-CoA:phospholipid acyltransferase.

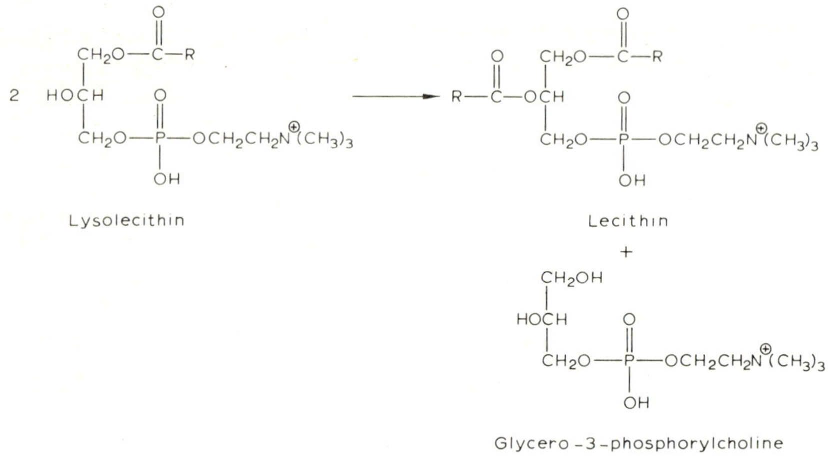

Nitrogen bases

The exchange of bases has been mentioned in the previous section of this chapter. This reaction can account for a considerable amount of intracellular lipid modification. The results of studies on phospholipid metabolism in rat liver mitochondria show that an appreciable exchange of serine and choline can occur in the absence of *de novo* phospholipid synthesis[80].

Exchange of entire molecules

Recent experimentation has established that entire phospholipid molecules as well as elements thereof can be relocated from one region of the cell to another, a process which is considered in Chapter 9.

Rates of phospholipid turnover in vivo

The bulk of this chapter is devoted to the reactions by which animal cells synthesize, degrade, and exchange their phospholipids. It will be instructive to consider briefly some of the available data pertaining to the rates at which these various pathways operate in the living cell.

Cells growing and dividing in a normal way must experience a net increase in phospholipids. This is brought about by a relatively greater rate of biosynthesis than degradation.

Superimposed, and having no effect on the net change in lipid content, are the exchange processes.

Various investigators have examined the dynamics of lipids in growing cells, most frequently using radioisotopes. The pattern often observed is as follows. All phospholipids become labeled from isotopically labeled precursors. However, the specific radioactivities of the phospholipids vary markedly and the time curve of labeling for one lipid may differ in shape from that of another, even though the precursor employed might be common to all the lipids, as would be $^{32}P_i$ (refs. 81, 82). Two likely explanations for the apparently different rates suggest themselves. First, the apparent differences could be misleading because of differences in the pool sizes of certain biosynthetic intermediates. While this undoubtedly has some effect, most workers believe that differences in turnover rates among the various phospholipids are the main contributing factor.

In tissues not undergoing significant growth, the same overall pattern is observed. Freysz et al.[83] used $^{32}P_i$ to measure the half-lives of various phospholipids of adult rat brain. In whole brain the values varied from 12.5 days (phosphatidylinositol) to 40 days (sphingomyelin). Variations were also found in isolated neurons and glial cells.

Summary and conclusions

Each membrane phospholipid is metabolized at a rate geared in some unknown way to the needs of the cell. There appears to be no requirement for a simultaneous turnover of all phospholipid types. Nor must membrane lipids and proteins turn over at the same

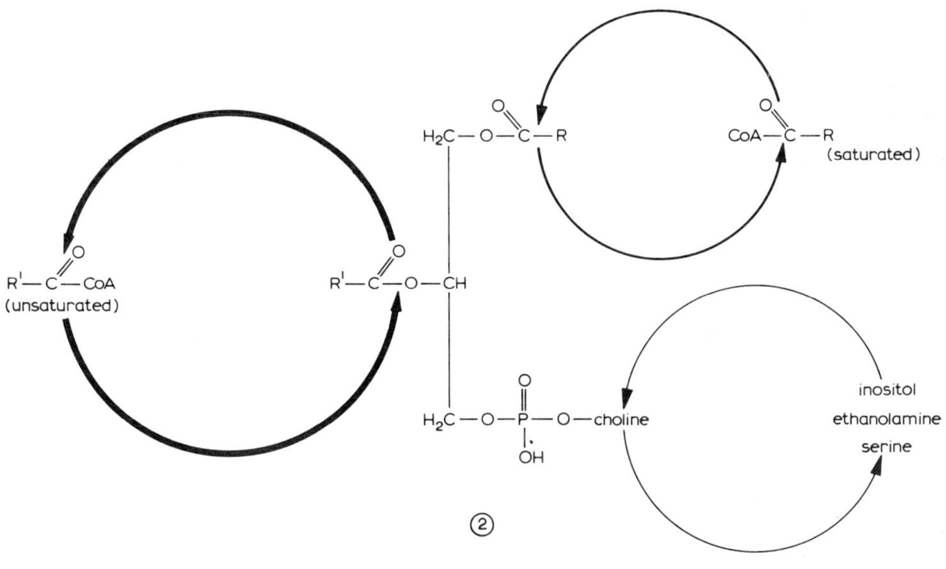

rate; lipids are as a rule more quickly replaced[84]. They may be replaced by the removal of intact lipid molecules or by the turnover of sub-molecular components. Phosphoglycerides, which are the lipids primarily involved in this partial turnover, undergo the

References p. 94

exchange reactions summarized in Fig. 2. The precise turnover rates *in vivo* are not known, but might be approximated by the size of the arrows, the more prominent representing a higher rate. The turnover process provides the cell a means of introducing structural changes, thereby modifying its membrane properties to meet changing environmental conditions. Why certain lipids are replaced more rapidly than others remains to be discovered.

4. DEGRADATION

Introduction

Enough data have now been assembled for us to recognize which metabolic pathways are available for the enzymatic degradation of phospholipids. Many important details remain to be discovered, and studies on the control of these pathways are only beginning.

In this survey of phospholipid catabolism, emphasis will be placed on the enzymes known to function within animal cells during the normal physiological process of lipid turnover. Only where it is needed for comparison will information on extracellular enzymes, e.g. those prepared from snake venoms and bacterial toxins, be introduced.

Substrate specificities of degradative enzymes have seldom been determined under conditions approaching those found within the cell, but our present knowledge suggests that most phospholipases will attack several types of phospholipid. For this reason, the discussion is organized so as to group information concerning individual enzymes rather than considering in sequence the stepwise degradation of specific lipids.

Hydrolysis of fatty acid esters

Phospholipase A_2

Because of its early discovery as an active principle of snake venoms, this enzyme, originally called lecithinase A and more recently designated phosphatide acyl-hydrolase (EC. 3.1.1.4), has been extensively studied. It carries out the reaction:

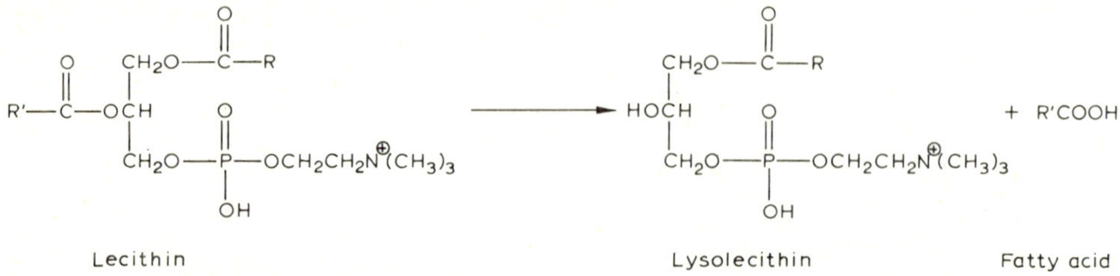

The specificity of venom phospholipase A has been studied extensively, notably by Hanahan and associates[85]. The enzyme deacylates at the 2-position of the glycerol moiety regardless of whether the fatty acid bound there is unsaturated or saturated.

Enzymes catalyzing this reaction are widely distributed among animal tissues[86], and although the degree of activity varies widely, it is likely that the enzyme is almost ubiquitous. Studies on rat liver have shown that phospholipase A_2 is localized predominantly in lysosomes[87] and in the outer membranes of mitochondria[88]. $Ca^{2\oplus}$ is required for activity. De Haas et al.[89] have purified the porcine pancreatic enzyme and found it to have a molecular weight of about 14 000, less than half that estimated for the *Crotalus adamanteus* venom enzyme[90]. The enzyme is remarkably heat-stable; activity persists in the pancreatic protein even after boiling for 5 min at pH4.

Phospholipase A_2 is normally present in porcine pancreas in zymogen form[91]. The inactive form was discovered after it was observed that phospholipase A_2 activity increases markedly upon storage of pancreas homogenates at room temperature. Trypsin was shown to activate the zymogen by cleaving off a heptapeptide. Because it is likely that phospholipase A_2 is secreted by the pancreas for eventual utilization in intestinal digestion, zymogen production may be peculiar to this tissue.

Phospholipase A_2 shows a rather pronounced specificity as to phospholipid type when studied with purified lipid substrates. On the other hand, the specificity is reduced in the presence of lipid mixtures, and a variety of evidence indicates that the affinity of a lipid for the enzyme depends upon the overall charge of the micelle in which it is contained[92]. The charge effect is not the sole determinant of specificity, however. The snake venom enzyme preferentially hydrolyzes lecithin while pancreatic phospholipase A_2 rapidly attacks very acidic lipids, such as phosphatidic acid, phosphatidylglycerol, and diphosphatidylglycerol, and hydrolyzes lecithin rather slowly[89]. Even among venom enzymes, differences can be noted. For example, the phospholipase A_2 of *Crotalus atrox* is much less active towards choline plasmalogen than towards phosphatidylcholine, but this kind of selectivity is not exhibited by the phospholipase A_2 of *Naja naja*[93].

It is impossible at this time to interpret the data on specificity obtained *in vitro* in a way that would allow us to understand the reactivity *in vivo*. It is likely that all the common phosphoglycerides are hydrolyzed to some extent. A certain degree of specificity is probably expressed *in vivo*, however, as evidenced by the preferential hydrolysis of mitochondrial phosphatidylethanolamine by endogenous phospholipase A_2 (ref. 94).

Phospholipase A_1

In 1967, Waite and van Deenen[95] published an extensive study of rat liver phospholipases, in which the presence of two different phospholipase A activities was clearly demonstrated. In addition to phospholipase A_2 (which had until that time customarily been called phospholipase A) they characterized a second lipolytic reaction which led to the removal of the 1-acyl ester of phospholipids. The new enzyme, which is of wide occurrence, is referred to as phospholipase A_1. In rat liver it exists mainly in the microsomal fraction while phospholipase A_2, as mentioned earlier, is localized in mitochondria[95]. Phospholipase A_1 is different from pancreatic lipase, which has also been shown to attack the 1-acyl ester of phospholipids[96].

Subsequent studies using rat liver have demonstrated that phospholipase A_1 also resides in the plasma membrane[97] and in lysosomes[87]. With the apparent exception of the lysosomal enzyme, $Ca^{2\oplus}$ is needed for activity.

The intracellular variation in the specificity of phospholipase A almost certainly has some significant physiological meaning although speculation as to the consequences of these localizations seems premature.

Lysophospholipase

This enzyme, also termed lysolecithin acyl-hydrolase (EC 3.1.1.5) and sometimes phospholipase B, catalyzes the following reaction:

$$\text{Lysolecithin} + H_2O \rightarrow \text{glycerylphosphorylcholine} + \text{fatty acid}$$

No positional specificity is indicated, and studies of liver lysophospholipase's substrate specificity suggest that both 1-acyl and 2-acyl lysophosphatidylcholine are hydrolyzed by the same enzyme[98].

The degradation of phosphatidylcholine to yield glycerylphosphorylcholine and fatty acids has been observed in many animal tissues[99, 100]. Intestinal mucosa is a particularly rich source. Typically, little or no lysophosphatide accumulates. In rat-liver microsomes endogenous phosphatidylethanolamine is deacylated at an even faster rate than endogenous phosphatidylcholine. Most investigators assume that the hydrolysis is due to combined action by phospholipase A and lysophospholipase, with the former enzyme being rate-limiting[101]. A different reaction sequence is seen in purified mitochondria, in which limited lysophospholipase activity results in an accumulation of lysophosphatides[94].

A further clarification of the rat-liver system stems from the work of Waite and van Deenen[95]. The deacylation of lysophosphatides was observed to be much more temperature-sensitive than is the phospholipase A activity. This was considered additional proof that lysophospholipase is distinct from the phospholipases. Furthermore, the phospholipase A activity, localized in particulate cell fractions, was easily separated from the lysophospholipase, which is mostly contained in the supernatant resulting from centrifugation at 100 000 $\times$ g for 1 h.

Ceramidase

The amide linkage of intact sphingomyelin is not cleaved enzymatically. Prior to deacylation, it is necessary to remove the phosphorylcholine moiety as described later. The resulting ceramide (*N*-acyl-sphingosine) can be deacylated by a particulate enzyme isolated from brain[59].

Ceramides containing either sphingosine or dihydrosphingosine are reactive. The reaction rate depends upon the nature of the bound fatty acid, being highest with oleic acid, less with palmitic or stearic, and insignificant with acetic or lignoceric. As previously mentioned, the formation of ceramide from sphingosine and free fatty acid as also catalyzed by the enzyme.

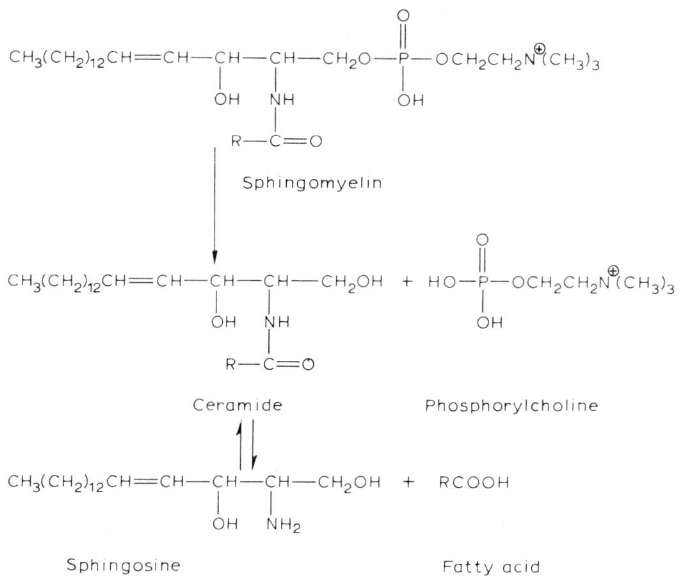

Scission of ether bonds

Degradation of the alk-1′-enyl bond of plasmalogens

Warner and Lands[102] described a microsomal enzyme from rat liver which is capable of hydrolyzing the ether bond of lyso-choline plasmalogen but not intact choline plasmalogen*. The analogous ethanolamine-containing phosphatides were not attacked. No divalent cation was required but, in a later study[103], the enzyme complex was shown to contain phospholipids which were essential for the activity towards the phospholipid substrate.

Brain tissue contains a cleavage enzyme possessing a different specificity[104]. It can rupture the ether linkage of intact ethanolamine plasmalogen or its lyso derivative. The conversion of ethanolamine plasmalogen to 2-acyl-*sn*-glycero-3-phosphorylethanolamine and fatty aldehyde requires $Mg^{2\oplus}$ and is most active at approximately pH 7.4.

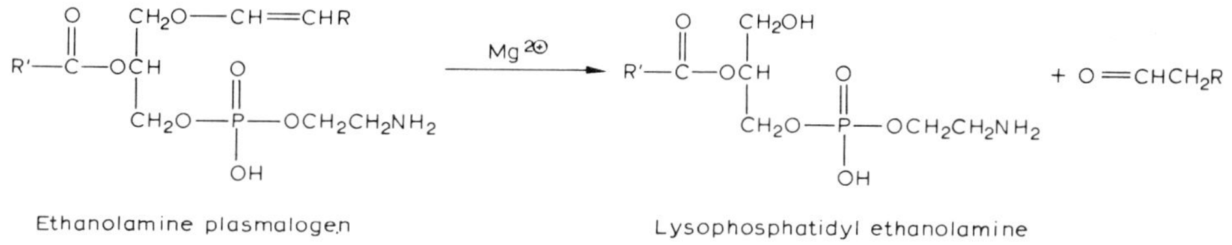

Kapoulas et al.[105] reported a third ether-splitting enzyme in cell-free preparations of *Tetrahymena pyriformis*. This enzyme, which presumably acts upon 1-*O*-alkyl glycerol as its natural substrate, can also cleave the alk-1′-enyl ether bond of free 1-*O*-alk-1′-enyl glycerol.

*Catabolism of ether-containing phospholipids has recently been fully reviewed in a monograph edited by Snyder[28].

References p. 94

It is clear from these varied findings that the alk-1′-enyl ether bond can be cleaved in intact or in partly degraded plasmalogen molecules. No conclusive evidence is yet available on which degradative method is active *in vivo*.

Degradation of 1-alkyl-2-acyl-glycerophosphatides

There have been no reports of enzymes capable of cleaving the alkyl ether linkage of intact glycerol ether-containing phospholipids. It would appear necessary to disassemble the molecule to the level of a free glycerol ether before the ether bond can be severed.

The principal details of the etherase reaction were revealed by the studies of Tietz et al.[106]. These workers incubated rat liver microsomes with radioactive glycerol ethers, and the data led them to propose the following sequence of events:

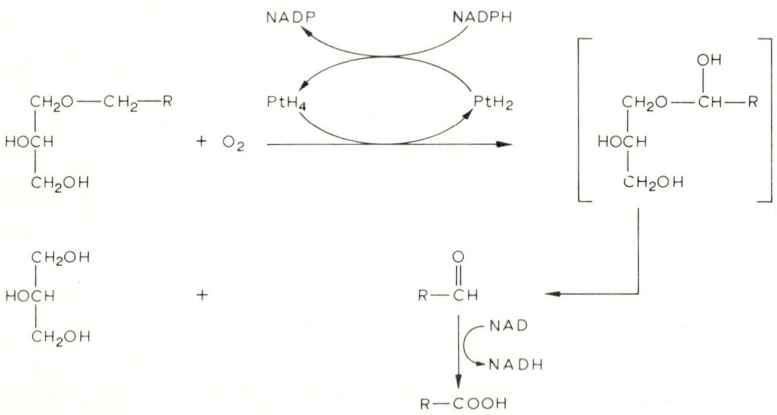

Molecular oxygen and a tetrahydropteridine cofactor are required for the reaction. The fully reduced pteridine could be regenerated by the addition of a NADPH-generating system to the incubation mixture. The existence of a hemiacetal intermediate rather than an α, β-unsaturated glyceryl ether was postulated because chemical degradation of the reaction mixture revealed no aldehydogenic compounds, thus ruling out the presence of the α,β-unsaturated derivative in more than trace amounts. Although the proposed hemiacetal has never been demonstrated, the other intermediate, fatty aldehyde, was shown to accumulate when NAD was omitted from the test system.

Additional properties of the rat liver system were defined by Pfleger et al.[107]. The etherase is only loosely associated with the microsomal membranes and can be separated by repeated centrifugation. Fatty aldehydes produced by the reaction can be reduced to fatty alcohols by an NADPH-requiring enzyme.

The liver enzyme described above would seem to catabolize dietary ether lipids since the tissue itself is essentially devoid of ether analogs. The protozoan *Tetrahymena pyriformis*, which is rich in glycerol ether phospholipids, can also degrade the ether bond, presumably as a part of the normal metabolic turnover of lipids[105]. The *in vitro* system differs in certain respects from the liver enzyme, notably in lacking any demonstrable need for a pteridine cofactor. Interestingly, cleavage of the ether bond by intact Tetrahymena cells is not initiated until the fed substrate exceeds a certain threshold concentration[108].

Cleavage of the phosphate-glycerol linkages

Phospholipase C

Although the official name for this enzyme, phosphatidylcholine cholinephosphohydrolase (EC 3.1.4.3), suggests that its primary substrate is lecithin, it can, under the

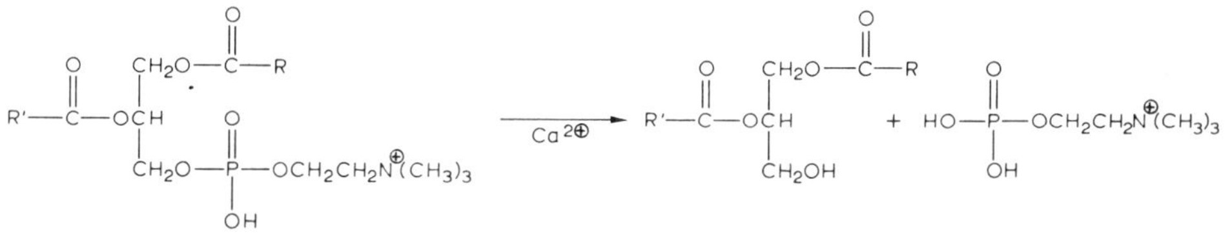

proper conditions, also attack phosphatidylethanolamine, sphingomyelin, the ethanolamine and 2-aminoethylphosphonate analogs of sphingomyelin[109], choline and ethanolamine plasmalogens[110], and the phosphonic analog of phosphatidylcholine[111].

Most of our data concerning phospholipase C comes from studies of enzymes from toxins of *Clostridium perfringens, Bacillus cereus*, and other bacteria. Enzymes possessing this activity are not as well studied in animal tissues. Interpretations are often complicated by the concurrent action of other phospholipases in the systems under examination. Heller and Shapiro[112] isolated a preparation from rat-liver particulate matter (mitochondria or lysosomes) which could hydrolyze sphingomyelin to ceramide and phosphorylcholine. The reaction proceeds more rapidly with the *erythro* isomer than with the *threo* form. Unlike the bacterial enzymes, this phospholipase was much more reactive towards sphingomyelin than towards phosphatidylcholine. The enzyme was later shown to be localized primarily in lysosomes[113]. Gatt[59] has studied a similar enzyme in brain. Although the crude system degraded both sphingolipids and glycerophosphatides, purification resulted in the separation of a phospholipase C type sphingomyelinase, which is inactive towards glycerophosphatides, and a phospholipase A_1. Brain phospholipase C is further distinguished from those of bacterial origin by showing no dependence on $Ca^{2\oplus}$ for full activity.

Atherton and Hawthorne[114] have described the properties of a phospholipase C type enzyme occurring in intestinal mucosa and a number of other animal tissues. The enzyme exhibits a very pronounced specificity, attacking only phosphatidylinositol and di- and triphosphoinositides among all the more common glycerophosphatides examined. Sphingomyelin was not tested. The enzyme, termed phosphoinositide inositolphosphohydrolase, is localized in the 105 000 × *g* supernatant fraction and depends upon the presence of $Ca^{2\oplus}$ for activity.

Phosphatidate phosphohydrolase

An intermediate step in the formation of several important lipids is the hydrolysis of phosphatidic acid to diglyceride and inorganic phosphate. The enzyme responsible for the reaction is *sn*-3-phosphatidate phosphohydrolase (EC 3.1.3.4), more commonly known as phosphatidic acid phosphatase.

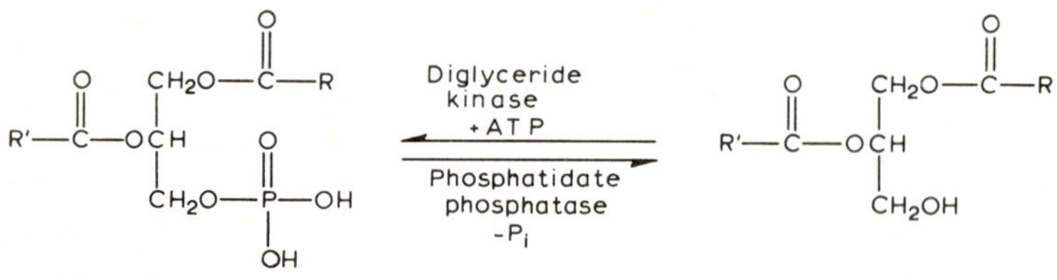

Phosphatidic acid 1,2-diacyl-*sn*-glycerol

In what may serve as an entirely separate function, the enzyme is sometimes coupled in its activity with another enzyme, diglyceride kinase, whereby a diglyceride molecule is phosphorylated with ATP acting as the donor.

The likelihood of at least two more or less independent functions for phosphatidate phosphatase is suggested by the finding that enzymatic activity occurs in various particulate fractions of the cell and that in these fractions certain properties of the enzyme differ significantly[115]. Some of the apparent differences in pH optima and kinetic parameters might be attributed to variations in the extent to which the enzyme is embedded within membranes. Coleman[116] has pointed out that much of the activity reported to be present in microsomes and certain other cell fractions could be due to contamination by fragments of plasma membrane, which is rich in phosphatidate phosphohydrolase.

Cleavage of the phosphate-base linkages (Phospholipase D)

Enzymes of this category have not been found in animal cells but are present in higher plants (Chapter 10). Although animal lipids are not the normal substrate for phospholipase D, they are susceptible to attack, and the enzyme is frequently used for the preparation of phosphatidic acid from animal phosphoglycerides.

Dephosphorylation of phosphoinositides

Tri- and diphosphoinositides are found principally in tissues of the nervous system. It has been recognized for several years that brain tissue contains polyphosphoinositide-specific phosphomonoesterases. These enzymes are so potent that special precautions must be taken to avoid appreciable degradation of the lipids during lipid extraction by classical methods. Lee and Huggins[117] succeeded in purifying the active triphosphoinositide phosphomonoesterase 50-fold from kidney cortex. The enzyme degraded triphosphoinositide in the presence of divalent cations, liberating inorganic phosphate and phosphatidylinositol.

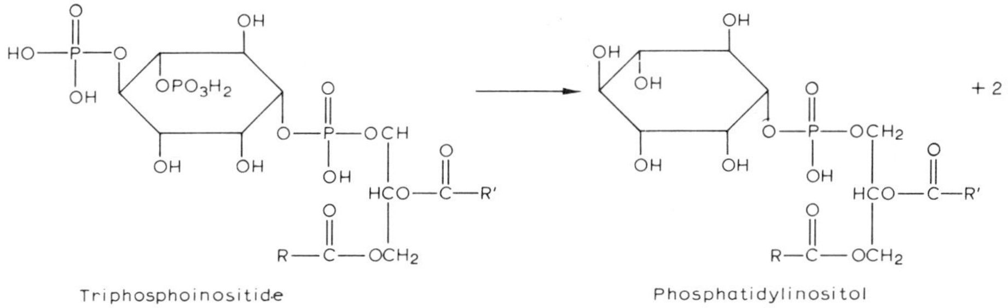

The mode of attack was shown by Chang and Ballou[118] to be the selective removal of the phosphate at position 5, followed by dephosphorylation at position 4.

Degradation of sphingosine bases

The enzymatic details of sphingosine-base catabolism have been determined principally through the study of dihydrosphingosine degradation. The compound is first converted to dihydrosphingosine-1-phosphate by a soluble kinase[119]. This activation is

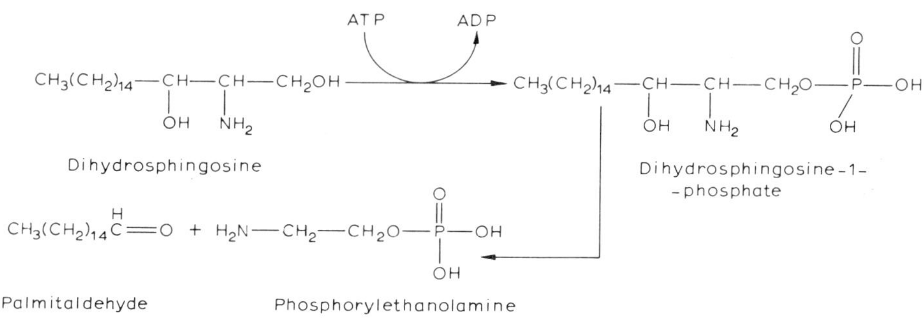

followed by an aldolase type reaction which produces phosphorylethanolamine and palmitaldehyde[56]. The latter enzyme is mainly microsomal in origin.

Summary and conclusions

In this section I have outlined the reactions which have been implicated in phospholipid degradation by animal tissues. With certain lipids, there is little doubt regarding the catabolic sequence, e.g. sphingomyelin clearly is degraded by (*1*) the removal of the phosphorylcholine moiety, (*2*) the hydrolysis of the amide linkage, and (*3*) the scission of sphingosine to yield palmitaldehyde and ethanolamine phosphate. On the other hand, phosphoglycerides can be dismantled by any one of several pathways. We do not know at present which sequence of catabolic steps is most prevalent *in vivo*. Assorted indirect evidence of one type or another suggests a scheme along the lines illustrated in Fig. 3, where the bolder arrows represent predominating reactions. Thus in my hypothetical tissue, the 2-acyl group is most frequently the first to be removed, and hydrolysis by a phospholipase C type attack is infrequent. There will undoubtedly prove to be many exceptions to this catabolic scheme.

References p. 94

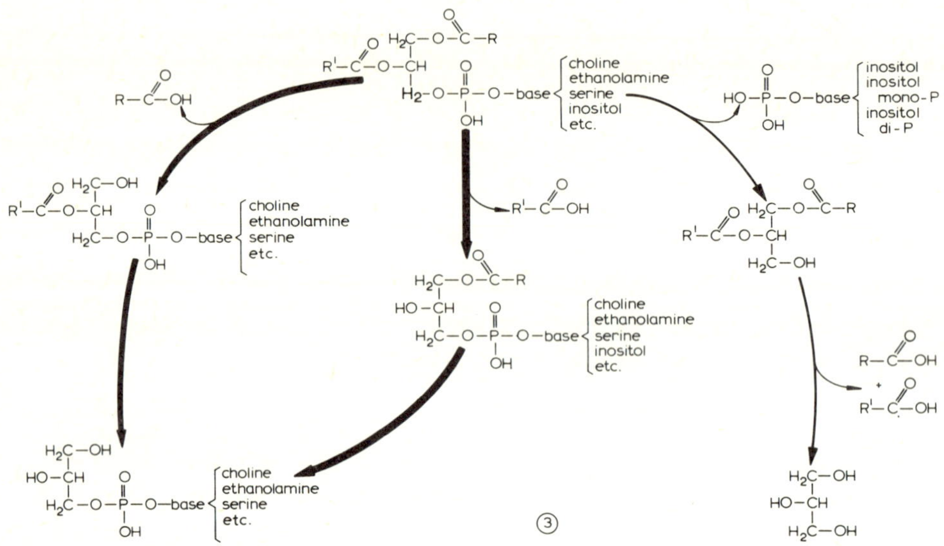

REFERENCES

1 G.A. Thompson Jr., in M. Florkin and E.H. Stotz (Eds.), *Comprehensive Biochemistry*, Vol. 18, Elsevier, Amsterdam, 1969, p. 157.
2 A. Kornberg and W.E. Pricer Jr., *J. Biol. Chem.*, 204 (1953) 345.
3 K. Lippel, J. Robinson and E.G. Trams, *Biochim. Biophys. Acta*, 206 (1970) 173.
4 H. van den Bosch and P.R. Vagelos, *Biochim. Biophys. Acta*, 218 (1970) 233.
5 W.E.M. Lands and P. Hart, *J. Lipid Res.*, 5 (1964) 81.
6 E. Sánchez de Jiménez and W.W. Cleland, *Biochim. Biophys. Acta*, 176 (1969) 685.
7 F. Possmayer, G.L. Scherphof, T.M.A.R. Dubbelman, L.M.G. van Golde and L.L.M. van Deenen, *Biochim. Biophys. Acta*, 176 (1969) 95.
8 L.E. Hokin and M.R. Hokin, *J. Histochem. Cytochem.*, 13 (1965) 113.
9 D.A. White and J.N. Hawthorne, *Biochem. J.*, 120 (1970) 533.
10 A.K. Hajra and B.W. Agranoff, *J. Biol. Chem.*, 243 (1968) 3542.
11 L.E. Puleo, G.A. Rao and R. Reiser, *Lipids*, 5 (1970) 770.
12 H. Okuyama and W.E.M. Lands, *Biochim. Biophys. Acta*, 218 (1970) 376.
13 B.W. Agranoff and A.K. Hajra, *Proc. Natl. Acad. Sci. (U.S.)* 68 (1971) 411.
13a G.A. Rao, M.F. Sorrels and R. Reiser, *Biochem. Biophys. Res. Commun.*, 44 (1971) 1279.
14 J. Wittenberg and A. Kornberg, *J. Biol. Chem.*, 202 (1953) 431.
15 A. Kornberg and W.E. Pricer, *Federation Proc.*, 11 (1952) 242.
16 E.P. Kennedy and S.B. Weiss, *J. Biol. Chem.*, 222 (1956) 193.
17 P. Bjørnstad and J. Bremer, *J. Lipid Res.*, 7 (1966) 38.
18 B. de Kruyff, L.M.G. van Golde and L.L.M. van Deenen, *Biochim. Biophys. Acta*, 210 (1970) 425.
19 H. Paulus and E.P. Kennedy, *J. Biol. Chem.*, 235 (1960) 1303.
20 J.R. Carter and E.P. Kennedy, *J. Lipid Res.*, 7 (1966) 678.
21 B.W. Agranoff, R.M. Bradley and R.O. Brady, *J. Biol. Chem.*, 233 (1958) 1077.
22 H.H. Bishop and K.P. Strickland, *Can. J. Biochem.*, 48 (1970) 269.
23 J.L. Harwood and J.N. Hawthorne, *Biochim. Biophys. Acta*, 171 (1969) 75.
24 J.-S. Tou, M.W. Hurst, C.G. Huggins and W.E. Floor, *Arch. Biochem. Biophys.*, 140 (1970) 492.
25 A. Sheltawy and R.M.C. Dawson, *Biochem. J.*, 100 (1966) 12.
26 E.A. Dennis and E.P. Kennedy, *J. Lipid Res.*, 11 (1970) 394.

27 E.A. Dennis and E.P. Kennedy, *J. Lipid Res.*, 13 (1972) 263.
28 F.L. Snyder (ed.), *Ether Lipids – Chemistry and Biology,* Academic Press, New York, 1972.
29 F. Snyder, R.L. Wykle and B. Malone, *Biochem. Biophys. Res. Commun.*, 34 (1969) 315.
30 R.L. Wykle and F. Snyder, *Biochem. Biophys. Res. Commun.*, 37 (1969) 658.
31 A.K. Hajra, *Biochem. Biophys. Res. Commun.*, 37 (1969) 486.
32 A.K. Hajra, *Biochem. Biophys. Res. Commun.*, 39 (1970) 1037.
33 F. Snyder, W.T. Rainey Jr., M.L. Blank and W.H. Christie, *J. Biol. Chem.*, 245 (1970) 5853.
33a S.J. Friedberg, A. Heifetz and R.C. Greene, *Biochemistry*, 11 (1972) 297.
34 M. Kates, M.K. Wassef and E.L. Pugh, *Biochim. Biophys. Acta*, 202 (1970) 206.
35 E.E. Hill and W.E.M. Lands, *Biochim. Biophys. Acta*, 202 (1970) 209.
36 G.A. Thompson Jr., *Biochim. Biophys. Acta*, 152 (1968) 409.
37 R. Wood and K. Healy, *J. Biol. Chem.*, 245 (1970) 2640.
38 W. Stoffel and D. LeKim, *Z. Physiol. Chem.*, 352 (1971) 501.
39 A. Poulos and G.A. Thompson Jr., *Lipids*, 6 (1971) 470.
40 R.L. Wykle, M.L. Blank and F. Snyder, *FEBS Letters*, 12 (1970) 57.
41 J.Y. Kiyasu, R.A. Pieringer, H. Paulus and E.P. Kennedy, *J. Biol. Chem.*, 238 (1963) 2293.
42 J.B. Davidson and N.Z. Stanacev, *Biochem. Biophys. Res. Commun.*, 42 (1971) 1191.
43 N.Z. Stanacev, L. Stuhne-Sekalec, K.B. Brookes and J.B. Davidson, *Biochim. Biophys. Acta*, 176 (1969) 650.
44 J.B. Davidson and N.Z. Stanacev, *Can. J. Biochem.*, 48 (1970) 633.
45 N.Z. Stanacev, Y.Y. Chang and E.P. Kennedy, *J. Biol. Chem.*, 242 (1967) 3018.
45a K.Y. Hostetler, H. van den Bosch and L.L.M. van Deenen, *Biochim. Biophys. Acta*, 260 (1972) 507.
46 J. Bremer, P.H. Figard and D.M. Greenberg, *Biochim. Biophys. Acta*, 43 (1960) 477.
47 K.D. Gibson, J.D. Wilson and S. Udenfriend, *J. Biol. Chem.*, 236 (1961) 673.
48 D. Rehbinder and D.M. Greenberg, *Arch. Biochem. Biophys.*, 109 (1965) 110.
49 G.A. Scarborough and J.F. Nyc, *J. Biol. Chem.*, 242 (1967) 238.
50 J.A. Balint, D.A. Beeler, D.H. Treble and H.L. Spitzer, *J. Lipid Res.*, 8 (1967) 486.
51 B. Lombardi, P. Pani, F.F. Schlunk and C. Shi-Hua, *Lipids*, 4 (1969) 67.
52 M.R. Steiner and R.L. Lester, *Biochemistry*, 9 (1970) 63.
53 G. Hübscher, *Biochim. Biophys. Acta*, 57 (1962) 555.
54 H.D. Crone, *Biochem. J.*, 104 (1967) 695.
54a J.N. Kanfer, *J. Lipid Res.*, 13 (1972) 468.
54b P. Morell and P. Braun, *J. Lipid Res.*, 13 (1972) 293.
54c W. Stoffel, *Ann. Rev. Biochem.*, 40 (1971) 57.
55 E.E. Snell, S.J. Dimari and R.N. Brady, *Chem. Phys. Lipids*, 5 (1970) 116.
56 W. Stoffel, *Chem. Phys. Lipids*, 5 (1970) 139.
57 K.A. Karlsson, *Chem. Phys. Lipids*, 5 (1970) 6.
58 M. Sribney, *Biochim. Biophys. Acta*, 125 (1966) 542.
59 S. Gatt, *Chem. Phys. Lipids*, 5 (1970) 235.
60 M. Sribney and E.P. Kennedy, *J. Biol. Chem.*, 233 (1958) 1315.
61 Y. Fujino, Mo. Nakano, T. Negishi and S. Ito, *J. Biol. Chem.*, 243 (1968) 4650.
62 R.O. Brady, R.M. Bradley, O.M. Young and H. Kaller, *J. Biol. Chem.*, 240 (1965) PC 3693.
62a H. Diringer, W.D. Marggraf, M.A. Koch and F.A. Anderer, *Biochem. Biophys. Res. Commun.*, 47 (1972) 1345.
63 B.K. Tam and P.B. McCay, *J. Biol. Chem.*, 245 (1970) 2295.
64 Y. Akamatsu and J.H. Law, *J. Biol. Chem.*, 245 (1970) 701.
65 M.I. Gurr and P. Brawn, *European J. Biochem.*, 17 (1970) 19.
66 R.J. Goss, *Advan. Cell Biol.*, 1 (1970) 233.
67 R.C. Reitz, M. El-Sheikh, W.E.M. Lands, I.A. Ismail and F.D. Gunstone, *Biochim. Biophys. Acta*, 176 (1969) 480.
68 W.E.M. Lands and P. Hart, *J. Biol. Chem.*, 240 (1965) 1905.
69 H. Eibl, E.E. Hill and W.E.M. Lands, *European J. Biochem.*, 9 (1969) 250.
70 M. Waite, P. Sisson and E. Blackwell, *Biochemistry*, 9 (1970) 746.
71 W.E.M. Lands and I. Merkl, *J. Biol. Chem.*, 238 (1963) 898.
72 H. van den Bosch, L.M.G. van Golde, A.J. Slotboom and L.L.M. van Deenen, *Biochim. Biophys. Acta*, 152 (1968) 694.

72a H. Okuyama and W.E.M. Lands, *J. Biol. Chem.*, 247 (1972) 1414.
73 M.E. de Tomás and R.R. Brenner, *Biochim. Biophys. Acta*, 202 (1970) 184.
74 K. Waku and W.E.M. Lands, *J. Biol. Chem.*, 243 (1968) 2654.
75 K. Waku and W.E.M. Lands, *J. Lipid Res.*, 9 (1968) 12.
76 E. Mulder and L.L.M. van Deenen, *Biochim. Biophys. Acta*, 106 (1965) 348.
77 J.A. Glomset, *J. Lipid Res.*, 9 (1968) 155.
78 J.F. Erbland and G.V. Marinetti, *Biochim. Biophys. Acta*, 106 (1965) 128.
79 H. van den Bosch, H.A. Bonte and L.L.M. van Deenen, *Biochim. Biophys. Acta*, 98 (1965) 648.
80 M.L. Williams and F.L. Bygrave, *European J. Biochem.*, 17 (1970) 32.
81 M.I. Gurr, C. Prottey and J.N. Hawthorne, *Biochim. Biophys. Acta*, 106 (1965) 357.
82 S.S. Tsao and W.E. Cornatzer, *Lipids*, 2 (1967) 424.
83 L. Freysz, R. Bieth and P. Mandel, *J. Neurochem.*, 16 (1969) 1417.
84 T. Omura, P. Siekevitz and G.E. Palade, *J. Biol. Chem.*, 242 (1967) 2389.
85 D.J. Hanahan, H. Brockerhoff and E.J. Barron, *J. Biol. Chem.*, 235 (1960) 1917.
86 J.J. Gallai-Hatchard and R.H.S. Thompson, *Biochim. Biophys. Acta*, 98 (1965) 128.
87 M. Waite, G.L. Scherphof, F.M.G. Boshouwers and L.L.M. van Deenen, *J. Lipid Res.*, 10 (1969) 411.
88 M. Waite, *Biochemistry*, 8 (1969) 2536.
89 G.H. de Haas, N.M. Postema, W. Nieuwenhuizen and L.L.M. van Deenen, *Biochim. Biophys. Acta*, 159 (1968) 103.
90 M.A. Wells and D.J. Hanahan, *Biochemistry*, 8 (1969) 414.
91 G.H. de Haas, N.M. Postema, W. Nieuwenhuizen and L.L.M. van Deenen, *Biochim. Biophys. Acta*, 159 (1968) 118.
92 A.D. Bangham, *Advan. Lipid Res.*, 1 (1963) 65.
93 G. Colacicco and M.M. Rapport, *J. Lipid Res.*, 7 (1966) 258.
94 P. Bjornstad, *J. Lipid Res.*, 7 (1966) 612.
95 M. Waite and L.L.M. van Deenen, *Biochim. Biophys. Acta*, 137 (1967) 498.
96 L. Sarda, M.F. Maylié, J. Roger and P. Desnuelle, *Biochim. Biophys. Acta*, 89 (1964) 183.
97 J.D. Newkirk and M. Waite, *Biochim. Biophys. Acta*, 225 (1971) 224.
98 H. van den Bosch, A.J. Aarsman, A.J. Slotboom and L.L.M. van Deenen, *Biochim. Biophys. Acta*, 164 (1968) 215.
99 A.F. Robertson and W.E.M. Lands, *Biochemistry*, 1 (1962) 804.
100 A. Ottolenghi, *J. Lipid Res.*, 5 (1964) 532.
101 P. Bjornstad, *Biochim. Biophys. Acta*, 116 (1966) 500.
102 H.R. Warner and W.E.M. Lands, *J. Biol. Chem.*, 236 (1961) 2404.
103 J.S. Ellingson and W.E.M. Lands, *Lipids*, 3 (1968) 111.
104 G.B. Ansell and S. Spanner, *Biochem. J.*, 94 (1965) 252.
105 V.M. Kapoulas, G.A. Thompson Jr. and D.J. Hanahan, *Biochim. Biophys. Acta*, 176 (1969) 250.
106 A. Tietz, M. Lindberg and E.P. Kennedy, *J. Biol. Chem.*, 239 (1964) 4081.
107 R.C. Pfleger, C. Piantadosi and F. Snyder, *Biochim. Biophys. Acta*, 144 (1967) 633.
108 V.M. Kapoulas, G.A. Thompson Jr. and D.J. Hanahan, *Biochim. Biophys. Acta*, 176 (1969) 237.
109 I. Arakawa, M. Sugita and T. Hori, *J. Japan. Biochem. Soc.*, 40 (1968) 154.
110 G.B. Ansell and S. Spanner, *Biochem. J.*, 97 (1965) 375.
111 E. Baer and N.Z. Stanacev, *Can. J. Biochem.*, 44 (1966) 893.
112 M. Heller and B. Shapiro, *Biochem. J.*, 98 (1966) 763.
113 S. Fowler, *Biochim. Biophys. Acta*, 191 (1969) 481.
114 R.S. Atherton and J.N. Hawthorne, *European J. Biochem.*, 4 (1968) 68.
115 B. Sedgwick and G. Hübscher, *Biochim. Biophys. Acta*, 106 (1965) 63.
116 R. Coleman, *Biochim. Biophys. Acta*, 163 (1968) 111.
117 T.-C. Lee and C.G. Huggins, *Arch. Biochem. Biophys.*, 126 (1968) 214.
118 M. Chang and C.E. Ballou, *Biochem. Biophys. Res. Commun.*, 26 (1967) 199.
119 C.B. Hirschberg, A. Kisic and G.J. Schroepfer Jr., *J. Biol. Chem.*, 245 (1970) 3084.

Chapter 5

SPECIFICITY OF ENZYMES INVOLVED IN THE METABOLISM OF PHOSPHOLIPIDS

R.M.C. DAWSON

Department of Biochemistry, A.R.C. Institute of Animal Physiology, Babraham, Cambridge (Great Britain)

1. TWO TYPES OF SPECIFICITY: CHEMICAL AND PHYSICAL

The previous chapter contains many references to the specificities of some of the enzymes participating in the synthesis of phospholipids or their breakdown. For example, the unsaturation or chain length of the fatty acids can make an appreciable difference to the rate of phospholipid biosynthesis *in vitro* and changes in the polar part of phosphoglyceride molecules can completely alter their susceptibility to attack by phospholipases. However, in such systems it is clearly necessary to distinguish two types of specificity, one due to the actual chemical changes in the structure of the substrate and the other due to the fact that the lipid substrate and also any lipoidal products tend to form polymolecular aggregates in an aqueous environment. The first type of specificity will influence the enzyme substrate reaction in the same way as when a completely soluble substrate interacts with the active centre of an enzyme, where closeness of fit, multiple attachment by various types of binding, and matching of polarity are of paramount importance in determining how readily enzyme substrate complexes are formed and the rate of the catalysed reaction. The second type of specificity will depend on the physical nature of the substrate including the size and type of polymeric aggregates it forms, the packing and arrangement of the molecules it contains and any electrostatic fields produced by the combined effects of the polar groups oriented in an orderly manner at the interface. Such physical factors can markedly control the ability of enzyme molecules both to approach the substrate and to orientate and form the correct stereochemical relationship to individual substrate molecules so that the enzyme substrate complex can be formed. Very often the way the substrate is presented to the enzyme can determine absolutely whether the enzymatically catalyzed reaction can occur or not. In reactions involving lipids the ease with which new substrate molecules can migrate from the interior of a lipid particle to the lipid–water interface undergoing enzymatic catalysis, or the removal of the lipoidal products of the reaction from the interface into either the lipid or water phase are additional physicochemical parameters which markedly influence the rate and extent of the reaction.

Unfortunately when an examination is made of the way that variations in the chemical structure of the lipid substrate affect the rate of an enzymatic reaction, the

References p. 114

chemical change can itself affect the thermodynamically stable form of the polymolecular aggregate of the lipid formed in an aqueous environment. A difference in rate can therefore arise both from the immediate effect on the enzyme-active-centre substrate interaction at a molecular chemical level and from physical changes in the lipid–water interface. In nearly all investigations on specificity performed up to the present, little attempt has been made to distinguish between the two types of rate control which are possible. If we define true specificity as interaction at a molecular level it will be necessary for many of the results presented in the literature to be re-examined. For example it was believed until recently that phospholipase C (EC 3.1.4.3) only attacked choline-containing phospholipids and was completely without action on analogous structures containing ethanolamine. It is now known that this is due to distinct differences in the double electrical layer surrounding the substrate particles and that if appropriate steps are taken to modify this, phosphatidylethanolamine can be degraded by the enzyme at a rate that is little different from that of phosphatidylcholine[1].

It is therefore appropriate to consider some of the ways in which the physical nature of the substrate can affect enzymatic reactions involving phospholipids.

2. PHYSICAL FACTORS AFFECTING ENZYMATIC REACTIONS INVOLVING PHOSPHOLIPIDS

Physical form of lipid participating in reactions

It is clear that some lipid substrates participating in enzymatic reactions involved in phospholipid synthesis need to be present as monomers or perhaps as loose aggregates of a few molecules rather than existing as polymolecular structures such as micelles or liquid crystals. The acylation of *sn*-glycero-3-phosphate by liver microsomes and the subsequent acylation of the 1-acyl-*sn*-glycero-3-phosphate to phosphatidic acid show kinetics which suggest that, above certain concentrations of palmitoyl-CoA, the rate is independent of substrate concentration[2]. The acylation enzyme appears to use free molecules of palmitoyl-CoA and once the critical micellar concentration is exceeded no further increase in reaction rate occurs. Thus Michaelis kinetics will be observed only below the critical micellar concentration and in the calculation of equilibrium constants the substrate concentration should also be below this. The acyl-CoA-lysophospholipid transferase activity of microsomes is unaffected by concentrations of various double bond positional isomers of octadecenoyl-CoA between 10 and 40 μM[65]. Although the critical micellar concentrations of these substrates are unknown it is again possible that the enzyme prefers free molecules to micelles. In such cases measurements of the specificity of an enzyme would have to include observations over a range of acyl-CoA concentrations, since merely measuring activity at saturation conditions could be measuring differences between critical micellar concentrations of differing substrates.

Often however, a reaction involving a lipid substrate will only proceed at a significant rate providing the lipid can form a definite lipid–water interface rather than be present in an aqueous environment as monomers or small loose aggregations of a few molecules only. Thus the phospholipase A_2 of *Crotalus adamanteus* venom scarcely attacks a water-soluble phosphatidylcholine (diacetyl) whereas the much more insoluble (didecanoyl) phosphatidylcholine which forms micelles is readily attacked[3]. A comparison of two isomers showed that 1-oleoyl-2-butyroyl lecithin which forms a turbid suspension in water was attacked more readily by venom phospholipase A_2 than 1-butyroyl-2-oleoyl lecithin which forms a clear solution[83]. Probably the best-studied reaction of this type is that of pancreatic lipase (which can also act as a phospholipase A_1)[4] on triglycerides. Short-chain triglycerides appear not to be attacked significantly until their solubility product is exceeded and an interface is formed[5,6]. The organized structure necessary for lipase action need not be a large particle since the addition of NaCl which causes short-chain triglycerides to aggregate and form small molecular clusters, enhances the activity[7]. It has been suggested[6] that the connection between hydrophobicity of the substrate and lipase activity could be due to three possible factors: (*1*) unfolding of the enzyme with exposure of the active centre at a certain critical interfacial tension, (*2*) hydrophobic binding of substrate to the active centre of the lipase, (*3*) shielding of the hydrophilic esters by organized water against the attack of a weakly nucleophilic enzyme.

Ratio between substrate mass and area of interface available for enzymatic attack

It has long been recognized that the state of dispersion of lipid substrates can affect the rate of an enzymatic reaction simply by presenting the enzyme with a far greater area of substrate surface for interaction. Thus, for example, lipoprotein lipase shows a high degree of dependence on the particle size or degree of dispersion of the substrate[96]. Substances such as fatty acids tend to disperse more readily at a high pH and as the degree of unsaturation of the fatty acid chains increases or its length decreases. Degree of unsaturation and chain length also influence, in the same way, the dispersion of neutral glycerides such as diglycerides and ceramides which participate in reactions involved in phospholipid synthesis. Phospholipids themselves can exist in an aqueous environment in two general forms, one the classical micellar form and the other bimolecular leaflets organized into liquid crystals of various sizes and shapes[8] (Chapter 6, p. 123). In both polymolecular aggregates the polar region of the phospholipid is oriented towards the water (and the enzyme) and the hydrophobic residues come into close contact through Van der Waal's forces. Most natural phospholipids form bimolecular leaflets in water even at high dilutions. This is because the bulk of them are zwitterions at a physiological pH with little electrostatic repulsion between the polar head groups (e.g. phosphatidylcholine, phosphatidylethanolamine); moreover the matching cross-sectional areas of the polar head group and the hydrocarbon chains make a molecule of roughly cylindrical shape which tends to stabilize the structure. On the other hand pure phospholipids which are highly negatively charged (e.g. triphosphoinositide) or have wedge-shaped molecules (e.g. lysolecithin) tend to form true micelles spontaneously with a diameter (generally 45–50 Å) depending on the length and characteristics of the fatty acid chains. The charge on the polar head group needed to bring about micelle formation needs to be appreciable since even acidic phosphoglycerides such as phosphatidylserine are believed to exist in aqueous suspension as bimolecular leaflets (Chapter 7, p. 157). Within the hydrated liquid crystalline structure itself, phase changes are possible depending on the temperature and the nature of the fatty acids in the phospholipid (Chapter 6, p. 123).

The bimolecular layers of natural phospholipids form liquid crystals which are smectic i.e. laminated bimolecular sheets of phospholipid intercalated with sheets of water. The mass forms a large particle consisting of concentric layers of bimolecular leaflets rather like the skins of an onion bulb. Because of the barriers to enzyme permeability caused by the hydrophobic regions of the bilayer, only the outermost layer of the particle will be initially exposed to a water-soluble enzyme attacking the particle. Thus, unless the enzymatic reaction destroys the bilayer or an exchange can occur between any lipoidal products and new substrate molecules from leaflets within the phospholipid particle the reaction will quickly stop. It seems likely that as the hydrocarbon chains become more saturated and pass at a given temperature from a liquid crystalline to a fully crystalline phase (Chapter 6, p. 120), this would hinder such interchanges which need to occur before the enzymatic reaction can continue.

If the phospholipid exists in a liquid crystalline form at room temperature, i.e. it contains a reasonable degree of unsaturation, the coarse dispersion which is formed by shaking it with water can be broken down by ultrasonication. Thus ovolecithin shaken with water, hydrates and breaks up into asymmetric particles of average molecular weight 10^8 and when these are sonicated they break down into smaller particles of molecular weight $2 \cdot 10^6$ (ref. 10). Although it is known that some hydrolysis of esters bonds and autoxidation can occur if the sonication is prolonged or performed in air, the resulting sonicated liposomes retain the bimolecular layer structure and are in fact vesicles consisting of single or at the most a few layers of bimolecular leaflets. There is still controversy as to whether the simple size reduction is attended by any other rearrangement of the molecules during sonication, since nuclear magnetic resonance studies of two groups are in disagreement[9,130].

The size reduction caused by sonication causes an enormous increase in the area of substrate surface available for enzymatic attack and this often results in a marked increase in the rate of phospholipase action, e.g. coarse lecithin particles are not hydrolysed by cobra venom phospholipase A_2 at a measurable rate whereas after sonication the dispersed liposomes are readily attacked. Of course, it is entirely possible that some change in the molecular organisation of the ovolecithin bilayers also occurs during sonication (as is suggested by certain studies[10]) and that this also assists hydrolysis.

Sometimes a marked increase in the rate of the enzymatic hydrolysis can be brought about by the introduction of detergents which penetrate into the bilayers, break up the phospholipid particles and similarly expose more substrate molecules to phospholipase attack. However, the action of these is somewhat more complicated since their introduction into the phospholipid leaflet can alter the packing of phospholipid molecules at the interface or alter the electrostatic field surrounding the surface. These in turn can either enhance or inhibit enzymatic activity. Thus, although ovolecithin particles can readily be dispersed by adding 0.25% sodium dodecylsulphate to the suspension, this, unlike sonication, results in no enhancement of cobra venom phospholipase A_2 action[11]. On the other hand, dodecylsulphate markedly stimulates the activity of phospholipase D towards the same substrate, although because of the large amount of calcium necessary as a cofactor there is not a corresponding reduction in the size of the substrate particles[12].

Influence of molecular interfacial packing of substrate on rate of enzymatic reactions

The influence of the interfacial tension, and consequently the area per molecule which lipid substrates occupy at the lipid–water interface can at present only be measured using monomolecular layers of substrate oriented at the air–water interface. So far investigations have been limited to the phospholipases, although it seems likely that similar effects may apply to the lipid substrates involved in enzymatic reactions synthesizing phospholipids. An enzyme in *Penicillum notatum* which can completely deacylate phosphatidylcholine, attacked monolayers of ^{32}P-labelled yeast lecithin with increasing rapidity as the film was compressed until at a pressure of 30 dynes/cm activity virtually ceased[13]. At this surface pressure each orientated yeast lecithin molecule would occupy an area of approximately 72 $Å^2$ at the interface. Similar results have been observed for the hydrolysis of monolayers of [*Me*-^{14}C]choline-labelled lecithin from yeast by phospholipase D (ref. 14), and yeast [^{32}P]lecithin by phospholipase C (ref. 1). It seems likely that these results can be correlated with the ability of part of the enzyme molecule (presumably the active centre) to penetrate between the orientated phospholipid molecules at the interface (Chapter 7, p. 162). Above a pressure of about 30 dynes/cm it is unlikely that proteins would be able to penetrate into yeast lecithin monolayers unless the electrostatic conditions at the interface were altered[15]. Snake venom phospholipase A_2 showed a similar cut-off point in its attack on lecithin monolayers since the pressure at which activity stopped depended on the degree of unsaturation of the fatty acid chains and their length[17, 18]. Insufficient experimental evidence is available at present to directly correlate enzymatic activity with the ability of the enzyme to penetrate between the phosphatidylcholine molecules of various fatty acid composition at the interface. It has recently been concluded by Quinn and Dawson[15, 19] that whether protein penetration into a phospholipid monolayer occurs or not at a given film pressure depends on the work required for penetration to occur rather than the size of the ‘gaps’ between the phospholipid molecules at the interface. However, the extent and amount of penetration will clearly depend on the size of these ‘gaps’ which at a given pressure will depend on the fatty acid composition of the film. Often it is found that the action of a phospholipase towards phospholipid monolayers is negligible at low surface pressures[13, 14, 17] when the packing density of the molecules is low and penetration would be easy. At low film pressures whole protein molecules are able to penetrate into phospholipid films and considerably unfold at the air–water interface[20]. This partially denatured enzyme protein at the interface may protect the phospholipid molecules from attack. Alternatively, the greater exposure of substrate hydrophobic residues at the interface, with the fatty acids flopping over rather than being vertically orientated as at higher surface pressures, may cause hydrophobic reactions with the protein which prevent the correct stereochemical alignment of the active centre of the enzyme and the vulnerable ester bond of the substrate.

Lipid solvent addition

Organic solvents, particularly diethylether, can greatly increase the rate at which some phospholipases attack a lecithin substrate. Ketones such as methyl isobutyl ketone[21, 14] can also stimulate the hydrolysis but many other fat solvents have a limited or negligible

References p. 114

effect. Snake venom phospholipase A_2 attacks phosphatidylcholine suspensions maximally when just sufficient diethylether has been added to saturate the aqueous system[22, 23] and it has the effect of clearing the turbidity in the suspension. It is likely that the ether penetrates into the lecithin particles and orientates at the lipid–water interface producing a surface dilution of the lecithin polar groups, thus allowing more ready access of the enzyme to the susceptible fatty acyl ester bond. This would be equivalent to the enhanced rate of enzymatic digestion of expanded lecithin monolayers discussed in the previous section. Increased dispersion of the phosphatidylcholine particles may occur although this does not appear to be the sole reason for stimulation since unlike diethyl ether, methyl isobutyl ketone does not produce an equivalent clearing, and furthermore diethylether can stimulate the enzymatic breakdown of ultrasonically dispersed lecithin preparations[11]. When cobra venom phospholipase A_2 is attacking ovolecithin there is evidence that ether assists the hydrolysis by removing the fatty acid formed from the surface of the lecithin particles[23]. Although this fatty acid could be taken into the bulk aqueous phase it seems more probable that there is a rearrangement of the molecular organisation of the lecithin particle with a loosening up of the structure so that fatty acids formed in the reaction can be repositioned in the interior. The hydrolysis of lecithin particles by phospholipase D is also markedly stimulated by saturating the incubation medium with diethylether. With the insoluble plant plastid enzyme it has been suggested that the solvent causes coalescence of the phospholipid and plastid[21] but soluble phospholipase D is also stimulated[14] and it is likely that the same physicochemical factors are operating as occur with phospholipase A.

The effect of the electrostatic field around the phospholipid–water interface

There is now good evidence that the electrical double layer which occurs adjacent to charged phospholipid surfaces can have a pronounced effect on enzymatic activity[8]. It was originally shown that a lecithinase present in *P. notatum* could only attack phosphatidylcholine providing a sufficient quantity of long chain anion had been added to the substrate to produce a negative ζ-potential of a certain magnitude[25]. Since then many other examples have been found where introduction of a charged amphipathic substance can markedly stimulate lipolytic activity towards phosphoglycerides. For example, pancreatic phospholipase A can be stimulated by the introduction of phosphatidic acid into the substrate[3] and phospholipase C by the production of a positive ζ-potential around the substrate either by the addition of long-chain cations or by adding water-soluble polyvalent cations such as $Ca^{2\oplus}$ or $Mg^{2\oplus}$ (ref. 1). The effect is not confined to hydrolytic enzymes since an anionic amphipathic substance is necessary for the introduction of the cyclopropane ring into a phosphatidylethanolamine substrate[26].

The addition of charged amphipathic substances to an organized phospholipid structure can have a dispersive effect, reducing the interfacial packing and increasing the hydration of the particle. It could therefore be argued that this would account for the initiation or stimulation of enzymatic activity. However, that this is not the sole effect is shown by experiments with unimolecular films in which it is possible to compress lecithin films to the same area per substrate molecule appertaining before the introduction of the charged amphipathic substance and still obtain initiation of activity. Furthermore, the dispersion of lecithin particles by a variety of long chain anions could not be correlated with the initiation of phospholipase activity whereas the ζ-potential could[27]. Often the activation occurs in the absence of any increased dispersion and frequently there is a massive increase in the size of the substrate particles due to aggregation. Nevertheless, it is possible that sometimes the reported stimulatory effect of anionic or cationic detergents on enzymatic reactions involving phospholipids can be partly or solely attributed to the increased dispersion.

The precise mechanism by which a change in the electrical field surrounding the surface of the phospholipid assists or inhibits the enzymatic activity is not understood, but it is possible that it affects the adsorption of enzyme to the surface and its penetration into the substrate. It is noteworthy that the penetration of non-enzyme proteins into phospholipid monolayers can only occur at high packing densities (compressed films) when the lipid polar groups possess a net charge or with zwitterionic phospholipids when a certain proportion of a charged amphipathic substance has been introduced into the film[19]. This effect is not entirely because the film bears a charge opposite to that of the net charge on the protein molecule and the sign of the charge can be of great importance. With unimolecular films of phospholipids, electrostatic requirements for phospholipase action are only observed at high pressures[13, 1, 14] conditions under which monolayer penetration by non-enzyme proteins would only occur with charged films[19]. At low film pressures (below 30 dynes/cm for yeast lecithin) neither phospholipase action nor penetration of model proteins is dependent on electrostatic conditions at the interface.

The electrostatic conditions at an interface can affect the activity of an enzyme by altering the surface pH[8]. Thus a negatively charged phospholipid will attract hydrogen ions from the bulk phase so the pH of the surface phase adjacent to the interface will be lowered. Alternatively, positively charged groups in the interface will increase the surface pH. In practice this could mean that the measured bulk pH optimum for an enzyme attacking a highly charged phospholipid such as phosphatidylinositol or phosphatidylserine might be 2–3 pH units higher than when it is hydrolysing zwitterionic phospholipids such as phosphatidylcholine or phosphatidylethanolamine. There is evidence also that a highly charged phospholipid micelle such as triphosphoinositide can prevent the attack of its specific phosphomonoesterases and phosphodiesterases possibly by adsorption with reversible partial denaturation on the highly charged surface. The reaction can be very greatly stimulated by the addition of cetyltrimethylammonium bromide to the system which reduces the ζ-potential of the substrate particles and at the same time aggregates them into semicrystalline masses[28, 29].

3. MEASUREMENT OF SPECIFICITY

The measurement of the substrate specificity of an enzyme involves measuring the reaction rate under standard conditions in which nothing is varied apart from the substrate itself. Enough has been written in the preceding sections to indicate that with enzymes synthesizing or catabolizing phospholipids, the *apparent specificity* measured by such a procedure may not give a picture of the true specificity of the enzyme because of the changing physical factors and variation in the structure and molecular organisation of the substrate. *True specificity* is defined in this instance as the ability of the active centre of the enzyme to bind with a molecule of the lipid substrate in such a way that an enzyme–substrate complex is formed and the reaction catalysed. To obtain some idea of the true specificity of a lipid substrate the latter must be presented to the enzyme in various ways. These would include large particles, sonicated particles, particles dispersed with non-ionic detergents and anionic detergents, and particles containing added cationic and anionic amphipathic substances over a concentration range. In the latter case, observations ought to be made over a pH range to compensate for any changes in surface pH. The addition of diethylether to saturate the incubation medium should be tried and ideally the substrate should be presented to the enzyme as a unimolecular film with varying degrees of compression. Because of phase transition changes which occur with phospholipid when the hydrophobic groups are varied the reaction rates should be compared over a range of temperatures. It seems clear that with phospholipid metabolism in plants temperature changes may cause a marked change in the specificity of enzyme reactions (Chapter 10, p. 281). In addition it seems likely that phospholipid-synthesizing enzymes may show a different specificity when substrates are presented individually and mixed[126].

Perhaps the most important specificity to be considered is that of *physiological specificity* or the specificity of the enzyme acting within the cell and when it is often present as a constituent of a cell membrane. Here it is unlikely that the substrates are presented as polymolecular phospholipid aggregates, as they are in the test-tube, but rather they would exist together with other lipids as lipoprotein complexes. There is a good deal of evidence that in such complexes the lipids are held in a much looser form than in the tight packing existing in polymolecular aggregates of pure lipid. Thus exchange of whole phospholipid molecules is much easier between lipoprotein complexes than it is between lipid particles or micelles. Consequently it is likely that the activation energy for a given enzymatic reaction would be less when the substrate lipid is present in a lipoprotein environment since less energy would be required for the formation of the enzyme–substrate complex. In this connection it is significant that some of the highest rates at which phospholipases A, C and D hydrolyse phospholipids have been obtained with lipoprotein substrates[1, 30, 87, 88]. Furthermore with such lipoprotein substrates the apparent specificity observed with pure phospholipids generally disappears as it often does when the latter are presented to the enzyme in varying physicochemical forms. Sometimes, however, when a phospholipase is present in a membrane structure it attacks the phospholipids in that membrane as lipoprotein complexes more slowly than exogenous added phospholipids[89]. Presumably this protects the membrane from too rapid self-autolysis. On the other hand soluble phospholipases in cells can apparently sometimes only attack membrane-bound substrate[24].

The determination of physiological specificity is at present almost an idealistic concept and we can only make informed guesses as to its nature from the lipid composition of tissues and membranes and the behaviour of the isolated enzymes towards pure lipid substrates presented in a variety of physical states. Nevertheless much valuable progress has already been made in understanding how the specificity of enzymatic reactions involving phospholipids might assist the control of their fatty acid composition in various tissues and membranes[102, 103] (Chapter 4, p. 67). However, the ease with which the fatty acid composition of liver lecithin can be altered by dietary means would suggest that the control due to enzyme specificity is only partial[127]. The pioneering work of the late Professor Hübscher and his colleagues[24] on the specificity of soluble phosphatidate phosphohydrolase decomposing membrane-bound phosphatidic acids of various fatty acid composition points the way to how further valuable information in this field might be obtained.

Recorded specificities of enzymes forming and breaking down phospholipids

In the following table are recorded some of the experimental results reported on the specificity of phospholipid metabolising enzymes together with some of the physico-chemical parameters which are known to affect these enzymes.

TABLE

SPECIFICITY AND PHYSICOCHEMICAL DETERMINANTS OF SOME ENZYMES CONCERNED IN PHOSPHOLIPID METABOLISM

Enzyme or system	*Specificity*	*Physicochemical factors affecting activity in bulk system*	*References*
Hydrolytic enzymes			
Phospholipase A_1 (microsomal, lysosomes)	Specifically deacylates PC or PE at the 1-acyl position. The enzyme in adrenal medulla lysosomes appears to deacylate PC and PE at the same rate but with the purified enzyme from brain, PE is hydrolysed more slowly than PC.	With purified brain enzyme Triton X-100 + Na taurocholate increases the rate of PE hydrolysis but not that of PC.	71, 72, 85, 86.
Phospholipase A_2 (venom, pancreas, intestine mitochondria)	Specificity can vary somewhat with source of enzyme even among snake venoms (see Chapter 4, p. 87). Removes a fatty acyl group on the glycerol carbon atom adjacent to the phosphoryl substituent in the phosphoglyceride structure. Thus with naturally occurring PC, PE, PS, PA, PG, CL, PI it removes the fatty acid esterified in the 2-position of the glycerol. CP is attacked but less readily than PC. An esterified fatty acid in the 1-position is not essential but the phospholipid must be the *sn*-3 enantiomorph. If the phosphorylated base is in the 2-position of the glycerol in synthetic PC only the fatty acid with the correct steric configuration can be hydrolysed. *Sn*-3-PCs containing fatty acids with greatly varying chain length are susceptible but certain water-soluble short chain PCs are hydrolysed only very slowly. (1-oleoyl-2-butyroyl)-PC which forms a turbid emulsion is readily attacked, (1-butyroyl-2-oleoyl)-PC which forms a clear solution is attacked more slowly. The pancreas enzyme has a preference for acidic *sn*-3-phosphoglycerides and hydrolyses fatty acids in the glycerol-C-2-position regardless of chain length or degree of unsaturation. Lyso-PC can be attacked: for specificity see ref. 94. When attacking a mixture of species in egg PC the rate of hydrolysis by venom enzyme is in the order (1 unsaturated–2 saturated) > (1 unsaturated–2 unsaturated) > (1 saturated–2 polyunsaturated) > (1 saturated–2 monounsaturated) > (1 saturated–2 saturated).	Diethyl ether activates PC hydrolysis by venom enzyme, as does sonication of the PC substrate. Pancreas enzyme attacks PC much more rapidly if the latter is treated with deoxycholate or mixed with acidic phospholipids e.g. PA or PS; PE is readily attacked. PC hydrolysis by intestinal enzyme activated by fatty acid but not by anionic phospholipids or dodecylsulphate.	31, 32, 33, 34, 35, 54, 56, 70, 82, 83, 3, 92, 93, 94.

Table, continued

Enzyme or system	*Specificity*	*Physicochemical factors affecting activity in bulk system*	*References*
Lysophospholipase, phospholipase B. (EC 3.1.1.5); lysolecithin acyl-hydrolase	1-acyl-*sn*-glycero-3-phosphorylcholine and 1-acyl-*sn*-glycero-3-phosphorylethanolamine attacked. Both 1- and 2-acyl lyso-PC can be deacylated but for both positional isomers the unsaturated analogues are degraded more rapidly. Compounds lacking the free hydroxyl group of lyso-PC are also hydrolysed e.g. acyl ethylene glycolphosphorylcholine and 1-acyl-propanediol-3-phosphorylcholine. Monoacyl derivatives of *sn*-glycero-1-phosphorylcholine, and *sn*-glycero-2-phosphorylcholine are hydrolysed.	When lyso-PC is incorporated in concentrations of up to 20 moles % in PC liposomes its hydrolysis is greatly suppressed and is not improved by sonication.	46, 47.
Phospholipase C (EC 3.1.4.3); phosphatidylcholine cholinephosphohydrolase	The enzyme from *Cl. perfringens* attacks choline-containing phospholipids e.g. PC, most readily but PE can be attacked under appropriate conditions (mixed with lecithin or a long-chain cation). CP is attacked less readily than PC in a mixture; it is reported that EP can be attacked providing it is mixed with lyso-PC. 1,3-diacyl-*sn*-glycero-2-phosphorylcholine and lyso-PC and phosphono analogues of PC are attacked but not glycerylphosphorylcholine. It has been reported that the *Cl. perfringens* enzyme attacks rat liver PC species at differing rates. Thus oleoyl-containing PCs are hydrolysed > linoleoyl > arachidonoyl. Although SM, ceramide-phosphorylethanolamine and ceramide-2-aminoethyl-phosphonate are attacked by *Cl. perfringens* extracts, recent evidence suggests that there may be a different enzyme that hydrolyses SM in preference to PC. The enzyme from *B. cereus* readily attacks PE, EP, PG, CL as well as the choline-containing phospholipids. The latter enzyme attacks short-chain PCs which are not hydrolysed appreciably by the *Cl. perfringens* enzyme.	Diethyl ether activates PC hydrolysis by both enzymes. PC hydrolysis by *Cl. perfringens* enzyme requires positive ζ-potential brought about either by introduction of long-chain cations or by addition of $Ca^{2\oplus}$ or $Mg^{2\oplus}$. The hydrolysis of PC in the presence of $Ca^{2\oplus}$ can be greatly enhanced by the addition of Na deoxycholate at 1.25 mg/ml; higher concentrations inhibit.	1, 35, 36, 48, 57, 97, 99, 100, 101, 104, 123, 125.
Phospholipase D (EC 3.1.4.4); phosphatidylcholine phosphatidohydrolase	PC, PE, PS, PG, CL, *O*-amino acylesters of PG are hydrolysed. Lyso-PC is attacked slowly compared with PC while CP is not hydrolysed. The enzyme can also transfer a phosphatidyl unit from PC to a variety of water-soluble primary alcohols, e.g. propanol, ethanol, ethylene glycol, glycerol.	PC hydrolysis usually stimulated by saturation of incubation medium with certain linear aliphatic ethers (e.g. diethyl ether), ketones (e.g. di-*n*-propylketone) or esters (e.g. ethyl butyrate) or by sonication, but not with enzyme from cotton seed. Lyso-PC hydrolysis inhibited by diethyl ether. Anionic detergents, e.g. Na dodecyl sulphate also stimulate PC hydrolysis as does the addition of long-chain anions, e.g. PA, TPI. It has been reported that cholesterol can enhance activity.	12, 21, 37, 56, 79, 80, Chapter 10 (p. 275)

Table, continued

Enzyme or system	*Specificity*	*Physicochemical factors affecting activity in bulk system*	*References*
General deacylase (potatoes)	Phospholipids, galactolipids, mono- and diglycerides and some esters of long-chain fatty acids attacked. Lysophospholipids attacked most readily. Acyl transfer reactions can occur in the presence of methanol.	PC, PE hydrolysis stimulated 30–40-fold by Triton X-100 or by the addition of free fatty acids, especially *cis*-unsaturated.	38, Chapter 10 (p. 277)
Sphingomyelinase	Spleen enzyme attacks SM and dihydro-SM but not sphingosylphosphorylcholine, *O*-acetyl-SM or PC. Spleen, brain and liver enzyme less active towards *threo*-SM than *erythro*-SM. *Cl. perfringens* enzyme hydrolyses PC and lyso-PC at 10% of the rate of SM while PS, PE and PI are not hydrolysed.	Purified liver enzyme only attacks SM if Triton X-100 or low concentrations of cetyltrimethylammonium bromide are added.	118–120, 104.
Hydrolysis of alkenyl ether in plasmalogens	An enzyme in liver microsomes attacks 1-(alk-1′-enyl)-*sn*-glycero-3-phosphorylcholine giving a long-chain aldehyde and GPC. 1-(alk-1′-enyl) glycerylphosphorylethanolamine, CP and EP not hydrolysed. An enzyme in brain produces long-chain aldehyde from EP more readily than from 1-(alk-1′-enyl)-*sn*-glycero-3-phosphorylethanolamine	The EP needs to be 'solubilized' with 4 mg/ml Na deoxycholate.	53, 55.
Phosphatidate phosphohydrolase (EC 3.1.3.4.)	Soluble enzyme attacks membrane-bound PA preferentially, lyso-PA not attacked. Unsaturated membrane-bound PA hydrolysed most readily. Dimyristoyl and palmitoyl–oleoyl PA attacked readily; dilauroyl and distearoyl poorly. Particulate enzyme most active towards distearoyl and dioleoyl PA. Intestinal enzyme will attack PA and lyso PA but not PE, PC or glycerophosphate. A chicken liver enzyme shows similar specificity. An enzyme preparation from pig brain was found to attack short-chain alcohol dihydrogen phosphates as well as PA but this is now thought to be due to a different enzyme.	Brain enzyme markedly inhibited by many detergents.	24, 110, 112.

Table, continued

Enzyme or system	*Specificity*	*Physiochemical factors affecting activity in bulk system*	*References*
Phosphatidylglycero-phosphatase	Phosphatidylglycerophosphate dephosphorylated but not PA or glycerophosphate.		124.
Acyl transferases			
Phosphatidic acid synthetase system (mammalian); Acid CoA-ligase (AMP) (EC 6.2.1.3) Acyl-CoA: L-glycerol-3-phosphate acyltransferase (EC 2.3.1.15)	The ligase operative in synthesizing naturally occurring PA can use fatty acids from C_6 to C_{20}. The acyl transferase acts only with CoA derivatives of chain length above C_{10} and with brain and liver microsomes, acyl CoA's containing saturated acids with 15–18 carbon are the best substrates. In rat liver microsomes (with CoA and ATP) and in *E. coli* unsaturated fatty acids preferentially esterified in the 2-position of glycero-3-phosphate and saturated fatty acids in the 1-position. (With guinea pig microsomes a more random acylation has been observed.) The first acylation probably occurs mainly in the 1-position when palmitoyl-CoA is the substrate and the 2-position with oleoyl-CoA. In this first acylation with liver microsomes it has been reported that activity with palmitoyl-CoA > linolenoyl-CoA > palmitoleoyl-CoA > oleoyl-CoA > stearoyl-CoA. In *E. coli* it is known that two enzymes are involved in the complete acylation of glycero-3-phosphate. More recently a non-selectivity for different acyl groups has been reported by other workers when the system is producing lyso-PA + PA in the presence of albumin. Only small differences in the acylation rates of 1-palmitoyl-glycero-3-phosphates using unsaturated fatty acids have been noted. Dihydroxyacetone phosphate may also act as acceptor for the first acylation and again more saturated fatty acid is incorporated into the 1-position.	Albumin and serum β-lipoproteins activate the acylation probably by preventing any inhibition caused by acyl-CoA derivatives acting as detergents.	40, 41, 42, 43, 84, 105–109, 114, 133, 134,
Phosphatidylcholine: cholesterol acyl transferase	Only the 2-acyl group of PC is transferred and the nature of the fatty acid appears to be of little consequence. Triglyceride and 1 acyl-glycero-3-phosphorylcholine are very poor substrates. The cholesterol contained in all plasma lipoproteins can act as a substrate but that in a small PC-rich high-density lipoprotein fraction is preferred.	Added PC can increase the rate of the transfer to cholesterol in plasma lipoproteins. Maximal activity with synthetic dispersions of PC and cholesterol in 4:1 molar ration in the presence of specific protein cofactor.	90, 91, 98.

Table, continued

Enzyme or system	*Specificity*	*Physicochemical factors affecting activity in bulk system*	*References*
Acyl-CoA:lysophospholipid transferase	**See Chapter 4, p. 82.** Lyso-PC and lyso-PE acylated but with different specificity for various acyl-CoA derivatives. The fatty acid in lyso-PC affects transferase activity very little but the latter must be the *sn*-3 isomer. Marked specificity for esterifying saturated fatty acids at 1-hydroxy position and unsaturated fatty acids at 2-hydroxy position. 1-(alk-1′-enyl)-*sn*-glycero-3-phosphorylcholine and 1-alkyl GPC not acylated by liver enzyme but erythrocyte and sarcoplasmic reticulum enzyme acylate; with the latter the transfer is highly specific for polyunsaturated acyl-CoA. When 16 positional isomers of *cis*-octadecenoic acid were used, acyl-CoA:1-acyl-GPC-acyl transferase (I) of liver preferred the 9 and 12 isomers while acyl-CoA:2-acyl-GPCacyl transferase (II) preferred the 8, 10, 12, 13 and 15 isomers with 12-octadecenoate showing the fastest rate. With enzyme I the introduction of double bonds increased activity but *trans* isomers of fatty acid transferred at slightly slower rate than *cis* acids. With enzyme II maximum activity was seen with $C_{18:0}$ in a series of saturated fatty acids from $C_{12}-C_{20}$. Addition of *cis* double bonds decreased rate but addition of *trans* double bonds increased rate. With I and II utilizing isomeric *cis cis* $C_{18:2}$ CoA thiol esters, acyl transfer was relatively fast when the double bonds were near the methyl end and slower when they were near the carboxyl end.	Specificity of liver microsome preparation can change as it ages (different enzymes?). The pattern of specificity for various acyl-CoA esters was not changed in the range 15–40°. (Specificity of Enzyme I may be affected more by solvent structure of solvated acyl-CoA whereas Enzyme II is not.)	65–69, 95, 122, 132.
Synthetic enzymes			
For phosphatidic acid synthesis see above			
CTP:cholinephosphate cytidylyltransferase (EC 2.7.7.15)	No nucleotide apart from CTP and *deoxy*-CTP can be used. A mixture of this enzyme and the following enzyme shows high specificity for phosphorylcholine or phosphorylethanolamine although some phosphorylated short-chain unnatural bases can react to a very limited extent with mixed liver enzymes. Phosphonate analogue of phosphorylcholine can act as substrate.		51, 58, 75, 78.
CTP:ethanolaminephos-	Only CTP and deoxy-CTP can act as substrates. See previous entry		58, 78.

Table, continued

Enzyme or system	*Specificity*	*Physiochemical factors affecting activity in bulk system*	*References*
Choline phosphotransferase; CDPcholine-1:2-diglyceride cholinephosphotransferase (EC 2.7.8.2)	A 1,2-diacyl-*sn*-glycerol is essential. PA, PC or triglyceride will not substitute. In plants and rat liver microsomes no specificity for various species of 1,2-diacyl-*sn*-glycerol. It is probable that the enzyme can use (*O*-alk-1′-enyl) acylglycerol or *O*-alkyl-acylglycerol as substrate (although *O*-alk-1′-enyl glycerol cannot be used) forming CP or the glycerol ether analogue. *Deoxy*-CDPcholine as active as CDPcholine, while UDPcholine, ADPcholine, GDPcholine will not act as substrates. CDPethanolamine is used by a separate enzyme for PE synthesis. For specificity of CDP esters of unnatural bases see next entry.	The activity of the enzyme is greatly stimulated by dispersion of the diglyceride with Tween 20 (polyoxyethylene sorbitan monolaurate) at 1 mg/ml; higher concentrations are inhibitory. Bile salts and digitonin stimulate but less so than Tween 20.	39, 44, 49, 50, 52, 58, 59, 126.
Ethanolamine phosphotransferase; CDPethanolamine 1,2-diglyceride ethanolaminephosphotransferase (EC 2.7.8.1)	*Deoxy*-CDPethanolamine much less active as substrate compared with CDPethanolamine. A number of CDP esters of unnatural bases can function, some of them more effectively with a brain enzyme (containing previous enzyme as well) than ethanolamine itself (e.g. L-2-aminopropan-1-ol). Increasing the distance between the N and O atoms of the bases reduces activity. A liver enzyme preparation consisting of a mixture of this and the previous enzyme was less active with CDP esters of unnatural bases. Rat liver microsomes show no specificity for various species of 1,2-diacyl-*sn*-glycerol of differing degrees of unsaturation but it has recently been reported that with mixed diglycerides an hexaenoic-containing diglyceride is used preferentially.		58, 59, 78, 126.
CDPcholine:ceramide choline phosphotransferase	CDPcholine specific and UDPcholine, ADPcholine and GDPcholine inactive. Originally reported that ceramide must have *trans* double bond in the sphingosine moiety while the hydroxyl group on carbon 3 must have *threo* relationship to the amino group on carbon 2. Recent work suggests that the enzyme in liver microsomes can only use *threo*-ceramides whereas that in liver mitochondria can use *erythro* as well. See however next column. Ceramides of dihydrosphingosine are inactive but derivatives of sphingosine containing a triple bond rather than a double bond are active.	Ceramides with unsaturated acyl fatty groups more reactive as easier to disperse. In the absence of detergent (Tween 20) activity falls with increasing chain length so that *N*-octanoyl-DL-*threo-trans*-sphingosine is quite inactive. Tween 20 allows considerable activity with this substrate but not with a C_{12} fatty acid. Recent work suggests *N*-acyl-*erythro*-sphingosines will act as substrates if these are dispersed by a combination of Tween 20, Triton X-100 and ultrasonication; see however previous column.	63–64, 117.

References p. 114

Table, continued

Enzyme or system	*Specificity*	*Physicochemical factors affecting activity in bulk system*	*References*
CTP:phosphatidic acid cytidylyl transferase	CTP active but not CMP, GTP, ITP, UTP. Of a series of saturated PA's tested from C_{10}–C_{22}, didecanoyl PA was the most active, but less so than natural unsaturated PA (see next column). Lyso-PA was poor substrate.	Synthetic saturated PA's are poor substitutes for unsaturated PA possibly because they are less readily dispersed since Tween 20 and other detergents render them more effective substrates. PA must not be added as Mg salt even though $Mg^{2\oplus}$ is required in the reaction.	60–62, 77.
CDPdiglyceride; inositol phosphatidyltransferase	*Myo*inositol and to a lesser extent *myo*-inosose-2, *scyllo*inosose and DL-*epi*-inosose are active. Many other cyclitols and sugars are inactive. DL-CDPdidecanoin is a better substrate than DL-CDP-dipalmitin or a CDPdiglyceride prepared from egg lecithin. DL-CDP-diglycerides containing C_{20} or C_{22} fatty acids are inactive.		60, 116.
Phosphatidylglycerol phosphate synthetase (CDPdiglyceride + glycerophosphate → phosphatidylglycerol phosphate + CMP)	Stereospecific for *sn*-glycero-3-phosphate.		76
Diglyceride kinase	For brain microsomes, diglyceride (from cabbage PA) is better substrate than diglyceride prepared from egg PC. 1-palmitoyl-2-oleoyl glycerol is even poorer. With the enzyme in *E. coli*, 1 monopalmitin can also act as substrate forming lyso-PA.	Dispersion of the diglyceride with detergents such as Cutscum has been reported as an absolute requirement for the *E. coli* enzyme whereas with the brain enzyme Na deoxycholate stimulated while Cutscum and Triton X did not.	113, 121, 128.

Table, continued

Enzyme or system	*Specificity*	*Physiochemical factors affecting activity in bulk system*	*References*
Phosphatidylcholine synthesis by methylation of PE (PE: *N*-methyltransferase + phosphatidyl-*N*-monomethylethanolamine: *N*-methyltransferase).	In rat liver the degree of unsaturation of the acceptor PE has little effect on the rate of methylation. The first transferase has an absolute dependence on PE and the second on *N*-monomethylethanolamine-or *NN*-dimethylethanolamine-phospholipids.	In *Neurospora* attempts to obtain Michaelis constants have given variable results presumably due to the degree of phospholipid dispersion.	45, 115.
Cyclopropanephospholipid synthetase. This enzyme forms a cyclopropane ring by catalyzing the reaction between *S*-adenosylmethionine and a mono-unsaturated fatty acid esterified in a phosphoglyceride.	PE, PG, PA, PS can act as substrates as can the diether analogue of PE. There is an exclusive requirement for a 3-phosphoglyceride and a slight preference for a monoenoic fatty acid esterified in the 1-position. With *Clostridium butyricum* esterified hexadec-9-enoic was a better substrate than hexadec-7-enoic, while esterified octadec-11-enoic was better than octadec-9-enoic.	The reaction rate with a pure PE substrate is markedly enhanced by adding an anionic detergent or an acidic phospholipid.	73–74, 26, 131.

References p. 114

REFERENCES

1 A.D. Bangham and R.M.C. Dawson, *Biochim. Biophys. Acta*, 59 (1962) 103.
2 W.L. Zahler, R.E. Barden and W.W. Cleland, *Biochim. Biophys. Acta*, 164 (1968) 1.
3 L.L.M. van Deenen and G.H. de Haas, *Biochim. Biophys. Acta*, 70 (1963) 538.
4 G.H. de Haas, L. Sarda and J. Roger, *Biochim. Biophys. Acta*, 106 (1965) 638.
5 L. Sarda and P. Desnuelle, *Biochim. Biophys. Acta*, 30 (1958) 513.
6 H. Brockerhoff, *Arch. Biochem. Biophys.*, 134 (1969) 366.
7 B. Entressangles and P. Desnuelle, *Biochim. Biophys. Acta*, 159 (1968) 285.
8 R.M.C. Dawson, in D. Chapman (Ed.), *Biological Membranes*, Academic Press, New York, 1968, p. 2C
9 B. Sheard, *Nature*, 223 (1969) 1057.
10 L. Saunders, *Biochim. Biophys. Acta*, 125 (1966) 70.
11 R.M.C. Dawson, *Biochim. Biophys. Acta*, 70 (1963) 697.
12 R.M.C. Dawson and N. Hemington, *Biochem. J.*, 102 (1967) 76.
13 A.D. Bangham and R.M.C. Dawson, *Biochem. J.*, 75 (1960) 133.
14 R.H. Quarles and R.M.C. Dawson, *Biochem. J.*, 113 (1969) 697.
15 P.J. Quinn and R.M.C. Dawson, *Biochem. J.*, 115 (1969) 65.
16 R.M.C. Dawson, *Biochem. J.*, 98 (1966) 35C.
17 G. Colacicco, *Nature*, 233 (1971) 202.
18 D.O. Shah and J.H. Schulman, *J. Colloid Interface Sci.*, 25 (1967a) 107.
19 P.J. Quinn and R.M.C. Dawson, *Biochem. J.*, 119 (1970) 21.
20 P.J. Quinn and R.M.C. Dawson, *Biochem. J.*, 116 (1970) 671.
21 M. Kates, *Can. J. Biochem. Physiol.*, 35 (1957) 127.
22 W.L. Magee and R.H.S. Thompson, *Biochem. J.*, 77 (1960) 526.
23 R.M.C. Dawson, *Biochem. J.*, 88 (1963) 414.
24 M.P. Mitchell, D.N. Brindley and G. Hübscher, *European J. Biochem.*, 18 (1971) 214.
25 A.D. Bangham and R.M.C. Dawson, *Biochem. J.*, 72 (1959) 486.
26 P.J. Thomas and J.H. Law, *J. Biol. Chem.*, 241 (1966) 5013.
27 H. Hauser and R.M.C. Dawson, *Biochim. Biophys. Acta*, 137 (1967) 518.
28 W. Thompson and R.M.C. Dawson, *Biochem. J.*, 91 (1964) 237.
29 R.M.C. Dawson and W. Thompson, *Biochem. J.*, 91 (1964) 244.
30 E. Condrea, I. Fabian and A. de Vries, *Experientia*, 20 (1964) 557.
31 P.P.M. Bonsen, G.H. de Haas, W.A. Pieterson and L.L.M. van Deenen, *Biochim. Biophys. Acta*, 270 (1972) 364.
32 N.H. Tattrie and J.R. Bennett, *Can. J. Biochem.*, 41 (1963) 1983.
33 L.L.M. van Deenen, G.H. de Haas and C.H.Th. Heemskerk, *Biochim. Biophys. Acta*, 67 (1963) 295.
34 G.H. de Haas, F.J.M. Daemen and L.L.M. van Deenen, *Biochim. Biophys. Acta*, 65 (1962) 260.
35 L.L.M. van Deenen, in R.M.C. Dawson and D.N. Rhodes (Eds.), *Metabolism and Physiological Significance of Lipids*, Wiley, London, 1964, p. 155.
36 F. Haverkate, U.M.T. Houtsmuller and L.L.M. van Deenen, *Biochim. Biophys. Acta*, 63 (1962) 547.
37 R.M.C. Dawson, *Biochem. J.*, 102 (1967) 205.
38 T. Galliard, *Biochem. J.*, 121 (1971) 379.
39 K.A. Devor and J.B. Mudd, *J. Lipid Res.*, 12 (1971) 403, 412.
40 F. Possmayer, G.L. Scherphof, T.M.A.R. Dubbelman, L.M.G. van Golde and L.L.M. van Deenen, *Biochim. Biophys. Acta*, 176 (1969) 95.
41 W.E.M. Lands and P. Hart, *J. Biol. Chem.*, 240 (1965) 1905.
42 W. Stoffel, M.E. de Tomas and H.G. Schiefer, *Z. Physiol. Chem.*, 348 (1967) 882.
43 A.K. Hajra, *Biochem. Biophys. Res. Commun.*, 33 (1968) 929.
44 J.B. Mudd, L.M.G. van Golde and L.L.M. van Deenen, *Biochim. Biophys. Acta*, 176 (1969) 547.
45 R.L. Lyman, S.M. Hopkins, G. Sheehan and J. Tinoco, *Biochim. Biophys. Acta*, 176 (1969) 86.
46 R.M.C. Dawson, *Biochem. J.*, 64 (1956) 192.
47 H. van den Bosch, A.J. Aarsman, A.J. Slotboom and L.L.M. van Deenen, *Biochim. Biophys. Acta*, 164 (1968) 215.

48 E. Baer and N.Z. Stanacev, *Can. J. Biochem.*, 44 (1966) 893.
49 F. Snyder, M.L. Blank and B. Malone, *J. Biol. Chem.*, 245 (1970) 4016.
50 S.B. Weiss, S.W. Smith and E.P. Kennedy, *J. Biol. Chem.*, 231 (1958) 53.
51 L.F. Borkenhagen and E.P. Kennedy, *J. Biol. Chem.*, 227 (1957) 951.
52 E.P. Kennedy and S.B. Weiss, *J. Biol. Chem.*, 222 (1956) 193.
53 H.R. Warner and W.E.M. Lands, *J. Biol. Chem.*, 236 (1961) 2404.
54 M.M. Rapport and R.E. Franzl, *J. Biol. Chem.*, 225 (1957) 851.
55 G.B. Ansell and S. Spanner, *Biochem. J.*, 94 (1965) 252.
56 W.E.M. Lands and P. Hart, *Biochim. Biophys. Acta*, 98 (1965) 532.
57 H.R. Warner and W.E.M. Lands, *J. Am. Chem. Soc.*, 85 (1963) 60.
58 E.P. Kennedy, L.F. Borkenhagen and S.W. Smith, *J. Biol. Chem.*, 234 (1959) 1998.
59 B. de Kruyff, L.M.G. van Golde and L.L.M. van Deenen, *Biochim. Biophys. Acta*, 210 (1970) 425.
60 H. Paulus and E.P. Kennedy, *J. Biol. Chem.*, 235 (1960) 1303.
61 J.R. Carter and E.P. Kennedy, *J. Lipid Res.*, 7 (1966) 678.
62 B.W. Agranoff, R.M. Bradley and R.O. Brady, *J. Biol. Chem.*, 233 (1958) 1077.
63 M. Sribney and E.P. Kennedy, *J. Biol. Chem.*, 233 (1958) 1315.
64 Y. Fujino, Mo. Nakano, T. Negishi and S. Ito, *J. Biol. Chem.*, 243 (1968) 4650.
65 R.C. Reitz, M. El-Sheikh, W.E.M. Lands, I.A. Ismail and F.D. Gunstone, *Biochim. Biophys. Acta*, 176 (1969) 480.
66 P. Jezyk and W.E.M. Lands, *J. Lipid Res.*, 9 (1968) 525.
67 R.C. Reitz and W.E.M. Lands, *J. Biol. Chem.*, 243 (1968) 2241.
68 W.E.M. Lands and I. Merkl, *J. Biol. Chem.*, 238 (1963) 898.
69 M.E. de Tomas and R.R. Brenner, *Biochim. Biophys. Acta*, 202 (1970) 184.
70 D.J. Hanahan, H. Brockerhoff and E.J. Barron, *J. Biol. Chem.*, 235 (1960) 1917.
71 M. Waite and L.L.M. van Deenen, *Biochim. Biophys. Acta*, 137 (1967) 498.
72 J.D. Newkirk and M. Waite, *Biochim. Biophys. Acta*, 225 (1971) 224.
73 A.E. Chung and J.H. Law, *Biochemistry*, 3 (1964) 967.
74 J.G. Hildebrand and J.H. Law, *Biochemistry*, 3 (1964) 1304.
75 K.S. Bjerve, *Biochim. Biophys. Acta,* 270 (1972) 348.
76 J.Y. Kiyasu, R.A. Pieringer, H. Paulus and E.P. Kennedy, *J. Biol. Chem.*, 238 (1963) 2293.
77 G.L. Petzold and B.W. Agranoff, *J. Biol. Chem.*, 242 (1967) 1187.
78 T. Chojnacki and G.B. Ansell, *J. Neurochem.*, 14 (1967) 413.
79 K.R. Bruckdorfer, P.A. Edwards and C. Green, *European J. Biochem.*, 4 (1968) 506.
80 C. Long, R. Odavic and E.J. Sargent, *Biochem. J.*, 102 (1967) 216.
81 G.A. Scarborough and J.F. Nyc, *Biochim. Biophys. Acta*, 146 (1967) 111.
82 J.H. Moore and D.L. Williams, *Biochim. Biophys. Acta*, 84 (1964) 41.
83 Ph.R. Bird, H.G. de Haas, C.H.Th. Heemskerk and L.L.M. van Deenen, *Biochim. Biophys. Acta*, 98 (1965) 566.
84 L.N.W. Daae, *Biochim. Biophys. Acta,* 270 (1972) 23.
85 A.D. Smith and H. Winkler, *Biochem. J.*, 108 (1968) 867.
86 S. Gatt, *Biochim. Biophys. Acta*, 159 (1968) 304.
87 E. Condrea, A. de Vries and J. Mager, *Biochim. Biophys. Acta*, 58 (1962) 389.
88 E. Condrea, C. Klibansky, R. Keret and A. de Vries, *Nature*, 200 (1963) 1096.
89 M. Waite, G.L. Scherphof, F.M.G. Boshouwers and L.L.M. van Deenen, *J. Lipid Res.*, 10 (1969) 411.
90 J.A. Glomset, *J. Lipid Res.*, 9 (1968) 155.
91 S.N. Shah, W.J. Lossow and I.L. Chaikoff, *Biochim. Biophys. Acta*, 84 (1964) 176.
92 G.H. de Haas, N.M. Postema, W. Nieuwenhuizen and L.L.M. van Deenen, *Biochim. Biophys. Acta*, 159 (1968) 103.
93 P.V. Subbaiah and J. Ganguly, *Biochem. J.*, 118 (1970) 233.
94 G.H. de Haas and L.L.M. van Deenen, *Biochim. Biophys. Acta*, 106 (1965) 315.
95 A.E. Brandt and W.E.M. Lands, *Biochim. Biophys. Acta*, 144 (1967) 605.
96 H.B. Eiber, A.N. Payza and B. Goldberg, *Biochim. Biophys. Acta*, 116 (1966) 256.
97 A. Ohsaka and T. Sugahara, *J. Biochem.*, 64 (1968) 335.

98 C.J. Fielding, V.G. Shore and P.E. Fielding, *Biochim. Biophys. Acta*, 270 (1972) 513.
99 T. Takahashi and H.H.O. Schmid, *Chem. Phys. Lipids*, 2 (1968) 220.
100 E. Graf and Y. Stein, *Biochim. Biophys. Acta*, 116 (1966) 166.
101 G.B. Ansell and S. Spanner, *Biochem. J.*, 97 (1965) 375.
102 W.E.M. Lands, M.L. Blank, L.J. Nutter and O.S. Privett, *Lipids*, 1 (1966) 224.
103 L.M.G. van Golde, G.L. Scherphof and L.L.M. van Deenen, *Biochim. Biophys. Acta*, 176 (1969) 635.
104 I. Pastan, V. Macchia and R. Katzen, *J. Biol. Chem.*, 243 (1968) 3750.
105 E. de Sánchez Jimenez and W.W. Cleland, *Biochim. Biophys. Acta*, 176 (1969) 685.
106 H.M. Abou-Issa and W.W. Cleland, *Biochim. Biophys. Acta*, 176 (1969) 692.
107 H.J. Fallon and R.G. Lamb, *J. Lipid Res.*, 9 (1968) 652.
108 R.G. Lamb and H.J. Fallon, *J. Biol. Chem.*, 245 (1970) 3075.
109 D.R. Husbands and W.E.M. Lands, *Biochim. Biophys. Acta*, 202 (1970) 129.
110 J.M. Johnston and J.H. Bearden, *Biochim. Biophys. Acta*, 56 (1962) 365.
111 B.W. Agranoff, *J. Lipid Res.*, 3 (1962) 190.
112 S.W. Smith, S.B. Weiss and E.P. Kennedy, *J. Biol. Chem.*, 228 (1957) 915.
113 M.R. Hokin and L.E. Hokin, *J. Biol. Chem.*, 234 (1959) 1381.
114 B. Sedgwick and G. Hübscher, *Biochim. Biophys. Acta*, 144 (1967) 397.
115 D. Rehbinder and D.M. Greenberg, *Arch. Biochem. Biophys.*, 109 (1965) 110.
116 J.A. Benjamins and B.W. Agranoff, *J. Neurochem.*, 16 (1969) 513.
117 M. Sribney, *Arch. Biochem. Biophys.*, 126 (1968) 954.
118 P.B. Schneider and E.P. Kennedy, *J. Lipid Res.*, 8 (1967) 202.
119 Y. Barnholz, A. Roitman and S. Gatt, *J. Biol. Chem.*, 241 (1966) 3731.
120 M. Heller and B. Shapiro, *Biochem. J.*, 98 (1966) 763.
121 R.A. Pieringer and R.S. Kunnes, *J. Biol. Chem.*, 240 (1965) 2833.
122 H. van den Bosch, L.M.G. van Golde, A.J. Slotboom and L.L.M. van Deenen, *Biochim. Biophys. Acta*, 152 (1968) 694.
123 G.H. de Haas, P.P.M. Bonsen and L.L.M. van Deenen, *Biochim. Biophys. Acta*, 116 (1966) 114.
124 Y. Chang and E.P. Kennedy, *J. Lipid Res.*, 8 (1967) 447.
125 T. Hori, I. Arakawa, M. Sugita and O. Itasaka, *J. Biochem.*, 64 (1968) 533.
126 H. Kanoh, *Biochim. Biophys. Acta*, 218 (1970) 249.
127 C. Pudelkewicz and R.T. Holman, *Biochim. Biophys. Acta*, 152 (1968) 340.
128 E.G. Lapetina and J.N. Hawthorne, *Biochem. J.*, 122 (1971) 171.
129 W.G. Fiscus and W.C. Schneider, *J. Biol. Chem.*, 241 (1966) 3324.
130 D. Chapman, D.J. Fluck, S.A. Plenkett and G.G. Shipley, *Biochim. Biophys. Acta*, 163 (1968) 255.
131 H. Goldfine and C. Panos, *J. Lipid Res.*, 12 (1971) 214.
132 K. Waku and Y. Nakazawa, *J. Biochem.*, 68 (1970) 459.
133 T.K. Ray, J.E. Cronan, R.D. Mavis and P.R. Vagelos, *J. Biol. Chem.*, 245 (1970) 6442.
134 R.G. Lamb and H.J. Fallon, *J. Biol. Chem.*, 245 (1970) 3075.

Chapter 6

PHYSICAL CHEMISTRY OF PHOSPHOLIPIDS

D. CHAPMAN

Department of Chemistry, Sheffield University, Sheffield (Great Britain)

The classical physical-chemical approach to the study of many molecules is to examine their properties in dilute solution, preferably in the form of monomers. This attitude is adopted so as to limit the many possible interactions and hence to reduce the number of variables in the system. The behaviour of phospholipids in a number of situations, for instance in water or in biological membrane systems is, however, perhaps best appreciated by consideration of their solid-state properties[1]. This is because the mesomorphic or liquid crystalline behaviour of phospholipids is of considerable importance to their behaviour in water and in biological membrane systems. In this chapter we first consider what we know of the pure anhydrous phospholipids, then the hydrated material and then discuss how this behaviour is affected by molecules such as cholesterol. We shall also point out the relationship which exists between the solid state and monolayer properties.

1. SOLID STATE BEHAVIOUR

No complete X-ray structure of a phospholipid fully esterified and containing fatty acid chains has been published. Preliminary X-ray studies, however, have been reported on a number of related derivatives e.g. the structure of *sn*-glycero-3-phosphorylcholine cadmium chloride trihydrate has been obtained[2]. This has shown that this molecule exists as a dipolar ion with the positive charge on the quaternary nitrogen neutralized by the negative charge on the phosphoric acid residue. The choline residue exists in the *gauche* conformation rather than the more extended form. The crystal structure of *sn*-glycero-3-phosphorylcholine has also been reported[3].

It is too early to say just how relevant these structures may be to the lipid organization in membranes. The most significant points seem to be (*a*) that the choline residue exists in the *gauche* conformation rather than the extended zig-zag form and (*b*) that the conformation of the glycerol residue can be *gauche gauche*.

Single crystal X-ray studies of some *rac*-phosphatidylethanolamines, e.g., 1,2-dilauroylphosphatidylethanolamine, are at present in progress (unpublished studies). An examination of 1,2-dilauroylphosphatidylethanolamine, as yet incomplete, suggests the following important structural features: (*a*) the two hydrocarbon chains of the fatty acid residue present in each lipid molecule are pointing in the same direction and are parallel to each other and (*b*) the polar portion of the molecule containing the phosphate and amino groups appears to be in line with the hydrocarbon chains.

The fact that the hydrocarbon chains in this structure are parallel to each other is not surprising. Long-chain fatty acids, esters and glycerides all pack in this way[4]. This is because, beyond a certain chain length, the long hydrocarbon chains tend to dominate the crystallization processes. This in turn is related to the favourable dispersion interaction forces which arise from parallel packed chains.

In related lipid molecules, while the hydrocarbon chains within the molecule lie parallel to each other, the chains do not always point in the same direction. The single crystal X-ray studies of triglycerides, e.g., trilaurin[5] and tricaprin[6], show that the two hydrocarbon chains on adjacent substituted positions, i.e. substituted at the 1- and 2-positions on the glycerol molecule, point in opposite directions. The different situation which appears to exist with the diacylphosphatidylethanolamines is important because the solid state structure can be regarded as being built up from 'infinite' sheets of phospholipid bilayers with the polar groups organized in sheets and with all the hydrocarbon chains of adjacent lipid molecules lying parallel, i.e., it has a somewhat similar arrangement to the organization of the lipid which has been suggested to occur in many biological membrane systems.

Other attempts to deduce information about the solid state properties have been made using electron and X-ray powder diffraction methods. Although studies using electron diffraction, have given only limited information[7] X-ray powder diffraction methods have provided more, e.g., the way in which the hydrocarbon chains are packed in the crystal lattice, and the angle of tilt of the chains to the basal planes has been determined. The diffraction data provide information about the long and short X-ray spacings. The values of the X-ray short spacings suggest that the hydrocarbon chains are packed in similar arrangements to those adopted by many other long chain molecules[8]. The long spacings of an homologous series of phospholipids are shown in Fig. 1.

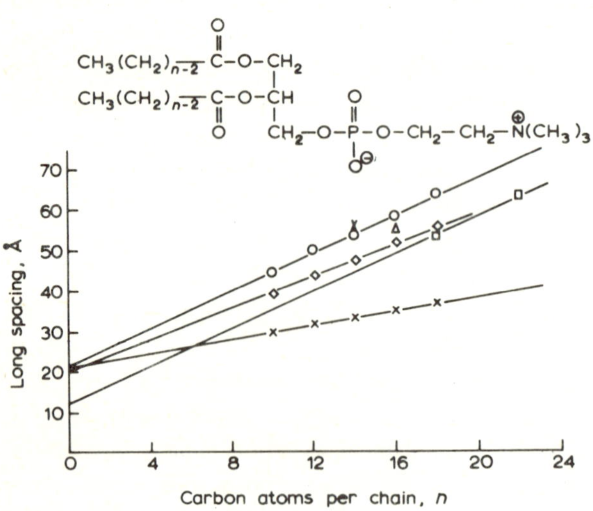

Fig. 1. Long spacing ~ chain length of fully saturated 1,2 diacyl-*sn*-phosphatidyl cholines. ○, α 1 monohydrate; △, α 2 monohydrate; X-β′ monohydrate; ◇, liquid crystalline (lamellar) at a temperature just above T_1°; x,liquid crystalline at 175°; □, crystalline, anhydrous β form.

Some impression of the polar group organization present in phospholipid molecules can be obtained by a consideration of their capillary melting points. For many years the melting points of both natural and synthetic phospholipids have been measured and recorded and these are summarized in Table 1. An examination of the melting points of

TABLE 1

MELTING POINT DATA OF PHOSPHOLIPIDS

	M. p. (°C)	*Comments*
1,2-diacylphosphatidylcholine	230	M. p. practically independent of chain length
1,2-diacylphosphatidylethanolamine	196	M. p. practically independent of chain length
1,3-diacylphosphatidylethanolamine	198	M. p. practically independent of chain length
1,2-diacylphosphatidylserine	160	Shorter chain lengths, lower m. p.
1,2-diacylphosphatidylinositol	136	Insufficient data to determine m. p. chain length dependency
1,2-diacylphosphatidylglycerol	67	Insufficient data to determine m. p. chain length dependency
1,2-diacylphosphatidic acid	71	Shorter chain lengths, lower m. p.
1,3-diacylphosphatidic acid	70	Shorter chain lengths, lower m. p.
bis-phosphatidic acid	70	Insufficient data to determine m. p. chain length dependency
Sphingomyelins	210	Insufficient data to determine m. p. chain length dependency
Lysophosphatidylcholine	240	Insufficient data to determine m. p. chain length dependency
Lysophosphatidylethanolamine	205	Insufficient data to determine m. p. chain length dependency

the different classes of phospholipids reveals a large variation. The phosphatidylglycerols, the phosphatidic acids and the bis phosphatidic acids have melting points which are chain length-dependent, and which do not exceed 71°. For these phospholipids the shorter, or more unsaturated the hydrocarbon chains, the lower the melting point. On the other hand, the phosphatidylethanolamines, phosphatidylcholines and sphingomyelins have melting points of about 197°, 230° and 210°, respectively, and these do not appear to be significantly affected by the type of fatty acyl residues present. Intermediate between these extremes are the phosphatidylserines, the melting points of which appear to be chain length-dependent. Insufficient data are available to know whether the melting points of the phosphatidylinositols are chain length-dependent.

The low melting points of the phosphatidic acids are of the same order as those of the corresponding chain length fatty acids, and from this we may infer that the forces of interaction between the molecules in the crystal are principally due to hydrogen bonds, and are consequently weak. The phospholipids with high melting points, such as the phosphatidylcholines, are more akin to the soaps (fatty acid salts) in their melting behaviour. The soaps are linked together by an ionic network which holds the structure together

References p. 141

until high temperatures (> 300°). It therefore seems likely that an ionic network exists in the crystal structure of these phospholipids. An ionic network would lead us to expect that the stronger the base, the higher the melting point and this is indeed found (see Table 1), in order of decreasing basicity choline > ethanolamine > serine, the same order as decreasing melting point.

An examination of a given class of these *high* melting phospholipids reveals that the melting points are mainly independent of the fatty acid residues present e.g., 1-stearoyl-2-oleoyl- and 1,2-distearoyl-*sn*-glycero-3-phosphorylcholine both have melting points of 230–231°. The melting points of these phospholipids are not markedly affected by the number of acyl chains present. The melting points of the lyso-derivatives are slightly higher than those of the corresponding diacyl phospholipids. It is thus quite clear that the polar head group is the major factor controlling phospholipid melting points.

2. THERMOTROPIC MESOMORPHISM

Lecithins and phosphatidylethanolamines

In addition to the capillary melting point, other phase changes have been shown to occur with phospholipids at lower temperatures, e.g., when pure dimyristoyl-phosphatidylethanolamine containing two fully saturated chains is heated from room temperature up to the capillary melting point, a number of thermotropic phase changes occur (i.e., phase changes caused by the effect of heat). This was first shown by infrared spectroscopic techniques[9], then by thermal analysis techniques[10]. Optical studies show that dimyristoyl-phosphatidylethanolamine at room temperature is birefringent under crossed polars. On heating, three processes occur: first some loss of birefringence at the first transition temperature ~120°; a small increase at 135° and a pronounced overall loss of birefringence near the capillary melting point of 200°. Above 120°, pressure on a coverglass with a needle causes the material to flow. When the temperature of the phospholipid reaches ~120° the i.r. absorption spectrum undergoes a remarkable change. Above this temperature the spectrum loses all the fine structure and detail which was present at lower temperatures.

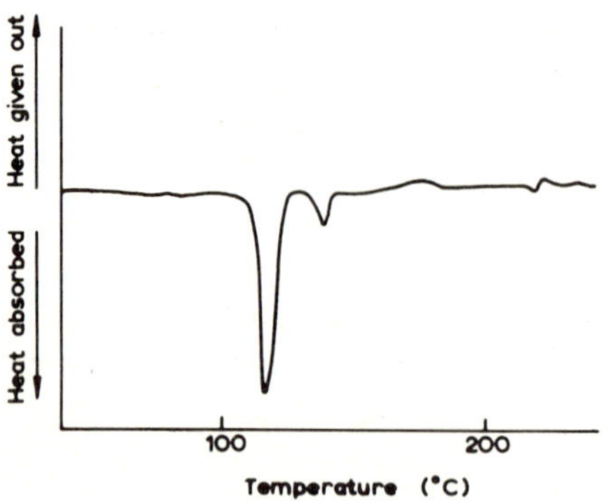

Fig. 2. Differential thermal analysis heating curve of 1,2-dimyristoyl-phosphatidylethanolamine (Chapman and Collin[10]).

At the high temperature the spectrum becomes similar to that obtained with a phospholipid dissolved in a solvent such as chloroform.

Differential thermal analysis (d.t.a.) shows that a marked endothermic transition (absorption of heat) occurs at this transition temperature[10]. An additional heat change occurs at ~135° and only a small heat change is involved near the capillary melting point of the lipid (see Fig. 2). This behaviour is similar to that which occurs with liquid crystals, such as ρ-azoxyanisole or cholesteryl acetate which form nematic and cholesteric liquid crystalline phases.

The X-ray long spacings show a dramatic reduction to some two-thirds of their original value at the first transition temperature with a further small reduction at the second transition temperature. The X-ray short spacings change at the first transition point from sharp diffraction lines to a diffuse spacing at ~4.6 Å.

Wide-line proton magnetic resonance (pmr) studies of various anhydrous phosphatidylethanolamines at liquid nitrogen temperature in the solid state show[11] broad resonance lines about 15 gauss wide (i.e., $\Delta\nu \approx 60$ kHz). As the temperature increases, a gradual narrowing of the line-width occurs until, at a particular transition temperature a considerable fall in the line-width to about 0.1 gauss is the onset of a mesomorphic condition. In this mesomorphic state, considerable molecular motion of the lipid chains occurs, and this motion almost averages to zero the magnetic dipole–dipole interactions. This averaging process involves translational modes as well as rotational oscillation of the methylene groups. These movements are particularly marked at the ends of the chains and also involve a marked departure from the all-*trans* character of the hydrocarbon chains*. This is shown by the infrared spectra of the lipids in this mesomorphic phase[8].

A similar behaviour has also been observed for anhydrous 1,2-distearoyl-*sn*-glycero-3-phosphorylcholine by proton magnetic resonance spectroscopy[12]. Reorientation of the $N^{\oplus}(Me)_3$ protons occurs at low temperatures and results in a reduction of line-width near −150° which is not observed for the phosphatidylethanolamines in the same temperature range.

The main conclusions from these various studies are that (*a*) even with the fully saturated phospholipid at room temperature, some molecular motion occurs in the solid. This is evident from the p.m.r. spectra and from the i.r. spectra taken at liquid nitrogen and at room temperatures; (*b*) when the phospholipid is heated to a higher temperature, it reaches a transition point, a marked endothermic change occurs and the hydrocarbon chains in the lipid 'melt' and exhibit a very high degree of molecular motion. This is evident both in the appearance of the i.r. spectrum and also in the narrow n.m.r. line-width. On the one hand the broad diffuse appearance of the i.r. spectrum is consistent with the chains flexing and twisting and with a 'break-up' of the all-planar *trans* configuration of the chains. (The i.r. spectrum can be regarded as giving a 'rapid snapshot' showing the condition and organization of the phospholipid molecule within a period of 10^{-15} sec.) The n.m.r. spectrum, on the other hand, is affected by slower motions. It

*This observation of greater chain movement at the methyl end of the chain of the lipid in its lipid crystalline form, deduced by Chapman and Salsbury[11] in 1966, was later confirmed by means of other techniques, *e.g.* ESR studies using spin labels.

References p. 141

provides an impression of any molecular motion, i.e., rotation, translation or diffusion, of the phospholipid molecule occurring within a much longer time period, some 10^{-7} sec. That the phase transition is concerned primarily with the hydrocarbon chains of the phospholipid is confirmed by the X-ray data.

When phospholipids contain shorter chain lengths, or unsaturated bonds, these marked endothermic phase transitions occur at low temperatures. The temperature at which these transitions occur parallels the behaviour of the melting point of the related fatty acids. The transition temperatures are high for the fully saturated long-chain phospholipids. They are lower when there is a *trans* double bond present in one of the chains and lower still when there is a *cis* double bond present. This variation of transition temperature also confirms that this phase transition is primarily associated with a 'melting' of the hydrocarbon chains of the phospholipid, while this in turn is a reflection of the dispersion forces between the chains.

Only one main 'melting of the chains' occurs even when there are two different types of chain present in the phospholipid. The transition temperatures for different phospholipid classes vary even though they contain exactly the same fatty acid residues. While natural phospholipids from erythrocytes or mitochondrial membranes contain large amounts of unsaturated *cis* double bonds and, therefore, in the dry condition have endothermic transition temperatures either near or below room temperatures, the highly saturated derivatives exhibit transition temperatures much higher than room temperature. With mixtures of phospholipids we might expect the transition temperatures to occur over a wider range of temperature than with a single phospholipid. This is true for the d.t.a. heating curves of egg phosphatidylcholine.

When a crystalline phospholipid approaches the first thermotropic transition temperature we can imagine that the molecular motion of the chains increases until, at the first transition point, lateral expansion of the crystal lattice is forced to take place. This expansion allows even greater chain mobility possibly involving co-operative coiling

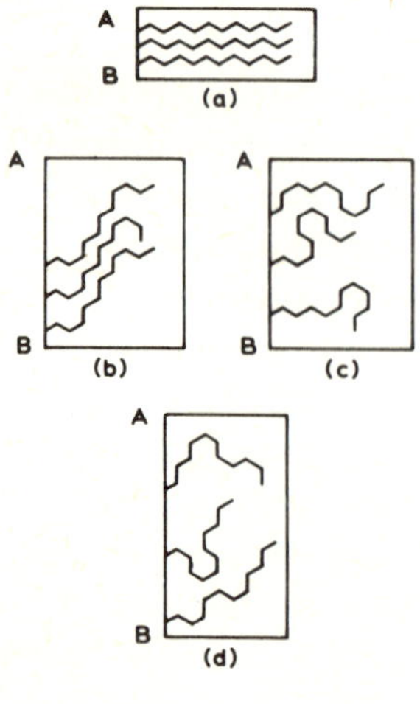

Fig. 3. Typical configuration of sets of chains on a two-dimensional hexagonal lattice as the interchain separation is increased. Separation increases from (a) to (d) (Whittington and Chapman[13]).

of the chains with rotation about C–C bonds occurring. Such a coiling of chains could explain the observed reduction in long spacing. A similar coiling of chains probably takes place with polar long-chain molecules in a monolayer at an air–water interface dependent upon the density of the chain packing.

A very simple theoretical system has been devised[13] to provide some insight for such situations. The model system is restricted to a set of simple chains in two dimensions and with the chains confined to lie on a two-dimensional hexagonal lattice. The end-to-end distance of each chain (corresponding in a real system to the distance from the polar group to the methyl group) and other properties were determined as a function of the density of packing of the chains using two simple potential functions and using the Monte Carlo computational method. Typical configurations of sets of simple chains on a two-dimensional hexagonal lattice are shown in Fig. 3. At the highest density only the fully extended configuration occurs. At lower densities other configurations are allowed. As the density of the chains decreases, the end-to-end distance of the chains suddenly falls, consistent with the occurrence of a co-operative phase transition.

An important conclusion which we obtain from these phase transition studies is that, near to the endothermic transition temperature, a given phospholipid can be in a highly mobile condition with its hydrocarbon chains flexing and twisting. The more unsaturated the chain, the lower the temperature will be at which this occurs. This is a fundamental property of the phospholipid and we can expect this chain mobility to occur in whatever situation the phospholipid occurs unless, for special reasons, this motion is somehow inhibited. We can envisage that inhibition of chain motion by interaction with other molecules would provide one of these special reasons, e.g. water or protein. In other circumstances, due to less perfect packing arrangements, we might expect, at a particular temperature, even greater mobility of the chains of the lipid and indeed of the whole lipid molecules themselves.

3. PHOSPHOLIPID–WATER SYSTEMS AND LYOTROPIC MESOMORPHISM

Phospholipids exhibit interesting behaviour in the presence of water. As with their thermotropic mesomorphism, where they do not pass directly from the crystalline state to a liquid, the phospholipids, in general, do not pass directly from the crystalline state to a solution in the presence of water. Various hydrated phases are encountered before 'solution' of the phospholipids in water occurs. Such behaviour is called 'lyotropic mesomorphism'. The lyotropic phases exhibit thermotropic mesomorphism in other words, the particular phase obtained is a function both of water content and of temperature[14].

The thermotropic mesomorphism of soaps is not unlike that of the phospholipids, and so it is to be expected that their lyotropic behaviour is similar. Indeed, the structural knowledge of the soap–water systems has been of help in undertaking the study of lyotropic mesomorphism of the phospholipids[15].

The soaps, phospholipids and similar molecules are often described as amphipathic or amphiphilic (Greek: *amphi*, dual; *pathi*, sympathy; *phile*, liking) because they contain both water-soluble and oil-soluble regions. Dervichian[16] has discussed the conditions governing the formation of lyotropic mesophases in detail, but it is useful to mention some of the points here. As we have seen, an increase in temperature melts the hydrocarbon chain region and, in all of the thermotropic mesophases, the polar and hydrocarbon parts of the molecules are separated into different regions. When water diffuses into the lattice it does so into the polar (ionic) region, but it *cannot* do this until a particular temperature (T_c line, Krafft point) is reached at which the hydrocarbon chains 'melt'. If the temperature is higher than this particular point, there is simultaneous dissociation of the ionic lattice by the penetration of water and the melting of the hydrocarbon chain region. The temperature of transition (T_c line) depends upon the nature of the hydrocarbon chains and of the polar region of the molecule, the amount of water present and on any solutes dissolved in the water.

Once the water has penetrated into the lattice of the amphiphile and the sample is then cooled to below the T_c line, the hydrocarbon chains arrange themselves into an orderly crystalline lattice, but the water is not necessarily expelled from the system. Lyotropic phases containing crystalline paraffin chain regions are called *gels*; these gels may or may not be metastable. If they are metastable they transform over a period of time to a suspension of microcrystals of the amphiphile in water, the *coagel* state. The coagel is a stable state, its structure is independent of the thermal history of the sample.

The structure of the lyotropic phases

From the general behaviour of amphiphile water systems it is known that the water and hydrocarbon chain region are separated by an interface of polar groups. The phases containing rigid, crystalline hydrocarbon chain regions can only form lamellar structures (i.e. gels and coagels). The lyotropic mesophases containing hydrocarbon chains may have structures other than lamellar. Larson[17] considers that, in view of the liquid-like nature of these mesophases, which implies freedom for molecular motion, it is reasonable to assume that all amphiphilic molecules, in one phase, are arranged in the same way, so that they are energetically equivalent. If this is so only three types of polar group interfaces are possible: spherical shells, cylindrical shells of infinite length and infinite planes. More structural types are possible, as water can either be inside or outside the spherical or cylindrical interface. In the phospholipid–water systems, diffraction patterns corresponding to all three types of polar-group interfaces have been obtained. A lamellar phase (infinite plane) an hexagonal phase (infinite cylinders) and a cubic phase (spherical shells) have been identified. Hexagonal phases exist with water inside and outside the cylinders, depending upon the phospholipid, whilst the cubic phase has, as yet, only been identified with water inside the spheres. In Fig. 4 diagrammatic representations of the structures of some lyotropic mesophases are given. Luzzati et al.[18] describe the cubic phase as being composed of identical rod-like elements of finite length, linked three by three into two distinct three-dimensional networks which are unconnected, but mutually interwoven. They refer to the presence of other more complicated phases which may exist over very narrow concentration ranges in the lecithin–water systems.

The structure of the gel (Fig. 4a) is lamellar with the lipid layers separated by water layers and with the hydrocarbon chains packed in a crystalline lattice. The gel has been analysed for lecithin–water systems. Gulik-Krzywicki et al.[19] studied a mitochondrial lipid–water system and have described some very complex gel-like phases which will be briefly discussed later.

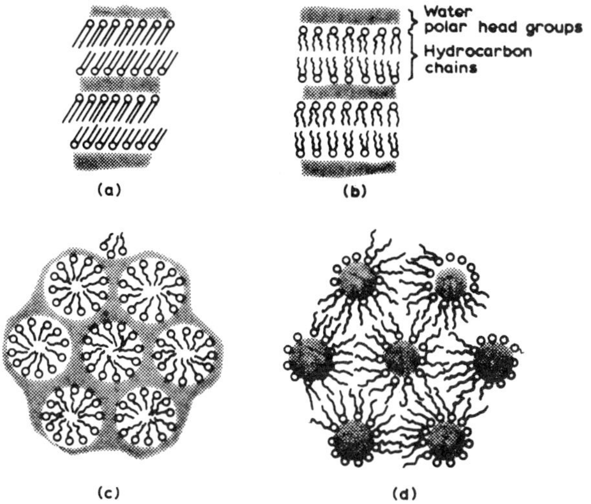

Fig. 4. Diagrammatic representation of the structures of some lyotropic mesophases. (a) gel; (b) lamellar (neat); (c) hexagonal I; (d) hexagonal II.

The lamellar phase (Fig. 4b) is made up of alternating layers of phospholipid and water. The lipid layers are filled with liquid-like hydrocarbon chains. Many phospholipid–water systems exhibit the lamellar lyotropic phase[20].

The hexagonal phases (Fig. 4c) consist of two-dimensional arrays of cylinders. In hexagonal I (type I, paraffin-in-water; type II, water-in-paraffin), the cylinders, filled by liquid paraffin chains and covered by the polar groups, are separated from each other by water. This phase has been described[21, 22] for the egg-yolk lysolecithin–water system. The inverse phase, formed by cylinders of water, covered by polar groups, and embedded in a liquid paraffin matrix, is called hexagonal II.

The cubic phase has been described for only one phospholipid system as yet, the phosphatidylcholine–water system, and the structure may either be body-centred or face-centred. Until recently the structure was thought to be composed of a lattice of water spheres embedded in a matrix of lipid. However, Luzzati et al.[18] now interpret the X-ray diffraction data in terms of a more complex structure formed of rods, of water and polar head groups, of finite length, linked three-by-three, to form two interwoven three-dimensional networks; it is both lipid continuous and water continuous.

The lyotropic phases obtained by mixing natural lipids with water have been studied[23, 23] by X-ray diffraction. Amongst the nerve lipid fractions examined was lecithin, which, on addition of two parts of water to one of lecithin, gives rise to a phase with a

References p. 141

lamellar structure, liquid-like hydrocarbon chains and a long spacing of 69 Å, some 60% greater than that of the dry material. In 1967, nearly a quarter of a century later, three papers by independent groups were published; two were concerned with egg lecithin[21, 22] whilst the other covered an homologous series of 1,2-diacyl-*sn*-glycero-3-phosphorylcholine[20].

3-sn-Phosphatidylcholine

The phase diagram of the 1,2-dipalmitoyl-*sn*-glycero-3-phosphorylcholine–water system is shown in Fig. 5. On addition of water, the transition temperature, T_c, of the

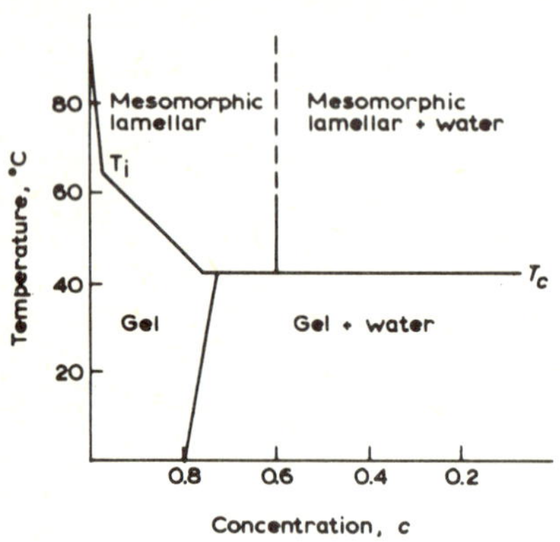

Fig. 5. Phase diagram of the 1,2-dipalmitoyl-*sn*-glycero-3-phosphorylcholine–water system. (Chapman et al.[20]).

phospholipid is lowered to a limiting value, T_c. This transition temperature is the minimum temperature required for water to penetrate between the layers of the lipid molecules. Above the T_c line the phosphatidylcholine –water system exists in a mesomorphic lamellar phase (Fig. 5), in which the hydrocarbon chains are in a liquid-like state. The composition of the system at maximum hydration is ~40 wt % water. Addition of more than 40 wt % water gives rise to a two-phase system consisting of fragments of the lamellar phase at maximum hydration dispersed in the excess water. The macroscopic appearance of these fragments has been the subject of many investigations.

When the phosphatidylcholine–water system is cooled below the T_c line, the hydrocarbon chains adopt an ordered packing. The structure of this phase, the gel, is lamellar with the hydrocarbon chains packed in a hexagonal subcell with the chain axes inclined at 58° to the lipid–water interface (Fig. 4).

The phase diagrams of the different chain length lecithin–water systems are essentially equivalent and are disposed along the temperature axis according to the melting temperature (T_c) of the hydrocarbon chains. Fig. 6 shows differential scanning calorimetric curves of a 1,2-distearoyl-*sn*-glycero-3-phosphorylcholine–water system obtained over a range of water concentrations. In the range $1.0 \geqslant c \geqslant 0.8$, the temperature of the endothermic transition T_c is observed to decrease steadily to a limiting value (T_c). In this concentration range, despite the fact that an appreciable amount of water is present, no transition is

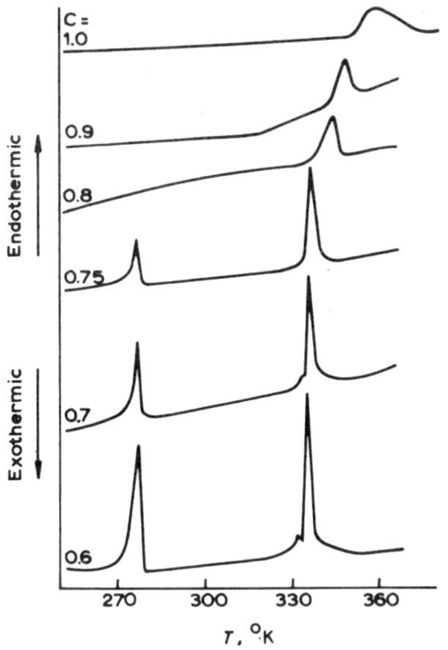

Fig. 6. Differential scanning calorimetry of 1,2-distearoyl-*sn*-glycero-3-phosphorylcholine in increasing amounts of water. (Concentration C = 1.0 means no H_2O present.)

observed in the heating or cooling curves due to any melting of ice or freezing of the water present. When the water content is greater than 20% ($c < 0.8$) the lipid endothermic transition temperature (T_c) remains constant and a peak at 0°, due to the melting of ice, can now be observed in the heating curve. As the concentration of water in the mixture further increases, so does the size of this peak. Quantitative studies are interpreted as

TABLE 2

HEATS AND TEMPERATURES OF TRANSITION FOR 3-*sn*-PHOSPHATIDYLCHOLINE–WATER SYSTEMS

1,2-Diacyl-phosphatidyl choline	*Limiting transition temperature (T_c) (°C)*	*Heat absorbed at T_c (kcal · mole^{-1}) phospholipid*	*Entropy change at T_c (cal · deg · mole^{-1})*	*Temperature of 'pre-transition' peak (°C)*
Dibehenoyl	75	14.9	42.8	75
Distearoyl	58	10.70	32.4	56
Dipalmitoyl	41	8.65	27.6	35
Dimyristoyl	23	6.65	22.4	14
Dilauroyl	~0 (under the ice peak)			
Dioleoyl	−22	7.6	30.3	
Egg yolk[a]	−15/−7			

[a]The egg-yolk lecithin transition is broad in water due to the heterogeneous fatty acid composition. The lower temperature is from the heating curve, the higher from the cooling curve.

showing that a proportion of the water is bound to the lecithin in a fixed ratio of 1:4 by weight (10 moles water/mole lecithin). This bound water is due to the formation of a hydrate structure associated with the polar group. The amount of bound water is independent of the fatty acid composition of the phospholipid, but is dependent upon the nature of the hydrophilic group. All the different lecithins studied behave similarly. The heat absorbed at T_c for lecithin–water systems with $c < 0.7$ (Table 2) is chain length-dependent. For the series of fully saturated lecithins a difference of 2 kcal·mole^{-1} occurs for a chain length difference of two methylene groups. Extrapolation to zero heat at this transition shows that the C_7 and shorter chain length lecithins have a negative heat of transition. A small 'pre-transition' peak is observed when $c < 0.8$ (Fig. 6). The temperature interval between this peak and the main endothermic peak (T_c) increases as the chain length of the lecithin becomes shorter (Table 2).

Egg-yolk lecithin

A complete binary phase diagram (Fig. 7) of the egg-yolk lecithin–water system has

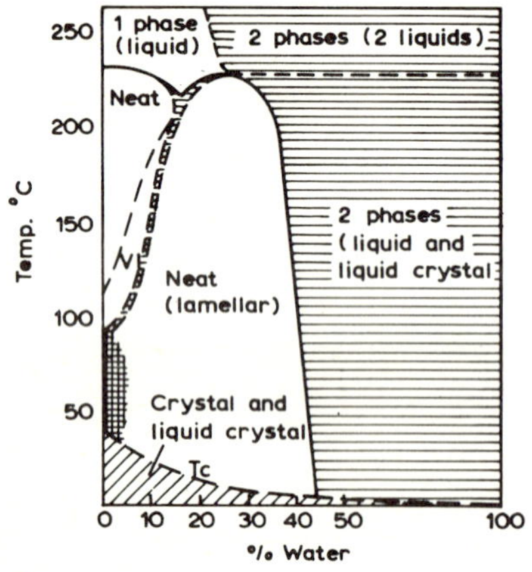

Fig. 7. Swelling of pure egg lecithin in water: the binary phase diagram for egg lecithin and water as a function of temperature. V.I. represents 'viscous isotropic' phase (face-centred cubic by X-ray diffraction). T_c is the ill-defined boundary of crystal–liquid-crystal phase transition. The cross-hatched area between 0 and 5% water from 45 to 90° represents a poorly defined area in which lamellar liquid crystalline phase may coexist with another liquid crystalline phase (Small[22]).

been described by Small[22]. The phase diagram is essentially the same as those for the lecithins of discrete chain length (cf. Fig. 5). The T_c line is rather broad and ill defined, and this observation is supported by other work on phospholipids derived from an heterogeneous mixture of fatty acids. The T_c line determined by Reiss-Husson[21] differs somewhat from that found by Small in that the latter shows a continuous decrease with increasing water content, whereas that of the former reaches a level value at about 20% water ($c = 0.8$); an observation in accord with the behaviour of discrete chain length lecithins. Calorimetric work from our laboratory shows the T_c line to level off at 15° (heating) –7° (cooling). This range of temperature gives an indication of the onset and the finish of the chain melting.

The maximum degree of hydration of the lamellar liquid crystalline phase of the egg-yolk lecithin–water system at 25° has been reported as 45% (ref. 22) and 40% (ref. 21) similar to that observed for the discrete chain length lecithins. The lamellar repeat distance, d_1 (Fig. 4b), increases to 64 Å at maximum hydration of this phase, whilst the thickness of the lipid layer falls from 43 Å (ref. 22) 36 Å (ref. 21) at $c = 0.85$ to 38 Å (ref. 22) 31 Å (ref. 21) at $c = 0.60$. This constant difference of 5 Å in the values of d_1 calculated by these workers undoubtedly reflects the assumptions involved (see original papers) in their calculations. The reduction in the thickness of the lipid layer, as the weight fraction of water is increased is also true of the lamellar liquid crystalline phase of many soaps and detergents.

In this egg-yolk lecithin–water phase diagram, Small[22] has delineated the boundaries of a viscous isotropic phase. This phase is thought to consist of spherical aggregates of the polar heads of the lecithin molecules surrounding small islands of water packed in a face-centred cubic lattice. This cubic phase can only contain up to 15 wt.% water. Both Reiss-Husson[21] and Chapman et al.[20] noted the presence of a cubic phase in lecithins but did not define the maximum water content with which this phase could exist. The latter authors, however, did define the minimum water content necessary for this phase, one molecule of water per molecule of lecithin.

The conformation of the choline head group. The conformation of the phosphorylcholine head group in the lecithin–water system is of interest in order to be able to understand certain properties of the system. It has been concluded, from a comparison of measured and theoretical electrokinetic potentials of lecithin bimolecular leaflets, that the head groups are oriented in a plane parallel to the leaflet[25]. Monolayer experiments have been variously interpreted as indicating that the head groups are parallel to the surface. From model membrane experiments some workers have concluded that the phosphate and trimethylammonium groups are in a plane parallel to the leaflet. The question of the conformation and orientation of the head group thus remains open.

Other phospholipid–water systems

The only phospholipid–water systems to be studied using chemically distinct phospholipids are some of the lecithin–water systems described above, where not only are the polar groups all identical, but also the fatty acid chains are chemically homogenous. Comparison of these systems with the egg-yolk lecithin–water system shows similar behaviour, with the exception of a change in the temperature of the crystal–liquid crystal transition (T_c line), and also, perhaps, of the detailed molecular structure of the gel phase.

(1) Nerve lipid extracts

In 1941 Bear, Palmer and Schmitt[23] studied the behaviour of various lipid fractions of beef spinal cord. The major lipids studied were the 'total lipid' extract and the 'cephalin' fraction. (The 'cephalin' fraction studied contained anionic lipids such as phosphatidylserine; it was not a pure phosphatidylethanolamine fraction.) These systems

References p. 141

adopt a lamellar structure in water, the uptake of water is almost indefinite, and the hydrocarbon chains are in a liquid-like condition. The large water uptake between the lipid bilayers was shown to be due to ionic charges on the bilayers. The addition of electrolytes to the aqueous phase reduced the interleaflet water content by an amount depending upon the charge and concentration of the electrolyte, calcium being much more effective at producing dehydration than either sodium or potassium.

(*2*) *Egg-yolk phosphatidylethanolamine*

The phase diagram of the egg-yolk phosphatidylethanolamine–water system[21] is comprised of a zone where the hydrocarbon chains of the lipid are crystalline (a gel or coagel), a dispersion zone (see p. 128) existing when more than 20 wt. % of the mixture is water, and two liquid crystalline lyotropic phases, lamellar and hexagonal.

The hexagonal phase of type II (cylinders of water in a lipid matrix, see p. 125) exists at 55° when the water content is between 4 and 21%. At maximum water content the diameter of the water cylinders is 32.2 Å and the surface area per phospholipid molecule at these cylinders is 41.3 Å. Reiss-Husson[21] finds the hexagonal phase to coexist with a lamellar phase at 25° and 35°. Papahadjopoulos and Miller[26] report that egg-yolk phosphatidylethanolamine, when dispersed in an excess of 145 m*M* KCl solution at pH 6.0, gives only a lamellar phase.

(*3*) *Sphingomyelin*

Reiss-Husson[21] has examined the effect of water on sphingomyelin extracted from beef brain. At 25° the hydrocarbon chains are crystalline, but at 40° a lamellar lyotropic phase is formed. This lamellar phase can incorporate up to 40 wt % water ($c = 0.6$) before a dispersion zone occurs (see p. 128). This behaviour is similar to that observed for 1,2-dipalmitoyl-phosphatidylcholine (see p. 126). At maximum hydration the thickness of the lipid bilayer is some 48 Å whilst the water layer thickness is 30 Å.

(*4*) *Brain lipid extract*

With the exception of the pioneer work of Bear et al.[23] all the phospholipid–water systems so far discussed have contained only one species of phospholipid, although the particular phospholipid may have contained a heterogeneous distribution of fatty acyl residues. Luzzati and Husson[27] examined an ethereal lipid extract of human brain containing approximately 52% ‘cephalin’, 35% lecithin and 13% phosphoinositides. It is not stated whether the cephalin is solely phosphatidylethanolamine or whether phosphatidylserines are present. However, the lipid does contain a fair percentage of ionic material and we might therefore expect, on the basis of the work of Bear et al., to find the lipid bilayers in any lamellar phase strongly repelling each other in the presence of water. This was, in fact, found by Luzzati and Husson. For the system with $c = 0.39$ at 22° the repeat distance d is 99.7 Å of which 60.4 Å is the water layer thickness. Compare this behaviour with that of an uncharged phospholipid such as lecithin where the water layer thickness at maximum hydration is only 30 Å. No mention is made as to whether more than 60% water can be incorporated between the brain lipid bilayers but, by comparison with the mitochondrial lipid–water system (following section), we should expect the lipid to be capable of greater water uptake.

It is seen that two lyotropic liquid crystalline phases exist, lamellar at high water content and hexagonal type II (see p. 125) at low water content. This was the first reported observation of a non-lamellar phase for phospholipid–water systems. The lipids, although a heterogeneous mixture, behave as a single component in these lyotropic, liquid crystalline phases. In the coagel region the hydrocarbon chains are crystalline in contrast to the other phases where the chains are liquid-like.

(5) *Mitochondrial lipid extract*

A detailed examination of the phase behaviour in water of a lipid extract from beef heart mitochondria has been made[19]. Molecular species analysis of this extract is given as lecithin 34 ± 5%, phosphatidylethanolamine (with some plasmalogen) 29 ± 5%, phosphatidylinositol 10 ± 2%, cardiolipins 20 ± 5%, cholesterol 2 ± 1%, and neutral lipids 5 ± 2%.

The phase diagram of the mitochondrial lipid–water system above about 5° (Fig. 8) bears a strong resemblance to the brain lipid–water system. The hexagonal phase is of

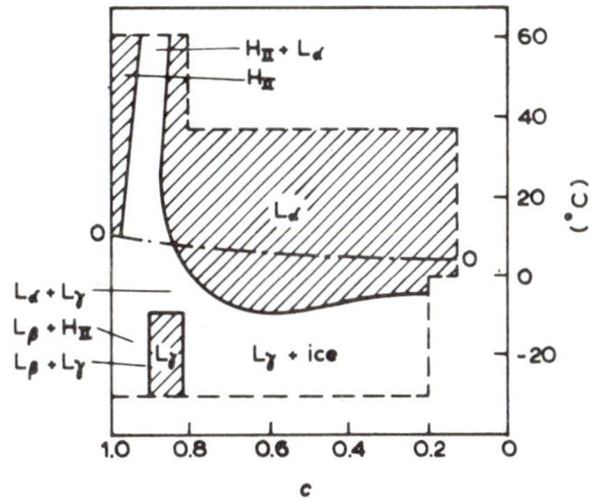

Fig. 8. Phase diagram of a mitochondrial lipid–water system. The dashed line represents the limits of exploration. The hatched areas are one-phase regions. Above the line 0–0 the conformation of the paraffin chains is 'liquid' (Gulik-Krzywicki et al.[19]).

type II, and exists when the water content of the system is low. Four different lamellar phases are noted, L_β, L_γ and two forms of L_α – a high-temperature form existing above the line 0–0, and a low-temperature form.

The high temperature L_α phase is of the type shown in Fig. 8, the hydrocarbon chains being liquid-like. Water uptake by this phase is continuous within the range studied ($c > 0.15$; i.e. up to 85 wt % water). The thickness of the lipid layer (d_l) and the surface area per molecule (S) at the lipid–water interface are both almost concentration-independent, being 40 Å and 60 Å respectively. At $c \approx 0.15$ the lipid bilayer thickness is 40 Å but the thickness of the water layer separating the lipid leaflets is some 250 Å. It is remarkable that the overall structure can be maintained over such large distances.

When the temperature of the L_α phase is lowered to below the line 0–0 (Fig. 8) the X-ray diffraction pattern in the high angle region, in addition to showing the diffuse 4.5 Å band (characteristic of liquid hydrocarbons), also exhibits a sharp reflection at 4.2 Å (characteristic of ordered paraffin chains). The intensity of this reflection increases, with respect to that of the 4.5 Å band, as the temperature is lowered. Gulik-Krzywicki et al.[19] interpret this gradual disorder–order transition as being due to domains of crystalline paraffin chains growing within the bilayer leaflets as the temperature is lowered, such that at any particular temperature below 0–0 part of each hydrocarbon chain is 'fluid' and part is crystalline. The surface area per molecule at the lipid–water interface at 58 Å^2 is only slightly less than that of 60 Å^2 found for the high-temperature L_α phase.

The lamellar L_β phase, a gel with rigid hydrocarbon chains, differs in structure from the gel shown in Fig. 4 in two respects. Firstly, the hydrocarbon chains are perpendicular to the lipid–water interface, and, secondly, a thin disordered region in the middle of the paraffin layer is postulated in order to allow for the unequal hydrocarbon chain lengths of the lipids.

The fourth lamellar phase, L_γ, appears to be radically different from all the other lamellar phases examined. Pure L_γ phase exists only at low temperatures and in the concentration range $0.9 > c > 0.8$ and yet it gives rise to a long spacing, d, of about 120 Å. Gulik-Krzywicki et al.[19] propose a structure comprised of alternative L_α (low temperature form) and L_β leaflets, separated as usual by water layers. Chemical fractionation of the lipids is thought to occur, thereby giving rise to different chemical compositions of the L_α and L_β layers. The overall structure proposed is thus:

$$\text{–water–}L_\beta\text{–(water–}L_\alpha\text{–water–}L_\beta)_n\text{–water–}$$

and this regularity must be maintained over a reasonable distance in order to give rise to a sharp diffraction pattern. The fractionation of the lipids, when the water content is low, appears to be perfectly feasible in view of the known fractionation of mixed lipid systems on drying. However, the structure proposed for the L_γ phase implies a greater attraction between L_α and L_β than between L_α and L_α or L_β and L_β leaflets.

Interactions with metal ions

Some phospholipids, e.g. phosphatidylserines, phosphatidylinositols and phosphatidylethanolamines, have a high affinity for $Ca^{2\oplus}$ and $Mg^{2\oplus}$ and this is related to interactions with their polar groups[28] (Chapter 7, p. 156). Electrostatic interactions of the polar head of the phospholipids can affect their thermotropic mesomorphism. Urbina and Chapman[29] have investigated the interaction of different divalent cations on the endothermic phase transition of dipalmitoyl-phosphatidylcholine (DPL) and ox-brain phosphatidylserine (PS). From monolayer and liposome experiments uranyl ions ($UO_2^{2\oplus}$) are known to interact stoichiometrically with the polar groups of phosphatidylcholine (PC) and to bind in the membrane surface of different biological membranes, presumably to phosphate groups. Infrared spectroscopy has confirmed that the strong binding of this ion is specifically to the phosphate groups[29]. It is also known that PS forms a stoichiometric complex with $Ca^{2\oplus}$ and other divalent ions.

These interactions have very strong effects on the thermotropic mesomorphism of both lipids. Fig. 9 shows a titration of dipalmitoyl-lecithin with $UO_2^{2\oplus}$ monitored by its

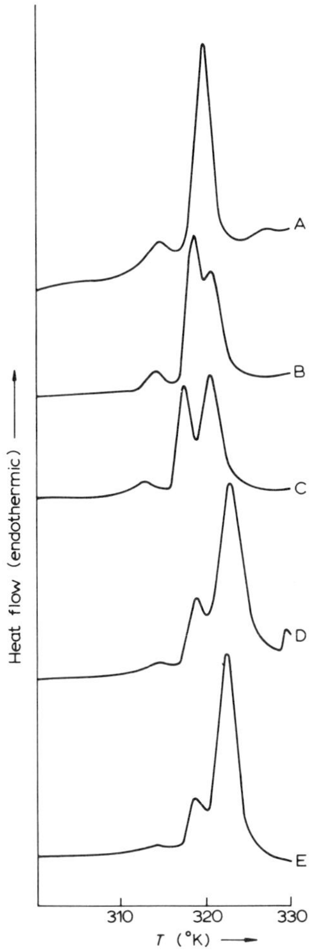

Fig. 9. Differential scanning calorimetry of dipalmitoyl-lecithin/water = 50/50 mixtures in the presence of

A: 0.0
B: 0.1
C: 0.2
D: 0.4
E: 0,8 equivalents of $UO_2^{2\oplus}$ per equivalent of lipid phosphorus (Urbina and Chapman 1971, ref. 29).

effect on the transition temperature of the lipid. (The first peak on these diagrams is thought to be associated with a modification of the polar head-group. The second peak is associated with melting of the lipid chains.) The original transition temperature is 314.5°K (41.5°) which is finally shifted in the 0.8 : 1 ($UO_2^{2\oplus}$: lipid P equivalents) complex to 319°K (46°). With PS, containing a net negative charge, the binding of $Ca^{2\oplus}$ at 1:1 proportion shifts the transition from 290°K observed with the free PS to 295°K. These effects have been confirmed by the effect on the mobility of the spin label 12 NS in the hydrocarbon phase of PS. This shows that the ion-binding dramatically reduces the reorientation of the label in the hydrocarbon phase of the lipid.

Monolayer studies

Pethica and Anderson[30] using distearoyl-phosphatidylcholine prepared by Malkin, studied its surface pressure and surface potential characteristics on substrates of varying ionic strength and pH. All monolayer states are possible with saturated lecithin (Fig. 10)

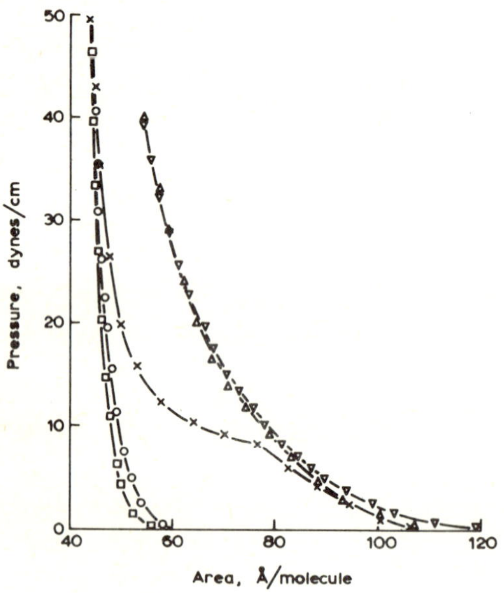

Fig. 10. Pressure-area curves for saturated 1,2-diacyl-phosphatidylcholines on 0.1 *M* NaCl at 22°: □, C_{22}; ○, C_{18}; ×, C_{16}; △, C_{14}; ▽ ,C_{10} (Phillips and Chapman[31]).

and also with the phosphatidylethanolamine homologues[31]. It is apparent that if the hydrocarbon chains are sufficiently long, condensed monolayers are formed whereas with shorter chains liquid-expanded films occur (Chapter 7, p. 147). These two limiting states are sufficiently well defined that at any particular temperature only one of the homologues studied ($C_{16:0}$) exhibits the transition state between the two. Variations in hydrocarbon chain length which do not give rise to a change in monolayer state do not have a significant effect on the pressure-area curves. Temperature changes can give rise to the above physical states in a monolayer of a single lecithin (Fig. 11). Obviously, a sufficiently low temperature causes the film to become completely condensed, whereas at higher temperatures it is fully expanded. Monolayers in the two limiting states are more or less invariant with temperature and it is only the sensitivity of the phase transition to temperature which leads to the variety of isotherms depicted in Fig. 11.

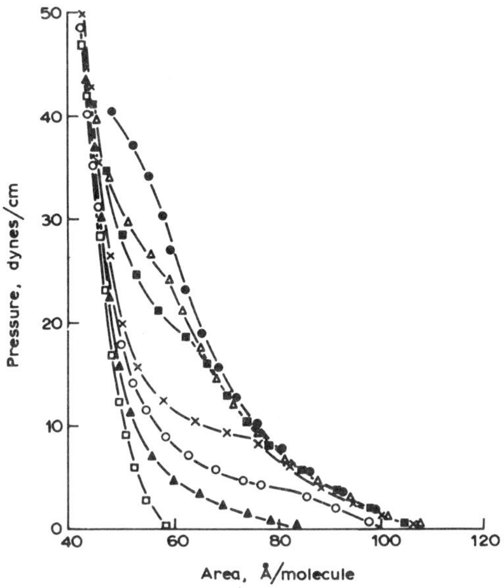

Fig. 11. Pressure-area curves for 1,2-dipalmitoyl-*sn*-glycero-3-phosphorylcholine on 0.1 *M* NaCl at various temperatures: ●, 34.6°; △, 29.5°; ■, 26.0°; x, 21.1°; ○, 16.8°; ▲, 12.4°; □, 6.2° (Phillips and Chapman[31]).

The molecules in a completely condensed phosphatidylethanolamine (PE) film are much more closely packed than those in the equivalent lecithin monolayer. In fact, condensed lecithins and PE give coherent films over the surprisingly wide range of about 14 $Å^2$/molecule. This is much larger than the equivalent range of about 4 $Å^2$/molecule for stearic acid and about 7 $Å^2$/molecule for a condensed diglyceride. It presumably arises from steric factors associated with the large polar groups on lecithins and PE. The larger limiting area found for the lecithin as opposed to the PE reflects the greater space required by the bigger and probably more hydrated choline groups.

The binding of cations to phospholipid monolayers and its relevance to the excitation mechanism in living cells has been considered by a number of workers (see also Chapter 7, p. 157). The phosphatidylserine molecule with its two anionic functional groups was thought particularly relevant. Bangham and Papahadjopoulos[32] followed the $Ca^{2\oplus}$ binding by surface potential (ΔV) and pressure (π) changes, and also by monitoring β-particle emission from $^{45}Ca^{2\oplus}$. ΔV–pH curves, both in the absence and presence of $Ca^{2\oplus}$, showed involvement of the $Ca^{2\oplus}$ with both anionic groups. The ΔV–$\log_{10}$ ($Ca^{2\oplus}$) concentration curve for the substrate (145 m*M* NaCl, pH 7.4) showed two linear regions, 0.1–1 m*M* and 1–10 m*M* $Ca^{2\oplus}$. They propose a transition to a second complex at 1 m*M* $Ca^{2\oplus}$. Electrophoresis results help to confirm that this ΔV discontinuity is due to ion binding and not just to changes in the permanent dipoles. The $Ca^{2\oplus}$ studies indicated a ratio of 0.9 equivalents $Ca^{2\oplus}$ to 1 equivalent phosphorus at 1 m*M* bulk concentration. This has been confirmed by other workers[33] at 0.1 m*M* bulk $Ca^{2\oplus}$, although no results are given for the $Ca^{2\oplus}$ binding at concentrations greater than 1 m*M*.

Phospholipid–cholesterol interactions

Many biological membranes contain appreciable proportions of cholesterol in addition to phospholipids, protein, etc. The interaction of cholesterol with phospholipids, is of considerable biological importance and many studies have been made to gain an understanding of the function of cholesterol in biological membranes.

Lecithin–cholesterol interaction

Leathes[34] studied monolayers of lecithin and cholesterol and concluded that the cholesterol has a condensing effect on the lecithin; in other words, the lecithin occupies a smaller area at the air–water interface in the presence of cholesterol than it does in its absence. Adam and Jessop[35] after further studies concluded that the condensing effect of cholesterol was 'due to mechanical obstruction of the tilting oscillations of the smaller molecules (lecithin), by the bulky and massive large molecules (cholesterol).' As we shall see later, this viewpoint is currently subscribed to, although it has been extended. Since this early work many monolayer experiments have been carried out on lecithin–cholesterol systems. The major conclusions from these investigations are that cholesterol condenses lecithin monolayers except, perhaps, for very short chain length or highly unsaturated lecithins in very expanded films. Standish and Pethica[36] studied mixtures of two disaturated phospholipids, dimyristoyl *rac*–PC and distearoyl *rac*-PE, each in admixture with cholesterol. In brief, only the dimyristoyl PC system showed non-ideal mixing, and, in addition, there were small shifts of 50 mD in the net dipole moment associated with the phospholipid molecule.

To what extent monolayer experiments at the air–water interface can be extrapolated to biological membranes is unknown. Chapman and Penkett[37] have shown by nuclear magnetic resonance spectroscopy that cholesterol inhibits the hydrocarbon chain motion of egg-yolk lecithin in water at 33.5°. This is in agreement with the monolayer results, and the experimental conditions were more akin to those pertaining to the biological condition in that the work was done on the bulk phase, not a monolayer.

Studies of the phase diagram of the egg lecithin–cholesterol–water system at room temperature have indicated that cholesterol, which alone is insoluble and does not swell in water, can be incorporated into the structure of an aqueous lamellar phase when associated with lecithin, providing that the molar ratio of cholesterol to lecithin is less than 1:1, that the hydrocarbon chains of the lecithin are in a liquid-like state, and that some water is present in the system[38]. This association is not a molecular association in the usual sense, much less the formation of a chemical compound. The situation is one of a mutual solution of the hydrocarbon parts of the lecithin and cholesterol molecules, occurring as a consequence of the orientation of the molecules by their contact with water and of the liquid-like nature of the paraffinic regions of the molecules. The structure of the lamellar phase is one of bilayers of lecithin and cholesterol with their hydrophilic groups oriented so as to be adjacent to the water layers. The lecithin molecules occupy a smaller area at the lipid–water interface in the presence of cholesterol than they do in its absence – a result in agreement with the above-mentioned monolayer work. The addition of cholesterol to the lecithin–water system does not give rise to any new phase with a basically different structure. Lecithin forms a lamellar phase in water (see p. 126) as does lecithin–cholesterol in water. Similar observations have been made on the lecithin–bile salt–water system on addition of cholesterol[39].

Until recently all the reported work on lecithin–cholesterol interactions had been concerned with systems where the lecithin was in a liquid-crystalline state so that the hydrocarbon chains were 'liquid-like'. The effect of temperature on the lecithin–cholesterol system had not been studied – a surprising omission in view of the thermotropic and lyotropic behaviour of phospholipids. In fact, disregard of the mesomorphism of phospholipids, and even lipids in general, has led to misleading statements about their behaviour.

Recent studies have been made on lecithin–cholesterol–water interactions by differential scanning calorimetry (DSC) and X-ray diffraction[40] in which the 1,2-dipalmitoyl-3-*sn*-phosphatidylcholine-cholesterol–water system was studied as a function of both temperature and of concentration of components. This particular lecithin was used because it exhibits a crystalline to liquid-crystalline phase change in the presence of water at a convenient temperatue (41°) for investigating the effect of cholesterol on both the gel and mesomorphic lamellar states. The DSC curves between 280 and 360°K for a series of 1,2-dipalmitoyl-3-*sn*-phosphatidylcholine-cholesterol mixtures, each containing 50% by weight of water show that the addition of cholesterol to the lecithin in water lowers the transition temperature between the gel and the lamellar liquid crystalline phase, and decreases the heat absorbed at the transition (Fig. 12). No transition at all is observed with

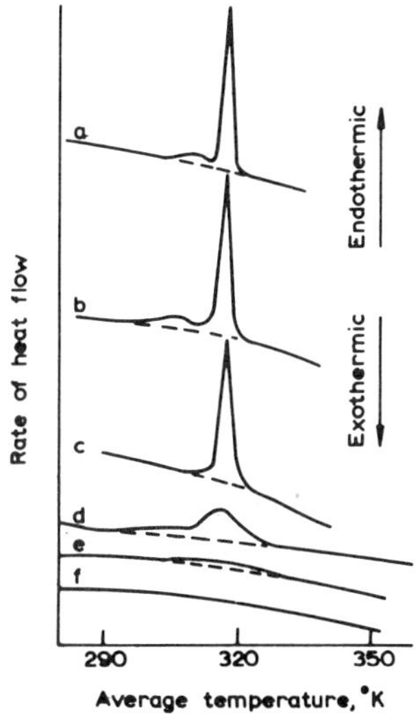

Fig. 12. Differential scanning calorimetry of 50 wt % dispersions in water of 1,2-dipalmitoyl-3-*sn*-phosphatidylcholine–cholesterol mixtures containing (a) 0.0 mole %, (b) 5.0 mole %, (c) 12.5 mole %, (d) 20.0 mole %, (e) 32.0 mole %, and (f) 50.0 mole % cholesterol (Ladbrooke et al.[40]).

an equimolar ratio of lecithin with cholesterol in water. This ratio corresponds to the maximum amount of cholesterol which can be introduced before cholesterol precipitation occurs. The addition of cholesterol also causes an increase in the amount of bound water. These effects are not specific for dipalmitoyl lecithin; unsaturated lecithins and the lipid extract of human erythrocyte ghosts exhibit similar behaviour. The addition of cholesterol to the lecithin–water system in the gel phase (see p. 137) causes a reduction in the cohesive forces between adjacent hydrocarbon chains of the lecithin; this leads to a fluidization of these chains. The ordered array of hydrocarbon chains in the gel phase is disrupted by cholesterol.

These observations[40] are entirely consistent with and complementary to the lecithin–cholesterol results referred to earlier. The additional finding is that cholesterol removes the T_c transition between the crystalline and liquid crystalline states in the presence of an excess of water. Thus we have the following situation:

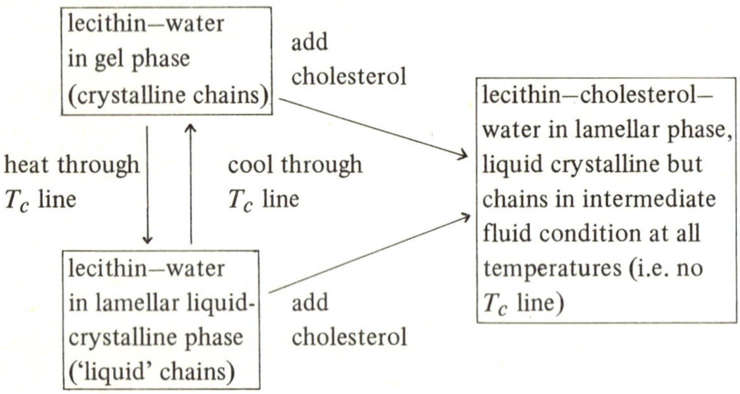

The hydrocarbon chains of the lecithin in water have less thermal notion in the presence of cholesterol than they have in the absence of cholesterol above the T_c line, and they have more thermal motion than in the absence of cholesterol below the T_c line. The requirement for water is, of course, essential.

Other observations are entirely in agreement with the interpretations advanced above for the lecithin–cholesterol interactions in the presence of water. High resolution nuclear magnetic resonance spectroscopy shows that the motion of the hydrocarbon chains of lecithin in the presence of cholesterol is less than in its absence above the T_c line (Penkett, unpublished observation). Monolayer experiments using pure 1,2-dipalmitoyl-*sn*-glycero-3-phosphorylcholine show that cholesterol removes the phase transition of the lecithin, that it condenses expanded monolayers at low pressure and that it causes an expansion of the condensed monolayer at high pressure.

Recently Oldfield et al.[41] have used deuterated lipids and examined them using deuteron resonance (DMR) spectroscopy. In the gel phase the lipid hydrocarbon chains are relatively rigid and the DMR spectrum shows only a very broad ill-defined resonance. On heating a transition to the liquid-crystalline phase occurs and at 30° a well-defined quadrupole splitting is observed. This splitting is approximately 27 ± 1 kHz. The static values of 270 kHz and 135 kHz expected for methylene deuterons are not observed and the result is therefore consistent with some motion in the chains. A central line of peak to peak width 4.2 ± 0.5 kHz also occurs and this may correspond to the relatively mobile terminal methyl groups as well as some terminal methylene groups. In presence of cholesterol the splitting observed at 30° increases to 49.4 ± 1.5 kHz showing that cholesterol has introduced a restriction on the freedom of motion of the lipid chains. The central component with width 4.5 ± 1 kHz is however, still present. Spin label studies using ESR spectroscopy are consistent with these conclusions[42].

'Solutions' of phospholipids in water

Do the phospholipids form true solutions in water and, if not, what is the structure of the particles dispersed in water? Let us consider what occurs as water is added to a lecithin above its crystal–liquid crystal transition temperature, T_c. As we have seen (p. 126), water is taken up between the lipid bilayers until a limiting concentration is reached. On addition of further water a two-phase system, of water and the lamellar phase at maximum hydration, then results. The macroscopic appearance of this lamellar phase dispersed in excess water has been the subject of many investigations[43].

When a particle of water-free lecithin above its limiting transition temperature T_c (see p. 127) is placed in contact with water, cylindrical growths appear from the surface of the particle. These growths (myelin tubes) maintain a constant diameter as they grow. The myelin figures are formed by macroscopic, coaxial, cylindrical sheets fitting into each other. The macroscopic sheets are themselves formed by a very large number of bilayers of lipid separated by water (i.e. lamellae) and which are rolled one inside the other.

Break up of these myelin tubes gives rise to globules[44]. These globules (liposomes or Bangham globules) have molecular weights of the order of 10^8 and are composed of spherically concentric lamellae. The globules are closed so that the water between the bilayers is isolated from that of the continuous aqueous phase[45]. The layered structure may thus be envisaged as being somewhat similar to that of an onion. Globules, formed of charged phospholipids, have been shown to act as almost perfect osmometers when alkali metal salts, glucose, sucrose or mannitol are used as solutes.

The phospholipid systems, such as mitochondrial lipid extract or lecithin with ionic impurities, which continue to swell in water almost without limit, do not behave like the uncharged phospholipids. The indefinite uptake of water between the lamellae is a direct consequence of the charged leaflets repelling each other. When salt is added to the system the repulsion forces between the leaflets are lessened and an equilibrium interlamellar separation is reached. Hence, colloidal systems become unstable on the addition of electrolyte. In the absence of salt the bilayers maintain their interaction over many hundreds of Ångström units. However, a dilution must be reached at which the bilayers cease to have any significant interaction with each other, and a dispersion of individual bilayers results. These bilayers may then be considered as very large planar 'micelles'. The sheets will, of course, not roll up on themselves to form 'Swiss roll' type structures because they are charged. However, they might form individual membranes separating an aqueous compartment from the bulk solution.

References p. 141

The effect of ultrasonic irradiation on phospholipid dispersions in water is of great importance, not only because of the interest of the system itself, but also because many physical studies are being carried out on natural membranes dispersed in water by sonic methods. Saunders and his colleagues have made extensive studies on sonicated phospholipid–water systems[46] by ultracentrifugation, viscosity and light scattering methods. They have demonstrated that the extremely large particles in coarse egg lecithin dispersions are disrupted by ultrasonic irradiation to give smaller, more symmetrical aggregates of molecular weight of about $2 \cdot 10^6$. Similar behaviour for phosphatidylserine dispersions has been noted. The effect of ultrasonic irradiation on lecithin sols was thought to be an expulsion of water from the aggregates and a rearrangement into folded laminae, a 'Swiss roll' type of structure.

Recently, some studies have been made on the effect of ultrasonic irradiation on egg-yolk lecithin–water dispersions[47]. By a combination of electron microscopy, X-ray diffraction and NMR spectroscopy, it was confirmed that the coarse dispersions, before sonication, consist of large particles (0.5 μ to 20 μ in diameter) with extensive lamellar ordering. The effect of sonication is to reduce the overall size of the particles to about 250 Å and that, during the breakdown of the particles, lamellar order is retained (i.e. no phase change occurs).

With egg lecithin–water dispersions no transition from the dispersed particles to free molecules is observed[48] even at dilutions of $5 \cdot 10^{-7}$ g · cm^{-3}. However, the di-C_2 and di-C_4 lecithin homologues are readily soluble in water[49] while the di-C_7 material is reported to be slightly soluble. Lecithins which are soluble fail to give a stable monolayer at the air–water interface and no stable film of the di-C_7 lecithin was formed[49]. It is of interest to note that for saturated lecithins with chain lengths of less than C_7, a negative heat of T_c is predicted. The heat absorbed at the T_c transition is a measure of the chain–chain interaction energy, and so we might expect that lecithins with a negative heat of T_c would disperse in water to form true solutions. The soaps, with one hydrocarbon chain per molecule, are known to dissolve, at high dilution, as individual molecules. Thus, we expect that the phospholipids with two hydrocarbon chains will only dissolve in water if the hydrocarbon chain length is short. The degree of unsaturation of the lipid should not have much effect upon its solubility in water.

Are the phospholipids with only one hydrocarbon chain per molecule soluble in water? Lysolecithin, derived enzymatically from egg lecithin, gives clear solutions in water consisting of spherical micelles each of which contains about 180 molecules[48]; the critical micellar concentration being 10^{-4} g · cm^{-3}.

Non-aqueous solvent systems

An early definition of lipids was those substances which are insoluble in water but soluble in 'fat solvents' such as ether, chloroform and benzene. Phospholipids, in contrast to triglycerides, are usually considered to be acetone insoluble; lecithins are considered to be soluble in alcohol, whilst phosphatidylethanolamines or phosphatidylserines are not. Sphingomyelins and cerebrosides are insoluble in diethyl ether. These generalizations, although useful, are not fully correct. The solubility of a given lipid in solvents depends markedly on the chain length and the degree of unsaturation (in other words on the hydrophobic–hydrophilic balance). The lecithins with hydrocarbon chain length of less than 10 carbon atoms are very soluble in acetone whereas the longer-chain members are not. Similarly, the solubility of lecithins in alcohol decreases considerably as the length of chain increases. This effect of chain length and unsaturation on lipid solubility has to be remembered when separation and analysis of different lipid classes are being carried out. The solubility of a given lipid may also be affected by the presence of other types of lipid.

In polar solvents such as ethanol, phospholipids are monomeric; in non-polar solvents such as benzene, however, they form micellar aggregates of large molecular weight (~50 000). The polar groups are considered to be in the interior and the hydrocarbon chains on the exterior of the micelle (inverted micelle). This is the opposite situation to that which occurs in water.

REFERENCES

1 D. Chapman and D.F.H. Wallach in D. Chapman (Ed.), *Biological Membranes*, Academic Press, London, 1968, p. 125.
2 M. Sundaralingam and L.H. Jensen, *Science*, 150 (1965) 1035.
3 S. Abrahamsson and I. Pascher, *Acta Crystallogr.*, 21 (1966) 79.
4 D. Chapman, *The Structure of Lipids by Spectroscopic and X-ray Techniques*, Methuen, London, 1965.
5 K. Larsson, *Arkhiv. Kemi*, 23 (1964) 1.
6 L.H. Jensen and A.J. Mabis, *Nature*, 197 (1963) 681.
7 D.F. Parsons and S.C. Nyburg, *J. Apply. Physiol.*, 37 (1966) 3920.
8 D. Chapman, P. Byrne and G.G. Shipley, *Proc. Roy. Soc. (London) Ser. A.*, 290 (1966) 115.
9 P. Byrne and D. Chapman, *Nature*, 202 (1964) 987.
10 D. Chapman and D.T. Collin, *Nature*, 206 (1965) 189.
11 D. Chapman and N.J. Salsbury, *Trans. Faraday Soc.*, 62 (1966) 2607.
12 N.J. Salsbury and D. Chapman, *Biochim. Biophys. Acta.*, 163 (1968) 314.
13 S.G. Whittington and D. Chapman, *Trans. Faraday Soc.*, 62 (1966) 3319.
14 R.M. Williams and D. Chapman, in R.T. Holman (Ed.), *Progress in the Chemistry of Fats and Other Lipids*, Vol. 11, Pergamon, Oxford, 1970, p. 3.
15 V. Luzzati, in D. Chapman (Ed.), *Biological Membranes*, Academic Press, London, 1968, p. 71.
16 D.G. Dervichian, *Mol. Cryst.*, 2 (1967) 55.
17 K. Larsson, *Z. Phys. Chem.*, 56 (1967) 173.
18 V. Luzzati, T. Gulik-Krzywicki and A. Tardieu, *Nature*, 218 (1968) 1031.
19 T. Gulik-Krzywicki, E. Rivas and V. Luzzati, *J. Mol. Biol.*, 27 (1967) 303.
20 D. Chapman, R.M. Williams and B.D. Ladbrooke, *Chem. Phys. Lipids*, 1 (1967) 445.
21 F. Reiss-Husson, *J. Mol. Biol.*, 25 (1967) 363.
22 D.M. Small, *J. Lipid Res.*, 8 (1967) 551.
23 R.S. Bear, K.J. Palmer and F.O. Schmitt, *J. Cell. Comp. Physiol.*, 17 (1941) 355.
24 K.J. Palmer and F.O. Schmitt, *J. Cell. Comp. Physiol.*, 17 (1941) 385.
25 T. Hanai, D.A. Haydon and J. Taylor, *J. Theoret. Biol.*, 9 (1965) 278.
26 D. Papahadjopoulos and N. Miller, *Biochim. Biophys. Acta*, 135 (1967) 624.
27 V. Luzzati and F. Husson, *J. Cell. Biol.*, 12 (1962) 207.
28 H. Hauser, D. Chapman and R.M.C. Dawson, *Biochim. Biophys. Acta*, 183 (1969) 320.
29 J. Urbina and D. Chapman, unpublished observations.

30 B.A. Pethica and P.J. Anderson, *1st Intern. Conf. Biochem. Problems of Lipids*, Paleis der Academien, Brussels, 1953, p. 129.
31 M.C. Phillips and D. Chapman, *Biochim. Biophys. Acta*, 163 (1968) 301.
32 A.D. Bangham and D. Papahadjopoulos, *Biochim. Biophys. Acta*, 126 (1966) 181.
33 E. Rojas, J.Y. Lettvin and W.F. Pickard, *Nature*, 209 (1966) 886.
34 J.B. Leathes, *Lancet*, (1925) 853.
35 N.K. Adam and G. Jessop, *Proc. Roy. Soc. (London) Ser. A*., 120 (1928) 473.
36 M.M. Standish and B.A. Pethica, *Biochim. Biophys. Acta*, 144 (1967) 659.
37 D. Chapman and S.A. Penkett, *Nature*, 211 (1966) 1304.
38 M. Bourgès, D.M. Small and D.G. Dervichian, *Biochim. Biophys. Acta*, 137 (1967) 157.
39 M. Bourgès, D.M. Small and D.G. Dervichian, *Biochim. Biophys. Acta*, 144 (1967) 189.
40 B.D. Ladbrooke, R.M. Williams and D. Chapman, *Biochim. Biophys. Acta*, 150 (1968) 333.
41 E. Oldfield, D. Chapman and W. Derbyshire, *FEBS Letters*, 16 (1971) 102.
42 A.S. Waggoner and L. Stryer, *Proc. Natl. Acad. Sci. U.S.*, 67 (1970) 579.
43 A.D. Bangham, M.M. Standish and J.C. Watkins, *J. Mol. Biol.*, 13 (1965) 238.
44 C.P. White, *Med. Chron.*, 47 (1908) 403.
45 A.D. Bangham, J. de Gier and G.D. Greville, *Chem. Phys. Lipids*, 1 (1967) 225.
46 D. Attwood and L. Saunders, *Biochim. Biophys. Acta*, 98 (1965) 344.
47 D. Chapman, D.J. Fluck, S.A. Penkett and G.G. Shipley, *Biochim. Biophys. Acta*, 163 (1968) 255.
48 N. Robinson, *Trans. Faraday Soc.*, 56 (1960) 1260.
49 L. Saunders, *Biochim. Biophys. Acta*, 125 (1966) 70.

Chapter 7

PHOSPHOLIPIDS AS MODEL MEMBRANES: MONOLAYERS, BILAYERS AND VESICLES

D. PAPAHADJOPOULOS

Department of Experimental Pathology, Roswell Park Memorial Institute, Buffalo, N.Y. (U.S.A.)

It might be that the function of phospholipids as components of biological membranes can be understood only in relation to the other membrane components including proteins, carbohydrates, cholesterol, glycolipids etc. The diversity in chemical composition and functional specialization observed with biological membranes presents a formidable obstacle to assigning a specific functional role to each of the membrane components. Although the mechanism of their participation at the molecular level is still not clear, phospholipids have been implicated in a number of vital membrane functions. These include the phenomena of chemical and electrical excitation, active transport of $Na^{\oplus}$ and $K^{\oplus}$, oxidative phosphorylation, electrical insulation, selective permeability properties, phagocytosis–pinocytosis, and other more general functions relating to membrane stability.

The study of the properties of phospholipids as model membranes can be rationalized as a shortcut approach necessitated by the complexity of biological membranes. The simplicity of the three major model membrane systems (monolayers, bilayers and vesicles) allows for the detailed examination of structure–function relationships and the application of physicochemical techniques whose interpretation would be much more difficult with biological membranes. The volume of recent work on model membranes indicates the attractiveness of the system as an experimental approach to the study of biological membranes. Phospholipid membranes, as will be discussed later in this chapter, exhibit some characteristics which are similar to those of biological membranes, such as water permeability, interfacial tension, electrical capacitance. In contrast to biological membranes, they appear to be virtually impermeable to ions and to other water-soluble compounds. However, their permeability can be increased after interacting with proteins, polypeptides, hormones, drugs, divalent metals, etc. Undoubtedly, the most interesting future research with phospholipid model membranes involves a study of how their permeability properties can be modified by the other membrane components.

Historically, research with phospholipids as model membranes can be separated into two periods. The early period involved mostly work with monolayers at the air–water and oil–water interphase. Although this technique is limited in terms of applicability to transport phenomena, it provides extremely useful information in terms of molecular packing and interactions at an interface. The most recent period of phospholipid model membranes started with the publication (in 1962) by Mueller, Rudin, Tien, and Wescott[1] of the first report on the properties of 'bimolecular lipid membranes' or 'black films'. This system is ideally suited for studies on the electrical parameters of model membranes. During the last few years numerous investigators have studied the basic physicochemical properties of the system as well as its application to biological membrane problems. An alternative system of phospholipid membranes was introduced by Bangham, Standish and Watkins[2]. This involves liquid-crystalline phospholipid particles (liposomes or vesicles) suspended in water and is particularly suited for measurements of diffusion rates. Each of the above model membrane systems offers mutually complementing advantages and often can be used in conjunction.

This chapter deals exclusively with the salient results concerning phospholipid monolayers, bilayers and vesicles and more specifically with their characteristics as models for biological membranes. For a wider coverage of the subject the reader is referred to a number of recent reviews and monographs (Monolayers[3,4], Techniques and theoretical concepts[5,6], Bilayer membranes[7–9], General reviews of phospholipid model membranes[10–16, 186]). Other model membrane systems not reviewed here, include millipore filters impregnated with phospholipids[17] and the three-phase system developed by the late Professor Schulman and colleagues[18].

1. MONOMOLECULAR FILMS (METHODOLOGY AND INTRINSIC PROPERTIES)

Phospholipids are amphipathic molecules of approximately cylindrical shape with the polar group located at one end of the long axis of the cylinder. They can be spread at an air–water interface to form an insoluble monolayer, in which molecules are oriented vertically to the plane of the surface, with the polar groups facing the water phase and the hydrocarbon chains pointing into the air. The characteristics of such monolayers are usually described by experimentally measured properties such as: surface pressure, surface potential and surface viscosity[6]. Surface pressure (π) is defined as the difference between the surface tension of the clean surface of the water (γ_0) and the surface tension of the film-covered surface (γ). The presence of phospholipids, and in general of other surfactants, usually lowers the surface tension of the water and thus surface pressure has a positive value expressed as dynes/cm. The magnitude of π varies with the surface concentration of the monolayer molecules (area per molecule) and it depends on the kinetic energy of the molecules in the film, the cohesion between the hydrocarbon chains and electrostatic interactions between the polar groups and between these and water. Surface pressures can be measured using either the Wilhelmy (vertical plate) or the Langmuir (horizontal barrier) techniques.

Surface potential (ΔV) is defined as the potential difference between the clean water and film-covered surfaces, and usually amounts to a few hundred millivolts. It arises from the components vertical to the surface of the permanent dipoles of the monolayer and the water phase molecules, as well as from the diffuse double-layer potential for charged molecules. The observed value is therefore the result of several partially compensating effects, and the interpretation in terms of specific molecular orientation is somewhat equivocal[6].

ΔV is measured either by the 'ionizing electrode' method (usually ^{226}Ra or ^{241}Am or by the 'vibrating plate' method[6]. In both, the circuit is completed with a calomel electrode and the potential measured with a high impedance electrometer.

Surface viscosity is another useful property, but very little work has been done involving phospholipids. The subject has been discussed in detail in a review by Joly[19] and details on methodology are also given by Gaines[6]. The rheological properties of phospholipid monolayers are an important subject for further study especially in relation to interactions with bivalent metals, proteins, cholesterol, etc. In the absence of an apparatus for measuring surface viscosity, the gross difference between a solid and a liquid film can be established by the simple test of sprinkled talcum powder[6].

The design of the trough for the containment of the monolayer can vary widely depending on the particular purpose for which it is used. A large rectangular trough with a moving barrier for sweeping the surface and compressing the films is best suited for accurate measurements of area per molecule at different values of surface pressure. This is the classical technique in which the phospholipid is spread initially at very large areas per molecule (zero pressure). Under these conditions the large intermolecular distances are favorable for evaporation of the solvent. Smaller, circular troughs have been used more recently by a number of investigators in studies of monolayer interaction with proteins and other biologically important compounds injected in the subphase under the film[20,21]. Here the phospholipid solution is added to the surface in sufficient amounts to obtain the desired surface pressure. Undoubtedly some solvent is trapped when the lipid film is spread on a limited area. However, these troughs provide uniform mixing (obtained easily with a magnetic stirrer) and can be designed for small bulk-phase volumes often necessitated by the scarcity of the biological materials to be studied. However, small rectangular troughs with movable barrier have also been used for such interactions[22,23]. A simple diagrammatic design for a surface trough is shown in Fig. 1.

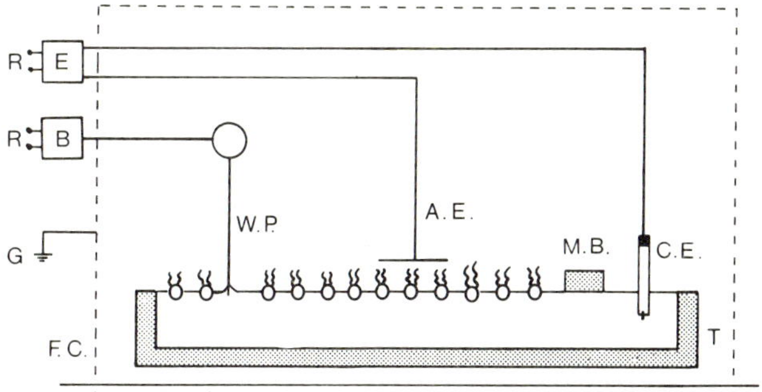

Fig. 1. Schematic representation of a surface balance assembly: T, trough; M.B., moving barrier; W.P., Wilhelmy plate; A.E., air electrode; C.E., calomel electrode; E, high-impedance electrometer; B, null-recording automatic balance; R, recorder; G, earth; F.C., Faraday cage.

Early work with phospholipid monolayers is complicated by the lack of purified preparations. Dervichian (1964) discussed the early results obtained with egg lecithin[24]. However, even purified phospholipids extracted from natural sources contain a variety of fatty acyl chains which complicate the interpretation of the surface properties and auto-

References p. 165

oxidation of the unsaturated components also causes problems. In general, natural phospholipids spread at the air–water interface at ambient temperature behave as liquid-expanded monolayers with a limiting area per molecule of 50–60 Å^2, obtained near the 'collapse pressure' which is approx. 47 dynes/cm. The surface potential of egg PC is 400–450 mV near the collapse pressure[24–26].

The most definitive data on the surface properties of phospholipids have been obtained with synthetic preparations, with defined hydrocarbon acyl chains. 1,2-Distearoyl lecithin at room temperature, low ionic strength and neutral pH, gave isotherms of a condensed film, with a limiting area of approx. 40 Å^2/molecule at high surface pressures (approx. 35 dynes/cm)[27]. This limiting area correlated well with the expected packing of two hydrocarbon chains per molecule (20 Å^2 per chain) found in earlier studies. The introduction of double bonds (by substitution of stearoyl with oleoyl acyl chains) or shorter chain length (by substitution of stearoyl with lauroyl) produced monolayers with expanded film characteristics[28]. Varying the head group (comparison of PA, PE, PS and PC containing the same acyl chains) produced a smaller, but significant change in the isotherms, at least at higher film pressures[28]. Thus for the distearoyl compounds, PA and PE generally reached smaller areas per molecule compared to PS and PC. A schematic diagram of the isotherms obtained with solid and expanded films is shown in Fig. 2.

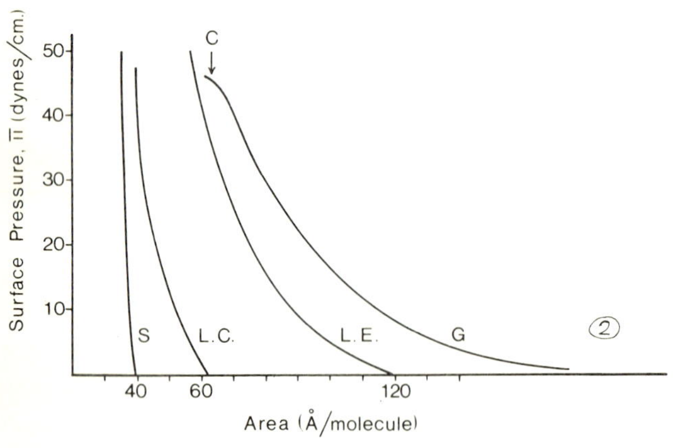

Fig. 2. Typical isotherms for monolayers in different physical states. S, solid; L.C., liquid condensed; L.E., liquid expanded; G, gaseous; C, collapse (point of maximum compression). Curves L.C. and L.E. correspond approximately to the behaviour of distearoyl PC and stearoyl–oleoyl PC, respectively.

In a comparative study of PE containing unsaturated acyl chains with either *cis* or *trans* double bonds, both liquid-condensed and liquid-expanded isotherms with transitions depending on the temperature have been observed[29]. Saturated phospholipid films can exhibit expanded or condensed monolayers depending on the temperature and the length of the acyl chain, a subtraction of two methylene groups being equivalent to a 20° temperature increase[30, 31]. Intermediate states with a phase transition along the isotherm were also observed. Thus at 25°, dipalmitoyl PC shows an intermediate region while PC having longer straight chains would be fully condensed and PC with shorter chains would be (liquid) expanded[4]. Such transitions correspond well with the hydrocarbon chain melting for the transition from crystalline to liquid-crystalline state[32] observed with Differential Thermal Analysis (DTA).

The importance of the liquid-expanded state to the integrity of biological membranes is indicated by the following observations: Yeast cells grown anaerobically cannot synthesize unsaturated fatty acids. Under these conditions the phospholipids contain fatty acids of shorter chain length ($C_{10:0}$, $C_{12:0}$ and $C_{14:0}$) at the 2-ester position, where unsaturated fatty acids are usually located[33]. Comparison of the data on the isotherms of different lecithins[28] indicates that $C_{18:0}$, $C_{12:0}$–PC gives an expanded film, similar to that given by the $C_{18:0}$, $C_{18:1}$–PC. Films of PE containing a high percentage of cyclopropane fatty acids (isolated from *E. coli*) are also expanded[34], and very similar to films composed of $C_{18:0}$, $C_{18:1}$–PE. The effects of shorter chains and double bonds in relation to areas per molecule are shown in Table 1.

TABLE 1

CROSS SECTIONAL AREAS OF SYNTHETIC LECITHINS AT THE AIR–WATER INTERFACE. PRESSURE ARBITRARILY CHOSEN AT 21 DYNES/cm

Fatty acyl groups		*Molecular cross sectional area in Å²*
Saturated		
C_{24}:0	C_{24}:0	40
C_{18}:0	C_{18}:0	40
C_{14}:0	C_{14}:0	60
C_{10}:0	C_{10}:0	75
Unsaturated		
C_{18}:0	C_{18}:1	65
C_{18}:1	C_{18}:1	75
C_{18}:0	C_{18}:2	80
C_{18}:2	C_{18}:2	90
C_{10}:0	C_{18}:3	85

Mycoplasma laidlawii supplies another example of the importance of the liquid-expanded state of the lipids on morphology and osmotic fragility of living cells. These microorganisms can incorporate into their membrane lipids, large amounts of saturated fatty acids (up to 80%) present in the growth medium[35]. Under these conditions the cells

References p. 165

change from filamentous to rounded swollen structures, and, depending on the particular fatty acid accumulated and the growth temperature, they cease to grow and lyse. The onset of these events can be correlated with the phase transition (observed by Differential Scanning Calorimetry; DSC) of the membrane lipids from liquid-crystalline to solid[36] as well a the phase-transitions observed by DSC with living *Mycoplasma laidlawii* cells[37]. Finally it should be noted that the function of dipalmitoyl PC as a component of lung surfactant could be related with the observed 'intermediate regions' between expanded and condensed films of this lipid at ambient temperatures[4].

2. BILAYER FILMS (METHODOLOGY AND INTRINSIC PROPERTIES)

The formation of bilayer membranes is a process analogous to that for black soap films[179] and to the usual soap bubbles formed from an aqueous soap solution in air. However, the bilayer membranes are formed within an aqueous phase from a solution in an organic solvent. The orientation of the phospholipid molecules would thus be opposite to the soap films, with the hydrocarbon chains in the middle and the polar groups facing the water. The first such membranes (black lipid films) were formed by applying the membrane-forming solution (brain lipids dissolved in *n*-tetradecane and α-tocopherol) with a small brush over a circular hole (0.5–5 mm^2) in a polyethylene septum separating two identical salt solutions[1]. More recent work has indicated that the only prerequisites for a membrane-forming solution, are the presence of an amphipathic compound (such as PC) and a neutral hydrocarbon solvent such as *n*-decane[38].

Following the application of the membrane-forming solution, a thick film occupies the area of the 'hole' which can be visualized by the interference colors under low magnification. Within a few minutes the film thins to an optically 'black' membrane which occupies most of the area of the hole, surrounded by a torus of the membrane-forming solution in contact with the plastic septum. The thickness of the resulting membranes as judged by capacitance and optical reflectance measurements is compatible with a bilayer of the amphipathic molecules present. A schematic drawing of a simple experimental setup and the gross morphology of a bilayer film is shown in Fig. 3. The formation of thin lipid films of bimolecular thickness have been considered thermodynamically[39,40]. Bilayer membranes can also be formed without a plastic frame in the form of spheres suspended in a sucrose gradient[41]. Most of the surface of such spheres is made up of the thin lipid films with a lens of the membrane-forming solution situated as a cap on the upper area. The methodology of bilayer membranes, the variety of compounds used for their formation and their physicochemical properties have been reviewed[11,7,15].

Theoretical considerations of the thinning process show that, as the hydrocarbon solvent drains from the membrane, the minimum thickness of a stable film will be approximately twice the length of the solute chains[42,43]. This is in agreement with the experimentally determined thickness of thin lipid films composed of a series of amphipathic molecules of different chain length by electrical capacitance measurements[44]. For egg PC the estimated[38] thickness is 48 ± 1 Å, a value which represents the average thickness of the hydrocarbon region. Estimates of thickness obtained by the method of light reflectance

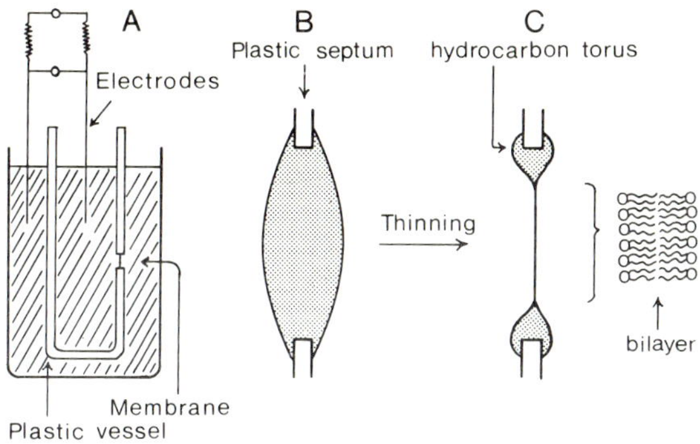

Fig. 3. Schematic representation of the formation of a bilayer membrane.

A, A simple experimental design showing a plastic vessel with a circular hole, held inside an external glass vessel. The bilayer membrane covers the hole and separates the inside from the outside aqueous phase.

B, Illustration of a drop of the membrane-forming solution covering the hole in the plastic septum before thinning.

C, Illustration of final bilayer membrane showing the hydrocarbon torus at the periphery of the hole, held on the plastic septum.

include both the hydrocarbon and the two polar group regions[7,15]. The thicknesses for PC membranes obtained by the latter technique vary from 62 ± 2 Å to 77 ± 10 Å. However, the lower value has been obtained more recently by consideration of the optical anisotropy of the bilayer membrane[45].

Despite the use of pure phospholipids and the experimentally determined thickness indicating a bilayer structure, the composition of these membranes is complicated by the presence of considerable amounts of hydrocarbon solvent[46]. However, it is possible that the residual solvent is not uniformly distributed within the membrane but rather present in small islands or lenses in equilibrium with the bilayer membrane. Occasional areas of considerably larger thickness have been observed by electron microscopy of fixed bilayer membranes[47]. The rest of the membrane appears as a trilamellar structure with a thickness of 60 ± 20 Å.

One of the most interesting properties of phospholipid bilayer membranes, at least from a historical point of view, is their low interfacial tension. Values have been reported by several groups and there is a general agreement for a value of approx. 1 dyne/cm for egg PC. Lower values have been obtained for membranes containing cholesterol. These results are within the range of the interfacial tension found with biological cells[48]. The surface tension at the clean oil–water interface is much higher, approx. 50 dynes/cm for water:*n*-octane[5]. However, the interfacial tension of the membrane-forming solution is usually quite low, ranging from 0.72 dynes/cm for PC-cholesterol-methyl oleate[49] to 3.72 dynes/cm for glycerol monoleate-*n*-decane[50]. These low values indicate that the oil–water interface is stabilized by the polar groups of the phospholipid molecules. This property of phospholipids was not known and consequently not taken into consideration in the early proposal[51] that proteins are needed in order to stabilize the oil–water interface of biological membranes. Thermodynamic considerations indicate that the surface energy of a bilayer must be somewhat less than the total surface energy of the two phospholipid monolayers. The difference between these two values is the free energy of formation for the bilayer and has been calculated as 0.00394 erg/cm^2 for glycerol monoleate[50] and 1.0–3.4 erg/cm^2 for PC-cholesterol membranes[49, 52]. This energy difference (free energy of formation) is presumably due mostly to the Van der Waals interactions between the molecules of the two monolayers.

The electrical capacitance of bilayer membranes was one of the first parameters to be determined and has since been studied by several investigators[7, 8]. Reported values range from 0.33 to 1.3 μF/cm^2 which compare well with a range of 0.5 to 1.3 μF/cm^2 obtained from different cell membranes[53]. Most of the values for bilayers, however, fall at the low end of the range, and a careful study[54] of the capacitance of PC membranes indicates a value of 0.38 ± 0.01 μF/cm^2. It was concluded that the capacitance is mostly contributed by the thin hydrocarbon layer, with negligible contribution from the head-groups. In agreement with this conclusion, it has been shown that neutral PC and acidic PS exhibit identical capacitance[55].

The great disparity between the phospholipid bilayer membranes and biological membranes is indicated by the former's very high electrical d.c. resistance. Thus, the reported values for bilayers[7] range from 10^6–10^9 $\Omega \cdot$ cm^2 while those for biological membranes[53] range from 10^2–10^5 $\Omega \cdot$ cm^2. However, it should be noted that the lower resistance values (10^5–10^6 $\Omega \cdot$ cm^2) are consistently obtained from bilayers formed from solutions containing chloroform-methanol[15]. This phenomenon appears to be related to the effect of general anaesthetics on the ionic permeability of phospholipid vesicles[57, 58, 140] Conduction across 'border leaks' may also contribute substantially to the measured resistance It can therefore be concluded that the true electrical resistance of phospholipid membranes[54] is more than 10^8 $\Omega \cdot$ cm^2. Such high electrical resistance is compatible with the very low permeability to cations observed with phospholipid vesicles[57, 58, 140].

In conclusion it should be noted that many of the properties of bilayer phospholipid membranes (electrical capacitance, permeability to water, overall thickness, trilamellar structure under electron microscopy), are similar to those of biological membranes. In contrast, bilayers seem to be rather impermeable to ions and other water-soluble neutral molecules. However, their permeability can be increased by the introduction of peptides, drugs, proteins etc. and the resulting modified phospholipid membranes can offer biologically important models for membrane function, as will be discussed later. Bilayer membranes are ideally suited for electrical measurements. However, due to the relatively small surface area and their relative instability they present difficulties for measuring diffusion rates of slowly diffusing solutes.

3. PHOSPHOLIPID VESICLES (LIPOSOMES)

When dried preparations of naturally occurring phospholipids are allowed to come in contact with an aqueous solution, they interact with water to form a variety of liquid-crystalline structures. These spontaneous molecular rearrangements are analogous to the lyotropic mesomorphism of soaps of long-chain aliphatic and other amphipathic molecules.

The particular liquid-crystalline phase assumed, depends on the lipid–water ratio and the temperature. Several recent reviews[24, 11, 59, 32] have dealt with the detailed physicochemical properties and liquid-crystalline structures of phospholipids in water. It is clear from these studies that when phospholipids are equilibrated with relatively large amounts of water (more than 50% of the dry weight), they assume a unique structure of lamellar configuration, which is usually referred to as 'smectic mesophase'. This structure can be visualized as stacked lamellae of phospholipid molecules in a bilayer configuration, with the water taking up the space between the lamellae and interacting with the polar groups of phospholipids. The tendency of phospholipids to assume this hydrated lamellar configuration is related to the transition temperature (melting of hydrocarbon chains). Thus, natural phospholipids, which are usually above this transition at ambient temperatures, will form such structures spontaneously. Synthetic phospholipids with saturated acyl chains will also form similar structures, if the temperature is raised close to the transition point (the major endothermic peak observed by DTA) of the dry compounds[60].

The lamellar configuration of phospholipids dispersed in excess water was visualized by electron microscopy[61, 62] and X-ray diffraction[63] long before the description of the system as model membranes[2]. The concentric lamellae seen under the electron microscope (Fig. 4) were usually referred to as 'myelin figures'. Bangham and colleagues[2] demonstrated experimentally that each lamella is formed of a continuous permeability barrier. Thus the water space inside the multilamellar particles constitutes a separate aqueous phase and the diffusion of trapped molecules or ions depends on the permeability properties of the phospholipid bilayers which separate the inside aqueous space from the suspending medium.

The original preparation of multilamellar vesicles[2] was made by simple mechanical shaking of a dry phospholipid sample with a salt solution in a round-bottom flask. This method produces a heterogeneous mixture of particles of varying size. The efflux rates of the incorporated isotopic tracers can be determined by dialysis or passage through a gel-filtration column. Alternatively the system can be adapted for the use of ion-specific electrodes or a spectrophotometric assay[64]. The original phospholipid used with this method was egg PC, but most of the phospholipids studied have been found to produce similar closed multilamellar vesicles[16, 65, 73]. Cholesterol can be incorporated up to a molar ratio of approx. 1:1.

Exposure of the coarse phospholipid suspensions to ultrasonic radiation tends to break the large multilamellar particles into smaller ones with fewer lamellae and eventually produces a homogeneous population of unilamellar vesicles (Fig. 5) of 200–500 Å diameter[65, 66]. Light scattering[67] and sedimentation velocity studies[66] of ultrasonicated aqueous dispersions of egg PC indicate an aggregate weight of $1.5–2.1 \cdot 10^6$. This corresponds to approx. $1.9–2.7 \cdot 10^3$ molecules per particle, and is in general agreement with the electron microscopic pictures of the vesicles. The minimum energy configuration suggested for smectic films[68] is a biconcave prolate or oblate spheroid. The same conclusion concerning the shape of the vesicles was also derived from consideration of the internal volume per vesicle by disaccharide capture[69]. It has also been concluded on the basis of paramagnetic studies with spin-labelled PC that approx. 65% of the phospholipid molecules in a vesicle are present in the external monolayer[69].

References p. 165

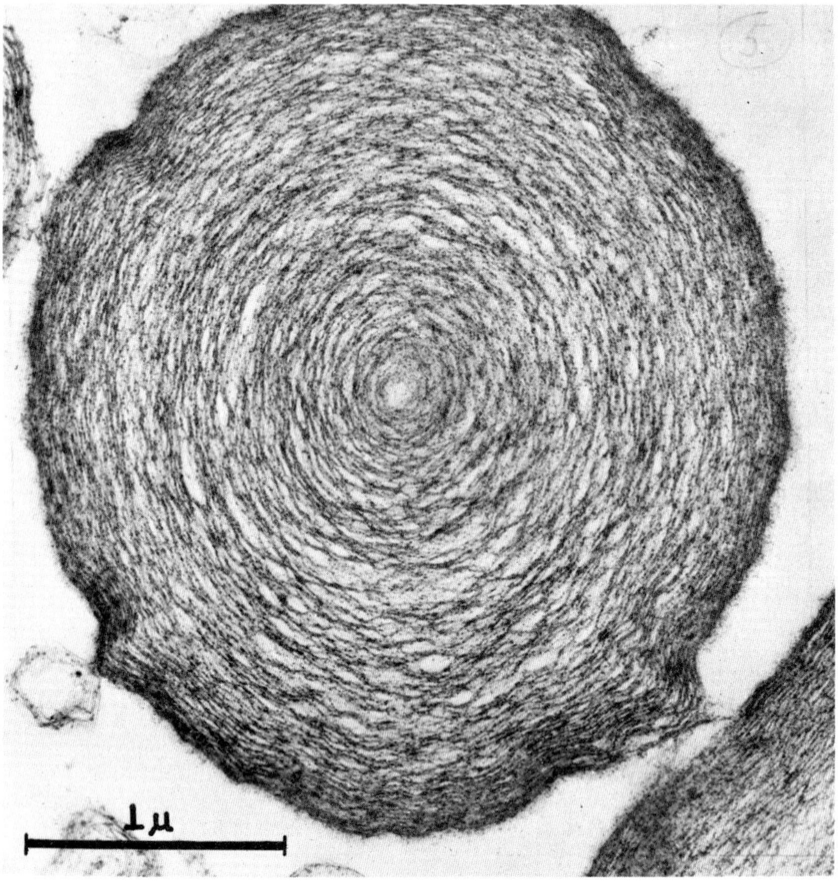

Fig. 4. Electron micrograph of a multilamellar phospholipid vesicle. The phospholipid (bovine heart cardiolipin) was dispersed by shaking in 10 m*M* NaCl solution at pH 7.4. Fixed in 10% OsO_4, dehydrated in acetone and embedded in Epon A12. Magnification: marker, 10^4 Å (1.0 μ). The preparation was made in this laboratory by Dr. H. Kimelberg and the electron micrograph by Mr. J.J. Black. Distortions in the contour of each lamella and the distance between lamellae could be the result of fixation and dehydration. From ref. 16.

Phospholipids containing unsaturated double bonds are quite susceptible to autoxidation, particularly during sonication[70]. Oxidation can have profound effects on the permeability properties and an atmosphere of nitrogen or argon should be used during the preparation and sonication of phospholipids. The use of a bath-type sonicator with controlled temperature[71] tends to minimize the thermal degradations likely to be produced by the probe-type sonicators (Fig. 6). Sonication of acidic phospholipids in 0.1 *N* NaCl produces populations of small vesicles which chromatograph on sepharose 4B as a single peak with constant captured volume[140] On the other hand PC, which is zwitterionic, is more resistant to sonication and shows a more complex pattern on passage through sepharose[66, 69]. The time period for complete sonication of PC was found to increase with increasing salt concentration[67].

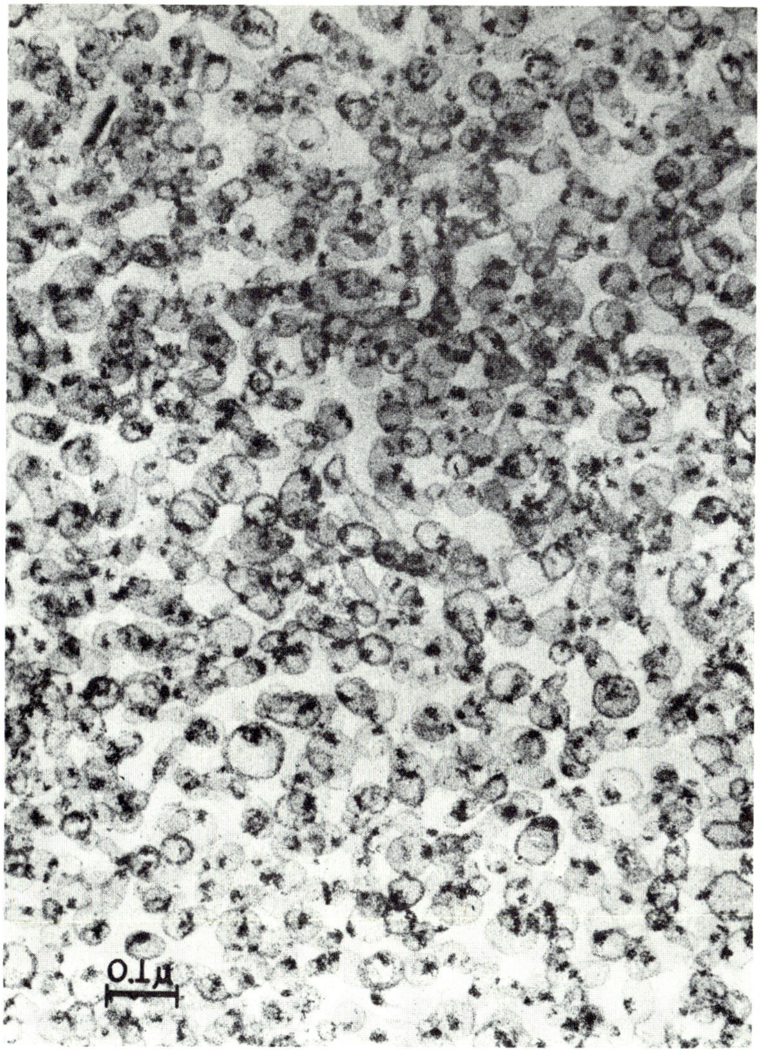

Fig. 5. Electron micrograph of unilamellar phospholipid vesicles. The phospholipid was beef heart cardiolipin and was dispersed in 10 m*M* NaCl at pH 7.4 by ultrasonication at room temperature under nitrogen for 30 min in a bath-type sonicator. Other details as in Fig. 5. Magnification: marker, 100 nm (0.1 μ). From ref. 16.

The thickness of the lipid layer in a fully hydrated smectic mesophase varies with lipids of different chain length. A value of 40 Å was calculated by X-ray diffraction for human brain lipids and also for egg-yolk PC[59]. The thickness of the diglyceride portion in egg PC liquid crystals was calculated[72] as 30 Å. The area occupied by each molecule in the above system of fully hydrated PC was found to be approx. 72 $Å^2$. This corresponds with the area per molecule of PC on monomolecular films when compressed to 20–25 dynes/cm[25, 26]. The thickness of the water compartment between lipid lamellae varies depending on the surface charge. Of all phospholipids tested only PE was found not to form

References p. 165

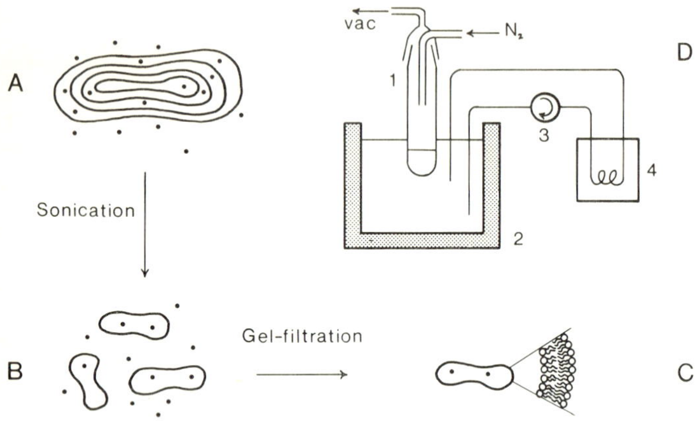

Fig. 6. Schematic representation of the formation of unilamellar phospholipid vesicles by ultrasonication.

A, multilamellar vesicle obtained by shaking dry phospholipid in aqueous solution. Black dots indicate the partial entrapment of a solute within the aqueous interlamellar spaces.

B, Formation of unilamellar vesicles after exposure to ultrasonic waves. Solute molecules still inside and outside the vesicles.

C, Vesicles after passage through a gel-filtration column. Inset indicates the bilayer structure. The non-trapped solute molecules have been removed from the bulk-phase during the passage through the column. Vesicles are now ready for determination of diffusion rates.

D, Diagram of sonication procedure in a bath-type sonicator. The phospholipid sample is evaporated to dryness in glass tube (*1*) under vacuum. The initial coarse suspension is made after the addition of the aqueous salt solution, by shaking under nitrogen. The tube is then positioned in the sonicating bath (*2*). The temperature of the bath is controlled by circulating cold water (*4*) through a peristaltic pump (*3*).

closed vesicles when suspended in salt solutions at neutral pH[65, 73]. This behavior could be related to the formation of salt linkages between vicinal amino and phosphate groups[74], which would tend to reduce the water of hydration around the polar groups. The formation of hexagonal patterns has also been reported for PE[75, 76]. However, PE tends to form swollen closed particles when suspended in low ionic strength or high pH, or in mixtures with PC[65, 73].

Phospholipid vesicles are osmotically sensitive to salts and various sugars. They tend to swell and rupture in hypotonic solutions and shrink in hypertonic solutions, and these properties can be used for measurements of the water permeability coefficient[77, 78]. Other methods and techniques involving phospholipid vesicles are summarized in a recent review[16].

In conclusion, it should be emphasized that phospholipid vesicles (or liposomes) have certain advantages and disadvantages when compared to the bilayer membranes (black lipid films). One of the advantages is the absence of extraneous hydrocarbon or other materials needed for the formation of black films. Thus, the membrane composition can be precisely and easily defined. In addition the surface area available for diffusion measurements is greater by orders of magnitude. Finally, the vesicles are very stable and amenable to study by various biochemical and biophysical methods including centrifugation, gel filtration, electrophoresis, spectrophotometry, fluorescence, nuclear magnetic resonance, electron spin resonance and calorimetric techniques. The disadvantages derive from the fact that it is technically difficult if not impossible to perform electrical measurements, and from uncertainties concerning the surface area and geometry of the multilamellar vesicles (liposomes). However, the use of sonicated preparations with fairly uniform size distribution of unilamellar vesicles has largely eliminated the uncertainty concerning surface area and geometry. Values for fluxes and permeability coefficients can thus be calculated without large errors.

4. PERMEABILITY PROPERTIES AND INTERACTIONS WITH IONS

Although monomolecular films have been used for permeability studies[79] this technique is limited to volatile compounds. The high electrical resistance of bilayer membranes (in the range of $10^8\ \Omega \cdot cm^2$) indicates a very low permeability to ions[15]. This has been verified by measurements of the diffusion rates for monovalent ions through phospholipid vesicles. Based on a conservative estimate of two lamellae per vesicle the diffusion rate for $Na^{\oplus}$ or $K^{\oplus}$ through PC was calculated as $4 \cdot 10^{-16}$ equiv./cm^2/sec or $2.8 \cdot 10^{-12}$ cm/sec and the electrical resistance for these ions as $5.8 \cdot 10^8\ \Omega \cdot cm^2$ (refs. 73, 58). The above figures are probably minimal values and more recently the permeability coefficient of $K^{\oplus}$ through vesicles containing 4% PA in PC was calculated as $0.3 \cdot 10^{-12}$ cm/sec[57, 140]. Rates for isotopic diffusion of $Cl^{\ominus}$ through phospholipid vesicles were much faster than for $Na^{\oplus}$ or $K^{\oplus}$ (ref. 2) but the ratio of $Cl^{\ominus}/K^{\oplus}$ varies for different phospholipids from 110 for 10% stearylamine in PC to 0.5 for PI[73].

Determinations of ionic transference numbers with bilayer membranes composed of either mixtures of phospholipids[80] or purified neutral (PC and PE) or acidic (PS) phospholipids[8] indicate that these membranes are cation perm-selective. The result is contrary to the higher anionic diffusion rates observed with vesicles. However, more recent studies with spherical bilayers of large surface area[41] have resolved the discrepancy by permitting a direct comparison of the isotopic diffusional rates with the transference numbers conductimetrically. It was thus established that although the observed $Na^{\oplus}$ isotopic flux can account for the calculated values of conductance, the $Cl^{\ominus}$ isotopic flux was larger by a factor of 10^3. The authors concluded that the $Cl^{\ominus}$ diffusion does not contribute appreciably to the steady-state electric current and proposed a carrier-type mechanism involving PC–heavy-metal complex which facilitates the $Cl^{\ominus}$ diffusion across the bilayer. The rates of inversion of orientation of molecules in PC vesicles, determined recently by electron spin resonance[69] are too fast compared to the rate of cation transport, but could account for the $Cl^{\ominus}$ diffusion rates.

After considering the repulsive forces generated by the presence of ionized headgroups, it has been suggested[81] that ionized phospholipids would tend to disrupt the bimolecular leaflet in favor of aggregates (micelles) of radially oriented molecules. However, recent studies with membranes composed of acidic phospholipids both as bilayers and vesicles at physiological ionic strength, have indicated high electrical resistance and very low permeability to ions[15, 82, 16, 140] in spite of one negative charge per molecule or per 60–70 Å^2. In agreement with these experimental findings, Ohki and Aono[83] have recently calculated the energy requirement for a bilayer–micelle transformation and concluded that even for phospholipids carrying one charge per molecule, the most stable configuration (in 0.1 *N* NaCl) is that of a bilayer. Higher surface charge (1.25–2.0 charges per molecule) would favor the formation of 'cylindrical aqueous pore' type of configuration.

The early studies involving vesicles and bilayer membranes indicated no appreciable discrimination between $Na^{\oplus}$ and $K^{\oplus}$ although the discrimination between anions and cations was considerable[2,73,80]. Since most cell membranes exhibit considerable discrimination between $Na^{\oplus}$ and $K^{\oplus}$, it was concluded that the property of monovalent cation selectivity does not reside in the membrane lipids. However, it is well known that monolayers of fatty acid soaps exhibit considerable specificity for monovalent cations[84]. Furthermore, Eisenman[85] predicted on the basis of carboxyl and phosphate group field strengths that $Na^{\oplus}$–$K^{\oplus}$ discrimination in biological membranes could depend on such groups. A recent re-evaluation of the subject with vesicles composed of purified phospholipids revealed that certain acidic phospholipids (such as PS and PG) do indeed discriminate between $Na^{\oplus}$ and $K^{\oplus}$ (ref. 86) with a maximal $K^{\oplus}/Na^{\oplus}$ ratio for diffusion rates of approx. 10. The same phospholipids also give maximal activation to $K^{\oplus}$–$Na^{\oplus}$ dependent ATPase isolated from rabbit kidney and partially freed from lipid[182]. It thus appears possible that the selectivity of biological membranes for $Na^{\oplus}$ and $K^{\oplus}$ is at least partly a function of the membrane phospholipids, although the mechanism is not at present clear. Other compounds such as the antibiotic valinomycin[87] confer high cation selectivity ratios to normally non-selective phospholipid membranes. Long-chain alcohols also increase the $K^{\oplus}/Na^{\oplus}$ selectivity ratio[88], but such compounds are not normally components of biological membranes.

Bivalent and multivalent metal ions have profound effects on the properties of phospholipid membranes. The early work on the binding of metal ions on aqueous phospholipid suspensions was based on charge reversal and flocculation[89]. However, the lack of purified preparations renders the interpretation difficult in terms of specific interactions. Binding has also been followed by measurements of the electrophoretic mobility of purified phospholipids dispersed in water[90,11,26]. By the use of monomolecular films it was established that the binding of metal ions generally increases the surface potential and decreases the surface pressure[27,26,25,20]. The order of binding efficiency on distearoyl PC was $UO_2^{2\oplus} \gg Al^{3\oplus} \gg Cs^{\oplus} > Cu^{2\oplus} \geqslant Mg^{2\oplus} \geqslant Ca^{2\oplus} > Na^{\oplus}$. As could be expected, anionic phospholipids bind with much higher affinity and there is some evidence that bivalent and monovalent metals do not bind appreciably to purified PC. Calcium binding on anionic phospholipid monolayers has also been followed by the method of surface radioactivity which has shown that $Ca^{2\oplus}$ can be displaced partially by sodium or potassium[90,91]. The affinity constants of PS and PA for $Ca^{2\oplus}$ have been determined by titrimetric techniques[92,93], and are of the order 10^4. This value is similar to that for ATP–$Ca^{2\oplus}$ complex and a polymeric chelation arrangement has been proposed to account for some of the effects of $Ca^{2\oplus}$ on phospholipid membranes[26]. The ionization of several phospholipids has been followed by pH titrations of monomolecular films, following changes in π and ΔV[26].

The effects of metal-ion binding in terms of permeability of membranes have been observed with both vesicles and bilayers. Ferric ions have been shown[94] to lower the resistance of PC bilayers by a factor of 10^3 at concentrations below 10^{-5} *M*. Under the influence of ferric and thorium ions, bilayer membranes switch from cation to anion permselectivity[88]. However, other metal ions such as $Cd^{2\oplus}$ and $Cu^{2\oplus}$ appear to increase the resistance of PC membranes. The effect of $Ca^{2\oplus}$ is of special interest, because, of its involvement in the excitation phenomena of nerve membranes. By the use of both vesicles and bilayer membranes it was established that calcium ions, present on both sides of a membrane composed of pure acidic phospholipids (PS), produce an increase in resistance. In contrast, the addition of small amounts of $Ca^{2\oplus}$ (1 m*M* $CaCl_2$ in 100 m*M* NaCl) only to one side of the membrane produces a decrease in resistance. If the concentration of $Ca^{2\oplus}$ on one side is increased above a threshold level (1 to 5 m*M* depending on the pH) the membranes become unstable and burst[95]. Similar instability has been observed when there is a pH gradient across PS membranes. However, it is possible to form stable PS membranes with $Ca^{2\oplus}$ present only on one side, if the pH of the other side is lowered to pH 3.0. None of these phenomena is observed with membranes composed of pure PC or with mixtures of up to 50% PS in PC. The above results are reminiscent of the 'asymmetric' effect of $Ca^{2\oplus}$ on biological membranes[96,97], and it has been proposed that the PS–$Ca^{2\oplus}$ complex and the resulting stable vs. unstable membrane could be implicated in the excitation phenomena of nerve membranes[98]. Other models involving phospholipids and phase transitions within the membrane have also been proposed to account for the electrochemical events observed during the generation of an action potential across a nerve membrane[99–101].

5. IONOPHORES AND EXCITABILITY

The most exciting results with model membranes were curiously the first to be reported. This was the discovery by Mueller and colleagues that lipid bilayer membranes could be made electrically excitable[1,102]. The phenomenon was observed following the addition to the aqueous phase bathing the membrane of a substance isolated from the culture-fluid of *Aerobacter cloacae*. This substance, usually referred to as EIM, has been partially purified and shown to be a complex of a basic protein and ribonucleic acid; both components appear to be essential for activity[103]. The addition of EIM lowers the electrical resistance of the lipid membranes by a factor of 10^5 and makes them cation-selective. For small steps of current the membranes behave as a passive resistance–capacitance system, but above a threshold value of imposed current, the resistance rises to an appreciably higher level and falls to the original level upon removal of current. Mueller and Rudin[102] describe these properties as an 'active-tunnel diode' analogue. Addition of protamine plus EIM produces characteristic action potentials and also repetitive firing. The molecular details of these phenomena which are interpreted as a field-dependent gating system are as yet not well understood, although they are analogous to the characteristics of some electrically excitable biological membranes. More recently alamethicin, a cyclic peptide of 18 amino acids and known sequence has been shown to produce similar electrical phenomena[104].

Phospholipid membranes with low electrical resistance and high selectivity between cations or anions can also be produced by the addition of small amounts of a variety of compounds which are generally referred to as ion-carriers or 'ionophores'. Most of these compounds exhibit antibiotic activity and have large ring structures[183]. They fall into several broad categories. The depsipeptides such as valinomycin, enniatins, gramicidins; the

References p. 165

macrotetralide actins such as nonactin, monactin etc.; the cyclic polyethers; the polyene cyclic lactones such as nystatin, amphotericin B, filipin; others are monocarboxylic polyethers such as nigericin, monensin, and large cyclic peptides such as alamethicin. The action of some of these antibiotics as ion-carriers was first observed with mitochondria[105] and later with erythrocyte membranes, vesicles and bilayers. The addition of valinomycin at concentrations as low as 10^{-10} to 10^{-7} g/ml to the aqueous phase of bilayer membranes has been shown to produce a selectivity of 300 to 1 for $K^{\oplus}$ over $Na^{\oplus}$. The permeability series for different cations[87] was $H^{\oplus} > Rb^{\oplus} > K^{\oplus} > Na^{\oplus} \simeq Li^{\oplus}$.

Studies of solutions of some of these compounds in organic solvents have demonstrated stoichiometric complexes with $K^{\oplus}$ or other metals. The conformation of some of the cation complexes has been deduced by ORD, NMR and X-ray diffraction studies. As an example, the conformation of valinomycin[106] indicates a large ring structure with most carbonyl oxygens directed inward while the periphery is covered by non-polar groups. It is thus apparent that a small metal ion can coordinate with the polar groups lining the inner cavity, while the whole complex has considerable solubility in an oil phase or the interior of a phospholipid membrane. Such an arrangement is thus suitable for the transport of metal ions through a medium of low dielectric. There is considerable evidence that the mechanism of cation transport for many of the ionophores such as valinomycin[107] and the family of macrotetralide actins[108] involves a lipophilic mobile carrier, with the ionophore-cation complex shuttling back and forth through the membrane.

The ion conduction induced by several other antibiotics appear to follow an alternative mechanism, perhaps involving the formation of channels or pores across the membranes. The inducement of discreet fluctuations of conductance in black lipid films by gramicidin (also observed with EIM[109]) has led Hladky and Haydon[110] to conclude that the drop of resistance is due to the formation of pores with high conductivity involving more than one molecule of the ionophore. Similar conclusions were arrived at earlier, concerning nystatin and amphotericin B, which induce anion transport. Amphotericin B produces a relatively large increase in the osmotic water permeability compared to the permeability coefficient for the diffusion of water through bilayer membranes[111], suggesting that the antibiotic induced the formation of aqueous pores in the range 7 to 10.5 Å. Bilayer membrane resistance varies with the 10th power of nystatin concentration indicating that perhaps 10 molecules of the compound form a pore through the bilayer membrane[112].

The sensitivity of *Mycoplasma laidlawii* to the polyene antibiotics depends on the presence of sterols[113]. The same sensitivity has been observed with all three membrane models. Filipin and nystatin penetrate monolayer films[114] and produce marked instability in bilayer membranes[115] only in the presence of cholesterol. Later studies on the electrical resistance of bilayers[116] and the permeability of liposomes[117] have confirmed and extended these observations on the presence of cholesterol and polyene antibiotic activity. Other studies with monolayers have shown that several of the cyclic polypeptides form insoluble films at the air-water interphase[118]. The presence of $K^{\oplus}$ in the subphase has a spectacular effect on the surface potential of valinomycin films (namely an increase of 500 mV compared to the same system with $Na^{\oplus}$)[119].

Other interesting aspects of the cyclic peptide antibiotics have been seen from studies with phospholipid vesicles. Alamethicin in a ratio of 1 to 600 molecules of PS suppresses completely the NMR signal of the phospholipid hydrocarbon chains, indicating reduced molecular motion[120]. The large number of phospholipid molecules affected by alamethicin indicates a co-operative phenomenon. Studies on the action of valinomycin on the $K^{\oplus}$ permeability of phospholipid vesicles indicated a molecular ratio of 10^6 lipid molecules per valinomycin molecule[57]. This implies the presence of only a few (and perhaps one) valinomycin molecule per phospholipid vesicle. More importantly, valinomycin induces, under these conditions, a 100-fold increase in $K^{\oplus}$ permeability, without an appreciable effect on the enthalpy of activation for the process ($\Delta H^* = 15.2$ kcal/mole). In order to account for the increase in permeability without concomitant charge in ΔH^*, it has been suggested[121] that an increase in entropy occurs, $\Delta S^* = 35$ cal/degree.

Uncouplers of mitochondrial oxidative phosphorylation have also been examined with bilayer membranes.2,4-Dinitrophenol (DNP) lowers the electrical resistance of PC membranes by a value sufficient to account for its effect in uncoupling mitochondria[122]. A variety of uncouplers decrease bilayer resistance with a dependence on the bulk pH characteristic for each compound[123—125]. Diffusion potential measurements indicate that the transported ion in the presence of most uncouplers is $H^{\oplus}$ or $OH^{\ominus}$ or both, evidence supporting the chemiosmotic hypothesis[126] for oxidative phosphorylation.

6. INFLUENCE OF ACYL CHAIN LENGTH, UNSATURATION AND CHOLESTEROL ON PERMEABILITY

The influence of chain length and unsaturation on the permeability of phospholipid membranes was studied recently by De Gier et al.[162] with several preparations of synthetic PC containing acyl chains of varying length and number of double bonds, and using the rate of swelling of phospholipid vesicles as a measure of permeability[77]. The presence of short chains (C-10 instead of C-18) and also double bonds, markedly increased the permeability of PC vesicles. An interesting recent report[128] emphasizes the role of the 'essential' fatty acids, linoleic and linolenic on the permeability of phospholipid vesicles composed of 2% dicetyl phosphate in different PCs isolated from several sources including 'essential' fatty acids, linoleic and linolenic, on the permeability of phospholipid vesicles diffusion could be correlated with the linoleic plus linolenic content of the different PC preparations, but not with total unsaturation. More recently, it has been reported that the unsaturation of the membrane lipids in *E. coli* K12 has an influence on the transport rates for β-glycosides in intact cells[181]. The effect of unsaturation and chain length has not been studied extensively with bilayer phospholipid membranes. One of the few existing reports[129] indicates that synthetic (dimyristoyl and dipalmitoyl) PC do not form stable membranes at 36°. More recently it has been reported that stable bilayer membranes with high electrical resistance can be formed with saturated branched-chain (diphytanoyl) PC[130].

Cholesterol is present in appreciable amounts in most plasma membranes, although not in mitochondria and bacteria. The phospholipid to cholesterol molar ratio varies considerably but in brain myelin and erythrocytes of several species, the ratio is close to

References p. 165

unity. The role of cholesterol in biological membranes is not known although recent evidence suggests that it is important in relation to the permeability and osmotic fragility of red cell membranes[131]. Evidence obtained from all three membrane model systems indicates that cholesterol has a strong influence on the packing of phospholipid molecules and also on the permeability properties of phospholipid membranes and these results can be extrapolated to biological membranes. The effect of cholesterol on the physical properties of phospholipid monolayers and liquid crystals has recently been reviewed[32] (see also Chapter 6, p. 136).

In monomolecular films the presence of cholesterol reduces the area occupied by each phospholipid molecule[132, 28]. However, the condensation effect varies with the liquidity of the PC films and is apparent only with PC films which are neither fully condensed nor fully expanded[32, 133]. More recently it was reported[134] that cholesterol lowers the viscosity of PC monolayers. The mechanism for the condensation effect can be explained in terms of a phase-transition involving the more expanded component (phospholipid) and cholesterol. This interpretation is supported by observations that cholesterol abolishes the solid to liquid-crystalline phase transition as indicated by differential scanning calorimetry[135]; cholesterol also restricts the motions of the CH_2 groups of the phospholipid hydrocarbon chains as shown by the broadening of the PMR spectrum[136]. It thus appears that the addition of cholesterol to phospholipids produces a new phase which is more condensed when compared to the pure phospholipids above their transition temperature and more fluid when compared to the pure phospholipids below their transition temperature. X-ray diffraction of phospholipid-cholesterol dispersions in water indicates that the mixed phase is also lamellar and liquid-crystalline up to a molar ratio of 1 to 1. Additional cholesterol does not incorporate into the mixed crystal, but crystallizes out as a pure component[137].

Permeability studies with phospholipid vesicles and bilayers support and extend the above physical studies. Cholesterol decreases the permeability of phospholipid vesicles to $Cl^{\ominus}$ (ref. 73) and also to glucose[64, 127]. Furthermore, it appears to decrease the water permeability[138] and increase the electrical capacitance and resistance of PC bilayers[139, 55, 140]. The cholesterol effect on the fluidity of the phospholipid membranes is also reflected in the activity of certain ionophoric antibiotics. Cholesterol diminishes the increase in permeability and electrical conductance produced by valinomycin[141] and the macrotetralide actins[108]. In contrast, the presence of highly unsaturated fatty acyl chains increases the valinomycin effect[141]. In summary, it appears that cholesterol has a condensing effect on phospholipid molecules above their transition temperature. As a result of this interaction, the mixed membranes have increased stability, lower permeability to ions, water etc. and can withstand wide variations in temperature without the appearance of any phase transitions. It can thus be concluded that the role of cholesterol in biological membranes[32] is to control the fluidity of the hydrocarbon chains of phospholipids, providing a coherent structure stable over a wide range of temperatures while permitting some latitude in the fatty acid content of the membrane lipids.

7. INTERACTIONS OF PHOSPHOLIPID MEMBRANES WITH PROTEINS

Proteins are normally present in all isolated biological membranes and it is thus of great interest to establish the principles of phospholipid–protein interactions and show how the presence of proteins modifies the structure and permeability properties of model membranes. In contrast to the numerous studies with phospholipid membranes, relatively little work has been done on the recombination of phospholipids with membrane proteins since the latter have not yet been isolated and characterized in a form useful for detailed work. The technical difficulties of extracting and purifying lipid-free proteins without interfering with their conformation have yet to be solved. The existing extraction procedures involve the use of either detergents, such as Triton, cholate, deoxycholate, Lubrol, or polar organic solvents such as *n*-butanol, pyridine, 2-chloro-ethanol, etc. or aqueous systems incorporating various 'chaotropic' agents, urea, low pH[184]. Proteins obtained from a variety of membranes have been shown by gel electrophoresis to be quite complex mixtures with many components[142, 143]. A certain percentage of proteins can be extracted from erythrocytes by judicious control of the ionic strength and bivalent metal concentration of the aqueous extracting solution[144–146].

With the availability of purified membrane protein, the study of their interaction with phospholipids to form lipoprotein model membranes is the obvious next step. The only two purified membrane proteins studied so far with model membranes are a bacterial ATPase and the protein 'spectrin' isolated from erythrocyte ghosts. The ATPase was purified from *Streptococcus fecalis* 'ghosts' by a mild salt treatment[147] and its effect on PC bilayer membranes[148] was to increase the conductance by a magnitude dependent on the presence of $Mg^{2\oplus}$ and ATP. An earlier study with a less characterized (possibly vesicular) preparation of ATPase from rat brain synaptic vesicles[149] also indicated a drop in resistance of bilayer membranes dependent on the ATP concentration, the observed zero-voltage current-flow being inhibited by ouabain. Further work is needed before the similarities of this model system to the $Na^{\oplus}$–$K^{\oplus}$ active transport ATPase can be established[182].

Spectrin, when added to multilamellar vesicles containing either negative or positive charges produced a 2- to 5-fold increase in the permeability to glucose[150]. The binding of this protein above its isoelectric point to negatively charged phospholipid vesicles probably therefore involves hydrophobic associations. More recently the interaction of spectrin was studied with unilamellar phospholipid vesicles at different pH and $Ca^{2\oplus}$ concentrations[151]. When spectrin was added to both acidic (PS) and neutral (PC) phospholipid vesicles at a pH below its isoelectric point (pH 3.5) it increased their permeability to $Na^{\oplus}$ 100- to 1000-fold: the effect at neutral pH (7.4) was minimal. However, in the presence of 0.5 m*M* $Ca^{2\oplus}$ and neutral pH, spectrin produced a substantial increase (2 to 3 orders of magnitude) in the $Na^{\oplus}$ permeability of acidic phospholipid membranes. The biological relevance of these studies is that the binding of spectrin on erythrocyte membranes depends on $Ca^{2\oplus}$ or other bivalent metals. Moreover, the $Na^{\oplus}$ permeability rates of phospholipid membranes in the presence of spectrin are comparable to those of erythrocytes and other biological membranes[152]. A similar synergistic effect with $Ca^{2\oplus}$ and a protein isolated from nerve membranes has been reported[153].

References p. 165

Other work on the interaction of phospholipids with total membrane proteins or protein fractions has been carried out. Recombination of erythrocyte lipids with the delipidized erythrocyte proteins obtained by butanol extraction only occurred after the sialic acid was removed from the proteins and then only at low pH (3.5–4.5) when the lipid vesicles and the proteins carried opposite charges[154]. However, it was concluded that although charge attractions are essential for the recombination, the final complex includes hydrophobic associations between lipids and proteins. The latter conclusion was supported by a parallel study on the effect of proteins on the PMR spectra of the lipids[155]. In earlier studies with PC bilayer membranes it was reported that the addition of the erythrocyte apoproteins caused a marked increase in the intensity of reflected light (indicating an increase in thickness) but had no effect on the electrical properties of the membranes[156].

The recombination of membrane lipids with proteins has also been studied following solubilization with detergents and polar organic solvents. Upon dilution of the solubilizing agent (cholate or deoxycholate) phospholipids and proteins from mitochondria recombine to form membranous structures exhibiting some of the original enzymatic activities[157]. Mycoplasma membranes solubilized by sodium dodecyl sulphate reaggregate into closed membranous vesicles upon prolonged dialysis of the detergent but only in the presence of $Mg^{2\oplus}$ (ref. 158). The re-aggregated membranes may, however, contain different protein species and ratios of lipid to protein, depending on the $Mg^{2\oplus}$ concentration and time of dialysis[159]. Similar recombinations have been obtained after separation of lipids and proteins using 2-chloroethanol following a slow re-introduction of water[180]. The permeability properties of these artificial lipoprotein membranes have not yet been studied[184].

For lack of well-characterized membrane proteins, several investigators have used pure soluble proteins most of which are not normally associated with membranes (such as albumin, cytochrome *c*, haemoglobin, lysozyme, ribonuclease etc.). Although the relevance of such studies to the biological membrane structure and function is debatable, the approach has the advantage of a better defined system and it has already provided some insight into the mechanism of phospholipid–protein interactions[160, 161, 16]. The early studies on lipid monolayer penetration by proteins[163, 164] have been greatly extended and refined more recently[165, 161, 174]. The main conclusions from these and other studies are the following: proteins will interact with lipid films at low surface pressure (π = 2–10 dynes/cm) by a process of 'free penetration' where whole protein molecules will occupy the interstitial spaces at the air–water interphase between lipids. The process is characterized by large increases in surface pressure (Fig. 7 (1)). At higher initial surface pressures, only part of the protein molecule occupies space at the air–water interphase (Fig. 7 (2)). In either case, the adsorption of proteins at the interface might be accompanied by conformational changes. However, at least for the cases of albumin and ribonuclease, the adsorbed molecules do not appear to be completely 'unfolded' as judged by the area per molecule occupied at the interface[165, 166]. In support of this last point, i.r. and deuterium-exchange studies of collapsed films of poly-benzyl-aspartate indicate the prevalence of helical configuration[167]. Furthermore CD studies with polylysine adsorbed on the PS vesicles indicate that the binding induces the appearance of α-helix configuration[168]. The

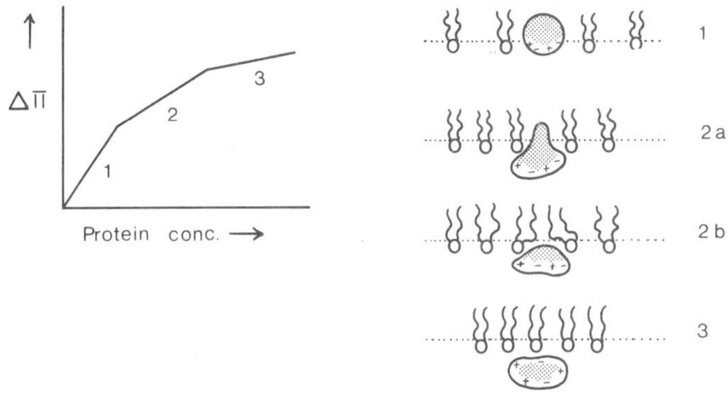

Fig. 7. Schematic representation of different forms of phospholipid–protein interactions. Plot of the increase in surface pressure ($\Delta\pi$) of phospholipid monolayers vs. protein concentration is shown on the left. Adapted from Quinn and Dawson[165]. The four figures on the right are illustrations of some of the possible types of interactions corresponding to the three stages of penetration shown in diagram on the left. The protein molecules are shown as encircled areas. Shaded areas indicate non-polar clusters within the protein molecule. Polar groups are represented by plus and minus signs. The conformation changes within the protein molecules shown for each type of interaction are hypothetical schematic representations, shown only for purposes of clarity. The conformational state of both phospholipids and proteins in such lipid-protein complexes is still largely unknown.

(*1*) 'Free penetration' of whole protein molecules within the available space between phospholipid molecules at the air–water interface.

(*2a*) Penetration of part of the protein molecule within the phospholipid film, at relatively high initial film pressures. Adapted from Kimelberg and Papahadjopoulos[174].

(*2b*) Rearrangement of the orientation of the phospholipid acyl chains to accomodate hydrophobic associations with the protein at the air–water interface[174].

(*3*) Simple adsorption of the protein by electrostatic and other polar interactions below the π interface.

CD spectrum of cytochrome *c* bound to PI liquid crystals, although different from that of the native molecule in solution does not indicate gross conformational changes[169]. Charge–charge interactions between phospholipids and proteins promote both binding and penetration as observed by an increase in π or by film expansion, even at high initial film pressures (π = 12–35 dynes/cm). Finally, binding of proteins to phospholipid films by polar group interactions can be observed without concomitant increase in surface pressure of the films (Fig. 7 (3)). All the above interactions depend on the chemistry and configuration of both lipids and proteins, and they appear to involve both polar and non-polar (hydrophobic) associations.

X-ray diffraction studies with cytochrome *c* and lysozyme complexes with acidic phospholipids indicate the existance of a lamellar phase. At high protein to lipid ratios, the repeat distance of 110–120 Å is in agreement with a structure involving one phospholipid bilayer sandwiched between two protein monolayers[65,170,169]. With lysozyme, appreciable thinning of the phospholipid bilayer and conformational changes in the protein molecule (by CD studies) were also observed[169]. The conclusion to be drawn from these studies is that soluble basic proteins will bind strongly to acidic phospholipids to form lipoprotein membranous structures which bear an overall similarity to the Danielli–

Davson–Robertson model for biological membranes. However, the interactions may also involve hydrophobic associations which might deform the packing and orientation of the phospholipid molecules within the bilayer structure to a certain degree[174, 169, 185]. In substantial agreement with this is the recent report[171] indicating that the binding of cytochrome *c* affects the phase-transitions (observed by differential scanning calorimetry) of acidic phospholipids.

Functional studies involving the influence of proteins on the permeability and electrical resistance of phospholipid membranes provide further evidence on the mechanism of phospholipid-protein interactions. Early studies with PC bilayers[139] and various phospholipid vesicles[73] showed that albumin had no effect on their permeability properties. However, it was shown more recently[172] that albumin will interact with negatively charged phospholipid vesicles at low pH (below its isoelectric point, $\simeq 4.8$) producing a substantial increase in the diffusion rate of glucose. It was concluded that the interaction depends on the 'unfolding' of the albumin molecule (induced by low pH) and involves both electrostatic and non-polar associations between lipid and protein. In a study of the effect of various soluble basic proteins on the permeability of phospholipid vesicles it was observed that although all such proteins (cytochrome *c*, lysozyme, ribonuclease, polylysine) bind strongly to acidic phospholipids (PS), their effects on permeability vary by orders of magnitude[173]. It was subsequently established by a combination of the techniques of monolayer penetration and vesicle permeability, that the observed increase in the permeability of phospholipid membranes can be correlated with the degree of penetration of the proteins into the monomolecular films[174]. Thus, proteins which tend to have minimal effects on permeability (such as ribonuclease and albumin at pH 7.4) do not penetrate into the monolayers (as judged by zero increase in surface pressure) when injected into the bulk-phase under the monolayer at an initial film pressure of $\geqslant 23$ dynes/cm. In contrast, proteins that produce a substantial increase in permeability (more than 2 orders of magnitude for cytochrome *c*, lysozyme, and albumin at low pH) penetrate into the PS films even at very high initial film pressures $\simeq 40$ dynes/cm. It is worth mentioning that the cut-off point of $\simeq 23$ dynes/cm represents a packing of approx. 67 $Å^2$ per phospholipid molecule which is similar to the area per phospholipid molecule in the liquid-crystalline lamellar phase obtained by X-ray diffraction[59, 72].

Other physiologically interesting reactions of phospholipid membranes with proteins have also been studied. These include the transient impedance changes of bilayers induced by an antigen–antibody reaction or enzyme substrate reaction[175] and the impedance changes due to acetylcholinesterase and acetylcholine[176]. Permeability changes have also been observed during the reaction of complement with antigen on the surface of phospholipid liposomes[177]. Finally, the lytic peptide melittin increases the permeability of liposomes and also penetrates monolayers of the same phospholipids[178].

8. INTERACTIONS WITH PHARMACOLOGICALLY ACTIVE COMPOUNDS

Perhaps one of the most useful aspects of phospholipids as model membranes has been their successful use for the elucidation of pharmacological action[11, 16, 186,187, 188]. The mechanism of action for certain vitamins, hormones, drugs etc. that involve membrane sites can be studied in a model system without the complications of a more physiological system involving whole cells, tissue slices, etc. Among the agents studied already, mention should be made of steroids, general and local anaesthetics, the reaction with antibodies, insect juvenile hormones and also metal ions and the ionophoric antibiotics mentioned earlier.

9. CONCLUDING REMARKS

Membranes composed of pure phospholipids exhibit some characteristics similar to those of biological membranes, such as interfacial tension, electrical capacitance, water permeability etc. However, these model membranes appear to be relatively impermeable to ions and have limited permeability to sugars and other water-soluble compounds. Moreover, they do not exhibit other important biological functions such as electrical excitability, active transport etc. These shortcomings are recognized among workers in the field and numerous recent studies have emphasized the modification of pure phospholipid membranes by proteins, peptides and other compounds which tend to induce biologically important functions. Studies with such modified membranes are of great promise but the field is still relatively unexplored.

Of equal importance are studies with pure lipid membranes, where the biological function of each particular constituent can be studied and defined in molecular terms. In addition, these simple membranes can serve as models for pharmacological action. All three model membrane systems described here (monolayers, bilayers and vesicles) offer mutually complementing advantages. Comparative studies between different model membrane systems can offer insights not easily obtainable with any one particular system.

Acknowledgements

I thank Dr. H.K. Kimelberg for valuable comments on the manuscript, Mrs. R. Lazo for help with bibliography and diagrams, and Mrs. J. Rossi for typing. Part of the material presented was obtained during the tenure of an established investigatorship of the American Heart Association supported in part by the mid-Hudson AHA chapter, and grants from the AHA (68-698 and 69-730), NIH (1 RO1-GM16106) and American Cancer Society (P-403 D).

REFERENCES

1 P. Mueller, D.O. Rudin, H.T. Tien and W.C. Wescott, *Nature*, 194 (1962) 979.
2 A.D. Bangham, M.M. Standish and J.C. Watkins, *J. Mol. Biol.*, 13 (1965b) 238.
3 J. Mingins and M.M. Standish, *Ann. Rep. Chem. Soc.*, 63 (1966) 91.
4 D.A. Cadenhead, *Recent Progress in Surface Science*, Vol. 3, Academic Press, New York, 1970 p. 169.
5 J.T. Davies and E.K. Rideal, *Interfacial Phenomena*, Academic Press, New York, 1961.
6 G.L. Gaines Jr., *Insoluble Monolayers at Liquid-Gas Interfaces*, Interscience, New York, 1966.
7 H.T. Tien and A.L. Diana, *Chem. Phys. Lipids*, 2 (1968) 55.
8 F.A. Henn and T.E. Thompson, *Ann. Rev. Biochem.*, 38 (1969) 241.
9 A. Goldup, S. Ohki and D.F. Danielli, *Recent Progress in Surface Science*, Vol. 3, Academic Press, New York, 1970, p. 193.

10 L. Rothfield and A. Finkelstein, *Ann. Rev. Biochem.*, 37 (1968) 463.
11 A.D. Bangham, in J.A.V. Butler and D. Noble (Eds.), *Progress in Biophysics and Molecular Biology*, Vol. 18, Pergamon, Oxford, 1968, p. 29.
12 J.P. Reeves, in R.M. Dowben (Ed.), *Biological Membranes*, Little, Brown, Boston, 1969, p. 223.
13 N. Laksminarayanaiah, *Transport Phenomena in Membranes*, Academic Press, New York, 1969, p. 414.
14 J.A. Lucy, in D. Chapman (Ed.), *Biological Membranes*, Academic Press, New York, 1968, p. 233.
15 T.E. Thompson and F.A. Henn, in E. Racker (Ed.), *Membranes of Mitochondria and Chloroplasts*, Van Nostrand/Reinhold, New York, 1969, p. 1.
16 D. Papahadjopoulos, in L. Prince and D.F. Sears (Eds.), *Biological Horizons in Surface Chemistry*, Academic Press, New York, 1973, in the press..
17 J.M. Tobias, D.P. Agin and R. Pawlowski, *J. Gen. Physiol.*, 45 (1962) 989.
18 J.H. Schulman and H.L. Rosano, in V.K. La Mer (Ed.), *Retardation of Evaporation by Monolayers*, Academic Press, New York, 1962, p. 97.
19 M. Joly, in J.F. Danielli, K.G.A. Pankhurst and A.C. Riddiford (Eds.), *Recent Progress in Surface Science*, Vol. 1, Academic Press, New York, 1964, p. 1.
20 A.D. Bangham and D. Papahadjopoulos, *Biochim. Biophys. Acta*, 126 (1966) 181.
21 G. Colacicco, *J. Colloid Interface Sci.*, 29 (1969) 345.
22 A.D. Bangham and R.M.C. Dawson, *Biochem. J.*, 75 (1960) 133.
23 P.J. Quinn and R.M.C. Dawson, *Biochem. J.*, 115 (1969) 65.
24 D.G. Dervichian, in J.A.V. Butler and H.E. Huxley (Eds.), *Progress in Biophysics and Molecular Biology*, Vol. 18, Pergamon, Oxford, 1964, p. 263.
25 D.O. Shah and J.H. Schulman, *J. Lipid Res.*, 6 (1965) 341.
26 D. Papahadjopoulos, *Biochim. Biophys. Acta*, 163 (1968) 240.
27 P.J. Anderson and B.A. Pethica, in G. Popjak and E. Le Breton (Eds.), *Biochemical Problems of Lipids*, Butterworth, London, 1956, p. 24.
28 L.L.M. van Deenen, V.M.T. Houtsmuller, G.H. de Haas and E. Mulder, *J. Pharm. Pharmacol.*, 14 (1962) 429.
29 D. Chapman, N.F. Owens and D.A. Walker, *Biochim. Biophys. Acta*, 120 (1966) 148.
30 D.A. Cadenhead, R.J. Demchak and M.C. Phillips, *Kolloid-Z. Z. Polym.*, 220 (1967) 59.
31 M.C. Phillips and D. Chapman, *Biochim. Biophys. Acta*, 163 (1968) 301.
32 D. Chapman and D.F.H. Wallach, in D. Chapman (Ed.), *Biological Membranes*, Academic Press, New York, 1968, p. 125.
33 F. Meyer and K. Bloch, *J. Biol. Chem.*, 238 (1963) 2654.
34 L.L.M. van Deenen, *Progress in Chemistry of Fats and Other Lipids*, Vol. 8, Pergamon, Oxford, 1965, p. 1.
35 S. Razin, M.E. Tourtelotte, R.N. McElhaney and J.D. Pollack, *J. Bacteriol.*, 91 (1966) 609.
36 J.M. Steim, *Advan. Chem.*, 84 (1968) 259.
37 J.C. Reinert and J.M. Steim, *Science*, 168 (1970) 1580.
38 T. Hanai, D.A. Haydon and J. Taylor, *Proc. Roy. Soc. Ser. A.*, 281 (1964) 377.
39 H.T. Tien, *J. Phys. Chem.*, 72 (1968) 2723.
40 R.J. Good, *J. Colloid Interface Sci.*, 31 (1969) 540.
41 R. Pagano and T.E. Thompson, *J. Mol. Biol.*, 38 (1968) 41.
42 D.A. Haydon, *J. Am. Oil Chem. Soc.*, 45 (1968) 230.
43 J.F. Danielli, *Molecular Associations in Biology*, Academic Press, New York, 1968, p. 529.
44 J.L. Taylor and D.A. Haydon, *Discussions Faraday Soc.*, 42 (1966) 51.
45 R.J. Cherry and D. Chapman, *J. Mol. Biol.*, 40 (1969) 19.
46 F.A. Henn and T.E. Thompson, *J. Mol. Biol.*, 31 (1968) 227.
47 F.A. Henn, G.L. Decker, J.W. Greenawalt and T.E. Thompson, *J. Mol. Biol.*, 24 (1967) 51.
48 A. Frey-Wyssling, *Submicroscopic Morphology of Protoplasm*, Elsevier, Amsterdam, 1953, p. 166.
49 A. Moran and A. Ilani, *Chem. Phys. Lipids*, 4 (1970) 169.
50 D.A. Haydon and J.L. Taylor, *Nature*, 217 (1968) 739.
51 J.F. Danielli and E.N. Harvey, *J. Cell Physiol.*, 5 (1935) 483.
52 H.G.L. Coster and R. Simons, *Biochim. Biophys. Acta*, 163 (1968) 234.
53 K.S. Cole, *Membranes, Ions and Impulses*, Univ. of California Press, Berkeley, 1968, p. 12.
54 T. Hanai, D.A. Haydon and J. Taylor, *J. Theoret. Biol.*, 9 (1965b) 433.

55 S. Ohki, *Biophys. J.*, 9 (1969) 1195.
56 A.D. Bangham, M.M. Standish and N. Miller, *Nature*, 208 (1965a) 1295.
57 S.M. Johnson and A.D. Bangham, *Biochim. Biophys. Acta*, 193 (1969a) 82.
58 D. Papahadjopoulos and S. Ohki, in J.F. Johnson and R.S. Porter (Eds.), *Liquid Crystals and Ordered Fluids*, Plenum, New York, 1970, p. 13.
59 V. Luzzati, in D. Chapman (Ed.), *Biological Membranes*, Academic Press, New York, 1968, p. 71.
60 D. Chapman and D.J. Fluck, *J. Cell Biol.*, 30 (1966) 1.
61 W. Stockenius, *J. Biophys. Biochem. Cytol.*, 5 (1959) 491.
62 A.D. Bangham and R.W. Horne, *J. Mol. Biol.*, 8 (1964) 660.
63 K.J. Palmer and F.O. Schmitt, *J. Cell Comp. Physiol.*, 17 (1941) 385.
64 R.A. Demel, S.C. Kinsky, C.B. Kinsky and L.L.M. van Deenen, *Biochim. Biophys. Acta*, 150 (1968) 655.
65 D. Papahadjopoulos and N. Miller, *Biochim. Biophys. Acta*, 135 (1967) 624.
66 C.H. Huang, *Biochemistry*, 8 (1969) 344.
67 D. Attwood and L. Saunders, *Biochim. Biophys. Acta*, 116 (1966) 108.
68 J.L. Fergason and G.H. Brown, *J. Am. Oil Chem. Soc.*, 45 (1968) 120.
69 R.D. Kornberg and H.M. McConnell, *Biochemistry*, 10 (1971) 1111.
70 R.A. Klein, *Biochim. Biophys. Acta*, 210 (1970) 486.
71 D. Papahadjopoulos, *Biochim. Biophys. Acta*, 211 (1970) 467.
72 D.M. Small, *J. Lipid Res.*, 8 (1967) 551.
73 D. Papahadjopoulos and J.C. Watkins, *Biochim. Biophys. Acta*, 135 (1967) 639.
74 D. Papahadjopoulos and L. Weiss, *Biochim. Biophys. Acta*, 183 (1969) 417.
75 F. Reiss-Husson, *J. Mol. Biol.*, 25 (1967) 363.
76 D.O. Tinker and L. Pinteric, *Biochemistry*, 10 (1971) 860.
77 A.D. Bangham, J. de Gier and G.D. Greville, *Chem. Phys. Lipids*, 1 (1967) 225.
78 R. Rendi, *Biochim. Biophys. Acta*, 135 (1967) 333.
79 M. Blank, *J. Phys. Chem.*, 66 (1962) 1911.
80 T.E. Andreoli, J.A. Bangham and D.C. Tosteson, *J. Gen. Physiol.*, 50 (1967) 1729.
81 D.A. Haydon and J. Taylor, *J. Theoret. Biol.*, 4 (1963) 281.
82 W. Lesslauer, J. Richter and P. Lauger, *Nature*, 213 (1967) 1224.
83 S. Ohki and O. Aono, *J. Colloid Interface Sci.*, 32 (1970) 270.
84 D.F. Sears and J.H. Schulman, *J. Phys. Chem.*, 68 (1964) 3529.
85 G. Eisenman, in A. Kleinzeller and A. Kotyk (Eds.), *Membrane Transport and Metabolism*, Academic Press, New York, 1961, p. 163.
86 D. Papahadjopoulos, *Biochim. Biophys. Acta*, 241 (1971) 254.
87 P. Mueller and D.O. Rudin, *Biochem. Biophys. Res. Commun.*, 26 (1967) 398.
88 J. Gutknecht and D.C. Tosteson, *J. Gen. Physiol.*, 55 (1970) 359.
89 H.G. Bungenberg de Jong, in H.R. Kruyt (Ed.), *Colloid Science*, Vol. 2, Elsevier, Amsterdam, 1949, p. 248.
90 H. Hauser and R.M.C. Dawson, *European J. Biochem.*, 1 (1967) 61.
91 E. Rojas and J.M. Tobias, *Biochim. Biophys. Acta*, 94 (1965) 394.
92 H.S. Hendrickson and J.G. Fullington, *Biochemistry*, 4 (1965) 1599.
93 M.B. Abramson, R. Katzman, H. Gregor and R. Curci, *Biochemistry*, 5 (1966) 2207.
94 V.K. Miyamoto and T.E. Thompson, *J. Colloid Interface Sci.*, 25 (1967) 16.
95 D. Papahadjopoulos and S. Ohki, *Science*, 164 (1969) 1075.
96 J.F. Hoffman, *Circulation*, 26 (1962) 1201.
97 I. Tasaki, A. Watanabe and L. Lerman, *Am. J. Physiol.*, 213 (1967) 1465.
98 S. Ohki and D. Papahadjopoulos, in M. Blank (Ed.), *Surface Chemistry of Biological Systems*, Plenum, New York, 1970, p. 155.
99 J.M. Tobias, *Nature*, 203 (1964) 13.
100 I. Tasaki and I. Singer, *Ann. N.Y. Acad. Sci.*, 137 (1966) 792.
101 J.H. Wang, *Proc. Natl. Acad. Sci. U.S.*, 67 (1970) 916.
102 P. Mueller and D.O. Rudin, *J. Theoret. Biol.*, 18 (1968a) 222.
103 L.D. Kushnir, *Biochim. Biophys. Acta*, 150 (1968) 285.
104 P. Mueller and D.O. Rudin, *Nature*, 217 (1968b) 713.
105 B.C. Pressman, *Proc. Natl. Acad. Sci. (U.S.)*, 53 (1965) 1076.

106 V.T. Ivanov, I.A. Laine, N.D. Abdulaev, L.B. Senyavina, E.M. Popov, Yu. A. Ovchinnikov and M.M. Shemyakin, *Biochem. Biophys. Res. Commun.*, 34 (1969) 803.
107 B.C. Pressman, E.J. Harris, W.S. Jagger and J.H. Johnson, *Proc. Natl. Acad. Sci., (U.S.)* 58 (1967) 1949.
108 G. Szabo, G. Eisenman and S. Ciani, *J. Membrane Biol.*, 1 (1969) 346.
109 R.C. Bean, W.C. Shepherd and H. Chan, *J. Gen. Physiol.*, 52 (1968) 495.
110 S.B. Hladky and D.A. Haydon, *Nature*, 225 (1970) 451.
111 T.E. Andreoli, V.W. Dennis and A.M. Weigl, *J. Gen. Physiol.*, 53 (1969) 133.
112 A. Finkelstein and A. Cass, *J. Gen. Physiol.*, 52 (1968) 145s.
113 M.M. Weber and S.C. Kinsky, *J. Bacteriol.*, 89 (1965) 306.
114 R.A. Demel, L.L.M. van Deenen and S.C. Kinsky, *J. Biol. Chem.*, 240 (1965) 2749.
115 H. van Zutphen, L.L.M. van Deenen and S.C. Kinsky, *Biochem. Biophys. Res. Commun.*, 22 (1966) 393.
116 V.W. Dennis, N.W. Stead and T.E. Andreoli, *J. Gen. Physiol.*, 55 (1970) 375.
117 G. Sessa and G. Weissmann, *J. Biol. Chem.*, 243 (1968) 4364.
118 A.V. Few, *Kolloidchemie Makromolekularer Naturstoffe*, 18 (1958) 63.
119 G. Colacicco, E.E. Gordon and G. Berchenko, *Biophys. J.*, 8 (1968) 22a.
120 H. Hauser, E.G. Finer and D. Chapman, *J. Mol. Biol.*, 53 (1970) 419.
121 S.M. Johnson and A.D. Bangham, *Biochim. Biophys. Acta*, 193 (1969b) 92.
122 J. Bielawski, T.E. Thompson and A.L. Lehninger, *Biochem. Biophys. Res. Commun.*, 24 (1966) 948.
123 V.P. Skulachev, A.A. Sharaf and E.A. Liberman, *Nature*, 216 (1967) 718.
124 U. Hopfer, A.L. Lehninger and T.E. Thompson, *Proc. Natl. Acad. Sci. (U.S.)*, 59 (1968) 484.
125 E.A. Liberman, V.P. Topaly and A.Y. Silberstein, *Biochim. Biophys. Acta*, 196 (1970) 221.
126 P. Mitchell, *Biol. Rev.*, 41 (1966) 445.
127 J. de Gier, J.G. Mandersloot and L.L.M. van Deenen, *Biochim. Biophys. Acta*, 173 (1969) 143.
128 J.L. Moore, T. Richardson and H.F. DeLuca, *Chem. Phys. Lipids*, 3 (1969) 39.
129 C. Huang, L. Wheeldon and T.E. Thompson, *J. Mol. Biol.*, 8 (1964) 148.
130 W.R. Redwood, F.R. Pfeiffer, J.A. Weisbach and T.E. Thompson, *Biochim. Biophys. Acta*, 233 (1971) 1.
131 K.R. Bruckdorfer, R.A. Demel, J. de Gier and L.L.M. van Deenen, *Biochim. Biophys. Acta*, 183 (1969) 334.
132 L. de Bernard, *Bull. Soc. Chim. Biol.*, 40 (1958) 161.
133 D.A. Cadenhead and M.C. Phillips, *Advan. Chem. Ser.*, 78 (1968) 131.
134 P. Joos, *Chem. Phys. Lipids*, 4 (1970) 162.
135 B.D. Ladbrooke, R.M. Williams and D. Chapman, *Biochim. Biophys. Acta*, 150 (1968) 333.
136 D. Chapman and S.A. Penkett, *Nature*, 211 (1966) 1304.
137 D.M. Small and M.C. Bourges, *Mol. Cryst.*, 1 (1966) 541.
138 A. Finkelstein and A. Cass, *Nature*, 216 (1967) 717.
139 T. Hanai, D.A. Haydon and J. Taylor, *J. Theoret. Biol.*, 9 (1965a) 422.
140 D. Papahadjopoulos, S. Nir and S. Ohki, *Biochim. Biophys. Acta*, 266 (1972) 561.
141 J. de Gier, C.W.M. Haest, J.G. Mandersloot and L.L.M. van Deenen, *Biochim. Biophys. Acta*, 211 (1970) 373.
142 E.D. Kiehn and J.J. Holland, *Proc. Natl. Acad. Sci. (U.S.)*, 61 (1968) 1370.
143 J. Lenard, *Biochemistry*, 9 (1970) 1129.
144 S.P. Burger, T. Fujii and D.J. Hanahan, *Biochemistry*, 7 (1968) 3682.
145 S.L. Marchesi, E. Steers, V.T. Marchesi and T.W. Tillack, *Biochemistry*, 9 (1970) 50.
146 J. TH. Hoogeven, R. Juliano, J. Coleman and A. Rothstein, *J. Membrane Biol.*, 3 (1970) 156.
147 A. Abrams and C. Baron, *Biochemistry*, 6 (1967) 225.
148 W.R. Redwood, H. Muldner and T.E. Thompson, *Proc. Natl. Acad. Sci. (U.S.)*, 64 (1969) 989.
149 M.K. Jain, A. Strickholm and E.H. Cordes, *Nature*, 222 (1969) 871.
150 C. Sweet and J.E. Zull, *Biochem. Biophys. Res. Commun.*, 41 (1970) 135.
151 R. Juliano, H. Kimelberg and D. Papahadjopoulos, *Biochim. Biophys. Acta*, 241 (1971) 894.
152 W.D. Stein, *The Movement of Molecules Across Cell Membranes*, Academic Press, New York, 1967.
153 P. Calissano and A.D. Bangham, *Biochem. Biophys. Res. Commun.*, 43 (1971) 504.

154 R.F.A. Zwaal and L.L.M. van Deenen, *Chem. Phys. Lipids*, 4 (1970) 311.
155 V.B. Kamat, D. Chapman, R.F.A. Zwaal and L.L.M. van Deenen, *Chem. Phys. Lipids*, 4 (1970) 323.
156 A.H. Maddy, C. Huang and T.E. Thompson, *Federation Proc.*, 25 (1966) 933.
157 D.E. Green, D.W. Allman, E. Bachmann, H. Baum, K. Kopaczyk, E.F. Korman, S. Lipton, D.H. MacLennan, D.G. McConnell, J.F. Perdue, J.S. Rieske and A. Tzagoloff, *Arch. Biochem. Biophys.*, 119 (1967) 312.
158 D.M. Engelman and H.J. Morowitz, *Biochim. Biophys. Acta*, 150 (1968) 376.
159 S. Razin, Z. Ne'eman and I. Ohad, *Biochim. Biophys. Acta*, 193 (1969) 277.
160 R.M.C. Dawson, in D. Chapman (Ed.), *Biological Membranes*, Academic Press, New York, 1968, p. 203.
161 G. Colacicco, *Lipids*, 5 (1970) 636.
162 J. de Gier, J.G. Mandersloot and L.L.M. van Deenen, *Biochim. Biophys. Acta*, 150 (1968) 666.
163 R. Matalon and J.H. Schulman, *Discussions Faraday Soc.*, 6 (1949) 27.
164 D.D. Eley and D.G. Hedge, *Discussions Faraday Soc.*, 21 (1957) 221.
165 P.J. Quinn and R.M.C. Dawson, *Biochem. J.*, 116 (1970) 671.
166 A. Khaiat and I.R. Miller, *Biochim. Biophys. Acta*, 183 (1969) 309.
167 B.R. Malcolm, *Biopolymers*, 9 (1970) 911.
168 G.G. Hammes and S.E. Schullery, *Biochemistry*, 9 (1970) 2555.
169 T. Gulik-Krzywicki, E. Shechter, V. Luzzati and M. Faure, *Nature*, 223 (1969) 1116.
170 G.G. Shipley, R.B. Leslie and D. Chapman, *Nature*, 222 (1969) 561.
171 D. Chapman and J. Urbina, *FEBS Letters*, 12 (1971) 169.
172 C. Sweet and J.E. Zull, *Biochim. Biophys. Acta*, 173 (1969) 94.
173 H. Kimelberg and D. Papahadjopoulos, *J. Biol. Chem.*, 246 (1971a) 1142.
174 H. Kimelberg and D. Papahadjopoulos, *Biochim. Biophys. Acta*, 233 (1971b) 805.
175 J. del Castillo, A. Rodriguez, C.A. Romero and V. Sanchez, *Science*, 153 (1966) 185.
176 J. del Castillo, A. Rodriguez and C.A. Romero, *Ann. N.Y. Acad. Sci.*, 144 (1967) 803.
177 J.A. Haxby, C.B. Kinsky and S.C. Kinsky, *Proc. Natl. Acad. Sci. (U.S.)*, 61 (1968) 300.
178 G. Sessa, J.H. Freer, G. Colacicco and G. Weissman, *J. Biol. Chem.*, 244 (1969) 3575.
179 K.J. Mysels, K. Shinoda and S. Frankel, *Soap Films: Studies of their Thinning*, Pergamon, Oxford, 1959, p. 1.
180 P. Zahler and E.R. Weibel, *Biochim. Biophys. Acta*, 219 (1970) 320.
181 G. Wilson, S.P. Rose and C.F. Fox, *Biochem. Biophys. Res. Commun.*, 38 (1970) 617.
182 H. Kimelberg and D. Papahadjopoulos, *Biochim. Biophys. Acta*, 282 (1972) 277.
183 S.C. Kinsky, *Ann. Rev. Pharmacol.*, 10 (1970) 119.
184 S. Razin, *Biochim. Biophys. Acta*, 265 (1972) 241.
185 R.P. Rand, *Biochim. Biophys. Acta*, 241 (1971) 823.
186 A.D. Bangham, *Ann. Rev. Biochem.*, 41 (1972) 753.
187 D. Papahadjopoulos, *Biochim. Biophys. Acta*, 265 (1972) 169.
188 S.C. Kinsky, *Biochim. Biophys. Acta*, 265 (1972) 1.

Chapter 8

PHOSPHOLIPIDS IN BIOLOGICAL MEMBRANES AND THE STUDY OF PHOSPHOLIPID–PROTEIN INTERACTIONS

J.B. FINEAN

Department of Biochemistry, University of Birmingham, Birmingham B15 2TT (Great Britain)

Tissue phospholipid is located mainly in the cell membranes and constitutes a substantial component of all membranes so far analysed. One of the major functions of phospholipid is thus to contribute to the structure and function of cell membranes.

Phospholipid alone does not, however, constitute a cell membrane: it forms a part of the lipid phase which may also include cholesterol and glycolipid and, in special cases, other lipid-type molecules such as chlorophyll in chloroplasts. The lipid phase is invariably associated with protein.

The protein of the membranes amounts to a major proportion of the total cell protein so that membranes form the main site of phospholipid–protein interaction in biological systems. This is the reason for including the topic in the present chapter but to complete the picture it will be necessary to refer also to phospholipid–protein interactions in situations where the lipoprotein does not form a clearly defined membrane structure.

1. GENERAL CHARACTERISTICS OF MEMBRANE LIPID

As the prime concern of this book is with the role of the phospholipid molecule the first point that will be considered in this chapter is the prominence in analytical terms of phospholipid in membrane composition. Detailed analytical figures will be found in Chapter 16 (p. 441) but for the present purpose it is necessary to have an overall view of the composition of membranes.

Although probably all membranes contain phospholipid the amount of phospholipid in different types of membrane preparations shows considerable variations[1–18]. The total lipid content (40–80% of dried weight) and the proportion of phospholipid are relatively high in mammalian cell membrane preparations but vary appreciably from one type of membrane to another[1–13]. Lipid content is also fairly high (30–40%) in plant membranes[17] and in yeast protoplasts[18] but the proportion of phospholipid is considerably reduced as compared with mammalian material. The membranes of microorganisms vary much more in their content of lipid and this can be modified by growth conditions[14]. There is, nevertheless, a substantial proportion of phospholipid which is, however, variable in its composition.

A further point concerning the nature of the lipid phase which is of considerable structural importance in lipoprotein systems is the nature of the hydrocarbon chains. In most membrane systems a substantial proportion of the fatty acid chains are unsaturated and chains which are branched or which incorporate ring structures occur in a small number of bacteria. Gas-liquid chromatography has facilitated the detailed analysis of the fatty acid chains of lipids in biological systems and revealed great complexity of fatty acid composition, but the most significant feature from the structural viewpoint may well be the average level of unsaturation in the lipid phase as a whole.

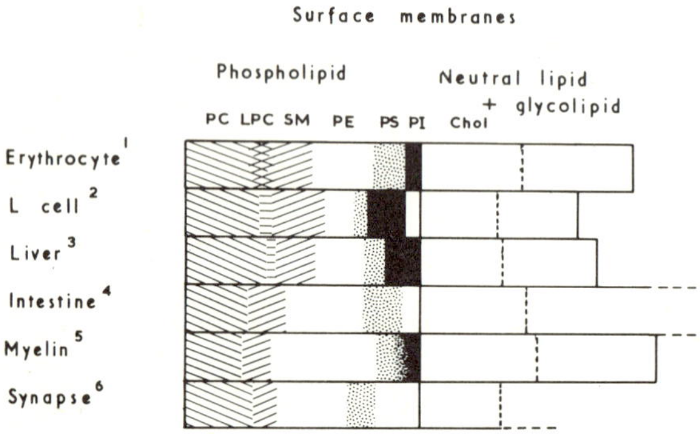

Fig. 1. Comparison of lipid compositions of a variety of cell surface membranes. Abbreviations: PC, phosphatidylcholine; LPC, lysophosphatidylcholine; SM, sphingomyelin; PE, phosphatidylethanolamine; PS, phosphatidylserine; PI, phosphoinositides; Chol, cholesterol. Sources of data indicated by reference numbers.

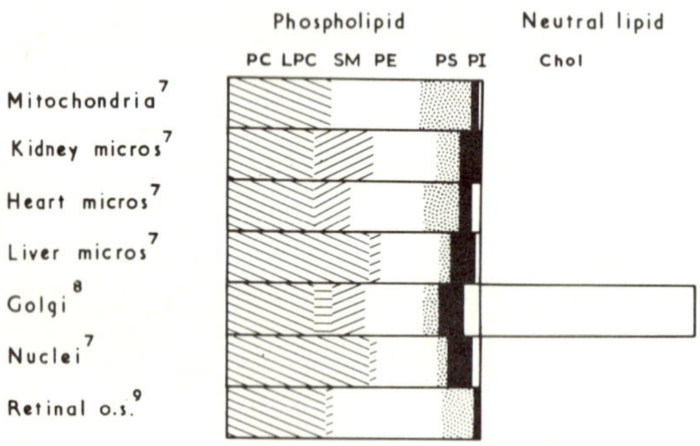

Fig. 2. Comparison of lipid compositions of a variety of membrane preparations from cell organelles. Abbreviations as in Fig. 1 (micros, microsomes; o.s. outer segments).

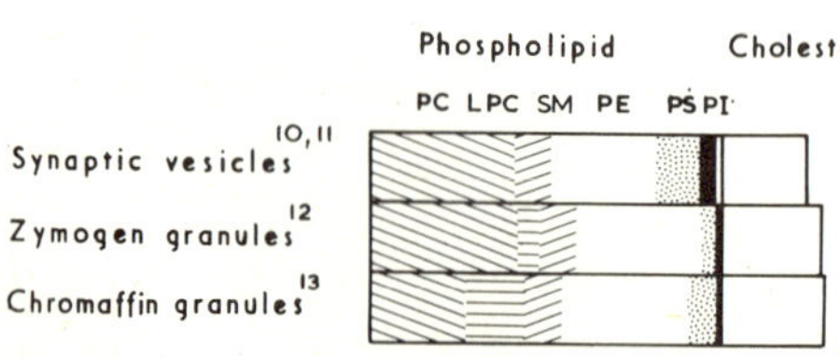

Fig. 3. Lipid compositions of vesicle and granule preparations. Abbreviations as in Fig. 1.

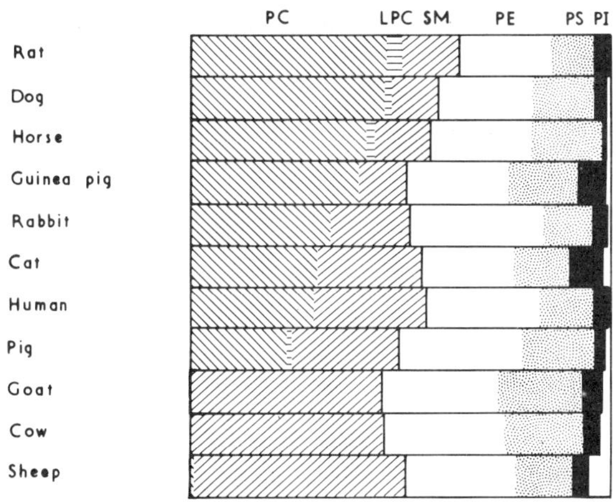

Fig. 4. Phospholipid distribution in erythrocytes from a variety of mammalian species. Abbreviations as in Fig. 1. Data taken mainly from Rouser et al.[7] and Nelson[1].

Some analyses of lipid phases in a variety of mammalian membranes assembled in Figs. 1, 2 and 3 emphasise the dominant position of phospholipid. From the point of view of the overall ionic status of the lipid phase one may group together phosphatidylcholine, lysophosphatidylcholine and sphingomyelin, all of which contain the phosphorylcholine end group with its balanced charges. Together these molecules constitute from 40–60% of the total phospholipid in these membranes and the analyses of erythrocytes from various animal species (Fig. 4) have suggested that in some respects at least sphingomyelin and phosphatidylcholine may substitute for one another[19]. The highly acidic phosphatidylserine, phosphatidic acid and phosphoinositides, taken together, account for up to 25% of the total phospholipid and the remainder is largely phosphatidylethanolamine and ethanolamine plasmalogen which have a small net negative charge at pH7. Thus the contribution of phospholipid to the net charge on the lipid phase in these membranes is a substantial negative one.

In some membranes, and particularly in membranes of cytoplasmic organelles in mammalian cells,the contribution of other lipids such as glycolipids and cholesterol is very small so that the negative charge arising from the phospholipid will not be diluted. In cell surface membranes and in membranes of synaptic vesicles and some secretory granules there are substantial proportions of cholesterol, which may have a considerable structural significance.

A variety of studies (see Chapter 6, p. 136) have indicated that cholesterol can be accommodated in a phospholipid phase up to a limiting proportion which is determined largely by the degree of unsaturation of the fatty acid chains of the phospholipid. The addition of cholesterol to monolayers of phospholipid with a moderate content of unsaturated fatty acid chains has a condensing effect[20,21] and also modifies the nuclear magnetic resonance spectrum in such a way as to suggest that cholesterol inhibits the movement of hydrocarbon chains[22]. If cholesterol in membranes also serves to reduce the fluidity of the lipid hydrocarbon phase then surface membrane lipid will be much less fluid than that in membranes of the cytoplasmic organelles. It can also be inferred from these data that the cholesterol will probably be dispersed among at least some of the phospholipid molecules so that it will also serve to dilute (or disperse) the charge arising from the phospholipid end groups. The glycolipid of surface membranes may play a similar role. There is, however, no direct evidence that lipid molecules are homogeneously distributed in membranes. Indeed, there are already indications that there may be specific localisations of some lipid components in relation to their interaction with protein.

2. MEMBRANE EXTRACTION AND SOLUBILISATION

Information on the nature and significance of lipid–protein associations in biological membranes and other lipoprotein systems has been derived both directly and indirectly from a wide variety of observations.

The nature of the lipid–protein interaction is undoubtedly an important factor in determining the conditions required for extraction of lipid or protein components from membrane systems but the interpretation of extraction data in such terms is by no means free from uncertainties. Solubilisation of up to 50% of the total protein content of erythrocyte membrane preparations (with minimum haemoglobin content) has been achieved by aqueous extraction under conditions which would be expected to affect only polar interactions. This amount of protein has been removed in a single step using a procedure in which levels of ions have been reduced as far as possible during isolation of the membranes and subsequent extraction[23]. In other work a similar level of protein extraction from erythrocyte membranes has been achieved in two steps, the first using EDTA in the aqueous extraction medium[24,25] and the second using extraction at high salt (NaCl or KCl) concentrations. This would appear to distinguish association by ion bridges and association by direct electrostatic binding but such associations are not necessarily directly with the lipid component of the membrane. In many cases the possibility that substantial amounts of lipid could have been extracted along with the protein appears to have been overlooked. However, in one study of extraction from erythrocyte ghosts by hypertonic (up to 1.2 *M*) NaCl solutions the proportion of the total lipid extracted was considerably greater than that of the protein[26]. The ghosts were reduced in diameter and the surface carried bead-like protuberances which may have contained lipid. Appreciable levels of protein can be extracted from other membranes under similar conditions but the data are generally less extensive than they are for the erythrocyte membrane. In the exceptional case of the halophilic bacteria which grow in high salt media the whole of the membrane may be solublilised in dilute salt solution[27] but here the lipoprotein system as a whole is solubilised and it is not clear to what extent the lipid–protein associations of the original membrane are modified.

A similar uncertainty exists in relation to attempts to solubilise membranes by the addition of detergent to the extraction medium. Extensive 'solubilisation' of a variety of membranes has been achieved through the use of detergents. This treatment has been considered to solubilise pre-formed lipoprotein 'subunits' of membranes[28–33] but recent work has emphasised that in many cases the detergent has modified the lipid–protein associations to produce a number of molecular complexes which incorporate the detergent molecules[34–38]. In some situations it has been possible to dialyse away most of the detergent to obtain relatively pure protein components but at the other extreme interaction with sodium dodecyl sulphate (SDS) has led to the formation of strong SDS-polypeptide complexes which have proved suitable for molecular weight determinations by polyacrylamide gel electrophoresis[39–41].

The significance of such studies from the point of view of the nature of lipid–protein interactions in membranes is that detergent molecules are probably capable of modifying the interaction with lipid of that portion of the membrane protein (about 50%) which is not extractable under conditions which simply modify the ionic environment. It may be that the detergent molecules displace lipid molecules from sites of interaction with the protein although such lipid may still maintain a general association with the new molecular complex. The variety of detergent molecules used in such studies may lead to a variation in the precise molecular associations produced so that only general conclusions can be drawn.

Organic solvents have also been used extensively to extract from, or to solubilise, lipoprotein systems. Addition of pyridine to aqueous extraction media has been shown to solubilise some of the protein that is resistant to purely aqueous extraction[42], but in general, studies involving organic solvents have been directed towards isolation of lipid components. The basis for the breakdown of lipoprotein systems by organic solvents is very complex. In early studies of the lipids of membranes it was customary to treat with ethanol to break down lipid–protein associations and then to extract differentially with other lipid solvents[43,44]. The mode of action of alcohol is not established. More complete one-step lipid extractions were achieved with alcohol–ether mixtures but this method has now been largely superseded by chloroform-methanol extraction of the wet tissue or membrane preparation[45]. It would appear that the water-miscible solvents such as ethanol and acetone are alone not sufficiently good lipid solvents to be effective for tissue extraction whilst the very good lipid solvents such as chloroform, benzene and ether are effective only when the lipid is dissociated from the protein or when they are associated with another more polar solvent. Chloroform alone will extract some lipid from wet lipoprotein systems and there is X-ray diffraction and electron-microscope evidence[38] that chloroform alone will destroy the ordered structure of myelin perhaps through penetration into non-polar regions of the myelin layer.

A further point of potential structural importance is that chloroform-methanol mixtures extract some protein along with the lipid from the wet tissues. The lipid and protein thus extracted are apparently associated in such a way that the molecular complex (proteolipid) has the external characteristics of a lipid and it is regarded by many as a macromolecule which may have specific biological functions. The amount of protein extracted in this way is particularly high in the case of brain white matter and spinal cord, and myelin fractions isolated from such tissues may dissolve completely in 2:1 chloroform methanol mixtures. Although the proportion of lipid firmly associated with the protein in this proteolipid complex may be varied by modification of the extraction

References p. 198

procedure it has nevertheless been suggested that the protein involved is a specific protein and that a part at least of the lipid–protein interaction is a constant feature which is of fundamental importance in determining the role of the proteolipid in the activity of the nervous system[47].

Another organic solvent which is very effective in the extraction of membrane lipoprotein is *n*-butanol[48,49]. When *n*-butanol is mixed briefly but vigorously with a comparable volume of a suspension of erythrocyte ghosts at 0° and then centrifuged to form two phases, 90 to 95% of the membrane protein is solubilised in the water phase along with a trace of lipid and some carbohydrate[49]. The bulk of the lipid is retained in the butanol. If the extraction is carried out at low ionic strength in the presence of EDTA the extracted protein can subsequently be fractionated by polyacrylamide gel electrophoresis and analytical ultracentrifugation to give three well-defined fractions, one equivalent to the EDTA extracted component of Marchesi et al.[50], one which includes the sialoprotein, and a major component with a pronounced tendency to aggregation[51]. Again it is difficult to establish the mechanism of action of the solvent in solubilising the protein and lipid components.

Studies of the effects of *n*-butanol on the organised structure of myelin have demonstrated that it very rapidly disrupts the membrane structure; in fact, it is the most effective member of the short-chain alcohol series in this respect. This could be attributed to its retention of a limited degree of miscibility with water together with its effectiveness as a lipid solvent. The lipid component is ultimately to be found mainly in the butanol phase and the protein mainly in the water and it may be that this separation takes place immediately when the membrane structure is disrupted and that the lipid and protein are never associated except in the membrane. If this is so it must be assumed that butanol saturated with water is able to penetrate the structure and solubilise the lipid whilst simultaneously water containing a little butanol can take up the protein in an apparently undenatured form[51]. A saturated solution of butanol (8%) in water has been used to extract from liver microsomes the lipid-dependent enzyme system, phosphatidic acid phosphatase[52]. When the concentration of butanol was increased a two-phase system was formed and the enzyme activity was lost, presumably because lipid and protein components became separated.

The situation may be somewhat different when membranes are dissolved in 2-chloroethanol. In this case the lipid and protein are both taken into the organic solvent[53] but whether they remain associated in this phase is not established. Certainly they can be readily separated by gel filtration and from the type of gel that is effective it has been suggested that the phospholipid is in monodisperse form in the solvent[54].

3. STUDIES OF INTACT MEMBRANE SYSTEMS

Information on the nature and strength of lipid–protein interaction in intact biological systems may be obtained from studies of the spatial distribution and orientation of molecular components and also from a comparison of the physical and chemical characteristics of these components in the biological system and in isolation. Thus techniques such as X-ray diffraction and electron microscopy may provide information on spatial relationships which will indicate to what extent specific molecular interactions are possible. Spectroscopic techniques will provide information on the physical states of molecular components in the intact biological system which can be compared with data obtained from such molecules in isolation. This may provide an indication of the kind of interaction present in the biological system. Similar information may be obtained by comparing activities (e.g. enzyme activities) of specific molecules when incorporated in the biological system with activities observed in isolation.

X-ray diffraction

Attempts to establish the spatial distribution of molecular components in lipoprotein membranes by X-ray diffraction analysis have so far tended to emphasise a layered structure. In the few cases in which it has been possible to calculate a most probable electron density profile through the membrane a most striking common feature has been the location of a trough of low electron density at the centre of the membrane with regions of high electron density to either side[55].

The most extensive and detailed diffraction studies have been of the nerve myelin sheath. This is a multilayered membrane system derived from Schwann cell (for peripheral nerves) or glial cell (for central nervous system) plasma membranes. It was long ago demonstrated by Schmitt and co-workers[56,57] that well-defined lamellar-type diffraction patterns could be recorded from the intact nerves maintained in a physiologically significant condition. Such patterns from a variety of nerves, both

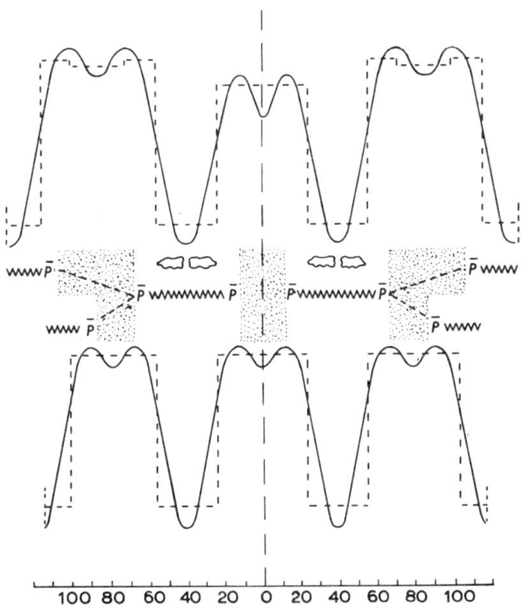

Fig. 5. Electron-density profiles for myelin layers in rat sciatic (upper) and rat optic (lower) nerves[55,60]. Locations of cholesterol molecules and phospholipid molecules are indicated schematically. Dotted regions indicate the probable locations of non-lipid components.

References p. 198

peripheral and central, were refined and a variety of modifications were studied by Finean and co-workers during the 1950s (ref. 58). A significant interpretation of the patterns in terms of an electron density profile was eventually obtained from a consideration of swelling phenomena[59–61]. Despite some ambiguities in the initial analysis (Worthington and Blaurock[62]) this electron density profile has proved to be essentially correct (Blaurock[63]). Recent improvements in the X-ray diffraction data have added further detail to this profile[64–66].

The general molecular distribution first put forward to account for this type of electron density profile[60] (Fig. 5) related the deep troughs to lipid hydrocarbon regions and the sharp peaks to phosphate groups of phospholipid and any associated metallic ions. Non-lipid components were suggested to be located mainly outside the lipid layer and the asymmetry of the units was considered to be due principally to different thicknesses of non-lipid components located to either side of the central lipid-containing layer. A quantitative scaling of electron densities (Worthington and Blaurock[62]) supported this suggestion but indicated that the region of liquid-like hydrocarbon was relatively narrow (~20 Å).

The higher resolution electron density profiles now available from the improved X-ray diffraction patterns[64] permit further speculations concerning the finer details of the molecular distribution. The most interesting features are the 'shoulders' that appear on the sides of the central troughs in each of the membrane pairs that form the structural unit of myelin (Fig. 6). These shoulders constitute an additional asymmetric feature in each membrane unit. It has been suggested that the shoulders indicate the location of cholesterol molecules within the lipid region (Caspar et al.[64]) but the apparent asymmetrical distribution leads one to suspect that additional factors may be involved. A further possibility is that the shoulders reflect to some extent the penetration of protein into the central region and as there is evidence of an asymmetric

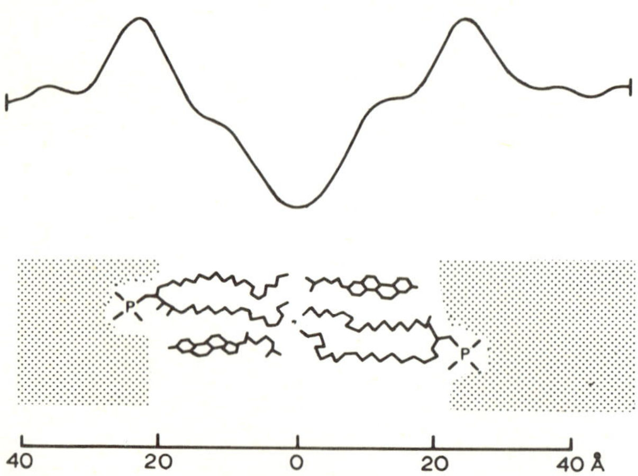

Fig. 6. Electron density profile for myelin layer in rabbit sciatic nerve[64]. Locations of cholesterol molecules and phospholipid molecules are indicated schematically and dotted regions represent protein plus water.

distribution of non-lipid components a difference in the extent of protein penetration from either side of the lipid layer would be readily understandable. Such an asymmetric penetration of protein may, of course, in turn lead to an asymmetric distribution of associated lipid components.

An alternative interpretation of the myelin diffraction pattern based on an assignment of phases from a heavy metal labeling technique[67] would require a somewhat different molecular organisation and has so far gained little support.

The membrane systems featured in the retinal outer segments in the eye have also yielded extensive diffraction data and again it has been possible to maintain the isolated retina in a physiologically significant state during the recording of the data. The main feature of these data (Blaurock and Wilkins[68]; Gras and Worthington[69]) was a series of diffractions related to a lamellar repeating structure of approx. 300 Å. Seven or eight reflections were included in an analysis in terms of an electron density profile through the membrane repeat. In the initial analysis the involvement of a sandwich-type structure of the Danielli–Davson kind was assumed but the resulting phase selection was later supported by a study of swelling and shrinking in solutions of different tonicities. Blaurock and Wilkins[68] calculated an electron density profile which featured membrane pairs at about 85 Å centre-to-centre separation each characterised by a centrally located deep trough (Fig. 7). Such pairs formed the retinal disc and the discs were stacked in parallel array perpendicular to the axis of the retinal rod or cone. The lipid layer was suggested to be no more than 26 Å thick and the liquid hydrocarbon region only 18 Å thick. Gras and Worthington[69] related their measured intensities to a five-parameter electron density strip model and concluded that the lowest density region (corresponding to liquid hydrocarbon) was only 16 Å thick. An additional feature of the diffraction data from retinal photoreceptors is a side-spacing of about 50 Å (Blasie et al.[70,71]; Blaurock and Wilkins[68]). This feature probably reflects a subdivision of a substantial proportion of the membrane substance and has been related in particular to the distribution of rhodopsin in the retinal discs.

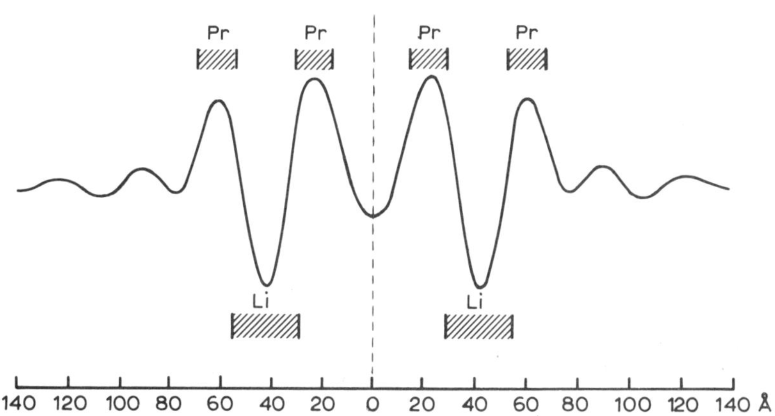

Fig. 7. Electron density profile of repeating layer in retinal outer segments of frog[68]. Pr, protein; Li, lipid.

A third membrane system which occurs naturally in a multilayered form is the grana of chloroplasts. This, too, has been studied by X-ray diffraction techniques (Kreutz[72, 73]) and has been observed to feature both lamellar reflections which can be related to an arrangement of protein and lipid layers and a very prominent side-spacing which suggests the subdivision of a substantial proportion of the substance of the membrane within the plane of the layers.

Other cell membranes need to be isolated and concentrated in order to obtain samples suitable for X-ray diffraction studies but isolation techniques have improved dramatically in recent years and significant X-ray diffraction data have now been reported for a variety of membrane preparations. Probably the most readily available cell membrane preparation is that of erythrocyte ghosts and these have provided the material for much of the exploratory work. When such ghosts are prepared by lysis of the erythrocytes at a variety of osmotic strengths they show wide variations in chemical composition[74, 75], permeability[76] and enzymatic activities. They also show variations in diffraction characteristics[75]. In general, when the ghosts are freshly prepared at about neutral pH the pellet obtained by ultracentrifugation is highly hydrated (more than 95% water) and of a very loose consistency. Such material has been studied by X-ray diffraction and Wilkins, Blaurock and Engleman[77], and Engleman[78] have recorded continuous diffraction which has provided a basis for comparison with similar data from water-dispersed lipid and from which it has been possible to conclude that the membrane probably features a lipid bilayer with parameters similar to those of the isolated lipid. Sharp diffraction bands had been recorded from such preparations but only after removal of a substantial proportion of the water content (Finean et al.[79]). Clear patterns were recorded when the water content was reduced to between 50 and 30% but two different patterns were reported for membranes prepared at very low osmolarities (below 5 imosm) and those prepared at higher molarities[75]. The low osmolarity preparations yielded a lamellar-type pattern which was interpreted in terms of a low-density-centre electron density profile but from the membranes prepared at higher osmolarities the authors were unable to obtain a diffraction pattern that they could consider to represent the intact, hydrated membrane. More recent studies[80] have shown that more compact ultracentrifuge pellets containing only 60 to 70% water may be obtained following lysis in distilled water or by adding very low (0.01%) concentrations of detergent (Triton X100) to the washing medium. Such pellets give well-defined low angle X-ray diffraction patterns without further reduction in water content and the patterns are very similar to those recorded earlier from 'condensed' samples.

Detailed low angle X-ray diffraction patterns with reflections out to the seventh order have also been recorded from 'condensed' samples of preparations of brush border membranes from intestinal epithelium and these, too, have been interpreted in terms of an electron density step profile model derived from the appearance of the membranes in electron micrographs of sections through the 'condensed' sample[81]. Several other membrane preparations including liver plasma membranes, kidney plasma membranes[82], muscle microsomes[83] (largely from sarcoplasmic reticulum) and outer and inner membranes of mitochondria[84] have yielded less detailed though still useful diffraction data which provide evidence of the wider significance of the layered membrane model. Characteristic membrane features have also been identified in an electron density profile calculated from diffraction data obtained from a compact preparation of virus particles[85]. Several additional membrane preparations have also been studied by the 'continuous diffraction' approach and these, too, appear to feature lipid bilayers in their structure[77].

In general then the X-ray diffraction patterns obtained from a variety of hydrated membrane preparations have provided strong support for the suggestion that the lipid component of membranes forms a centrally located bilayer which provides a narrow (20 to 30 Å) layer of liquid hydrocarbon at the centre of the membrane. The possibility of penetration of protein into this layer is recognised in all such membrane studies and from considerations of molecular densities and structural parameters it has been possible to suggest tentatively (Wilkins et al.[77]) that in retinal outer segment membranes this penetration amounts to about 20 to 30% occupation of the central region by non-lipid material. Estimates of a similar magnitude have also been made from a consideration of the area occupied by granular particles seen in freeze-etched erythrocyte membranes (Branton[86]) and from studies of the modification of erythrocyte membranes and of muscle microsomal membranes by phospholipase C (Coleman et al.[87], Finean et al.[88]). This latter study also involved the recording of low angle X-ray diffraction patterns from control and phospholipase C modified erythrocyte membrane preparations. It was observed that although the membrane area appeared to be reduced by about 50% as a result of displacement of lipid the thickness of the membrane layer as reflected in the low-angle diffraction pattern showed a small decrease. This suggested to the authors that the distribution of the non-lipid components was such that they could undergo a 50% lateral contraction without increase in thickness and hence that any layer of such material overlying the lipid would initially form some kind of open structure through which much of the lipid would be exposed directly to the external environment. Such an arrangement might provide an explanation of the apparently ready access of lipolytic enzymes to the membrane lipid and also for the ready exchangeability of some of the membrane lipids. A condensation of the non-lipid layers would also provide a plausible interpretation of the changes in X-ray diffraction intensities which are observed when erythrocyte membranes are modified by phospholipase C[89].

Spectroscopic and optical studies

It is only during the last few years that techniques such as circular dichroism, optical rotatory dispersion and nuclear magnetic resonance and electron spin resonance spectroscopy have been applied intensively to the problems of membrane structure. Interpretations were initially based solely on experience gained from studies of biologically important molecules such as protein and nucleic acids dispersed in relatively dilute solution and little was known of the effects of molecular interactions on such data. Membrane and other lipoprotein systems which involve lipid–lipid, protein–protein and lipid–protein interactions in structures containing probably no more

References p. 198

than 30% water presented new problems of interpretation which were not at first fully appreciated and are still not fully resolved.

The first clear indication of the type of configuration featured in membrane protein came from infra-red absorption spectra of dried membrane preparations. In studies of erythrocyte membranes, Maddy and Malcolm[90,91] noted the lack of any indication of a β-configuration of the membrane protein and suggested that the spectrum favoured the presence of an appreciable proportion of α-helix. Chapman et al.[92] supported these observations relating to the protein configuration and failed to demonstrate any β-configuration even after the preparation had been heated to 150°. They also compared the infra-red spectrum of the membranes with that of the total lipid extract and concluded that the hydrocarbon chains in the membranes displayed less all-*trans* character than those in the isolated lipid perhaps because of a greater freedom of rotation or perhaps because of alternative configurations adopted when in association with the protein. An absence of protein β-configuration was also noted in the infra-red spectra of plasma membranes and endoplasmic reticulum of Ehrlich ascites carcinoma cells[93].

Studies of optical rotatory dispersion and circular dichroism of membrane preparations in aqueous dispersion[93–95] apparently confirmed the absence of protein β-configuration in several membranes and provided more positive and quantitative assessment of the α-helix configuration. Both techniques have long been used for the study of the configuration of protein molecules in solution where it has been possible to use reference polymers (e.g. poly-L-lysine) of known configuration and to obtain an accurate assessment of the proportions of α-helix and random coil in globular protein molecules. In the case of the optical data from membrane preparations the interpretation is complicated by a reduced amplitude and by the fact that the critical features of the data curves all show a small shift towards the red end of the spectrum as compared with reference polymers in aqueous solution.

Several alternative explanations of these effects have now been proposed.

Urry and co-workers[96,97] contend that the reduced magnitudes and the red shift of ORD are primarily artifacts arising from light scattering and absorption flattening effects by the particulate membrane preparation. They also suggest that the circular dichroism will be further distorted by a differential scattering of left and right circularly polarised light. This kind of explanation has drawn support from the work of Steim and Fleischer[98] on mitochondrial structural protein and from Wrigglesworth and Packer[99] on mitochondria and more recently from work on intact and sonicated erythrocyte ghosts by Schneider, Schneider and Rosenheck[100] and by Glaser and Singer[101]. On the other hand Ottaway and Wetlaufer[102] have concluded from a detailed theoretical analysis that the contribution to circular dichroism measurements by light scattering from particulate material is small.

In some of the earliest studies[93,95] the light-scattering effects were considered to be small and the red shift was considered to be due to protein–protein and/or lipid–protein interactions. Evidence that protein–protein interactions can lead to lowered amplitudes and spectral shifts as compared with synthetic polymers used as reference systems is available[103, 104] but the possibility that in membranes these effects are due largely to the interaction of predominantly non-polar segments of protein α-helix with lipid hydrocarbon chains has been favoured by several authors and has provided the basis for new speculations on membrane structure (Fig. 9).

The initial estimations of α-helix content of erythrocyte membrane protein were in the range of 25–30%. The content of β-configuration was considered to be low and of random coil relatively high. Corrections for light-scattering effects tended to increase the estimated proportion of α-helix to about 50% (Schneider et al.[100]) but recent, more sophisticated curve-fitting computer analysis, has indicated that in mycoplasma membranes β-configuration is higher than would be expected from the appearance of the curves and the α-helix and random coil correspondingly lower. Wallach, Graham and Fernback[106] have also concluded that there is appreciable β-configuration in mitochondrial membrane protein. One can probably conclude that the optical studies have demonstrated that membrane protein contains appreciable proportions of α-helix and other coiled chain

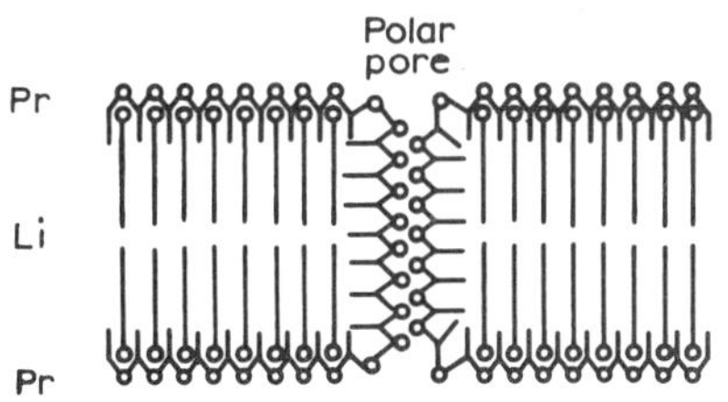

Fig. 8. Schematic drawing of proposed pore structure in a cell membrane (After Stein and Danielli[205]). Pr, protein; Li, lipid.

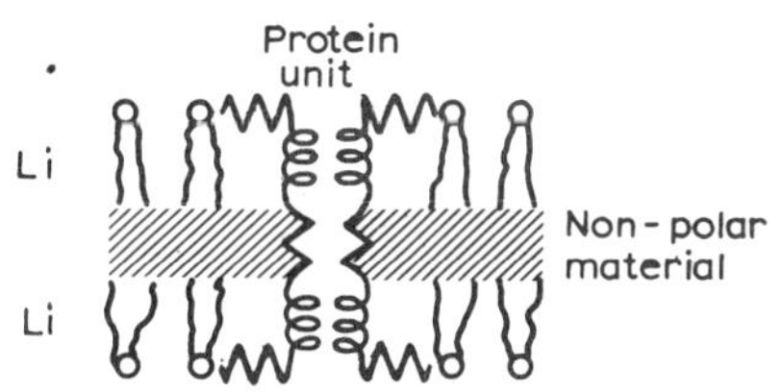

Fig. 9. Schematic drawing of a cell membrane (After Lenard and Singer[95]). Li, lipid.

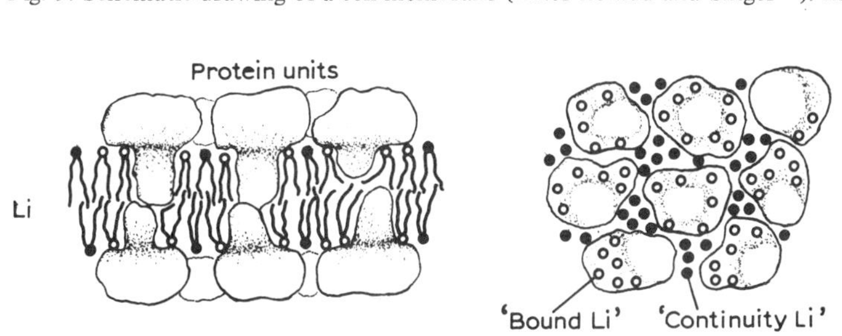

Fig. 10. Model of a cell membrane based on scheme proposed by Zahler and Weibel[206]. Li, lipid. Protein units penetrate into or through the central layer so as to occupy about 30% of central region. Protein, in either compact (as illustrated) or extended form, occupies only about 50% of external layers. Remainder is occupied mainly by water.

configurations but that there may be more β-configuration present than is at first apparent from the spectra. The overall configuration is, nevertheless, essentially that of a globular-type protein. There is as yet no final agreement on the origin of the low amplitude and the red shift of the spectral data from membrane protein as compared with that of reference polymers in solution and any structural suggestions based on these observations must be regarded as very speculative.

Nuclear magnetic resonance absorption spectra of lipoprotein systems[107,108] also present some unsolved problems of interpretation. The proton resonance spectrum (60 MHz) of intact erythrocyte ghosts[107,109] is weak and ill-defined. Membranes disrupted by sonication give a sharp $N(CH_3)_3^{\oplus}$ resonance absorption peak but the peak corresponding to the protons in the lipid hydrocarbon chains remains weak and broad. This peak can be made sharp by sonication with deoxycholate and is also sharp in sonically dispersed lipid extract although it remains broadened when the lipid is only crudely dispersed in water. The explanation given by Chapman and co-workers[107,109] for the broadening of the absorption bands in the spectrum of the intact membranes is that it is due to 'magnetic' inhomogeneity effects which become averaged out in the small particles produced by sonic fragmentation. An alternative explanation considered by Sheard and Bradbury[110] is that in the larger particles there is insufficient segmental motion to average the proton dipolar interactions which are responsible for the line broadening but that particles below a certain critical size adopt a more mobile structure. Neither explanation is at present accepted as completely satisfactory. The continued restriction in freedom of movement of hydrocarbon chains in the membrane fragments produced by sonication as compared with the isolated and sonicated phospholipid may be due to lipid–protein interaction in the membrane. Information on the mobilities of specific regions in lipid molecules may also be obtained from studies of the magnetic resonance absorption spectra of other nuclei such as ^{13}C and ^{31}P. The ^{13}C NMR spectrum of phosphatidylcholine in water has already provided an assessment of relative mobilities along the alkyl chains and in the glycerol residue and choline group but the spectra of erythrocyte ghost preparations were initially too broad for accurate assignments to be made[110a]. However, the ^{31}P resonance signals in phospholipid of sarcoplasmic reticulum preparations have been identified and reported to indicate that the motion of the phosphate group is considerably restricted as compared with other regions of the molecule which can be characterised by proton resonance absorption[110b].

A different approach in the application of nuclear magnetic resonance techniques to the study of lipoproteins has been to observe modifications of spectra of molecules artificially incorporated into the lipoprotein structure. The introduction of low concentrations of a lipid soluble anaesthetic such as benzyl alcohol[111,112] into erythrocyte membranes results initially in a pronounced immobilisation of the benzyl alcohol molecules but this effect decreases as the concentration of benzyl alcohol molecules is increased. This can be interpreted as indicating that the membrane first exerts an immobilising influence on the anaesthetic molecules but that as their concentration increases they introduce disorder into the lipid phase of the membrane. At concentrations greater than 60 m*M* the mobility again starts to decrease and this has been accounted for in terms of the exposure of previously inaccessible sites on the membrane protein as the membrane structure begins to break down[113]. The binding of the benzyl alcohol to membrane protein has been confirmed in studies of the isolated protein components. Similar effects have been observed with other anaesthetic molecules introduced into a variety of membranes but there have been appreciable differences in the magnitudes of the effects observed[112].

Other spectroscopic probe techniques which have now been extended to the study of biological membranes include electron spin resonance[114, 115] and fluorescence[116–118] spectroscopy. Both ESR probes and fluorescence probes are very sensitive to the environment in which they become located and both generally involve small molecular groupings which can be incorporated into larger molecules in order to specifically vary their location in the membrane structure. Thus nitroso groups which have a pronounced electron spin resonance absorption have been attached at various points along fatty acid molecules and introduced into a variety of membranes. Stearic acid with a nitroso group attached to C_{12} has been incorporated *in vivo* into the membrane of *Mycoplasma laidlawii* (Tourtelotte et al.[119]) and the ESR spectra have indicated that although the motion of the spin label may be slightly restricted as compared with the same label in an isolated lipid phase the environment is still that of a semi-viscous, hydrophobic liquid and there is no evidence of non-polar interaction with protein. Fatty acid type molecules with a spin label at a variety of points along the hydrocarbon chain have also been introduced into *Mycoplasma laidlawii* membranes by an exchange process (Rottem et al.[120]) and it has been noted that molecular motion is increased as the label is moved away from the –COOH end of the fatty acid molecule. Spin-label derivatives of 5- and 12-ketostearic acids have also been introduced into the membranes of non-myelinated nerves and into erythrocyte membranes by the exchange procedure (Hubbell and McConnell[121, 122]). In the nerve membranes the molecular motion of the probe was not significantly different from that in the isolated lipid phase but in the erythrocyte membrane there were strong indications of restricted movement. The 12-labelled stearic acid has also been incorporated *in vivo* into Neurospora mitochondria (Keith et al.[123,124]) and again the mobility of the spin label was not significantly different from that in the isolated lipid phase with no pronounced restriction of movement such as is found in a fatty acid–bovine serum albumin complex.

In general the use of spin-labelled fatty acids in the study of membranes would seem to indicate that at least at the position of C_{12} the lipid hydrocarbon forms a highly mobile liquid-like environment in a variety of membranes but where there is evidence of some restricted mobility this is greatest towards the polar regions of the system. There appears to be little evidence of a pronounced restriction of mobility by non-polar association with protein.

Some spin-labelled steroids in which the label is rigidly attached to the ring structure are partially immobilised by incorporation into nerve membranes[122]. The –OH group of these molecules has been considered to be firmly anchored in the polar regions of the membrane. Such probes have been recommended as being particularly useful in

Referecens p. 198

detecting conformational changes in membranes and have been used along with a variety of smaller spin-labelled molecules to study the perturbing effects of other small molecules such as the anaesthetics[125]. The results of such studies have in general supported the conclusions of experiments in which the NMR spectra of the perturbing agents themselves were used to detect changes in fluidity and ultimately breakdown of the membrane structure.

Spin labels have also been covalently linked to such molecules as maleimides (Chapman et al.[126]) which react specifically with protein reactive groups such as –SH and $-NH_2$. They have provided information on the availability of such groups in membranes and also served as conformation-sensitive markers for use in attempts to isolate protein components from membranes (Schneider and Smith[127]). Such probes have also indicated that the conformation of erythrocyte membrane protein is not significantly changed when a high proportion of the membrane lipid is modified by phospholipase C treatment (Simpkins et al.[128]).

Some fluorescent probes such as retinol (Radda and Smith[129]) will also provide information on the fluidity of the lipid phase in membranes and other lipoprotein systems. Such studies have suggested that the fluidity of the lipid phase in erythrocyte membranes is similar to that in isolated lipid. This result would appear to be at variance with the conclusion drawn from some ESR studies but as there is no absolute quantitation of the data the magnitude of the disagreement is not apparent. Other fluorescent probes such as 1-analine-naphthalene-8-sulphonate (ANS)[130] have been found to show a tendency to become oriented at lipid–protein interfaces. Gulik-Krzywicki et al.[131] introduced ANS into a variety of lipid–water and simple lipid–protein model systems and reported that the fluorescence properties of ANS were markedly different in the two types of system. They have suggested that this probe clearly distinguishes a situation in which purely electrostatic interaction is involved from one in which hydrophobic interaction is also possible. They emphasise that the distinction is unlikely to be so clear-cut in membranes where a variety of protein molecules may be involved. In membrane studies so far the main use of ANS has been to detect energy-dependent changes, particularly in mitochondria and mitochondrial particles (Chance[132], Brocklehurst et al.[133]).

The potential variety of fluorescent probe molecules that can be synthesised is very great and they may be expected to provide valuable information on both the mobility and the polarity of the environment in which they become located. The initial problem is to determine a priori the precise location of the probe in the lipid–protein system but it is anticipated that this will become apparent from the study of a wide range of probes which have different physical and chemical characteristics. Meanwhile such probes will continue to provide a very sensitive means of carrying out studies to compare systems of varying compositions and conformations.

Electron microscopy

Although the electron microscope can now resolve detail down to the atomic level and is theoretically capable of providing a direct visualisation of lipid–protein interactions, in practice there is difficulty in preserving lipid–protein interactions through the imposed preparative conditions and also in identifying individual types of molecule in the electron micrographs. A variety of preparative procedures have been applied and although each has its peculiar limitations, together they have provided valuable information on the nature of lipoprotein systems.

Thin-sectioned material has generally been fixed, stained and embedded in such a way that it is difficult to decide directly to what the intensity distribution in the final image should be attributed. In osmium tetroxide-fixed preparations osmium undoubtedly makes a major contribution to the densities seen in the final image but although a good deal is known concerning the chemistry of the initial interaction of osmium tetroxide with lipid and protein components there is considerable doubt as to the ultimate location of the osmium in relation to the lipoprotein architecture when the tissue has been further processed and finally presented for viewing in the electron beam. The weight of present evidence[55, 134, 135], both circumstantial and direct from X-ray diffraction studies, strongly suggests that the highest densities are eventually to be found at the regions of interaction between lipid and protein components. Furthermore X-ray diffraction studies of nerve myelin have indicated that treatment with osmium tetroxide approximately doubles the electron density of the system[136] but additional changes undoubtedly take place during further processing and viewing[137].

The general location of high densities in pairs of parallel sheets in membrane systems studied at low resolution by sectioning techniques may provide an indication of the location of lipoprotein interfaces as well as a convenient means of identifying a membrane structure. However, when viewed at high resolution these lines of high density can be seen to consist of dense granules which may have more specific structural significance. Such dense granules could conceivably correspond to protein molecules or lipoprotein particles featured in a variety of membrane models[138, 139] but at present they are not directly identifiable as such and cannot be taken as positive support for any particular membrane model.

The negative staining technique in which non-fixed membrane preparations are dried down from a dispersion in a solution containing a heavy metal salt has also provided intriguing detail at high resolution. This, too, has led to speculative interpretations in terms of protein or lipoprotein particulate structures in membranes but such interpretations have been criticised on the grounds that they appear to ignore the possibility of molecular re-arrangements which are known from other studies to take place when non-fixed lipoprotein systems are dried[134, 140]. However, strikingly regular features have been observed consistently in a variety of membrane preparations and in some cases corresponding features have been observed in electron micrographs of sectioned material. Such images may at least provide a useful means of distinguishing different membranes and at best reveal important structural components which may have been submerged in the intact structure.

One of the most striking examples of such a granular structure is the so-called 'stalk and knob' feature which is characteristic of negatively stained preparations of mitochondrial inner membrane[141]. These appear to cover the outer surface (facing the

References p. 198

matrix) of tube-like projections which have been suggested to represent individual cristae in the intact mitochondrion. As a result of experiments in which a granular fraction has been recovered following treatment which resulted in the membranes losing their knob-like projections it is now generally accepted that the knobs represent ATPases involved in the mitochondrial electron transport system[142, 143]. Similar experiments have led to the identification of granules seen to be associated with negatively stained preparations of intestinal brush border membrane preparations as specific degradative enzymes[144]. Such images of negatively stained membrane preparations have been considered to support the theory that membranes in general consist of a two-dimensional assemblage of lipoprotein subunits but they have cast no light on the details of the lipid–protein association.

During the last few years there has been a rapid increase in the application of freeze-etching (or freeze-cleaving) techniques to the study of membrane structure. There was initially a strong difference of opinion as to the location of the cleavage plane with respect to the molecular organisation of the membrane but several ingenious experiments by Branton and coworkers[145, 146] in particular have indicated that the cleavage is most likely to occur along a non-polar plane and hence, in terms of the sandwich-type structure, to expose the inside of the membrane. The cleaved surface of myelin has been reported to be smooth but all other membranes so far studied by this method have revealed a scattering of granules on an otherwise smooth surface. In the case of freeze-etched erythrocyte ghosts it has been calculated[147] that the granules occupy about 30% of the exposed surface. In some experiments it has been possible to study both of the complementary surfaces exposed by freeze-cleaving but although a qualitative complementarity has been seen a quantitative matching does not seem to have been possible. The granules are generally considered to be protein particles which penetrate into or through the centre of the membrane and the smooth regions are presumably related to non-polar regions of the lipid phase. In terms of lipid–protein interaction these observations suggest that protein molecules could come into association with non-polar regions of lipid.

In general, although there appears to be considerable doubt about the precise molecular significance of the fine detail in electron micrographs of all types of preparation there is nevertheless a definite trend in the minds of microscopists towards interpretations which involve interactions between lipid and protein at more than one level in the membrane. Completely segregated lipid and protein layers or complete segmentation into discrete lipoprotein sub-units are likely to be the exception rather than the rule in membrane structure.

4. LIPID-DEPENDENT ENZYME ACTIVITIES

Many membrane-located enzyme activities have been observed to decline or to be lost completely when membrane lipid is extracted or modified but some have been fully or partially restored by the addition of new lipid[148, 149]. A partial list of such enzyme activities is given in Table 1. The implication of such a lipid dependence would appear

TABLE 1

LIPID-DEPENDENT ENZYME SYSTEMS

EC No.	*Enzyme activity*	*Lipid requirement demonstrated*	*Reference*
1.1.1.30	D(–) β-Hydroxybutyrate dehydrogenase	Phosphatidylcholine	Sekuzu et al.[150, 152] Jurtshuk et al.[151, 153]
	Mitochondrial electron transport	Total mitochondrial lipid	Fleischer et al.[154] Green and Fleischer[155] Green et al.[33] Racker and Bruni[156]
3.6.1.3	$Na^{\oplus}/K^{\oplus}$ ATPase	Phosphatidylserine	Fenster and Copenhauer[157] Wheeler and Whittam[158]
		Lecithin > lysolecithin > phosphatidic acid	Tanaka and Strickland[163]
1.3.99.3	Stearyl coenzyme A desaturase	Phospholipid, triglyceride and fatty acid	Jones et al.[160]
3.1.3.9	Glucose-6-phosphatase	Phosphatidylethanolamine	Duttera et al.[161]
3.1.3.4	Phosphatidic acid phosphatase	Total microsomal lipid	Coleman and Hubscher[52]
2.7.7.15	Phosphorylcholine cytidylyl transferase	Lysolecithin ≫ lecithin	Fiscus and Schneider[162]
3.6.1.3	$Ca^{2\oplus}/Mg^{2\oplus}$ ATPase	Lysolecithin, lecithin, neutral detergents	Martonosi et al.[165]
	$Ca^{2\oplus}$ transport	Lysolecithin, lecithin, phosphatidic acid	Martonosi et al.[165]
2.4.1.–	UDPgalactose: lipo-polysaccharide α-3-galactosyl transferase	Phosphatidylethanolamine	Weiser and Rothfield[166]
	Deoxycorticosterone 11-β-hydroxylase	Total mitochondrial lipid	Williamson and O'Donnell[167]
1.6.99.3	NADH-cytochrome *c* reductase	Lecithin:lysolecithin, 1:1.	Jones and Wakil[167a]
6.2.1.–	Acyl-CoA synthetase	Phosphatidylcholine	Sartorelli et al.[167b]

to be that there is some interaction between lipid and the residual enzyme system which is critical either for maintaining a specific configuration of the enzyme protein or for facilitating interaction between enzyme protein and the substrate. In the case of the reactivation of β-hydroxybutyrate dehydrogenase by an unsaturated phosphatidylcholine there is additional evidence of direct interaction between the depleted mitochondrial enzyme and the reactivating lipid.

The demonstrated requirements are predominantly for polar lipids and in a few cases the requirement appears to be for a particular polar lipid or for a restricted group of lipids. Such a specificity would seem to imply that the interaction between the enzyme protein and the lipid is an electrostatic one and that in the intact membrane many enzyme proteins are normally associated electrostatically with specified lipid molecules. The conclusion is, however, a tentative one, for the experimental demonstration is of a reactivation which is not necessarily a reconstitution as is frequently assumed. Furthermore, although the activating lipid is presented to the depleted enzyme in the form of a polar molecular aggregate and the initial interaction would therefore be expected to be an electrostatic one, there may well be additional steps in the activating process which would further modify the nature and extent of interaction between lipid and protein components of the enzyme system. It would be of interest in this respect to establish the conditions necessary to remove this reactivating lipid so as to again render the system inactive.

Despite these reservations concerning present conclusions relating to the detailed nature of the lipid–protein interactions essential for some membrane-located enzyme activities the reactivation experiments still provide important sources of information on functionally significant lipid–protein interactions which should be of significance in relation to membrane structure.

5. SOLUBLE LIPOPROTEINS

In discussing lipid–protein interaction in soluble lipoproteins there is a clear distinction to be made between low density (below 1.063: includes very low density and low density fractions) lipoproteins of blood and of egg yolk which contain relatively little (up to 25%) protein and the high density lipoproteins in which the protein content approaches 50% (ref. 168).

In the case of the low density lipoproteins the protein component is considered to be largely confined to a surface layer but there is insufficient to form a condensed layer over the whole surface. Optical and spectroscopic studies (Scanu and Hirz[169]) have indicated that this protein includes extended β-configuration (antiparallel pleated sheets), α-helix and random coil configurations. The β-configuration is probably dominant but the overall conformation appears to be dependent on the proportion of lipid in the particle (Dearborn and Wetlaufer[170]) and is sensitive both to temperature changes and to removal of lipid[169,171]. There is a suggestion that flexibility or perhaps adaptability of the protein conformation in the lipoprotein particle may be significant in relation to lipid transport[169].

It has been suggested that phospholipid (present to the extent of about 25%) together with free cholesterol (Schumaker and Adams[172]; Gotte[173]) complete the surface layer. This phospholipid is readily accessible to phospholipases and the whole surface is visualised as a mosaic of lipid and protein, both components presenting largely ionised groups to the external environment and anchored to the predominantly neutral lipid core through non-polar interaction. The extent and nature of any interaction between the protein and phospholipid at the surface is not evident in this general picture. Nuclear magnetic resonance studies of the blood lipoproteins (Steim et al.[174], Chapman and co-workers[175,109]) have suggested that the lipid in the particles is in a state of mobility approaching that of free lipid in organic solvents. The line widths of the hydrocarbon chain proton absorption bands are such as to suggest that there is no restriction of mobility such as might arise from extensive non-polar association with protein. The absorptions specific to groups on the phospholipids also suggest complete freedom from restrictive interactions. On the other hand, proton resonance absorptions associated with groups on the protein do not show up until the lipoproteins have been treated with detergents[176]. This could mean that the mobilities of such groups are significantly restricted in the lipoprotein structure. Thus, although the protein conformation would appear to be dependent on the association with lipid, there is no evidence of restriction of mobility of the lipid such as is found in the membrane lipoprotein structures.

The high density lipoprotein of blood plasma has also been studied by spectroscopic and optical methods[172–176]. The protein has a much higher content of helical configuration (55%–70%) and the general conformation is probably less affected by the lipid component than it is in the low density lipoproteins although the lipid may have some stabilising effect. Nuclear magnetic resonance studies again show that the lipid is in a very mobile state which is not significantly affected by association with protein. Lipoprotein reconstituted from the water-soluble apoprotein of HDL and phospholipid displays density and spectroscopic characteristics very similar to those of the original high density lipoprotein[175]. In some electron spin resonance studies a spin label was attached to –SH groups of the high density lipoprotein and observed to distinguish two attachment sites which showed appreciably different constraints[173]. The high constraint was greatly reduced by delipidation.

Thus, although the lipid components of both low density and high density lipoproteins exhibit great mobility which precludes any appreciable restriction through association with protein, there are nevertheless certain features of the protein structure which seem to be sensitive to changes in the lipid phase.

6. LIPID–PROTEIN INTERACTIONS *IN VITRO*

The interaction of individually well-characterised lipid and protein components has been studied in a variety of physical situations using a variety of analytical techniques.

The interaction of protein with lipid monolayers at an air–water interface was studied first by Doty, Matalon and Schulman[178,179] and later by Eley and Hedge[180]. Doty, Matalon and Schulman maintained the lipid monolayer at a constant surface area established initially at a pressure of 15–20 dynes/cm and then noted the increase in surface pressure when soluble protein such as albumin or haemoglobin was injected into the aqueous substrate. Under these conditions they established that the primary

References p. 198

interaction was an electrostatic one and that the adsorption of protein was reversible. At this pressure the protein was unable to penetrate the lipid layer to any appreciable extent but when the interaction was carried out at constant surface pressure extensive penetration did occur. In experiments involving the interaction of proteins with monolayers of synthetic phosphatidylethanolamines Eley and Hedge established the monolayer at a low pressure of 2 dynes/cm so that the protein was able to penetrate into the phospholipid layer. They concluded that there was extensive penetration by protein side-chains although the initial interaction was ionic. Such penetration would involve some unfolding of the protein chains which appears not to have been completely reversible. Thus the monolayer studies have so far emphasised a primary electrostatic interaction between phospholipid monolayers and protein substrate followed at low surface pressures by a penetration of some of the protein side chains deeper into the lipid layer as the polypeptide chains unfold.

Observations on the modification of properties of lipid bilayers by protein have been largely confined to the effects of excitability inducing material (EIM) and of some cyclic polypeptide antibiotics on membrane permeability and conductance. EIM is most readily prepared from egg white which has been inoculated with *A. cloacae* and although it is not finally characterised it is at present considered to be a nucleoprotein. When a phospholipid-containing bilayer was exposed to this material there was a dramatic change in conductance which displayed the characteristics of an action potential[181]. A similar effect[182] can be obtained with alamethicin, a cyclic polypeptide of nineteen amino acids, and a variety of similar cyclic antibiotics have been observed to greatly modify the permeability to ions both of experimental bilayers and cell membranes[183–188]. A common structural feature of many of these antibiotics is that the non-polar portions of the amino acids may be oriented so as to form a predominantly non-polar external surface which might enable the antibiotic to penetrate into the non-polar regions of lipid bilayer or membrane. The interior of the ionic molecule would then be polar and capable of accommodating metallic cations. In the case of alamethicin the effect on the membrane properties is reported[182] to vary with the sixth power of the concentration which would suggest a hexamolecular cooperative effect. It was initially suggested that such molecules might provide polar pores through the non-polar regions of the membrane although not all aspects of the experimental observations are readily reconciled with this idea.

When phospholipid is dispersed in an aqueous medium it forms aggregates in which the ionic groups are directed towards the water phase and the hydrocarbon chains are accommodated inside the aggregate to form a non-polar environment. The aggregate may be a multilamellar structure in which each layer is a bimolecular leaflet or it may be reduced to a single layered vesicle and even to a fragment of a bilayer or a spherical 'micelle' depending on the nature of the phospholipid and the techniques used for dispersion. Always, the aggregate presents a polar surface to the aqueous environment and any interaction with protein in solution must initially at least involve electrostatic forces. Phospholipid alone under these conditions will invariably form an aggregate mostly with a negative surface charge and will interact most readily with protein which has a net positive charge. In some very early X-ray diffraction studies the addition of basic proteins such as histones to aqueous dispersions of brain 'cephalin' produced an insoluble lipoprotein complex which the diffraction data suggested to feature bimolecular layers of lipid 'cemented' together by monolayers of protein[189]. Electron micrographs which appear to demonstrate such a structural arrangement were obtained by Stoeckenius[190] from dispersions of brain phospholipid to which soluble protein had been added.

More extensive X-ray diffraction studies of such systems have been made by Gulik-Krzywicki et al.[191]. Cytochrome *c*, lysozyme and ribonuclease were allowed to interact with a variety of phospholipid systems in aqueous dispersion and several different lipoprotein phases were identified. Most were lamellar in which one or more layers of protein molecules were intercalated between bilayers of lipid but despite the predicted close-packing of protein molecules within the layers there were no indications of a corresponding diffraction. Optical and fluorescent probe studies of these same systems[131] have distinguished situations in which the interaction between lipid and protein is purely electrostatic and also some in which there is extensive interpenetration of the two components so that appreciable non-polar interaction is achieved. The circular dichroism measurements indicated appreciable changes in protein conformation although protein could be recovered in an active state.

A further series of studies which has emphasised the importance of electrostatic conditions on the surface of the phospholipid aggregate in relation to interaction with protein involved the interaction of phospholipases with phospholipid substrates (Dawson[192]). Dispersions of lecithin alone were hydrolysed at a very slow rate by phospholipase B but the rate rose markedly when an anionic long-chain compound was added so as to give the lipid aggregate a small net negative charge. An uncharged lecithin aggregate also remained little affected by phospholipase C until the surface charge was changed but in this case a net positive charge was required and this could be achieved either by introducing a long-chain cation or by adding divalent metallic ions to the bulk phase. In both cases it would appear that electrostatic forces were primarily involved in the initial approach of the enzyme protein to the lipid substrate. Relationships between surface charge on lipid aggregates and the action of other phospholipases appeared to be less simple. In order for phospholipase D to attack lecithin effectively certain long-chain anions such as dodecyl sulphate, phosphatidic acid and triphosphoinositide (but not all acidic phospholipids) were required but there was also an absolute requirement for high concentrations of calcium which would counter the charge effect of the anions. This would suggest that there may be several distinct steps in the approach and interaction of the enzyme with the lipid aggregate. These phenomena are discussed further in Chapter 5 (p. 97).

The substrate for triphosphoinositide phosphoesterases is naturally highly negatively charged and the hydrolysis of the pure substrate normally continues for a very short time. The hydrolysis can be maintained by the addition of substances which partially neutralise the negative charge. A suggested explanation is that the high negative charge on the lipid aggregate leads to an unfolding of enzyme protein over the surface. This limits the hydrolysis.

Re-aggregation of dispersed membrane components

Most of the re-aggregation experiments fall into three categories. One is the restoration of an extracted component to the residual structure as, for example, the addition of lipid to a membrane system that has been partially extracted with an organic solvent. As discussed earlier the criterion for significant re-aggregation may be the reactivation of an enzyme system but this does not necessarily imply reconstitution in the sense that the original membrane structure has been completely restored and in most studies additional evidence of reconstitution has not been provided.

A second category of re-aggregation experiments is one in which membranes which have been 'solubilised' by, for example, detergent treatment are made to re-aggregate. In many instances this has been achieved by dialysing away the detergent and the re-aggregation to form a membrane-like structure has been confirmed by electron microscopy frequently using only the negative staining procedure. In many instances no attempt has been made to test the restoration of any function so the significance of the aggregation is not established. Furthermore, since the method of solubilisation of membranes by detergents is not well understood the experiments throw little light on the nature of the lipid–protein interaction.

The third category includes those experiments in which the lipid and protein components have been separated before re-aggregation. Probably the most extensively studied system from this point of view is the electron transport system of the mitochondrion. The individual complexes of this system have been separated and freed of phospholipid and then the active system reformed by adding back the phospholipid together with any other essential component such as coenzyme Q which may also have been extracted[32, 193]. The complexes can be reactivated individually but it would appear that in order to completely restore the whole electron transport system the lipid-depleted complexes must be recombined with lipid simultaneously[194, 195]. The reconstituted system is then seen by electron microscopy to be in the form of membranous vesicles. Green and co-workers have suggested that the initial separation of the complexes by deoxycholate involves the breaking of predominantly non-polar protein–protein and lipid–lipid bonds but there is some uncertainty concerning the nature of the lipid–protein associations. Cardiolipin, which is specifically required for the reactivation of some of the complexes would be expected to interact ionically. Cytochrome *c* which appears to link the activities of complexes III and IV also forms a predominantly polar interaction with phospholipid *in vitro* but the interaction of phospholipid and 'structural protein' *in vitro* is non-polar[196]. There is, however, considerable doubt as to the origin and significance of structural protein from mitochondria[194].

In terms of the tripartite units suggested by Green and co-workers[31] to be the fundamental structural units of mitochondrial inner membranes the main (and perhaps the only) lipid–protein interactions are located in the basepieces each of which is suggested to contain at least one of the electron transport complexes together with essential phospholipid. Such basepieces are considered to be associated in two dimensions to form the membrane continuum in which coenzyme Q and cytochrome *c* can act as mobile elements linking the activities of the individual complexes[195]. The physical distinction between this continuum of lipoprotein basepieces and the more traditional view of a membrane as a predominantly continuous lipid layer with associated structural and functional proteins appears to have become largely a matter of emphasis. However, the precise nature of the lipid–protein interactions is still not established.

In a few experiments lipid and protein components of membranes have been separated without the involvement of detergents and attempts have then been made to re-aggregate them in a meaningful way. Ji and Benson[198] prepared chloroplast protein by sequential extraction of sonically dispersed chloroplasts with 60 to 100% acetone. Recombination with chloroplast lipid was then effected by equilibration of a dispersion of protein in methanol with lipid in methanol across a dialysis membrane followed by precipitation by dialysis against aqueous ethanol. The composition of the precipitate was similar to that of the intact chloroplast but other characteristics of significance in relation to membrane structures were not demonstrated.

A recombination in an organic solvent was also attempted by Zahler and Weibel[199] who solubilised erythrocyte membranes in 2-chloroethanol, separated the protein by gel filtration and then brought both protein and lipid together again in 2-chloroethanol. In 2-chloroethanol the protein was observed by optical studies to be in a predominantly α-helical configuration[200] and the phospholipid was considered to be in a monodisperse form. The recombined lipoprotein was precipitated by dialysis against aqueous buffer and a main component similar in density and composition to the intact erythrocyte membrane was separated on a density gradient. This component showed membrane characteristics by electron microscopy and nuclear magnetic resonance studies also indicated a significant recombination but membrane activities were not demonstrated.

Zwaal and Van Deenen[201] separated erythrocyte membrane protein and lipid by the butanol extraction procedure of Maddy[49] and then attempted re-aggregation in aqueous media. In this case re-aggregation could be accomplished only after removal of sialic acid residues from the protein and then only at pH 3.5 to 4.5 when the lipid and protein were of opposite net charge. The resulting precipitate had a phospholipid : protein ratio similar to that of the original erythrocyte ghosts and showed membrane-like trilamellar characteristics in electron micrographs. Nuclear magnetic resonance[202] and electron spin resonance studies[203] showed the lipid in the re-aggregated lipoprotein to be in a physical state comparable to that in the erythrocyte membrane but restoration of physiological or biochemical activities was not demonstrated. There was no breakdown of the re-aggregated lipoprotein when the pH was subsequently raised to pH 7: neither could any breakdown be effected with strong salt solution. From this it was concluded that non-polar interactions had taken place after the initial ionic association at pH 3.5 to 4.5.

7. CONCLUSIONS

The probability that in many membranes the interaction between lipid and protein involves both polar and non-polar elements appears to be clearly indicated even though some of the supporting evidence may be interpreted in alternative ways. The data do not,

References p. 198

however, permit a quantitative assessment of the relative significance of the two types of interaction and any conclusion that one or the other is dominant should still be viewed with reserve.

Any spatial models proposed for membranes should make provision for both polar and non-polar interactions and a variety of published models do make such a provision. Several recent reviews have systematically assessed the basis for, and the criticisms of, these models and at least two[134, 135] have concluded that the sandwich-type structure still provides the most satisfactory general explanation of the available experimental data. There is general agreement that the lipid component of most membranes is predominantly in bilayer form.

The sandwich-type model was first proposed in 1935 by Danielli and Davson[204] but initially they made no specific provision for non-polar interaction. A later version, introduced by Stein and Danielli[205] specifically to make provision for ion permeabilities, also introduced the possibility of non polar interaction between the lipid hydrocarbon chains and protein which penetrated through the membrane to form protein-lined pores (Fig. 8) . In this version of the sandwich structure, however, the protein was indicated as being entirely in an extended β-type configuration and this is not supported by more recent optical data.

To conform with the optical data from erythrocyte membranes Lenard and Singer[95] introduced a symbol for α-helix into this kind of molecular scheme (Fig. 9) and also made provision for further non-polar material in the centre of the sandwich, perhaps to allow for more non-polar interaction between lipid and protein. The necessity for any additional centrally located protein component is made doubtful by the recent quantitative estimates of the amount of protein accommodated in the central lipid layer. The data of Branton[86] reveal (presumed) non-lipid granules occupying 30% of the central interface of the lipid layer, some of which are considered to represent cylindrical components penetrating from the predominantly non-lipid regions. The observations of Coleman et al.[87] suggest that the extent of interruption of the lipid layer in erythrocyte membranes is no more than 30%. Taking these suggestions together it would seem that the most satisfactory explanation would be a centrally located lipid bilayer with protein penetrating into or through it so as to occupy up to 30% of the total area.

Of the published molecular models, that of Zahler and Weibel[206] is probably the one that can most readily be adapted to represent the quantitative data available on erythrocyte ghosts. In this model (Fig. 10), mushroom-shaped or dumb-bell-shaped protein moieties are arranged with their narrower shafts or stalks penetrating into or through a lipid bilayer and their more bulbous heads located outside the lipid to form predominantly non-lipid layers. If the density of the lipid is taken to be 1.05 and the protein which penetrates into it is at the maximum density of 1.33 then from the known protein:lipid ratio of about 60 to 40 by weight it can be calculated that about two-thirds by weight of the protein lies outside the lipid layer.

The absence of a thickness increase during the shrinkage of erythrocyte membranes following phospholipase C treatment would suggest that initially this protein does not form a fully condensed layer or layers but one that can contract by at least 50%.

The two extreme structures that one can consider are mushrooms with fully condensed heads (density 1.33) floating independently on the sea of lipid or mushrooms with more expanded heads making more protein:protein contacts and forming a kind of open patchwork on the lipid surface. Most probably there would be considerable variation in the extent of penetration of individual protein molecules into the lipid layer and the configuration of the exposed protein may also vary. If the undersides of the mushroom heads were to provide sites for ionic interaction with lipid then this would allow for only a very limited ionic interaction and non-polar lipid–protein and lipid–lipid interactions would predominate. Much of the lipid would of necessity remain free from direct association with the protein. A proportion of exposed lipid, held in the structure solely by non-polar lipid–lipid interaction could account for accessibility of lipid to such agents as lipolytic enzymes and for exchangeability of lipid. Such exposed lipid might include mainly the less highly charged phospholipids and cholesterol which are known to be fairly readily extracted from membranes and which may be regarded as 'continuity lipid' to distinguish it from that which is bound directly to the protein. The proportion of such 'continuity lipid' may vary in different types of membranes.

The clustering of lipid molecules around predominantly non-polar protein stalks and their simultaneous binding to the polar undersides of the protein heads would tend to make the clean separation of protein and lipid difficult both in aqueous meaia and in organic solvents and would also encourage conformational changes in the protein when isolated in either polar or non-polar media. The difficulties encountered in interpreting the action of solvents on membranes would certainly be understandable in these circumstances. So, too, would be the problems of reconstituting a membrane from separated lipid and protein constituents. Reaggregation through predominantly polar or non-polar forces might be readily effected and this would probably form a layered structure with membrane-like morphology, but the restoration of the specific penetrations of protein moieties into the interior of the lipid would be much less certain and on these deeper interactions many of the membrane functions may depend.

The ability of some detergent molecules to fragment biological membranes without irreversibly inactivating membrane proteins or changing their configurations could be due to a limited interaction with lipid sufficient simply to break the continuity of the membrane. Functionally significant reconstitution might then be possible simply by removal of the detergent by dialysis. Some membrane-located enzyme activities are, however, modified even by traces of a mild detergent which must therefore be assumed to interact with the immediate environment of the active protein though not necessarily with the protein itself. Strongly ionic detergents, such as sodium dodecyl sulphate which modify the protein configuration probably displace closely associated lipid and interact directly with the protein to produce irreversible changes.

A model of this kind also has interesting implications in relation to membrane biogenesis and membrane fusion. It might be assumed that some of the membrane proteins have specific requirements for lipid which become directly associated with them. They may acquire this lipid immediately after synthesis in order to preserve a critical configuration or they may at a later stage select lipid molecules from an existing general lipid phase in order to acquire a functionally critical configuration. The lipid associated directly with the protein might be genetically prescribed but any further associations required to incorporate the active lipoproteins into a continuous membrane system may be relatively unspecific. Consequently a specific functional lipoprotein may be accommodated in a variety of lipid structures and perhaps transferred from one to another. It may also be able to migrate independently within the plane of the membrane. In situations where the 'continuity lipid' is similar in two membranes fusion may be readily effected, perhaps aided by the incorporation into these regions of further elements which tend either to introduce a slight instability or to make the two systems more compatible. On the other hand, in situations where lipid-dependent proteins are dominant and where protein–protein interactions play a more prominent role in establishing the overall molecular organisation a departure from the lamellar theme in which lipid is largely responsible for the continuity may be expected.

The present situation with respect to opinions on the molecular organisation of membranes is that there is widespread agreement on general features and differences in proposed models are largely a matter of emphasis. Thus the suggestion that membranes are an assembly of 'nesting subunits'[31] appears to differ fundamentally from the idea that membranes feature a predominantly continuous lipid layer with which active proteins associate and into or through which they may penetrate. However, when the penetration amounts to 30% and the proteins carry with them a substantial amount of prescribed lipid, such lipoproteins might be considered to become equivalent to nesting subunits cemented together by continuity lipid. They nevertheless form what amounts to a predominantly continuous phase of lipid which serves to maintain the orientation of the active lipoproteins and to complete the barriers required to compartmentalise the cell.

Models of necessity oversimplify and they also tend to overemphasise certain features in order to stress their importance. Qualifications made in the accompanying text are frequently overlooked.

REFERENCES

1 G.J. Nelson, *Biochim. Biophys. Acta*, 144 (1967) 221.
2 D.B. Weinstein, J.B. Marsh, M.C. Glick and L. Warren, *J. Biol. Chem.*, 244 (1969) 4103.
3 T.K. Ray, V.P. Skipski, M. Barclay, E. Essner and F.M. Archibald, *J. Biol. Chem.*, 244 (1969) 5528.
4 G.G. Forstner, K. Tanaka and K.J. Isselbacher, *Biochem. J.*, 109 (1968) 51.
5 L.C. Mokrasch, in A. Lajtha (Ed.), *Handbook of Neurochemistry*, Vol. I, Plenum, New York, 1969, p. 171.
6 C. Cotman, M.L. Blank, A. Moehl and F. Snyder, *Biochemistry*, 8 (1969) 4606
7 G. Rouser, G.J.M. Nelson, S. Fleischer and G. Simon, in D. Chapman (Ed.), *Biological Membranes*, Academic Press, London, 1968, p. 5.
8 T.W. Keenan and D.J. Morré, *Biochemistry*, 9 (1970) 19.
9 N.C. Nielsen, S. Fleischer and D.G. McConnell, *Biochim. Biophys. Acta*, 211 (1970) 10.
10 V.P. Whittaker, *Ann. N.Y. Acad. Sci.*, 137 (1966) 982.

11 E.G. Lapetina, E.F. Soto and E. De Robertis, *J. Neurochem.*, 15 (1968) 437.
12 D.A. White and J.N. Hawthorne, *Biochem. J.*, 120 (1970) 533.
13 H. Blaschko, H. Firemark, A.D. Smith and H. Winkler, *Biochem. J.*, 104, (1967) 545.
14 J.A.F. op den Kamp and L.L.M. van Deenen, in E. Tria and A.M. Scanu (Eds.), *Structural and Functional Aspects of Lipoproteins in Living Systems*, Academic Press, London, 1969, p. 227.
15 P.F. Smith, *Advan. Lipid Res.*, 6 (1968) 69.
16 S. Razin, *Ann. Rev. Microbiol.*, 23 (1969) 317.
17 M. Kates, *Advan. Lipid Res.*, 8 (1970) 225.
18 R.P. Longley, A.H. Rose and B.A. Knights, *Biochem. J.*, 108 (1968) 401.
19 L.L.M. van Deenen, *Progr. Chem. Fats Lipids*, 8 (1965) 1.
20 L. de Bernhard, *Bull. Soc. Chim. Biol.*, 40 (1958) 161.
21 L.L.M. van Deenen, V.M.T. Houtsmuller, G.H. de Haas and E. Mulder, *J. Pharm. Pharmacol.*, 14 (1962) 429.
22 M.C. Phillips, V.B. Kamat and D. Chapman, *Chem. Phys. Lipids*, 4 (1970) 409.
23 D. Mazia and A. Ruby, *Proc. Natl. Acad. Sci. (U.S.)*, 61 (1968) 1005.
24 V.T. Marchesi and E. Steers Jr., *Science*, 159 (1968) 203.
25 S.A. Rosenberg and G. Guidotti, *J. Biol. Chem.*, 244 (1969) 5118.
26 C.D. Mitchell and D.J. Hanahan, *Biochemistry*, 5 (1966) 51.
27 A.D. Brown, *Biochim. Biophys. Acta*, 75 (1963) 425.
28 S. Razin, H.J. Morowitz and T.M. Terry, *Proc. Natl. Acad. Sci. (U.S.)*, 54 (1965) 219.
29 D.M. Engelman, T.M. Terry and H.J. Morowitz, *Biochim. Biophys. Acta*, 135 (1967) 381.
30 D.M. Engelman and H.J. Morowitz, *Biochim. Biophys. Acta*, 150 (1968) 385.
31 D.E. Green and J.F. Perdue, *Proc. Natl. Acad. Sci. (U.S.)*, 55 (1966) 1295.
32 D.G. McConnell, A. Tzagaloff, D.H. MacLennan and D.E. Green, *J. Biol. Chem.*, 241 (1966) 2373.
33 D.E. Green, D.W. Allmann, E. Bachmann, H. Baum, K. Kopaczyk, E.F. Korman, S.H. Lipton, D.H. MacLennan, D.G. McConnell, J.F. Perdue, J.S. Rieske and A. Tzagaloff, *Arch. Biochem. Biophys.*, 119 (1967) 312.
34 D. Chapman, V.B. Kamat, J. de Gier and S.A. Penkett, *J. Mol. Biol.*, 31 (1968) 101.
35 S. Razin and V. Barash, *FEBS Letters*, 3 (1969) 217.
36 W.S. Bont, P. Emmelot and H. Vaz-Dias, *Biochim. Biophys. Acta*, 173 (1969) 389.
37 P. Emmelot and H. Vaz-Dias, *Biochim. Biophys. Acta*, 203 (1970) 172.
38 J. Philippott, *Biochim. Biophys. Acta*, 225 (1971) 201.
39 A.L. Shapiro, E. Viñuela and J.V. Maizel, *Biochem. Biophys. Res. Commun.*, 28 (1967) 815.
40 K. Weber and M. Osborn, *J. Biol. Chem.*, 244 (1969) 4406.
41 A.K. Dunker and R.R. Rueckert, *J. Biol. Chem.*, 244 (1969) 5074.
42 P.P. Blumenfeld, *Biochem. Biophys. Res. Commun.*, 30 (1968) 200.
43 M.A. Macheboeuf and G. Sandor, *Bull. Soc. Chim. Biol.*, 14 (1932) 1168.
44 D.G. Dervichian, *Progr. Biophys. Mol. Biol.*, 14 (1964) 263.
45 J. Folch and M. Lees, *J. Biol. Chem.*, 191 (1951) 807.
46 M.G. Rumsby and J.B. Finean, *J. Neurochem.*, 13 (1966) 1501.
47 E. De Robertis, *Science*, 171 (1971) 963.
48 R.K. Morton, *Nature*, 166 (1950) 1092.
49 A.H. Maddy, *Biochim. Biophys. Acta*, 117 (1966) 193.
50 S.L. Marchesi, E. Steers, V.T. Marchesi and T.W. Tillack, *Biochemistry*, 9 (1970) 50.
51 A.H. Maddy and P.G. Kelly, *Biochem. J.*, 122 (1971) 62P.
52 R. Coleman and G. Hubscher, *Biochim. Biophys. Acta*, 73 (1963) 257.
53 D.F.H. Wallach and P.H. Zahler, *Proc. Natl. Acad. Sci. (U.S.)*, 56 (1966) 1552.
54 P.H. Zahler and D.F.H. Wallach, *Biochim. Biophys. Acta*, 135 (1967) 371.
55 J.B. Finean, *Quart. Rev. Biophys.*, 2 (1969) 1.
56 F.O. Schmitt, R.S. Bear and G.L. Clark, *Radiology*, 25 (1935) 131.
57 F.O. Schmitt, R.S. Bear and K.J. Palmer, *J. Cell. Comp. Physiol.*, 18 (1941) 39.
58 J.B. Finean, *Int. Rev. Cytol.*, 12 (1961) 303.
59 J.B. Finean, *Circulation*, 26 (1962) 1151.
60 J.B. Finean and R.E. Burge, *J. Mol. Biol.*, 7 (1963) 672.

61 M.F. Moody, *Science*, 142 (1963) 1173.
62 C.R. Worthington and A.E. Blaurock, *Nature New Biol.*, 218 (1968) 87.
63 A.E. Blaurock, *J. Mol. Biol.*, 56 (1971) 35.
64 D.L.D. Caspar and D.A. Kirschner, *Nature*, 231 (1971) 46.
65 C.R. Worthington and G.I. King, *Nature*, 234 (1971) 143.
66 A.E. Blaurock, *Neurosci. Res. Program Bull.*, 9 (1971) 530.
67 C.K. Akers and D.F. Parsons, *Biophys. J.*, 10 (1970) 101.
68 A.E. Blaurock and M.H.F. Wilkins, *Nature*, 223 (1969) 906.
69 W.J. Gras and C.R. Worthington, *Proc. Natl. Acad. Sci. (U.S.)*, 63 (1969) 233.
70 J.K. Blasie, M.M. Dewey, A.E. Blaurock and C.R. Worthington, *J. Mol. Biol.*, 14 (1965) 143.
71 J.K. Blasie, C.R. Worthington and M.M. Dewey, *J. Mol. Biol.*, 39 (1969) 407.
72 W. Kreutz, *Z. Naturforsch.*, 18b (1963) 1098.
73 W. Kreutz, *Z. Naturforsch,*, 19b (1964) 441.
74 J.T. Dodge, C. Mitchell and D.J. Hanahan, *Arch. Biochem. Biophys.*, 100 (1963) 119.
75 S. Knutton, J.B. Finean, R. Coleman and A.R. Limbrick, *J. Cell Sci.*, 7 (1970) 357.
76 T.A. Bramley, R. Coleman and J.B. Finean, *Biochim. Biophys. Acta*, 241 (1971) 752.
77 M.H.F. Wilkins, A.E. Blaurock and D.M. Engleman, *Nature New Biol.*, 230 (1971) 72.
78 D.M. Engleman, *J. Mol. Biol.*, 58 (1971) 153.
79 J.B. Finean, R. Coleman, W.G. Green and A.R. Limbrick, *J. Cell Sci.*, 1 (1966) 287.
80 R. Freeman and J.B. Finean, in preparation.
81 A.R. Limbrick and J.B. Finean, *J. Cell Sci.*, 7 (1970) 373.
82 J.B. Finean, R. Coleman and W.A. Green, *Ann. N.Y. Acad. Sci.*, 137 (1966) 287.
83 J.B. Finean, R. Coleman, S. Knutton, A.R. Limbrick and J.E. Thompson, *J. Gen. Physiol.*, 51 (1968) 19s.
84 J.E. Thompson, R. Coleman and J.B. Finean, *Biochim. Biophys. Acta*, 150 (1968) 405.
85 S.C. Harrison, D.L.D. Caspar, R.D. Camerini-Otero and R.M. Franklin, *Nature New Biol.*, 229 (1971) 197.
86 D. Branton, *Ann. Rev. Plant Physiol.*, 20 (1969) 209.
87 R. Coleman, J.B. Finean, S. Knutton and A.R. Limbrick, *Biochim. Biophys. Acta*, 219 (1970) 81.
88 J.B. Finean, T.A. Bramley and R. Coleman, *Nature*, 229 (1971) 114.
89 R. Freeman and J.B. Finean, in preparation.
90 A.H. Maddy and B.R. Malcolm, *Science*, 150 (1965) 1616.
91 A.H. Maddy and B.R. Malcolm, *Science*, 153 (1966) 213.
92 D. Chapman, V.B. Kamat and R.J. Levene, *Science*, 160 (1968) 314.
93 D.F.H. Wallach and P.H. Zahler, *Proc. Natl. Acad. Sci. (U.S.)*, 56 (1966) 1552.
94 B. Ke, *Arch. Biochem. Biophys.*, 112 (1965) 554.
95 J. Lenard and S.J. Singer, *Proc. Natl. Acad. Sci. (U.S.)*, 56 (1966) 1828.
96 D.W. Urry and T.H. Ji, *Arch. Biochem. Biophys.*, 128 (1968) 802.
97 D.W. Urry, L. Masotti and J. Krivacic, *Biochem. Biophys. Res. Commun.*, 41 (1970) 521.
98 J.M. Steim and S. Fleischer, *Proc. Natl. Acad. Sci. (U.S.)*, 58 (1967) 1292.
99 J.M. Wrigglesworth and L. Packer, *Arch. Biochem. Biophys.*, 128 (1968) 790.
100 A.S. Schneider, M.J.T. Schneider and K. Rosenheck, *Proc. Natl. Acad. Sci. (U.S.)*, 66 (1970) 793.
101 M. Glaser and S.J. Singer, *Biochemistry*, 10 (1971) 1780.
102 C.A. Ottaway and D.B. Wetlaufer, *Arch. Biochem. Biophys.*, 139 (1970) 257.
103 J.Y. Cassim and J.T. Yang, *Biochem. Biophys. Res. Commun.*, 26 (1967) 58.
104 W.B. Gratzer, G.H. Beaven, H.W.E. Rattle and E.M. Bradbury, *European J. Biochem.*, 3 (1968) 276.
105 G.L. Choules and R.F. Bjorklund, *Biochemistry*, 9 (1970) 4759.
106 D.F.H. Wallach, J.M. Graham and B.R. Fernbach, *Arch. Biochem. Biophys.*, 131 (1969) 322.
107 D. Chapman, in D. Chapman (Ed.), *Biological Membranes*, Academic Press, London, 1968, p. 125.
108 H. Lecar, G. Ehrenstein and I. Stillman, *Biophys. J.*, 11 (1970) 140.
109 D. Chapman and N.J. Salsbury, *Recent Progr. Surface Sci.*, 3 (1970) 121.

110 B. Sheard and E.M. Bradbury, *Progr. Biophys. Mol. Biol.*, 20 (1970) 187.
110a J.C. Metcalfe, N.J.M. Birdsall, J. Feeney, A.G. Lee, Y.K. Levine and P. Partington, *Nature*, 233 (1971) 199.
110b D.G. Davies and G. Inesi, *Biochim. Biophys. Acta*, 282 (1972) 180.
111 J.C. Metcalfe and A.S.V. Burgen, *Nature*, 220 (1968) 587.
112 J.C. Metcalfe, P. Seeman and A.S.V. Burgen, *Mol. Pharmacol.*, 4 (1968) 87.
113 C.M. Colley, S.M. Metcalfe, B. Turner, A.S.V. Burgen and J.C. Metcalfe, *Biochim. Biophys. Acta*, 233 (1971) 720.
114 H.M. McConnell and B.G. McFarland, *Quart. Rev. Biophys.*, 3 (1970) 91.
115 W.L. Hubbell and H.M. McConnell, *J. Am. Chem. Soc.*, 93 (1971) 314.
116 A.S. Waggoner and L. Stryer, *Proc. Natl. Acad. Sci. (U.S.)*, 67 (1970) 579.
117 D.F.H. Wallach, E. Ferber, D. Selin, E. Weidekamm and H. Fischer, *Biochim. Biophys. Acta*, 203 (1970) 67.
118 G.K. Radda, *Biochem. J.*, 122 (1971) 385.
119 M.E. Tourtelotte, D. Branton and A. Keith, *Proc. Natl. Acad. Sci. (U.S.)*, 66 (1970) 909.
120 S. Rottem, W.L. Hubbell, L. Hayflick and H.M. McConnell, *Biochim. Biophys. Acta*, 219 (1970) 104.
121 W.L. Hubbell and H.M. McConnell, *Proc. Natl. Acad. Sci. (U.S.)*, 61 (1968) 12.
122 W.L. Hubbell and H.M. McConnell, *Proc. Natl. Acad. Sci. (U.S.)*, 63 (1969) 16.
123 A.D. Keith, A.S. Waggoner and O.H. Griffith, *Proc. Natl. Acad. Sci. (U.S.)*, 61 (1968) 819.
124 A. Keith, B. Bulfield and W. Snipes, *Biophys. J.*, 10 (1970) 618.
125 W.L. Hubbell, J.C. Metcalfe, S.M. Metcalfe and H.M. McConnell, *Biochim. Biophys. Acta*, 219 (1970) 415.
126 D. Chapman, M.D. Barratt and V.B. Kamat, *Biochim. Biophys. Acta*, 173 (1969) 154.
127 H. Schneider and I.C.P. Smith, *Biochim. Biophys. Acta*, 219 (1970) 73.
128 H. Simpkins, E. Panko and S. Tay, *J. Membrane Biol.*, 5 (1971) 334.
129 G.K. Radda and D.S. Smith, *FEBS Letters*, 9 (1970) 287.
130 R.B. Freedman and G.K. Radda, *FEBS Letters*, 3 (1969) 150.
131 T. Gulik-Krzywicki, E. Schechter, M. Iwatsubo, J.L. Ranck and V. Luzzati, *Biochim. Biophys. Acta*, 219 (1970) 1.
132 B. Chance, *Proc. Natl. Acad. Sci. (U.S.)*, 67 (1970) 560.
133 J.R. Brocklehurst, R.B. Freedman, D.J. Hancock and G.K. Radda, *Biochem. J.*, 116 (1970) 721.
134 W. Stoeckenius and D.M. Engleman, *J. Cell Biol.*, 42 (1969) 613.
135 R.W. Hendler, *Physiol. Rev.*, 51 (1971) 66.
136 J.B. Finean, in R.J.C. Harris (Ed.), *The Interpretation of Ultrastructure*, Academic Press, New York, 1962, p. 89.
137 H. Fernandez-Moran and J.B. Finean, *J. Biophys. Biochem. Cytol.*, 3 (1957) 725.
138 G. Vanderkooi and D.E. Green, *Proc. Natl. Acad. Sci. (U.S.)*, 66 (1970) 615.
139 F.S. Sjostrand, in A.J. Dalton and F. Haguenau (Eds.), *The Membranes*, Academic Press, New York, 1968, p. 151.
140 J.B. Finean and M.G. Rumsby, *Nature*, 197 (1963) 1326.
141 H. Fernández-Morán, T. Oda, P.V. Blair and D.E. Green, *J. Cell Biol.*, 22 (1964) 63.
142 Y. Kagawa and E. Racker, *J. Biol. Chem.*, 241 (1966) 2475.
143 E. Racker and L.L. Horstman, *J. Biol. Chem.*, 242 (1967) 2547.
144 T. Oda and S. Seki, *Proc. 6th Intern. Congr. Electron Microscopy, Kyoto*, 1965, Biol. Vol., p. 387.
145 C.F. Johnson, *Proc. 6th Intern. Congr. Electron Microscopy, Kyoto*, 1965, Biol. Vol., p. 389.
146 D.W. Deamer and D. Branton, *Science*, 158 (1967) 655.
147 P. Pinto da Silva and D. Branton, *J. Cell Biol.*, 45 (1970) 598.
148 D.J. Triggle, *Recent Progr. Surface Sci.*, 3 (1970) 273.
149 W.J. Lennarz, *Ann. Rev. Biochem.*, 39 (1970) 359.
150 I. Sekuzu, P. Jurtshuk Jr. and D.E. Green, *Biochem. Biophys. Res. Commun.*, 6 (1961) 71.
151 P. Jurtshuk Jr., I. Sekuzu and D.E. Green, *Biochem. Biophys. Res. Commun.*, 6 (1961) 76.
152 I. Sekuzu, P. Jurtshuk Jr. and D.E. Green, *J. Biol. Chem.*, 238 (1963) 975.
153 P. Jurtshuk Jr., I. Sekuzu and D.E. Green, *J. Biol. Chem.*, 238 (1963) 3595.

154 S. Fleischer, G. Brierley, H. Klouwen and D.B. Slautterback, *J. Biol. Chem.*, 237 (1962) 3264.
155 D.E. Green and S. Fleischer, *Biochim. Biophys. Acta*, 70 (1963) 554.
156 E. Racker and A. Bruni, in L. Bolis and B.A. Pethica (Eds.), *Membrane Models and the Formation of Biological Membranes*, Wiley, New York, 1968, p. 138.
157 L.J. Fenster and J.H. Copenhaver Jr., *Biochim. Biophys. Acta*, 137 (1967) 406.
158 K.P. Wheeler and R. Whittam, *Nature*, 225 (1970) 449.
159 T. Noguchi and S. Freed, *Nature New Biol.*, 230 (1971) 148.
160 P.D. Jones, P.W. Holloway, R.O. Peluffo and S.J. Wakil, *J. Biol. Chem.*, 244 (1969) 744.
161 S.M. Duttera, W.L. Byrne and M.C. Ganoza, *J. Biol. Chem.*, 243, (1968) 2216.
162 W.G. Fiscus and W.C. Schneider, *J. Biol. Chem.*, 241 (1966) 3324.
163 R. Tanaka and K.P. Strickland, *Arch. Biochem. Biophys.*, 111 (1965) 583.
164 L.A. Pesch and J. Peterson, *Biochim. Biophys. Acta*, 96 (1965) 390.
165 A. Martonosi, J. Donley and R.A. Halpin, *J. Biol. Chem.*, 243 (1968) 61.
166 M.M. Weiser and L. Rothfield, in L. Bolis and B.A. Pethica (Eds.), *Membrane Models and the Formation of Biological Membranes*, Wiley, New York, 1968, p.
167 D.G. Williamson and V.J. O'Donnell, *Biochemistry*, 8 (1969) 1289.
167a P.D. Jones and S.J. Wakil, *J. Biol. Chem.*, 242 (1967) 5267.
167b L. Sartorelli, L. Galzigna, C.R. Rossi and D.M. Gibson, *Biochem. Biophys. Res. Commun.*, 26 (1967) 90.
168 S. Margolis, in E. Tria and A.M. Scanu (Eds.), *Structural and Functional Aspects of Lipoproteins in Living Systems*, Academic Press, London, 1969, p. 369.
169 A. Scanu and R. Hirz, *Nature*, 218 (1968) 200.
170 D.G. Dearborn and D.B. Wetlaufer, *Proc. Natl. Acad. Sci. (U.S.)*, 62 (1969) 179.
171 A.M. Scanu, H. Pollard, R. Hirz and K. Kothary, *Proc. Natl. Acad. Sci. (U.S.)*, 62 (1969) 171.
172 V.N. Schumaker and G.H. Adams, *Ann. Rev. Biochem.*, 38 (1969) 113.
173 A.M. Gotto Jr., *Proc. Natl. Acad. Sci. (U.S.)*, 64 (1969) 1119.
174 J.M. Steim, O.J. Edner and F.G. Bargoot, *Science*, 162 (1968) 909.
175 D. Chapman, R.B. Leslie, R. Hirz and A.M. Scanu, *Nature*, 221 (1969) 260.
176 R.B. Leslie, D. Chapman and A.M. Scanu, *Chem. Phys. Lipids*, 3 (1969) 152.
177 A. Scanu, in E. Tria and A.M. Scanu (Eds.), *Structural and Functional Aspects of Lipoproteins in Living Systems*, Academic Press, London, 1969, p. 425.
178 P. Doty and J.H. Schulman, *Discussions Faraday Soc.*, 66 (1949) 21.
179 R. Matalon and J.H. Schulman, *Discussions Faraday Soc.*, 66 (1949) 27.
180 D.D. Eley and D.G. Hedge, *Discussions Faraday Soc.*, 21 (1956) 221.
181 P. Mueller and D.O. Rudin, *Nature*, 213 (1967) 603.
182 P. Mueller and D.O. Rudin, *Nature*, 217 (1968) 713.
183 B.C. Pressman, *Proc. Natl. Acad. Sci. (U.S.)*, 53 (1965) 1076.
184 J.B. Chappell and A.R. Crofts, *Biochem. J.*, 95 (1965) 393.
185 H.A. Lardy, S.N. Graven and S. Estrada-o, *Federation Proc.*, 26 (1967) 1355.
186 P. Mueller and D.O. Rudin, *Biochem. Biophys. Res. Commun.*, 26 (1967) 398.
187 A.D. Bangham, *Progr. Biophys. Mol. Biol.*, 18 (1968) 29.
188 A. Goldup, S. Ohki and J.F. Danielli, *Recent Progr. Surface Sci.*, 3 (1970) 193.
189 K.J. Palmer, F.O. Schmitt and E. Chargaff, *J. Cell. Comp. Physiol.*, 18 (1941) 43.
190 W. Stoeckenius, *J. Biophys. Biochem. Cytol.*, 5 (1959) 491.
191 T. Gulik-Krzywicki, E. Shechter, V. Luzzati and M. Faure, *Nature*, 223 (1969) 1116.
192 R.M.C. Dawson, in D. Chapman (Ed.) *Biological Membranes*, Academic Press, London, 1968 p. 203.
193 D.E. Green and A. Tzagoloff, *J. Lipid Res.*, 7 (1966) 587.
194 A. Tzagoloff, D.H. MacLennan, D.G. McConnell and D.E. Green, *J. Biol. Chem.*, 242 (1967) 2051.
195 D.E. Green and D.H. MacLennan, *Bioscience*, 19 (1969) 213.
196 D.E. Green and S. Fleischer, in R.M.C. Dawson and D.N. Rhodes (Eds.), *Metabolism and Physiological Significance of Lipids*, Wiley, London, 1964 p. 581.
197 D.M. Kaplan and R.S. Criddle, *Physiol. Rev.*, 51 (1971) 249.
198 T.H. Ji and A.A. Benson, *Biochim. Biophys. Acta*, 150 (1968) 686.

199 P. Zahler and E.R. Weibel, *Biochim. Biophys. Acta*, 219 (1970) 320.
200 P.H. Zahler and D.F.H. Wallach, *Biochim. Biophys. Acta*, 135 (1967) 371.
201 R.F.A. Zwaal and L.L.M. van Deenen, *Chem. Phys. Lipids*, 4 (1970) 311.
202 V.B. Kamat, D. Chapman, R.F.A. Zwaal and L.L.M. van Deenen, *Chem. Phys. Lipids*, 4 (1970) 323.
203 K.V. Berger, M.D. Barratt and V.B. Kamat, *Biochem. Biophys. Res. Commun.*, 40 (1970) 1273.
204 J.F. Danielli and H. Davson, *J. Cell Comp. Physiol.*, 5 (1935) 495.
205 W.D. Stein and J.F. Danielli, *Discussions Faraday Soc.*, 21 (1956) 238.
206 P. Zahler, in *Modern Problems of Blood Preservation*, Fischer, Stuttgart, 1969, p. 1.

Chapter 9

PHOSPHOLIPIDS IN SUBCELLULAR ORGANELLES AND MEMBRANES

W.C. McMURRAY

Department of Biochemistry, Health Sciences Centre, University of Western Ontario, London 72, Ontario (Canada)

1. INTRODUCTION

In the phylogeny of organisms the membrane systems of the cell have acquired progressively complex forms and functions. With the evolution of eukaryotic cells this differentiation has progressed to the point of development of discrete organelles, each composed of a characteristic arrangement of membranes adapted to its functional role. As the external plasmalemma or plasma membrane of even the simple cells provides a protective barrier between the external environment and the internal milieu of the cell, so the selective permeability properties of organellar membranes form a series of microenvironments or compartments of metabolism within the eukaryotic cell. Moreover, the concept of membrane functions essentially in terms of a series of permeability barriers, is rapidly broadening to include the role of membranes as catalytic surfaces. Thus the membrane can constitute a uniquely ordered system for metabolic reactions, providing a hydrophobic environment where the components of multienzyme complexes may interact with a high degree of efficiency.

The phospholipids are ubiquitous and essential components of these membrane systems. Many of the permeability and electrochemical properties of biological membranes may be mimicked by artificial membranes prepared from pure phospholipids, and the conformation and enzymatic activity of membrane proteins are known to be markedly modified by the presence of the phospholipids. The first two parts of this Chapter attempt to explore some of the characteristics of the phospholipids in intracellular membranes and organelles, which may lead to an understanding of their contribution to the diverse functions of the varied membranes of the cell. The later three sections relate to questions of the renewal of membranes with particular reference to the phospholipid components. The approach has been admittedly critical and argumentative, with no attempt to encompass all aspects of the subject. Since models of membrane structure, as well as micro-organism, plant, and nerve-cell phospholipids are covered in other sections these topics are largely omitted here. Most of the subject material will relate to mammalian systems, and in particular to that much belaboured organ, the liver of the rat.

2. PHOSPHOLIPID COMPOSITION OF MEMBRANES

With the advent of new techniques for subcellular fractionation an ever-expanding literature has arisen in recent years relating to the components of various membrane systems. Recognition of specific sites of localization of characteristic 'marker' enzymes for different fractions has aided workers whose bent is chiefly morphological to apply biochemical criteria in the identification and the assessment of contamination of isolated fractions[1]. Conversely, most biochemists have learned the value of the techniques of electron microscopy as a correlate to the standard methods of chemical analysis. Hopefully this marriage of cellular chemistry and ultrastructure will continue to be felicitous and fruitful, although its offspring in the growth of new journals and burgeoning publications may bear ominous similarities to the population explosion, to the despair of the would-be reviewer! This brief summary of the field will attempt to present a comparative overview of the analytical data. A more comprehensive survey of the values for phospholipid and fatty acid content of cell membranes will be found in Chapter 16 (p. 441).

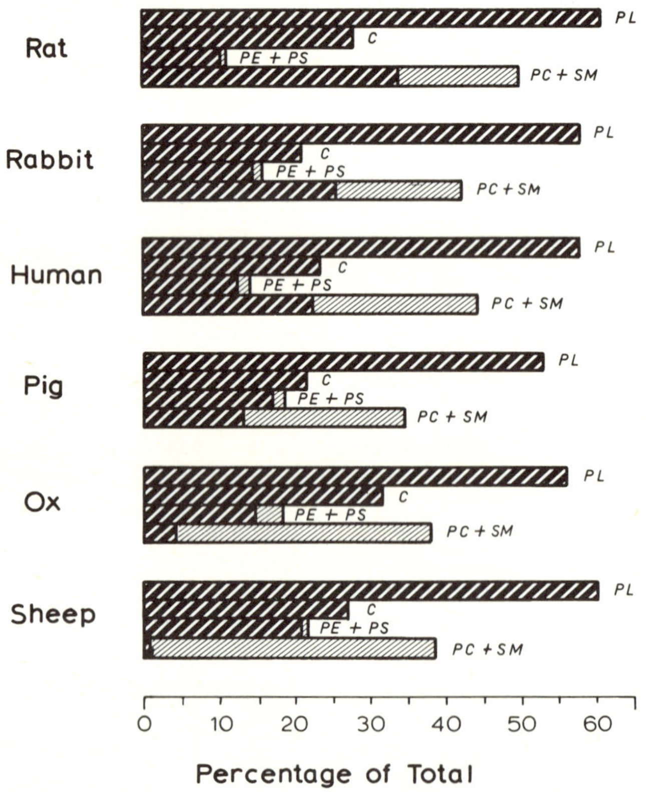

Fig. 1. Cholesterol and phospholipid composition of erythrocyte ghosts from several species[2]. Values for each fraction were calculated as the percentage of the total lipids. PL, phospholipids; C, cholesterol; PE + PS, 'cephalins' – PE, phosphatidylethanolamine (heavy cross-hatching), PS, phosphatidylserine (light cross-hatching); PC + SM, choline-containing phospholipids – PC, phosphatidylcholine (heavy cross-hatching), SM, sphingomyelin (light cross-hatching).

The phospholipid composition of membranes may vary considerably with species and tissue of origin. This variation is illustrated by erythrocyte ghosts from a number of mammalian species (Fig. 1). As noted by De Gier and Van Deenen[2] the phospholipids constitute half or more of the total erythrocyte membrane lipids, with the remainder contributed by cholesterol. Comparison with Table 9 of Chapter 16 suggests that the phosphatidylserine figures are all too low. Total 'cephalins' are a fairly constant proportion of the total lipids, although the fatty residues in the 1-position may vary from predominantly alkenyl ethers in the human[3] to mainly alkyl ethers in the ox[4]. The total choline phospholipids are also constant, but the proportion contributed by lecithin or sphingomyelin varies over a wide range. Significant alterations in the fatty acid composition of erythrocyte membranes from various species are also known, reflecting the differences in dietary habits and alterations effected by rumen microorganisms, as well as species differences of biosynthetic capability[5].

Organ-specific variations are illustrated best by the comprehensive data of Getz et al[6]. (Fig. 2). It is noteworthy that in spite of the many metabolic differences between

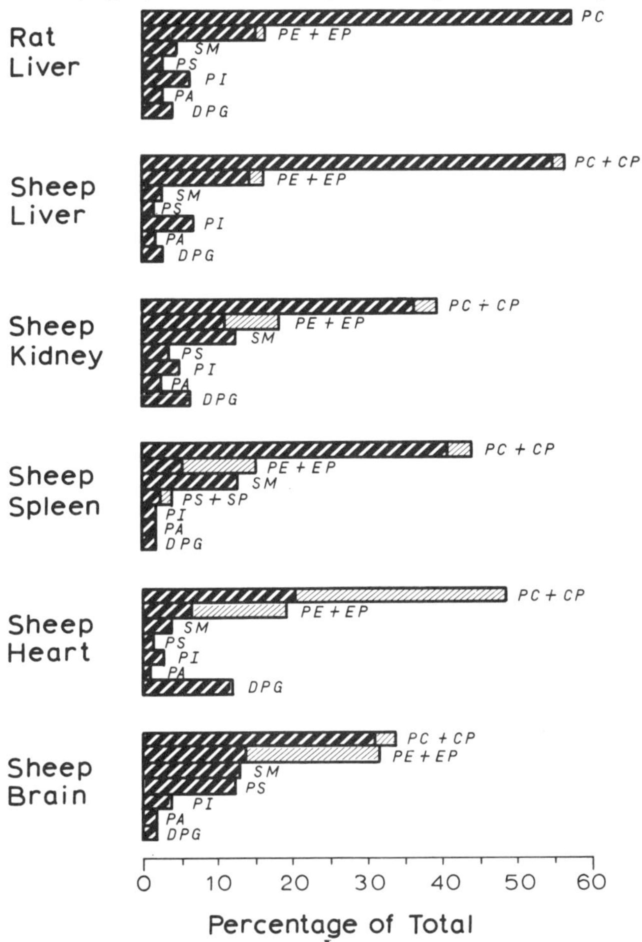

Fig. 2. Phospholipid composition of rat and sheep tissues[6]. Values are given as percentages of the total P in the phospholipid fraction. PI, phosphatidylinositol; PA, phosphatidic acid; DPG, diphosphatidylglycerol; CP, EP and SP, choline-, ethanolamine- and serine-plasmalogens (light cross-hatching). Other abbreviations as in Fig. 1.

species, differences of phospholipid composition among various tissues were greater than between homologous tissues of different species (compare rat liver and sheep liver). Choline phospholipids predominate in most tissues, constituting over half of the total phospholipid fraction; however, the amount present as sphingomyelin may vary greatly, and the proportion of phosphoglyceride in the plasmalogenic form is also tissue specific. It is characteristic of heart for example that the plasmalogens account for more than half of both choline and ethanolamine phosphoglycerides, while in brain the ethanolamine plasmalogens predominate and choline plasmalogens are only found in traces. High values for sphingomyelin and phosphatidylserine in brain are related to enrichment of these phospholipids in myelin; the high respiratory capacity of cardiac muscle is reflected in its large proportion of diphosphatidylglycerol, a specific mitochondrial component. In view of animal and tissue variations, values for phospholipid composition of subcellular membrane fractions are difficult to compare, unless determined for the same species and organ. Accordingly the phospholipid distribution for a number of organelles is illustrated for rat liver (Figs. 3 and 4).

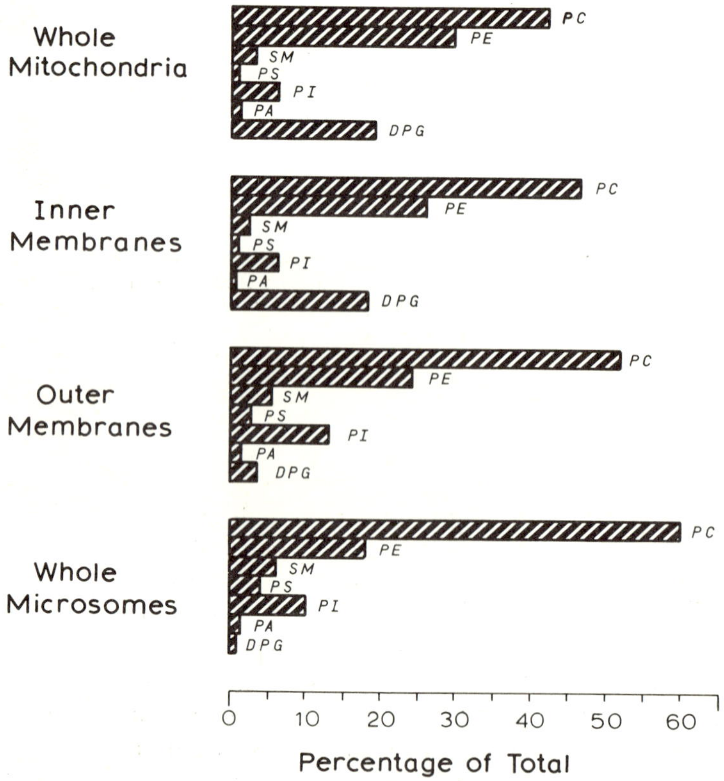

Fig. 3. Phospholipid composition of rat liver mitochondrial membranes and microsomes[30]. Abbreviations as in Figs. 1 and 2.

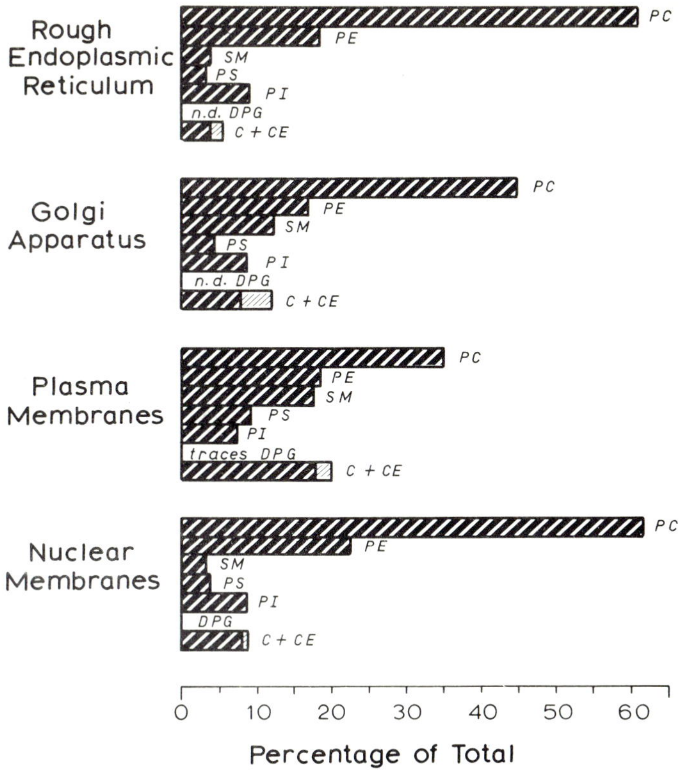

Fig. 4. Phospholipid and sterol composition of rough endoplasmic reticulum[49], Golgi apparatus[49], plasma membranes[56], and nuclear membranes[65] from rat liver. Individual phospholipid values are given as percentages of the total P of the phospholipid fraction. C, cholesterol; CE, cholesterol esters (light cross-hatching). Sterols are given as percentages of total lipids by weight. Other abbreviations as in Figs. 1 and 2.

Mitochondria

Isolated mitochondria from a variety of sources are rich in phospholipid, which may account for as much as 90% of the total lipids[7]. Relative to the whole tissue of origin mitochondria generally contain less lecithin and more phosphatidylethanolamine when the values are expressed in terms of the percentage contribution of each component in the total phospholipids. The concentrations of sphingomyelin and phosphatidylserine in mitochondria are quite low. The salient feature of the mitochondrial phospholipids is the high concentration of diphosphatidylglycerol, which is found in only small quantities in other membranes. Diphosphatidylglycerol was reported to be greatly enriched in mitochondrial fractions from a wide variety of tissues by Getz et al[6]. Recent investigations have shown that the diphosphatidylglycerol content of mitochondria may vary considerably depending upon their state of differentiation. Thus, Chan and Lester[8] found that the diphosphatidylglycerol concentration in muscle mitochondria of the tobacco horn worm increased three-fold during maturation from the larval stage to the adult moth. Conversely Bergelson et al.[9] reported that mitochondria of hepatoma cells exhibit dedifferentiation since the content of diphosphatidylglycerol is less than half of that found for normal liver mitochondria. Repression of the yeast respiratory chain by glucose or anaerobiosis also produces a reduction of the diphosphatidylglycerol content[10,11]. In the mitochondrial membrane it is probable that the diphosphatidylglycerol exists in the form of salts or chelation complexes with cations, particularly the divalent cations $Ca^{2\oplus}$ and $Mg^{2\oplus}$ for which it has a high affinity[12]. There is considerable evidence that the mitochondrial phospholipids play a physiological role in the passive binding of cations[13,14], and the contribution of the strongly anionic diphosphatidylglycerol molecule undoubtedly is significant in this connection.

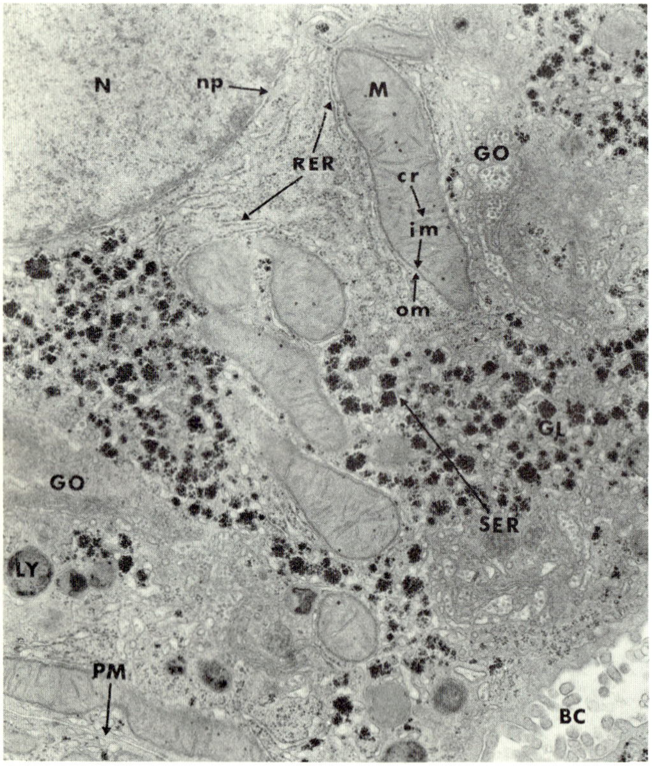

Fig. 5. Electron micrograph of a rat liver cell (× 12 500) illustrating the localization of the subcellular membranes. Mitochondria (M) show the typical double membrane structure, with the outer membrane (om) forming an envelope, the inner membrane (im) invaginating to produce the cristae (cr). Elements of both rough (RER) and smooth (SER) endoplasmic reticulum are evident, the latter lying amidst the dense glycogen particles (GL). Smooth membranes of the Golgi apparatus (GO) lie outside the glycogen-rich areas. Plasma membrane (PM) and bile canaliculi (BC) are seen at the bottom, a portion of a nucleus (N) with double membrane and nuclear pores (np) at the top left. LY, lysosome. Electron micrograph kindly supplied by Dr. K.R. Porter.

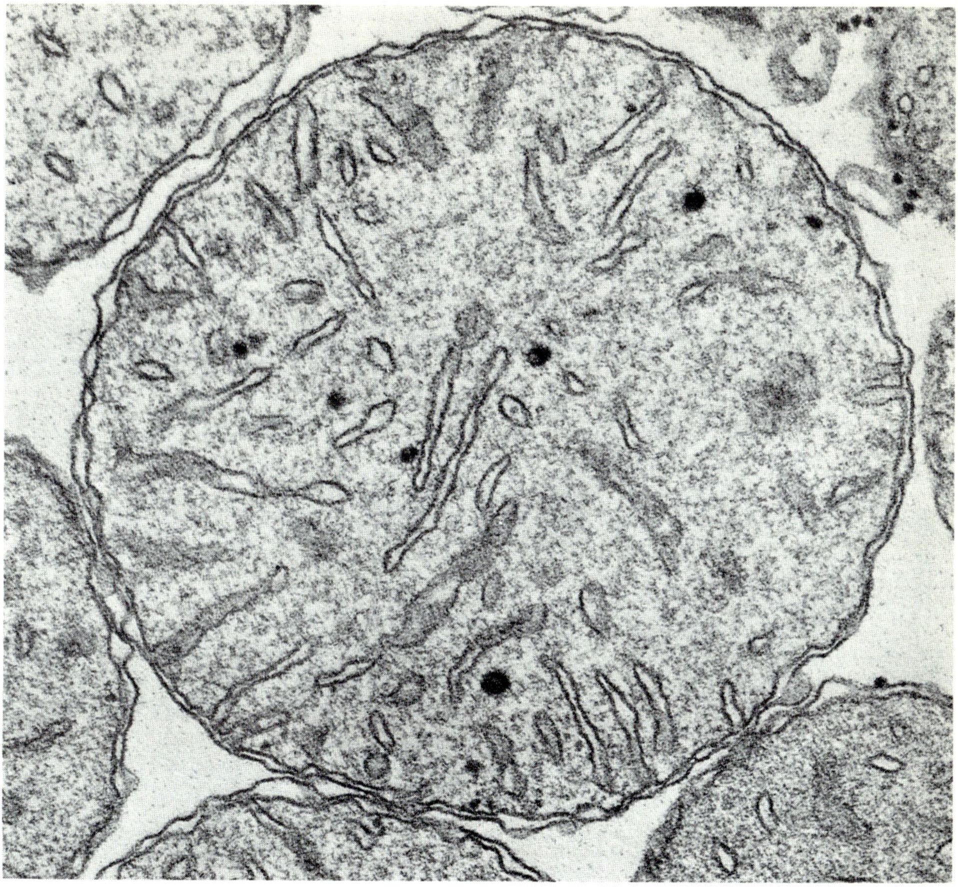

Fig. 6. Isolated rat liver mitochondrion (× 110 000) in the 'orthodox' conformation, showing the typical double membrane structure seen *in vivo*. (Reproduced from *J. Cell Biol.* 37 (1968) 365, with the kind permission of Dr. C.R. Hackenbrock and the publishers.)

In recent years a number of different procedures (treatment with phospholipase A[15], or with digitonin[16], osmotic lysis[17], sonication[18]) has been used to disrupt mitochondria, followed by density gradient or differential centrifugation to separate the fragments arising from the outer membrane envelope, and from the inner membrane-containing components of the cristae mitochondriales. The latter may be recognized morphologically by the regular array of projecting subunits on the membrane fragments which are visualized on electron microscopy after negative staining. It is generally accepted that the inner membrane is the seat of the respiratory chain and may be characterized by its content of cytochromes b, c_1, a and a_3, and by the 'marker' enzymes cytochrome oxidase, succinate:cytochrome *c* oxidoreductase, and the NADH:cytochrome *c* oxidoreductase which is inhibited by rotenone[15–20]. Smooth membranes derived from the mitochondrial lysates were assumed to derive from the outer envelope, although they could not be distinguished on morphological grounds from smooth membranes of contaminating microsomes. It was observed that the smooth mitochondrial fragments were enriched in monoamine oxidase[21] and an NADH:cytochrome *c* oxidoreductase that was insensitive to rotenone[18], and these enzymes were proposed as markers for the outer membrane. Green et al.[22] have criticized this conclusion on the basis that the rotenone-insensitive NADH:cytochrome *c* oxidoreductase simply represented microsomal contamination, while the monoamine oxidase assay used was inadequate to estimate activity at its true site in the inner membrane. Space does not permit a full discussion of this interesting controversy. However, it would appear that the criticisms have been fully answered, and that the original assignment of marker enzymes is probably correct[23–26].

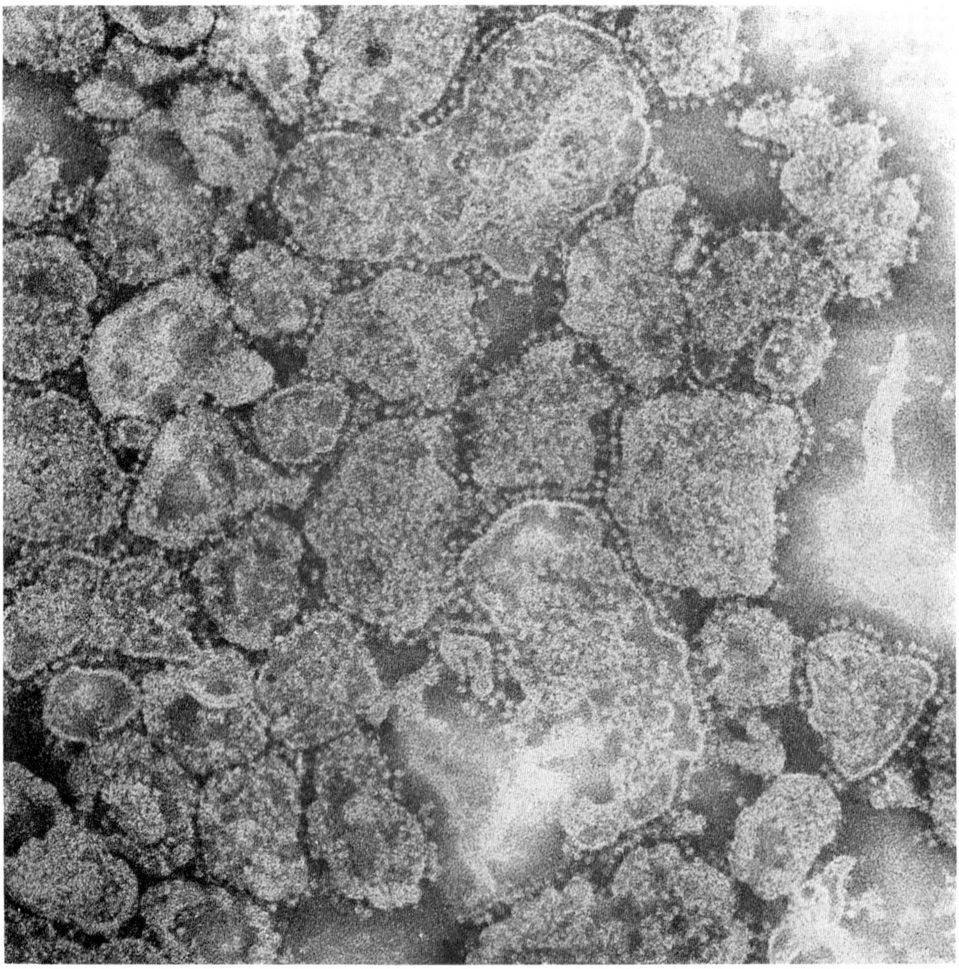

Fig. 7. Submitochondrial vesicles (× 140 000) negatively stained with phosphotungstic acid. Projecting subunits of the inner membrane are clearly visible. (The electron micrograph was kindly supplied by Drs. Akitsugo Saito and Sidney Fleischer.)

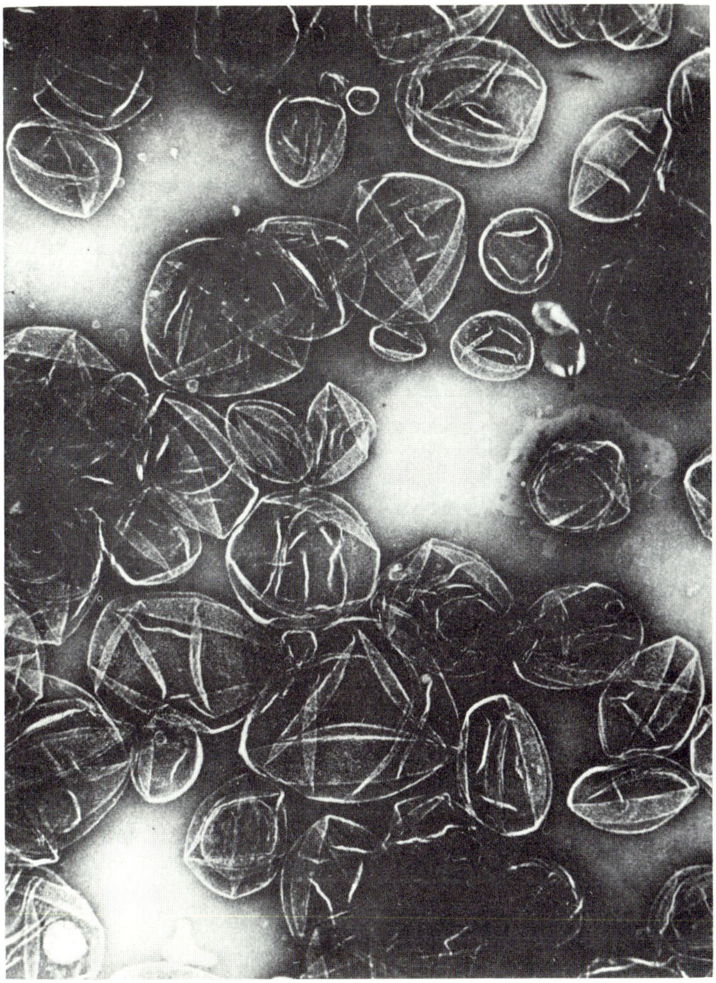

Fig. 8. Purified mitochondrial outer membrane fraction (× 35 000) fixed with OsO_4, then negatively stained with phosphotungstic acid. (Reproduced from Ref. 27 with kind permission of Dr. D.F. Parsons and the publishers.)

Phospholipid composition of the isolated membrane fractions was first reported by Parsons et al.[27] for guinea pig liver mitochondria, and yielded similar patterns to those depicted for rat liver (Fig. 3). Thus the inner membranes showed a similar pattern to that for whole mitochondria with high concentrations of diphosphatidylglycerol, while the outer membranes showed greater similarity to the microsome fraction with low concentrations of diphosphatidylglycerol but higher concentrations of phosphatidylinositol, phosphatidylserine and sphingomyelin than the whole mitochondria. The ratio lecithin : phosphatidylethanolamine for outer membranes is generally high, approaching that found in microsomes (~3.0), while that for inner membranes approaches the lower ratios of whole mitochondria (~1.5) (refs. 28–30). Since the convoluted inner membrane has a much larger surface area[31], and accounts for 25–29% of the total mitochondrial volume as compared to 6–10% for the outer envelope[32], it is not surprising that the phospholipid composition of whole mitochondria more closely resembles that for the inner membrane than for the outer membrane. The outer membranes are three-fold richer in total phospholipid and six-fold richer in cholesterol than inner membranes based upon protein[33,34], which accounts for the different sedimentation behaviour of the membranes in density gradients (outer membrane density = 1.13 versus inner membrane density = 1.21) (ref. 17). Parkes and Thompson[35] investigated the fatty acid composition of mitochondrial phospholipids and concluded that the patterns were basically similar for inner and outer membranes from guinea pig liver, in confirmation of the findings for rat liver by Stoffel and Schieffer[29]. The most highly unsaturated phospholipids were phosphatidylethanolamine which contained nearly 20% of total fatty acids as arachidonate, and diphosphatidylglycerol in which over 70% of the fatty acyl groups were accounted for as linoleate. Although the membranes differ in possessing different amounts and proportions of individual phospholipids, there seems to be a distinctive fatty acid pattern for each phospholipid which is approximately the same for all membranes[35].

Endoplasmic reticulum

The microsome fraction contains a high concentration of phospholipids, particularly lecithin, and may account for as much as one-half of the total cellular content of phospholipids[36]. Although the microsomes derive principally from the endoplasmic reticulum by pinching off elements of this tubular network of membranes during homogenization of the cells, the resulting vesicles are undoubtedly as heterogeneous in origin as they are in size[37]. Elements of 'rough' endoplasmic reticulum are identifiable by the presence of attached ribosomes in electron micrographs and by the high RNA content of the fractions, but the smooth membrane fractions may derive from a number of cellular membranes (including the plasma membrane) in addition to the agranular endoplasmic reticulum, depending upon the tissue of origin. Rough membrane vesicles are denser than smooth membranes owing to the contribution of ribosomal protein and hence the lower lipid : protein ratios of the former. It is possible to isolate fractions which are enriched in rough or smooth membrane components by density gradient centrifugation procedures[38], although it is not possible to determine how much of the smooth membrane fraction could have arisen by removal of ribosomes from the rough endoplasmic reticulum during preparation[37]. Both the rough and smooth fractions from liver contain the typical marker enzymes for endoplasmic reticulum, glucose-6-phosphatase and NADPH: cytochrome *c* oxidoreductase[39].

Glaumann and Dallner[40] compared the lipid composition of rough and smooth rat liver microsome fractions, isolated by a gradient centrifugation technique that depended upon cation-binding properties of the fractions[38]. Rough and smooth membrane fractions each accounted for approximately half of the total microsomal phospholipid. When the ribosomes and adherent proteins of the lumen were removed, rough and smooth membranes exhibited similar phospholipid : protein ratios. Moreover, the quantitative distribution of individual phospholipids showed no significant differences between the

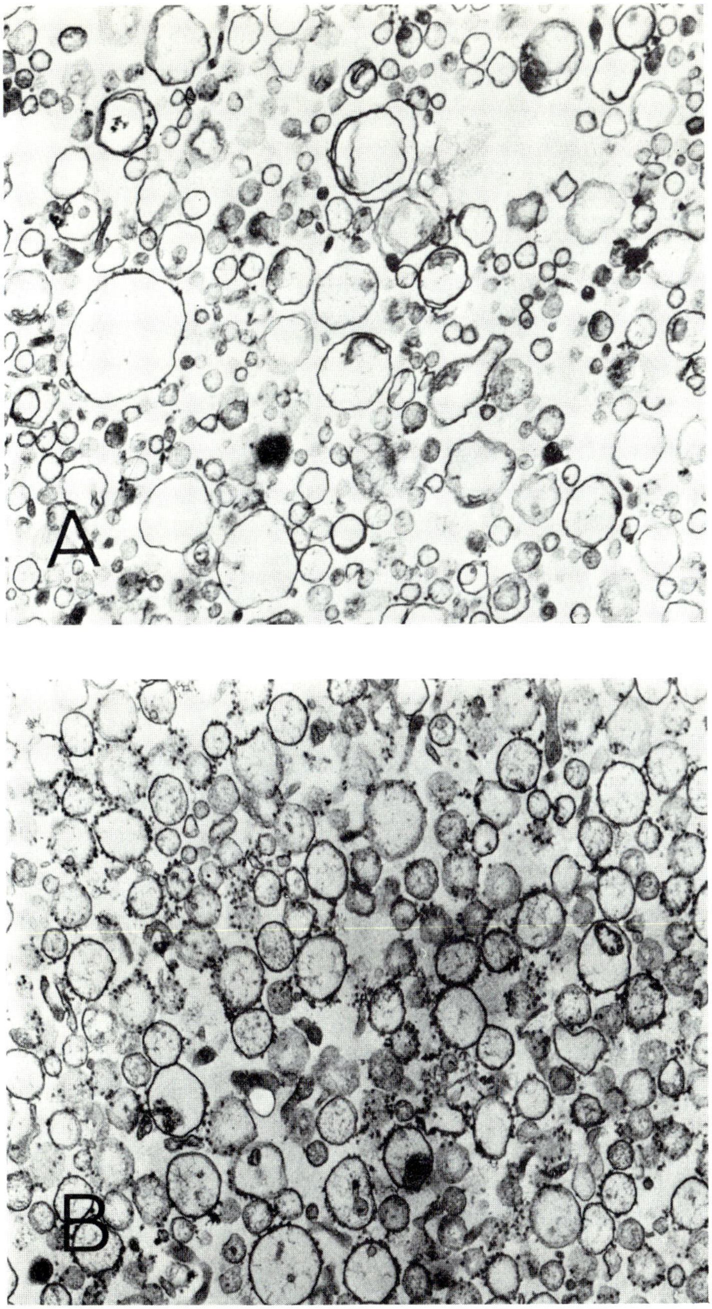

Fig. 9. Thin sections of endoplasmic reticulum subfractions (× 35 000) isolated by density gradient centrifugation. The light sub-fraction A (density 1.000-1.094) shows a preponderance of 'smooth' vesicles; the heavy microsome fraction B (density 1.142-1.192) contains mainly 'rough' vesicles covered with ribosomes. (Reproduced from Ref. 27 with kind permission of Dr. D.F. Parsons and the publishers.)

fractions. (The similarity in the phospholipid patterns for whole microsomes and rough endoplasmic reticulum may be seen by comparing Figs. 3 and 4). A similar fractionation of liver microsomes from the embryonic chick was conducted by Pollak and Ward[41]. At various stages of development the smooth membrane fractions had quite a different phospholipid composition, with approximately double the sphingomyelin proportion found for rough membranes[42]. Recently, Meldolesi et al.[43,44] have also reported the presence of a high sphingomyelin content in smooth membranes isolated from the microsome fraction of guinea pig pancreas. In this instance, the smooth membranes appeared to be derived principally from the Golgi apparatus however, rather than from the endoplasmic reticulum.

Golgi apparatus

Quite recently there have been major advances in the separation and characterization of the membrane system which makes up the Golgi apparatus. Morré et al.[45], Leelavathi et al.[46], and Fleischer and Fleischer[47] have described procedures for the isolation from rat liver of the smooth membrane fraction which appears to correspond to the strongly osmiophilic network of Golgi membranes visualized by impregnation techniques in tissue sections. Phosphatases specific for thiamine pyrophosphate and nucleoside diphosphates are concentrated several-fold in Golgi-rich membranes[48], but the best marker would seem to be the activity of UDPgalactose: *N*-acetylglucosamine galactose transferase; isolated Golgi membranes showed high recovery[46] and one hundred-fold enrichment[47] with respect to the latter enzyme, but contained low quantities of enzymes characteristic of endoplasmic reticulum (glucose-6-phosphatase, rotenone-insensitive NADH:cytochrome *c* oxidoreductase) or plasma membrane (5′-nucleotidase)[46–48].

Keenan and Morré[49], and Yunghans et al.[50] have described the phospholipid composition of isolated Golgi fractions as intermediate between that of the endoplasmic reticulum and the plasma membrane. The Golgi membranes contain a larger proportion of sphingomyelin than the endoplasmic reticulum with correspondingly lower lecithin content. Cholesterol and cholesterol ester concentrations are also higher in the Golgi fraction. Based on the transition in staining properties from the proximal face of Golgi lamellae adjacent to endoplasmic reticulum to the distal face of the stacked membranes next to the secretory vesicles, it has been postulated that the Golgi may represent a transitional membrane undergoing differentiation to form new plasma membrane[51]. In addition to the transitional phospholipid distribution noted above, the fatty acid composition of the Golgi membrane showed a progression toward that characteristic of plasma membranes, with more saturated residues in the lecithin and a greater proportion of long-chain residues in the sphingomyelin than was observed for the endoplasmic reticulum membranes[49].

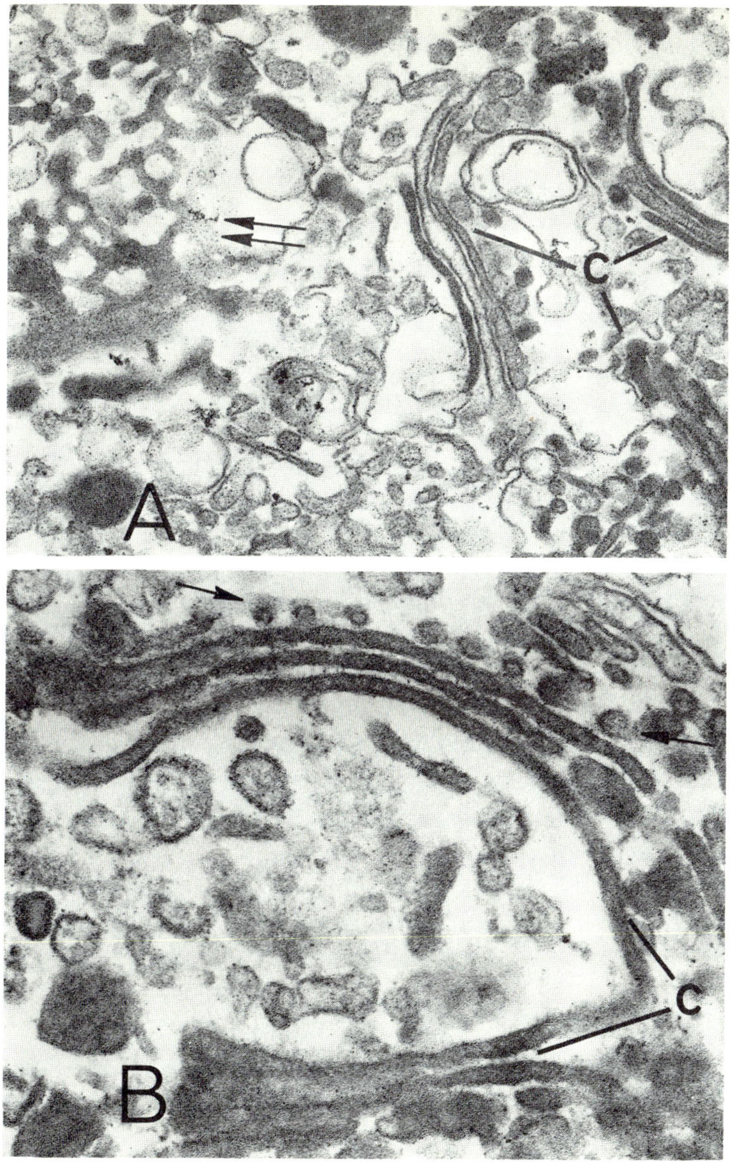

Fig. 10. Isolated Golgi fractions, (A–× 47 000; B–× 81 000). The tubular structures of the densely stained cisternae (c) are noted by the arrows. (Reproduced from Ref. 46 with kind permission of Dr. D.S. Feingold and the publishers.)

References p. 246

Plasma membranes

Until recently investigations of the phospholipids of cell surface membranes were mostly limited to erythrocyte ghosts, which in non-nucleated red cells could be isolated free of other cytomembrane systems. In 1960 Neville[52] described a procedure for the

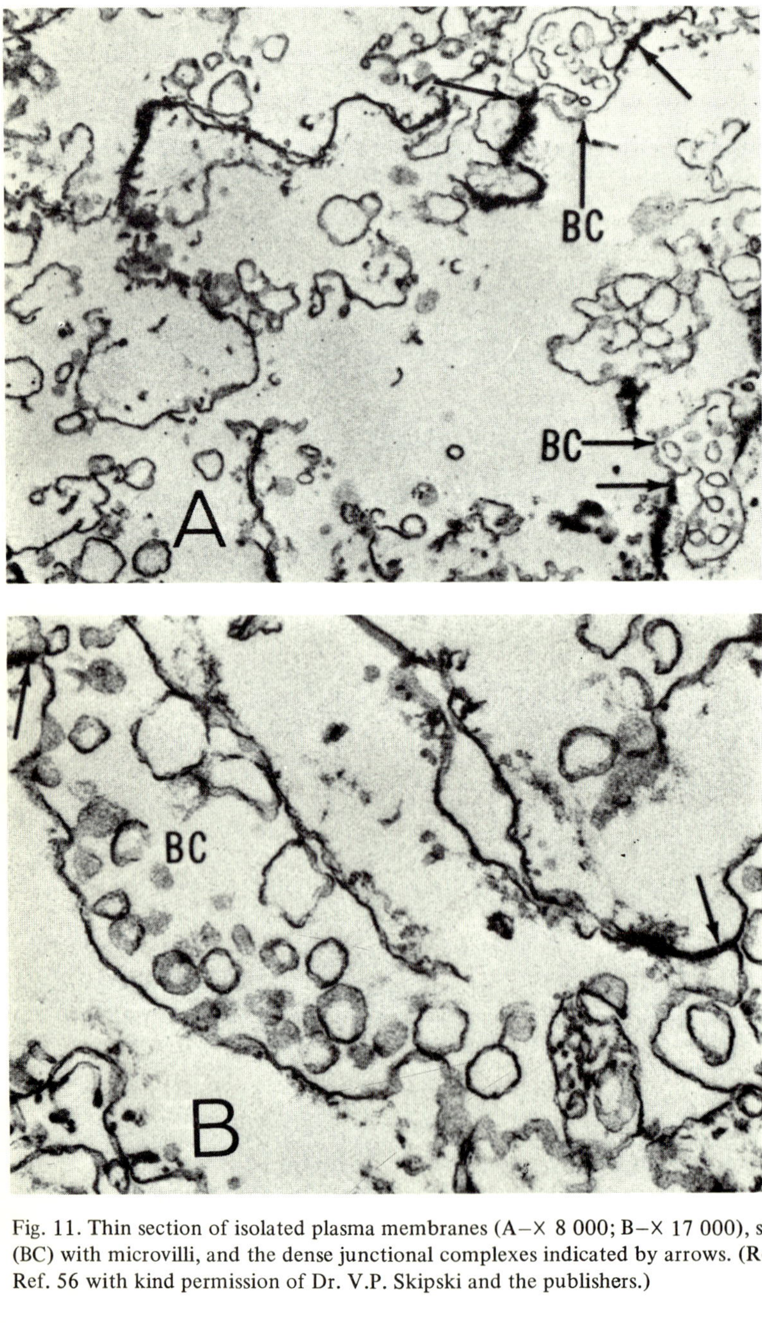

Fig. 11. Thin section of isolated plasma membranes (A–× 8 000; B–× 17 000), showing bile canaliculi (BC) with microvilli, and the dense junctional complexes indicated by arrows. (Reproduced from Ref. 56 with kind permission of Dr. V.P. Skipski and the publishers.)

separation of plasma membranes from rat liver, and variations and improvements upon this method have provided the basis for analytical investigations over the past decade. The surface membranes are characterized by enrichment with respect to marker enzymes such as 5′-nucleotidase and a $Na^{\oplus}/K^{\oplus}$-activated ATPase[53]. Evans[54] has sub-fractionated the plasma membranes into a light vesicular fraction which is enriched in 5′-nucleotidase, and a heavy fraction containing membrane strips with junctional complexes and bile canaliculi which is enriched in the $Na^{\oplus}/K^{\oplus}$-activated ATPase.

A complete lipid analysis for rat liver plasma membranes has been reported by Pfleger et al.[55] and by Ray et al.[56]; the phospholipid pattern is illustrated in Fig. 4. In comparison with other cytomembranes the plasma membrane fraction shows the greatest proportion of sphingomyelin and phosphatidylserine, and the lowest content of lecithin. It has been suggested that the consequent preponderance of saturated and longer chain fatty acyl substituents of plasma membranes contributes to tighter packing of phospholipid molecules in the membrane bilayer, and hence to increased stability and impermeability by comparison with other cytomembranes[49]. It is noteworthy also that plasma membranes exhibit the greatest enrichment with respect to cholesterol[57]; the molar ratio cholesterol : total phospholipid approaches 1:1, as noted earlier for erythrocyte membranes. Studies of artificial membranes suggest that the presence of high concentrations of cholesterol would enhance the close packing of phospholipids to stabilize the plasma membrane[5] further.

Similar compositional analyses have been reported for plasma membranes derived from HeLa cells[58], and for L cells grown in culture[59]. The results agree with those reported for liver in showing enrichment of surface membranes with respect to cholesterol and sphingomyelin compared to the rest of the cell. Weinstein et al.[59] observed that, although the fatty acid composition of L-cells was dependent upon lipids added to the culture medium, the proportion of long-chain fatty acids in the surface membranes was selectively increased. The degree of unsaturation, however, was similar to that of the whole cell.

Nuclear membranes

Despite their prominence both with respect to size and function the nuclei have received surprisingly little attention from phospholipid biochemists until comparatively recently. Analytical difficulties have arisen owing to the uncertain degree of contamination of isolated nuclei by tightly adherent tags of cytoplasmic material. The presence of ribosomes on the outer nuclear envelope and the continuity of this membrane with elements of the rough endoplasmic reticulum have made it difficult to determine the existence of true differences of composition between nuclear membranes and the endoplasmic reticulum[60]. The isolation of nuclear membrane fragments has been reported from several laboratories, but there is conflicting evidence regarding the occurrence of various enzyme markers among the different preparations. Thus the presence of NADH:cytochrome *c* oxidoreductase has been verified for several nuclear membrane preparations, but NADPH:cytochrome *c* oxidoreductase, which was reported

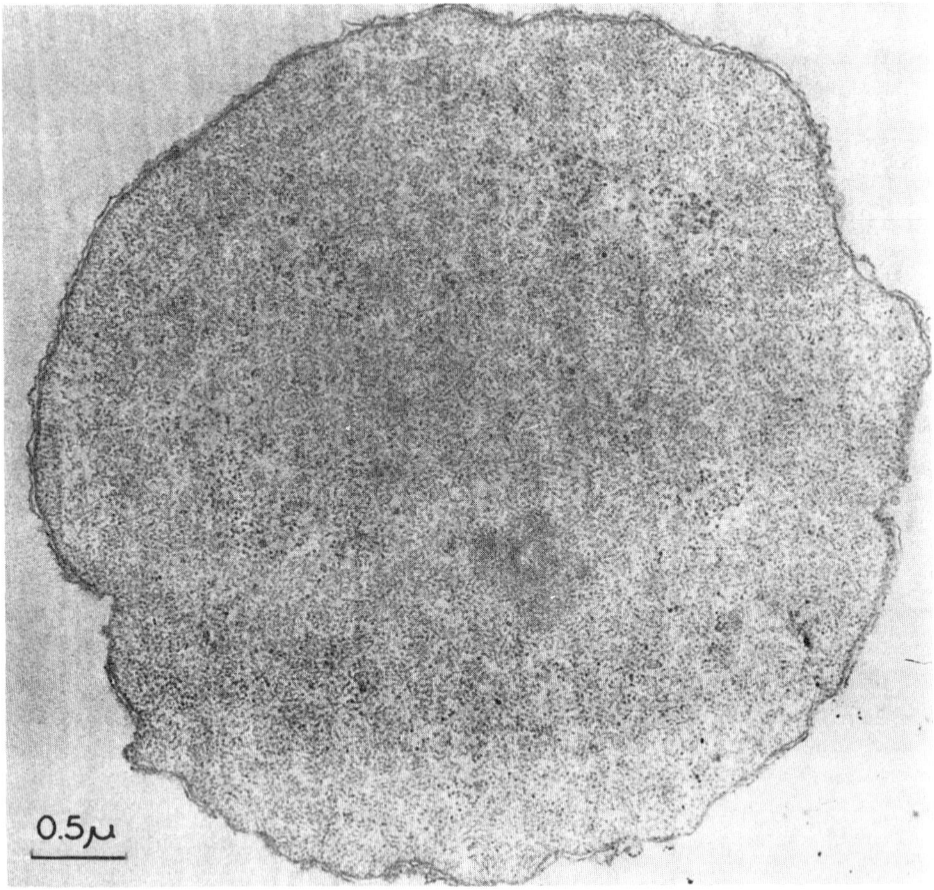

Fig. 12. Isolated rat liver nucleus showing the inner and outer membranes as seen *in vivo*. (Electron micrograph reproduced from Ref. 63 with kind permission of Dr. C.B. Kasper and the publishers.)

to be an integral component of nuclear membranes by Franke et al.[61] for pig and rat liver, was detected in only very small amounts for nuclear membranes from bovine liver by Berezney et al.[62] Kashnig and Kasper[63] on the other hand, found similar high levels of glucose-6-phosphatase and NADH:cytochrome *c* oxidoreductase, with about one-half the specific activity of microsomes in their rat liver nuclear membranes, while Franke et al.[61] detected little or no glucose-6-phosphatase in their preparations. Obviously, resolution of this controversy must occur before a clear assessment of cross-contamination of isolated nuclear membranes by other cytomembranes can be carried out based upon enzyme markers.

The phospholipid composition of nuclei was first examined by Gurr et al.[64] who demonstrated that preparations isolated from media containing citric acid contained considerably less total phospholipid than nuclei isolated in sucrose media. Electron-microscopic examination revealed that the citric acid nuclei had lost their outer envelopes, while sucrose nuclei resembled those seen *in situ* with an intact double

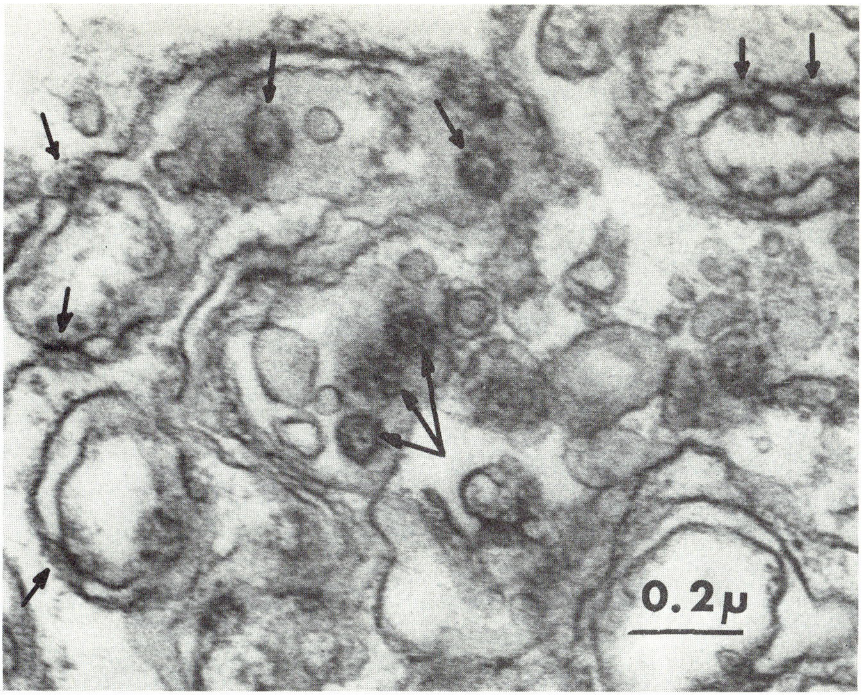

Fig. 13. Isolated nuclear membrane fraction. Both transverse and tangential sections of the nuclear pores are evident (arrows). (Reproduced from Ref. 63 with kind permission of Dr. C.B. Kasper and the publishers.)

membrane structure. The pattern of phospholipid distribution of rat liver nuclear membranes was shown by Kleinig[65] to resemble closely that of the rough endoplasmic reticulum (Fig. 4); Keenan et al.[66] have arrived at the same conclusion with respect to bovine nuclear preparations, although the latter workers reported somewhat higher values for the sphingomyelin and phosphatidylserine content of nuclear membranes. The cholesterol content of the nuclear membranes from rat liver described by Franke et al.[61] (shown in Fig. 4) was much lower than the value given by Keenan et al.[66] who claimed that 16% of total bovine nuclear membrane lipids was contributed by cholesterol. Kashnig and Kasper[63] also reported higher sterol content for their preparations, with cholesterol : phospholipid mole ratios of 0.30 and 0.42 as compared to the value of 0.10 reported by Kleinig[65] for the same tissue.

An interesting comparison between the nuclear membrane and plasma membrane is provided by the experiments of Zentgraf et al.[67] with hen erythrocytes. As noted by these workers the erythrocyte of the mature bird provides a convenient system for such studies since it contains these two membrane systems alone, and contamination with other organelles is essentially negligible. The isolated nuclear membranes contained an enrichment of NADH:cytochrome *c* oxidoreductase, but had little or no 5′-nucleotidase or $Na^{\oplus}/K^{\oplus}$-ATPase activities. The latter enzymes were enriched in the plasma membrane fractions. Neither membrane fraction contained detectable glucose-6-phosphatase. The lipid composition of the nuclear membrane fraction[68] was similar to that observed for bovine liver fractions by Keenan et al.[66]. The plasma membrane fraction showed much higher contents of sphingomyelin and cholesterol than those of mammalian preparations; lecithin, phosphatidylethanolamine and cholesterol each constituted 30% of the total phospholipids, and the cholesterol : phospholipid molar ratio was 1:1. The fatty residues were found to be characteristic of the individual phospholipid species. About 70% of the lecithin and phosphatidylethanolamine molecules were diacyl substituted; 20–25% of lecithin molecules were saturated ether analogues (1-alkyl, 2-acyl) and a similar proportion of phosphatidylethanolamine molecules were plasmalogens (1-alkenyl, 2-acyl). These proportions were observed to be identical for plasma membrane and nuclear membrane components. The fatty acid distributions were also shown to be distinctive for different phospholipids, but as with mitochondrial and microsomal membranes, there were no striking differences between membranes.

General comments

The findings discussed above have been chosen where possible to compare the compositions of homologous cell fractions. Ideally, the comparisons would be most reliable when conducted with the same animals using the same analytical procedures, and with appropriate controls to evaluate artefacts arising from hydrolytic or oxidative reactions during preparation. Thus for example, it would seem premature at present to comment extensively upon reported values for the lysophospholipid composition of membrane fractions. Also conspicuous by their absence are values for membrane phospholipids of the lysosomes, owing to a paucity of information on this subject. The results that have appeared show a phospholipid distribution for lysosomal fractions that is similar to that of mitochondria (i.e. high diphosphatidyl glycerol content (see Chapter 16 p. 459), but contamination by light mitochondria and ingested mitochondrial fragments must be assessed before attributing this composition to lysosomes proper.

Patton[69] has noted an interesting correlation between the contents of sphingomyelin and cholesterol in a variety of endocytoplasmic membranes. This relationship may be seen by comparing the percentage contribution of these two lipids in a variety of liver membrane fractions, from the mitochondrion which contains the lowest concentrations of sphingomyelin and cholesterol, to the plasma membrane which has the highest proportions of both lipids (see Fig. 4). Although the two lipids do not always exist in a simple stoichiometric ratio (e.g. molar ratios of cholesterol: sphingomyelin may be greater than three in erythrocyte membranes), the consistent parallel relationship within a given species would appear to be more than coincidental.

The intracellular membrane systems may appear to be compositionally distinct with respect to phospholipid distribution as well as being morphologically distinct, but there are obviously areas of continuity in both parameters. Of particular interest are the parallels between endoplasmic reticulum, outer mitochondrial membranes, and nuclear membranes. Phospholipid compositions of these membranes are strikingly similar, as are some of their enzymatic activities. Each of these three membranes contains an

NADH:cytochrome *c* oxidoreductase which differs from the enzyme of mitochondrial respiration in its insensitivity to rotenone; associated with this reductase activity is an appreciable content of a cytochrome of the b_5-type. Endoplasmic reticulum and nuclei also contain cytochrome P-450 (ref. 70), a component of NADPH-catalysed hydroxylations which is absent from mitochondrial outer membranes[71], as well as from other cell membranes (Golgi, plasma membranes)[70]. Nuclei are also unique in containing cytochrome *c*-dependent NADH oxidase activity that is distinct from mitochondrial respiratory activity[72]. Recent reports have indicated that isolated nuclear membranes contain significant quantities of cytochromes of the a–a_3 type[73], as well as cytochrome oxidase activity[74], which cannot be accounted for by mitochondrial contamination. Possibly, as in the mitochondrion, these oxidative reactions are localized in the inner of the two nuclear membranes. It will be of some interest to establish whether the nuclear membranes may be separable, and to what extent their compositions and functions may differ.

The postulated flow and transformation of membranes in the direction endoplasmic reticulum–Golgi apparatus–plasma membrane[51], and the parallel progression of phospholipid distribution patterns[49], has been commented upon earlier. The formation of membrane blebs and migration of whole segments of membranes from nuclei and endoplasmic reticulum have been inferred repeatedly from microscopic studies. Other morphological evidence has indicated membrane transfer from the Golgi apparatus to lysosomes and plasma membrane, and from the plasma membrane back into the cytoplasm during the formation of various endocytotic vesicles and vacuoles. (See reviews by Whaley et al.[75] and by Morré et al.[76].) Thus cellular membranes may be seen to exist in a continual state of transfer, exchange, and transformation of components among different loci of the cytoplasm, the balance of which would be regulated by the morphogenetic, secretory, ingestive, and synthetic processes of the cell. Despite the discrete profiles of protein and lipid compositions of membrane loci, undoubtedly future studies will reveal that these loci should be considered less as discrete organelles and more as components within a continuum of an integrated and interconnected multi-functional membrane system.

3. FUNCTIONS OF MEMBRANE PHOSPHOLIPIDS

Mitochondrial membranes

The dependence of the reactions of the respiratory chain upon the presence of mitochondrial phospholipids has been well-documented, and has provided a model system for the study of protein–lipid interactions in the morphological and functional assembly of membrane systems[77–79]. In early studies treatment by crude phospholipase preparations indicated the phospholipid-dependence of mitochondrial oxidations, although products of the enzymatic degradation such as fatty acids and lysophospholipids were not always excluded as inhibitory agents[80,81]. Recently more definitive investigations have been conducted with purified phospholipases under controlled conditions, and with adequate procedures for removal of degradation

References p. 246

products[79]. These studies, and others in which phospholipids were removed by extraction with 90% acetone[82], established conclusively the dependence of electron transport enzyme systems upon the presence of the membrane phospholipids. Succinoxidase activity, and segments of the multi-enzyme system, namely, succinate–CoQ reductase, $CoQH_2$ –cytochrome *c* reductase, and cytochrome oxidase, were almost totally inactivated in the phospholipid-depleted mitochondria[83]. Restoration of these activities was effected by the addition of micellar suspensions of mitochondrial phospholipids, and was accompanied by binding of the phospholipids to the lipid-depleted membranes. The reactivation of electron transport followed parallel with the phospholipid uptake until a saturating value was reached when the phospholipid content of the reactivated particles approximated that of the original mitochondria.

Several different individual phospholipids have been shown to be capable of restoring respiratory activity[84]. In terms of the amount of added phospholipid-P which was required to produce maximal reactivation of succinate: cytochrome *c* oxidoreductase the following order of effectiveness was observed: diphosphatidylglycerol > total mitochondrial phospholipid > phosphatidylethanolamine >> lecithin. However, saturating concentrations of each phospholipid produced similar maximum activities of the enzyme. Fleischer et al.[84] showed that the differential effects of individual phospholipids were not due to specific activations, but were related to the extent to which each was bound by depleted mitochondria. Thus half-maximal reactivation was attained with approx. 5 μg of diphosphatidylglycerol-P/mg protein, and 60–70% of the added phospholipid was bound to the particles; with lecithin half-maximal reactivation required the addition of approx. 50 μg P/mg protein, but only 8% of the added phospholipid was bound. In all cases the extrapolated value for 100% activation was approx. 11 μg lipid-P rebound/mg protein, compared with 13 μg P/mg protein in native mitochondria.

The kinetics of phospholipid rebinding, and effects of alteration of the constituent fatty acids, were studied by De Pury and Collins[85]. Lecithin, labelled *in vivo* with ^{32}P, was isolated from livers of normal rats and rats which had been fed a diet deficient in essential fatty acids. Micellar preparations of each phospholipid were then incubated with liver mitochondria that had been extracted with aqueous acetone. Normal lecithin was bound more slowly and was correspondingly less effective in restoring succinate: cytochrome *c* oxidoreductase activity to the depleted particles than the lecithin from deficient animals in which the essential fatty acids (linoleic and arachidonic acids) were replaced by the oleic and palmitoleic series. These findings were interpreted to indicate that there was no special requirement for essential fatty acids in the electron transport chain, but that both types of lecithin were equally effective once they were bound to mitochondrial protein.

The studies of Widmer and Crane[86] demonstrating the existence of a complex form of cytochrome *c* in mitochondria which could be released in a lipid-soluble state, focussed attention upon the interaction of phospholipids with this electron transport carrier. Cytochrome *c* was shown to react with mitochondrial phosphatidylethanolamine to form lipid complexes, which were soluble in diethyl ether or iso-octane[87,88]. Ability to form the complexes was greatest for highly unsaturated species of phosphatidylethanolamine, and the complex formed with the dimyristoyl derivative was insoluble in iso-octane[88]. Acidic phospholipids such as diphosphatidylglycerol or phosphatidylinositol also formed iso-octane-insoluble complexes with cytochrome *c*, with 8–10 moles of phospholipid bound per mole of cytochrome *c*, corresponding to the freely available basic sites of the protein[89]. Phosphatidylethanolamine, at pH values where it is dissociated to the acidic form, was bound in the molar ratio 24:1 suggesting neutralization of both external and internal basic sites of the cytochrome *c* molecule, and the interaction was prevented when free NH_2 groups of the cytochrome *c* had been destroyed by treatment with nitrous acid[88]. Presumably both hydrophobic and electrostatic forces are involved in these interactions. Thus lecithin alone failed to form a complex with cytochrome *c*, but did react when the addition of small amounts of acidic lipids had fully neutralized the external basic sites[89]. Moreover, formation of cytochrome *c* lipid complexes is retarded in salt solutions[90]. By contrast the binding of phospholipids to lipid-depleted mitochondria is unaffected by high ionic strength of the medium, indicating that electrostatic interactions are probably of less importance than for the binding to cytochrome *c*[91].

The dual nature of phospholipid–cytochrome *c* interactions was underlined by the studies of Quinn and Dawson[92] in which changes in surface pressure of phospholipid monolayers were followed simultaneously with appearance of surface radioactivity after the introduction of ^{14}C-labelled cytochrome *c* into the sub-phase. The rate of penetration of cytochrome *c* into the phospholipid film was strongly affected by the ionic strength and pH of the sub-phase, but there was no evidence for any stoichiometric relationship. At high lipid film pressures cytochrome *c* failed to penetrate the monolayer, but adsorbed to the film by initial electrostatic interaction followed by formation of non-ionic bonds which were not readily disrupted by introduction of *M*-sodium chloride into the sub-phase. Compared with monolayers of phosphatidylethanolamine or lecithin which bound only a small percentage of the added cytochrome *c*, diphosphatidylglycerol monolayers were capable of taking up the cytochrome *c* quantitatively from the sub-phase[93]. Cytochrome *c* bound to films of diphosphatidylglycerol was almost completely displaced by high salt concentrations, and in this respect the binding of cytochrome *c* resembled that found in the mitochondrion. It seems not unreasonable to conclude that diphosphatidylglycerol is a likely candidate as a site for complex formation with cytochrome *c* in the inner mitochondrial membrane[93].

The lipid–cytochrome *c* complex, containing a mixture of mitochondrial phospholipids, is readily reduced by ascorbic acid, and the reduced form is functional as a substrate for cytochrome oxidase[94]. By contrast the complex studied by Kimelberg et al.[95] which was prepared by interaction of a mixture of lecithin and diphosphatidylglycerol with cytochrome *c*, was reduced by ascorbate only in the presence of a lipophilic electron carrier such as phenazine methosulfate, or a lytic detergent such as Triton X-100 (ref. 96). In this multi-laminated preparation, the cytochrome *c* appears to be trapped within a lipophilic barrier, analogous to the situation found with 'inside-out' submitochondrial particles where the cytochrome *c* may be released only after disruption of the membrane with phospholipase[97].

Preparations of cytochrome oxidase containing low quantities of phospholipids have been 'solubilized' in dispersed form by the use of various detergents. Such lipid-depleted forms of the enzyme require the addition of phospholipids for the maximum rate of oxidation of ferro-cytochrome *c*[98–100]. However, Sun and Crane[94] showed that the reduced form of the iso-octane-soluble phospholipid–cytochrome *c* complexes could be rapidly oxidized by lipid-depleted cytochrome oxidase at rates comparable to those for oxidation of free ferro-cytochrome *c* by lipid-containing oxidase preparations. In view of the ability of cytochrome *c* to act as a mediator in the transport of phospholipids to mitochondrial membranes, as noted later in this chapter, it might be argued that the lipid complex form of reduced cytochrome *c* was not the true substrate, but rather was acting to reconstitute the active lipoprotein form of the oxidase by transferring the phospholipids to the lipid-depleted enzyme.

However, further studies in Crane's laboratory have established multiple roles for phospholipids in the cytochrome oxidase reaction, and suggest that both the substrate and enzyme must exist as lipoprotein complexes for maximal reactivity[101,102]. Thus complete removal of the phospholipids as well as the detergent used in isolation of the enzyme, produced an aggregated form of the oxidase which reacted poorly with lipid–cytochrome *c*. Addition of phospholipids, or of low concentrations of detergents (Triton X-100, deoxycholate or Emasol 1130) reorganized the aggregated enzyme into membrane vesicles and restored the activity toward reduced lipid–cytochrome *c*[101]. The results suggest three functions for mitochondrial phospholipids in this region of the electron transport chain: (*1*) a specific role in producing a micellar form of cytochrome *c* for lipophilic interaction with the enzyme; (*2*) a nonspecific requirement, which may be satisfied *in vitro* by detergents, in maintaining suitable conformation of the oxidase in a dispersed state such that catalytically active copper atoms and SH groups are in an exposed functional orientation; and (*3*) a unique requirement for 1–2 molecules of diphosphatidylglycerol which are very tightly bound to 'lipid-free' cytochrome oxidase, and cannot be replaced by detergents[102].

An extensive literature exists on the reconstitution of the mitochondrial electron transport system from its components (for reviews see refs. 78, 103–108). Frequently the subunit complexes that have been studied in recombination experiments have contained bound phospholipid, and each lipoprotein complex may be considered as a functional system, catalysing reactions in a segment of the respiratory chain. Recent experiments in Racker's laboratory in which elements of the chain have been resolved as lipid-free protein components have emphasized the important role of phospholipids in reconstituting the form and function of mitochondrial membranes. Thus the recombination of succinate dehydrogenase with cytochrome *b* to produce a functional complex capable of reducing coenzyme Q showed an absolute and selective requirement for phospholipids; detergents failed to replace the requirement for a mixture of mitochondrial phospholipids. Lecithin or phosphatidylethanolamine were effective, while diphosphatidylglycerol did not satisfy the requirements for reconstitution but produced a variable inhibition by inactivating the cytochrome *b*[109]. The effects of the phospholipids in this system are complicated and undoubtedly include factors such as the stabilization of soluble succinic dehydrogenase as noted by Cerletti et al.[110]. However, it is noteworthy that the complex reconstituted in the absence of phospholipid is capable of reducing phenazine methosulfate but not coenzyme Q, indicating that one of the functions of phospholipids in this region of the chain may be to provide the appropriate environment for interaction with hydrophobic carriers such as the quinones[109].

In further studies by Yamashita and Racker[111] the entire succinoxidase system was reconstituted from the purified cytochrome components in the presence of succinate dehydrogenase, coenzyme Q, and phospholipids. The oxidation of succinate by the reconstituted system occurred at a rate comparable to that observed with sub-mitochondrial particles, and showed similar sensitivity to respiratory inhibitors (antimycin A, trifluorothienyl butadione, cyanide). Several neutral and acidic phospholipids were effective in the reconstitution, including some phospholipids not normally found in significant amounts in mitochondrial membranes such as phosphatidylserine. As had been observed for the span of the respiratory chain between succinate and coenzyme Q, the diphosphatidylglycerol preparation used in these studies was less effective and resulted in inactivation of the restored activity with time.

The study of phospholipid requirements and the reconstitution of activity from components has proven more difficult for the mitochondrial NADH oxidase system, possibly owing to the greater lability of the NADH dehydrogenase. For example, treatment of mitochondria with aqueous acetone solutions to extract the phospholipids results in inactivation of NADH oxidation that is not reversed by restoration of phospholipid to the depleted preparations under conditions where the succinoxidase activity is reactivated[79]. Controlled incubation with phospholipase A demonstrated that sub-mitochondrial particles lost NADH oxidase activity at early stages of digestion when succinoxidase activity was not greatly affected[112]. Reactivation of NADH oxidation by addition of phospholipids could be observed only if the inhibitory fatty acids and lysophospholipids produced by the digestion were removed by washing with serum albumin solutions[79]. Digestion with low levels of phospholipase A results in preferential hydrolysis of phosphatidylethanolamine and lecithin of mitochondrial membranes; considerably higher levels of phospholipase A are required for digestion of the diphosphatidylglycerol[113]. Parallel with the diphosphatidylglycerol hydrolysis, the NADH dehydrogenase of the membranes was released in soluble form[114].

The mechanism of the coupling of electron transport reactions to the production of ATP in oxidative phosphorylation remains obscure. However, it is clear that the chemical coupling of oxidative phosphorylation is associated with a coupling in the morphological sense between the respiratory chain of the inner mitochondrial membrane and the projecting sub-unit structures of the F_1-ATPase[106,107]. It is also apparent that phospholipids play a vital role in the structural requirements for coupling of oxidative phosphorylation in the interactions of the components.

Investigations of the properties of F_1 have shown that the enzyme behaves in a very different manner when attached to the mitochondrial membrane, than when free in solution. Solubilization of the enzyme renders it susceptible to inactivation by cold exposure, due to dissociation into inactive subunits, and the soluble enzyme is no longer sensitive to the effects of the oxidative phosphorylation inhibitors, oligomycin or rutamycin. The ATPase protein may be rebound to preparations of the inner membrane devoid of F_1, when it regains cold-stability and oligomycin-sensitivity. This modification of the properties of an enzyme upon binding to a membrane have been termed 'allotopy' by Racker[115].

When lipid-depleted components of the inner membrane are used for the reconstitution, restoration of the allotopic properties of F_1 may be shown to depend on the presence of phospholipids. Thus Kagawa and Racker[116] showed that cholate-treated fragments of the inner membrane (CF_O) were capable of binding F_1 to form an amorphous preparation with projecting subunits, but with little ATPase activity. On the addition of phospholipids to this CF_O–F_1 complex the amorphous structure becomes organized into vesicular membranes with projecting subunits, similar in appearance to inner mitochondrial membranes on negative staining, and simultaneously ATPase activity, which is sensitive to oligomycin, is fully restored to the membranes[117]. Kopaczyk et al.[118] have also shown that addition of phospholipid to soluble preparations of the ATPase complex could produce a highly active membranous form of the enzyme which was inhibited by oligomycin.

Further resolution of the membrane complex has been obtained by extraction under alkaline conditions, which releases a protein necessary for the conferral of oligomycin sensitivity upon membrane-bound F_1. This oligomycin sensitivity conferring protein (OSCP)[119] when added to the alkaline-extracted membrane proteins (ACF_O) in the presence of F_1, resulted in masking of the bound ATPase activity[120]. The further addition of mixed mitochondrial or soya bean phospholipids to the complex (OSCP–ACF_O–F_1) restored the ATPase activity which was sensitive to inhibition by rutamycin. The effects of individual phospholipids were variable. Phosphatidylinositol alone was as effective as the mixture of phospholipids in unmasking the ATPase, while lecithin and phosphatidylethanolamine only partially reactivated the enzyme; in each case the unmasked activity was susceptible to inhibition by rutamycin. On the other hand the addition of strongly acidic phospholipids (diphosphatidylglycerol, phosphatidylserine, or phosphatidylinositol diphosphate) caused full reactivation of the ATPase but the allotopic property of sensitivity to rutamycin was lost.

A different approach to the role of phospholipids in the coupling phenomenon was followed by Berezney et al.[121] who studied the effects of sequential hydrolysis of the phospholipids of submitochondrial particles with phospholipase A and removal of lysophospholipids by washing with serum albumin[113]. Partial digestion, in which lecithin and phosphatidylethanolamine but not diphosphatidylglycerol were hydrolysed, reduced the oligomycin-sensitive ATPase by more than half; complete digestion of phospholipids caused simultaneous disappearance of the projecting subunits and loss of 97% of the oligomycin-sensitive ATPase. When the endogenous lysophospholipids produced by digestion were not removed by washing, or when exogenous lysophospholipids were added to the lipid-depleted membranes, the ATPase was active but was insensitive to rutamycin. Interestingly, the highly lipid-depleted preparations retained a membranous structure, as had been observed previously by Fleischer et al.[122] for solvent-extracted mitochondria. Of even greater interest was the observation that disappearance of projecting subunits was not associated with solubilization of F_1 from the membrane fragments; moreover, addition of intact phospholipids reactivated the oligomycin-sensitive ATPase of the lipid-depleted preparations to 30% of control values although there was no reappearance of the projecting subunits. These results suggest that upon removal of the phospholipids the projecting F_1 subunits are no longer visible because they collapse and become buried in the membrane in an inactive form. The activity and allotopic properties of F_1 are restored by phospholipid addition even though the subunits remain buried in the membrane[121].

Microsomal and other membranes

Recent investigations have established that components of electron transport systems in the endoplasmic reticulum are also dependent upon the phospholipids in the membranes. Jones and Wakil[123] extracted chicken liver microsomes with aqueous acetone, and observed that the consequent removal of 80% of the total phospholipid resulted in a parallel reduction of the activity of NADH: cytochrome *c* oxidoreductase. The activity of NADPH: cytochrome *c* oxidoreductase was not similarly affected, nor was the reduction of ferricyanide by NADH. Restoration of NADH : cytochrome *c* oxidoreductase was obtained by addition of phospholipid micelles to the depleted preparations. Most effective in the reconstitution were mixed micellar preparations containing lecithin with either lysolecithin or oleic acid added to facilitate formation of the dispersions. The addition of 1-oleoyl, 2-butyryl-glycero-3-phosphorylcholine which produced micellar preparations with similar properties to lecithin–lysolecithin sols was effective in restoring the activity in the absence of lysophospholipid or fatty acid.

Oxidation of NADPH coupled either to the oxygen-dependent desaturation of stearoyl CoA to oleoyl CoA[124], or to the hydroxylation and demethylation reactions of liver microsomal enzymes have also been shown to possess a rather specific requirement for phospholipids[125]. Investigation of the lipid requirement for fatty acid, hydrocarbon, and drug hydroxylations by Strobel et al.[126] showed that the transfer of electrons from NADPH to the haemoprotein designated P-450 was completely dependent on microsomal phospholipid, and that lecithin, but not lysolecithin or phosphatidylethanolamine, was effective.

There is an extensive literature on the role of phospholipids in membrane transport reactions. Permeability properties of several membrane systems can be shown to bear close relationships to their phospholipid composition. Artificial membranes composed solely of bimolecular layers of phospholipid mimic many of the passive permeability phenomena seen with biological membranes (see Chapter 7, p. 143, for details). Thus uncouplers of oxidative phosphorylation produce large increases in proton conductance of phospholipid bilayers[127], consistent with Mitchell's hypothesis[128] that uncouplers can act as lipid-soluble proton carriers across the mitochondrial inner membrane barrier to prevent the formation of $H^{\oplus}$ gradients during electron transport. Valinomycin and other ionophores produce selective transport of $K^{\oplus}$ and other cations across artificial phospholipid membranes, in similar fashion to their effects on cation movement across mitochondrial and other biological membranes[129]. In various species of mammals permeability of erythrocyte membranes to both ionic and non-ionic permeants follows a striking correlation with the ratio of lecithin : sphingomyelin[5,130]. Since the lecithin species generally contain a greater proportion of unsaturated fatty acids than do the sphingomyelins it has been inferred that permeability may be partially dependent upon the degree of unsaturation of membrane phospholipids. The observations of De Gier et al.[131] with artificial vesicular membranes prepared from pure phospholipids (liposomes) are consistent with this interpretation. The rate of penetration of liposomal vesicles by non-electrolytes or cations increased in proportion to increased unsaturation of the phospholipid fatty acids.

Reference was made previously to the modulatory effects of phospholipids upon the mitochondrial ATPase that is involved in oxidative phosphorylation. Phospholipids also play an important part in the function of other subcellular ATPases which have been implicated in membrane transport reactions. The plasma membrane ATPase, which is activated by sodium and potassium ions, and undoubtedly participates in translocation of these cations, is dependent on phospholipid in a number of tissues[132]. A detailed consideration of this enzyme system is presented in connection with the discussion of excitable tissues in Chapter 15 and will not be included here.

Another active-transport ATPase is that of the sarcoplasmic reticulum, which appears to be involved with the uptake of calcium ions within a permeability barrier isolated from the muscle cytoplasm and the myofibrils. Uptake of $Ca^{2\oplus}$ is also concomitant with ATP hydrolysis in isolated muscle microsomal vesicles. Treatment of the microsomes with phospholipase C inactivates both the ATPase and $Ca^{2\oplus}$ transport parallel with the hydrolysis of the microsomal lecithin to diglycerides + phosphorylcholine[133]. The ATPase activity, but not $Ca^{2\oplus}$ transport was reactivated by detergents; reconstitution of both ATPase and $Ca^{2\oplus}$ transport was obtained only upon restoration of the phospholipid content of depleted membranes by binding added lecithin, lysolecithin or phosphatidic acid. Consistent with this lack of specificity of phospholipid requirement for reconstitution are the observations of Yu et al.[134], who confirmed that the ATPase and $Ca^{2\oplus}$ uptake were lost on treatment of muscle microsomes with phospholipase C, but not with phospholipase A or phospholipase D (the latter two enzymes would convert lecithin to lysolecithin or to phosphatidic acid respectively). The $Ca^{2\oplus}$-activated ATPase is inactivated by phospholipase A only if the lysolecithin produced is washed out of the membranes with serum albumin[135]

Duttera et al.[136] have studied the effects of phospholipases on liver microsomes, and showed a phospholipid requirement for the glucose-6-phosphatase activity. Digestion with phospholipase C caused inactivation of the enzyme which was reversed by addition of phosphatidylethanolamine but not lecithin. Lysolecithin produced restoration of activity at low concentrations, but was inhibitory at high concentrations. The latter observation undoubtedly accounts for the fact that digestion with phospholipase A produced irreversible inactivation of the enzyme unless serum albumin was also added. Restoration of normal microsomal membrane appearance in electron micrographs also occurred in reconstituted preparations[137]. The anomalous specificity for phosphatidylethanolamine in restoring enzyme activity was studied by Cater et al.[138], and was attributed to temperature-sensitivity of the lipid depleted enzyme. Thus when phospholipase C digestion was performed at 37° reversal of enzyme inactivation required phosphatidylethanolamine, not lecithin, despite the fact that loss of glucose-6-phosphatase activity with time paralleled the course of lecithin digestion, not the slower hydrolysis of phosphatidylethanolamine. Treatment with phospholipase C at 5° produced a partially inactivated enzyme that could be reactivated either by lecithin or phosphatidylethanolamine. As noted previously for succinate dehydrogenase[110] and the F_1–ATPase[120], these observations underline the importance of the lipid–protein interactions in membranes in providing a stabilizing environment for enzyme molecules which may be susceptible to inactivation in their soluble lipid-free forms.

4. INTRACELLULAR SYNTHESIS AND TURNOVER

The enzymatic steps involved in the biosynthesis of phospholipids have been largely elucidated, and are discussed in detail in an earlier section (Chapter 4, p. 67). Study of the process of turnover, or rate of renewal of phospholipids in the living cell, has presented considerable difficulty, however. The problems have arisen because different portions of phospholipid molecules may be renewed at different rates, and different molecular species of a given phospholipid may exhibit varying metabolic stability dependent upon their fatty acid composition. Further complications are introduced by the reutilization of isotopically labelled phospholipid precursors in the cell, and by the possibility that intact phospholipid molecules may undergo exchange-transfer between membranes. Early studies on the metabolism of phospholipids in relation to the renewal of membrane components must accordingly be re-evaluated in the light of these limitations.

Virtually all organs and tissues exhibit some capacity for neoformation of phospholipids from simple low molecular weight precursors (inorganic phosphate, glycerol, acetate, choline, ethanolamine, serine, inositol), although the rates may vary greatly[139]. Myelin phospholipids of adult animals, for example, show (but see Chapter 14, p. 393) little incorporation of isotopically labelled precursors[140,141], and the phospholipids of mature erythrocytes appear to undergo negligible renewal as a result of net biosynthesis *in situ*[142–144]. In other cellular membranes the phospholipids are continually being synthesized and broken down yielding a dynamic steady state as first enunciated for proteins by Schoenheimer[145].

The rate of phospholipid replacement is very high in the liver, an organ which not only provides the phospholipids needed for synthesis of its own subcellular membranes, but also acts as the source of phospholipids secreted in the bile and the circulating plasma lipoproteins. Following the administration of radioactive precursors *in vivo* the microsomal phospholipids of the liver have the highest specific radioactivities initially[146–148]. After the injection of [^{32}P] orthophosphate values for the specific radioactivity of individual phospholipid fractions were in the order: phosphatidic acid > phosphatidylethanolamine > phosphatidylcholine > phosphatidylinositol > phosphatidylserine = diphosphatidylglycerol; this relation held for all cell fractions examined, and the specific activity of each phospholipid was greatest initially in the microsomal fraction[149]. Isolated liver microsomes contain at high activity a number of enzyme reactions catalysing the biosynthesis of glycerides and phospholipids including acylation of 3-glycerophosphate[150] and of lysophosphatides[151], methylation of phosphatidylethanolamine to lecithin[152], and the formation of phosphatidylinositol[153], phosphatidylethanolamine and lecithin *de novo* from cytidine nucleotides[154, 155]. In the latter case the enzyme was almost exclusively localized in the microsomal fraction. This predominance of microsomal enzyme systems suggested a special role for the endoplasmic reticulum in synthesis of the major glycerophospholipids parallel to its importance in cellular protein biosynthesis.

Several reports have been at odds with an exclusive role for the endoplasmic reticulum in phospholipid synthesis. McCaman and Cook[156] reported significant activity of CDPcholine : 1,2-diglyceride cholinephosphotransferase in the mitochondria from beef heart and rat brain. In rat liver mitochondria, activity was reported to be localized in the outer membrane[29,157], and experiments on [^{14}C] choline incorporation *in vivo* showed lecithin of the outer membrane to be labelled more rapidly than that of the inner membrane[158]. Experiments *in vitro* demonstrated incorporation of [^{14}C] choline into lecithin of isolated rat liver mitochondria by a reaction that was stimulated by CTP and ATP, but was independent of the presence of added microsomes[159]. Kaiser[160] studied the mitochondrial system further, and reported that labelled ethanolamine, serine, and 3-glycerophosphate were also incorporated into phospholipids *in vitro* by energy-requiring reactions. Incorporation of the nitrogenous bases was not enhanced by addition of microsomes; [^{14}C] glycero-3-phosphate was incorporated into lecithin and phosphatidylethanolamine indicating total *de novo* synthesis of glycerophospholipids by the mitochondria. Such studies suggested that endoplasmic reticulum might not be the sole site of phospholipid synthesis in liver, but, that the mitochondrion might possess some degree of autonomy in forming its own membrane phospholipids, as it does with polypeptide synthesis[161].

In evaluating the findings on incorporation of radioactive precursors by isolated mitochondria the following complications must be considered and accounted for: (*1*) Extent of contamination of mitochondrial fractions by microsomes; (*2*) Adequacy of the incorporation system for measurement of pathways of phospholipid synthesis *de novo*; (*3*) Extent of operation of exchange reactions catalysing replacement of portions of phospholipid molecules without synthesis *de novo*; (*4*) The exchange of whole phospholipid molecules between different membranes.

The experiments reported by McMurray and Dawson[30] examined some of these criteria. With regard to point (*1*), the microsomal contamination of liver mitochondrial preparations was assessed by the activity of the marker enzyme for endoplasmic reticulum, NADPH : cytochrome *c* oxidoreductase[18]. Incorporation studies were then conducted *in vitro* in the presence of fixed amounts of mitochondria with variable known increments of microsomes based on units of enzyme marker. In this way it was possible to establish the relation between phospholipid labelling and microsomal concentration, and, by extrapolating to zero microsome concentration, to estimate the contribution of pure mitochondria to the incorporation in the absence of microsomal contamination. For the incorporation of [^{32}P] orthophosphate into lecithin, phosphatidylethanolamine, and phosphatidylinositol, and of CDP[^{14}C]choline into lecithin, the labelling was linearly related to microsomal content and extrapolated to zero at zero microsome concentration. These data suggested that mitochondria lacked essential reactions for *de novo* synthesis of these phospholipids in the absence of microsomes. The findings with CDP[^{14}C]choline incorporation have been confirmed by Zborowski and Wojtczak[162] and by Sarzala et al.[163], indicating that the terminal step in lecithin biosynthesis is a microsomal reaction and is essentially absent from pure liver mitochondria. Porcellati et al.[164] have made parallel observations on the biosynthesis of ethanolamine phospholipids in brain preparations Phospholipid labelling from [^{14}C]phosphorylethanolamine in the presence of CTP was carried out by microsomes in the presence of the soluble cytoplasmic fraction; when CDP[^{14}C]ethanolamine was the labelled precursor, incorporation into ethanolamine lipids was observed with microsomes alone[165]. Presumably, the cytosol is necessary for the biosynthesis of CDPethanolamine from CTP and phosphorylethanolamine[155], the rate-limiting step in the sequence. In neither case was there any labelling with brain mitochondria.

In consideration of points (*2*) and (*3*) the nature of the labelled substrate is of importance. For example, it is difficult to envisage how [^{32}P]orthophosphate could be incorporated into lipid phosphodiester linkages, or how [^{14}C]glycero-3-phosphate could be incorporated into the glycerol moiety of those phospholipids, other than by *de novo* synthesis. On the other hand, it is well established, mainly through the important work of Hübscher, that ^{14}C-labelled choline[166], ethanolamine[167], serine or inositol[168] may be incorporated into the corresponding phospholipids in the absence of *de novo* synthesis, in liver microsomes and mitochondria, by way of exchange reactions. Although such exchange reactions are generally energy-independent, they may be stimulated to variable degrees by nucleotides (ATP, CTP, or CMP), and by divalent cations ($Ca^{2\oplus}$, $Mn^{2\oplus}$) (refs. 166–169). Phospholipid base exchanges in mammalian tissues appear to be analogous to the transphosphatidylation reactions catalysed by purified preparations of phospholipase D from plants[170–171].

The incorporation of [^{14}C]choline by rat liver mitochondria *in vitro* was only partially dependent on energy-yielding reactions of respiration; the labelling was stimulated by cytidine nucleotides, but CMP was more effective than CTP[30]. Roughly half the [^{14}C]choline recovered in the lipid fraction was incorporated by a CoA-dependent reaction into a component that chromatographed with fatty acyl esters of choline, derivatives that may be mistaken for phosphatidylcholine[172,173]. These, and parallel experiments with [^{14}C]inositol and ethanolamine[30], indicated that such precursors could undergo extensive exchanges and other side-reactions yielding a grossly misleading estimate of the extent of *de novo* synthesis of the phospholipids by mitochondria.

Williams and Bygrave[174] in a comprehensive study of the conditions for optimal incorporation of [^{32}P]orthophosphate, concluded that previous investigations had been conducted under sub-optimal conditions for *de novo* phospholipid biosynthesis. Maximal incorporation into major phospholipids (lecithin, phosphatidylethanolamine, phosphatidylinositol) occurred only in the presence of microsomes, cytosol and mitochondria. In this system, which utilized NADH and fructose-1,6-diphosphate as energy-producing substrates, some incorporation was observed in the presence of antimycin A, or in the absence of mitochondria, indicating that anaerobic glycolytic reactions could provide some of the energy requirement for phospholipid synthesis, as shown earlier for brain preparations[175]. Requirement for the cytosol fraction was more stringent, and the extent of labelling was not saturated by increasing concentrations of cytosol protein. Reactions catalysed by the soluble fraction thus appear to be rate-limiting as shown also by Porcellati et al.[164]; even with whole liver cells addition of the cytosol fraction was found to be essential for optimal incorporation of inorganic ^{32}P (ref. 176). Williams and Bygrave[174] could find no [^{32}P]orthophosphate incorporation into major glycerophospholipids by mitochondria that could not be accounted for by contamination with microsomes or bacteria, and concluded that mitochondria are not capable of *de novo* synthesis of these phospholipids. However, the universality of this generalization may not be valid. In *Tetrahymena pyriformis* the formation of lecithin from CDPcholine and diglyceride has been reported to be localized chiefly in the mitochondrial fraction[177], as had been reported to be the case previously for chicken liver mitochondria, for the biosynthesis of both lecithin[178] and sphingomyelin[179]. Recent experiments reported by Ostrow[180] showed significant activity of this enzyme system in mitochondria from the yeast, *Saccharomyces cerevisiae*, with similar high activities in isolated outer membranes, and in microsomal membranes. Swartz and Mitchell[181] showed considerable activity of the enzyme also in a mitochondrial fraction from rat retina, but obviously the extent of contamination by microsomes must be evaluated in these systems before rigorous conclusions may be drawn. Similar criticism may be applied to the observation of [^{14}C] glycero-3-phosphate incorporation into lecithin reported by Kaiser[160], since the contribution from microsomal contamination was not documented in these experiments. Sauner and Lévy[182] could demonstrate little [^{14}C] glycerol incorporation into mitochondria, under conditions where microsome-containing systems were active.

The *de novo* synthesis of lecithin and phosphatidylethanolamine from inorganic ^{32}P requires the co-operation of three subcellular fractions: (*1*) Mitochondria: (*a*) provide the major means for formation of γ-[^{32}P] ATP by oxidative phosphorylation; and (*b*) contribute to the formation of phosphatidic acid by acylation of glycero-3-phosphate[162, 183]. (*2*) The cytosol: (*c*) is essential for phosphorylation of the *N*-containing bases to produce ^{32}P-labelled phosphorylcholine and phosphorylethanolamine[176]; (*d*) promotes the synthesis of labelled cytidine diphosphate derivatives of choline and ethanolamine[155]; and (*e*) can provide diglycerides by action of the soluble phosphatidic acid phosphohydrolase on particulate phosphatidic acid[184]. (*3*) The endoplasmic reticulum: would then be responsible for the terminal step of the sequence, (*f*) reaction of the labelled cytidine nucleotides with diglyceride acceptors to form the appropriate phospholipids[154, 155].

The incorporation of serine lipids of mammalian liver occurs predominantly by the reaction:

phosphatidylethanolamine + serine → phosphatidylserine + ethanolamine

which is reported to be localized in the microsomal fraction[168]. Bygrave and Bücher[185] described two systems in rat liver mitochondria for serine incorporation. One system was greatly stimulated by $Ca^{2\oplus}$ and was energy-independent, characteristic of the exchange reaction. The second system was dependent on the presence of oxidizable substrate (succinate) and $Mg^{2\oplus}$ and was stimulated by CMP. In locust flight muscle mitochondria incorporation of serine occurred primarily by the second pathway[186]. In *Tetrahymena pyriformis* this activity sedimented primarily with mitochondria[187].

There seem to be two independent systems for phosphatidic acid biosynthesis in rat liver, one in the microsomal fraction, the other in the mitochondria. It was concluded that microsomal contamination could not possibly account for the labelling of phosphatidic acid by isolated mitochondria[30]. Upon fractionation of mitochondrial membranes the biosynthetic activity segregated together with monoamine oxidase[188], and it was concluded that both enzymes were localized in the outer membrane[162,163]. The mitochondrial enzyme leads to a preponderance of lysophosphatidic acid, while the microsomal enzyme in the same conditions gives rise to the fully acylated product[189]. Acylation of lysophospholipids has been reported to occur in extracts prepared from sonicated mitochondria[190], and this activity also seems to occur in the outer membrane[191].

When isolated mitochondria are incubated with [^{32}P] orthophosphate under optimal conditions for oxidative phosphorylation, or with γ-[^{32}P] ATP, the phosphatidic acid and phosphatidylinositol phosphate fractions become strongly labelled[192–194]. Labelling of the latter component probably arises from contamination of mitochondrial fractions with fragments of plasma membrane, since the latter seems to be the principal site for the phosphatidylinositol kinase[195]. A third phospholipid fraction was also observed to be rapidly labelled with ^{32}P by guinea pig liver mitochondria[196], and was identified by Hajra and Agranoff[197] as acyl dihydroxyacetone phosphate. Further studies by Hajra[198] established that isolated mitochondria were capable of synthesizing this novel phospholipid by the acylation of dihydroxyacetone phosphate with fatty acyl-CoA derivatives; the acylation was preferential for saturated fatty acyl-CoA's. The acyl dihydroxyacetone phosphate could then be reduced by a NADPH-dependent reaction in the mitochondrial fraction to form lysophosphatidic acid[199]; further acylation to produce phosphatidic acid was observed under appropriate conditions. This route to phosphatidic acid provides an interesting alternative to the direct acylation of glycero-3-phosphate, since it could lead to greater selectivity of positional distribution of fatty acids in phospholipids.

Until recently the mechanism of renewal of the polyglycero-phospholipids in mitochondria has been poorly understood. In bacteria the formation of diphosphatidylglycerol occurs by a sequence of reactions of two moles of CDPdiglyceride with glycero-3-phosphate, producing phosphatidylglycerol as an intermediate[200]. Isolated mitochondria from mammalian tissues or yeast catalyse the synthesis of CDPdiglyceride from CTP and phosphatidic acid[201–203]. In the presence of added CDPdiglyceride, mitochondria from brain[204,205] and heart[206] were able to incorporate labelled glycero-3-phosphate into phosphatidylglycerol, but there appeared to be no further condensation of phosphatidylglycerol with a second molecule of CDPdiglyceride to form labelled diphosphatidylglycerol.

A possible alternative to the cytidine pathway was provided by the interesting observation that diphosphatidylglycerol could be formed in the presence of purified phospholipase D by the transfer of a phosphatidic acid moiety from one molecule of phosphatidylglycerol to a second acceptor molecule of phosphatidylglycerol[207]. However, this transphosphatidylation reaction has not been demonstrated in mitochondria, and might not be anticipated to be of major physiological significance for mammalian mitochondria, since phospholipase D activity has thus far been demonstrated to be mainly confined to plant tissues[208].

The recent important study by Davidson and Stanacev[209] appears to have settled some of the previous inconsistencies, and establishes the ability of mitochondria to synthesize diphosphatidylglycerol. The results confirm that ^{3}H-labelled glycero-3-phosphate may be incorporated only as far as phosphatidylglycerol by guinea pig liver mitochondria in the presence of an exogenous source of CDPdiglyceride; no significant incorporation into diphosphatidylglycerol was observed under any conditions with this system. When the system was supplemented with the necessary cofactors for the endogenous production of CDPdiglyceride within the mitochondria, there was significant incorporation of glycero-3-phosphate into diphosphatidylglycerol. As noted by Davidson and Stanacev[209] either the synthetic reaction occurs within a barrier that is impermeable to exogenous CDPdiglyceride, or the enzyme is highly specific for the fatty acid substituents found in the CDPdiglyceride produced *in situ*. In this connection there is another inconsistency which may bear a relation to the selectivity of the diphosphatidylglycerol synthetic system. Paradoxically, the observations of Gray[210] demonstrated that mitochondrial diphosphatidylglycerol possesses a very different fatty acid composition from its postulated precursor, phosphatidylglycerol. In view of the fact that CDPdiglyceride molecules involved in the synthesis of these two lipids may exist in two different compartments, or may be acted upon selectively based upon structural differences, this paradox may turn out to be more apparent than real.

5. EXCHANGE OF PHOSPHOLIPIDS BETWEEN MEMBRANES

The ability of phospholipids to undergo exchanges from sites of binding in the membranes to lipoprotein complexes has been recognized for some time[211]. Thus, the apparent ability of the erythrocyte to incorporate [^{32}P] orthophosphate *in vivo* into lecithin and sphingomyelin could be attributed to an exchange of these labelled lipids from serum α- and β-lipoproteins formed in the liver[143,212]. The exchange reaction was studied *in vitro* using either labelled lipoprotein with unlabelled erythrocytes, or labelled erythrocytes with unlabelled lipoprotein, and was shown to be reversible. The rate of exchange was substantial (10% of total phospholipids per hour). Studies of this type indicated the difficulties that might be encountered in determining the metabolic stability and turnover of membrane components through use of radioactive isotopes. Obviously, lipid components of the membrane such as lecithin and sphingomyelin, must have been less firmly bound than heretofore assumed in order to exchange so readily with extracellular lipid. Estimates of turnover might bear little or no relation to rates of true renewal of molecules in such circumstances.

Tarlov[213] first clearly demonstrated the exchange of lipoprotein phospholipids with intracellular membranes. Labelled lecithin, lysolecithin and sphingomyelin of soluble plasma lipoproteins underwent exchange with the mitochondrial bound lipids in a reversible process that was not energy-dependent and occurred at a similar rate to that observed previously for erythrocyte lipids. Other investigators at roughly the same time were studying somewhat similar phenomena, but chiefly from the stand-points either of gross uptake or excretion of lipids by cells[214–216].

The important study of Wirtz and Zilversmit[217] demonstrated that an exchange of phospholipid could also occur *in vitro* between two different membrane systems. In these experiments rat liver subcellular fractions were labelled *in vivo* by [^{3}H] leucine, and either [^{32}P]orthophosphate or [^{14}C]glycerol; in this way the proteins were labelled with ^{3}H, the phospholipids with ^{32}P or ^{14}C. Upon incubation of the isolated labelled mitochondria with unlabelled microsomes a substantial portion of the ^{32}P or ^{14}C was found to be transferred to the unlabelled fraction, but not the ^{3}H. It was thus possible to apply a correction from the ^{3}H radioactivity radioactivity data for the degree of cross-contamination of the fractions. The exchange was shown to be reversible, and time- and temperature-dependent. After 90 min at 37°, 30 to 40% of the particulate phospholipid had exchanged. No requirement for energy was demonstrated, but the exchange was promoted by a soluble factor present in the 105 000 × *g* supernatant fraction. The exchange was more rapid for phosphatidylcholine than for phosphatidylethanolamine; diphosphatidylglycerol of the labelled mitochondria did not exchange with the microsomes.

The experiments of McMurray and Dawson[30] were also consistent with the existence of a mechanism for exchange of lecithin molecules between liver microsomes and mitochondria *in vitro* . Thus, although the mitochondria had been shown to lack the enzymes for lecithin biosynthesis, incubation of mitochondria together with microsomes and CDP[^{14}C] choline resulted in substantial labelling of mitochondrial lecithin. Previously labelled microsomes exchanged ^{14}C-labelled lecithin with unlabelled mitochondria, without transfer of the microsomal marker enzyme, NADPH : cytochrome *c* oxidoreductase. Lecithin exchange also occurred in the reverse sense without transfer of marker enzymes of mitochondrial outer or inner membranes (monoamine oxidase and cytochrome oxidase respectively) to the microsomal fraction. The exchange was enhanced by a soluble factor which appeared to be protein since the activity was labile to heating and non-dialysable. The rate of exchange of lecithin was such that isotopic equilibration was approximated after 2 h incubation; phosphatidylinositol exchange was more rapid, with isotopic equilibration attained after 1 h. Exchanges of phosphatidylethanolamine and phosphatidic acid were appreciably slower than for lecithin.

The exchange of phospholipids between rat liver organelles was confirmed also by Akiyama and Sakagami[218], although the results differed in one respect from those already cited, namely in a failure to show any great temperature dependence of the process. Other parameters of the system seemed compatible with those studied previously, including the requirement for the soluble 105 000 × *g* supernatant fraction. A transfer of labelled phospholipid was also demonstrated from microsomes to the supernatant fraction; incubation of labelled supernatant fraction with mitochondria resulted in phospholipid transfer to the mitochondria reinforcing the suggestion that a soluble factor in the supernatant could play an intermediary carrier role in lipid exchange between membranes.

The generality of the exchange phenomenon was underlined by the studies of Ben Abdelkader and Mazliak[219] with plant tissues. These experiments, conducted with sub-

References p. 246

cellular fractions from potato tubers and cauliflowers, exhibited strong similarities to the results obtained in animal tissues, and demonstrated that the exchange process could occur between organelles from two different species. In some cases the results suggested, however, that there was a degree of interspecies specificity with respect to the intermediary function of the soluble supernatant factors; thus more of the labelled phospholipid appeared to be blocked in the supernatant fraction when it was incubated with heterologous labelled organelles than with those from the same species. The data are complex and the interpretations are difficult, but the results suggested also some selectivity in the mechanism for renewal of phospholipids of varying fatty acid composition. It is of interest in this connection that Wirtz et al.[220] could detect no difference in the rate of exchange of different molecular species of lecithin characterized by varying degrees of unsaturation, between rat liver microsomes and mitochondria.

The soluble stimulatory factor has been isolated from the 105 000 $\times$ *g* supernatant fraction and partially characterized. Preliminary studies showed that the heat-labile factor was probably a polypeptide since it was non-dialysable, precipitable by acetone or ammonium sulphate, and the activity was lost following trypsin digestion[221]. Ability to stimulate phospholipid exchange was not affected by treatment of the supernatant factor with organic solvent to remove bound phospholipids. On Sephadex chromatography the factor migrated with the slow, low molecular weight components. Rechromatography of the slow-moving fraction yielded a highly purified preparation, which on the basis of protein content was more than 100 times as active in promoting phospholipid exchange as the original supernatant[222,223]

In view of the low molecular weight of the active polypeptide fraction, and the suggestion[224] that phospholipids and cytochrome *c* could act in a reciprocal manner in transfer as a lipoprotein complex from microsomes to mitochondria, the effect of cytochrome *c* upon lecithin transfer was investigated. Cytochrome *c* from horse heart was capable of stimulating lecithin exchange between rat liver microsomes and mitochondria by a process that was independent of respiratory reactions. However, relatively large amounts of cytochrome *c* (1 mg/ml) were required, well in excess of the amounts present in the supernatant fraction, and it seems unlikely that cytochrome *c* could play a physiological role in the exchange process[223].

It has been established that both the outer and inner membrane phospholipids of mitochondria participate in the exchange with those from labelled microsomal membranes[182,225]. The specific activities of the outer membrane phosphatidylethanolamine and phosphatidylinositol were intermediate between values for the microsomes and inner membrane; with the conditions used lecithin had virtually reached equilibration between inner and outer membranes, but values for the outer membrane were marginally higher. The results were consistent with prior exchange between microsomal phospholipids and those of the outer mitochondrial membrane, followed by exchange between the latter and the inner membrane. The role of the outer membrane as an intermediary between microsomes and inner membrane is not obligatory, however, since mitochondria stripped of outer membranes are capable of exchanging phospholipids with microsomes. In addition to providing confirmation of the observations of Sauner and Lévy[182], Blok et al.[225] have shown that an exchange of phospholipids can occur *in vitro* between isolated inner and outer membranes of mitochondria independent of the presence of microsomes. Moreover, when pulse-labelled mitochondria in which the phospholipids had been much more highly labelled in the outer than in the inner membrane were incubated in nonradioactive medium the differences between the specific activities of the phospholipids of the two membranes were eliminated. Thus it is apparent that phospholipid exchange may occur readily between membranes despite the strictures imposed by their *in situ* organization.

The mechanism of the exchange process and the function of the supernatant factor remain to be elucidated. Although the factor is not lipoprotein in nature, it may form a carrier complex with phospholipids while acting as an intermediary between the two membranes. Direct evidence for such a complex is lacking, although a soluble form of labelled phospholipid is released from microsomes during incubation[218,219]; this solublized phospholipid then might exchange with phospholipids of mitochondria, blood plasma, or chylomicrons[226,227]. One may calculate, however, that only a small percentage of the total soluble phospholipid could be bound to the infinitesimal amounts of the stimulatory protein that are required to effect the exchange process[222,223]. If the factor plays a transporting role it must be as a transient intermediate rather than as a bulk carrier of phospholipid.

The nature of the soluble form of phospholipid produced as an intermediate in the exchange between membranes is unknown. Previous observations and theoretical considerations based on the lack of exchangeability between lecithin molecules of monolayers, made it appear unlikely that phospholipid liposomes could play such a role[139]. However, recent experiments have shown that particulate preparations of pure phospholipid are capable of exchange with membrane phospholipid[223,227]. Sonicated, optically clear dispersions of mixed microsomal phospholipids containing ^{14}C-labelled lecithin exchanged the radioactivity with mitochondria at a similar rate to that observed for labelled microsomes of similar specific radioactivity. Moreover, the exchange of lecithin was dependent to a similar extent upon the presence of the supernatant protein factor. Possibly membrane and micellar phospholipids exist in a dynamic equilibrium within the cell, with only a small energy barrier to the exchange of lipid molecules in different phases. The protein factor might further lower the barrier by hastening formation of a collision complex at the membrane surface[139].

Relevance of the exchange process observed *in vitro* to transfer of phospholipid molecules between membranes *in vivo* must be considered as not proven at present, although there is much suggestive evidence. The rapid labelling of mitochondrial, nuclear, and other non-microsomal phospholipids both *in vivo* and *in vitro*, despite the exclusive localization of many pathways of biosynthesis in the endoplasmic reticulum, is one example of a paradoxical observation which could be explained by physiological operation of the exchange–transfer process. It is well-established that phospholipids such as lecithin, phosphatidylethanolamine and phosphatidylinositol are more rapidly labelled initially in the endoplasmic reticulum, consistent with the latter being the site of biosynthesis[146,149]. After a lag the phospholipids of other membranes increase in specific activity to reach

References p. 246

isotopic equilibration with those in the endoplasmic reticulum. Phospholipids of the outer mitochondrial membrane *in vivo* were intermediate in specific activity between those of the endoplasmic reticulum and the inner membrane[30, 225], as was observed for the exchange process *in vitro*[182]. Moreover, the degree of equilibration of microsomal and mitochondrial phosphatidylethanolamine in rat liver *in vivo* was considerably less than for lecithin[228], which is consistent with the observed difference in the exchange rates for these phospholipids *in vitro*[30, 217].

The interesting experiments of Wirtz and Zilversmit[229] on labelling by [^{32}P]orthophosphate *in vivo* provide additional strong confirmation for the existence of exchange–transfer of phospholipid in liver cells of intact animals. The animals were pretreated either with phenobarbital which caused a reduction of the phospholipid specific activity of the microsomes by specifically increasing the endoplasmic reticulum phospholipid pool, or with carbon tetrachloride which selectively increased the microsomal phosphatidylethanolamine specific activity without affecting the pool size. In both cases changes in specific activity of mitochondrial phospholipids paralleled those in the microsomes; the ratios of mitochondrial: microsomal specific activities for lecithin and for phosphatidylethanolamine in control, phenobarbital-, and CCl_4-treated animals were identical, underlining the dependence of mitochondrial labelling upon that in the endoplasmic reticulum. If the process of mitochondrial phospholipid incorporation had occurred by some mechanism that was independent of the biosynthetic reactions occurring in the endoplasmic reticulum, it is difficult to conceive why changes of microsomal phospholipid pool sizes or specific activities should result in changes of phospholipid labelling in the mitochondria.

Pulse-labelling experiments with isolated hepatic cells or with liver slices were also consistent with transfer of phospholipids *in situ*. In the experiments of Jungalwala and Dawson[176], liver cells were incubated *in vitro* with [^{32}P] orthophosphate under conditions where microsomal phospholipids became more strongly labelled than mitochondria in short-term incubations. After washing to remove the radioactive intracellular fluid, the cells were reincubated in unlabelled phosphate-buffered medium. Although there was no further incorporation of label into microsomal phospholipids, those of the mitochondria showed progressive increase of labelling until isotopic equilibration was approached. Similar results were obtained in pulse-chase experiments with [^{14}C]choline and [^{14}C]ethanolamine incorporation in rat liver slices (Wong and McMurray, unpublished work). The evidence thus far regarding the participation of phospholipid exchange–transfer *in vivo* is admittedly all of the circumstantial variety. However, as observed by Thoreau, 'Some circumstantial evidence is very strong, as when you find a trout in the milk'.

6. MEMBRANE BIOGENESIS

The renewal of membrane components is a continuing process in cells. The assessment of rates of renewal of the membrane phospholipids is generally based upon determination of the incorporation and loss of radioactivity from these components at intervals following administration of appropriate labelled precursors to animals *in vivo*. Thus for example [*Me*-^{14}C] choline would appear to be an ideal specific precursor for determination of the rates of *de novo* synthesis of lecithin in the microsomes and other subcellular fractions[230]. The experiments of Treble et al.[231] cast serious doubt on this assumption. These workers injected rats with [^{32}P] orthophosphate and ^{3}H-labelled choline simultaneously and determined the incorporation into lecithin and phosphorylcholine, the precursor of lecithin via the pathway for *de novo* synthesis. The ratio of ^{3}H:^{32}P in liver lecithin was several-fold greater than in phosphorylcholine, indicating that the choline moiety of lecithin underwent more rapid renewal than could be accounted for by the cytidine pathway. The results suggested the existence *in vivo* of a rapid choline exchange reaction, analogous to the base exchange described previously *in vitro*, which was roughly four times more active than the reactions for synthesis of lecithin *de novo*.

Heterogeneous rates of renewal have also been demonstrated for various moieties of brain phospholipids following the intracranial injection of different labelled precursors. For the total phospholipids of whole rat brain, Lapetina et al.[232] observed the following values for the biological half-life ($T_{½}$): glycerol, 5.1 days; P_i, 9.2 days; acetate, 38 days. For the lecithin fraction of rat brain particles Abdel-Latif and Smith [233] obtained these values for the $T_{½}$: glycerol, 2.9 days; choline, 11.7 days; P_i, 18.6 days. Moreover, turnover of a given precursor may vary greatly with its position in the phospholipid molecule. As an example, the incorporation of labelled glycerol into polyglycerophospholipids of *Hemophilus parainfluenza* membranes[234] shows two distinct components: (*1*) The unacylated glycerol moieties of phosphatidylglycerol and diphosphatidylglycerol turn over at a similar rapid rate to the phosphate moieties; (*2*) The acylated glycerol moieties of these and the other membrane phospholipids by contrast are quite stable metabolically.

One of the basic assumptions in estimating turnover rates from the rate of decay of radioactivity following the labelling of the phospholipids is that the precursors and the breakdown products are not being continually reincorporated into the phospholipids to a significant degree during the decay period. With the extended times for which the decay is usually measured it is assumed that the pool size of non-radioactive precursors becomes large enough to dilute non-metabolized residual radioactive precursors from the injected dose, as well as the radioactive breakdown products, to the point where their reutilization becomes negligible. Unfortunately, data are not generally included on the turnover of precursor pools, together with the results on phospholipid turnover. In some known instances the radioactivity of these pools may remain quite high over prolonged periods. Thus McMurray and Dawson[30] found that within one day after injection of a single dose of [^{32}P]orthophosphate in rats, the liver phospholipids attained isotopic equilibrium with the acid-soluble phosphate pool of the tissue. The rate of decline of phospholipid-specific radioactivity in the ensuing 21 days was identical with the fall in the precursor acid-soluble phosphate-specific activity. These results suggest that values for $T_{½}$ of ^{32}P in phospholipids do not reflect turnover of the latter, but rather the turnover of phosphate in the whole organism, possibly via equilibration with the large phosphate reservoir of bone.

Recycling of components such as the fatty acyl moieties may also obscure the true

References p. 246

turnover of phospholipid molecules. Various transacylation systems, as well as deacylation and reacylation of lysophospholipids may account for much of the apparent metabolic stability of fatty acyl moieties in phospholipids[235]. Thus, turnover data from [^{14}C] acetate incorporation studies indicate the existence of a metabolically static fatty acid pool in liver, heart, kidney and adipose tissue, which is in equilibrium with phospholipid fatty acyl groups via rapid deacylation–reacylation reactions[236]. The common usage of turnover in relation to biogenesis has become so frequently qualified by these discrepancies that it seems more appropriate to refer to the isotopic data in terms of 'apparent turnover' and 'apparent T½' (ref. 237).

Apart from the difficulties occasioned by attempts to evaluate renewal of membrane phospholipids in terms of rates of replacement of moieties of those molecules is the complication of metabolic heterogeneity among members of phospholipid species differing with respect to their fatty acid substituents. The first indication of this was the observation by Collins[238] that phospholipid metabolism was increased in the livers of rats deficient in the essential fatty acids, and the inference that arachidonoyl-containing phospholipids are more metabolically inert than others. These conclusions have been strikingly confirmed. Incorporation of [^{32}P] orthophosphate, ^{14}C-labelled glycerol, choline or ethanolamine is most rapid for the phospholipid species with monoenoic and dienoic unsaturated fatty acids in position 2, reflecting a predominant involvement of diacylglycerols containing oleic and linoleic acids in *de novo* pathways of biosynthesis[239–242]. Molecules containing arachidonoyl moieties on the other hand appear to be synthesized mainly by the acylation of endogenous lysophospholipids[243, 244]; consequently the tetraenoic phospholipids exhibit a slower rate of renewal of the glycerolphosphate-base backbone[245], while continual equilibration of the acyl groups of these phospholipids with metabolically stable fatty acid pools in the tissues would contribute to the long apparent half-life of these molecules[236].

Fletcher and Sanadi[246] first reported values for the T½ of mitochondrial phospholipids and proteins. In these studies [^{14}C] acetate or [^{35}S] methionine were administered to rats by intraperitoneal injection and the rate of decay of radioactivity in components of liver mitochondria was measured in the period 14–28 days. A similar logarithmic decay was observed for the phospholipids, cytochrome *c*, a soluble and an insoluble protein fraction. The data suggested that turnover was identical for each of these components and that the mitochondrion was renewed as a unit with a T½ of about 10 days. Since the mixed liver proteins yielded values for T½ of 5–6 days, mitochondria appeared to turn over independently of the renewal of the whole cell. Bailey et al.[236] and Taylor et al.[247] obtained results somewhat at variance with these findings, which they construed as contradictory to the concept of turnover of the mitochondrion as a unit. Incorporation of [^{14}C] acetate and [^{32}P]orthosphate into phospholipids, and of [^{14}C]methionine into proteins for shorter time intervals ($<$ 14 days) yielded non-linear decay curves on semi-log plots which were interpreted in terms of multiple exponential functions. Decay of phospholipid labelling in the first 72 h indicated the presence of two rapidly metabolized components with half-lives of 4 h and 1.9 days; later decay was attributed to another first-order component with T½ of 10 days, corresponding to the half-life for the insoluble proteins. Thus part at least of the mitochondrial phospholipid appeared to be renewed more rapidly than the proteins.

The labelling of diphosphatidylglycerol was slower with inorganic ^{32}P but more rapid with [^{14}C] acetate than the other phospholipids. This finding was confirmed by McMurray and Dawson[30] who showed that incorporation of ^{32}P into diphosphatidylglycerol attained maximal values much later than other phospholipids (2 days in comparison with 6 h), and showed some persistence of labelling above the values for the acid-soluble phosphate pool. Gross et al.[237] confirmed the slower turnover of diphosphatidylglycerol with ^{32}P for a number of tissues, with very long half-lives for brain and heart; in liver and kidney half-lives were more nearly equal for all the phospholipids and corresponded closely to values for the mitochondrial DNA (range $T_{1/2}$, 9.4–11.5 days).

The inner and outer mitochondrial membranes exhibit clear differences with respect to turnover of their phospholipids and proteins[248–252]. *In vivo* the rate of labelling of outer membrane phospholipids from [^{32}P] orthophosphate was significantly greater than that found for inner membranes of rat liver mitochondria[30]. Similarly the maximum degree of incorporation of ^{14}C-labelled acetate or choline in lecithin was attained more rapidly in the outer than the inner membranes[248,249]. These findings could be interpreted in terms of more rapid and direct transfer of labelled phospholipids from the sites of biosynthesis in the endoplasmic reticulum to the outer membrane by the exchange process. Results from studies of protein turnover also indicate that a difference exists between renewal rates of the two membranes[250]. Values for the half-lives with [^{35}S] methionine labelling indicated the presence of at least two distinct components in liver mitochondria: a relatively short-lived fraction ($T_{1/2}$ = 6.7 days) exhibiting first-order kinetics in the period 2–15 days after the initial injection which was probably contributed chiefly by outer membrane proteins ($T_{1/2}$ = 4.2 days); a longer-lived fraction observed after 15 days ($T_{1/2}$ = 12.2 days) which corresponded to turnover of the inner membrane ($T_{1/2}$ = 12.6 days) and soluble matrix proteins ($T_{1/2}$ = 11.6 days). The difference between turnover rates of the proteins of the two membranes has been confirmed using a double-labelling technique which allows correction for recycling of radioactivity[251]. The more metabolically stable components of the inner membrane are associated with the proteins that are not readily solubilized by KCl extraction[252]. An examination of turnover of haemoproteins using the non-reutilizable precursor, [5-^{3}H] aminolaevulinic acid, also showed differences in the components of the membranes in rat liver mitochondria[253]. Thus the haem groups of cytochrome oxidase, and of cytochromes *b* and *c* of the inner membrane turned over with identical $T_{1/2}$ values (5–6 days), while cytochrome b_5 of the outer membrane was renewed at a significantly faster rate ($T_{1/2}$ = 4.4 days)[254]. Proteins of the matrix compartment, notably the inducible enzymes, alanine aminotransferase and ornithine-ketoacid aminotransferase ($T_{1/2}$ = 0.75 days), and 5-aminolaevulinate dehydratase ($T_{1/2}$ = 0.05 days), also appear to turnover independently of the inner membrane[255].

Omura et al.[256] have investigated the renewal of phospholipids in rat liver microsomes. With [^{14}C] acetate the turnover of total microsomal phospholipids ($T_{1/2}$ = 2.3–3.3 days) was more rapid than for total microsomal protein ($T_{1/2}$ = 3.8–4.0 days) measured with [^{14}C] leucine labelling. Undoubtedly there was some recycling of fatty acyl groups by the acylation reactions discussed above, since turnover of the glycerol moiety of phospholipids appeared to be considerably faster ($T_{1/2}$ = 1.2 days). [^{14}C] leucine may also be extensively reutilized; the non-reutilizable precursor, [*guanido*-^{14}C] arginine, gives much lower values for half-lives of microsomal protein (T½ = 2.0–2.1 days)[257], and shows also distinctly faster turnover than [^{14}C] leucine in liver mitochondria[253]. Despite the uncertainties surrounding the values for turnover constants of individual constituents of the membrane, the evidence to date appears to favour the existence of independent renewal processes for these consituents[256], rather than turnover of the membrane as a unit. The heterogeneity of turnover rates is most strikingly illustrated by microsomal cytochrome b_5. Thus the turnover of the haem prosthetic group of the molecule (T½ = 1.7 days measured with [5-^{3}H] aminolaevulinate) is appreciably faster than for the apoprotein (T½ = 3.5 days estimated with the non-reutilizable [*guanido*-^{14}C] arginine)[258]. Recent experiments by Kadenbach[259] indicate similar independence of the renewal of prosthetic group and protein moiety of cytochrome *c*; kinetic studies suggested that only the apoprotein of cytochrome *c* is synthesized in the endoplasmic reticulum, with attachment of the haem portion occurring after transfer of the apoprotein to the mitochondria as had been suggested earlier by Halder et al.[260]. A similar dissociation between renewal of the protein and prosthetic group of cytochrome oxidase may be indicated from studies of the adaptation of anaerobic yeast to oxygen, since the oxidase activity appears considerably later than the production of haem a[261].

Further data consistent with the independent or ‘mosaic’ assembly of components during membrane biogenesis are afforded by investigation of the effects of phenobarbital injection. Administration of this barbiturate to rats induces a proliferation of the smooth membranes of endoplasmic reticulum in liver cells, which is coincident with an increase in the activities of some microsomal enzymes, but not all[262]. Thus the amount of NADPH-cytochrome *c* reductase was markedly elevated by phenobarbital injection, while cytochrome b_5 was not greatly affected, and the synthesis of some microsomal proteins was actually decreased[257,263]. Differential effects were also observed in the effects on membrane phospholipids. The latter were increased in phenobarbital-treated animals, to a greater extent than the increase of microsomal protein, so that phospholipid : protein ratios were enhanced[262]. The enrichment of the endoplasmic reticulum with respect to phospholipid was particularly marked for lecithin. This selective increase in lecithin biosynthesis was accompanied by an increase in the *S*-adenosylmethionine : phosphatidylethanolamine methyl transferase activity; by contrast lecithin biosynthesis from the CDPcholine pathway was not affected[264]. Thus it may be seen that subcellular membranes may alter in response to inductive stimuli by selective alterations in the biosynthesis of individual protein and phospholipid constituents of the membrane. Whether or not this method of independent replacement of membrane components plays the cardinal role in formation of new membranes during normal growth and development, it must be recognized that the composition of mature, differentiated membranes may respond to environmental changes by synthesis or by exchange–transfer of their constituents in the absence of total cell growth.

Changes in the lipid composition and enzyme content of microsomal membranes have also been studied during cellular differentiation. In the early embryonic period a great increase is noted in the quantity of rough-surfaced endoplasmic reticulum in rat liver; this is followed post-natally by a major increase of smooth-surfaced elements[265]. In liver microsomes a progressive decrease in the ratio RNA:protein, and an increase in the phospholipid:protein ratio, are noted in the neonatal period indicating that membrane proliferation exceeds the rate of ribosome formation after birth. The finding of a significant amount of non-membranous non-ribosomal protein in chick embryonic tissue prior to membrane production, and its disappearance in later stages of differentiation, suggested that maturation of the endoplasmic reticulum could occur through addition of lipids to protein precursors of the membranes[41]. Lipid-poor osmiophilic granules, termed reticulosomes, were observed in the undifferentiated liver and were isolated from chick embryos by Pollak and Shorey[266], and shown to contain considerable glucose-6-phosphatase activity. On incubation with sonicated phospholipids the reticulosomes formed a complex with a phospholipid: protein ratio similar to isolated microsomes, and with similar membranous structure as seen by electron microscopy[267]. The experiments of Tzur and Shapiro[268] had indicated that synthesis of microsomal phospholipids could be regulated by the availability of proteins capable of forming lipoprotein complexes. In the developing rat hepatocyte the most rapid labelling of protein and phospholipid components occurs in the rough-surfaced elements of the endoplasmic reticulum, suggesting that newly formed membrane may be assembled first in the rough membranes and subsequently transferred to the smooth endoplasmic reticulum[264]. Constitutive enzymes of the endoplasmic reticulum (glucose-6-phosphatase, NADPH-cytochrome *c* reductase) were produced at different stages of development in rat liver and at different rates of synthesis[269]. As in the labelling experiments, increases in enzyme activity of the rough membranes occurred sooner and faster than in the smooth membranes.

Taken together the findings suggest that the rough endoplasmic reticulum plays a primary biosynthetic function and may have a secondary role in assembly of lipoprotein constituents of other subcellular membranes[270,271]. Since the catalytic proteins of the membranes may occur in varying proportions and arise in an asynchronous manner during development and adaptation, it seems highly improbable that membrane formation could occur by a simultaneous, single-step process. With the evidence available at present the hypothesis advanced by Dallner et al.[269] is more tenable. The membrane components, proteins and lipids, turn over independently of one another and are assembled in a multi-step mechanism. Some interdependence between the protein and lipid components may exist to the extent that availability of one or the other would be required to form lipoprotein complexes for transfer from the sites of biosynthesis to newly formed membranes. The dynamic equilibrium between synthesis and breakdown of each of the components of the membrane 'mosaic' may be controlled to allow for net synthesis during growth, or for modification of composition and function in response to adaptive stimuli. To date, most attention has been given to the mechanisms of biosynthesis and their regulation. It now appears quite possible that the biosynthesis of membrane components may proceed at a fairly constant rate during the cell cycle[272] and periods of accelerated growth of cell surfaces[273]. It may well be that the principal regulatory systems for membrane renewal will be found to lie, not in the positive control of biosynthetic pathways but in the negative control of degradative reactions.

References p. 246

REFERENCES

1 C. de Duve, in D.B. Roodyn, (Ed.), *Enzyme Cytology*, Academic Press, New York, 1967, p. 1.
2 J. de Gier and L.L.M. van Deenen, *Biochim. Biophys. Acta,* 49 (1961) 286.
3 J.W. Farquhar, *Biochim. Biophys. Acta*, 60 (1962) 80.
4 D.J. Hanahan, J. Ekholm and C.M. Jackson, *Biochemistry*, 2 (1963) 630.
5 L.L.M. van Deenen, *Progr. Chem. Fats Lipids*, 8 (1965) 1.
6 G.S. Getz, W. Bartley, D. Lurie and B.M. Notton, *Biochim. Biophys. Acta*, 152 (1968) 325.
7 S. Fleischer, G. Rouser, B. Fleischer, A. Casu and G. Kritchevsky, *J. Lipid Res.*, 8 (1967) 170.
8 S.K. Chan and R.L. Lester, *Biochim. Biophys. Acta*, 210 (1970) 180.
9 L.D. Bergelson, E.V. Dyatlovitskaya, T.I. Torkhovskaya, I.B. Sorokina and N.P. Gorkova, *Biochim. Biophys. Acta*, 210 (1970) 287.
10 F. Paltauf and G. Schatz, *Biochemistry*, 8 (1969) 335.
11 S. Jakovcic, G.S. Getz, M. Rabinowitz, H. Jakob and H. Swift, *J. Cell Biol.*, 48 (1971) 490.
12 H. Nielsen, *Biochim. Biophys. Acta*, 231 (1971) 370.
13 A. Scarpa and G.F. Azzone, *Biochim. Biophys. Acta*, 173 (1969) 78.
14 W.E. Jacobus and G.P. Brierley, *J. Biol. Chem.*, 244 (1969) 4995.
15 E. Bachmann, D.W. Allmann and D.E. Green, *Arch. Biochem. Biophys.*, 115 (1966) 153.
16 M. Lévy, R. Toury and J. André, *Biochim. Biophys. Acta*, 135 (1967) 599.
17 D.F. Parsons, G.R. Williams and B. Chance, *Ann. N.Y. Acad. Sci.* 137 (1966) 643.
18 G.L. Sottocasa, B. Kuylenstierna, L. Ernster and A. Bergstrand, *J. Cell Biol.*, 32 (1967) 415.
19 C. Hoppel and C. Cooper, *Biochem. J.*, 107 (1968) 367.
20 C. Schnaitman and J.W. Greenawalt, *J. Cell Biol.*, 38 (1968) 158.
21 C. Schnaitman, V.G. Erwin and J.W. Greenawalt, *J. Cell Biol.*, 32 (1967) 719.
22 D.E. Green, D.W. Allmann, R.A. Harris and W.C. Tan, *Biochem. Biophys. Res. Commun.*, 31 (1968) 368.
23 D.S. Beattie, *Biochem. Biophys. Res. Commun.*, 31 (1968) 901.
24 J.W. Greenawalt and C. Schnaitman, *J. Cell Biol.*, 46 (1970) 173.
25 J.M. Smoly, B. Kuylenstierna and L. Ernster, *Proc. Natl. Acad. Sci. (U.S.)*, 66 (1970) 125.
26 L. Ernster and B. Kuylenstierna, in E. Racker (Ed.), *Membranes of Mitochondria and Chloroplasts*, Van Nostrand Reinhold, New York, 1970, p. 172.
27 D.F. Parsons, G.R. Williams, W. Thompson, D. Wilson and B. Chance, in E. Quagliariello, S. Papa, E.C. Slater and J.M. Tager (Eds.), *Mitochondrial Structure and Compartmentation*, Adriatica, Bari, 1967, p. 29.
28 M. Lévy and M.-T. Sauner, *Compt. Rend. Soc. Biol.*, 161 (1967) 277.
29 W. Stoffel and H.-G. Schiefer, *Z. Physiol. Chem.*, 349 (1968) 1017.
30 W.C. McMurray and R.M.C. Dawson, *Biochem. J.*, 112 (1969) 91.
31 P. Baudhuin and J. Berthet, *J. Cell Biol.*, 35 (1967) 631.
32 G. Brunner and Th. Bücher, *FEBS Letters*, 6 (1970) 105.
33 D.F. Parsons and Y. Yano, *Biochim. Biophys. Acta*, 135 (1967) 362.
34 M. Lévy and M.T. Sauner, *Chem. Phys. Lipids*, 2 (1968) 291.
35 J.G. Parkes and W. Thompson, *Biochim. Biophys. Acta*, 196 (1970) 162.
36 G.S. Getz, W. Bartley, F. Stirpe, B.M. Notton and A. Renshaw, *Biochem. J.*, 83 (1962) 181.
37 E. Reid, in D.B. Roodyn (Ed.), *Enzyme Cytology*, Academic Press, New York, 1967, p. 321.
38 G. Dallner, *Acta Pathol. Microbiol. Scand. Supp.*, 166 (1963).
39 G. Dallner, A. Bergstrand and R. Nilsson, *J. Cell. Biol.*, 38 (1968) 257.
40 H. Glaumann and G. Dallner, *J. Lipid Res.*, 9 (1968) 720.
41 J.K. Pollak and D.B. Ward, *Biochem. J.*, 103 (1967) 730.
42 D.B. Ward and J.K. Pollak, *Biochem. J.*, 104 (1967) 861.
43 J. Meldolesi, J.D. Jamieson and G.E. Palade, *J. Cell Biol.*, 49 (1971a) 109.
44 J. Meldolesi, J.D. Jamieson and G.E. Palade, *J. Cell Biol.*, 49 (1971b) 130.
45 D.J. Morré, R.L. Hamilton, H.H. Mollenhauer, R.W. Mahley, W.P. Cunningham, R.D. Cheetham and V.S. Lequire, *J. Cell Biol.*, 44 (1970) 484
46 D.E. Leelavathi, L.W. Estes, D.S. Feingold and B. Lombardi, *Biochim. Biophys. Acta*, 211 (1970) 124.

47 B. Fleischer and S. Fleischer, *Biochim. Biophys. Acta*, 219 (1970) 301.
48 R.D. Cheetham, D.J. Morré and W.N. Yunghans, *J. Cell Biol.*, 44 (1970) 492.
49 T.W. Keenan and D.J. Morré, *Biochemistry*, 9 (1970) 19.
50 W.N. Yunghans, T.W. Keenan and D.J. Morré, *Exptl. Mol. Pathol.*, 12 (1970) 36.
51 S.N. Grove, C.E. Bracker and D.J. Morré, *Science*, 161 (1968) 171.
52 D.M. Neville Jr., *J. Biophys. Biochem. Cytol.*, 8 (1960) 413.
53 H.M. Berman, W. Gram and M.A. Spirtes, *Biochim. Biophys. Acta*, 183 (1969) 10.
54 W.H. Evans, *Biochim. J.*, 116 (1970) 833.
55 R.C. Pfleger, N.G. Anderson and F. Snyder, *Biochemistry*, 7 (1968) 2826.
56 T.K. Ray, V.P. Skipski, M. Barclay, E. Essner and F.M. Archibald, *J. Biol. Chem.*, 244 (1969) 5528.
57 R. Coleman, *Chem. Phys. Lipids*, 2 (1968) 144.
58 H.B. Bosmann, A. Hagopian and E.H. Eylar, *Arch. Biochem. Biophys.*, 128 (1968) 51.
59 D.B. Weinstein, J.B. Marsh, M.C. Glick and L. Warren, *J. Biol. Chem.*, 244 (1969) 4103.
60 G.P. Georgiev, in D.B. Roodyn (Ed.), *Enzyme Cytology*, Academic Press, New York, 1967, p. 27.
61 W.W. Franke, B. Deumling, B. Ermen, E.-D. Jarasch and H. Kleinig, *J. Cell Biol.*, 46 (1970) 379.
62 R. Berezney, L.K. Funk and F.L. Crane, *Biochim. Biophys. Acta*, 203 (1970) 531.
63 D.M. Kashnig and C.B. Kasper, *J. Biol. Chem.*, 244 (1969) 3786.
64 M.I. Gurr, J.B. Finean and J.N. Hawthorne, *Biochim. Biophys. Acta*, 70 (1963) 406.
65 H. Kleinig, *J. Cell. Biol.*, 46 (1970) 396.
66 T.W. Keenan, R. Berezney, L.K. Funk and F.L. Crane, *Biochim. Biophys. Acta*, 203 (1970) 547.
67 H. Zentgraf, B. Deumling, E.-D. Jarasch and W.W. Franke, *J. Biol. chem.*, 246 (1971) 2986.
68 H. Kleinig, H. Zentgraf, P. Comes and J. Stadler, *J. Biol. Chem.*, 246 (1971) 2996.
69 S. Patton, *J. Theoret. Biol.*, 29 (1970) 489.
70 S. Fleischer, B. Fleischer, A. Azzi and B. Chance, *Biochim. Biophys. Acta*, 225 (1971) 194.
71 G. Brunner and F.L. Bygrave, *European J. Biochem.*, 8 (1969) 530.
72 R. Penniall, W.D. Currie, N.R. Connell and W.R. Bibb, *Biochem. Biophys. Res. Commun.*, 17 (1964) 752.
73 K. Ueda, T. Matsuura, N. Date and K. Kawai, *Biochem. Biophys. Res. Commun.*, 34 (1969) 322.
74 R. Berezney, L.K. Funk and F.L. Crane, *Biochem. Biophys. Res. Commun.*, 38 (1970) 93.
75 W.G. Whaley, M. Dauwalder and J.E. Kephart, in J. Reinert and H. Ursprung (Eds.), *Origin and Continuity of Cell Organelles*, Springer , New York, 1971, p. 1.
76 D.J. Morré, H.H. Mollenhauer and C.E. Bracker, in J. Reinert and H. Ursprung (Eds.), *Origin and Continuity of Cell Organelles*, Springer, New York, 1971, p. 82.
77 D.E. Green and S. Fleischer, *Biochim. Biophys. Acta*, 70 (1963) 554.
78 D.E. Green and A. Tzagoloff, *J. Lipid Res.*, 7 (1966) 587.
79 S. Fleischer and B. Fleischer, in R.W. Estabrook and M.E. Pullman (Eds.), *Methods in Enzymology*, Vol. X, Academic Press, New York, 1967, p. 406.
80 H.L. Tookey and A.K. Balls, *J. Biol. Chem.*, 220 (1956) 15.
81 E. Petrushka, J.H. Quastel and P.G. Scholefield, *Canad. J. Biochem. Physiol.*, 37 (1959) 989.
82 R.L. Lester and S. Fleischer, *Biochim. Biophys. Acta*, 47 (1961) 358.
83 G.P. Brierley, A.J. Merola and S. Fleischer, *Biochim. Biophys. Acta*, 64 (1962) 218.
84 S. Fleischer, G. Brierley, H. Klouwen and D.B. Slautterback, *J. Biol. Chem.*, 237 (1962) 3264.
85 G.G. de Pury and F.D. Collins, *Chem. Phys. Lipids*, 1 (1966) 1.
86 C. Widmer and F.L. Crane, *Biochim. Biophys. Acta*, 27 (1958) 203.
87 M. Reich and W.W. Wainio, *J. Biol. Chem.*, 236 (1961) 3058.
88 M.L. Das, H. Hiratsuka, J.M. Machinist and F.L. Crane, *Biochim. Biophys. Acta*, 60 (1962) 433.
89 M.L. Das, E.D. Haak and F.L. Crane, *Biochemistry*, 4 (1965) 859.
90 M.L. Das and F.L. Crane, *Biochemistry*, 3 (1964) 696.
91 G. Lenaz, A.M. Sechi, G. Parenti-Castelli and L. Masotti, *Arch. Biochem. Biophys.*, 141 (1970) 79.
92 P.J. Quinn and R.M.C. Dawson, *Biochem. J.*, 113 (1969) 791.
93 P.J. Quinn and R.M.C. Dawson, *Biochem. J.*, 115 (1969) 65.
94 F.F. Sun and F.L. Crane, *Biochim. Biophys. Acta*, 172 (1969) 417.
95 H.K. Kimelberg, C.P. Lee, A. Claude and E. Mrena, *J. Membrane Biol.*, 2 (1970) 235.
96 H.K. Kimelberg and C.P. Lee, *J. Membrane Biol.*, 2 (1970) 252.
97 K.S. Ambe and F.L. Crane, *Science*, 129 (1959) 98.

98 M. Reich and W.W. Wainio, *J. Biol. Chem.*, 236 (1961) 3062.
99 G.P. Brierley and A.J. Merola, *Biochim. Biophys. Acta*, 64 (1962) 205.
100 D.G. McConnell, A. Tzagoloff, D.H. MacLennan and D.E. Green, *J. Biol. Chem.*, 241 (1966) 2373.
101 T.F. Chuang, F.F. Sun and F.L. Crane, *Bioenergetics*, 1 (1970) 227.
102 Y.C. Awasthi, T.F. Chuang, T.W. Keenan and F.L. Crane, *Biochem. Biophys. Res. Commun.*, 39 (1970) 822.
103 L.R. Fowler and S.H. Richardson, *J. Biol. Chem.*, 238 (1963) 456.
104 Y. Hatefi, in M. Florkin and E.H. Stotz (Eds.), *Comprehensive Biochemistry*, Vol. 14, Elsevier, Amsterdam, 1966, p. 199.
105 F.L. Crane, J.W. Stiles, K.S. Prezbindowski, F.J. Ruzicka and F.F. Sun, in J. Järnefelt (Ed.), *Regulatory Functions of Biological Membranes*, Elsevier, Amsterdam, 1968, p. 21.
106 D.H. MacLennan, in F. Bronner and A. Kleinzeller (Eds.), *Current Topics in Membranes and Transport*, Vol. 1, Academic Press, New York, 1970, p. 177.
107 E. Racker, in E. Racker (Ed.), *Membranes of Mitochondria and Chloroplasts*, Van Nostrand Reinhold, New York, 1970, p. 127.
108 W.W. Wainio, *The Mammalian Mitochondrial Respiratory Chain*, Academic Press, New York, 1970.
109 A. Bruni and E. Racker, *J. Biol. Chem.*, 243 (1968) 962.
110 P. Cerletti, M.A. Giovenco, M.G. Giordano, S. Giovenco and R. Strom, *Biochim. Biophys. Acta*, 146 (1967) 380.
111 S. Yamashita and E. Racker, *J. Biol. Chem.*, 244 (1969) 1220.
112 S. Minakami, F.J. Schindler and R.W. Estabrook, *J. Biol. Chem.*, 239 (1964) 2042.
113 R. Berezney, Y.C. Awasthi, L.K. Funk and F.L. Crane, *Bioenergetics*, 1 (1970) 445.
114 Y.C. Awasthi, F.J. Ruzicka and F.L. Crane, *Biochim. Biophys. Acta*, 203 (1970) 233.
115 E. Racker, *Federation Proc.*, 26 (1967) 1335.
116 Y. Kagawa and E. Racker, *J. Biol. Chem.*, 241 (1966) 2467.
117 Y. Kagawa and E. Racker, *J. Biol. Chem.*, 241 (1966) 2475.
118 K. Kopaczyk, J. Asai and D.E. Green, *Arch. Biochem. Biophys.*, 126 (1968) 358.
119 D.H. MacLellan and A. Tzagoloff, *Biochemistry*, 7 (1968) 1603.
120 B. Bulos and E. Racker, *J. Biol. Chem.*, 243 (1968) 3901.
121 R. Berezney, Y.C. Awasthi and F.L. Crane, *Bioenergetics*, 1 (1970) 457.
122 S. Fleischer, B. Fleischer and W. Stoeckenius, *J. Cell. Biol.*, 32 (1967) 193.
123 P.D. Jones and S.J. Wakil, *J. Biol. Chem.*, 242 (1967) 5267.
124 P.D. Jones, P.W. Holloway, R.O. Peluffo and S.J. Wakil, *J. Biol. Chem.*, 244 (1969) 744.
125 D.G. Williamson and V.J. O'Donnell, *Biochemistry*, 8 (1969) 1289.
126 H.W. Strobel, A.Y.H. Lu, J. Heidema and M.J. Coon, *J. Biol. Chem.*, 245 (1970) 4851.
127 U. Hopfer, A.L. Lehninger and T.E. Thompson, *Proc. Natl. Acad. Sci. (U.S.)*, 59 (1968) 484.
128 P. Mitchell, *Biol. Rev.*, 41 (1966) 445.
129 P.J.F. Henderson, J.D. McGivan and J.B. Chappell, *Biochem. J.*, 111 (1969) 521.
130 B. Deuticke and W. Gruber, *Biochim. Biophys. Acta*, 211 (1970) 369.
131 J. de Gier, C.W.M. Haest, J.G. Mandersloot and L.L.M. van Deenen, *Biochim. Biophys. Acta*, 211 (1970) 373.
132 R. Tanaka, T. Sakamoto and Y. Sakamoto, *J. Membrane Biol.*, 4 (1971) 42.
133 A. Martonosi, J. Donley and R.A. Halpin, *J. Biol. Chem.*, 243 (1968) 61.
134 B.P. Yu, F.D. de Martinis and E.J. Masoro, *J. Lipid Res.*, 9 (1968) 492.
135 W. Fiehn and W. Hasselbach, *European J. Biochem.*, 13 (1970) 510.
136 S.M. Duttera, W.L. Byrne and M.C. Ganoza, *J. Biol. Chem.*, 243 (1968) 2216.
137 B.F. Trump, S.M. Duttera, W.L. Byrne and A.U. Arstila, *Proc. Natl. Acad. Sci. (U.S.)*, 66 (1970) 433.
138 B.R. Cater, J. Poulter and T. Hallinan, *FEBS Letters*, 10 (1970) 346.
139 R.M.C. Dawson, in P.N. Campbell and G.D. Greville (Eds.), *Essays In Biochemistry*, Vol. 2, Academic Press, London, 1966, p. 69.
140 A.N. Davison, in R.M.C. Dawson and D.N. Rhodes (Eds.), *Metabolism and Physiological Significance of Lipids*, Wiley, London, 1964, p. 527.
141 M.E. Smith, *Advan. Lipid Res.*, 5 (1967) 241.
142 A.A. Buchanan, *Biochem. J.*, 75 (1960) 315.

143 C.F. Reed, *J. Clin. Invest.*, 47 (1968) 749.
144 C.M. Redman, *J. Cell Biol.*, 49 (1971) 35.
145 R. Schoenheimer, *The Dynamic State of Body Constituents*, Harvard Univ. Press, Cambridge, 1942.
146 G.L. Ada, *Biochem. J.*, 45 (1949) 422.
147 M. Griffiths and N. Pace, *Proc. Soc. Exptl. Biol. Med.*, 83 (1953) 771.
148 M. Pascaud, *Biochim. Biophys. Acta*, 84 (1964) 528.
149 M.I. Gurr, C. Prottey and J.N. Hawthorne, *Biochim. Biophys. Acta*, 106 (1965) 357.
150 W.E.M. Lands and P. Hart, *J. Lipid Res.*, 5 (1964) 81.
151 W.E.M. Lands, *J. Biol. Chem.*, 235 (1960) 2233.
152 J. Bremer and D.M. Greenberg, *Biochim. Biophys. Acta*, 46 (1961) 205.
153 H. Paulus and E.P. Kennedy, *J. Biol. Chem.*, 235 (1960) 1303.
154 G.F. Wilgram and E.P. Kennedy, *J. Biol. Chem.*, 238 (1963) 2615.
155 W.C. Schneider, *J. Biol. Chem.*, 238 (1963) 3572.
156 R.E. McCaman and K. Cook, *J. Biol. Chem.*, 241 (1966) 3390.
157 H.G. Schiefer and W. Stoffel, *Z. Physiol. Chem.*, 349 (1968) 14.
158 F.L. Bygrave and Th. Bücher, *FEBS Letters*, 1 (1968) 42.
159 F.L. Bygrave and W. Kaiser, *European J. Biochem.*, 4 (1968) 578.
160 W. Kaiser, *European J. Biochem.*, 8 (1969) 120.
161 M. Ashwell and T.S. Work, *Ann. Rev. Biochem.*, 39 (1970) 251.
162 J. Zborowski and L. Wojtczak, *Biochim. Biophys. Acta*, 187 (1969) 73.
163 M.G. Sarzala, L.M.G. van Golde, B. de Kruyff and L.L.M. van Deenen, *Biochim. Biophys. Acta*, 202 (1970) 106.
164 G. Porcellati, M.G. Biasion and G. Arienti, *Lipids*, 5 (1970) 725.
165 G. Porcellati, M.G. Biasion and M. Pirotta, *Lipids*, 5 (1970) 734.
166 R.R. Dils and G. Hübscher, *Biochim. Biophys. Acta*, 46 (1961) 505.
167 C. Artom, *Federation Proc.*, 20 (1961) 280.
168 G. Hübscher, *Biochim. Biophys. Acta*, 57 (1962) 555.
169 R.R. Dils and G. Hübscher, *Biochim. Biophys. Acta*, 32 (1959) 293.
170 R.M.C. Dawson, *Biochem. J.*, 102 (1967) 205.
171 S.F. Yang, S. Freer and A.A. Benson, *J. Biol. Chem.*, 242 (1967) 477.
172 E.P. Kennedy, *Can. J. Biochem. Physiol.*, 34 (1956) 334.
173 J.F. Berry and V.P. Whittaker, *Biochem. J.*, 73 (1959) 447.
174 M.L. Williams and F.L. Bygrave, *European J. Biochem.*, 17 (1970) 32.
175 W.C. McMurray, K.P. Strickland, J.F. Berry and R.J. Rossiter, *Biochem. J.*, 66 (1957) 621.
176 F.B. Jungalwala and R.M.C. Dawson, *Biochem. J.*, 117 (1970) 481.
177 J.D. Smith and J.H. Law, *Biochim. Biophys. Acta*, 202 (1970) 141.
178 T. Chojnacki and T. Korzybski, *Acta Biochim. Polon.*, 10 (1963) 455.
179 M. Sribney, *Arch. Biochem. Biophys.*, 126 (1968) 954.
180 D. Ostrow, *Federation Proc.*, 30 (1971) 1226 (Abstr.).
181 J.G. Swartz and J.E. Mitchell, *J. Lipid Res.*, 11 (1970) 544.
182 M.-T. Sauner and M. Lévy, *J. Lipid Res.*, 12 (1971) 71.
183 W.C. McMurray, K.P. Strickland, J.F. Berry and R.J. Rossiter, *Biochem.J.*, 66 (1957) 634.
184 M.P. Mitchell, D.N. Brindley and G. Hübscher, *European J. Biochem.*, 18 (1971) 214.
185 F.L. Bygrave and Th. Bücher, *European J. Biochem.*, 6 (1968) 256.
186 F.L. Bygrave and W. Kaiser, *European J. Biochem.*, 8 (1969) 16.
187 E.A. Dennis and E.P. Kennedy, *J. Lipid Res.*, 11 (1970) 394.
188 E.H. Shephard and G. Hübscher, *Biochem. J*, 113 (1969) 429.
189 L.N.W. Daae and J. Bremer, *Biochim. Biophys. Acta*, 210 (1970) 92.
190 P.R. Turkki and J.L. Glenn, *Biochim. Biophys. Acta*, 152 (1968) 104.
191 M. Waite, P. Sisson and E. Blackwell, *Biochemistry*, 9 (1970) 746.
192 J. Garbus, H.F. DeLuca, M.E. Loomans and F.M. Strong, *J. Biol. Chem.*, 238 (1963) 59.
193 T. Galliard and J.N. Hawthorne, *Biochim. Biophys. Acta*, 70 (1963) 479.
194 T. Galliard, R.H. Michell and J.N. Hawthorne, *Biochim. Biophys. Acta*, 106 (1965) 551.
195 R.H. Michell, J.L. Harwood, R. Coleman and J.N. Hawthorne, *Biochim. Biophys. Acta*, 144 (1967) 649.

196 A.K. Hajra, E.B. Seguin and B.W. Agranoff, *J. Biol. Chem.*, 243 (1968) 1609.
197 A.K. Hajra and B.W. Agranoff, *J. Biol. Chem.*, 243 (1968) 1617.
198 A.K. Hajra, *J. Biol. Chem.*, 243 (1968) 3458.
199 A.K. Hajra and B.W. Agranoff, *J. Biol. Chem.*, 243 (1968) 3542.
200 N.Z. Stanacev, Y.Y. Chang and E.P. Kennedy, *J. Biol. Chem.*, 242 (1967) 3018.
201 G.L. Petzold and B.W. Agranoff, *J. Biol. Chem.*, 242 (1967) 1187.
202 M.L. Vorbeck and A.P. Martin, *Biochem. Biophys. Res. Commun.*, 40 (1970) 901.
203 D. Mangnall and G.S. Getz, *Federation Proc.*, 30 (1971) 1226 (Abstr.).
204 N.Z. Stanacev, D.C. Isaac and K.B. Brookes, *Biochim. Biophys. Acta*, 152 (1968) 806.
205 F. Possmayer, G. Balakrishnan and K.P. Strickland, *Biochim. Biophys. Acta*, 164 (1968) 79.
206 N.Z. Stanacev, L. Stuhne-Šekalec, K.B. Brookes and J.B. Davidson, *Biochim. Biophys. Acta*, 176 (1969) 650.
207 N.Z. Stanacev and L. Stuhne-Sekalec, *Biochim. Biophys. Acta*, 210 (1970) 350.
208 R.H. Quarles and R.M.C. Dawson, *Biochem. J.*, 112 (1969) 787.
209 J.B. Davidson and N.Z. Stanacev, *Canad. J. Biochem.*, 49 (1971) 1117.
210 G.M. Gray, *Biochem. Biophys. Acta*, 84 (1964) 35.
211 C.E. Rowe, *Biochem. J.*, 76 (1960) 471.
212 T. Sakagami, O. Minari and T. Orii, *Biochim. Biophys. Acta*, 98 (1965) 111.
213 A. Tarlov, *Federation Proc.*, 27 (1968) 458.
214 M. Dobiasova and N.S. Radin, *Lipids*, 3 (1968) 439.
215 G.H. Rothblat, M.K. Buchko and D. Kritchevsky, *Biochim. Biophys. Acta*, 164 (1968) 327.
216 J.T. Buckley, T.J. Delahunty and D. Rubinstein, *Canad. J. Biochem.*, 46 (1968) 341.
217 K.W.A. Wirtz and D.B. Zilversmit, *J. Biol. Chem.*, 243 (1968) 3596.
218 M. Akiyama and T. Sakagami, *Biochim. Biophys. Acta*, 187 (1969) 105.
219 A. Ben Abdelkader and P. Mazliak, *European J. Biochem.*, 15 (1970) 250.
220 K.W.A. Wirtz, L.M.G. van Golde and L.L.M. van Deenen, *Biochim. Biophys. Acta*, 218 (1970) 176.
221 K.W.A. Wirtz and D.B. Zilversmit, *Biochim. Biophys. Acta*, 193 (1969) 105.
222 K.W.A. Wirtz and D.B. Zilversmit, *FEBS Letters*, 7 (1970) 44.
223 P.J. Stewart and W.C. Murray, *Proc. Can. Federation Biol. Soc.*, 14 (1971) 92.
224 B. Kadenbach, in L. Ernster and Z. Drahota (Eds.), *Mitochondria, Structure and Function*, Academic Press, New York, 1969, p. 179.
225 M.C. Blok, K.W.A. Wirtz and G.L. Scherphof, *Biochim. Biophys. Acta*, 233 (1971) 61.
226 D.B. Zilversmit, *J. Lipid Res.*, 12 (1971) 36.
227 D.B. Zilversmit, *J. Biol. Chem.*, 246 (1971) 2645.
228 O. Stein and Y. Stein, *J. Cell Biol.*, 40 (1969) 461.
229 K.W.A. Wirtz and D.B. Zilversmit, *Biochim. Biophys. Acta*, 187 (1969) 468.
230 P. Nagley and T. Hallinan, *Biochim. Biophys. Acta*, 163 (1968) 218.
231 D.H. Treble, S. Frumkin, J.A. Balint and D.A. Beeler, *Biochim. Biophys. Acta*, 202 (1970) 163.
232 E.G. Lapetina, G. Rodriguez de Lores Arnaiz and E. de Robertis, *Biochim. Biophys. Acta*, 176 (1969) 643.
233 A.A. Abdel-Latif and J.P. Smith, *Biochim. Biophys. Acta*, 218 (1970) 134.
234 D.C. White and A.N. Tucker, *J. Lipid Res.*, 10 (1969) 220.
235 W.E.M. Lands, *Ann. Rev. Biochem.*, 34 (1965) 313.
236 E. Bailey, C.B. Taylor and W. Bartley, *Biochem. J.*, 104 (1967) 1026.
237 N.J. Gross, G.S. Getz and M. Rabinowitz, *J. Biol. Chem.*, 244 (1969) 1552.
238 F.D. Collins, *Biochem. Biophys. Res. Commun.*, 9 (1962) 289.
239 G.A.E. Arvidson, *European J. Biochem.*, 4 (1968) 478.
240 G.A.E. Arvidson, *European J. Biochem.*, 5 (1968) 415.
241 J. Tinoco, G. Sheehan, S. Hopkins and R.L. Lyman, *Lipids*, 5 (1970) 412.
242 T. Akino and T. Shimojo, *Biochim. Biophys. Acta*, 210 (1970) 343.
243 H. Kanoh, *Biochim. Biophys. Acta*, 176 (1969) 756.
244 L.M.G. van Golde, G.L. Scherphof and L.L.M. van Deenen, *Biochim. Biophys. Acta*, 176 (1969) 635.
245 M.A. Trewhella and F.D. Collins, *Lipids*, 4 (1969) 304.
246 M.J. Fletcher and D.R. Sanadi, *Biochim. Biophys. Acta*, 51 (1961) 356.

247 C.B. Taylor, E. Bailey and W. Bartley, *Biochem. J.*, 105 (1967) 605.
248 D.S. Beattie, *J. Membrane Biol.*, 1 (1969) 383.
249 F.L. Bygrave, *J. Biol. Chem.*, 244 (1969) 4768.
250 G. Brunner and W. Neupert, *FEBS Letters*, 1 (1968) 153.
251 B. de Bernard, G.S. Getz and M. Rabinowitz, *Biochim. Biophys. Acta*, 193 (1969) 58.
252 D.S. Beattie, *Biochem. Biophys. Res. Commun.*, 35 (1969) 721.
253 V. Aschenbrenner, R. Druyan, R. Albin and M. Rabinowitz, *Biochem. J.*, 119 (1970) 157.
254 R. Druyan, B. de Bernard and M. Rabinowitz, *J. Biol. Chem.*, 244 (1969) 5874.
255 R.W. Swick, A.K. Rexroth and J.L. Stange, *J. Biol. Chem.*, 243 (1968) 3581.
256 T. Omura, P. Siekevitz and G.E. Palade, *J. Biol. Chem.*, 242 (1967) 2389.
257 I.M. Arias, D. Doyle and R.T. Schimke, *J. Biol. Chem.*, 244 (1969) 3303.
258 K.W. Bock and P. Siekevitz, *Biochem. Biophys. Res. Commun.*, 41 (1970) 374.
259 B. Kadenbach, *European J. Biochem.*, 12 (1970) 392.
260 D. Haldar, K. Freeman and T.S. Work, *Nature*, 211 (1966) 9.
261 W.L. Chen and F.C. Charalampous, *J. Biol. Chem.*, 244 (1969) 2767.
262 S. Orrenius and J.L.E. Ericsson, *J. Cell Biol.*, 28 (1966) 181.
263 Y. Kuriyama, T. Omura, P. Siekevitz and G.E. Palade, *J. Biol. Chem.*, 244 (1969) 2017.
264 D.L. Young, G. Powell and W.O. McMillan, *J. Lipid Res.*, 12 (1971) 1.
265 G. Dallner, P. Siekevitz and G.E. Palade, *J. Cell Biol.*, 30 (1966) 73.
266 J.K. Pollak and C.D. Shorey, *Biochem. J.*, 93 (1964) 36C.
267 J.K. Pollak, K. Ward and C.D. Shorey, *J. Mol. Biol.*, 16 (1966) 564.
268 R. Tzur and B. Shapiro, *J. Lipid Res.*, 5 (1964) 542.
269 G. Dallner, P. Siekevitz and G.E. Palade, *J. Cell. Biol.*, 30 (1966) 97.
270 C.A. Ashley and T. Peters Jr., *J. Cell Biol.*, 43 (1969) 237.
271 J.R. Tata, *Biochem. J.*, 116 (1970) 617.
272 A.M.H. Warmsley, B. Phillips and C.A. Pasternak, *Biochem. J.*, 120 (1970) 683.
273 L. Warren and M.C. Glick, *J. Cell Biol.*, 37 (1968) 729.

Chapter 10

PHOSPHOLIPID METABOLISM IN PHOTOSYNTHETIC PLANTS

T. GALLIARD

Agricultural Research Council, Food Research Institute, Colney Lane, Norwich, NOR 70F (Great Britain)

This chapter will cover the phospholipid metabolism of higher plants and also of those micro-organisms which have photosynthetic capabilities, i.e. algae and photosynthetic bacteria while yeasts are dealt with separately in Chapter 11. It is not intended, however, to limit discussion to the biochemistry of photosynthetic tissues. On the contrary, a significant amount of our limited knowledge of lipid biochemistry in plants has come from studies on non-photosynthetic tissues of higher plants. Metabolism of the fatty acid moieties of plant lipids has been extensively reviewed recently[1–3] and will only be discussed here in relation to the involvement of phospholipids in fatty acid metabolic pathways. In photosynthetic tissues, phospholipid metabolism is closely associated with that of the phosphorus-free polar lipids of chloroplasts, i.e. galactosyl diglycerides and sulpholipids, and the reader is referred to recent reviews on these lipids[2–4]

1. COMPOSITION OF PLANT PHOSPHOLIPIDS

In describing the phospholipid composition of plants, it is convenient to distinguish between the lipid content of chloroplast-containing tissues and non-photosynthetic tissues. Table 1 shows the concentration and composition of the phospholipid fraction from a wide range of plants and photosynthetic micro-organisms and the following summary relates to the information presented. The total lipids may account for 10% of the dry weight of green leaves and up to 40% of the dry weight of isolated chloroplasts. Phospholipid accounts for approximately 14% of the lipids in chloroplast lamellae, the remainder being galactolipids (40%), sulpholipid (7%), chlorophyll, carotenoids and quinones (37%) (ref. 5). Phosphatidylglycerol is the major phospholipid of chloroplasts. Purification of chloroplast lamellae results in a further increase in the proportion of phosphatidylglycerol. Phosphatidylcholine and phosphatidylinositol are present only in small amounts (3% and 2% respectively of the total lipids), and no other phospholipids are observed in purified lamellae[5,6].

In etiolated leaves and in non-photosynthetic tissues from higher plants, phosphatidylcholine and phosphatidylethanolamine are normally the most abundant phospholipids. All green algae so far studied contain the same phospholipids as higher plants, although relative proportions of individual lipids vary with the species and, in a given species, with

TABLE 1

PHOSPHOLIPID COMPOSITION OF VARIOUS PLANT TISSUES

Plant tissue		*Phospholipid conc. (µmoles/g fresh wt. unless stated otherwise)*	*PC*	*PE*	*PG*	*PI*	*CL*	*PS*	*PA*	*Other*	*Reference*
Leaves:[a]											
Barley[b], *Hordeum sp.*		–	52	1	23	14					145
Maize (corn), *Zea mais*	light-grown	1.3	15	28	22		18			unidentified (27)	146
Maize (corn), *Zea mais*	dark-grown	0.7	48	31	7		11			unidentified (7)	
Maize (corn), *Zea mais*		1.7	26	14	28	7	14			*lyso* PC (10)	147
Paspalum dilatatum		1.4	28	13	35	7	7			*lyso* PC (10)	147
Perennial ryegrass, *Lolium perenne*		3.0	44	18	25	7	7				147
Cock's foot, *Dactylis glomerata*		3.4	33	24	33	3	8				147
Xanthium orientale		1.0	51	15	25	5	4				147
Lettuce, *Lactuca sativa*		0.8	38	26	12	7	16				147
Elder, *Sambucus nigra*	green	4.4	43	11	39	7			0		148
Elder, *Sambucus nigra*	yellow	3.3	55	16	21	12			0		
Tomato, *Solanum lycopersicum*		2.2	51	21	20	5	5				147
Tobacco[b], *Nicotiana tabacum*		–	47	8	22	22					145
Squash, *Cucurbita pepo*		3.7	43	27	24	4	2				147
Pumpkin, *Cucurbita pepo*		3.9	36	23	28	13					54
Camellia japonica		4.7	43	27	24	4	2				147
Poplar, *Populus italica*		4.6	40	18	24	11	8				147
Lucerne (alfalfa), *Medicago sativa*		2.8	27	18	24	10	21				147
Lucerne (alfalfa), *Medicago sativa* var. Vernal[c]	grown at 15°	–	44	36	12	10					

Plant tissue		*Phospholipid conc. (μmoles/g fresh wt. unless stated otherwise)*	*PC*	*PE*	*PG*	*PI*	*CL*	*PS*	*PA*	*Other*	*Reference*
Lucerne (alfalfa), *Medicago sativa* var. Vernal[c]	grown at 30°	–	12	7	19	63					
Lucerne (alfalfa), *Medicago sativa* var. Caliverde[d]	grown at 15°	–	35	14	13	38					118
Lucerne (alfalfa), *Medicago sativa* var. Caliverde[d]	grown at 30°	–	8	1	25	66					
White clover, *Trifolium repens*		3.9	36	22	29	7	7				147
White clover, *Trifolium repens*	light-grown	–	32	23	45		tr				149
White clover, *Trifolium repens*	dark-grown	–	42	21	33		4				
Sweet clover, *Melilotus alba*		–	30	9	25	18		6			145
Runner bean, *Phaseolus multiflorus*		2.7	45	17	22	9		6		PA (1); unidentified (6)	150
Rowan, *Sorbus aucuparia*		5.5	40	31	18	7	5				147
Rose, *Rosa cv.*		3.1	48	18	19	9	6				147
Spinach, *Spinacia oleracea*		2.3	47	17	26	9	tr				148
Sugar beet, *Beta vulgaris*		4.7	47	23	19	11					147
Sugar beet, *Beta vulgaris*		22[e]	36	12	17	7	4	2	4	5 unidentified (total 18)	151
Pine, *Pinus radiata*		2.3	35	12	24	16	13				147
Ginkgo biloba		3.7	48	15	23	7	7				147
Leaf stalk:											
Sugar beet, *Beta vulgaris*		14[e]	32	16	8	7	3	2	10	4 unidentified (total 22)	151
Cotyledon:											
Pea, *Pisum sativum*	dry seed	4.1[f]	52	14	1	18	1	0.4	3	phosphatidylmethanol (0.5)	
Pea, *Pisum sativum*	11 days dark-grown	1.7[f]	54	14	1	11	2	1	4	phosphatidylmethanol (2–5)	37
Pea, *Pisum sativum*	11 days light-grown	1.9[f]	62	16		11	2	1	3	phosphatidylmethanol (2)	

Table 1, continued

Plant tissue		*Phospholipid conc. (μmoles/g fresh wt. unless stated otherwise)*	*PC*	*PE*	*PG*	*PI*	*CL*	*PS*	*PA*	*Other*	*Reference*
Root:											
Sugar beet, *Beta vulgaris*		3.5[e]	35	17	6	5	2	12	12	4 unidentified (total 21)	151
Parsnip, *Pastinaca sativa*		0.9	38	20	8	13	22				147
Tuber:											
Potato, *Solanum tuberosum*, var. Majestic		0.7	55	28	2	12	1	1	1		89
Potato, *Solanum tuberosum*, var. Bintje		0.5[g]	50	35	3	7	2	2	tr		137
Potato, *Solanum tuberosum*, var. Netted Gem		3.0[e; g]	42	32		26					152
Fruit:											
Apple, *Pyrus malus*, var. Cox's Orange Pippin	pre-climacteric	0.5	46	31	7	13	2	1	1		77
Apple, *Pyrus malus*, var. Cox's Orange Pippin	post-climacteric	0.5	54	25	2	15	1	1	1		
Apple, *Pyrus malus*, var. Golden Delicious	post-climacteric	–	35	42	tr	3	3	tr	17		153
Pollen:											
Pine, *Pinus ponderosa*		–	60	18	6	13	2	1			154
Bud:											
Cotton, *Gossypium sp.*		–	68	6		26					155

Table 1, continued

Plant tissue		*Phospholipid conc. (µmoles/g fresh wt. unless stated otherwise)*	*PC*	*PE*	*PG*	*PI*	*CL*	*PS*	*PA*	*Other*	*Reference*
Seed:											
Castor bean, *Ricinus communis*		1.5[g]	13	83		tr				4	156
Soya bean, *Glycine max.*		1.2[g]	45	15		25	4	1	8	phytoglycolipid (3–4)	157
Groundnut, *Arachis hypogaea*		0.7[g]	44	18		24			4	phytoglycolipid (9)	157
Pine, *Pinus sp.*		1.2[g]	54	11		30				phytoglycolipid (4)	157
Cotton, *Gossypium sp.*		2% of seed oil	33	22		37				8	158
Sunflower, *Helianthus annuus*		–	51	23		22			4		159
Cocoa, *Theobroma sp.* (3 varieties)		4[g]	36-40	11-17		25-29		2-3	1-3	*lyso* PC (4-9); unidentified (8-11)	160
Callus:											
Kalanchoe crenata	light-grown	2.8[e, g]	48	17	8	11			17		161
Kalanchoe crenata	dark-grown	2.6[e, g]	54	16	4	13			14		
Pteridophyta:											
Blechnum fluviatile		2.3	35	18	26	6	7				147
Moss:											
Furoria and *Leptobryum pyriforme*		2.3	56	18	18	7	2				147
Liverwort:											
Marchantia berteroana		0.7	27	14	24	11	24				147

References p. 284

Table 1, continued

Plant tissue		*Phospholipid conc. (μmoles/g fresh wt. unless stated otherwise)*	*PC*	*PE*	*PG*	*PI*	*CL*	*PS*	*PA*	*Other*	*Reference*
Green algae:											
Mesotaenium caldariorum		4.4	25	18	38	18	tr				147
Chlorella vulgaris		26[e]	46	9	27	14	1				26
Chlorella pyrenoidosa		16[h]	22	11	55	12	tr				38
Scenedesmus obliquus		–[b]	36	6	41	15	tr	1			145
Euglena gracilis	light-grown[i]	20μg/10^{6}cells	47	15	10	9	14	4		traces of PA, *lyso* PC, *lyso* PE	62
Euglena gracilis	dark-grown[i]	19μg/10^{6}cells	56	17	tr	10	14	3			62
Euglena gracilis	auxotrophic	–	72	10	8	tr	tr	8			162
Thermophilic alga:											
Cyanidium caldarum	grown at 45°	–	40-50	24-32	6-27	7-11					163
Blue-green algae:											
Anacystis nidulans		12[e, g]	0	0	100	0					5, 164
Anabaena variabilis		–	0	0	100						7
Photosynthetic bacteria:											
Rhodospirillum rubrum		87[e]	6	57	29	tr	8			*O*-ornithyl esters in PG fraction	165, 166
Rhodopseudomonas spheroides		–[e]	4	36	54		6			*O*-ornithyl esters in PG fraction	165
Rhodopseudomonas capsulatus		106[e, g]	0	76	22	0	0				166
Rhodomicrobium vannielli		50[e, g]	26	5	10		7			*O*-ornithyl esters of PG (47)	167
Chromatium strain D		–	0	47	53	0	tr				165
Chromatium strain D		–[b]	0	58	32	0	8			*lyso* PG (2)	168
Chromatophores		–	23	35	34		tr				169

Table 1, continued

Plant tissue	*Phospholipid conc. (μmoles/g fresh wt. unless stated otherwise)*	*PC*	*PE*	*PG*	*PI*	*CL*	*PS*	*PA*	*Other*	*Reference*
Subcellular particles:										
Chloroplasts:										
Maize, *Zea mais*	–	22	tr	65	10	tr	0			101
Tobacco, *Nicotiana tabacum*	–	23	15	58	4					170
Tobacco[j], *Nicotiana tabacum*	–	24	10	42	10			14		170
Spinach, *Spinacea oleracea*	5.3	47	17	26	9	tr				148
Spinach, lamellae, *Spinaca oleracea*	72[e, g]	24	0	65	11					5, 6
Spinach, lamellae, *Spinaca oleracea*	75[e, g]	19		67	14					164
Sugar beet, *Beta vulgaris*	6.9	38	15	38	9					148
Euglena gracilis	–	35	4	50	6	tr		3		62
Green pepper fruit, *Capsicum annuum*	–	21	8	60	5			6		171
Red pepper fruit, *Capsicum annuum* (plastids)	–	53	5	30	12			0		171
Mitochondria:										
Cauliflower inflorescence, *Brassica oleracea*	–	42	30	4	10	12	tr			101
Cauliflower inflorescence, *Brassica oleracea*	–	43	32	3	11	7	1	2		106
Potato tuber, *Solanum tuberosum*	–	43	30	3	7	8	3	5		106
Apple fruit, *Pyrus malus*	–	45	35	7	5	tr	3	5		100
Mungbean hypocotyl, *Phaseolus aureus*	–	44	29	3	9	11	tr			101
Grown gall of Scorsonera	–	44	30	3	11	10	2			101
Euglena gracilis	–	62	11	t	8	22		16	*lyso* PC (t) *lyso* PE (t)	62
Microsomal fraction:										
Cauliflower inflorescence	–	31	31	5	13	5	4	6		106
Potato tuber	–	45	33	1	16	1	1	3		106

[a] Listed in approximate taxonomic sequence.
[b] Composition calculated from ^{32}P-labelling pattern.
[c] Cold-resistant variety.
[d] Cold-sensitive variety.
[e] μmoles/g dry weight.
[f] μmoles/pea (average).
[g] Molar concentration calculated from weight concentration.
[h] μmoles/ml of cell volume.
[i] Grown on glutamate + lactate; DPG only 7% when grown on lactate.
[j] Isolated by non-aqueous technique.

References p. 284

growth conditions[2,4]. The variation in phospholipid composition with temperature is strikingly illustrated by lucerne grown at different temperatures (see Table 1) and will be discussed later. In the more primitive photosynthetic blue-green algae, choline-, ethanolamine- and inositol-containing lipids are absent. Phosphatidylglycerol is the major phospholipid, although some unidentified phospholipids have been observed in some species of the blue-green algae[5,7]. Photosynthetic bacteria show lipid distributions common to many bacteria in that phosphatidylglycerol and phosphatidylethanolamine are normally the major phospholipids and *O*-aminoacyl esters of phosphatidylglycerol are present in most species examined, sometimes as the major phospholipid; phosphatidylcholine, phosphatidylinositol and diphosphatidylglycerol are also found in some, but not all, species.

With the exception of chloroplasts, little information is available on the lipid composition of homogeneous preparations of subcellular fractions from plant cells. Mitochondria are rich in phosphatidylcholine and phosphatidylethanolamine; phosphatidylinositol and diphosphatidylglycerol are also present with the latter at a higher concentration in mitochondria than in whole tissue.

Although phosphatidylcholine, phosphatidylethanolamine and phosphatidylglycerol are, in general, the major phospholipids in higher plants, all tissues so far studied contain some phosphatidylinositol and diphosphatidylglycerol. Phosphatidylserine is found, if at all, only in small amounts. Phosphatidic acid also occurs only in trace amounts in plant tissues and its presence in significant amounts in *extracts* of plants is almost certainly a result of phospholipase D activity during extraction.

A phospholipid which has been identified in legume seeds (pea[8] and soyabean[9]) and also in wheat flour[10] is *N*-acyl phosphatidylethanolamine. Diatoms, a group of marine algae, contain phosphatidyl-*N*-monomethyl-ethanolamine[11]. A novel, nitrogen-free, phospholipid which contains only glycerol, fatty acid and phosphate moieties was isolated from spinach leaves[12]. Some plants contain a complex phosphorus-containing phytoglycolipid, the structure of which was recently elucidated[13].

Fatty acid distribution in plant phospholipids

Some typical examples of the fatty acid content of individual phospholipids in various types of plants and in different tissues of plants are listed in Table 2. Table 3 presents currently available information on the positional distribution of fatty acids between the *sn*-1 and -2 acyl groups for phospholipids.

The fatty acid composition of plant lipids in general is dependent upon environmental conditions, particularly light and temperature, and these aspects will be discussed in some detail later. At this point, some general observations can be made on the information presented in Tables 2 and 3.

Although a large number of different fatty acids have been identified in plants, only a few common fatty acids are found in phospholipids. Palmitic, linoleic and linolenic acids are, in general, the major constituents of the phospholipids of higher plants; stearic and oleic acids are usually present in smaller amounts and chain lengths other than 16 or 18 carbons are unusual in other than trace amounts. Hexadecenoic acid, as the 9-*cis* isomer

TABLE 2

EXAMPLES OF THE FATTY ACID COMPOSITION OF INDIVIDUAL PHOSPHOLIPIDS OF PLANT TISSUES

			Percentage fatty acid composition								Others	Reference
			14:0	*16:0*	*16:1^{-9} cis*	*16:1^{-3} trans*	*18:0*	*18:1*	*18:2*	*18:3*		
Leaves:												
Castor bean,												
Ricinus communis	photosynthetic	PC		38			3	3	24	33		173
		PE		36			2	1	21	40		
		PG	1	47		19	4	12	4	20		
		PI		54	4		5	8	6	34		
	etiolated	PC		15	2		9	11	37	24		173
		PE		21	4		5	8	41	18		
		PG	2	38	5		9	11	20	14	12:0(2)	
		PI	4	32	5		11	14	23	11	12:0(3)	
		CL	6	24			11	15	29	9	12:0(4)	
Root:												
Turnip,												
Brassica campestris		PC		10	1		1	11	20	58		18
		PE		13	1		1	11	25	50		
		PI		34	1		1	4	15	47		
		CL		11	1		1	9	21	57		
		total lipid		13	1		1	9	19	57		
Tuber:												
Potato,												
Solanum tuberosum		PC		17			3	5	67	8		89
		PE		19			3	6	64	8		
		PG		48			5	6	38	4		
		PI		43			3	2	43	8		
		CL		10			7	10	58	16		
		total lipid		20			3	4	60	15		

TABLE 2, continued

		Percentage fatty acid composition								Others	Reference
		14:0	16:0	$16:1^{-9}$ cis	$16:1^{-3}$ trans	18:0	18:1	18:2	18:3		
Bulb:											
Narcissus sp.	PC	3	23	1		1	10	61	3	12:0(3)	172
	PE	1	25	0		2	7	65	1	12:0(1)	
	PG	1	44	9		3	15	22	0	12:0(2)	
	PI	1	40	3		3	7	37	2	12:0(1); 16:2(0.6); 16:3(0.4)	
	CL	2	34	10		3	11	32	0	12:0(2); 16:2(1); 16:3(1)	
	total lipid		17	2		1	9	69	3		
Fruit:											
Apple, *Pyrus malus*	PC		23			3	8	57	10		77
	PE		23			2	2	65	9		
	PG		42			2	17	30	8		
	PI		39			3	5	39	14		
	CL		25			4	5	49	17		
	total lipid		25			4	7	54	11		
Pollen:											
Pine, *Pinus ponderosa*	PC		18			2	17	52	4	20:2(2); others (6)	154
	PE		21			0	8	50	8	20:2(4); others (8)	
	PG		38			2	10	38	3	– others (9)	
	PI		40			0	17	39	0	– others (4)	
	CL		22			5	14	29	2	20:2(5); others (22)	
	PS		41			4	22	18	0	20:2(8); others (8)	
Bud:											
Cotton, *Gossypium sp.*	PC		20			5	11	30	33		155
	PE		26			0	0	47	27		
	PI		37			6	9	9	39		

TABLE 2, continued

			Percentage fatty acid composition								Others	Reference
			14:0	16:0	$16:1^{-9}$ cis	$16:1^{-3}$ trans	18:0	18:1	18:2	18:3		
Moss:												
Hypnum cupressiforme		PC		17	7		3	11	27	27	20:4(8)	174
		PE		13	12		3	12	21	21	16:2(1); 20:4(17); 22:0(3)	
		PG		32		18	3	12	10	19	20:4(3)	
		CL		28	13		7	26	14	8	16:2(2); 20:4(4)	
		total lipid	1	14	5		2	7	20	24	20:4(12); 20:5(7); 22:0(5)	
Green alga:												
Chlorella vulgaris	auxotrophic	PC		16	3	0	5	9	44	18	16:2(3); 16:3(2)	173
		PE		29	6	0	3	9	33	15		
		PG		31	5	16[a]	6	10	25	5		
		PI		30	11	0	6	23	10	4	16:2(4)	
		CL		24	9	0	14	27	27	6		
Thermophilic alga:												
Cyanidium caldarum	grown at 20°	PC		32			10	11	16	31		119, 120
		PE		7			6	59	9	18		
	grown at 55°	PC		43			3	30	23	0		
		PE		27			8	50	15	0		
Blue-green algae:												
Anabaena variabilis		PG	8	31	9	0	6	21	12	6		128
Anacystis nidulans		PG	8	35	17	0	8	22	0	0		128
Photosynthetic bacteria:												
Chromatium		PG	2	31	16[b]		3	49[c]	0	0		14
		PE	2	10	15[b]		16	56[c]	0	0		
Chloroplast lamellae:												
Spinach, *Spinaca oleracea*		PG	1	11		32		2	4	47		6
		PC		12				9	16	58	16:3(4)	
		PI	4	34		1	2	7	15	26	16:2(6); 16:3(2)	

[a] Absent in dark-grown tissue. [b] Double-bond position not specified. [c] Probably vaccenic acid.

References p. 284

TABLE 3

DISTRIBUTION OF FATTY ACIDS BETWEEN 1- and 2-POSITIONS OF PLANT PHOSPHOLIPIDS

		Percentage fatty acid composition						*Reference*
		16:0	*16:1*[a]	*18:0*	*18:1*	*18:2*	*18:3*	
Phosphatidylcholine								
Leaves *(Phaseolus multiflorus)*	total	26.6	tr	6.0	3.7	37.7	25.7	150
	1-position	51.0	0	11.6	2.8	20.3	14.4	
	2-position	2.3	tr	0.4	4.6	55.3	37.1	
	% at 2-position	4	–	25	62	73	72	
Spinach chloroplasts	total	14.3	0.4	0.4	8.4	21.0	55.5	58
	1-position	26.3	1.2	1.1	6.8	19.0	45.6	
	2-position	3.7	3.4	0.8	8.6	23.8	57.9	
	% at 2-position	12	70	40	56	56	58	
Spinach microsomal fraction	total	20.2	0.4	0.8	12.8	25.6	4.0	58
	1-position	38.4	2.2	2.2	12.1	17.6	27.4	
	2-position	6.9	1.9	0.8	15.8	25.4	49.3	
	% at 2-position	15	45	29	56	59	64	
Cauliflower inflorescence	total	18.5		4.8	9.5	17.2	50.0	58
	1-position	42.0		11.3	29.0	7.1	11.7	
	2-position	15.2		4.4	13.5	21.2	45.8	
	% at 2-position	28		34	35	70	81	
Cotton buds	total	20.4		5.2	10.7	30.4	33.3	155
	1-position	23.7			19.3	32.7	24.5	
	2-position	4.0			20.3	38.8	38.8	
	% at 2-position	14			51	54	61	

TABLE 3, continued

		Percentage fatty acid composition							Reference
		16:0	*16:1*[a]	*18:0*	*18:1*	*18:2*	*18:3*		
Soyabean	total	17.2		3.8	22.6	47.8	8.6		175
	1-position	34.4		7.6	30.2	23.8	4.0		
	2-position	–		–	15.8	71.8	13.1		
	% at 2-position	0		0	34	75	77		
Chlorella vulgaris	total	18	3	2	5	44	26		176
	1-position	30	2	3	5	39	19		
	2-position	4	3	1	7	48	34		
	% at 2-position	12	60	25	58	55	64		
Phosphatidylethanolamine									
Chlorella vulgaris	total	35	2		3	39	19		176
	1-position	67	3		4	15	8		
	2-position	7	1		5	58	28		
	% at 2-position	9	25		6	80	78		
Phosphatidylinositol									
Soyabean	total	25.6		11.7	9.1	44.4	6.4	others 2.7	40
	1-position	71.9		17.3	4.6	6.7	0	others 0.5	
	2-position	1.3		0.5	9.7	74.3	12.5	others 1.4	
	% at 2-position	2		3	68	92	100		
Cauliflower inflorescence	total	14.5		2.2	13.1	16.5	49.7	others 4.0	40
	1-position	34.5		3.7	22.1	10.5	25.3	others 3.0	
	2-position	3.5		0.9	2.3	17.2	76.1	others 0.5	
	% at 2-position	9		22	12	66	77		

References p. 284

TABLE 3, continued

		Percentage fatty acid composition						Reference
		16:0	*16:1*[a]	*18:0*	*18:1*	*18:2*	*18:3*	
Phosphatidylglycerol								
Spinach leaves	total	20.0	31.7	0.6	2.6	8.0	37.1	14
	1-position	32.4	0.5	1.2	2.4	2.1	61.4	
	2-position	11.1	58.9	–	3.1	14.5	12.4	
	% at 2-position	25	99	0	54	87	17	
Chlorella vulgaris	total	34.8	22.0	4.2	9.0	9.6	19.1	14
	1-position	28.1	0	3.9	12.2	17.0	36.2	
	2-position	41.6	44.0	4.4	5.8	2.2	2.0	
	% at 2-position	58	100	53	32	11	5	

[a]16:1 in phosphatidylglycerol is *trans*-3-hexadecenoic acid; for other phospholipids 16:1 refers to *cis*-9-hexadecenoic acid.

or sometimes as the 7-*cis* isomer, is found widely in small amounts in phospholipids; the 3-*trans*-hexadecenoic acid isomer, however, occurs specifically at the 2-position of phosphatidylglycerol in actively photosynthesising tissues of higher plants and can represent 70% of the total fatty acids at the 2-position of this lipid in spinach chloroplasts[14]. Oxygenated, branched-chain and odd-numbered chain length fatty acids occur widely in the triglycerides of seed oils and in the waxy components on the surface of plant tissues but these are not normally found in phospholipids.

Phosphatidylcholine and phosphatidylethanolamine generally show similar fatty acid compositions. Phosphatidylinositol and diphosphatidylglycerol normally contain a higher proportion of saturated (mainly palmitic) fatty acids than phosphatidylcholine or phosphatidylethanolamine. The distribution of fatty acids between the 1- and 2-positions of the phospholipids shows a general pattern similar to that in animal lipids. The saturated fatty acids are located mainly at the 1-position, whereas the 2-position contains predominantly unsaturated fatty acids. However, linoleic and linolenic acids normally comprise 60–80% of the total fatty acid in plant phospholipids and the distribution of these polyunsaturated fatty acids between the two positions is similar. A significant amount of the linolenic acid may be at the 1-position (Table 3). With phosphatidylglycerol from green leaves, the predominant molecular species (47% total) is 1-linolenoyl-2(*trans*-3-hexadecenoyl)-*sn*-glycero-3-phosphoryl-1′-*sn*-glycerol[14].

2. SOME PRACTICAL CONSIDERATIONS IN BIOCHEMICAL STUDIES WITH PLANT LIPIDS

In working with plant lipids, the biochemist faces certain obstacles, some of which are unfamiliar to his colleagues studying mammalian or microbial systems. Problems arise in both the isolation of lipids from plant tissues, in the preparation of cell-free extracts and the limited growing season. Some of the important considerations will be briefly indicated.

Extraction of lipids

Many plants contain highly active lipolytic enzymes which are released when cells are ruptured. Phospholipase D and acyl hydrolase activities have been implicated in these effects. Such enzymes can be destructively active even at low temperatures and their effects may be increased by some organic solvents conventionally used for lipid extractions. The classical demonstration of these autolytic processes was the isolation[16], from cabbage leaves, of phosphatidic acid, an artifact produced by the action of phospholipase D on endogenous phospholipids. Precautions now taken to avoid such pitfalls include rapid blanching of the tissue before extraction[15] and initial homogenization with propan-2-ol, preferably hot, before conventional solvent extraction[17]. Even with such precautions, the lipids extracted under different solvent conditions may show some variation[18].

Preparation of cell-free extracts

In many types of plant cells, the bulk of the cell volume is occupied by one or more

References p. 284

vacuoles. These organelles, which frequently contain high concentrations of organic acids at low pH, are invariably ruptured during homogenization producing acidic mixtures, e.g. the pH of a water extract of lemon fruits is about 2.5. Phenolic compounds are commonly found in plants and enzymatic oxidation during extraction produces polyphenolic substances, frequently highly coloured, which inactivate many enzymes by a tanning process. The lipolytic enzymes mentioned above can attack lipoprotein membrane structures causing, for example, inactivation of chloroplasts and mitochondria. Careful control of pH, prevention of oxidation and removal of phenolics, fatty acids, etc. are required for satisfactory production of active extracts and functional sub-cellular organelles[19–24].

Other factors which require consideration include the gross contamination of many plants with micro-organisms; growth and environmental conditions (particularly light and temperature) during storage of plants; and varietal differences in a given species.

In the selection of plant material for particular studies, the above factors have led to a few favoured tissues appearing repeatedly in the literature. This situation is analogous to the popularity of rat liver and *E. coli* in other fields and the same danger of generalization is present.

3. BIOSYNTHESIS OF PHOSPHOLIPIDS

Definitive work on pathways of lipid biosynthesis in plants has developed more slowly than corresponding studies on microbial and mammalian systems. The enzymes concerned will not be discussed in detail when these do not differ significantly from the same enzymes previously defined in other organisms and dealt with in Chapter 4 (p. 67). Fig. 1 indicates the enzymatic pathways of biosynthesis and interconversions of phospholipids which have been demonstrated or indicated in plant tissues.

Acyl-CoA:sn-glycero-3-phosphoric acid O-acyltransferase (EC 2.3.1.15) (Fig. 1,1)

The central role of phosphatidic acid in the biosynthesis of plant lipids was indicated by the high specific radioactivity of this intermediate in the labelling studies of Kates who used [^{32}P]orthophosphate with leaves[25] and *rac*-glycero-3-[^{32}P]phosphoric acid with Chlorella[26]. The acyl transferase activity was demonstrated in microsomal preparations from a non-photosynthetic tissue (avocado mesocarp)[27] and from leaves[28, 29]. Radioactivity from ^{14}C- or ^{32}P-labelled glycerophosphate in these systems was incorporated mainly into phosphatidic acid. Fatty acyl-CoA esters served as acyl donors but the possibility that the immediate acyl donor is acyl carrier protein (ACP), as demonstrated in bacterial systems[30], has not been explored.

Diglyceride phosphokinase (ATP:1,2-diacyl-sn-glycerol 3-phosphotransferase) (Fig. 1,2)

This alternative route for phosphatidic acid formation was shown to occur in mitochondria from peanut cotyledons[31, 32]. It is unlikely to be a major route for *de novo* biosynthesis of phospholipids in plant tissue but may be involved in mobilization of materials from the neutral glyceride reserves of some seeds.

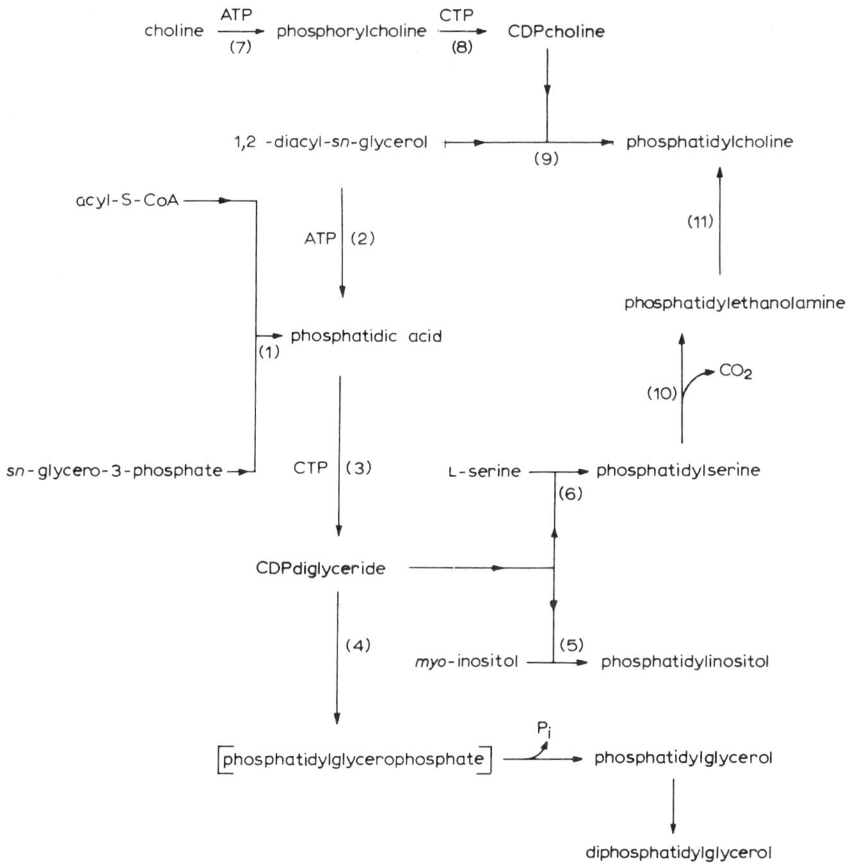

Fig. 1.

CTP:phosphatidic acid cytidyl transferase (Fig. 1,3)

The involvement of CDPdiglyceride in the biosynthesis of phosphatidylinositol, phosphatidylglycerol and diphosphatidylglycerol in plants was established after it had been shown[33, 34] that mitochondrial preparations from cauliflower inflorescence catalysed the conversion of CTP to CDPdiglyceride. Subsequent studies on a variety of plant tissues have shown that the enzyme occurs predominantly in microsomal fractions[35] but is also present in microsome-free mitochondrial preparations. It is, however, absent from chloroplasts. Using cauliflower mitochondria, no requirement for added phosphatidic acid could be demonstrated[35, 36] but this tissue is particularly rich in phospholipase D activity[37] which could generate phosphatidic acid from endogenous phospholipids. Recently, a requirement for exogenous phosphatidic acid was demonstrated with highly purified mitochondria from cauliflower[177].

Phosphatidylglycerol and diphosphatidylglycerol (Fig. 1,4)

Although phosphatidylglycerol is the major phospholipid in chloroplasts and is rapidly labelled from $^{14}CO_2$ during photosynthesis in algae[38], Sastry and Kates[29] were unable to demonstrate the conversion of phosphatidic acid to phosphatidylglycerol in cell-free homogenates of spinach leaves and the *de novo* biosynthetic pathway for phosphatidylglycerol in chloroplasts has not yet been demonstrated. However, the enzyme sequence: phosphatidic acid → [phosphatidylglycerophosphate] → phosphatidylglycerol → diphosphatidylglycerol does occur in cauliflower mitochondria[39]. Radioactivity from ^{14}C-labelled *sn*-glycero-3-phosphoric acid was incorporated into phosphatidylglycerol and diphosphatidylglycerol. When $HgCl_2$ (1 m*M*) was added to inhibit the dephosphorylation of phosphatidylglycerophosphate, the synthesis of these two phospholipids was inhibited and the label accumulated in phosphatidylglycerophosphate.

Phosphatidylinositol (Fig. 1,5)

Free inositol stimulates the incorporation of [^{32}P]phosphatidic acid into phosphatidylinositol in cell-free extracts of spinach leaves[29]. The formation of phosphatidylinositol from CDPdiglyceride and *myo*-inositol can also occur through an $Mn^{2\oplus}$-requiring system located in a particulate preparation from cauliflower inflorescence tissue[40].

Phosphatidylserine (Fig. 1,6)

The CDPglyceride:serinetransferase system which operates in bacteria (Chapter 11) has not been demonstrated in cell-free systems from plants or animals. However, labelling studies with *rac*-glycerol-3-[^{32}P]phosphate in Chlorella[26] and with L-[^{14}C]serine on excised tissue from tomato roots[41] and potato tubers[42], have indicated that this pathway probably exists in plants.

Phosphatidylcholine and phosphatidylethanolamine

These are the two major phospholipids of all non-photosynthetic tissues of higher plants and of all subcellular organelles, with the exception of chloroplasts, in photosynthetic tissues. Kinetic studies with many plants have shown that these lipids are metabolically active with respect to turnover of fatty acyl, glycerol, phosphate and base moieties. It is therefore surprising that there is a dearth of information covering their biosynthesis. By analogy with microbial and mammalian systems, there are two possible routes for *de novo* synthesis of these lipids: (a) via CDPcholine (or CDPethanolamine) and diglyceride (reactions (7)–(9) in Fig. 1) or (b) via phosphatidylserine (Fig. 1,10 and 11). The possibility of exchange reactions is discussed later.

Little information is available on the biosynthesis of phosphatidylethanolamine via CDPethanolamine in cell-free extracts of plants. Very recently, the individual enzymes for phosphatidylcholine synthesis have been identified in extracts of both leaves and non-photosynthetic tissues from higher plants.

ATP:choline phosphotransferase (EC 2.7.1.32) (Fig. 1, 7)

This is a soluble enzyme occurring widely in plants[43, 44]. In plants, a large proportion of organic phosphorus may occur in the form of phosphorylcholine, which may be an agent of phosphorus translocation[45].

CTP:cholinephosphate cytidylyltransferase (EC 2.7.7.15) (Fig. 1,8)

This enzyme has recently been demonstrated in cell-free extracts of onion stem[44] the highest specific activity being located in the dicytosome fraction. The enzyme, which had a pH optimum between 6 and 7, required a divalent cation for activity ($Mn^{2\oplus} > Mg^{2\oplus}$) and was inhibited by $Ca^{2\oplus}$ ions. Morré et al.[44] compared the enzyme activities in onion and in rat liver. In both, the highest specific activity was in the dicytosome (Golgi apparatus) fractions and, expressed as activity per mg protein, the specific activities in both tissues were of the same order of magnitude. These results supported the concept of the Golgi apparatus as a site of membrane biosynthesis[44].

CDPcholine:1,2-diacyl-sn-glycerol cholinephosphotransferase (EC 2.7.8.2) (Fig. 1,9)

Morré et al.[44] have shown that this enzyme is present in onion stem and is located almost entirely in the particulate fraction. It is also present in spinach leaves[46] and predominantly in the microsomal fraction. The spinach enzyme, pH optimum 8.0, required divalent cations ($Mn^{2\oplus} > Mg^{2\oplus}$) and was inhibited by sulphydryl reagents; $Mn^{2\oplus}$ ions protected the enzyme from this inhibition and, together with dithiothreitol, stabilized the enzyme which was otherwise inactivated at 30°. Since the spinach microsomal enzyme showed no specificity for particular species of 1,2-diacyl-*sn*-glycerol, it was concluded that the fatty acid distribution in the phosphatidylcholine of spinach leaves is not controlled by any specificity of the choline phosphotransferase[46].

Interconversions of nitrogenous phospholipids

An alternative route for the formation of phosphatidylethanolamine and phosphatidylcholine is from phosphatidylserine by decarboxylation (Fig. 1,10) and successive *N*-methylation with *S*-adenosylmethionine (Fig. 1,11) as described in Chapters 4 (p. 77) and 11. Labelling studies with excised tissues of higher plants, algae and photosynthetic bacteria have shown that the pathway is operative in these tissues. However, the importance of this pathway for phosphatidylcholine biosynthesis in higher plants is questionable, although it may be the mechanism used by photosynthetic bacteria and, possibly, by algae. The methyl group of [*Me*-^{14}C]methionine was rapidly incorporated into phosphatidyl-*N*-methylethanolamine, phosphatidyl-*N*,*N*-dimethylethanolamine and phosphatidylcholine of *Rhodopseudomonas spheroides*[47]. In cell-free extracts, *S*-adenosylmethionine was the methyl donor. Kinetic studies of [^{32}P]orthophosphate incorporation into the lipids of Chlorella[26] and the finding of phosphatidylmethylethanolamine in marine algae[11] suggests that the above pathway are operative in lower plants.

References p. 284

When uniformly labelled L-[^{14}C]serine was incubated with disks of potato tuber, the most heavily labelled lipid was phosphatidylethanolamine[42]. In the same system, [^{14}C]ethanolamine and [*Me*-^{14}C]methionine were incorporated into phosphatidyl-*N*-methylethanolamine and phosphatidylcholine. However, kinetic studies indicated that the formation of both phosphatidylethanolamine and phosphatidylcholine from their respective bases (presumably via CDP derivatives) was quantitatively a more important route for biosynthesis of these lipids than either the interconversion system or base exchange reactions[42]. A study of the incorporation of ^{14}C-labelled serine and ethanolamine into the lipids of excised roots from tomato plants provided evidence for their incorporation into the corresponding phospholipids and for the decarboxylation-methylation pathway from phosphatidylserine via phosphatidylethanolamine to phosphatidylcholine[41].

Although the relative importance of this pathway for the *de novo* biosynthesis of phospholipids in this tissue is not known, it is possibly the major route by which free ethanolamine and, perhaps free choline, are formed; i.e. the individual phospholipids act as intermediates in the conversion of serine to the free bases[41]. An indication if this process in cell-free preparations from etiolated pea seedlings has been obtained recently[48].

Transphosphatidylation reactions

Kinetic studies[38] on the incorporation of $^{14}CO_2$ into algae, suggested that phosphatidylglycerol (which was the most rapidly labelled lipid) was converted into other phospholipids by transphosphatidylation reactions in which the phosphatidyl residue was transferred from glycerol to inositol and subsequently to choline and ethanolamine. Support for such a process was obtained when it was found that certain plant tissues did, in fact, catalyse transphosphatidylation reactions. When various tissues were extracted in the presence of ethanol[49] or methanol[50], large amounts of phosphatidylethanol or phosphatidylmethanol respectively were formed. These lipids, which were not observed if the tissues were heated before solvent extraction, were shown to be artefacts produced by the action of phospholipase D on endogenous phospholipids[51]. Yang et al.[51] and Dawson[52] have studied the transphosphatidylation reaction of phospholipase D using a range of alcohols as acceptors of the phosphatidyl unit from phosphatidylcholine. Among active acceptors were glycerol and ethanolamine. It was suggested[51,52] that interconversion of phospholipids could be catalysed thus, for example:

$$\text{phosphatidylcholine + glycerol} \xrightarrow[\text{Ca}^{2\oplus}]{\text{phospholipase D}} \text{phosphatidylglycerol + choline}$$

However, there may be little physiological significane of this reaction in the intact cell because the product in the above reaction is phosphatidyl-*rac*-1′-glycerol[51], i.e. a mixture of diastereoisomers, whereas the natural product obtained from plants is phosphatidyl-1′-*sn*-glycerol[15]. Also, it is difficult to account for the different fatty acid distributions in individual phospholipids if such a process occurs *in vivo*, e.g. *trans*-3-hexadecenoic acid in photosynthetic tissues is found only in phosphatidylglycerol[53].

Although phospholipase D itself may not be involved in phospholipid interconversion *in vivo*, there is evidence that exchange reactions may occur in plants by a similar mechanism. A particulate preparation from etiolated pea seedlings catalysed the incorporation of ^{14}C-labelled serine, ethanolamine and choline into the corresponding phospholipids[48]. The reaction was stimulated by $Ca^{2\oplus}$ ions (as is phospholipase D) in contrast to the cytidylyltransferase enzyme in the *de novo* synthetic system, which is inhibited by $Ca^{2\oplus}$ ions[44]. Whether the reaction was a true exchange reaction of the type described in mammalian systems (Chapter 4, p. 78) or was a transphosphatidylation reaction involving different acceptor and donor base groups was not made clear. Competition experiments suggested that a single enzyme was involved in the transfer of serine, ethanolamine and choline into the lipids[48]. Again, the physiological significance of these exchange reactions must be questioned because the activity of the above reactions in cell-free extracts was low and pulse-chase experiments, in which [^{14}C]choline was administered to potato disks, indicated that the exchange of choline between an endogenous pool and phospholipid was slow compared with the rate of pulse labelling of phospholipid[42].

Fatty acyl transfer reactions

Evidence from labelling experiments with $^{14}CO_2$ using algae[38] and mature leaves[54] points to a turnover of fatty acid moieties independent of that of the whole phospholipid molecule. This phenomenom is also implicit in many studies *in vivo* on the biosynthesis of fatty acids in various plant tissues. Exchange of fatty acyl groups in preformed lipids would also provide a mechanism for introducing specific fatty acids to give the characteristic distributions which are observed in each phospholipid. The possible significance of this process is emphasised by the observation[46] that the choline phosphotransferase enzyme involved in the *de novo* biosynthesis of phosphatidylcholine in spinach leaf is not specific for diglyceride species and could not in itself explain fatty acid distribution in the product. Also, other routes of phospholipid formation involving transphosphatidylation would not confer fatty acid specificity on the products.

Homogenates of spinach leaves were found to catalyse the acylation of 1-acyl-*sn*-glycero-3-phosphorylethanolamine in the presence of oleic acid, coenzyme A and ATP, producing the diacyl phospholipid[55]. The acyl transferase enzyme had its highest specific activity in the microsomal fraction when 1-acyl-*sn*-glycero-3-phosphorylcholine was used as substrate[56]. Incorporation of linoleic acid from [^{14}C]linoleoyl-*S*-CoA into the endogenous lipids of a microsomal fraction (solubilized with sodium dodecyl sulphate) was highly specific for phosphatidylcholine with little incorporation into other phospholipids. A similar specificity for phosphatidylcholine was observed with a fatty acyl transferase system in a microsomal fraction from safflower seeds[57]. A significant observation was the positional specificity of the spinach microsomal acyl transferase[56]; when saturated fatty acyl-CoA derivatives were used the saturated acids were incorporated primarily into the 1-position of the *sn*-glycerol moiety of phosphatidylcholine and the unsaturated acids entered the 2-position. In the safflower system, oleic acid from oleoyl-*S*-CoA was incorporated into the 2-position of phosphatidylcholine[57]. This positional specificity of the acyl transferase correlates with the fatty acid distribution found in phosphatidylcholine from plant tissues[58] (Table 3). The specificity for phosphatidylcholine suggests that either the enzyme is specific for this lipid or that the transfer is limited by availability of endogenous substrate. It is assumed that the endogenous species to which the acyl group is transferred is the *lyso*-phospholipid as in animal systems (Chapter 4, p. 82). However, this mechanism would require a source of endogenous *lyso*-phosphatidylcholine acylated in the 1-position to accept mainly unsaturated fatty acids and another species acylated in the 2-position to receive the saturated acids. There is little evidence to date for specific phospholipases A_1 and A_2 in plants. In plant systems the possibility that acyl groups are transferred by *acyl exchange* reactions has not been excluded. An exchange would not require *lyso*-phospholipid substrates, but transfer could take place between the donor (e.g. acyl lipid or acyl-*S*-CoA) and the diacyl lipid; the specificity for fatty acid and position could be carried by the transferase or by the physicochemical environment of the membrane-bound lipids.

An acyl hydrolase enzyme which also catalyses acyl transfer reactions from acyl lipids, including *lyso*- and diacyl-phospholipids, to acceptor alcohols was demonstrated in potato tubers[59]; the role of this enzyme in the turnover of acyl groups in phospholipids is not known. Another rather unspecific acyl transfer system was found in homogenates of spinach leaves[60] in which fatty acyl groups were transferred between galactolipids and phospholipids.

In addition to transfer reactions of acyl or polar moieties of phospholipids, there is evidence for reactions involving the exchange of intact phospholipid molecules between subcellular organelles in plants; this will be discussed later.

4. PHOSPHOLIPID CATABOLISM

Enzymes attacking the polar group of phospholipids

Phospholipase C (phosphatidylcholine:choline phosphohydrolase EC 3.1.4.3)

This is not a major enzyme in higher plants although some evidence for its presence was obtained in plastid fractions from homogenates of carrot root and spinach leaves[61].

Phospholipase D (phosphatidylcholine:phosphatidohydrolase EC 3.1.4.4)

This is the only enzyme attacking the polar groups of phospholipids which has been well characterized in plants. This enzyme is present in a wide range of higher plants, and probably in *Euglena gracilis*[62], but its existence in other micro-organisms and in animals has not been established. The enzyme was first described in carrot root and cabbage leaves[63], although its occurence was, unwittingly, demonstrated in 1927 when phosphatidic acid was isolated from cabbage leaves[16].

Although phospholipase D was originally thought to be localized in plastids[64], there is good evidence that in most plants the enzyme is present mainly in non-particulate fractions of extracts[37, 65, 66]. It is completely absent from chloroplasts of maize and spinach leaves when these organelles are prepared under conditions which eliminate contamination by other subcellular fractions[67], whereas chloroplasts isolated by conventional differential centrifugation showed significant activity. Similar observations were made with mitochondria from cauliflower inflorescence and hypocotyls from *Phaseolus aureus*[67]. On the other hand, Heller et al.[68] found that although phospholipase D activity in mature peanuts was mainly soluble, in immature seeds a lower enzyme activity was present but this was associated with a particulate fraction.

The soluble form of phospholipase D has been partially purified from cabbage leaves[65, 69] and requires $Ca^{2\oplus}$ ions for activity. As described above, in addition to its hydrolytic activity phospholipase D also catalyses transphosphatidylation reactions which are also $Ca^{2\oplus}$-dependent. In addition to water, acceptors for the transphosphatidylation reaction are several primary alcohols[52], including methanol[50], ethanol, *n*-propanol, ethanolamine and glycerol[51, 52, 68], but not inositol or glycerol-1-phosphate[51].

Phospholipase D is inhibited by *p*-chloromercuribenzoate, but not by diisopropylphosphofluoridate and Yang et al.[51] have proposed that the phosphatidyl–enzyme complex contains a phosphothioester linkage. These authors also consider that, since $Ca^{2\oplus}$ ions are required for both hydrolase and transphosphatidylation activities of phospholipase D, the cation must play a role in substrate–enzyme interaction rather than acting to bind free phosphatidic acid[70]. Basic proteins are strongly inhibitory in the hydrolysis of phosphatidylcholine[69] and this could explain the inhibition obtained by a heat-labile, high molecular weight factor in the soluble fraction of some leaf extracts[64, 66].

Assuming that a phosphothioester is involved in the enzyme-substrate interaction, a general reaction for phospholipase D may be written:

$$\mathrm{R{-}O{-}\overset{\displaystyle O}{\overset{\|}{\underset{\underset{\displaystyle O^-}{|}}{P}}}{-}O{-}X} \;+\; \mathrm{enz{-}SH} \;\longleftrightarrow\; \mathrm{R{-}O{-}\overset{\displaystyle O}{\overset{\|}{\underset{\underset{\displaystyle O^-}{|}}{P}}}{-}S{-}enz} \;+\; \mathrm{X{-}OH}$$

$$\mathrm{R{-}O{-}\overset{\displaystyle O}{\overset{\|}{\underset{\underset{\displaystyle O^-}{|}}{P}}}{-}S{-}enz} \;+\; \mathrm{Y{-}OH} \;\longleftrightarrow\; \mathrm{R{-}O{-}\overset{\displaystyle O}{\overset{\|}{\underset{\underset{\displaystyle O^-}{|}}{P}}}{-}O{-}Y} \;+\; \mathrm{enz{-}SH}$$

where R represents the 1,2-diacyl-*sn*-glycerol residue of a phospholipid.

The substrate specificity of the enzyme is not unequivocally determined; the activity of the enzyme *in vitro* on a given substrate seems to depend on the physical state in which the substrate is presented and on the presence or absence of organic solvents and detergents. Diethylether stimulates the hydrolysis of phosphatidylcholine by phospholipase D from most plant tissues[71], but not from cottonseed[66], whereas this solvent inhibits the hydrolysis of *lyso*-phosphatidylcholine[65] by the cabbage enzyme. Diphosphatidylglycerol was

References p. 284

not hydrolysed by phospholipase D from cabbage[72] or peanuts[68] when conditions optimal for the hydrolysis of phosphatidylcholine were used; when ether was omitted, diphosphatidylglycerol was hydrolysed by the peanut enzyme[68]. Natural plant phospholipids which are attacked by the enzyme under varying conditions include phosphatidylcholine, phosphatidylethanolamine, phosphatidylserine[71], phosphatidylglycerol, diphosphatidylglycerol[68] and *O*-amino-acyl esters of phosphatidylglycerol[73]. No hydrolysis of phosphatidylinositol has yet been demonstrated with phospholipase D.

The effects of organic solvents and detergents on phospholipase D have been studied in some detail[68, 69, 71, 74]. The results in general suggest that the observed effects are due to the physical state of the substrates in an aqueous environment. Not only are effects on reaction rates observed, but pH optima and substrate specificities also depend on the solvent or detergent added. Similar effects are observed with the lipolytic acyl hydrolase described later.

It is unlikely that phospholipase D would be freely soluble in the cytoplasm of plant cells but, in common with other hydrolytic enzymes, it could be localized in organelles equivalent to lysosomes or in the vacuoles, the contents of which are released by rupture of the tonoplast membrane during homogenization. A comprehensive survey[37] of many plant tissues, has shown that in plants containing the enzyme, activity is highest in swollen storage tissues, e.g. the stalk of cabbage and celery, the root of carrot, the seeds of pea and marrow and the tuber of Jerusalem artichoke. This generalization does not hold for all plants because the phospholipase D is absent from potato tuber and in fruits (e.g. apple, lemon, tomato and rose)[37]. The physiological role of the enzyme is not known; its activity does not appear to be associated with utilization of the reserve materials in storage tissues and seeds[37].

Enzymes attacking fatty acyl ester bonds of phospholipids

Under certain conditions higher plants can lose polar lipids from tissues. Phospholipase D activity does not lead to a net loss of acyl ester content and the product, phosphatidic acid, does not accumulate in plant tissues. To account therefore for a net loss of acyl lipids, acyl hydrolases must be operative. Also, if the fatty acyl transfer reactions discussed previously involve the monoacyl phospholipid intermediates, then phospholipases A_1 and A_2 acting on endogenous diacyl phospholipids would be involved. Evidence for phospholipid acyl hydrolases acting *in vivo* is provided by the loss of polar lipids in senescing leaves[75, 76] and fruits[77], in stored edible tissues[78] and in germinating seeds[79, 37].

Enzymes that deacylate neutral glycerol esters, i.e. lipases, (glycerol ester hydrolase; EC 3.1.1.3) are well known in many seeds[80] and the enzymic deacylation of galactolipids has been established[81–83]. Surprisingly little attention has been paid to phospholipid acyl hydrolase enzymes in plants, in contrast to the wealth of information on phospholipases A and B in animals, reptiles and insects. Early work[84] reported that rice seeds contained enzymes which catalysed the deacylation of phosphatidylcholine and *lyso*-phosphatidylcholine and more recently[85–88] a similar system has been described in a particulate fraction from germinating barley. The barley enzyme(s) removes fatty acids from the 1- and

2-positions of phosphatidylcholine and *lyso*-phosphatidylcholine is formed only as a transient intermediate[88].

When potato tubers were homogenised in water at 0° and the extracts immediately analysed, the lipid content of the homogenate differed remarkably from that of the intact tissue[89] in that during homogenization a rapid breakdown of polar lipids (phospholipids and galactolipids) had occurred and corresponding amounts of free fatty acids and fatty acid hydroperoxides were formed[90]. A lipolytic acyl hydrolase was partially purified from potato tubers and the hydrolysis of both phospholipids and galactolipids was catalysed by the same enzyme preparation which also deacylated monoglycerides and diglycerides (but not triglycerides) and some artificial esters of long-chain fatty acids[59]. The specific activity of the partially purified enzyme towards galactolipids was seventy times greater than that of a purified homogeneous 'galactolipase' enzyme from *Phaseolus multiflorus* leaves[83]. Since the Phaseolus galactolipase also showed some activity towards phosphatidylcholine[76, 80] and had very similar properties to the potato enzyme and since crude extracts of barley deacylate both phospholipids and galactolipids[76], it is possible that the same lipolytic acyl hydrolase enzyme is concerned in these tissues and that a description such as 'galactolipase' or 'phospholipase B' is misleading. In a survey of various plants, no tissue has yet been found with activity comparable to that in potato tuber; leaves of potato and of *Phaseolus multiflorus* and germinating barley embryos showed relatively low levels of activity[76].

The potato enzyme deacylated most natural diacyl phospholipids at similar rates, but was more active towards monoacyl phospholipids and no *lyso*-intermediates were detected during the hydrolysis of phospholipids. In addition to its acyl hydrolase activity, the enzyme also catalyses acyl transfer reactions in the presence of alcohols. In common with most carboxyl ester hydrolases, it is inhibited by diisopropyl phosphofluoridate, but is insensitive to sulphydryl reagents. The barley enzyme was inhibited by very high concentrations of *p*-chloromercuribenzoate ($10^{-2}M$). Neither enzyme was affected by $Ca^{2\oplus}$ ions or by EDTA[59, 88]. It is probable that the hydrolysis of phospholipids proceeds through a monoacyl intermediate and that removal of the first fatty acid is rate-limiting.

When sonicated aqueous dispersions of phosphatidylcholine were used as substrates, purified preparations of the potato enzyme were inactive in the absence of surface-active agents. The most effective agents were *cis*-unsaturated free fatty acids and the activity in crude homogenates was due to the presence of endogenous fatty acids[91]. Triton X-100 was also effective but caused a marked shift upwards in the pH optimum, analogous to the effects shown by surface-active agents on phospholipase D[74]. Free fatty acids were also required for full activity of the barley enzyme on phosphatidylcholine[88].

The marked effect of surface-active agents was shown only when phosphatidylcholine and phosphatidylethanolamine were used as substrates. Although many studies on phospholipases are performed with these phospholipids, the 'liposome' structures which they form in aqueous media[92] may render the susceptible bonds unavailable to enzyme in the aqueous phase and surface-active agents or organic solvents are required to disperse the lipids into less ordered structures[91]. The lipolytic acyl hydrolase was active in the absence of surface-active agents, on *lyso*-phosphatidylcholine which forms a true micellar dispersion. However, the 'liposome' structure of phosphatidylcholine dispersions is a useful model

References p. 284

system in that, in some respects, it resembles the properties of phospholipids in natural membrane structures[93]. For example, the action of the lipolytic acyl hydrolase from potato on mitochondria resembled the action on phosphatidylcholine in that no hydrolysis of the mitochondrial phospholipids took place unless free fatty acid was added to the incubation mixture[91].

5. SUBCELLULAR LOCALIZATION OF PHOSPHOLIPID METABOLISM

On the assumption that phospholipids in plant cells are primarily, if not solely, constituents of lipoprotein membrane structures, metabolism of these lipids should be considered as a function of membrane synthesis, breakdown or turnover. Studies *in vitro* with isolated subcellular fractions have so far failed to show complete *de novo* synthesis of phospholipids in any single fraction and point to a cooperative effect between different fractions in both synthesis and exchange reactions. It might be expected, however, that chloroplasts, which in many respects behave as autonomous symbiants in plant cells, would be able to synthesise their lipids, particularly since the lipid composition of chloroplasts is vastly different from that of other organelles. There is no doubt that chloroplasts can synthesise fatty acids and that the synthesis is stimulated by light[94]. Extensive studies on fatty acid biosynthesis in isolated chloroplasts have shown that, although saturated and monounsaturated fatty acids are synthesised, no desaturation to the polyunsaturated fatty acids characteristic of most chloroplast lipids takes place[1]. Although it is unlikely that newly synthesised fatty acids in chloroplasts exist in the free form, very little attention has yet been paid to polar lipids formed in cell-free systems. A small incorporation of *sn*-glycero-3-[^{32}P]phosphate into phosphatidic acid by isolated chloroplasts from spinach leaves has been observed[29] but the possibility of contamination with other cell fractions was not examined. Chloroplasts free from other cell fractions incorporate *sn*-glycero-[3-^{14}C]phosphate into phosphatidic acid and 1,2-diacyl-*sn*-glycerol[95]. However, CTP:phosphatidic acid cytidylyltransferase could not be demonstrated in uncontaminated chloroplast preparations[35] which suggests that the major phospholipid, phosphatidylglycerol, is not formed in chloroplasts by the known *de novo* route. Failure to demonstrate total lipid synthesis in chloroplasts could be due to removal of essential components during isolation; e.g. fatty acid synthesis is more efficient when chloroplasts are isolated in an intact form[96]. Although the chloroplast lipids could be formed by transfer from other cell constituents (see later), this would be inconsistent with the concept of autonomy in chloroplasts.

Although the question of polyunsaturated fatty acid and phospholipid synthesis in isolated chloroplasts remains unsolved, there is evidence that chloroplasts can synthesise galactosyl diglycerides, the major lipids of chloroplasts. Studies[97, 98] have shown that chloroplasts can form galactolipids by galactosylation reactions involving UDPgalactose and 1,2-diacyl-*sn*-glycerol. Since 1,2-diacylglycerol is probably formed *de novo* in chloroplasts[95], a route for galactolipid biosynthesis exists in chloroplasts[99].

When mitochondria are isolated from various tissues from higher plants following administration of [^{14}C]acetate, the labelling pattern of mitochondrial lipids and constituent fatty acids, including polyunsaturated fatty acids is not selective and resembles the lipid composition of these organelles[100]. Isolated mitochondria on the other hand, do not synthesise polyunsaturated fatty acids and phosphatidic acid is the major phospholipid labelled from [^{14}C]acetate[100] or [^{32}P]orthophosphate[101]. In several plant tissues, CTP: phosphatidic acid cytidylyltransferase activity is present in mitochondria isolated free from other cell fractions[35] suggesting that they may have limited lipid-synthesising capabilities which operate in the formation, in particular, of diphosphatidylglycerol which is concentrated in the mitochondria of higher plants[101]. A possible explanation as to why isolated mitochondria produce mainly phosphatidic acid from simple precursors, but will form CDP-diglyceride from added CTP could lie in a deficiency of this nucleotide in isolated mitochondria.

From the earlier discussion of phospholipid biosynthesis in plants, it will be apparent that most of the later stages of *de novo* synthesis are found in microsomal fractions of plant extracts. Whether this phospholipid synthesis is a property of the endoplasmic reticulum in general, or of dictysomes[44] similar to the Golgi apparatus, is an interesting problem.

Many of the apparent anomalies observed from studies *in vivo* and *in vitro* in animals have been explained recently by the discovery of exchange processes between subcellular organelles involving transfer of intact phospholipid molecules[102–104]. Mazliak et al.[105] proposed that the replacement of mitochondrial phospholipids in higher plants depended on a contribution from other subcellular fractions and recently exchange processes between the phospholipid of mitochondrial and microsomal membranes in extracts of potato tubers and cauliflower inflorescence have been demonstrated[106]. A soluble factor in the post-microsomal supernatant was necessary for the exchange process[106]. A significant observation in this work was that this soluble factor showed some specificity for the plant species. It was observed that efficient exchange of lipids only occurred between subcellular fractions from the same tissue; e.g. when a [^{14}C]phospholipid-labelled microsomal fraction from potato tubers was incubated with cauliflower mitochondria in the presence of a supernatant fraction from cauliflower, most of the phospholipid leaving the microsomal fraction was transferred only to the supernatant and incorporation into the mitochondria was blocked. Similar results were obtained with labelled mitochondria and unlabelled microsomes from the different species[106].

Thus, from evidence at present available, it appears that mitochondria have the ability to synthesise at least some of the phospholipids of their membrane structures, but that they depend on either the endoplasmic reticulum or a more specialized membrane for the supply of other phospholipids, including phosphatidylcholine and phosphatidylethanolamine, and a transfer of lipids between membranes is mediated by a cytoplasmic 'carrier' system. Several potential sites exist for synthesis of the fatty acyl moieties of phospholipids; chloroplasts, mitochondria and the soluble fatty acid synthesis system of plants[11]. However, none of these systems *in vitro* produce the major polyunsaturated fatty acids of plants i.e. linoleic and linolenic acids. At present, the only known sites at which polyunsaturated fatty acids are formed in plants are located in membranes present in microsomal preparations. The fact that isolated mitochondria do not synthesise polyunsaturated fatty acids might support the hypothesis that these organelles have 'evolved' from bacteria-like

References p. 284

systems. In general, bacteria, including photosynthetic bacteria and primitive algae, do not contain polyunsaturated fatty acids or phosphatidylcholine. Is it possible that these 'symbiotic' subcellular organelles have not evolved a system for synthesising these components for which they are dependent on the 'host' cell and lipid transfer systems? It is interesting to note that the lipid which is characteristic of mitochondria, i.e. diphosphatidylglycerol, is more typical of procaryotic bacteria than of other membranes in higher organisms. The above comments on mitochondria could equally apply to chloroplasts which are also thought to be autonomous symbiants of procaryotic origin. Current work in several laboratories on the culture of isolated chloroplasts in defined media should clarify this point.

6. FUNCTION OF PHOSPHOLIPIDS

The function of phospholipids as components of lipoprotein membranes is discussed in other chapters. The discussion below will only relate to any special functions in plant cells or to certain functions which have been studied in more detail in these tissues.

Photosynthesis

The following facts, until recently, suggested the possibility that phosphatidylglycerol, the major lipid of chloroplasts, had a specific role in the photosynthetic process: (a) all photosynthetic organisms so far studied contain phosphatidylglycerol associated with the photosynthetic apparatus; (b) the fatty acids of phosphatidylglycerol in *Chlorella vulgaris* turned over more rapidly than those of any other lipid[26, 38, 107]; (c) *trans*-3-hexadecenoic acid, which is found only in phosphatidylglycerol[53] and occurs specifically on the 2-position of the glycerol moiety[15], is only present in tissues exposed to light, being replaced by palmitic acid in etiolated tissues[108]. However, recent observations have cast doubt on any essential role for this lipid in the photosynthetic process. Thus, although phosphatidylglycerol occurs in blue-green algae it does not contain *trans*-3-hexadecenoic acid[7]. Mature leaves, in contrast to actively growing photosynthetic algae, show a very low turnover of the fatty acids of phosphatidylglycerol compared with other lipids[54]. This suggests that phosphatidylglycerol is an essential component of the membrane structure in the photosynthetic apparatus, but that it probably does not have a specific role in the photochemical reactions. The proposal that the polyunsaturated fatty acids of chloroplast lipids are directly involved in the photosynthetic electron transport system[109] has also been countered by the observation that these fatty acids are absent in blue-green algae which, however, can carry out the Hill reaction[110]. It could be that phosphatidylglycerol is not essential for photosynthesis *per se* but that its presence in photosynthetic structures reflects the composition of the procaryotic organisms from which chloroplasts may have evolved. The occurence of polyunsaturated fatty acids in chloroplast lipids may be related to the ability of higher plants to function efficiently over a wide range of ambient temperatures as discussed below.

Chloroplast lipids may not be directly involved in the photochemical reactions but their presence is essential for the electron transport processes since re-addition of the lipid fraction to lipid-depleted chloroplast preparations restored much of the electron transport system[111].

The polar lipids of chloroplasts obviously play a major role in the structural organization of the photosynthetic apparatus. The specific roles of the individual lipids have not been proved but on the basis of the known morphology and properties of chloroplast lamellae, Benson[112–114] has suggested in some detail how the different lipids may be arranged in the membranes.

Effect of temperature on membrane structure and function

Although a high concentration of polyunsaturated fatty acids is characteristic of chloroplast lipids, the fatty acid composition of membrane lipids in many plants shows a predominance of the 'essential fatty acids', linoleic and linolenic acids. In contrast to animals, higher plants can function efficiently over a wide range of temperatures and some attention has been paid to the relationship between the efficiency of mitochondrial activity and physiological behaviour of plants with respect to temperature. Mitochondria from tissues which were resistant to chilling at 0–10° (cauliflower bud, turnip root, pea shoot) were more 'flexible', as measured by their ability to swell, and had a higher content of unsaturated fatty acids than tissues sensitive to chilling (tomato fruit, corn shoot, sweet potato root and bean shoots)[115]. Arrhenius plots of the respiration rates of mitochondria from chilling-resistant tissues[116] showed a linear decrease over the range 25° to 1°, whereas the curves for chilling-sensitive tissues showed a linear decrease only to 9–12°, at which point there occurred a marked deviation with a steeper gradient towards 1°.

Temperature is known to affect the fatty acid composition of plant tissues[117]. The phospholipid composition also varies with temperature. Table 1 shows the example of lucerne (alfalfa), which when grown at 15° shows a lipid composition similar to that observed for many leaves under natural conditions. However, plants grown at 30° contain much lower proportions of phosphatidylcholine and phosphatidylethanolamine and a very high proportion of phosphatidylinositol[118]. Since phosphatidylinositol has relatively saturated fatty acids, the shift to phosphatidylinositol at the higher temperature resulted at the same time in a more saturated fatty acid composition. A striking example of the effect of temperature on fatty acid composition of phospholipids was observed with the thermophilic alga *Cyanidium caldanum*[119, 120] (see Table 2). Although linolenic acid is present in the phospholipids of the organism cultured at 20° (30% of the fatty acids of phosphatidylcholine), it is completely absent when grown at 55°.

It is obvious that a change in the fatty acid composition of a tissue can be effected in two ways. Either the lipid composition may remain constant and the component fatty acids alter, or the lipid composition can change, altering the proportion of lipids containing a higher saturated fatty acid content (e.g. phosphatidylinositol) to those with more unsaturated fatty acids (e.g. phosphatidylcholine and phosphatidylethanolamine). The same principle applies to other conditions giving changes in fatty acid compositions, for example fatty acid changes during senescence of tissues are due more to losses of certain lipids than to actual changes in fatty acid composition of individual lipids[77]. Unfortunately, most studies on lipid changes in plant tissues have concentrated on fatty acid distribution and have neglected net changes of individual lipids.

Role of phospholipids in fatty acid metabolism

Nichols et al.[121] studied the incorporation of [^{14}C]acetate into the polar lipids of Chlorella and proposed that some lipids, including phosphatidylglycerol and phosphatidylcholine, in which the fatty acids had a higher rate of turnover, were involved in fatty acid metabolism, whereas others, including phosphatidylethanolamine and phosphatidylinositol, with low turnover rates, had structural roles. Support for these proposals has come from work on algae and higher plants in several laboratories. Acetate incorporation into lipids of blue-green algae[122], flax seeds[123] and mature leaves[54] has shown a rapid turnover of some phospholipids, indicative of a role in fatty acid conversions. Moreover, Gurr et al.[124–126], working with Chlorella, have recently demonstrated that exogenous [^{14}C]-oleic acid is incorporated rapidly and specifically into phosphatidylcholine and that, although once incorporated, the label in the phospholipid turns over slowly, the linoleic acid of phosphatidylcholine becomes labelled. Roughan[54] showed that most of the [^{14}C]-acetate fed to mature leaves appeared initially in phosphatidylcholine which had a higher specific radioactivity than other lipids and that the labelling within the phosphatidylcholine showed a sequence oleic → linoleic →linolenic acid with no change in the specific radioactivity. These results and those of others were interpreted as indicating that conversion of oleic acid to linoleic acid took place within the phospholipid molecule. Although there is a precedent for this type of phospholipid covalently bound fatty acid conversion in the formation of cyclopropane fatty acids in bacteria[127], Nichols et al.[121] and Gurr et al.[124, 125] have emphasised that their results do not distinguish between (a) direct desaturation of phospholipid-bound fatty acids and (b) the phospholipids acting as carriers of fatty acyl groups within the desaturation system and, in fact, the latter was considered most likely[121]. Vijay and Stumpf[57] have obtained proof that the desaturation of oleate to linoleate by a microsomal system from safflower seeds takes place with oleic acid bound as the CoA thioester and not as an oxygen ester in phospholipids. They also observed a very active acyl transfer enzyme which transferred oleic acid and pre-formed linoleic acid from CoA specifically to phosphatidylcholine.

Since *trans*-3-hexadecenoic acid occurs specifically bound to phosphatidylglycerol, two possible mechanisms to explain this were suggested: (a) that the biosynthetic system for phosphatidylglycerol had a specific requirement for *trans*-3-hexadecenoic acid[128] or (b) that a desaturation of the palmitic acid moiety on the 2-position of phosphatidylglycerol occurred[129]. The first possibility was excluded when it was found that the exogenous *trans*-3-hexadecenoic was incorporated into all the major lipid classes[128] but the second mechanism has not yet been proved.

Thus, available evidence points to an intimate involvement of phospholipids in fatty acid conversions, but perhaps they serve as carriers at the site of desaturation rather than as substrates.

Role of phospholipids in other cellular processes

In addition to their possible roles as co-factors or carriers in fatty acid conversions, phospholipids of plant cells, as in other organisms, are involved in the enzymatic activities of membrane-bound enzymes. For example, re-addition of the lipids to lipid-depleted mitochondria from cauliflower restored the activity of NADH:cytochrome *c* oxido-reductase[130].

The rapid turnover of the glycerol moiety of phosphatidylglycerol in *Chlorella vulgaris* grown in $^{14}CO_2$ (ref. 36) prompted the suggestion[131] that this lipid played a dynamic role in the carbohydrate economy of photosynthesis. However, objections to this possibility were raised when it was shown that rapid turnover of the glycerol moiety does not occur in mature leaves[54]. The probable functions of nitrogenous phospholipids in the biosynthesis of free ethanolamine and choline were discussed earlier.

7. SOME PHYSIOLOGICAL ASPECTS OF LIPID METABOLISM

Many physiological processes in plants involve the net synthesis or breakdown of membrane structures and subcellular organelles and these account for changes observed in the phospholipid content of tissues during development, some of which were mentioned earlier. Obviously, the lipid content and the biosynthetic or catabolic processes observed are related to particular stages in the life cycles of plant tissues and also to environmental conditions.

An unusual process which is proving to be useful in the study of lipid metabolism in plants is the so-called 'ageing' of some excised tissues. When thin slices made from bulky storage tissues (such as potato and artichoke tubers, carrot and beet roots, etc.) are allowed to incubate aerobically, marked changes take place. This 'ageing' process is associated with large increases in respiration[132], RNA and protein synthesis[133], ion-uptake[134], fatty acid[135, 136] and lipid biosynthesis[42, 137]. The development of all these processes is subject to inhibition by inhibitors of protein synthesis. The physiological significance of these changes is not fully understood, but they have much in common with processes of wound-healing and breaking of dormancy and involve active renewal of endoplasmic reticulum and mitochondrial membranes. In excised tissues from unripe fruit, ageing phenomena involving altered lipid metabolism have also been observed[142, 143] and these mimic many of the changes which occur *in vivo* during fruit ripening.

Many investigations have been made on the incorporation of labelled substrates into the lipids of a wide variety of plant tissues. With the exception of actively developing photosynthetic tissues (in which galactolipids, phosphatidylglycerol, sulpholipid and pigments are rapidly synthesised) most studies have indicated that, of all the lipids, phospholipids show the highest turnover. Phosphatidylcholine and phosphatidylethanolamine are particularly active in higher plants and, although the fatty acyl groups frequently show the highest turnover rates, the glycerol and phosphate moieties of the lipids are also involved. In contrast, triglycerides, sterol-containing lipids, cerebrosides and pigments generally have slow turnover rates. These generalizations hold for developing tissues as well as for

References p. 284

tissues which are not showing rapid net changes in lipid composition and suggest dynamic roles for phospholipids in all plant cells.

Currently, investigations of the physiological and biochemical effects of plant growth hormones are being actively pursued in many laboratories but little is known of their effects on lipid metabolism. The ageing phenomena described above are thought to be induced by changes in hormone balance. Indolacetic acid induced changes in the phospholipid metabolism of oat coleoptiles, possibly by causing enhanced synthesis of mitochondrial membrane phospholipids[144].

RELEVANCE OF PHOSPHOLIPID METABOLISM IN EDIBLE PLANTS

Enzymic changes causing changes in the lipid content of edible plants are of importance in agriculture and the food industry. Edible foods undergo physiological and biochemical changes which may be detrimental to the quality of the product. These changes may be of practical importance between harvesting and processing, or in the case of unprocessed foods, during the storage life.

As far as phospholipids are concerned, most problems arise from the presence of lipolytic and oxidative enzymes. Damage or bruising of plants can cause cell rupture and increased activity of the enzymes. A major cause of off-flavour in many edible tissues is lipoxygenase which is involved in the production of odourous volatile carbonyl compounds from the polyunsaturated fatty acids of phospholipids and galactolipids. Lipolytic acyl hydrolases probably contribute to these processes by liberating free fatty acids. Freezing of tissues does not by itself eliminate enzyme activity (the phospholipids in unblanched frozen peas are hydrolysed at $-20°$)[78]. Brief heating (blanching) before freezing does prevent enzymatic breakdown. The deteriorative changes during storage of unprocessed foods are, as yet, less easily prevented.

REFERENCES

1 P.K. Stumpf, *Ann. Rev. Biochem.*, 38 (1969) 159.
2 B.W. Nichols and A.T. James, *Progr. Phytochem.*, 1 (1968) 1.
3 C. Hitchcock and B.W. Nichols, *Plant Lipid Biochemistry*, Academic Press, London, 1971.
4 M. Kates, *Advan. Lipid Res.*, 8 (1970) 225.
5 C.F. Allen, O. Hirayama and P. Good, in T.W. Goodwin (Ed.), *Biochemistry of Chloroplasts*, Vol. 1, Academic Press, New York, 1966 p. 195.
6 C.F. Allen, P. Good, H.F. Davies, P. Chisum and S.D. Fowler, *J. Am. Oil. Chem. Soc.*, 43 (1966) 223.
7 B.W. Nichols, R.V. Harris and A.T. James, *Biochem. Biophys. Res. Commun.*, 20 (1965) 256.
8 R.H. Quarles, N. Clarke and R.M.C. Dawson, *Biochem. Biophys. Res. Commun.*, 33 (1968) 964.
9 R. Aneja, J.S. Chadha and J.A. Knaggs, *Biochem. Biophys. Res. Commun.*, 36 (1969) 401.
10 R.A. Bomstein, *Biochem. Biophys. Res. Commun.*, 21 (1965) 49.
11 M. Kates and B.E. Volcani, *Biochim. Biophys. Acta*, 116 (1966) 264.
12 H. Debuch and E. Rotsch, *Z. Physiol. Chem.*, 347 (1966) 79.
13 H.E. Carter, D.R. Strobach and J.N. Hawthorne, *Biochemistry*, 8 (1969) 383.
14 L.L.M. van Deenen and F. Haverkate, in T.W. Goodwin (Ed.), *Biochemistry of Chloroplasts*, Vol. 1, Academic Press, London, 1966, p. 117.

15 F. Haverkate and L.L.M. van Deenen, *Biochim. Biophys. Acta*, 106 (1965) 78.
16 A.C. Chibnall and H.J. Channon, *Biochem. J.*, 21 (1927) 233.
17 M. Kates and F.M. Eberhardt, *Can. J. Botany*, 35 (1957) 895.
18 M. Lepage, *Lipids*, 2 (1967) 244.
19 W.D. Bonner, *Methods Enzymol.*, 10 (1967) 126.
20 J.W. Anderson, *Phytochemistry*, 7 (1968) 1973.
21 W.D. Loomis and J. Battaile, *Phytochemistry*, 5 (1966) 423.
22 R.L. Bieleski, *Anal. Biochem.*, 9 (1964) 431.
23 M. Friedlander and J. Neumann, *Plant Physiol.*, 43 (1968) 1249.
24 H. Ikuma, *Plant Physiol.*, 45 (1970) 773.
25 M. Kates, *Biochim. Biophys. Acta*, 41 (1960) 315.
26 P.S. Sastry and M. Kates, *Can. J. Biochem.*, 43 (1965) 1445.
27 E.J. Barron and P.K. Stumpf, *Biochim. Biophys. Acta*, 60 (1962) 329.
28 G.M. Cheniae, *Plant Physiol.*, 40 (1965) 235.
29 P.S. Sastry and M. Kates, *Can. J. Biochem.*, 44 (1966) 459.
30 H. Goldfine, *Ann. Rev. Biochem.*, 37 (1968) 303.
31 M. Mazelis and P.K. Stumpf, *Plant. Physiol.*, 30 (1955) 237.
32 C. Bradbeer and P.K. Stumpf, *J. Lipid Res.*, 1 (1960) 214.
33 S. Sumida and J.B. Mudd, *Plant. Physiol.*, 43 (1968) 1162.
34 R. Douce, *Compt. Rend., Ser. D*, 267 (1968) 534.
35 J. Bahl, T. Guillot-Salomon and R. Douce, *Physiol. Vég.*, 8 (1970) 55.
36 S. Sumida and J.B. Mudd, *Plant Physiol.*, 45 (1970) 719.
37 R.H. Quarles and R.M.C. Dawson, *Biochem. J.*, 112 (1969) 787.
38 R.A. Ferrari and A.A. Benson, *Arch. Biochem. Biophys.*, 93 (1961) 185.
39 R. Douce and J. Dupont, *Compt. Rend., Ser. D*, 268 (1969) 1657.
40 S. Sumida and J.B. Mudd, *Plant Physiol.*, 45 (1970) 712.
41 C. Willemot and W.G. Boll, *Can. J. Botany*, 45 (1967) 1863.
42 W.J. Tang and P.A. Castelfranco, *Plant Physiol.*, 43 (1968) 1232.
43 K. Tanaka, N.E. Tolbert and A.F. Gohlke, *Plant Physiol.*, 41 (1966) 307.
44 D.J. Morré, S. Nyquist and E. Rivera, *Plant Physiol.*, 45 (1970) 800.
45 J.V. Maizel, A.A. Benson and N.E. Tolbert, *Plant Physiol.*, 31 (1956) 407.
46 K.A. Devor and J.B. Mudd, *J. Lipid Res.*, 12 (1971) 403.
47 A. Gorchein, A. Neuberger and G.H. Tait, *Proc. Roy. Soc. (London), Ser. B*, 170 (1968) 299.
48 S.L. Vandor and K.E. Richardson, *Can. J. Biochem.*, 46 (1968) 1309.
49 A.A. Benson, S. Freer and S.F. Yang, *Proc. 9th Intern. Conf. Biochem. Lipids,* Noordwijk aan Zee (The Netherlands) 1965.
50 R. Douce, M. Faure and J. Maréchal, *Compt. Rend., Ser. C*, 262 (1966) 1549.
51 S.F. Yang, S. Freer and A.A. Benson, *J. Biol. Chem.*, 242 (1967) 477.
52 R.M.C. Dawson, *Biochem. J.*, 102 (1967) 205.
53 R.O. Weenink and F.B. Shorland, *Biochim. Biophys. Acta*, 84 (1964) 613.
54 P.G. Roughan, *Biochem. J.*, 117 (1970) 1.
55 C.T. Bartels and L.L.M. van Deenen, *Biochim. Biophys. Acta*, 125 (1966) 395.
56 K.A. Devor and J.B. Mudd, *J. Lipid Res.*, 12 (1971) 412.
57 I.K. Vijay and P.K. Stumpf, *J. Biol. Chem.*, 246 (1971) 2910.
58 K.A. Devor and B.J. Mudd, *J. Lipid Res.*, 12 (1971) 396.
59 T. Galliard, *Biochem. J.*, 121 (1971) 379.
60 E. Heinz, *Biochim. Biophys. Acta*, 144 (1967) 333.
61 M. Kates, *Can. J. Biochem. Physiol.*, 33 (1955) 575.
62 R. Calvayrac and R. Douce, *FEBS Letters*, 7 (1970) 259.
63 D.J. Hanahan and I.L. Chaikoff, *J. Biol. Chem.*, 168 (1947) 233; *ibid.*, 169 (1947) 699.
64 M. Kates, *Can. J. Biochem. Physiol.*, 32 (1954) 571.
65 F.M. Davidson and C. Long, *Biochem. J.*, 69 (1958) 458.
66 H.L. Tookey and A.K. Balls, *J. Biol. Chem.*, 218 (1956) 213.
67 H. Clermont and R. Douce, *FEBS Letters*, 9 (1970) 284.
68 M. Heller, E. Aladjem and B. Shapiro, *Bull. Soc. Chim. Biol.*, 50 (1968) 1395.
69 R.M.C. Dawson and N. Hemington, *Biochem. J.*, 102 (1967) 76.

70 G.B. Ansell and J.N. Hawthorne, in *Phospholipids*, Elsevier, Amsterdam, 1964, p. 163.
71 M. Kates, in K. Bloch (Ed.), *Lipid Metabolism*, Wiley, New York, 1960, p. 165.
72 G.H. de Haas, P.P.M. Bonsen and L.L.M. van Deenen, *Biochim. Biophys. Acta*, 116 (1966) 114.
73 P.P.M. Bonsen, G.H. de Haas and L.L.M. van Deenen, *Biochim. Biophys. Acta*, 106 (1965) 93.
74 R.H. Quarles and R.M.C. Dawson, *Biochem. J.*, 112 (1969) 795.
75 S.R. Draper, *Phytochemistry*, 8 (1969) 1641.
76 T. Galliard, unpublished observations.
77 T. Galliard, *Phytochemistry*, 7 (1968) 1915.
78 B. Bengtsson and I. Bosund, *J. Food Sci.*, 31 (1966) 474.
79 R.C. Jordan and A.C. Chibnall, *Ann. Botany (London)*, 47 (1933) 163.
80 P. Desnuelle and P. Savary, *J. Lipid Res.*, 4 (1963) 369.
81 P.S. Sastry and M. Kates, *Biochemistry*, 3 (1964) 1280.
82 P.J. Helmsing, *Biochim. Biophys. Acta*, 144 (1967) 470.
83 P.J. Helmsing, *Biochim. Biophys. Acta*, 178 (1969) 519.
84 A. Contardi and A. Ercoli, *Biochem. Z.*, 261 (1933) 275.
85 L. Acker and H. Bücking, *Z. Lebensm. Untersuch. Forsch.*, 104 (1956) 423.
86 L. Acker and K. Müller, *Nahrung*, 9 (1965) 3.
87 L. Acker and J. Geyer, *Z. Lebensm. Untersuch. Forsch.*, 137 (1968) 231.
88 L. Acker and J. Geyer, *Z. Lebensm. Untersuch. Forsch.*, 140 (1969) 269.
89 T. Galliard, *Phytochemistry*, 7 (1968) 1907.
90 T. Galliard, *Phytochemistry*, 9 (1970) 1725.
91 T. Galliard, *European J. Biochem.*, 21 (1971) 90.
92 A.D. Bangham and R.W. Horne, *J. Mol. Biol.*, 8 (1964) 660.
93 L.L.M. van Deenen, *Progr. Chem. Fats Lipids*, 8 (1966) 1.
94 P.K. Stumpf, J. Brooks, T. Galliard, J.C. Hawke and R.D. Simoni, in T.W. Goodwin (Ed.), *Biochemistry of Chloroplasts*, Vol. II, Academic Press, New York, 1967, p. 213.
95 R. Douce, Thèse Doctorat, Paris (1970), quoted in R. Douce and T. Guillot-Salomon, *FEBS Letters*, 11 (1970) 121.
96 L.A. Appelqvist, P.K. Stumpf and D. von Wettstein, *Plant Physiol.*, 43 (1968) 163.
97 A. Ongun and J.B. Mudd, *J. Biol. Chem.*, 243 (1968) 1558.
98 J.B. Mudd, H.H.D.M. van Vliet and L.L.M. van Deenen, *J. Lipid Res.*, 10 (1969) 623.
99 R. Douce and T. Guillot-Salomon, *FEBS Letters*, 11 (1970) 121.
100 P. Mazliak and A. Ben Abdelkader, *Rev. Gén. Botan.*, 77 (1970) 53.
101 R. Douce, T. Guillot-Salomon, C. Lance and M. Signol, *Bull. Soc. Franc. Physiol. Végétale*, 14 (1968) 351.
102 K.W.A. Wirtz and D.B. Zilversmit, *J. Biol. Chem.*, 243 (1968) 3596.
103 W.C. McMurray and R.M.C. Dawson, *Biochem. J.*, 112 (1969) 91.
104 F.B. Jungalwala and R.M.C. Dawson, *European J. Biochem.*, 12 (1970) 399.
105 P. Mazliak, U. Stoll and A. Ben Abdelkader, *Biochim. Biophys. Acta*, 152 (1968) 414.
106 A. Ben Abdelkader and P. Mazliak, *European J. Biochem.*, 15 (1970) 250.
107 B.W. Nichols, A.T. James and J. Breuer, *Biochem. J.*, 104 (1967) 486.
108 B.W. Nichols, *Biochim. Biophys. Acta*, 106 (1965) 274.
109 J.A. Erwin, D. Hulanicka and K. Bloch, *Comp. Biochem. Physiol.*, 12 (1964) 191.
110 R.W. Holton, H.H. Blecker and M. Onore, *Phytochemistry*, 3 (1964) 595.
111 I. Shibuya and B. Maruo, *Nature*, 207 (1965) 1096.
112 A.A. Benson, *J. Am. Oil Chem. Soc.*, 43 (1966) 265.
113 T.E. Weier and A.A. Benson, in T.W. Goodwin (Ed.), *Biochemistry of Chloroplasts*, Vol. 1, Academic Press, New York, 1966, p. 91.
114 T.E. Weier and A.A. Benson, *Am. J. Botany*, 54 (1967) 389.
115 J.M. Lyons, T.A. Wheaton and H.K. Pratt, *Plant Physiol.*, 39 (1964) 262.
116 J.M. Lyons and J.K. Raison, *Plant Physiol.*, 45 (1970) 386.
117 T.P. Hilditch and P.N. Williams, in *The Chemical Constitution of Natural Fats*, 4th ed., Chapman and Hall, London, 1964.
118 P.J.C. Kuiper, *Plant Physiol.*, 45 (1970) 684.
119 M.G. Kleinschmidt and V.A. McMahon, *Plant Physiol.*, 46 (1970) 286.

120 M.G. Kleinschmidt and V.A. McMahon, *Plant Physiol.*, 46 (1970) 290.
121 B.W. Nichols, A.T. James and J. Breuer, *Biochem. J.*, 104 (1967) 486.
122 B.W. Nichols, *Lipids*, 3 (1968) 354.
123 C.D. Dybing and B.M. Craig, *Lipids*, 5 (1970) 422.
124 M.I. Gurr, M.P. Robinson and A.T. James, *European J. Biochem.*, 9 (1969) 70.
125 M.I. Gurr and P. Brawn, *European J. Biochem.*, 17 (1970) 19.
126 M.I. Gurr and B.W. Nichols, personal communication.
127 A.E. Chung and J.H. Law, *Biochemistry*, 3 (1964) 967.
128 B.W. Nichols, P. Harris and A.T. James, *Biochem. Biophys. Res. Commun.*, 21 (1965) 473.
129 C.T. Bartels, A.T. James and B.W. Nichols, *European J. Biochem.*, 3 (1967) 7.
130 A. Jolliot and P. Mazliak, *Compt. Rend., Ser. D*, 272 (1971) 246.
131 A.A. Benson, *Ann. Rev. Plant Physiol.*, 15 (1964) 1.
132 G.G. Laties, in B. Wright (Ed.), *Control Mechanisms in Respiration and Fermentation*, Ronald, New York, 1963, p. 129.
133 R.E. Click and D.P. Hackett, *Proc. Natl. Acad. Sci. (U.S.)*, 50 (1963) 243.
134 R.L. Bieleski and G.G. Laties, *Plant Physiol.*, 38 (1963) 586.
135 C. Willemot and P.K. Stumpf, *Can. J. Botany*, 45 (1967) 579.
136 C. Willemot and P.K. Stumpf, *Plant Physiol.*, 42 (1967) 391.
137 A. Ben Abdelkader, *Physiol. Végétale*, 6 (1968) 417.
138 M.E. Jackman and R.F.M. van Steveninck, *Austral. J. Biol. Sci.*, 20 (1967) 1063.
139 M. Cotte-Martinon, M. Diano, D. Meunier and G. Ducet, *Compt. Rend., Ser. D*, 269 (1969) 827.
140 A. Ben Abdelkader, P. Mazliak and A.M. Catesson, *Phytochemistry*, 8 (1969) 1121.
141 S.G. Lee and R.M. Chasson, *Physiol. Plantarum*, 19 (1966) 199.
142 T. Galliard, M.J.C. Rhodes, L.S.C. Wooltorton and A.C. Hulme, *Phytochemistry*, 7 (1968) 1453.
143 A.C. Hulme, M.J.C. Rhodes, T. Galliard and L.S.C. Wooltorton, *Plant Physiol.*, 43 (1968) 1154.
144 R.P. Zimmerer and R.H. Hamilton, *Plant Cell Physiol. (Tokyo)*, 6 (1965) 681.
145 A.A. Benson and B. Maruo, *Biochim. Biophys. Acta*, 27 (1958) 189.
146 O. Hirayama and T. Suzuki, *Agr. Biol. Chem. (Tokyo)*, 32 (1968) 549.
147 P.G. Roughan and R.D. Batt, *Phytochemistry*, 8 (1969) 363.
148 J.F.G.M. Wintermans, *Biochim. Biophys. Acta*, 44 (1960) 49.
149 A. Trémolières and P. Mazliak, *Physiol. Végétale*, 8 (1970) 135.
150 P.S. Sastry and M. Kates, *Biochemistry*, 3 (1964) 1271.
151 U. Beiss, *Landwirtsch. Forsch.*, 23/1 (1969) 198.
152 M. Lepage, *Lipids*, 3 (1968) 477.
153 P. Mazliak, A. Ben Abdelkader and A.-M. Catesson, *Physiol. Végétale*, 5 (1967) 237.
154 D.L. McIlwain and C.E. Ballou, *Biochemistry*, 5 (1966) 4054.
155 A.C. Thompson, R.D. Henson, J.P. Minyard and P.A. Hedin, *Lipids*, 3 (1968) 373.
156 M.M. Paulose, S. Venkob Rao and K.T. Achaya, *Indian J. Chem.*, 4 (1966) 529.
157 H. Wagner and P. Wolff, *Fette, Seifen, Anstrichmittel*, 66 (1964) 425.
158 B. Vijayalakshmi, S. Venkob Rao and K.T. Achaya, *Fette, Seifen, Anstrichmittel*, 71 (1969) 757.
159 L.A. Shustanova, A.U. Umaron and A.L. Markham, *Khim. Prir. Soedin.*, 3 (1970) 292.
160 J.G. Parsons, P.G. Keeney and S. Patton, *J. Food Sci.*, 34 (1969) 497.
161 D.R. Thomas and A.K. Stobart, *J. Exptl. Botany*, 21 (1970) 274.
162 F. Haverkate and L.L.M. van Deenen, *Proc. Koninkl. Ned. Akad. Wetensch., Ser. B*, 68 (1965) 141.
163 C.F. Allen, P. Good and R.W. Holton, *Plant Physiol.*, 46 (1970) 748.
164 O. Hirayama, *J. Biochem. (Tokyo)*, 61 (1967) 179.
165 F. Haverkate, F.A.G. Teulings and L.L.M. van Deenen, *Proc. Koninkl. Ned. Akad. Wetensch., Ser. B*, 68 (1965) 154.
166 O. Hirayama, *Agr. Biol. Chem. (Tokyo)*, 32 (1968) 34.
167 C.-E. Park and L.R. Berger, *J. Bacteriol.*, 93 (1967) 221.
168 S. Steiner, S.F. Conti and R.L. Lester, *J. Bacteriol.*, 98 (1969) 10.
169 A. Gorchein, *Proc. Roy. Soc. (London), Ser. B*, 170 (1968) 279.
170 A. Ongun, W.W. Thomson and J.B. Mudd, *J. Lipid Res.*, 9 (1968) 409.
171 J.A. McArthur, T.V. Marsho and D.W. Newman, *Plant Physiol.*, 39 (1964) 551.

172 B.W. Nichols and A.T. James, *Fette, Seifen, Anstrichmittel*, 66 (1964) 1003.
173 A.T. James and B.W. Nichols, *Nature*, 210 (1966) 372.
174 B.W. Nichols, *Phytochemistry*, 4 (1965) 769.
175 H. Debuch, *Z. Physiol. Chem.*, 306 (1957) 279.
176 R. Safford and B.W. Nichols, *Biochim. Biophys. Acta*, 210 (1970) 57.
177 R. Douce, *Compt. Rend. Ser. D*, 272 (1971) 3146.

Chapter 11

PHOSPHOLIPIDS IN MICRO-ORGANISMS*

RICHARD T. AMBRON** and RONALD A. PIERINGER***

Department of Biochemistry, Temple University School of Medicine, Philadelphia, Pa. (U.S.A.)

1. PHOSPHATIDIC ACID

Studies on the metabolism of phosphatidic acid in animals have firmly placed this phospholipid in the key position of being the precursor *de novo* of all other glyceride-containing lipids[1]. More recent investigations with micro-organisms have demonstrated that the metabolism of phosphatidic acid in micro-organisms is essentially the same, except for the one notable difference discussed below. The fundamental metabolic role of phosphatidic acid in micro-organisms was established by showing that glycerol-3-phosphate and not free glycerol served as the acceptor-substrate of the fatty acyl moiety in a fatty acyltransferase reaction catalyzed by particulate preparations from *E. coli*[2,3]. Thus lysophosphatidic acid and not monoglyceride is the first lipid formed *de novo* from a water-soluble, carbohydrate-derived metabolite (Fig. 1, reaction A and B)[2–6]. Lysophosphatidic acid is then acylated to phosphatidic acid[2,5,6]. These findings indicate

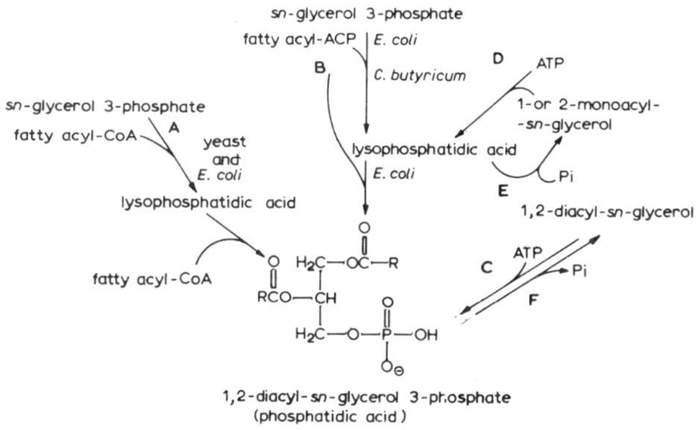

Fig. 1. Phosphatidic acid biosynthesis in bacteria.

---

*Supported by grants from the National Science Foundation, the N.I.A.I.D. (AI-05730), and the N.I.C.H.D. (1-PO1-HD-05874-01).

**Present adress: Department of Neurobiology and Behavior, Public Health Research Institute, New York, N.Y. 10016 (U.S.A.).

***Recipient of Career Development Award 1K4 NS 12,776-01AI from the National Institute of Neurological Diseases and Stroke.

that the monoglyceride and diglyceride phosphokinase pathways do not function in the synthesis *de novo* of lysophosphatidic and phosphatidic acids even though they are relatively active and stereospecific reactions in bacteria[7, 8, 9] (Fig. 1, C and D). The lack of participation of the glyceride phosphokinases in the net synthesis of phosphatidic acid has been corroborated by the *in vivo* studies of Chang and Kennedy[9], who detected a small but metabolically inactive pool of glyceride in logarithmically growing cells of *E. coli.*

Perhaps the greatest difference of the microbial pathway from the animal pathway for phosphatidic acid biosynthesis is the replacement of the coenzyme A moiety of the fatty acyl-CoA substrate with acyl carrier protein (ACP) in some micro-organisms. Goldfine, Ailhaud, and Vagelos[3] demonstrated that the acylation of glycerol 3-phosphate by a particulate enzyme preparation from *Clostridium butyricum* had a high requirement for the acyl carrier protein derivative (Fig. 1, B). Fatty acyl-CoA did not function as an acyl donor under conditions in which fatty acyl-acyl carrier protein was a suitable substrate. The acyltransferase of *C. butyricum* displayed an even higher degree of specificity in that fatty acyl-acyl carrier protein isolated from *E. coli* functioned as a substrate only about one-third as well as the fatty acyl-acyl carrier protein from *C. butyricum*[3]. The major lipid produced in the reaction between glycerol 3-phosphate and fatty acyl-acyl carrier protein in this bacterium was lysophosphatidic acid; no phosphatidic acid was detected. The nature of the fatty acyl-donating substrate that converts lysophosphatidic acid to phosphatidic acid in *C. butyricum* is still unknown[3].

In marked contrast to *C. butyricum* the glycerol 3-phosphate acyltransferase system of yeast requires the coenzyme A derivative of the fatty acid-donating substrate[4] (Fig. 1, A). The fatty acid synthetase of yeast differs from that in most bacteria as it is a multi-enzyme complex to which the fatty acyl intermediates are bound. Thus the 'acyl carrier protein' of yeast is a component of the multi-enzyme complex[10]. To test the possibility of a direct transfer of fatty acid moiety from the multi-enzyme synthetase to glycerol 3-phosphate, Kuhn and Lynen[4] charged purified fatty acid synthetase with palmitate in the complete absence of coenzyme A by using malonyl – and acetyl-pantetheine as fatty acid precursors instead of malonyl – and acetyl-CoA. Under these conditions the palmitate is generated on the synthetase complex and released as the free acid; whereas in the presence of CoA palmitoyl-CoA is the terminal product[10]. When radioactive glycerol 3-phosphate and the fatty acyltransferase enzyme were added to the fatty acid synthetase charged with palmitate, no radioactive glycerol-3-phosphate was incorporated into phospholipid unless CoA was added to the incubation mixture. Kuhn and Lynen[4] concluded from this experiment that palmitoyl-CoA is an obligatory intermediate in the transfer of palmitate from the synthetase to the acyltransferase system, and that 'palmitoyl-synthetase' is not the immediate donor of fatty acid to glycerol 3-phosphate.

The glycerol 3-phosphate: fatty acyltransferase of *E. coli* is probably the most extensively studied of all the microbial forms of this enzyme. In *E. coli* the acyltransferase which is a particulate enzyme, uses both fatty acyl-CoA[11, 12, 2, 5, 6] and fatty acyl-acyl carrier protein[12, 6] as substrates. With fatty acyl-CoA as the substrate both lysophosphatidic acid and phosphatidic acid are formed in the acyltransferase reaction (refs. 2, 5, 6, 11, 12). However initial studies by Ailhaud and Vagelos[12] using the saturated palmitoyl-acyl carrier protein as the substrate, showed that the major lipid formed was monoglyceride. Subsequent studies proved that lysophosphatidic acid was the initial product formed from glycerol 3-phosphate and fatty acyl-acyl carrier protein, but that it was actively dephosphorylated to monoglyceride by a relatively specific phosphatase in the crude particulate enzyme preparation[6] (Fig. 1, reaction E). The ability of acyl-CoA to convert lysophosphatidic acid to phosphatidic acid was shown not to be due to a greater inhibition of the phosphatase by the acyl-CoA over that of the acyl-acyl carrier protein substrate[6]. Both substrates yielded approximately equal amounts of monoglyceride[6]. Rather was the difference in the type of lipid formed due to the inability of the acyltransferase to accept a saturated fatty acyl-acyl carrier protein substrate in catalyzing the acylation of lysophosphatidic acid to phosphatidic acid[6]. When the coenzyme A derivative was the substrate the selectivity of acyltransferase of the second step in phosphatidic acid biosynthesis was less pronounced. In fact, Pieringer, Bonner, and Kunnes[2] found little difference either in the amount of glycerol 3-phosphate incorporated into phospholipid or in the ratio of the lysophosphatidic acid to phosphatidic acid products, when mixtures of oleoyl-CoA and palmitoyl-CoA in variable ratio concentrations were compared with oleoyl-CoA and palmitoyl-CoA taken separately over an entire concentration range. Using the same substrates, Ray et al.[15] obtained the same results. However, upon closer examination of the lysophosphatidic acid product these workers[15] found the oleoyl-CoA substrate yielded almost exclusively the 2-isomer (2-monoleoyl-*sn*-glycerol 3-phosphate) and the palmitoyl-CoA-yielded (92%) the 1-isomer (1-monopalmitoyl-*sn*-glycerol 3-phosphate). Investigation of other acyl-CoA substrates *in vitro* revealed a positional distribution pattern remarkably consistent with the distribution of the fatty acid moieties of phospholipids derived from whole cells of *E. coli*[15].

Through the use of temperature sensitive mutants of *E. coli* Vagelos and coworkers[15] have been able to demonstrate convincingly that the acylation of glycerol 3-phosphate is catalyzed by an enzyme quite different from the enzyme that catalyzes the acylation of lysophosphatidic acid. The lysophosphatidic acid: acyltransferase, which previously had been observed to have different pH optima and magnesium ion concentration requirements[6] from the other acyltransferase, was not thermolabile in three mutants of *E. coli* under conditions in which the glycerol 3-phosphate: acyltransferase activity deteriorated rapidly at 37° (Ref. 15). The thermolability was the same with various saturated and unsaturated fatty acyl-CoA substrates. In the case of the glycerol 3-phosphate:acyltransferase, this result suggests that one enzyme catalyzes acyl transfer to either position 2 or to position 1 of glycerol 3-phosphate.

The ability of the acyltransferase to incorporate oleic acid, as well as lauric, stearic, linoleic, and *trans*-vaccenic (fatty acids not usually found in the phospholipids of *E. coli*), suggests that it is not the acyltransferase that regulates the type of fatty acids found in *E. coli* grown under normal conditions[2]. Its specificity seems to be to direct a specific fatty acid to the 1- or to the 2-position of the glycerol-3-phosphate[15].

Although the dephosphorylation of lysophosphatidic acid and phosphatidic acid by

References p. 327

phosphatase activity in *E. coli* had been known for some time[7], the specificity and properties of this enzyme had not been studied. Van den Bosch and Vagelos[6] observed that the phosphatase of lysophosphatidic acid and phosphatidic acid was not the same as the phosphatidylglycerophosphate phosphatase observed by Chang and Kennedy[16]; nor was it the same as the alkaline phosphatase of *E. coli*. Thus phosphatidic acid seems to be attacked by a relatively specific phosphatase. Since phosphatidic acid and not diglyceride is the precursor *de novo* of all phospholipids of *E. coli* (a situation which does not occur in animals), the function of these phosphatase activities at first appears uncertain. However in Gram-positive bacteria this reaction would have great significance since diglyceride is the required precursor of the glycosyl diglycerides[17] and the glycerophosphoryl diglycosyl diglycerides[17, 18]. Further study may reveal that these glycolipids exist in Gram-negative bacteria as well; for example, they recently have been found in Pseudomonas[19]. The production of diglyceride by the phosphatase may provide some clue as to the need for the existence of the glyceride phosphokinase[7]. Under certain conditions of over-production of diglyceride and under-production of phosphatidic acid *de novo* the diglyceride phosphokinase[7] might be needed for the regulation of the proper levels of these metabolites. The partial inhibition of the glycerol 3-phosphate : fatty acyltransferase by ATP[5], the substrate of the glyceride phosphokinase, offers some intriguing possibilities on the regulation of these pathways. During the growth of *Haemophilus parainfluenza* White and Tucker[20] found the phosphate of phosphatic acid turned over about 2.5 times faster than the glycerol moiety. The most likely enzymes responsible for this difference in the turnover of these two moieties are the phosphatase and glyceride phosphokinase.

2. CDPDIGLYCERIDE

The occurrence and metabolic role of CDPdiglyceride in animal systems was first shown more than a decade ago by Agranoff, Bradley and Brady[21] and by Paulus and Kennedy[22]. In animal systems CDPdiglyceride is the direct precursor of phosphatidylglycerol, phosphatidylinositol and diphosphatidylglycerol, but not of phosphatidylethanolamine, phosphatidylcholine, or phosphatidylserine. However, in bacterial metabolism the role of CDPdiglyceride as an intermediary metabolite takes on added significance in that all of the common phospholipids of bacteria are formed *de novo* from this nucleotide-linked lipid intermediate. The studies of Carter[23] on *E. coli*, of McCaman and Finnerty[24] on *Micrococcus cerificans* and of Patterson and Lennarz[25] on *Bacillus PP* have shown that particulate-membrane fractions from each bacterium will catalyze the formation of CDP-diglyceride from CTP and phosphatidic acid.

$$\text{phosphatidic acid} + \text{CTP} \xrightarrow{Mg^{2\oplus}} \text{CDPdiglyceride} + \text{PPi}$$

All three enzyme systems displayed a requirement for bivalent cation (best fulfilled by $Mg^{2\oplus}$), and an exogenous supply of substrates. The *M. cerificans* and *Bacillus PP* CTP-phosphatidate cytidylyltransferase but not the *E. coli* enzyme, was stimulated by $K^{\oplus}$ ion and required the presence of nonionized detergent Cutscum or Triton-X-100. The activity of the particulate enzyme from *M. cerificans* was high and its use for the synthesis of CDP-diglyceride might be considered as an alternative to chemical procedures[26].

3. PHOSPHATIDYLGLYCEROPHOSPHATE, PHOSPHATIDYLGLYCEROL, AND DIPHOSPHATIDYLGLYCEROL

The pulse-chase studies of Kanfer and Kennedy[27] with ^{32}P proved the presence of phosphatidylglycerol in *Escherichia coli* and showed that the turnover of ^{32}P in this phospholipid occurred at a relatively high rate. Such results, which have been confirmed by others[28] suggested a dynamic metabolic role for phosphatidylglycerol. Subsequent work *in vitro*[29] demonstrated that particulate enzyme preparations of *E. coli* catalyzed the synthesis of phosphatidylglycerol in a two-step reaction in which phosphatidylglycerophosphate participated as an intermediate and CDPdiglyceride and *sn*-glycerol 3-phosphate were the substrates. This *E. coli* enzyme system which has been demonstrated recently in ghost membranes of Baccilli[25], proved to be analogous to the sequence of reactions catalyzed by enzymes in liver mitochondria[30]. In *E. coli* the two enzymes have been partially purified and separated from each other on the basis of differential extraction into solutions containing Triton-X-100. The phosphatidylglycerophosphate-synthesizing enzyme was extractable from particulate, cell-envelope fractions with buffer–Triton-X-100. But the phosphatidylglycerophosphate phosphatase was extractable from the same particulate fractions only when EDTA was also present in the Triton-X-100 solution[31, 32]. The partially purified phosphatase was relatively specific for phosphatidylglycerophosphate and was clearly different from the phosphatidic acid and alkaline phosphatases of *E. coli*[32].

Crude particulate enzyme fractions from *E. coli* also catalyzed the synthesis of the next higher homologue, diphosphatidylglycerol (cardiolipin)[33], in this series. Phosphatidylglycerol and CDPdiglyceride were the direct precursors of this lipid*. (See p. 331.)

The work of Ono and White[34] has uncovered a very active phospholipase D-type enzyme in *Haemophilus parainfluenza* that is apparently specific for diphosphatidylglycerol. The products of the reaction catalyzed by this membrane-bound enzyme were phosphatidylglycerol and phosphatidic acid. One feature of the enzyme that is of interest and remains to be studied is its stereospecificity. Which side of the internal glycerol of the diphosphatidylglycerol does the enzyme attack? It is possible that a phosphatidylglycerol might be produced in which the acylated and unacylated glycerol moieties are both linked at the *sn*-3-position through a phosphodiester bond. The stereochemistry of this phosphatidylglycerol would be different from that synthesized *de novo* in which the unacylated glycerol is bonded at the *sn*-1-position and the acylated glycerol is linked to the phosphodiester bond at the *sn*-3-position[30].

4. PHOSPHATIDYLETHANOLAMINE

Phosphatidylethanolamine, the presence of which in bacteria was shown by Law and coworkers[35], is found in high concentration (often the highest concentration of all phospholipids) in a great variety of bacteria (ref. 36 and references listed in 36). This predominance is maintained in many organisms at all stages of growth[36]. Turnover studies by pulse-chase labeling demonstrated phosphatidylethanolamine to be one of the most metabolically stable lipids found in bacteria[20,27,28,37]. In a more detailed study on *Haemophilus parainfluenzae*, White and Tucker[20] measured the turnover of the component parts of phosphatidylethanolamine during growth. They found that whereas the total amount of phosphatidylethanolamine doubled in 42 min, half the radioactivity in the ethanolamine moiety disappeared in 114 min (2.7 doublings), and half from the phosphate in 120 min (2.8 doublings). However, the glycerol did not turn over in 4.3 doublings. Thus although phosphatidylethanolamine may be classified in general as a relatively stable lipid, it is not completely static in that different parts of phosphatidylethanolamine do turn over, but to varying degress. This study, which showed the moieties of other phospholipids also turn over at different rates, emphasizes the need to recognize the variable turnover of component parts of lipids and stresses the importance of specific degradative enzymes in the overall metabolism of phospholipids even during growth.

Studies *in vitro* on specific hydrolytic enzymes have revealed the presence of phospholipase A_1, A_2, and C activities in *Escherichia coli*[38,39]. Proulx and Van Deenen[38] demonstrated that cell-free enzyme preparations of *E. coli* under the proper incubation conditions released diglyceride and phosphorylethanolamine from 1-([9, 10-$^3H_2$] stearoyl)-2-([1-^{14}C] linoleoyl)-*sn*-glycero-3-phosphorylethanolamine. The ratio of ^3H to ^{14}C in the diglyceride product was the same as in the phosphatidylethanolamine substrate[38] indicating that the diglyceride was produced directly from the substrate and not by a resynthesis mechanism. Phospholipase A_1 activity could be made almost free of A_2 activity if the incubation containing broken cells of *E. coli* was carried out at pH 5 (ref. 39). However, A_2 activity could not be separated from phospholipase A_1. Lysophosphatidylethanolamine, the product of phospholipase A activity, has been tested as a possible acceptor of fatty acids in reactions catalyzed by an acyltransferase of *E. coli*[40]. Taylor and Heath[40] found that 2-acyl-*sn*-glycero-3-phosphorylethanolamine incorporated β-hydroxydecanoate, β-hydroxylaurate or β-hydroxymyristate specifically at the C-1 position. The fatty acids functioned as substrates only when in the form of an acyl carrier protein derivative. The acyltransferase catalysing this reaction was freed from particulate-matter by sonication. These studies *in vitro* on the degradation and reacylation of phosphatidylethanolamine provide insight into the mechanism of turnover of the component parts of phospholipids.

The enzymatic formation of phosphatidylethanolamine in bacteria occurs by a pathway that is uniquely different from the pathway that exists in animal systems. In bacteria Kanfer and Kennedy[29] demonstrated phosphatidylserine to be an obligatory intermediate in the formation of phosphatidylethanolamine. The synthesis *de novo* of phosphatidylserine in bacteria is brought about by a direct interaction between CDPdiglyceride and L-serine[25,29] and the phosphatidylserine so produced is then decarboxylated to phosphatidylethanolamine and CO_2.

CDPdiglyceride + L-serine → phosphatidylserine + CMP
phosphatidylserine → phosphatidylethanolamine + CO_2

This sequence of reactions has been thoroughly studied in representatives of Gram-negative (*E. coli*)[29] and Gram-positive (Bacilli)[25] bacteria. CDPethanolamine does not function as a precursor of phosphatidylethanolamine in bacteria as it does in animal systems[29]: on the other hand L-serine will not interact with CDPdiglycerides in animals[41] and phosphatidylserine is formed by an exchange reaction (Chapter 4, p. 73).

This variance between the animal and bacterial systems for phosphatidylserine formation provided the basis of an interesting study by Dennis and Kennedy[41] on the pathway for biosynthesis of this phospholipid in protozoa. Their results demonstrated that *Tetrahymena pyriformis* formed phosphatidylserine in a manner completely analogous to the higher animal systems, and support the classification of protozoa as a lower form of animal life.

5. PHOSPHATIDYLCHOLINE

Phosphatidylcholine is a widely distributed and quantitatively important phospholipid in animals and higher plants, but is generally not present among the bacteria. Only among the orders Pseudomonadales (Athiorhodacae, Thiobacteriaceae) and Eubacteriales (Rhizobiacae) are there genera in which several species contain phosphatidylcholine. Geiger and Anderson[42] identified choline in hydrolysates of the lipids from *Agrobacterium tumefaciens*, but it was not until almost a quarter-century later that Kaneshiro and Marr[43] demonstrated phosphatidylcholine in this organism. In a subsequent investigation Law et al.[44] showed that the label from L-[*Me*-^{14}C] methionine could be incorporated *in vivo* into the phosphatidyl-*N*-methylethanolamine, phosphatidyl-*N*, *N*-dimethylethanolamine, and phosphatidylcholine of *A. tumefaciens*, suggesting the presence of a methyltransferase pathway. With the finding of phosphatidylcholine in *A. rhizogenes* and *A. radiobacter*[45] it appears that this lipid may be common to the genus Agrobacteriae. Numerous investigations into the lipids of the photosynthetic bacteria[46–51] have revealed that phosphatidylcholine is also a major constituent of the phospholipids among members of the genus Rhodopseudomonae. Several studies of *Rhodopseudomonas spheroides*, for instance, have indicated that phosphatidylcholine accounts for up to 23% of the lipid phosphorus of whole cells (Table 1). Lascelles and Szilagyi[47] showed that ^{32}P could be incorporated into the phospholipids of *R. spheroides* NICB 8253 and noticed that the turnover of phosphatidylcholine was quite slow in comparison with the other phospholipids. Similar results were obtained by Gorchein et al.[52] using the same strain. Wood et al.[49] investigated several species of Rhodopseudomonas and found phosphatidylcholine in *R. spheriodes*, *R. capsulata*, and *R. palustris*, but it was seemingly absent in *R. gelatinosa* and *R. rubrum* (Table 1). Why some members of a given genus have phosphatidylcholine as a major lipid constituent, while it is not present in others, has not yet been satisfactorily explained. Ikawa[53] has suggested that phosphatidylcholine represents a biochemically advanced form of phospholipid which might be associated with bacteria possessing efficient electron transport chains. Hagen et al.[54] on the other hand, suggest a correlation between bacterial phosphatidylcholine and the presence of intracytoplasmic membranes. The Athiorhodacae discussed above are known to possess complex internal membrane systems

TABLE 1

OCCURRENCE OF METHYL DERIVATIVES OF PHOSPHATIDYLETHANOLAMINE IN BACTERIA

	N-methyl *PE*	*N,N-dimethyl* *PE*	*PC*	*Ref.*
Rhodopseudomonas spheroides	–	–	19.2[a]	46
Rhodopseudomonas spheroides	–	–	22[a]	47
R. spheroides (Chromatophores)	–	–	63[b]	48
R. spheroides (Chromatophores)	–	–	23[a]	49
R. spheroides	–	–	+	49
R. capsulata	–	–	+	49
R. palustris	–	–	+	49
R. gelatinosa	–	–	–	49
R. rubrum	–	–	–	49
Rhodospirillum rubrum	–	–	30[a]	50
Rhodomicrobium vannielii	–	–	27	51
Thiobacillus thiooxidans	36–54[a, c]		?	57
Hyphomicrobium vulgare strain NQ 521	?	36[a]	29[a]	55
Hyphomicrobium vulgare strain ZV-580	?	+	32[a]	55
Hyphomicrobium strain H-526, M-552	?	+	33–36[a]	55
Nocardia coeliaca	–	–	25–40	56
Nocardia polychromogenes	–	–	–	
Nocardia asteroides	–	–	–	
Nocardia erythropolis	–	–	–	
Nocardia leishmanii	–	–	–	
Nocardia corallina	–	–	–	
Nocardia rubra	–	–	–	
Nocardia transvalensis	–	–	–	
Nocardia flava	–	–	–	
Nocardia madurae	–	–	–	
Nocardia lutea	–	–	–	
Clostridium butyricum ATCC 6015, 8260	+	–	–	45
Clostridium acetobutylicum	–	–	–	45
Clostridium propionicum	–	–	–	45
Clostridium histolyticum	–	–	–	45
Clostridium tetanomorphum	–	–	–	45
Proteus vulgaris	+			45
Agrobacterium tumefaciens (6 strains)	+	+	+	44
Brucella abortus Bang II	+	+	34.5[a]	63
Brucella abortus Bang 1119			37.3	
Ferrobacillus ferrooxidans	42	1[a]	1.5[a]	60
Hemophilus parainfluenzae	+	±	?	58,72
Ochromonas malhamensis ATCC 11532	+	+	+	59
Chlamydomonas sphagnophilia			+	68
Chlamydomonas reinhardii Y-1			+	69
Nitrosocystis oceanus			+	54
Nitrosocystis europaea			–	54

[a]% lipid phosphorus; [b]nmoles per mg chromatophore; [c]depends on age of culture.

called chromatophores and these authors found phosphatidylcholine in several strains of Hyphomicrobium and in *Nitrosocystis oceanus*, bacteria also known to contain elaborate intracellular membranes[55]. In spite of these attempts to find a rationale for the presence of phosphatidylcholine in bacteria, there is the nagging reality that some bacteria just do not fit into such classifications. For example, several species of photosynthetic bacteria[49] as well as *Nitrosocystis europaea*[54] have complex membranes but lack phosphatidylcholine. Similarly, phosphatidylcholine does not appear to be a prerequisite for aerobic metabolism since several genera among the Eubacteriales (Azotobacter and Bacillus) and Pseudomonadales (Pseudomonas) have electron transport chains while they do not contain phosphatidylcholine[53].

Yano et al.[56] found that of the ten species of Nocardia, which they investigated, only one, *N. coeliaca*, contained phosphatidylcholine (Table 1). Since this particular species is believed to lack chromogenesis [56] there may be some relationship between this process and the presence or absence of certain lipids. Approximately 87% of the fatty acids of the phosphatidylcholine of this organism were shown to be branched-chain. This is quite unlike the situation in Agrobacteria[14] and in *R. capsulata*[49] both of which have been shown to contain large amounts of 18-carbon monoenoic acids.

An unusual finding was made by Goldfine and Ellis[45] during their rather extensive investigations into the lipids of a variety of bacteria. Incorporation of L-[*Me*-^{14}C] methionine into the lipid fraction of several species of Clostridia revealed that in *C. butyricum* ATCC 6015 and 8260, only phosphatidyl-*N*-methylethanolamine, and not phosphatidyl-*N*,*N*-dimethylethanolamine or phosphatidylcholine, was present. Furthermore, four other Clostridial species (*acetabutyricum*, *propionicum*, *histolyticum* and *tetanomorphum*) lacked all three methyl-derivatives of phosphatidylethanolamine[45]. Whether the synthesis of only the phosphatidyl mono-methylethanolamine in certain strains of *C. butyricum* represents a vestigial pathway which has been lost in the other Clostridia or whether it represents the initial step in acquiring a means by which phosphatidylcholine can be synthesized remains to be seen. The same situation seems to exist in *Thiobacillus thiooxidans*[57] and *Proteus vulgaris*[45]. In *Ferrobacillus ferrooxidans*[60] phosphatidyl-*N*-methylethanolamine is the predominant lipid (see Table 1), but traces of phosphatidyl-*N*,*N*-dimethylethanolamine and phosphatidylcholine have been detected.

Several papers with contradictory results have been published on the lipid composition of *Brucella abortus.* Thiele et al.[63] reported that, in strains Scherle II and 1119, phosphatidylcholine represented 34.5 and 37.3% of the lipid phosphorus, respectively. In the former strain, C-19 cyclopropane fatty acids accounted for the majority of the fatty acids in position C-2 of the phosphatidylcholine, whereas hydroxy fatty acids predominated at position C-1. In contrast, Bobo and Eagon[62] found traces of phosphatidylcholine in this organism although some choline was present in the medium. As Goldfine and Ellis[45] have pointed out, many media (particularly yeast extracts) may contain choline (or phosphatidylcholine) in sufficient quantities to yield questionable results. This is especially true when attempts are made to quantify the phosphatidylcholine content of an organism in which it exists as a minor component. Under these circumstances it is imperative that a defined lipid-free medium be used.

There are numerous papers dealing with the lipid composition of the protozoa (for reviews see refs. 64, 65) but these have dealt with a relatively small number of species. Furthermore, in many of the earlier papers, identification of a given lipid was based solely on its R_F value, thus rendering any conclusions, tentative at best. In general, however, phosphatidylcholine appears to be widespread among the various classes of protozoa. In *Dictyostelium discoideum*, a Mycetozoan, phosphatidylcholine accounts for 37% of the lipid phosphorus; 50% of the total fatty acid associated with this phosphatidylcholine was Δ 5.11–18.2 (ref. 66). Among the true amoebas (Sarcolina), the fatty acid composition of Acanthamoeba has been studied by Korn[67] and its phospholipid composition by Hack et al.[68] who found phosphatidylcholine. The various photoauxotrophic protozoa have been studied extensively in many laboratories in attempts to elucidate the role of lipids in the photosynthetic process. *Ochromonas malhamensis*, a member of the Chrysomonads has been shown by Lust and Daniel[59] to possess the methyltransferase pathway for the formation of phosphatidylcholine. Evidently phosphatidylcholine is also present among the Phytomonads since it has been found in *Chlamydomonas sphagnophilia*[68] and *C. reinhardii* Y-1 (ref. 69) and among the Euglenoids since Hulanicka et al.[70] identified phosphatidylcholine in *Euglena gracilis* Z.

Much of the work on ciliate lipid composition has been performed on Tetrahymena. Thus, Hack et al.[68] reported the presence of phosphatidylcholine in *Tetrahymena pyriformis* and this was later confirmed by Erwin and Bloch[71] who found that it comprised 30% of the phospholipids. The fatty esters were mainly linoleic and α-linoleic acids. In their investigation of rumen Holotrich protozoa, Katz and Keeney[72] found comparable amounts of phosphatidylcholine (28%) containing significant amounts of branched-chain and unsaturated fatty acids. Dawson and Kemp[73], while studying the aminoethylphosphonates among the protozoa reported finding considerable amounts of phosphatidylcholine in *Entodinium caudatum* (26% of lipid phosphorus) and in a variety of rumen protozoa.

The significance of the studies on phosphatidylcholines of yeast and *Neurospora crassa* will be discussed in a later section, but in general, phosphatidylcholine is rather common in these relatively higher organisms and recently phosphatidylcholine was shown to account for up to 40% of the lipid contents of the Endomycetales, Cryptococceae, and Sporobolomyces[75]. In the yeast-like *Pullularia pullulans*, however, no phosphatidylcholine was detected[76].

The biosynthesis of phosphatidylcholine in bacteria was first investigated by Law et al.[44] in *Agrobacterium tumefaciens*, an organism known to contain both phosphatidylethanolamine and phosphatidylcholine[42,43,77]. When this organism was grown in the presence of L-[*Me*-^{14}C] methionine, hydrolysis of the labeled lipids resulted in the isolation of radioactive *N*-methylethanolamine, *N,N*-dimethylethanolamine, and choline. These results suggested that phosphatidylcholine was being synthesized via a stepwise methylation of phosphatidylethanolamine similar to that which was shown to occur in liver microsomes[78]. Kaneshiro and Law[79] mechanically disrupted 'log phase' cells of *A. tumefaciens* TT 105 and separated the cell-free extract into cell envelope, 100 000 × *g* particulate and supernatant fractions. All three of these fractions were able to incorporate label from *S*-adenyl-[*Me*-^{14}C] methionine into phosphatidylcholine, but only the particulate and supernatant fractions could be stimulated by the addition of exogenous phosphatidylethanolamine. The particulate fraction carried out all three transmethylation reactions with some accumulation of the intermediate phosphatidylmonomethyl- and dimethylethanolamines. In contrast, the main reaction catalysed by the soluble enzyme was the conversion of phosphatidylethanolamine to phosphatidyl-*N*-methylethanolamine, indicating that at least two enzymes are required for the entire methylation sequence. Purification of this soluble enzyme forty-fold established the fact that the *S*-adenosyl homocysteine markedly inhibited the first methyl transfer and that no co-factor seemed to be required. The pathway can be schematically indicated by the following:

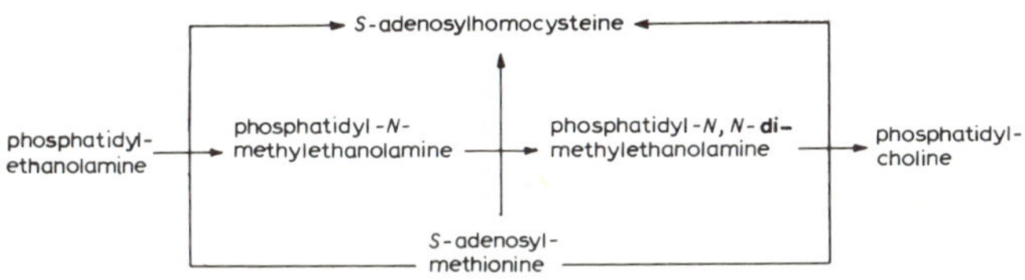

Sherr and Law[82] attempted to ascertain whether the CDPcholine pathway[80,81]; (Chapter 4, p. 70) also exists in *A. tumefaciens*. Although it was known that this organism could grow optimally in the absence of exogeneous choline, it was possible that choline released by degradative processes in the cell could re-enter phospholipid via CDP-choline. Growth of *A. tumefaciens* on [*Me*-^{14}C] choline resulted in a rapid incorporation of label into lipid. However, the label did not appear specifically in phosphatidylcholine but was incorporated equally well into all lipids. Apparently, a choline oxidase had been induced which resulted in label entering the C-1 unit pool and ultimately being incorporated into the methyl group of cyclopropane fatty acids. Despite the fact that there was an indication that some choline was directly incorporated into phosphatidylcholine *in vivo*, attempts at assaying for choline phosphotransferase *in vitro* proved to be fruitless, even under conditions when induction of the enzyme would be expected. It appears therefore, that the physiologically significant pathway for phosphatidylcholine synthesis in *A. tumefaciens* is via *N*-methylation of phosphatidylethanolamine. Since phosphatidylcholine has also been found in *A. rhizogenes* and *A. radiobacter*[45] the above statement is probably true of the Agrobacteriae in general.

An investigation of phosphatidylcholine biosynthesis in *Hyphomicrobium vulgare* NQ-521 (ref. 55) has demonstrated the presence of an active methyltransferase pathway *in vitro* in which there was a considerable accumulation of phosphatidyl-*N,N*-dimethylethanolamine. The cause of the seemingly anomalous behavior of the second enzyme was not determined. Wood et al.[49] have shown that phosphatidylcholine is a constituent of the phospholipid fraction in several species of photosynthetic bacteria. Although no investigations were made concerning the biosynthesis of the phosphatidylcholine, the fact that the fatty acid composition of this phospholipid was quite similar to that of phosphatidylethanolamine (and different from that of phosphatidylglycerol) indicates a metabolic relationship between these two lipids, probably via the methyltransferase pathway.

Two very interesting studies[45, 83] have revealed that in certain bacteria only the initial *N*-methyltransferase step is present. Thus, the incorporation of L-[*Me*-^{14}C]methionine *in vivo* into the lipids of *Clostridium butyricum* ATCC 6015 and 8260, and a strain of *Proteus vulgaris* resulted in the labeling of phosphatidyl-*N*-methylethanolamine, but with no indications of the presence of other more highly methylated derivatives. The presence of only phosphatidylmonomethylethanolamine in these organisms is quite puzzling and is made more so by the finding that four other strains of Clostridia (acetobutyricum, propionicum, histolyticum, tetanomorphum) completely lack all three methylated derivatives of phosphatidylethanolamine[45]. Nyc and his co-workers[84–86] have investigated two mutant strains of *Neurospora crassa* in which the methylated intermediates on the pathway to phosphatidylcholine accumulate[84]. Scarborough and Nyc[85] compared the ability of microsomal preparations from the two mutant strains (34486 and 47904) and the wild type (1A) to incorporate radioactivity from *S*-adenosyl-L-[*Me*-^{14}C]methionine into lipid. Microsomes from both normal and mutant 34486 cells transferred label into phosphatidylmono- and dimethylethanolamine and phosphatidylcholine, but the mutant had very low levels of phosphatidyl-*N*-methylethanolamine. The levels of endogenous phosphatidylmono- and dimethylethanolamine in the microsomes of the mutant were also low, leading to the conclusion that phosphatidyl-*N*-methylethanolamine was synthesized at a subnormal rate due to the presence of a defective phosphatidylethanolamine-*N*-methyl transferase. Microsomes from mutant 47904 on the other hand incorporated label primarily into phosphatidyl-*N*-methylethanolamine with virtually no counts in phosphatidyl-*N,N*-dimethylethanolamine or phosphatidylcholine. The monomethyl derivative also accumulated *in vivo*, indicating that the transfer of the second methyl group was impaired in this mutant. In view of the fact that the lesion in 47904 was known to be at a single genetic locus[87], the experiments described above suggested that a single enzyme converts phosphatidylmonomethylethanolamine to phosphatidylcholine. This was confirmed by Scarborough and Nyc[86] who were able to 'solubilize' the phosphatidyl-*N*-methylethanolamine *N*-methyltransferase from the microsomal fraction of a normal strain of *N. crassa* and demonstrate that both methylation steps were altered to the same extent under a variety of conditions.

Investigating the control of the methyltransferase pathway in yeast, Waechter et al.[88] grew *Saccharomyces cerevisiae* on a defined medium containing L-[^{14}C-*Me*]methionine in the presence or absence of choline. When no choline was present, label was incorporated into phosphatidylcholine > phosphatidyl-*N,N*-dimethylethanolamine > phosphatidyl-*N*-methylethanolamine and the kinetics of the incorporation indicated a precursor–product relationship. In the presence of 1 m*M* choline, however, the extent of incorporation into phosphatidylcholine was reduced to less than 5% despite the initial rate of incorporation of label appearing normal. When the incorporation of [^{14}C] serine into phosphatidylserine and phosphatidylethanolamine was measured in the presence and absence of choline, no significant difference could be detected. This indicated that the reduced synthesis of the methylated phosphatidylethanolamines in the presence of choline was not due to lack of endogenous substrates. The 90 000 × *g* particulate fraction obtained from cells grown in the presence of choline also displayed a greatly reduced transfer of [*Me*-^{14}C] from *S*-adenosyl methionine compared to particles obtained from cells grown in the absence of choline[89]. Taken collectively, these data clearly suggest that growth of *S. cerevisiae* in the presence of choline results in the repression of the methyltransferase pathway. A corollary of the conclusion is the necessity for a mechanism whereby exogenous choline could be taken up into lipid. In fact, labeled choline was incorporated directly into phosphatidylcholine by a mechanism which did not appear to be inducible since the extent of incorporation was the same in cells grown both in the presence and absence of choline. Although the reaction sequence probably involves CDPcholine, no direct evidence for the presence of this pathway was obtained.

A rather novel pathway for lecithin synthesis has also been reported in bakers yeast[90]:

2-lysophosphatidylcholine → phosphatidylcholine + glycerophosphorylcholine

However, because of the presence of an active phospholipase, it was not possible to recover the products in equimolar amounts. Thus, the relative contribution of this pathway, versus direct reacylation of the lyso compound, could not be assessed. However subsequently, Van den Bosch et al.[91] were able to show that when [1-^{14}C] acyl-*sn*-glycero-[3-^{32}P] phosphorylcholine was incubated with yeast supernatant fraction in the presence of ATP, CoA, and linoleic acid, phosphatidylcholine could be isolated which had a ^{14}C/^{32}P ratio twice that of the original substrate. This indicated that the exchange pathway shown above was solely responsible for the formation of the product. This reaction sequence may also explain the observation made by Hanahan and Jayko[92] that some phosphatidylcholines occurring in yeast have identical fatty acids in positions 1 and 2 (e.g., dipalmitoleoyl phosphatidylcholine).

Smith and Law[61] have recently reported the presence of choline phosphotransferase activity in the mitochondrial fraction obtained from *Tetrahymena pyriformis*. This enzyme required 15 m*M* $Mn^{2\oplus}$ or 10 m*M* $Co^{2\oplus}$ for maximum activity and incorporated label from CDP[1,2-^{14}C] choline into phosphatidylcholine. A microsomal fraction obtained from the same organism incorporated label from *S*-adenosyl-L-[*Me*-^{14}C] methionine primarily into phosphatidylcholine with only minor amounts of the methylated intermediates being formed. Thus, Tetrahymena seemingly can form phosphatidylcholine by the two pathways previously characterized in animal cells. An earlier report by Lust and Daniel[59] had shown that both the microsomal and mitochondrial fractions from the protozoan *Ochromonas malhamensis* 11532 could readily methylate phosphatidylethanolamine to phosphatidylcholine using *S*-adenosylmethionine. The addition of CDPcholine did not suppress the ^{14}C uptake into microsomal phosphatidylcholine from *S*-adenosyl-L-[*Me*-^{14}C] methionine. Unfortunately a similar experiment was not attempted using the mitochondrial fraction.

Relatively little is known concerning the catabolism of phosphatidylcholine in micro-organisms and of the information which is available, most involves inferences from chase experiments. Short et al.[60] grew *Ferrobacillus ferroxidans* in the presence of $NaH^{14}CO_3$ and [^{32}P] orthophosphate and then followed the loss of label during growth at constant pH. The results revealed that there was essentially no turnover of either ^{14}C or ^{32}P in phosphatidylmonomethylethanolamine or phosphatidylcholine. The most rapid turnover was observed in phosphatidylethanolamine and phosphatidylglycerol. The relative metabolic inertness of phosphatidylcholine in micro-organisms has also been observed by Yano et al.[56], who noticed the slow incorporation of [1-^{14}C] glycerol or [1-^{14}C] acetate into phosphatidylcholines of *Nocardia coeliaca*. Lascelles and Szilagyi[47] found the same metabolic inertness in Rhodopseudomonas. The significance of the differential rates of synthesis among the microbial phospholipids remains an important problem.

6. AMINO-ACYL PHOSPHATIDYLGLYCEROLS

The amino-acyl derivatives of phosphatidylglycerol were initially discovered by MacFarlane[93] in *Clostridium perfringens* and have subsequently been found in Mycoplasma[94, 95] and in a variety of Gram-positive bacteria[97, 98] (Table 2). Although a number of L-amino acids have been identified as components of this class of lipids, the L-lysyl and L-alanyl esters seem to predominate. Lennarz et al.[99] were the first to discover the mode of biosynthesis of these unusual lipids. They demonstrated that cell-free extracts of *Staphylococcus aureus* strain 81/80 could incorporate L-[^{14}C] lysine into lysyl-phosphatidylglycerol. The ini-

TABLE 2

OCCURRENCE OF AMINO ACYL-PHOSPHATIDYLGLYCEROL

Organism	*Amino acid derivative*	*Reference*
Mycobacterium 607	ornithine[a]	111
Streptococcus faecalis ATCC 9790	L-lysine[b], alanine, arginine (?)	112, 102
Clostridium perfringens	L-lysine, alanine	93
Bacillus megaterium	L-lysine	113
Bacillus cereus	L-ornithine	114
Staphylococcus aureus	L-lysine	108
Streptococcus faecalis 10CI	alanine, L-lysine	105
Mycoplasma laidlawii B	D-alanine, L-alanine	94, 95
Bacillus subtilis	L-lysine	121, 122

[a]UV-irradiated cultures or low pH only.
[b]Both 2′ and 3′.

tial reaction involved the incorporation of L-[^{14}C] lysine into lysyl-tRNA catalyzed by a soluble (100 000 X *g* supernatant) enzyme. The mechanism is believed to be the same as the amino acid activation which results in the formation of amino-acyl-tRNA's for participation in protein synthesis. The second step was catalyzed by a particulate enzyme and involved the transfer of lysine from the lysyl-tRNA^{Lys} to phosphatidylglycerol. A complete dependence on phosphatidylglycerol was reported, clearly indicating a net synthesis and not merely an exchange of lysine. The lysyl-tRNA^{Lys} requirement was not species-specific since the lysyl-tRNA prepared from *E. coli* was as active as that from *S. aureus.* In a later paper[100] a similar reaction was found to occur in *Bacillus megaterium* KM, *Bacillus cereus* ATCC 9139, and *Clostridium welchii* ATCC 3624. This last organism was able also to incorporate alanine from alanyl-tRNA into alanyl-phosphatidylglycerol.

Investigations into the lipid specificity illustrated that *sn*-2'deoxy-, but not the *sn*-3'deoxyphosphatidylglycerol was a substrate for the reaction[96]. This indicated that the 3'-position on the glycerol was the site of esterification. Bonsen et al.[109] had previously shown that the synthetic 1-L-phosphatidyl-3-L-lysyl-glycerol was chromatographically indistinguishable from the natural lipoamino acid from *S. aureus.* A possible migration of the lysyl group from the 2- to the 3-position during isolation was ruled out by the work of Molotkovski and Bergelson[101] who chemically synthesized both forms and found no such interconversion. This work provides solid evidence that the 3'-lysyl derivative is found in *S. aureus*, but does not preclude the presence of the 2'-derivative in other species. In fact, recently[102] a lysyl-phosphatidylglycerol was isolated from *Streptococcus faecalis* ATCC 9790 which migrated with standard 1,2-distearoyl-*sn*-glycero-3-phosphoryl-1'-(2'-*O*-L-lysyl)-glycerol. In addition, the 3'-derivative and a compound tentatively characterized as 2', 3'-dilysyl-phosphatidylglycerol were also found in this organism[102]. Conclusive evidence in support of the proposed 2'- and 2',3'-compound has not appeared.

Nesbitt and Lennarz[103] and Gould et al.[104] structurally modified the amino acyl-tRNA substrate in an attempt to discern what part of the molecule is recognized by the enzyme. *S*-*β*-aminoethyl cysteinyl-tRNA^{Lys}, an analog of lysyl-tRNA^{Lys} was an excellent substrate in amino acyl-phosphatidylglycerol synthesis whereas aminoethyl cysteinyl-tRNA^{Cys} was inactive. These results clearly show that the enzyme can recognize structural differences in the ribonucleotide substrate. Studies in which the amino acyl portion of the amino-acyl-tRNA was structurally modified indicated that the enzyme also recognized the amino acid component since substitution of the amino or methyl group of the alanyl moiety of alanyl-tRNA rendered the molecule inactive. There are some recent papers[95, 105] which suggest that other mechanisms may also exist for formation of the amino acyl-phosphatidylglycerols. Lennarz et al.[99] had shown that only L- and not D-lysine was incorporated into lysylphosphatidyl-glycerol. This is inconsistent with the data of Ito and Strominger[106] and Okuda et al.[107] which show that D-amino acids are incorporated into polymers via a mechanism not involving tRNA. Koostra and Smith[95] characterized a D-alanyl-phosphatidylglycerol in *Mycoplasma laidlawii* strain B. Their careful studies on the biosynthesis of this compound revealed an active ATP [^{32}P] pyrophosphate exchange in the product resulting from the incubation of D-alanine, ATP, and crude *Mycoplasma laidlawii* amino acyl synthetases. Such results indicate the probable formation of an AMP-

References p. 327

D-alanyl-enzyme complex. Labeled D-alanine from this complex could then be transferred to phosphatidylglycerol in a reaction not requiring tRNA. It was also demonstrated that the synthesis of L-alanyl-phosphatidylglycerol in *Mycoplasma laidlawii* proceeds via the reaction elucidated by Lennarz et al.[99]. The finding of two different mechanisms for the synthesis of these unusual amino acyl lipids underlines the probable importance of these compounds to the cell.

In the original study of Houtsmuller and Van Deenen[108] on the metabolism of phosphatidylglycerol and lysyl-phosphatidylglycerol in *Staphylococcus aureus*, they concluded that the relationship between the two lipids was dictated by the pH of the medium. Thus, when the pH was low (e.g. 4.8) there was an increase in the level of lysyl-phosphatidylglycerol and a concomitant decrease in phosphatidylglycerol. If the pH of the medium was maintained about 7.0, there was a significant decrease in the level of the amino acyl derivative and an increase in the level of phosphatidylglycerol. Subsequent studies, however, have revealed that the metabolism of these compounds is more complex than hitherto believed. When Houtsmuller and Van Deenen[97] extended their investigations into the mechanism responsible for the fluctuation of these compounds with pH, they discovered that, within experimental limits, the decrease in phosphatidylglycerol was only partially compensated by a rise in diphosphatidylglycerol and lysyl-phosphatidylglycerol. Furthermore, a similar pH effect is not found in all bacteria since the amino acyl phosphatidylglycerol of *Bacillius cereus* was not altered by changing the pH of the medium and the relationship between pH and lipid metabolism in *Bacillus megaterium* appeared to be more complex than in *S. aureus*. Kocun[105] studied the influence of pH on diphosphatidylglycerol, phosphatidylglycerol, and the alanyl and lysyl phosphatidylglycerols of *Streptococcus faecalis* ATCC 8049 and found that as the pH of the medium decreased with growth, the level of lysyl-phosphatidylglycerol showed an initial increase followed by an apparent leveling off. In contrast, the level of alanyl-phosphatidylglycerol greatly decreased to an eventual constant level. The fluctuation in diphosphatidylglycerol and phosphatidylglycerol during this period was slight in comparison to that of the alanyl derivative. The demonstration that the alanyl and lysyl-phosphatidylglycerols fluctuate in opposite directions with pH suggests that the metabolism of these compounds responds in different ways to change in the hydrogen ion concentration of the medium. Gould and Lennarz[110] carefully investigated the metabolism of phosphatidylglycerol and lysyl-phosphatidylglycerol in *S. aureus* under conditions of growth and rest at both pH 7.0 and 5.2. They found that whereas there were considerable changes in the levels of phosphatidylglycerol, the level of lysyl-phosphatidylglycerol was not greatly affected by incubation of the cells under these four conditions. The most significant decline in phosphatidylglycerol occurred in resting cells at pH 5.2 and this was due to both a decrease in synthesis and an increase in catabolism. Other factors such as $Mg^{2\oplus}$ ion concentration and cell density also seemed to play some role in the rather complex metabolism of this lipid. Although the level of lysyl-phosphatidylglycerol remained relatively constant, chase experiments revealed that the lysyl group was actively turning over. These studies indicate that under the described experimental conditions, the change in the ratio of phosphatidylglycerol to lysyl-phosphatidylglycerol is primarily due to fluctuation in the levels of phosphatidylglycerol.

An unusual observation was made by Khuller and Subrahmanyam[111] who noted that the ornithinyl ester of phosphatidylglycerol appeared in cells of Mycobacterium 607 grown under UV light. The mechanism by which the irradiation induced the formation of the lipid was not elucidated, but a similar process did not occur in *M. phlei.*

Speculations as to the function of these lipoamino acids have included suggested roles in amino acid transport, cell-wall biosynthesis[108] and maintenance of cell membrane charge. However, kinetic studies[110] indicate that the turnover of the lysyl moiety of the lysyl-phosphatidylglycerol in *S. aureus* is at least one order of magnitude lower than the rate of lysine uptake. Furthermore, this turnover is relatively unaffected by environmental conditions, including the presence of cell-wall inhibitors. These findings do not completely preclude a role for lysyl-phosphatidylglycerol in such cell processes, but it diminishes the probability that this lipid performs these functions in *S. aureus.* The suggested role of these derivatives in maintaining membrane charge under adverse environmental conditions (e.g. low pH) is appealing but has not been substantiated. Kocun's[105] finding that lysyl-phosphatidylglycerol increased in concentration with decreasing pH would seem to contradict this contention as would the work of Gould and Lennarz[110] which indicated minimal fluctuation of lysyl-phosphatidylglycerol with changes in pH. A more appealing function for the amino-acyl phosphatidylglycerols would be to stabilize the membrane, and some evidence in favor of this proposal has appeared. Van Iterson and Op den Kamp[130] grew *B. subtilis* at low pH under conditions in which the lysyl-phosphatidylglycerol accumulated. When these cells were treated with lysozyme, the resultant gymnoplasts (protoplasts) maintained the rod-shape of the normal cell and were quite resistant to osmotic lysis. This type of gymnoplast could not be produced from cells grown at neutral pH by merely increasing the magnesium concentration. The exact mechanism whereby the membrane is stabilized is not yet known, but it is presumed to involve the presence of lysyl-phosphatidylglycerol.

7. GLUCOSAMINYL PHOSPHATIDYLGLYCEROL

A derivative of phosphatidylglycerol containing a glucosamine moiety was discovered in *Bacillus megaterium*[113] and also in *Pseudomonas ovalis*[116]. In the former organism this lipid accumulated as the pH of the medium declined and reached a maximum at approximately 80% of the total phospholipids[117]. Isolation of the compound revealed a glucosamine:nitrogen:phosphorus:glycerol ratio of 1.0:1.1:1.1:2.0 and hydrolysis at pH 9 converted the phospholipid to phosphatidylglycerol and glucosamine. In a subsequent report, Op den Kamp et al.[118] established the structure of the lipid as 1-(1,2-diacyl-*sn*-glycero-3-phosphoryl)-2-*O*-(2-amino-2-deoxy-D-glucopyranosyl)-*sn*-glycerol. This structure was substantiated by Gurr et al.[119] who used some elegant chemistry to synthesize the above derivative and then showed that the synthetic and natural compounds were the same. It was not possible, however, to determine unequivocally whether the glucosamine is in an α- or β-linkage to the glycerol. This problem has recently been solved by MacDougall and Phizackerly[120] who converted the parent lipid to the *N*-acetyl derivative and then removed the fatty acids by mild alkaline hydrolysis. Treatment of the water-soluble derivative with calf brain β-*N*-acetyl-

References p. 327

hexosaminidase quantitively released the *N*-acetylglucosamine. The structure of the complete lipid is shown in Fig. 2. Substantial evidence is presented in this same paper[120]

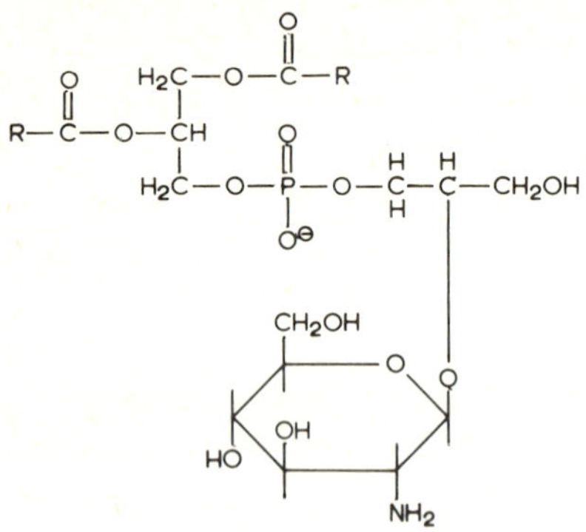

Fig. 2. Glucosaminyl phosphatidylglycerol. The 3′-derivative has also been detected (see text).

showing that the 3′-derivative of glucosaminyl phosphatidylglycerol is also present in *B. megaterium*, and that in this compound, too, the linkage between the glycerol and glucosamine is *β*. No information is available on the synthesis or metabolism of these compounds.

8. DIGLUCOSYL DERIVATIVES OF PHOSPHATIDYLGLYCEROL

In a recent study of the lipids of *Streptococcus hemolyticus* D-58, Ishizuka and Yamakawa[127, 128] isolated and characterized a diglucosyl derivative of phosphatidylglycerol. The two glucose moieties were linked by an α 1 → 2 bond and were presumably in a glycosidic linkage to the 3–OH of the terminal glycerol. The same structure was proposed by Fischer and Seyferth[129] for a lipid which they isolated from *Streptococcus lactis* and *Streptococcus faecalis* and also by Dos Santos Mota et al.[102] for a lipid purified from *S. faecalis* ATCC 9790. The data presented in support of the proposed structure, however, serve equally well for a glycerophosphate derivative of diglucosyl diglyceride, and this will be discussed more fully under the heading phosphoglycolipids.

9. DIETHER DERIVATIVES OF PHOSPHATIDYLGLYCEROL

The polar lipids of the halophile *Halobacterium cutirubrum* are unique in that they consist exclusively of di-*O*-phytanylglycerol ether derivatives[123]. The predominant phospholipids are the di-*O*-phytanyl analogs of phosphatidylglycerophosphate (PGP-diether) and phosphatidylglycerol (PG-diether). Joo and Kates[124] chemically synthesized the diastereoisomers of both phytanyl compounds and then compared the natural and synthetic products. The structure of PGP-diether was shown to be 2,3,-di-*O*-phytanyl-*sn*-glycero-1-phosphoryl-3′-*sn*-glycerol 1′-phosphate. Thus, these compounds are ether derivatives of *sn*-glycero-1-phosphate, rather than the usual *sn*-glycerol 3-phosphate, and consequently are the mirror images of the diacyl phospholipids found in other bacteria. Efforts directed toward an understanding of how the glycerol moieties are incorporated into lipid revealed that the glycerophosphate dehydrogenase and glycerokinase of *H. cutirubrum* form only the *sn*-glycerol 3-phosphate isomer[125]. Also, Kates et al.[126] observed that label from [1(3)-^{3}H] glycerol was retained in the glycerol moieties of the ethers, but that label from [2-^{3}H] glycerol was lost. These data effectively eliminate dihydroxyacetone phosphate and glyceraldehyde-3-phosphate as glycerophosphate precursors since these two compounds are interconverted via an isomerase and this reaction would result in the loss of label at carbon-3. Although attempts at determining the precursor of the glycerol moiety were not successful, the incorporation of label from [1(3)-^{3}H] glycerol into the phytanyl portions of the diether was believed to occur by the following pathway:

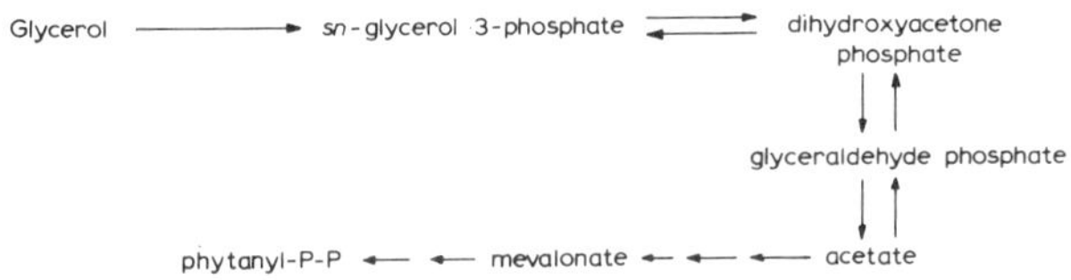

10. PHOSPHATIDYLINOSITOL MANNOSIDES

Chemical structures

In 1930 Anderson and his colleagues[131–134] noted the occurrence of inositol and D-mannose as components of a phospholipid fraction from a human Mycobacterium. Saponification of this material yielded a water-soluble phosphorus-containing polysaccharide which they named 'manninositose', containing inositol and mannose in a ratio of 2:1. Subsequent studies on Mycobacterial phospholipids have suggested that the phospholipid fraction investigated by Anderson actually contained a series of phospho-*myo*-inositol glycosides which differed in the ratio of mannose to inositol. Work by Vilkas and Lederer[135] demonstrated the existence in this fraction of a glycerophospholipid containing two mannoses and one *myo*-inositol moiety. Both mannose units were in α-glycosidic linkage to the inositol indicating the lipid structure to be phosphatidyl-*myo*-inositol dimannoside. Nojima[136, 137] discovered both a phosphatidyl-*myo*-inositol tetra- and penta-mannoside in *Mycobacterium tuberculosis* BCG and, when Vilkas[138] subjected crude Mycobacterium phospholipid to an alkaline hydrolysis, free *myo*-inositol and *myo*-inositol monomannoside were found, suggesting the prescence of phosphatidyl-*myo*-inositol and phosphatidyl-*myo*-inositol monomannoside. A trimannoside derivative was tentatively identified by Lederer[139] in *M. tuberculosis* and this finding was eventually substantiated by Lee and Ballou (140) in both *M. tuberculosis* and *M. phlei.* Thus, a family of phosphoinositides containing from one to five mannose units is present in the Mycobacteria.

Studies on the complete characterization of these interesting phospholipids were initiated by Ballou et al.[141] who revealed that the mono-, di-, and pentamannosyl phosphoinosides of *M. tuberculosis* BCG were related stereochemically to all other *myo*-inositol phospholipids in having the phosphatidyl group on the D-1 position of the *myo*-inositol ring. These studies were greatly facilitated through use of selective hydrolytic procedures permitting isolation of the individual glycoside moieties. In a series of elegant structural analyses[140, 142, 143] Lee and Ballou subsequently deduced the basic structure of these mannosides in *M. tuberculosis* and *M. phlei* to be shown in Fig. 3.

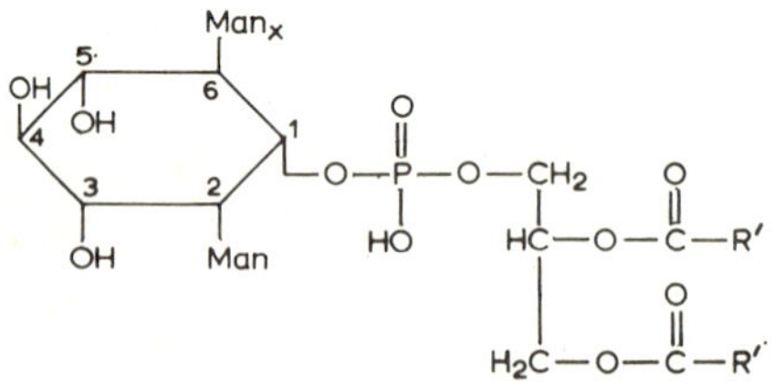

Fig. 3. Structures of phosphatidyl-*myo*-inositols. R′, R″ = fatty acid residues: X = 0 (monomannoside), X = 1 (dimannoside), X = 2 (trimannoside), X = 3 (tetramannoside), X = 4 (pentamannoside).

In the phosphatidyl-*myo*-inositol monomannoside the single mannose group was shown to be attached via a glycosidic linkage to the 2-hydroxyl group of the *myo*-inositol ring. In the diamannoside, one α-D-mannose unit was found at the 2-position and one at the 6-position of the inositol ring. The additional mannose units in the tri- and tetra-analogues are connected in α-1-6 linkages to the mannose at position 6. The penta-derivative has the terminal mannose connected to the oligosaccharide by an α-1,2-linkage. Analysis of the lipids of *M. tuberculosis* and *M. phlei* revealed no apparent difference in the structure of the glycosidic moiety of the phosphoinositides but a hexoside mannose was found in the latter which had an additional α-1,2-linked mannose on the mannotetraose chain.

Since these studies were conducted on the deacylated lipids, the position and number of the fatty acid residues was not known with certainty although two fatty acyl groups were believed to be present on the glycerol. Vilkas[144] had suggested earlier that multiple forms of the individual mannophosphoinositides might exist, but it was not until the work of Pangborn and McKinney[145] that the presence of variously acylated derivatives of the phosphatidyl-*myo*-inositol mannosides was demonstrated. Using a combination of solvent fractionation and TLC on silicic acid they were able to separate two phosphatidyl-*myo*-inositol dimannosides each containing two fatty acyl residues, two additional dimannosides containing three or four fatty acyl residues respectively, and two phosphatidyl-*myo*-inositol pentamannosides containing three and four fatty acyl groups, respectively. Palmitic and tuberculostearic (10 methyl octadecanoic) acids accounted for 90% of the fatty acids of these lipids. Three strains of *M. tuberculosis* were investigated (BCG, H37RV, and a human strain) with no apparent differences in the lipid composition. Similar findings were made subsequently by Brennan and Ballou[151] in *M. phlei.* In a brief paper on the phospholipids of *Mycobacterium* 607, Khuller and Subrahmanyan[150] were able to isolate 7 mannophosphoinositides. Four fatty acid groups each were found on hexa-, penta- and dimannophosphoinositides, while the tetra-, tri-, and another dimannoside each had three fatty acyl residues. Only the monomannoside, which was the predominant glycoside of the organism, possessed two fatty acids. The position of these additional fatty acid moieties has not yet been determined (but see later discussion on biosynthesis of phosphatidyl-*myo*-inositols).

Although the phosphatidyl-*myo*-inositol mannosides are generally associated with the Mycobacteria, there is a growing body of evidence to indicate that they may be present in other microorganisms as well. Asselineau[146] and Ferreira-Gomes et al.[147] have reported finding mannose- and inositol-containing phospholipids in *Corynebacterium diphtheriae* and Brennan[148] has found a diacyl phosphatidyl-*myo*-inositol di-mannoside in *Corynebacterium xerosis*, although the assignment of four fatty acids to the lipid must be viewed as tentative. When Brennan and Ballou[149] extended their investigation to the Propionibacteria, they found a phosphatidyl-*myo*-inositol monomannoside in *P. shermanii.* This lipid seemed to be the major phosphatidylinositol glycoside in the organism and this is in contrast to the Mycobacteriae in which the dimannoside generally predominates. Laneelle and Asselineau[153] also found evidence for a phosphatidyl-*myo*-inositol monomannoside among the lipids of *P. freudenreichii*. These findings are of taxanomic interest since they suggest a relationship between the Mycobacteriae, Corynebacteriae, and propionic acid bacteria. The Actinomycetales which are believed to be closely related to the Mycobacteria also possess phosphatidylinositol mannosides. Laneelle et al.[152] identified these compounds in *Nocardia brasiliensis* and *N. asteroides* and more recently Kataoka and Nojima[154] extended these studies to include *N. polychromogenes, Streptomyces griseus*, and *Microbisporangium chromogenes.*

Biosynthesis

The first report concerning the biosynthesis of these unique phospholipids was made by Hill and Ballou[155]. They disrupted *M. phlei* cells by sonication and found that the 100 000 $\times$ *g* pellet could incorporate label from GDP[^{14}C]mannose into lipid. The reaction was quite specific for the guanidine nucleotide sugar and was stimulated 10–20% by the addition of crude *M. phlei* phospholipids or Tween 80. Despite the relatively low incorporation of label, they were able to identify the product of the reaction as 1-phosphatidyl-L-*myo*-inositol-2-*O*-α-D-mannopyranoside. The anomalous migration of the product on chromatography however, indicated a more highly acylated species. There was virtually no incorporation of labeled mannose into any of the higher analogues – a surprising result since the monomannoside analogue is present in only minor quantities in the Mycobacteria. The scheme proposed[155] for the biosynthesis of the mannosides was the following:

GDPmannose + phosphatidyl-*myo* inositol → GDP + phosphatidyl-*myo*-inositolmannoside

This pathway also describes the biosynthesis of the monomannoside in *P. shermanii*[149] where the results were less equivocal since phosphatidyl-*myo*-inositol monomannoside is the

References p. 327

major phosphoinositide in this organism. In a subsequent paper[151] Brennan and Ballou demonstrated that a total cell-free extract of *M. phlei* was able to effect a transfer of mannose from GDP[^{14}C]mannose into two chromatographically distinct dimannosyl phosphatidyl-*myo*-inositols. One of these dimannosides contained three fatty acyl groups (dimannophosphoinositide B) and the other four fatty acyl groups (dimannophosphoinositide A). The relative proportions of the two products could be altered by the addition of nonionic detergent (Cutscum) or by using exogenous phospholipid acceptors. When a 100 000 × *g* particulate fraction was incubated with GDP[^{14}C]mannose and exogenous crude lipid acceptors, labeled mannose was also incorporated into another dimannoside (dimannophosphoinositide C, containing two fatty acid residues), and a tri- and penta-phosphatidyl-*myo*-inositol, in addition to dimannophosphoinositides A and B. The authors were able to prepare a 'solubilized' enzyme from acetone powders of the 100 000 × *g* particulate fraction and this system could be stimulated two-fold by the addition of *M. phlei* phosphatidyl-*myo*-inositol. The products consisted of approximately equal amounts of the three dimannosides and some trimannosyl phosphatidyl-*myo*-inositol. However, when the exogenous acceptor was yeast phosphatidyl-*myo*-inositol, the primary product was the dimannoside C. This suggested that although the yeast phosphatidylinositol was a suitable substrate for the glycosyl transferase, the product of the reaction was not a suitable substrate for the acylase and consequently dimannosides A and B were not formed. Actually, however, the direct synthesis of phosphatidyl-*myo*-inositol dimannoside from phosphatidyl-*myo*-inositol was never demonstrated since it was never shown that both mannose units were labeled, nor was phosphatidyl-*myo*-inositol monomannoside ever used as a substrate. Takayama and Goldman[156] observed that a cell-free particulate preparation from the H37Ra strain of *M. tuberculosis* was capable of synthesizing both phosphatidyl-*myo*-inositol mono- and dimannosides from GDP[^{14}C]mannose as indicated by the isolation of their respective deacylation products. Treatment of the ^{14}C-labeled deacylated dimannoside with β-glucosidase (which has α-mannoside activity) resulted in the selective removal of only the mannose attached to position 6 of the myoinositol ring. It was discovered that all of the radioactivity in the deacylated dimannosyl phosphatidylmyoinositol resided in this single mannose – the mannose attached to the 2-hydroxyl of the inositol was not labeled. Apparently there was a restriction on the sequential addition of the second mannose to position 6 following the incorporation of a mannose unit at position 2 of phosphatidyl-*myo*-inositol. This suggested to the authors that the phosphatidyl-*myo*-inositol monomannoside has to be modified, perhaps by acylation, before the second mannose can be transferred. Studying the problem of acylation, Brennan and Ballou[157] showed that the 100 000 × *g* particulate fraction from *M. phlei* was able to incorporate a variety of fatty acyl-CoA substrates into dimannophosphoinositides using *M. phlei* phospholipids as acceptors. Use of the lipid-depleted, 'solubilized' particulate fraction permitted study of the acylation of the dimannophosphoinositols by [^{14}C] palmitoyl-CoA. When dimannoside C (two fatty acids) was the exogenous substrate, the major product was dimannophosphatidylinositol B (three fatty acids). When dimannoside B was incubated under the same conditions, label was found in the substrate and, after short incubations, in dimannophosphoinositide A (four fatty acids). The appearance of label in the substrate indicated that deacylation–reacylation reactions were occurring, further complicating an already complex series of reactions. A summary of the interrelationships between the phosphoinositide mannosides is shown below[157].

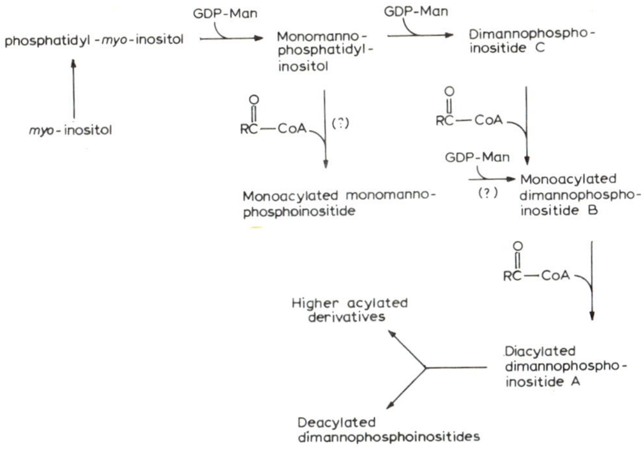

The elucidation of the synthesis of these compounds is additionally complicated by a consideration of some recent information which implies an alternative pathway for the synthesis of the phosphatidyl-*myo*-inositol monomannoside. The lipids of the propionic acid bacteria contain small amounts of phosphatidyl-*myo*-inositol monomannoside[148, 149], but they also possess a compound characterized as a diacyl *myo*-inositol mono-mannoside[158, 159] (Fig. 4). One fatty acid is esterified to position 6 of mannose and the

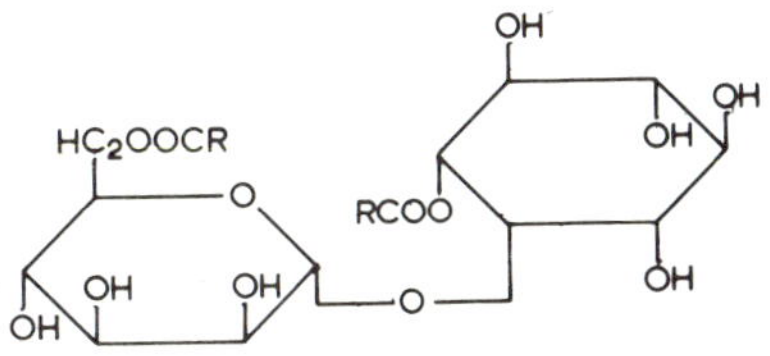

Fig. 4. Diacyl *myo*-inositol monomannoside.

other to position 1 of *myo*-inositol while the D-mannose is attached via an α-glycosidic linkage to the 2-hydroxyl of *myo*-inositol. Thus, the deacylated component obtained from the above lipid appears to be identical to the 2-*O*-α-D-mannopyranosyl-*myo*-inositol obtained by deacylation of phosphatidyl-*myo*-inositol monomannoside (except for the absence of the glycerolphosphate moiety in the former) from *M. phlei* and *M. tuberculosis*[142]. A structural relationship between this glycolipid and the phosphatidyl-*myo*-inositols is further indicated by the findings[151] that phosphatidyl-*myo*-inositol di-mannoside A (four fatty acids) has a fatty acyl group at position 6 on one of the mannose units. These facts have prompted Shaw[17] to propose an alternative route for the biosynthesis of the phosphatidylinositol monomannosides:

References p. 327

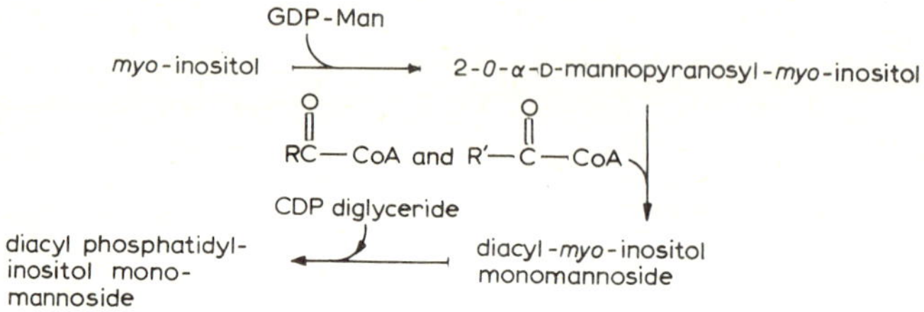

This scheme would result in a product having four acyl residues and it is significant that in their biosynthetic studies, Brennan and Ballou[149] observed a labeled product formed by *Propionibacterium shermanii* with endogenous acceptor that they believed was a more highly acylated phosphatidyl-*myo*-inositol monomannoside. It is also possible, of course, that the phosphatidyl-*myo*-inositol monomannoside which is synthesized from exogenous phosphatidyl-*myo*-inositol and GDP-mannose by *P. shermanii* could be further acylated and then hydrolyzed via phospholipase D to yield the diacyl *myo*-inositol monomannoside. The contribution of the various pathways to the overall metabolism of the phosphatidylinositol mannoside awaits further study.

Relatively little is known concerning the metabolism of the phosphatidylinositol oligomannosides. Akamatsu et al.[159] incorporated ^{32}P into the lipids of *M. phlei* during logarithmic growth. Chase experiments revealed that diphosphatidylglycerol was the most metabolically active major phospholipid while phosphatidylethanolamine and the total phosphatidylinositol oligomannoside fraction had relatively low turnovers. Ono and Nojima[160] found phospholipase activities seemingly associated with the membrane fraction of *M. phlei*. These enzymes readily degraded endogenous diphosphatidylglycerol, yet the phosphatidylinositol oligomannosides were not substrates. In a subsequent note[161] a phospholipase A was partially purified from homogenates of *M. phlei*. The substrate specificities were not mentioned, but the presence of this enzyme suggests that there may be turnover of the fatty acid moieties of the phospholipids in this organism.

Studies on the distribution of the phospholipids in *M. phlei* by Akamatsu et al.[162] indicated that whereas diphosphatidylglycerol was concentrated in the cell membrane, the phosphatidylinositol oligomannosides were mainly associated with the wall. A similar study by Goldman[163] led to the conclusion that the phosphatidylinositol pentamannoside was located predominantly in the wall whereas the dimannoside was found in the membrane.

11. ACYLATED DERIVATIVES OF GLYCEROPHOSPHORYL DIGLUCOSYL DIGLYCERIDE

Several years ago Pieringer[164] demonstrated that a 30 000 × *g* membrane fraction from *S. faecalis* ATCC 9790 could catalyze the synthesis of three glycolipids from 1,2-diglyceride and UDPglucose. Time-course studies suggested a precursor–product relationship between the three lipids and the first two products were identified as monoglucosyl diglyceride and diglucosyl diglyceride, respectively. The third product has recently been characterized by Ambron and Pieringer[18] as an acylated derivative of

3-*O*-(2-*O*-α-D-glucopyranosyl-6-*O*-phosphoryl-3′(1′)-*O*-*sn*-glycerol)-α-D-glucopyranosyl-*sn*-glycerol which is a new phosphoglycolipid. Fatty acid analysis led to the tentative assignment of four fatty acids to the compound. Periodate oxidation studies on the purified, intact lipid failed to release formaldehyde, indicating that at least one of the hydroxyls on the external glycerol is acylated. Diglucosyl diglyceride isolated from *S. faecalis* was shown to act as a direct precursor of this new lipid[18] thus establishing the following structure:

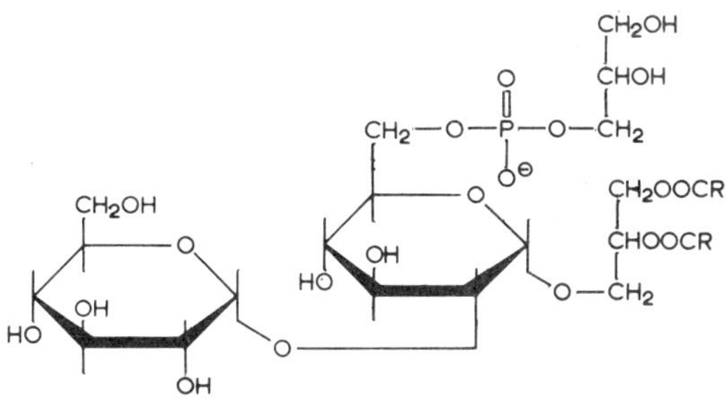

Fig. 5. Glycerophosphoryl diglucosyl diglyceride (two of the four fatty acids are not shown).

The nature of the endogenous glycerophosphate donor is at present unknown** (but see p. 331), although phosphoenol pyruvate stimulates the conversion of label from UDP-[U-^{14}C]glucose to the phosphoglycolipid[165]. While this work was in progress, Ishizuka and Yamakawa[127, 128], characterized a lipid from *Streptococcus hemolyticus* D-58 as a diglucosyl derivative of phosphatidylglycerol. Independently, Fisher and Seyferth[129] reported the presence of this same compound in both *S. faecalis* and *Streptococcus lactis*. The data presented in support of this proposed structure however, would also serve equally well for the acyl-glycerophosphoryl diglucoside structure proposed by Ambron and Pieringer[18].

Mild alkaline hydrolysis of the deacylated lipid was performed by each of the three groups with the result that glycerophosphate and diglucosylglycerol were isolated as the main products. Since it is known that under alkaline conditions phosphodiester bonds are cleaved primarily via a mechanism involving adjacent hydroxyl groups, phosphorylglycerol diglucose would also have to be a product if the structure proposed by Ishizuka and Yamakawa[127, 128], and Fisher and Seyferth[129] is correct. No such product was reported. In addition, when Ambron and Pieringer[18] subjected the glucose-labelled, deacylated lipid to periodate oxidation, borohydride reduction and acid hydrolysis, glycerophosphate was the only radioactive phosphorylated product, substantiating the structure of the compound shown in Fig. 5. Dos Santos Mota et al.[102] while studying the lipids of *S. faecalis* ATCC 9790 also isolated a phosphoglycolipid which had the same mobility on silicic acid paper as the lipid from *S. hemolyticus.* Treatment of this compound with phospholipase D resulted in the release of phosphatidic acid. Although this result would seem to indicate a phosphatidylglycerol diglucoside structure, neither the amount of phosphatidic acid released nor the identity of the other fragment produced during the hydrolysis was revealed. Since the lipid identified by Ambron and Pieringer has four acyl residues, it is possible that phosphatidic acid would result from treatment of this lipid with phospholipase D. The other product then would have to be diglucosyl diglyceride.

Recently, Fisher[166] published a brief communication in which the lipid in *S. lactis* was given the structure shown in Fig. 5 and it is probable that the lipid in *S. hemolyticus* is also the same. Shaw et al.[167] have re-examined the structure of the 'phosphatidylglucose' from *Mycoplasma laidlawii* and concluded that this compound also is the same as in Fig. 5 but with only two fatty acids. Thus, it appears that this phosphoglycolipid may exist in those organisms which contain diglucosyl diglycerides. It will be very interesting to see if there are such derivatives in bacteria which possess analogous diglucosyl diglycerides.

The similarity in composition and structure between the diglucosyl phosphoglycolipid and the intracellular 'membrane' teichoic acids of the group D Streptococci suggests that there may be some relationship between these two classes of compounds. Important in this regard is the recent paper of Wicken and Knox[168] who isolated the intracellular teichoic acid of *Lactobacillus fermenti* and found a galactosyl glucosyl-containing glycolipid bound to the polymer, apparently by a phosphodiester linkage. Preliminary evidence indicates that the glycolipid linked to the teichoic acid has the same structure as the glycolipid found among the lipids of the organism. Although more experiments must be conducted, it is possible that the function of the phosphoglycolipids of the type shown in Fig. 5 is to anchor the intracellular teichoic acids to the membranes of Gram-positive bacteria[165].

12. PHOSPHOSPHINGOLIPIDS AND UNUSUAL LIPIDS OF MICROORGANISMS

Sphingolipids are exceedingly rare components of the lipids of microorganisms. It is of considerable interest, therefore, that White and his coworkers have been able to isolate and identify three ceramide phosphates in the Gram-negative anaerobe *Bacteroides melaninogenicus*[169–172]. Approximately 50% of the total extractable lipid phosphorus of this organism was stable to mild alkaline hydrolysis. Analysis of these resistant compounds led to the discovery of ceramide phosphorylglycerol, ceramide phosphorylethanolamine[170], and ceramide phosphorylglycerophosphate[172]. All three sphingolipids were present in strains CR2A_1, CR2A_2, and Lev, whereas ceramide phosphorylglycerophosphate was absent in strain Dog. Strains Lev, Dog, and CR2A_2 also had similar quantities of ceramide phosphates which were tentatively identified as ceramide phosphate and ceramide phosphorylserine. Should further experimentation substantiate the identity of the latter lipids, then *B. melaninogenicus* will have ceramide analogues of all its major diacyl phospholipids except diphosphatidylglycerol (i.e. phosphatidic acid, phosphatidylglycerol, phosphatidylserine[169]. Treatment of the ceramide phosphates with methanolic 2 *N* HCl at 100° resulted in the release of the amide-linked fatty acids, the dihydrosphingosine moieties, and the respective organic phosphate derivatives. Characterization of the released long-chain bases via GLC and mass spectroscopy revealed 17-methyloctadecasphinganine* (63%), octadecasphinganine (21%), 15-methyl-hexadecasphinganine

*Sphinganine is D-erythro-2-amino-octadecane-1,3-diol. (see Chapter 2).

(12%), and (possibly) 16-methyl-heptadecasphinganine (4%). The presence of branched-chain sphinganines is yet another unusual feature of the *B. melaninogenicus* lipids since they are quite rare. Carter and Gaver[173] had previously found branched-chain sphingosines among the components of ceramide aminoethylphosphonates of Tetrahymena.

The fatty acid constituents of the diacyl phospholipids and phosphosphingolipids of *B. melaninogenicus* seemed to be the same. Both fractions contained normal saturated, normal monoenoic, and iso- and anteiso-branched fatty acids and in both the major residue was 15:0 branched. There were considerable differences in distribution, however, in that the amide-linked fatty acids had a greater proportion of saturated and monoenoic residues while the diacyl phospholipids had more branched chain fatty acids[170]. Nothing is known about the metabolism of these and it will be interesting to see if the relationship between ceramide phosphorylglycerol and ceramide phosphorylglycerophosphate is analogous to that of phosphatidylglycerol and phosphatidylglycerophosphate.

A sphingomyelin-like lipid has been isolated from avian *Mycoplasma gallisepticum*[179]. The lipid could be labelled *in vivo* with ^{32}P, gave a positive test for choline, and had an infrared spectrum similar to that of authentic sphingomyelin. Despite these results, the identity must be considered tentative since the purity of the compound was not certain.

Investigations into the lipid composition of the yeasts has resulted in the discovery of two unique ceramide phosphoglycolipids. The first was found by Wagner and Zofcsik[174] in *Candida utilis* and *Saccharomyces cerevisiae*. Called 'mycoglycolipid', this compound on hydrolysis yielded C_{18}-dihydrosphingosine, fatty acids (containing 20, 22, and 24 carbon atoms), inorganic phosphate, inositol and mannose. The structure proposed for the intact lipid can be expressed:

mannose-inositol-phosphate-ceramide.

The linkages in the glycoside portion were not established. A similar lipid was isolated from *S. cerevisiae* by Steiner et al.[175] but it appears to have a higher content of inositol and phosphate. Using [^{3}H]*myo*-inositol, [^{14}C] glucose and ^{32}P to label specific parts of the molecule *in vivo*, the authors were able to hydrolyze the molecule selectively and analyze the radioactive products. On the basis of the results from these experiments, the following scheme best describes the structure of the intact lipid:

mannose-(inositol-phosphate)$_2$-ceramide.

In view of the obvious similarity between this lipid and the mycoglycolipid shown above, it is possible that only one of these components actually exists, the other resulting from an error in quantitation of the compounds. *S. cerevisiae* is also known to contain phosphatidyl-*myo*-inositol and recently inositol di- and triphosphate were isolated among the hydrolysis products of the phospholipid fraction, suggesting the presence of diphosphoinositide and triphosphoinositide[176]. Tanner[180] incorporated [^{14}C]*myo*-inositol into *S. cerevisiae* and found labelled *O*-α-D-mannopyranosyl-*myo*-inositol, *myo*-inositol phosphate, glycerylphosphoryl-*myo*-inositol and two unidentified compounds, in the degradation products from the lipid fraction. It appears then, that *S. cerevisiae* possesses a minimum of five inositol-containing lipids and a study of the metabolic relationships between these compounds would certainly prove to be rewarding.

During the course of studies on the aminoethylphosphonates as components of rumen protozoan lipids, Dawson and Kemp[177] noted the presence of an alkali-stable lipid which released inorganic phosphate and ethanolamine upon acid hydrolysis. The infrared spectrum indicated the presence of a ceramide and thus the compound was tentatively identified as a ceramide ethanolamine phosphate. An added feature of the unusual lipid composition of these rumen protozoa was the subsequent discovery of a novel phospholipid with the structure phosphatidyl-*N*-(2-hydroxyethyl)alanine[178]. The identity of the compound was aided by NMR studies and substantiated by chromatography of the nitrogenous fragment with synthetic DL-*N*-(2-hydroxyethyl)alanine in 11 solvent systems. The authors suggest that this lipid may substitute for phosphatidylserine which is known to be absent from these organisms[177].

13. SOME FACTORS INFLUENCING PHOSPHOLIPID METABOLISM *in vivo*

A. Temperature shifts

The literature contains an abundance of studies concerning the relationship between the growth temperature and fatty acid composition of bacterial lipids (for a review, see ref. 181). In general, it appears that as the temperature decreases the degree of fatty acid unsaturation increases. This empirical observation is teleologically appealing since unsaturated fatty acids are known to have lower melting points than the corresponding saturated acids. Thus, by increasing the content of unsaturated fatty acids in response to lower temperatures, micro-organisms can be said to preserve the proper ‘fluidity’ in the hydrophobic portions of their complex lipids. Despite several proposals which argue that the temperature range for viability is set by the fatty acid composition[182–184], the true significance of these changes has not yet been determined. Marr and Ingraham[185] examined the effect of growth temperature on the fatty acid composition of exponentially growing *E. coli* ML-30. They noticed that as the temperature of growth was decreased in 5° increments from 35° to 10°, the level of octadecanoic acid increased while that of palmitic acid decreased. In contrast, the levels of palmitoleic and methylene hexadecanoic acid remained essentially unchanged. Since Law et al.[44] have subsequently observed an increase in the proportion of cyclopropane fatty acids in late exponential phase in *E. coli* grown at 37°, it appears that the results obtained by Marr and Ingraham were due specifically to the change in temperature. It should be mentioned, however, that many of these changes could also be induced by changes in the growth medium[185] emphasizing once again the balance which exists between the pathways of lipid metabolism and the growth conditions. In another experiment the temperature of growth was raised to 43° (ref. 185), and this resulted in the octadecanoic : palmitic ratio dropping from 0.78 to 0.25, suggesting that increases in temperature lead to increase in saturation[101]. In a brief communication, Haest et al.[191] substantiated the findings of Marr and Ingraham and additionally made the novel proposal that alterations in fatty acid composition as a result of temperature changes are efforts by the bacteria to compensate for changes in membrane permeability. Thus, an increase in temperature results in an increased permeability which can be compensated for by increasing the degree of saturation. This is supported by the work of Demel et al.[192] who recognized that the packing of phospholipid in molecules in monolayers is determined primarily by the nature of the hydrophobic groups, the smallest area being occupied by saturated fatty acids. Increasing fatty acid saturation with increasing growth temperature is apparently not a universal phenomenon, however, since increases in unsaturation have also been observed in micro-organisms under conditions of increasing temperature[186, 187].

Ng et al.[188] noticed that when *E. coli* ML-30 growing exponentially at 37° were suddenly shifted to 10°, a 4–5-h lag period occurred before growth was resumed. During this lag there was no net RNA, DNA or protein synthesis and no increase in cell number[187], but changes in the fatty acid composition, similar to those reported by Marr and Ingraham[185] took place. When glucose was withheld during the lag period, no changes in fatty acid composition were observed, although the addition of glucose at the end of the lag resulted in a resumption of growth[189]. This evidence that these cells could grow at 10° despite the fact that they possessed the lipid composition of normally grown cells clearly shows that lipid changes during the lag are not prerequisites for growth.

In view of the evidence that lag phase cells at 10° are not growing, this system affords an excellent opportunity to study the turnover rate of individual phospholipids. Such a study was recently conducted by Okuyama[190] who found that during the lag period there was a 20% increase in the total amount of phospholipid per cell. As would be expected, most of the increase was due to synthesis of phosphatidylethanolamine and phosphatidylglycerol. Unexpected, however, was the finding from chase experiments that phosphatidylethanolamine had a considerable turnover. This was surprising in view of previous studies which had indicated that phosphatidylethanolamine in *E. coli* was stable[27, 28]. Evidently, the metabolism of the phospholipids during the 4-h lag differs from that under normal conditions. In apparent agreement with this conclusion are data suggesting that there is also more extensive degradation of the phospholipids during the lag period at 10° than during growth at 37°. When cells homogeneously labeled with [1-^{14}C]acetate at 37° were chased at 10°, there was essentially no change in the radioactivity of the total lipid fraction. Analysis of the individual lipids, however, detected a decrease in the label present in the neutral (triglyceride, diglyceride, fatty acid) lipids and a concomitant increase in the phospholipid fraction. The author interpreted these results as indicating a transfer of acyl moieties from the neutral lipid to the phospholipid fraction. The mechanism whereby the deacylation–reacylation of the phospholipid is effected is not known, but the Lands cycle (phospholipase–acyltransferase) or the diglyceride kinase pathways may be involved[2].

De Siervo[193] grew *E. coli* B at 27° without any major influence on the growth rate although the growth yield seemed to be higher at this temperature than at 37°. The 10° fall in temperature resulted in an altered lipid pattern involving the metabolism of phosphatidylethanolamine, phosphatidylglycerol, and phosphatidylglycerophosphate. Of particular interest was the observation that phosphatidylglycerophosphate (tentative identification) accumulates at 27°, indicating that the dephosphorylation of this lipid

References p. 327

may be impaired at the lower temperature. Such results give credence to the idea that the complex changes which occur in the phospholipid metabolism of *E. coli* during the temperature shifts are merely manifestations of the effect of the temperature on the individual enzymes involved. Evidently this explanation cannot be extrapolated to all bacteria, however, since Joyce et al.[194] lowered the temperature of an exponentially growing culture of *S. aureus* U-71 from 37° to 25° without any noticeable changes in the fatty acid or phospholipid composition. Exponential growth continued at this lower temperature but with a three-fold increase in the doubling time. Perhaps there is a tighter coupling of lipid metabolism to the other metabolic processes in the cells of Gram-positive organisms.

A recent study of the relationship between increased growth temperature and the composition of the cellular membranes of *Tetrahymena pyriformis* has been made by Erwin[195]. Growth of this organism at 40° induced morphological abnormalities in the cell surface and resulted in a nutritional requirement for certain lipids. The requirement would seem to be for polyunsaturated fatty acids since either asolectin (a mixture of plant phosphatidylcholine, phosphatidylethanolamine, and phosphatidylinositol, with linoleic acid as the major fatty acid) or purified fatty acids from the cilia of this protozoan (60% unsaturated C_{18}) will support growth at this elevated temperature. The author believes that the increased temperature at least partially inhibits the biosynthesis of polyunsaturated fatty acids and probably also interferes with the formation of phospholipid since the relative phospholipid content of these abnormal cells was only one third that of the normal cells. Paradoxically, the synthetic phospholipid dipalmitoyl-*sn*-glycero-3-phosphoryl ethanolamine, despite the presence of only saturated fatty acids, also supports growth at 40°. Analysis of the fatty acid composition of cells grown with the synthetic phospholipid revealed that the fatty acid pattern was drastically changed: whereas 60% of the fatty acids in cells grown at 35° were unsaturated C_{18}, these accounted for less than 10% of the fatty acids in cells grown at 40° in the presence of the synthetic phospholipids. Conversely, the level of palmitic acid rose from 10% in the normal to over 70% in the high temperature cells. The evidence suggests that the synthetic phospholipids are able to ameliorate the condition of unfavourable temperature by being incorporated intact into the membranes of the organism. Such an exchange should prove to be valuable in the studies of ciliate protozoal membranes.

B. Induction of electron transport systems

One of the most profound alterations in the metabolism of facultative organisms occurs during the transition from the anaerobic state. A primary consequence of this change is the formation of a membrane-bound respiratory chain consisting of protein, lipid, cytochromes, and so forth, which enables the organism to derive energy via electron transport. Several studies have been made of this adaptive phenomenon in an effort to understand the mechanism by which the integration of these compounds is controlled. As would be expected, the bacteria, with their rapid growth rate, ease of handling, and relatively uncomplicated structure, have been invaluable in this regard. *Staphylococcus* *aureus*, when grown anaerobically, derives its energy from glycolysis and contains no detectable respiratory system[196]. The presence of oxygen during growth however, results in an increased growth rate and induces the formation of a membrane-bound electron-transport system consisting of flavoproteins, cytochrome b-b_1, cytochrome a, and cytochrome oxidase O[197, 198]. Frerman and White[198] observed that during the aerobic shift, cytochrome b_1 and cytochrome oxidase O appeared first, followed 4 h later by the appearance of cytochrome a. The rate of induction of each cytochrome was not constant so that the ratios of these constituents tended to vary throughout the transition. Concomitant with the formation of cytochromes was a 100-fold decrease in the critical oxygen tension*, but 66% of this decrease occurred only after cytochrome a had appeared. Previous studies on the lipids of *S. aureus*[199, 200] had established that phosphatidylglycerol, lysyl-phosphatidylglycerol, and diphosphatidylglycerol accounted for up to 98% of the total lipid phosphorus. During the induction of the cytochrome system in *S. aureus*, Frerman and White[198] observed a simultaneous 2-fold increase in the total phospholipid concentration. The increase was not uniform, however, since phosphatidylglycerol increased 2-fold and diphosphatidylglycerol 1.6-fold whereas the concentration of lysyl-phosphatidylglycerol remained relatively constant. As a consequence of this, the molar ratio between phosphatidylglycerol and lysyl-phosphatidylglycerol varied between 4.1:1 and 8.9:1. Since it is known that essentially all of the lipid in *S. aureus* is part of the same membrane as the respiratory pigments[200], these results indicate that the electron-transport membrane of this organism can function under conditions in which widely different molar proportions of the components of this system are present. Furthermore, such results imply that the membrane of *S. aureus* is a mosaic of regions which differ in composition and possibly function[200].

In contrast to the relatively minor changes in phospholipid composition, White and Frerman[201] noted dramatic changes in the fatty acid composition of complex lipids of *S. aureus* during the anaerobic transition. In an exhaustive study, 64 fatty acids from this organism were separated and identified via GLC as normal, iso(i)- and anteiso(a)-derivatives of saturated and monoenoic acids containing from 10 to 21 carbon atoms. Under conditions resulting in the formation of the respiratory chain, the 2-fold increase in phospholipid was paralleled by a 2-fold increase in the fatty acid dry weight. This increase, however, was not proportional for all fatty acids. In anaerobically growing *S. aureus*, C_{18} and C_{20} fatty acids comprised 59% of the total. These plus i-C_{11}, C_{14}, a-C_{15}, C_{16} and C_{19} also constituted 91% of the total fatty acid dry weight. In contrast, in aerobically growing cells, a-C_{15} and C_{20} comprised 58% of the total and these plus C_{16}, a-C_{17}, C_{17}, C_{18}, a-C_{18}, i-C_{19}, and C_{19} composed 91% of the total fatty acid. In all, there were major changes in 23 (of 64) of the fatty acids of the complex phospholipids (Table 3). A study of the fatty acids of the individual lipids revealed that in lysyl-phosphatidylglycerol, 14 fatty acyl residues showed significant decreases in concentration while 6 fatty acyl residues showed significant increases in concentration (Table 3). Also, whereas C_{18} was the major fatty acid in the anaerobic state, a-C_{15} assumed this position under

*The concentration at which the rate of oxygen utilization is directly proportional to the oxygen tension i.e. it provides a measure of the affinity of the electron transport chain for oxygen.

TABLE 3

CHANGES IN THE AMOUNTS OF FATTY ACID PER g (DRY WEIGHT) IN THE COMPLEX LIPIDS OF *Staphylococcus aureus* DURING THE SHIFT FROM ANAEROBIC TO AEROBIC GROWTH

Fatty acid	*LPG*	*PG*	*DPG*
10	N	D	D
11, i	D	I	I
11	D	I	D
11, =	I	D	D
12, i	D	I	D
12, a	D	I	D
12, =	I	I	D
13, i	D	D	I
13, a	D	I	I
13	N	I	N
13, =	D	D	I
14, i	D	D	I
14, a	D	I	D
14, =	I	I	D
15	D	N	I
15, =	I	I	D
16, a =	I	I	N
16, =	N	D	D
17, i	D	I	I
17, =	D	I	I
18	D	I	I
19	I	I	D
19, =	D	I	I

D indicates a 2- to over 200-fold decrease in the amount of fatty acid per g (dry weight) in the lipid; I indicates a 2- to over 120-fold increase in the molar content of fatty acid per g (dry weight) in the lipid, and N indicates a less than 2-fold change in the absolute molar content of the fatty acid per g (dry weight) in the lipid. LPG, PG, and DPG indicate lysyl phosphatidylglycerol, phosphatidylglycerol, and diphosphatidylglycerol. From White and Frerman[201], reproduced by permission of the authors and the American Society for Microbiology.

aerobic conditions. The fatty acids of phosphatidylglycerol were also considerably altered; 16 increased in concentration while 6 decreased, but a-C_{15} remained the major fatty acid. In diphosphatidylglycerol a total of 21 fatty acids were altered, 11 decreasing in concentration and the rest increasing. A comparison of the fatty acid composition of phosphatidylglycerol, lysyl-phosphatidylglycerol, and diphosphatidylglycerol revealed considerable differences in the proportions of the various residues, but relatively few instances in which a fatty acid was present in one of the phospholipids yet absent in another. Since these lipids are metabolically related, the authors interpreted these results as indicating that deacylation–reacylation reactions were occurring at the level of the intact lipid during the anaerobic–aerobic shift. It is also possible, however, that the differing proportions of fatty acids in these three phospholipids is a reflection of changes in the flux of the phosphatidylglycerol branch-point. If the fatty acid composition of the precursor phosphatidic acid is changing with time, then only if lysyl-phosphatidylglycerol and diphosphatidylglycerol are formed at the same rate would the fatty acid composition of both of these products be similar. The possibility of separate CDPdiglyceride pools must also be considered since the enzyme responsible for the formation of diphosphatidylglycerol from phosphatidylglycerol may have a different CDPdiglyceride specificity than the enzyme which forms phosphatidylglycerol from glycerol-3-phosphate. Regardless of the mechanism responsible for the alterations in fatty acid composition, the results cited above clearly indicate that dramatic alterations of membrane phospholipid metabolism occur during the formation of the electron transport system.

White and co-workers have also studied the phospholipid metabolism of *Hemophilus parainfluenzae* during an oxygen-induced modification of the respiratory chain. In this organism, an electron-transport system composed of dehydrogenases, cytochromes, cytochrome oxidases, quinone, and phospholipid[202, 203] is required for metabolic activity and has been shown to be associated with the membrane-wall complex. The predominant phospholipids of this complex are phosphatidylethanolamine (78%), phosphatidylglycerol (18%), and diphosphatidylglycerol (3%); phosphatidylmonomethylethanolamine, phosphatidylcholine, phosphatidylserine and phosphatidic acid have also been identified, but each accounts for less than 0.5% of the total phospholipid[72]. Extraction of these lipids with organic solvents results in disruption of electron transport, suggesting some role for phospholipids in this process[204]. If the oxygen tension of the medium is lowered, *H. parainfluenzae* responds by producing a more extensive cytochrome system which is capable of a much more rapid electron transport[203, 205]. Thus, during the transition from high to low oxygen tension, there is an average 50-fold increase in the content of cytochromes a_2, a, b_1, c_1, and O, with a concomitant 10-fold increase in oxygen utilization. Chase experiments conducted with ^{32}P during the period of rapid cytochrome synthesis demonstrated a marked decrease (2.3 and 2.0-fold, respectively) in the turnover of the phosphate moiety of phosphatidylethanolamine and diphosphatidylglycerol, but a 1.8-fold increase in the turnover of phosphatidylglycerol[206]. It is significant that these differential rates of turnover occurred despite the fact that there was no change in the bacterial doubling time. This is in contrast to the situation in *S. aureus* where the aerobic shift resulted in a more rapid growth rate[198]. Early in the period of rapid cytochrome synthesis in *H. parainfluenzae* there was a steady increase in the proportions of phosphatidylglycerol (from 17 to 23% of the total lipid phosphate) and diphosphatidylglycerol (from 2 to 4%) and a decrease in phosphatidylethanolamine (from 80 to 75%). The net result of these changes was an overall 5–10% increase in total phospholipid. As the steady-state level of the cytochromes was achieved, however, the total lipid phosphate, the rate of incorporation of labelled phosphate into phospholipids, and the proportions of phosphatidylglycerol and diphosphatidylglycerol all decreased[206]. During the period of cytochrome induction, major alterations were evidently occurring in pool sizes also, since considerably more time was required to saturate the phosphate pools entering phosphatidylethanolamine, phosphatidylglycerol, and diphosphatidylglycerol. This was especially true in the case of diphosphatidylglycerol wherein it took 7–9 times longer for ^{32}P to saturate the diphosphatidylglycerol precursor pool when comparing aerobic and non-aerated cells[206]. In the reverse situation, when cells grown without aeration (and therefore having a high cytochrome content) were suddenly shifted to high aeration, the turnover of phosphatidylglycerol and diphosphatidylglycerol increased 1.5 and 1.1-fold, respectively, but phosphatidylethanolamine turnover decreased 1.2-fold.

It is evident then, that profound changes in phospholipid metabolism occur during the proliferation of the *H. parainfluenzae* cytochrome system in response to lower oxygen tensions. The fact that both the rate of turnover and incorporation of phosphate into phospholipid can be rapidly altered despite a constant rate of growth certainly suggests that lipid metabolism is in some way coupled to cytochrome synthesis in this organism. It is also significant that most of the changes elicited by growth under low aeration are reversible. This implies that lipids must be modified at the membrane level. The recent finding by Ono and White[34] of a diphosphatidylglycerol-specific phospholipase D associated with the membrane of *H. parainfluenzae*, is very interesting in this regard. The enzyme requires $Mg^{2\oplus}$ for maximal activity and hydrolyzes diphosphatidylglycerol, phosphatidic acid and phosphatidylglycerol. The possible role of this phospholipase in the highly complex lipid metabolism of *H. parainfluenzae* is discussed in the section dealing with the metabolism of these lipids *in vitro*.

Rhodopseudomonas spheroides is a much studied photosynthetic bacterium which undergoes adaptive changes during the transition from growth in air to semi-anaerobic growth in the light. The pigments of this organism together with the enzymes and co-factors required for photophosphorylation are located in relatively homogeneous structures called 'chromatophores'. Work by Cohen-Bazire and Kunisawa[211], Gorchein[208, 209] and Gorchein et al.[46, 210] has revealed that the chromatophores arise as invaginations of the peripheral cytoplasmic membrane in response to the transition from the non-pigmented, aerobic state to (semi)-anaerobic growth in the light. During this shift, the membrane is modified by the addition of a number of constituents and then portions of the modified membrane are further changed by deposition of the photosynthetic pigmenta (carotenoids, bacterio-chlorophyll) to form the chromatophores. These can be isolated as homogeneous entities due to the presence of so-called 'weak points'[209, 212] at the junction of the chromatophore and external membrane. The isolated chromatophore vesicles have a high lipid content and resemble the cell membrane in containing phosphatidylethanolamine, phosphatidic acid, and a lipid tentatively identified as diphosphatidylglycerol[47, 207]. Gorchein[207] also reported the presence of a sulpholipid and a lipid containing ornithine. Lascelles and Szilagyi[47] studied the phospholipid metabolism of *R. spheroides* NCIB 8253 during the formation of pigmented chromatophores. They observed that cells grown anaerobically in the light had 50–70% more phospholipid than did non-pigmented cells grown with high aeration. Despite this increase, howevet, the proportion of the individual lipids did not appear to vary during the period of pigment formation. Essentially the same results were obtained by Gorchein[207, 46, 52] who additionally observed that if cells previously grown aerobically on ^{32}P and [*Me*-^{14}C] methionine were shifted to anaerobic growth in nonradioactive medium in the light the total radioactivity of the phospholipids remained constant, as did the ratio of ^{32}P to ^{14}C. Since very little label from the [*Me*-^{14}C] methionine was incorporated into fatty acyl moieties[46], these results lead to the conclusion that the non-fatty acid portions of the phospholipids were metabolically stable. This is probably true of the fatty acid residues also, since previous studies[213, 214] had shown that there were no major changes in the fatty acid composition of the phospholipids when comparing aerobic and photosynthesizing cells.

Lascelles and Szilagyi[47] found that there was a different rate of ^{32}P incorporation for each lipid during pigment formation such that phosphatidylglycerol was labeled more rapidly than either phosphatidic acid or phosphatidylethanolamine which in turn were more rapidly labeled than phosphatidylcholine. The same pattern was observed under conditions of aeration and also in a mutant (L-70) which was unable to form pigments, thus indicating that the effect was probably not related to light-induced chromatophore induction. Although the shift toward pigment formation resulted in increased amounts of phospholipid, there was no evidence to indicate that pigmented cells had more protein than non-pigmented cells. It was thus postulated that the membrane formed during the transition (i.e. the chromatophore membrane) represents a more lipid-rich structure than the membrane present under aerobic conditions. In contrast, Gorchein et al.[52] found that the specific activity of the phospholipids in the chromatophore was the same as that of the adapted cells. They concluded therefore, that during adaptation to anaerobic growth in the light, newly formed phospholipid was evenly distributed throughout the entire membrane system of this organism. Despite the disparity of these views it is clear that the apparent metabolic stability and constant proportion among the phospholipids of *R. spheroides* during the aerobic–anaerobic light shift is indicative of a different mode of membrane biosynthesis than that seen in *S. aureus* and *H. parainfluenzae.*

An interesting situation exists in the photosynthetic bacterium Chromatium strain D where growth under conditions of low incident light results in the proliferation of cells containing large numbers of photosynthetic vesicles (chromatophores)[215]. This phenomenon is reversible because growth under conditions of high intensity incident light leads to a significant reduction in the number of these vesicles. Steiner et al.[216] found that although the proportions of the major lipids phosphatidylethanolamine, phosphatidylglycerol, and diphosphatidylglycerol[217] remained approximately the same, cells grown under low-intensity light had more chromatophore material and more bacteriochlorophyll, but less phospholipid than did cells grown under high-intensity light. Thus the amount of pigment in the cell does not correlate with the amount of phospholipid. Evidently then, the formation of photosynthetic vesicles in this organism differs from that in *R. spheroides* in which pigment and phospholipid formation increased in parallel.

14. FUNCTIONAL ASPECTS OF PHOSPHOLIPIDS IN MICROBIAL MEMBRANES

The intense interest of investigators in membranes has resulted in a number of excellent recent studies which reveal in greater detail the functional importance of phospholipids in microbial membranes. Experiments on the assembly of sugar transport systems in bacteria have been particularly fruitful in demonstrating the involvement of phospholipids in a dynamic process. Using an unsaturated fatty acid auxotroph of *Escherichia coli* K-12, Fox[218] showed that β-galactosidase and galactoside acetyltransferase activities but not β-galactoside transport (all three are derived from the structural genes of the lac-operon) were inducible during unsaturated fatty acid starvation. The correlation between the absence of unsaturated fatty acids (a condition yielding lipids containing only saturated fatty acids[219]) and the lack of β-galactoside transport suggested that the synthesis of unsaturated lipid was required for the induction of β-galactoside transport. Also, this requirement indicated that the newly synthesized transport proteins resulting from induction appeared to associate preferentially with newly synthesized lipids. By controlling the type of exogenous unsaturated fatty acid incorporated into the lipids of the membrane of the auxotroph, Wilson and Fox[220] were able to study sugar transport in cells containing lipids of varying physical (melting or liquidity) characteristics. The rate of transport of either β-galactosides or β-glucosides when measured over a temperature range yielded biphasic Arrhenius plots. The point of intersection of the two-slopes of the Arrhenius plot, which is referred to as the transition temperature, was unique for each of the essential unsaturated fatty acids incorporated into the lipids of the auxotroph. The transition temperatures for β-glucoside and β-galactoside transport were identical for cells grown on the same unsaturated fatty acid. Since the two types of sugars have transport systems that are genetically and mechanistically different, the above observation supports the idea that the temperature-transport characteristics are mainly due to the physico-chemical properties of the fatty acid moiety of the lipids in the membrane.

In the biphasic Arrhenius plots which measured sugar transport as a function of temperature, the slopes of the lines above the transition temperature did change and were characteristic of each unsaturated fatty acid presented to the auxotroph[220]. The effect of fatty acid composition on transport was determined readily by assaying for transport at two different temperatures; one above and the other below the transition temperature. The temperatures selected for β-galactoside transport were 10° and 28°, respectively; and for β-glucoside transport 5° and 28°, respectively. When the auxotroph was first grown in the presence of one unsaturated fatty acid (oleate) and then induced in the presence of a second unsaturated fatty acid (linoleate) or if induced first and then grown further on these fatty acids the characteristics of transport were predominantly those of the fatty acid present during the induction period and not of the fatty acid present only during growth. Again it would seem that newly synthesized transport proteins associate and remain associated with the newly synthesized lipids formed during induction of transport protein[220].

In a glycerol-requiring mutant of *E. coli*.K-12 with a defect in *sn*-glycerol-3-phosphate formation from dihydroxyacetone phosphate, Hsu and Fox[221] observed that glycerol starvation resulted in the inhibition of synthesis of total cell protein, deoxyribonucleic acid and phospholipid. Under these conditions phospholipid synthesis ceased abruptly whereas total cell protein and DNA synthesis proceeded for a short time. Glycerol starvation produced no pronounced decrease in the amount of labeled protein (label derived from [^{14}C]isoleucine) into total membrane protein. However, omission of glycerol during induction of the lac-operon markedly reduced induction of lac transport but had much less effect on the induction of β-galactosidase or galactoside acetyltransferase activities. This reduction in transport was not improved if the mutant was grown further in the presence of glycerol, but in the absence of inducer. These results confirm the data obtained with the unsaturated fatty acid auxotroph, and show that synthesis of lipid and of transport protein must occur simultaneously if functional transport sites are to be made[221].

The amount of M-protein (the β-galactoside carrier of the lac-operon) formed in the presence of inducer, but in the absence of glycerol changed to the same extent as the amount of β-galactosidase protein. However, the change in β-galactoside transport was almost 10 times lower than the change in concentration of either of these two proteins. The M-protein induced during glycerol starvation showed no defect in its binding capacity for β-galactosides. These results argue against the idea that the lack of transport is due to the production of a defective M-protein. Further, these data show that the defect in transport induction during glycerol starvation is brought about by events subsequent to the translation step in protein synthesis.

A glycerol auxotroph of *Staphylococcus aureus*, which also displayed a mutation for temperature-sensitive inducibility of β-galactoside transport, has been isolated and studied by Mindich[222]. In this auxotroph glycerol starvation prevented induction of transport to a much greater extent than the induction of enzymes (6-phospho-β-galactosidase and the phosphotransferase which transfers phosphate from phosphoenol pyruvate to galactoside) associated with transport process. The permease of β-galactosides (lactose) in *S. aureus* involves this phosphotransferase system described by Kundig and Roseman[223, 224] rather than the lactose system mentioned above for *E. coli*. The *S. aureus* mutant, like the *E. coli* mutant[221], did incorporate protein into membrane but did not synthesize phospholipids and glycerides during induction carried out in the absence of glycerol from the growth medium. Another similarity is that once transport is induced in the *S. aureus* mutant (by growth in the presence of inducer and glycerol) removal of the glycerol from the medium results in only small losses of transport activity[222]. However, unlike the *E. coli* mutant, the *S. aureus* auxotroph which had failed to be induced for transport activity when grown in the absence of glycerol slowly regained partial transport activity when placed in a second incubation medium containing glycerol but no inducer. Thus, in the *S. aureus* double mutant β-galactoside transport is not irreversible in that there is a partial increase in efficiency of transport in the absence of protein synthesis. This variance with the lac system of *E. coli* may be due to the difference in the method of induction (*E. coli* system used a β-galactoside; *S. aureus* system used a shift in temperature from 34° to 44°) or to the difference in the genetics and mechanism of action of the β-galactoside permeases in the two bacteria.

β-Galactoside transport in *S. aureus* as mentioned above occurs by the phosphotransferase system[223, 224]. Studies by Kundig and Roseman on the phosphotransferase of *E. coli*[224] have shown that membrane-bound enzyme II, which catalyzes the phosphorylation of certain sugars from phosphohistidine-protein (P-HPr), contains phosphatidylglycerol as an essential, structural constituent. Phosphatidylglycerol has been directly implicated in

References p. 327

the transport of sugars mediated by the phosphotransferase system. Milner and Kaback[225] found that treatment of membrane preparations of *E. coli* ML 308-225 with phospholipase D resulted in the apparently specific hydrolysis of phosphatidylglycerol and a concomitant loss of α-methylglucoside. Transport activity was restored on resynthesis of phosphatidylglycerol in the membrane. Although no one specific phospholipid has been implicated in the transport studies of Fox and coworkers[218, 220, 221], and Mindich[222] the association of phosphatidylglycerol with the phosphotransferase system[224] implies that it may very well be involved in the systems they studied.

Phosphatidylethanolamine has been conclusively demonstrated to be required in the biosynthesis of the lipopolysaccharide located in the outer membrane of the cell envelope of Gram-negative bacteria[226]. This subject has been reviewed recently[226], and will only be summarized here. One portion of the lipopolysaccharide, which is referred to as the core region, is made up of a chain of glucose, galactose, and glucosamine residues. These monosaccharides are enzymatically linked together by the stepwise transfer of sugar residues from nucleoside diphosphate sugar substrates to the nonreducing end of an elongating chain. Two of the glycosyl transglycosyl transferases (UDP-glycoside : lipopolysaccharide glycosyl transferase) could be partially solubilized by sonication and were found to have an absolute requirement for phosphatidylethanolamine. The specificity for a phosphatidylethanolamine having a particular structure was rather high: choline could not be substituted for the ethanolamine moiety and unsaturated or cyclopropane fatty acid moieties could not be replaced by saturated fatty acids nor could they be partially removed (to yield lysophosphatidylethanolamine). The role of the phosphatidylethanolamine is to interact noncovalently with the elongating lipopolysaccharide and in so doing convert the lipopolysaccharide from an inactive to an active substrate. This interaction was promoted *in vitro* by gentle heating to 60°. Once formed, the binary complex is capable of binding the glycosyl transferase to form a ternary complex (lipopolysaccharide:phosphatidylethanolamine:enzyme), which in turn will accept the sugar from UDP-sugar substrate. These studies *in vitro*, which included the use of monolayer techniques, showed that the reconstitution of a fully operative transferase system specifically required that the components be added in the order listed above, that is; phosphatidylethanolamine, lipopolysaccharide, enzyme and UDP-sugar. Formation of the ternary complex did not occur if the order of addition was changed. This prescribed sequence is likely to be required in membrane assembly *in vivo*. The suggestion was made[236] that the arrangement of the molecules in the binary and ternary monolayers constructed *in vitro* are similar to their arrangement in the cell envelope, except that the natural membrane would be a bimolecular leaflet, the equivalent of two molecular monolayers.

The above examples of the participation of phospholipids in the function of microbial membranes provide an excellent foundation on which future studies in this area can be built.*

---

*While this chapter was being prepared, a proof ot a review by Goldfine[227] was made available to us. The reader should find the Tables in this review listing the lipid composition of many bacteria and spirochaetales particularly valuable.

REFERENCES

1 E.P. Kennedy, *Ann. Rev. Biochem.*, 26 (1957) 119.
2 R.A. Pieringer, H. Bonner Jr. and R.S. Kunnes, *J. Biol. Chem.*, 242, (1967) 2719.
3 H. Goldfine, G.P. Ailhaud and P.R. Vagelos, *J. Biol. Chem.*, 242 (1967) 4466.
4 N.J. Kuhn and F. Lynen, *Biochem. J.*, 94 (1965) 240.
5 M. Kito and L.I. Pizer, *J. Bacteriol.*, 97 (1969) 1321.
6 H. van den Bosch and P.R. Vagelos, *Biochim. Biophys. Acta*, 218 (1970) 233.
7 R.A. Pieringer and R.S. Kunnes, *J. Biol. Chem.*, 240 (1965) 2833.
8 W.E.M. Lands, R.A. Pieringer, P.M. Slakey and A. Zschocke, *Lipids*, 1 (1966) 444.
9 Y.-Y. Chang, and E.P. Kennedy, *J. Biol. Chem.*, 242 (1967) 516.
10 F. Lynen, D. Oesterhelt, E. Schweizer and K. Willecke, in F.C. Gran (Ed.) *Cellular Compartmentalization and Control of Fatty Acid Metabolism*, Academic Press, New York, 1968, p. 1.
11 R.A. Pieringer, *Federation Proc.*, 24 (1965) 476.
12 G.P. Ailhaud and P.R. Vagelos, *J. Biol. Chem.*, 241 (1966) 3866.
13 T. Kaneshiro and A.G. Marr, *J. Biol. Chem.*, 236 (1961) 2615.
14 J.G. Hildebrand and J.H. Law, *Biochemistry*, 3 (1964) 3.
15 T.K. Ray, J.E. Cronan Jr., R.D. Mavis and P.R. Vagelos, *J. Biol. Chem.*, 245 (1970) 6442.
16 Y.-Y. Chang and E.P. Kennedy, *J. Lipid Res.*, 8 (1967) 456.
17 N. Shaw, *Bacteriol. Rev.*, 34 (1970) 365.
18 R.T. Ambron and R.A. Pieringer, *J. Biol. Chem.*, 246 (1971) 4214.
19 S.G. Wilkinson, *Biochim. Biophys. Acta*, 187 (1969) 492.
20 D.C. White and A.N. Tucker, *J. Lipid Res.*, 10 (1969) 220.
21 B.W. Agranoff, R.M. Bradley and R.O. Brady, *J. Biol. Chem.*, 233 (1958) 1077.
22 H. Paulus and E.P. Kennedy, *J. Biol. Chem.*, 235 (1960) 1303.
23 J.R. Carter Jr., *J. Lipid Res.*, 9 (1968) 748.
24 R.E. McCaman and W.R. Finnerty, *J. Biol. Chem.*, 243 (1968) 5074.
25 P.H. Patterson and W.J. Lennarz, *J. Biol. Chem.*, 246 (1970) 1062.
26 B.W. Agranoff and W.D. Suomi, *Biochem. Prep.*, 10 (1963) 47.
27 J. Kanfer and E.P. Kennedy, *J. Biol. Chem.*, 238 (1963) 2919.
28 Y. Kanemasa, Y. Akamatsu and S. Nojima, *Biochim. Biophys. Acta*, 144 (1967) 382.
29 J. Kanfer and E.P. Kennedy, *J. Biol. Chem.*, 239 (1964) 1720.
30 J. Kiyasu, R.A. Pieringer, H. Paulus and E.P. Kennedy, *J. Biol. Chem.*, 238 (1963) 2293.
31 Y.-Y. Chang and E.P. Kennedy, *J. Lipid Res.*, 8 (1967) 447.
32 Y.-Y. Chang and E.P. Kennedy, *J. Lipid Res.*, 8 (1967) 456.
33 N.Z. Stanacev, Y.-Y. Chang and E.P. Kennedy, *J. Biol. Chem.*, 242 (1967) 3018.
34 Y. Ono and D.C. White, *J. Bacteriol.*, 104 (1970) 712.
35 J.H. Law, *Bacteriol. Proc.*, 129 (1961).
36 C.L. Randle, P.W. Albro and J.C. Dittmer, *Biochim. Biophys. Acta*, 187 (1969) 214.
37 G. Ames, *J. Bacteriol.*, 95 (1968) 833.
38 P.R. Proulx and L.L.M. van Deenen, *Biochim. Biophys. Acta*, 144 (1967) 171.
39 P.R. Proulx and C.K. Fung, *Can. J. Biochem.*, 47 (1969) 1125.
40 S.S. Taylor and E.C. Heath, *J. Biol. Chem.*, 244 (1969) 6605.
41 E.A. Dennis and E.P. Kennedy, *J. Lipid Res.*, 11 (1970) 394.
42 W.B. Geiger Jr. and R.J. Anderson, *J. Biol. Chem.*, 129 (1939) 519.
43 T. Kaneshiro and A.G. Marr, *J. Lipid Res.*, 3 (1962) 184.
44 J.H. Law, H. Zalkin and T. Kaneshiro, *Biochim. Biophys. Acta*, 70 (1963) 143.
45 H. Goldfine and M.E. Ellis, *J. Bacteriol.*, 87 (1964) 8.
46 A. Gorchein, A. Neuberger and G.H. Tait, *Proc. Roy. Soc., B.*, 170 (1968) 299.
47 J. Lascelles and J.F. Szilagyi, *J. Gen. Microbiol.*, 38 (1965) 55.
48 A. Gorchein, *Biochim. Biophys. Acta*, 84 (1964) 356.
49 B.J. Wood, B.W. Nichols and A.T. James, *Biochim. Biophys. Acta*, 106 (1965) 261.
50 A.A. Benson, F.G.M. Wintermans and R. Wiser, *Plant Physiol.*, 34 (1959) 315.
51 C.E. Park and L.R. Berger, *J. Bacteriol.*, 93 (1967) 221.
52 A. Gorchein, A. Neuberger and G.H. Tait, *Proc. Roy. Soc., B.*, 170 (1968) 311.
53 M. Ikawa, *Bacteriol. Rev.*, 31 (1967) 54.

54 P.O. Hagen, H. Goldfine and P.J.L. Williams, *Science*, 151 (1966) 1543.
55 H. Goldfine and P.O. Hagen, *J. Bacteriol.*.95 (1968) 367.
56 I. Yano, Y. Furukawa and M. Kusunose, *J. Bacteriol.*, 98 (1969) 124.
57 J.M. Shively and A.A. Benson, *J. Bacteriol.*, 94 (1967) 1679.
58 R.L. Lester and D.C. White, *J. Lipid Res.*, 8 (1967) 565.
59 G. Lust and L.J. Daniel, *Arch. Biochem. Biophys.*, 113 (1966) 603.
60 S.A. Short, D.C. White and M.I.H. Aleem, *J. Bacteriol.*, 99 (1969) 142.
61 J.D. Smith and J.H. Law, *Biochim. Biophys. Acta*, 202 (1970) 141.
62. R.A. Bobo and R.G. Eagon, *Can. J. Microbiol.*, 14 (1968) 503.
63 O.W. Thiele, D. Busse and K. Hoffmann, *European J. Biochem.*, 5 (1968) 513.
64 V.C. Dewey, in G.W. Kidder, M. Florkin and B.T. Schaer, (Eds.) *Chemical Zoology*, Vol I, Academic Press, New York, 1967, p. 161.
65 J.A. Erwin, in D.E. Buetow (Ed.), *Biology of Euglena*, Vol. II, Academic Press, New York, 1968, p. 133.
66 F. Davidoff and E.D. Korn, *Biochem. Biophys. Res. Commun.*, 9 (1962) 54.
67 E.D. Korn, *J. Biol. Chem.*, 238 (1963) 3584.
68 M.H. Hack, R.G. Yaeger and T.D. McCaffery, *Comp. Biochem. Physiol.*, 6 (1962) 247.
69 I. Goldberg and I. Ohad, *J. Cell Biol.*,.44 (1970) 563.
70 D. Hulanicka, J. Erwin and K. Bloch, *J. Biol. Chem.*, 239 (1964) 2778.
71 J. Erwin and K. Bloch, *J. Biol. Chem.*, 238 (1963) 1618.
72 D.C. White, *J. Bacteriol.*, 96 (1968) 1159.
73 R.M.C. Dawson, and P. Kemp, *Biochem. J.*, 105 (1967) 837.
74 H. Meyer and F. Meyer, *Biochim. Biophys. Acta*, 231 (1971) 93.
75 G.L. Graff, B. Vanderkelen, C. Guening and J. Humpers, *Compt. Rend. Soc. Biol.*, 162 (1968) 1635.
76 E. Merdinger, P. Kohn and R.C. McClain, *Can. J. Microbiol.*, 14 (1968) 1021.
77 R.J. Manasse and W.A. Corpe, *Can. J. Microbiol.*, 13 (1967) 1591.
78 J. Bremer and D.M. Greenberg, *Biochim. Biophys. Acta*, 46 (1961) 205.
79 T. Kaneshiro and J.H. Law, *J. Biol. Chem.*, 239 (1964) 1705.
80 E.P. Kennedy and S.B. Weiss, *J. Biol. Chem.*, 222 (1956) 193.
81 S.B. Weiss, S.W. Smith and E.P. Kennedy, *J. Biol. Chem.*, 231 (1958) 53.
82 S.L. Sherr and J.H. Law, *J. Biol. Chem.*, 240 (1965) 3760.
83 H. Goldfine, *Biochim. Biophys. Acta*, 59 (1962) 504.
84 M.O. Hall and J.F. Nyc, *J. Am. Chem. Soc.*, 81 (1959) 2275.
85 G.A. Scarborough and J.F. Nyc, *J. Biol. Chem.*, 242 (1967) 238.
86 G.A. Scarborough and J.F. Nyc, *Biochim. Biophys. Acta*, 146 (1967) 111.
87 N.H. Horowitz, *J. Biol. Chem.*, 162 (1946) 413.
88 C.J. Waechter, M.R. Steiner and R.L. Lester, *J. Biol. Chem.*, 244 (1969) 3419.
89 M.R. Steiner and R.L. Lester, *Biochemistry*, 9 (1970) 63.
90 R. Kokke, G.J.M. Hooghwinkel, H.L. Booij, H. van den Bosch, L. Zelles, E. Mulder and L.L.M. van Deenen, *Biochim. Biophys. Acta*, 70 (1963) 351.
91 H. van den Bosch, H.A. Bonte and L.L.M. van Deenen, *Biochim. Biophys. Acta*, 98 (1965) 648.
92 D.J. Hanahan and M.E. Jayko, *Biochem. Prep.*, 4 (1955) 12.
93 M.G. MacFarlane, *Nature*, 196 (1962) 136.
94 N. Shaw, P.F. Smith and W.L. Koostra, *Biochem. J.*, 107 (1968) 329.
95 W.L. Koostra and P.F. Smith, *Biochemistry*, 8 (1969) 4794.
96 W.J. Lennarz, P.P.M. Bonsen and L.L.M. van Deenen, *Biochemistry*, 6 (1967) 2307.
97 U.M.T. Houtsmuller and L.L.M. van Deenen, *Biochim. Biophys. Acta*, 106 (1965) 564.
98 W.J. Lennarz, in S.J. Wakil (Ed.), *Lipid Metabolism*, Academic Press, New York, 1970, p. 155.
99 W.J. Lennarz, J.A. Nesbit, III and J. Reiss, *Proc. Natl. Acad. Sci., (U.S.)*, 55 (1966) 934.
100 R.M. Gould and W.J. Lennarz, *Biochem. Biophys. Res. Commun.*, 26 (1967) 510.
101 J.G. Molotkovsky and L.D. Bergelson, *Chem. Phys. Lipids*, 2 (1968) 1.
102 J.M. dos Santos Mota, J.A.F. op den Kamp, H.M. Verheij and L.L.M. van Deenen, *J. Bacteriol.*, 104 (1970) 611.
103 J.A. Nesbitt, III and W.J. Lennarz, *J. Biol. Chem.*, 243 (1968) 3088.
104 R.M. Gould, M.P. Thorton, V. Liepkalns and W.J. Lennarz, *J. Biol. Chem.*, 243 (1968) 3096.
105 F.J. Kocun, *Biochim. Biophys. Acta*, 202 (1970) 277.

106 E. Ito and J.L. Strominger, *J. Biol. Chem.*, 237 (1962) 2696.
107 K. Okuda, I. Uemura, J.W. Bodley and T. Winnick, *Biochemistry*, 3 (1964) 100.
108 U.M.T. Houtsmuller and L.L.M. van Deenen, *Biochim. Biophys. Acta*, 84 (1964) 96.
109 P.P.M. Bonsen, G.H. de Haas and L.L.M. van Deenen, *Biochemistry*, 6 (1967) 1114.
110 R.M. Gould and W.J. Lennarz, *J. Bacteriol.*, 104 (1970) 1135.
111 G.K. Khuller and D. Subrahmanyam, *J. Bacteriol.*, 101 (1970) 654.
112 M.L. Vorbeck and G.V. Marinetti, *Biochemistry*, 4 (1965) 296.
113 J.A.F. op den Kamp, U.M.T. Houtsmuller and L.L.M. van Deenen, *Biochim. Biophys. Acta*, 106 (1965) 438.
114 U.M.T. Houtsmuller and L.L.M. van Deenen, *Biochim. Biophys. Acta*, 70 (1963) 211.
115 C.K. Huston, P.W. Albro and G.B. Grindey, *J. Bacteriol.*, 89 (1965) 768.
116 P.J.R. Phizackerly, J.C. MacDougal and M.J.O. Francis, *Biochem. J.*, 99 (1966) 21C.
117 J.A.F. op den Kamp and L.L.M. van Deenen, *Chem. Phys. Lipids*, 1 (1966) 86.
118 J.A.F. op den Kamp, W. van Iterson and L.L.M. van Deenen, *Biochim. Biophys. Acta*, 135 (1967) 862.
119 M.I. Gurr, P.P.M. Bonsen, J.A.F. op den Kamp and L.L.M. van Deenen, *Biochem. J.*, 108 (1968) 211.
120 J.C. MacDougall and P.J.R. Phizackerly, *Biochem. J.*, 114 (1969) 361.
121 J.A.F. op den Kamp, I. Redai and L.L.M. van Deenen, *J. Bacteriol.*, 99 (1969) 298.
122 D.G. Bishop, L. Rutberg and B. Samuelsson, *European J. Biochem.*, 2 (1967) 448.
123 M. Kates, L.S. Yengoyan and P.S. Sastry, *Biochim. Biophys. Acta*, 98 (1965) 252.
124 C.N. Joo and M. Kates, *Biochim. Biophys. Acta*, 176 (1969) 278.
125 M.K. Wassef, J. Sarner and M. Kates, *Can. J. Biochem.*, 48 (1970) 69.
126 M. Kates, M.K. Wassef and E.L. Pugh, *Biochim. Biophys. Acta*, 202 (1970) 206.
127 I. Ishizuka and T. Yamakawa, *J. Biochem., (Tokyo)*, 64 (1968) 13.
128 I. Ishizuka and T. Yamakawa, *Japan J. Exptl. Med.*, 39 (1969) 321.
129 V.W. Fischer and W. Seyferth, *Z. Physiol. Chem.*, 349 (1968) 1662.
130 W. van Iterson and J.A.F. op den Kamp, *J. Bacteriol.*, 99 (1969) 304.
131 R.J. Anderson, *J. Am. Chem. Soc.*, 52 (1930) 1607.
132 R.J. Anderson and A.G. Renfrew, *J. Am. Chem. Soc.*, 52 (1930) 1252
133 R.J. Anderson and E.G. Roberts, *J. Am. Chem. Soc.*, 52 (1930) 5023.
134 R.J. Anderson, W.C. Lothrop and M.M. Creighton, *J. Biol. Chem.*, 125 (1938) 299.
135 E. Vilkas and E. Lederer, *Bull. Soc. Chim. Biol.*, 42 (1960) 1013.
136 S. Nojima, *J. Biochem. (Tokyo)*, 46 (1959) 607.
137 S. Nojima, *J. Biochem. (Tokyo)*, 46 (1959) 499.
138 E. Vilkas, *Bull. Soc. Chim. Biol.*, 42 (1960) 1005.
139 E. Lederer, *Biochem. J.*, 81 (1961) 31P.
140 Y.C. Lee and C.E. Ballou, *Biochemistry*, 4 (1965) 1395.
141 C.E. Ballou, E. Vilkas and E. Lederer, *J. Biol. Chem.*, 238 (1963) 69.
142 C.E. Ballou and Y.C. Lee, *Biochemistry*, 3 (1964) 682.
143 Y.C. Lee and C.E. Ballou, *J. Biol. Chem.*, 239 (1964) 1316.
144 E. Vilkas, *Compt. Rend.*, 240 (1955) 1156.
145 M.C. Pangborn and J.A. McKinney, *J. Lipid Res.*, 7 (1966) 627.
146 J. Asselineau, *Biochim. Biophys. Acta*, 54 (1961) 359.
147 N. Ferreira-Gomes, T. Ioneda and J. Pudles, *Nature*, 211 (1966) 81.
148 P.J. Brennan, *Biochem. J.*, 109 (1968) 158.
149 P.J. Brennan and C.E. Ballou, *Biochem. Biophys. Res. Commun.*, 30 (1968) 69.
150 G.K. Khuller and D. Subrahmanyam, *Experientia*, 24 (1968) 851.
151 P.J. Brennan, and C.E. Ballou, *J. Biol. Chem.*, 242 (1967) 3046.
152 M.A. Laneelle, J. Asselineau and Castelnuous, *Ann. Inst. Pasteur*, 108 (1965) 69.
153 M.A. Laneele and J. Asselineau, *Compt. Rend.*, 266 D (1968) 1901.
154 T. Kataoka and S. Nojima, *Biochim. Biophys. Acta*, 144 (1967) 681.
155 D.L. Hill and C.E. Ballou, *J. Biol. Chem.*, 241 (1966) 895.
156 K. Takayama and D.S. Goldman, *Biochim. Biophys. Acta*, 176 (1969) 196.
157 P.J. Brennan and C.E. Ballou, *J. Biol. Chem.*, 243 (1968) 2975.
158 C. Prottey and C.E. Ballou, *J. Biol. Chem.*, 243 (1968) 6196.

159a Y. Akamatsu, Y. Ono and S. Nojima, *J. Biochem. (Tokyo)*, 61 (1967) 96.
159 N. Shaw and F. Dinglinger, *Biochem. J.*, 109 (1968) 700.
160 Y. Ono and S. Nojima, *Biochim. Biophys. Acta*, 176 (1969) 111.
161 Y. Ono and S. Nojima, *J. Biochem. (Tokyo)*, 65 (1969) 979.
162 Y. Akamatsu, Y. Ono and S. Nojima, *J. Biochem. (Tokyo)*, 59 (1966) 176.
163 D.S. Goldman, *Am. Rev. Resp. Dis.*, 102 (1970) 543.
164 R.A. Pieringer, *J. Biol. Chem.*, 243 (1968) 4894.
165 R.T. Ambron, *Ph. D. Thesis*, (1971), Temple University, Philadelphia, Pa.
166 W. Fischer, *Biochem. Biophys. Res. Commun.*, 41 (1970) 731.
167 N. Shaw, P.F. Smith and H.M. Verheij, *Biochem. J.*, 120 (1970) 439.
168 A.J. Wicken and K.W. Knox, *J. Gen. Microbiol.*, 60 (1970) 293.
169 V. Rizza, A.N. Tucker and D.C. White, *J. Bacteriol.*, 101 (1970) 84.
170 J.P. LaBach and D.C. White, *J. Lipid Res.*, 10 (1969) 528.
171 D.C. White, A.N. Tucker and C.C. Sweeley, *Biochim. Biophys. Acta*, 187 (1969) 527.
172 D.C. White and A.N. Tucker, *Lipids*, 5 (1970) 56.
173 H.E. Carter and R.C. Gaver, *Biochem. Biophys. Res. Commun.*, 29 (1967) 886.
174 H. Wagner and W. Zofcsik, *Biochem. Z.*, 346 (1966) 343.
175 S. Steiner, S. Smith, C.J. Waechter and R.L. Lester, *Proc. Natl. Acad. Sci. (U.S.)*, 64 (1969) 1042.
176 R.L. Lester and M.R. Steiner, *J. Biol. Chem.*, 243 (1968) 4889.
177 R.M.C. Dawson and P. Kemp, *Biochem. J.*, 105 (1967) 837.
178 P. Kemp and R.M.C. Dawson, *Biochem. J.*, 113 (1969) 555.
179 M.E. Tourtelotte, R.G. Jensen, G.W. Gander and H.J. Morowitz, *J. Bacteriol.*, 86 (1963) 370.
180 W. Tanner, *Arch. Mikrobiol.*, 64 (1968) 158.
181 M. Kates, *Advan. Lipid Res.*, 2 (1964) 17.
182 E.R.L. Gaughran, *J. Bacteriol.*, 53 (1947) 506.
183 L.V. Heilbrunn, *Am. J. Physiol.*, 69 (1924) 190.
184 J. Belehradek, *Protoplasma*, 121 (1931) 406.
185 A.G. Marr and J.L. Ingraham, *J. Bacteriol.*, 84 (1962) 1260.
186 S.K. Long and O.B. Williams, *J. Bacteriol.*, 79 (1960) 629.
187 E.A. Prill, P.R. Wenck and W.H. Peterson, *Biochem. J.*, 29 (1935) 21.
188 H. Ng, J.L. Ingraham and A.G. Marr, *J. Bacteriol.*, 84 (1962) 331.
189 M.K. Shaw and J.L. Ingraham, *J. Bacteriol.*, 90 (1965) 141.
190 H. Okuyama, *Biochim. Biophys. Acta*, 176 (1969) 125.
191 C.W.M. Haest, J. de Gier and L.L.M. van Deenen, *Chem. Phys. Lipids*, 3 (1969) 413.
192 R.A. Demel, L.L.M. van Deenen and B.A. Pethica, *Biochim. Biophys. Acta*, 135 (1967) 11.
193 A.J. de Siervo, *J. Bacteriol.*, 100 (1969) 1342.
194 G.H. Joyce, R.K. Hammond and D.C. White, *J. Bacteriol.*, 104 (1970) 323.
195 J.A. Erwin, *Biochim. Biophys. Acta*, 202 (1970) 21.
196 K.C. Strasters and K.C. Winkler, *J. Gen. Microbiol.*, 33 (1963) 213.
197 H.W. Taber and M. Morrison, *Arch. Biochem. Biophys.*, 105 (1964) 367.
198 F.E. Frerman and D.C. White, *J. Bacteriol.*, 94 (1967) 1868.
199 M. MacFarlane, *Biochem. J.*, 82 (1962) 40P.
200 D.C. White and F.E. Frerman, *J. Bacteriol.*, 94 (1967) 1854.
201 D.C. White and F.E. Frerman, *J. Bacteriol.*, 95 (1968) 2198.
202 D.C. White and L. Smith, *J. Biol. Chem.*, 239 (1964) 3956.
203 D.C. White, *J. Bacteriol.*, 83 (1962) 851.
204 D.C. White, *J. Biol. Chem.*, 240 (1965) 1387.
205 D.C. White, *J. Biol. Chem.*, 238 (1963) 3757.
206 D.C. White and A.N. Tucker, *J. Bacteriol.*, 97 (1969) 199.
207 A. Gorchein, *Proc. Roy. Soc. B.*, 170 (1968) 279.
208 A. Gorchein, *Proc. Roy. Soc. B.*, 170 (1968) 247.
209 A. Gorchein, *Proc. Roy. Soc. B.*, 170 (1968) 255.
210 A. Gorchein, A. Neuberger and G.H. Tait, *Proc. Roy. Soc. B.*, 170 (1968) 319.
211 G. Cohen-Bazire and R. Kunisawa, *J. Cell. Biol.*, 16 (1963) 401.
212 G. Cohen-Baziere and R. Kunisawa, in H. Gest, A. San Pietro and L.P. Vernon (Eds.) *Bacterial. Photosynthesis*, Antioch, Yellow Springs, 1963, p. 59.

213 G. Scheuerbrandt and K. Bloch, *J. Biol. Chem.*, 237 (1962) 2064.
214 A.R. Hands and W. Bartley, *Biochem. J.*, 84 (1962) 238.
215 R.C. Fuller, S.F. Conti, and D.B. Mellin, in H. Gest, A. San Pietro and L.P. Vernon (Eds.) *Bacterial Photosynthesis*, Antioch, Yellow Springs, 1963, p. 71.
216 S. Steiner, J.C. Burnham, S.F. Conti and R.L. Lester, *J. Bacteriol.*, 103 (1970) 500.
217 S. Steiner, S.F. Conti and R.L. Lester, *J. Bacteriol.*, 98 (1969) 10.
218 C.F. Fox, *Proc. Natl. Acad. Sci. (U.S.)*, 63 (1969) 850.
219 L. Kass, *J. Biol. Chem.*, 243 (1968) 3223.
220 G. Wilson and C.F. Fox, *J. Mol. Biol.*, 55 (1971) 49.
221 C.C. Hsu and C.F. Fox, *J. Bacteriol.*, 103 (1970) 410.
222 L. Mindich, *Proc. Natl. Acad. Sci. (U.S.)*, 68 (1971) 420.
223 W. Kundig and S. Roseman, *J. Biol. Chem.*, 246 (1971) 1393.
224 W. Kundig and S. Roseman, *J. Biol. Chem.*, 246 (1971) 1407.
225 L.S. Milner and H.R. Kaback, *Proc. Natl. Acad. Sci. (U.S.)*, 65 (1970) 683.
226 L. Rothfield and D. Romeo, *Bacteriol. Rev.*, 35 (1971) 14.
227 H. Goldfine, *Advan. Microbial Physiol.*, 8 (1972) 1.

NOTES ADDED IN PROOF

*(p. 293). New evidence indicates that the synthesis of diphosphatidyl glycerol in *E. coli* occurs by a condensation of two molecules of phosphatidylglycerol (C.B. Hirschberg and E.P. Kennedy, *Proc. Natl. Acad. Sci. (U.S.)*, 69 (1972) 348).

**(p. 313). The donor of the phosphatidyl group to diglycosyldiglyceride to form the compound in Fig. 5 is either diphosphatidylglycerol or phosphatidylglycerol (R.A. Pieringer, *Biochem. Biophys. Res. Commun.*, 49 (1972) 502).

Chapter 12

PHOSPHONOLIPIDS

H. ROSENBERG

Department of Biochemistry, John Curtin School of Medical Research, Institute of Advanced Studies, Australian National University, Canberra (Australia)

The term 'phosphonolipids' was originally suggested by Baer and Stanacev[1] to describe phospholipids carrying a covalent bond between the phosphorus atom and the carbon of the nitrogenous base. Interest in this class of compounds followed the isolation by Horiguchi and Kandatsu[2] of 2-aminoethylphosphonic acid (AEP):

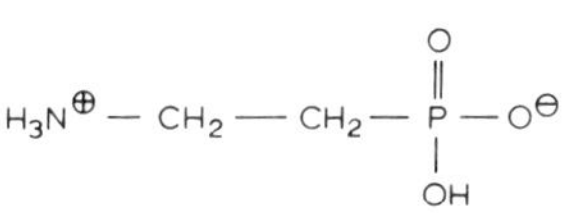

from rumen Protozoa and the subsequent demonstration that it and its derivatives were widely distributed in several phyla of the animal kingdom, predominantly in a bound form in lipids or other structural elements of the cells. Several reviews have already dealt with the distribution, chemistry and metabolism of compounds containing carbon–phosphorus bonds[3–7]. This chapter presents a picture of the metabolism and possible function of the phosphonolipids in the parts of the animal kingdom where these compounds uniquely occur.

1. PHOSPHONOLIPID CLASSES: DISCOVERY AND CHEMISTRY

Because AEP was originally isolated from hydrolysates of material extracted from tissues by organic solvents[2], it was soon realised that this compound, so similar in structure to *O*-phosphorylethanolamine, was probably an integral part of phospholipids[3, 5]. Following the finding of AEP in Protozoa[2], Kittredge et al.[8] were the first to demonstrate its presence in higher organisms, when they found AEP in hydrolysates of alcoholic extracts of the sea anemone, *Anthopleura elegantissima*, and suggested that it could be a constituent of lipids. From the same source, Rouser et al.[9] isolated the first pure phosphonolipid, ceramide-AEP (Fig. 1, I). The occurrence of a phosphonic analogue of a glyceride (for which 'phosphonoglyceride' appears to be a suitable name), in the Protozoan *Tetrahymena pyriformis*, was demonstrated by Liang and Rosenberg[10] and by Thompson[11] (Fig. 1, II), who also showed that a large proportion of this compound contained an alkyl ether bond (Fig. 1, III) predominantly in position 1 since, on hydrolysis, it yielded almost entirely chimyl alcohol. The alk-1′-enyl analogue (Fig. 1, IV), has been isolated from rumen Protozoa[12].

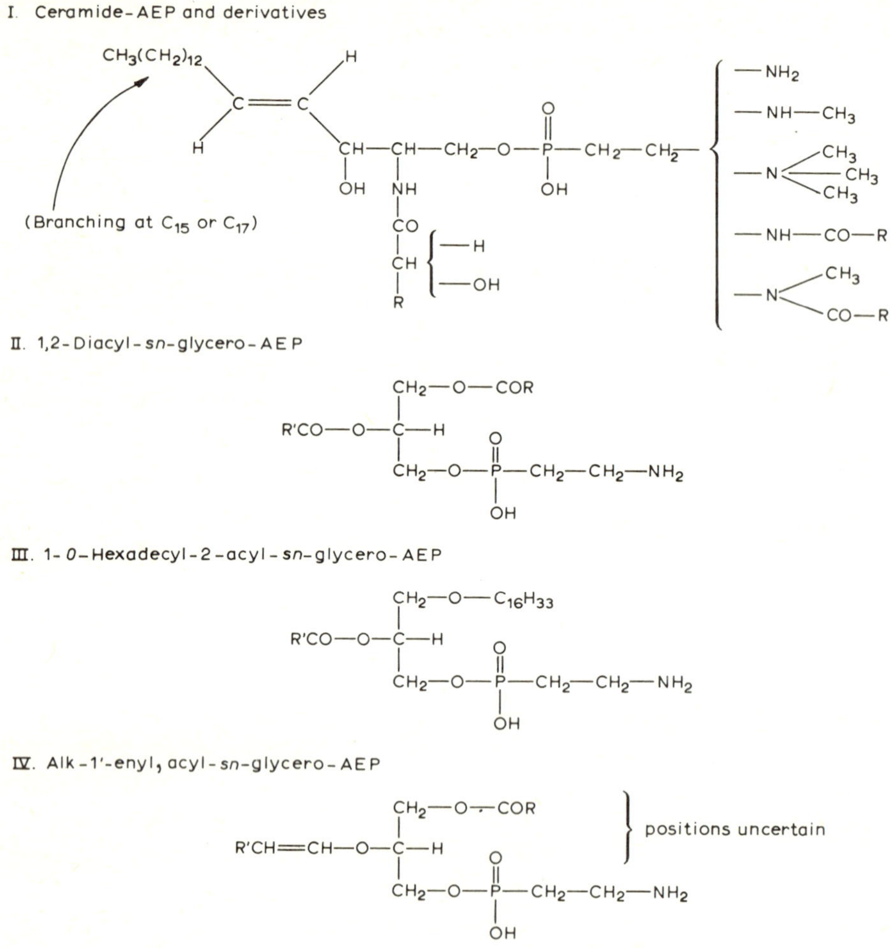

Fig. 1. Structures of known phosphonolipids.

Kittredge et al.[13] reported the presence of mono-, di- and tri-*N*-methyl-AEP in hydrolysates of alcoholic extracts of the sea anemone, and later reported[14] the isolation of lipids containing these bases. Hori and Arakawa[15] isolated from the mussel, Corbicula, the *N*-methylated and the *N*-acylated ceramide-AEP (Fig. 1, I); Ceramide(*N*-methyl)-AEP was found[16] in the shellfish Turbo and recently in Tetrahymena[79].

The presence of the compounds described above has since been confirmed in various species (see below). The cases given above refer only to the initial demonstration of new classes of phosphonolipids, and in species collected from their natural habitat.

2. DISTRIBUTION OF PHOSPHONOLIPIDS IN NATURE

A large proportion of the early work on distribution of the phosphonates relied on procedures which involved the total acid hydrolysis of either the whole tissue, or of material extracted by organic solvents. In the hydrolysates, AEP was directly identified or its presence was demonstrated by an indirect method[4,17]. This method is based on the assumptions that acid hydrolysis (usually 48 h at 105° in 6 *N* HCl) of biological material does not affect the highly stable C—P bond, but that it results in the *complete* mineralization of the phosphorus of *all* phosphate esters. The latter assumption is particularly important since the difference between total phosphorus (determined after combustion) and acid-labile phosphorus is taken to be carbon-bound phosphorus. Such methods cannot afford direct evidence of the presence of phosphonolipids, but the latter is suggested by the fact that AEP is obtained from the solvent-extractable fraction after liberation by hydrolysis[4]. It is also true that, whenever fractionation was attempted, phosphonolipids were usually the major phosphonate-containing fraction. An accurate picture of the distribution of phosphonolipids in various phyla cannot be given until careful isolation and analysis of the phosphonolipids is made. At present, such information is scarce and often the occurrence of phosphonolipids has been accepted on the basis of indirect, albeit convincing, evidence concerning hydrolytic fragments. The field could be greatly advanced by broader surveys, based on the more specific and direct methods of phosphonolipid analysis now available[73–76].

From the reviews quoted above[3–6] and from a recent survey by Quin and Shelburne[17] it is obvious that, while traces of 'phosphonates' can be detected by the indirect method in a variety of animals, their occurrence is beyond doubt only in the Protozoa, Coelenterata and Mollusca. Recently, however, phosphonolipids have been detected in Mycobacteria[18] thus including the Prokaryotes in the list. Horiguchi[3] quoted results indicating the presence of AEP in other bacteria, but the levels found there were generally below 0.2% of total phosphorus.

Table 1 shows the distribution of phosphonolipids in the animal kingdom. It does not include reports of phosphonolipids, where AEP is suspected to have been derived from the diet or the rumen. For instance, AEP-containing lipid has been detected in the liver of the goat[34], and in beef brain[35], and this probably represents AEP derived from the rumen and absorbed from the alimentary tract. This mechanism was demonstrated experimentally by administering AEP to rats and showing its incorporation into liver lipids[36]. Humans may consume AEP with dairy products and by ingestion of invertebrates, *e.g.* the amount of AEP ingested with a meal of edible molluscs could be quite high and could, predictably, lead to the deposition of some phosphonolipids in human tissues[38, 39]. This material, most likely of exogenous derivation, is outside the scope of the present chapter, as is the occurrence of phosphonolipids in insects experimentally fed AEP and its methylated derivatives[40, 41].

In deciding the origin of AEP-lipids in the three phyla where these compounds are widely distributed, one has no difficulty about Protozoa: AEP synthesis in a number of species has been demonstrated in axenic culture[2, 10]; the same applies to bacteria[18]. The

References p. 342

TABLE 1

DISTRIBUTION OF PHOSPHONOLIPIDS IN NATURE

Phylum and species	*Lipid species (see Fig. 1)*	*Ref.*
Prokaryotes:		
Bacteria		
Mycobacterium spp.	Unidentified phospholipid, probably phosphonoglyceride	18
Eukaryotes:		
Protozoa		
Tetrahymena pyriformis	1,2-Diacyl-*sn*-glycero-AEP	3[a], 10[a], 11, 19, 20, 21
	1-*O*-Alkyl-2-acyl-*sn*-glycero-AEP	11, 20, 22
	Ceramide-AEP (Branched LCB)	23
	Ceramide-(*N*-methyl)AEP	79
Entodinium caudatum	1,2-Diacyl-*sn*-glycero-AEP	12
	AEP-plasmalogen	12
	Ceramide-AEP	12
Monochrysis lutheri	Unidentified lipid	24
Coelenterata		
Anthopleura elegantissima	Ceramide-AEP	9, 14[a], 25
Anthopleura xanthogrammica	Ceramide-(*N*-monomethyl)AEP	13[a], 14[a]
	Ceramide-(*N*-trimethyl)AEP	13[a], 14[a]
Mollusca		
Corbicula sandai	Ceramide-AEP	26
Corbicula sandai	Ceramide-(*N*-hexadecanoyl)AEP	15
	Ceramide-(*N*-octadecanoyl)AEP	15
5 Other bivalves	Ceramide-AEP	26
Snails – 13 species	Ceramide-AEP	26
Cephalopods – 2 species	Ceramide-AEP	26
Bivalve, *Pinctada martensii*	2-OH-Ceramide-AEP	27
Monodonta labio (snail)	Ceramide-(*N*-monomethyl)AEP	28
Ostrea gigas (oyster)	Ceramide-AEP	29
Turbo cornutus	Ceramide-(*N*-monomethyl)AEP	16
Lymnea stagnalis	Ceramide-AEP	30
	Diacyl-*sn*-glycero AEP	30
Helix lactea	Ceramide-AEP	31
Haliotis midae	Ceramide-AEP	32[a]
Haliotis corrugata	Ceramide-AEP	33
Hinnites giganteum	Ceramide-AEP	33

[a]The lipid has not been fully identified, but is strongly implicated by circumstantial evidence.

possibility has been suggested that in Coelenterata the phosphonates are derived from unicellular organisms associated with the sea anemone[8]. However, since small fish are probably the main food and because of the high levels of AEP encountered in some coelenterates (in *Tealia felina*, about 50% of the total phosphorus[4]), it is unlikely that they are derived from exogenous sources. In Mollusca, AEP is present in many species in large amounts. Furthermore, AEP-lipids occur in the (vegetarian) garden snails and slugs, where short-term synthesis of phosphonolipids from [^{32}P] orthophosphate *in vivo* can be demonstrated[42]. Most importantly, ceramide-AEP has been shown to be synthesized *in vitro* by liver homogenates from the mussel, *Hyriopsis schlegelii*[43]. The single report[44] of phosphonolipids in a marine crab is not sufficient to include Arthropoda amongst the phyla containing C–P compounds. The total amount found is comparatively low ($<$ 0.2%) and the AEP may have originated from the diet which is likely to include both molluscs and coelenterates[44]. De Koning[44] could detect no AEP in the fresh-water crab, Potamon, and Hori[26] failed to find phosphonolipids in various Crustacea.

In conclusion, present evidence restricts the distribution of endogenously biosynthesized phosphonolipids to certain bacteria, and a wide variety of Protozoa, Coelenterata and Mollusca. A distinctive pattern of distribution can also be seen (Table 1): In Protozoa the phosphonoglycerides seem to predominate, although some ceramide-AEP is present. In the Coelenterata ceramide-AEP and its methylated derivatives predominate, apparently to the exclusion of phosphonoglycerides[25]. In Mollusca, while ceramide-AEP and its derivatives appear to be the main phosphonolipid species, some phosphonoglycerides also occur. The significance of these differences is not understood.

3. BIOSYNTHESIS

The biosynthesis of phosphonolipids is closely related to the formation of AEP itself. In Tetrahymena, [^{32}P] orthophosphate incorporation into AEP *in vivo* proceeds most rapidly into the lipid-bound, rather than into the free AEP molecule[45,46], thus suggesting that a lipid is involved somewhere in the pathway of AEP biosynthesis. In an attempt to elucidate the biosynthetic pathway, several workers traced the origin of AEP-carbon. Trebst and Geike[47], by careful analysis of the distribution of radioactivity from [^{14}C] glucose labelled in positions 1, 2 or 6, concluded that the most likely carbon precursor of AEP was phosphoenolpyruvate (PEP). On the basis of similar results Warren[48]

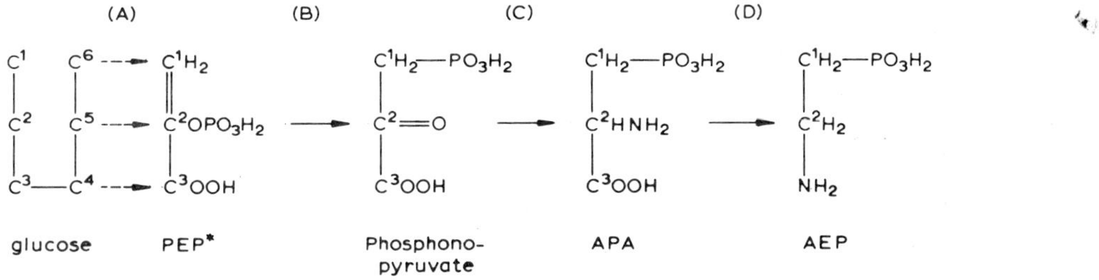

Fig. 2. The origin of AEP carbon.
*The numbering of carbon atoms in PEP is opposite to the usual convention. The reactions are: (A) Glycolysis; (B) Rearrangement (Warren[48], see Fig. 3); (C) Amination; (D) Decarboxylation (with the loss of carbon derived from glucose C^3 and C^4).

reached the same conclusion, having added the observation that carbons 3 and 4 from glucose did not contribute to AEP carbon, but contributed to 2-amino-3-phosphonopropionate. (This compound, sometimes referred to as alanine-3-phosphonic acid (APA) was previously discovered by Kittredge and Hughes[49], in Tetrahymena and in a zoanthid). All these observations were consistent with a rearrangement of PEP and a loss of the C_1 (carboxyl) at some stage (Fig. 2). Such a rearrangement was proposed simultaneously by Warren[48] and Liang and Rosenberg[46]. Horiguchi et al.[50] also proposed PEP or enoloxaloacetate as the likely precursors. Recently Horiguchi[78] demonstrated another pathway in homogenates of Tetrahymena. This is similar to that proposed by Warren, but the order of decarboxylation and deamination is reversed, so that 2-phosphonoacetaldehyde (PAAL), rather than APA, is an intermediate in the reaction (Fig. 3). However, the reaction may be more complex than originally envisaged (Horiguchi, personal commun.). The versions propose by Horiguchi and by Warren (Fig. 3) differ from that proposed by Liang and Rosenberg in

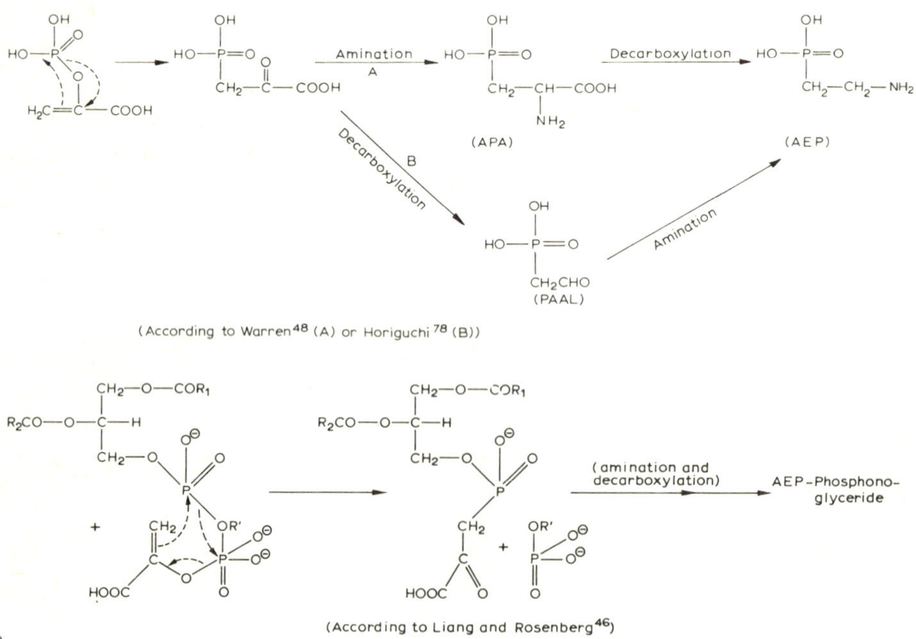

Fig. 3. Three schemes for PEP rearrangement.

the origin of the phosphorus moiety and in the involvement of a lipid molecule.

In addition to the proposed reactions shown in Fig. 3, the direct incorporation of AEP into phosphonoglycerides was established by the demonstration[10] that preparations from Tetrahymena catalysed the following reactions:

$$\text{CTP} + \text{AEP} \longrightarrow \text{CMP-AEP} + \text{PP}_i \quad (1)$$

$$\text{CMP-AEP} + \text{1,2-diacyl-}sn\text{-glycerol} \longrightarrow \text{phosphonoglyceride} + \text{CMP} \quad (2)$$

the intermediate, CMP–AEP having been isolated. There thus exists the 'classical' (see Chapter 4) pathway for phosphonoglyceride synthesis but, from the labelling kinetics with ^{32}P (refs. 45, 46), it cannot be of primary importance in Tetrahymena when *de novo* synthesis of AEP is taking place. Reactions (1) and (2) can, however, account for the observed incorporation of exogenously supplied AEP[51] into the phospholipids of Tetrahymena, and may well constitute the main pathway for phosphonolipid biosynthesis in molluscs, where labelling kinetics indicate that free AEP is a likely precursor of phosphonolipid[42]. The enzyme which catalyses reaction (1) above has also been demonstrated in rat liver, but the rates with AEP are insignificant compared to those with phosphorylethanolamine (about 1/100). On the other hand, in Tetrahymena this reaction proceeds at least as rapidly as CDPE synthesis, and the affinity for AEP is high[52].

The problem of phosphonolipid biosynthesis in Tetrahymena thus consists of the need to elucidate a pathway in which PEP gives rise to lipid-bound AEP, but not through free AEP. The scheme presented in Fig. 3 was one that satisfied such a condition. Another scheme involves phosphatidylenolpyruvate as an intermediate[5]. Smith and Law[51] have recently shown that APA was incorporated into the phosphonolipids of Tetrahymena, where it ended up as AEP. Neither APA-containing lipid nor free AEP was found in the cells. Incorporation of APA-carbon into AEP of the lipid fraction was more efficient than incorporation of AEP-carbon. These authors considered, but dismissed, a decarboxylation cycle involving APA, on the grounds that no precedent existed for such an exchange (i.e. serine phosphate for ethanolamine phosphate) and that no lipid-bound APA was found. They suggested that APA was decarboxylated to AEP which was, in turn,

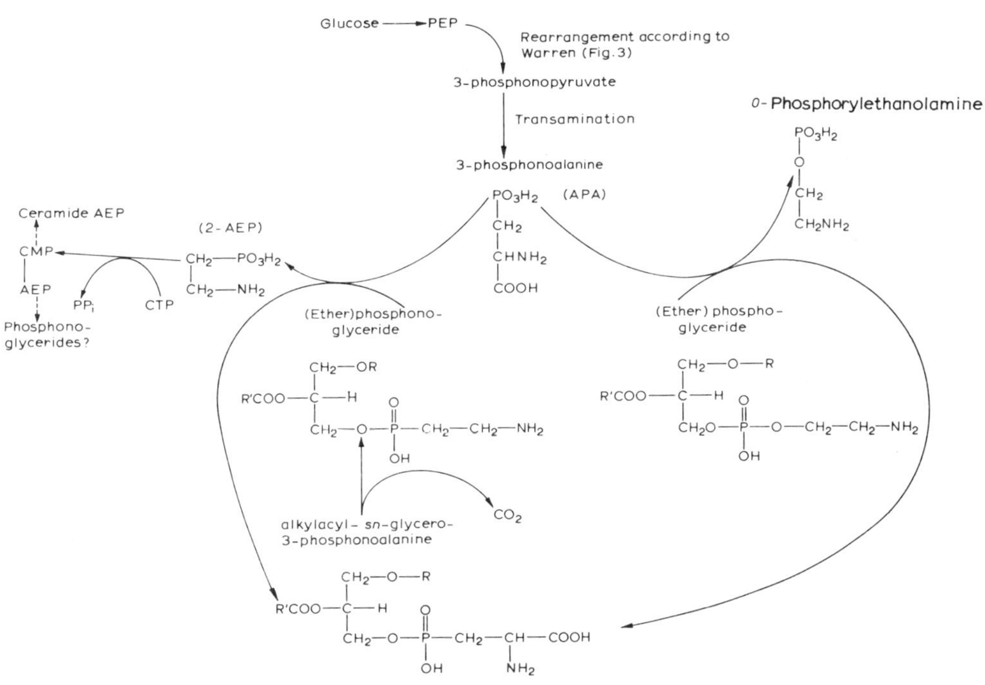

Fig. 4. Hypothetical phosphonoalanine decarboxylation cycle.

References p. 342

incorporated into phosphonolipid. This alternative scheme, however, fails to account for the known[45, 46] labelling kinetics, and is not likely to represent a predominant pathway in Tetrahymena *in vivo* where lipid-bound AEP does not appear to arise from the soluble species. Furthermore, it is subject to the same objections as a 'decarboxylase cycle' scheme. Firstly, no labelled (free) AEP was found in the cells which has incorporated label from APA. Secondly, there is no precedent for an 'aqueous phase' serine (or phosphoserine) decarboxylase: In fact, a decarboxylase cycle for serine via phosphatidylserine was recently described in Tetrahymena by Dennis and Kennedy[53]. Thompson[54] showed that the lipid moiety of phosphoglycerides behaved, in labelling experiments, as a precursor of phosphonoglycerides.

All these facts can be unified in a single hypothesis involving the operation of a 3-phosphonoalanine decarboxylase cycle, similar to the classical serine decarboxylase exchange reaction[55], except that the exchange occurs at the oxygen between the phosphorus and the C_3 of the glycerol. The scheme (Fig. 4) allows 3-phosphonoalanine to displace a phosphoethanolamine group from a phosphoglyceride in the *de novo* synthesis of phosphonolipids, or the AEP group of a phosphonoglyceride (most likely alkyl-acyl[11]) in the 'decarboxylation cycle' with the production of free AEP which may enter into CMP–AEP and hence into other lipid molecules, including (possibly) ceramide-AEP. It is clear that the scheme does not cover the problem of ceramide-AEP biosynthesis in coelenterates and molluscs, and that it is proposed primarily for Tetrahymena, where it is in accord with all the points mentioned above.

4. DEGRADATION

The enzymic degradation of phosphonolipids can be separated into two parts: The breakdown of the lipid molecule with the release, among other components, of 2-AEP or its derivatives, and the cleavage of the C–P bond itself. The degradation of the phosphonolipid molecule has been demonstrated in a number of biological systems. Thus Saito and Mukoyama[56] showed that clostridial phospholipase C hydrolysed ceramide-AEP. Hori et al.[57] also showed the enzymic breakdown of ceramide-AEP by bacterial enzymes from *Clostridium perfringens* and from *Bacillus cereus*. They made the interesting observation that a 2-hydroxy-group on the fatty acyl residue, a structure peculiar to some molluscs[27], rendered these ceramides completely resistant to attack[58]. Thompson[54] showed that phosphonoglycerides were broken down in Tetrahymena, but at a rate slower than the breakdown of phosphoglycerides. Thus, although these lipids may be degraded at slower rates, they are nevertheless degradable by known enzymes although it is possible that specific enzymes exist which can cleave off the aminophosphonate base in a more efficient manner.

The cleavage of the C–P bond of AEP may be ecologically important in the return of the phosphorus to the natural pool in a utilizable form. This reaction is brought about by a variety of micro-organisms. Thus Zeleznick et al.[59], Mastalerz et al.[60] and Harkness[61] showed that a number of bacterial strains could utilize organic phosphonates as a source of phosphorus, carbon, or both, although only Harkness used the naturally occurring AEP as a substrate. Details of the strains used have been summarised by Mastalerz[6]. Six strains of

B. cereus have also been shown to break down AEP[62]. The breakdown proceeded via a transamination step, and was shown[63] to follow the route:

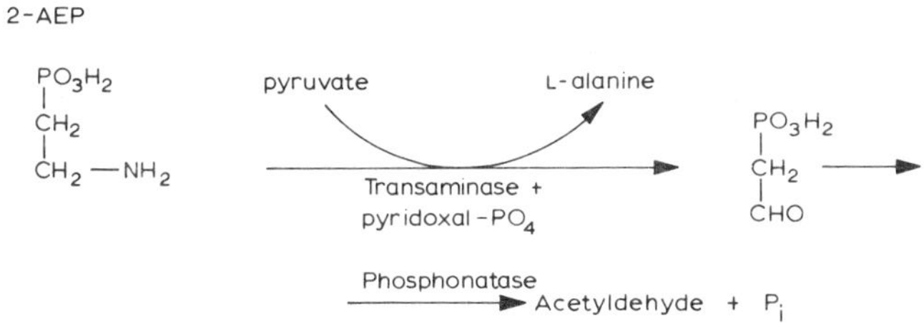

More recently, Lacoste and Neuzil[64], who studied AEP breakdown by Pseudomonas, showed that the first step also involved transamination, but were unable to detect 2-phosphonoacetaldehyde. It is possible that in this strain the breakdown of the intermediate is very rapid. The enzyme responsible for this reaction, 2-phosphonoacetaldehyde phosphonohydrolase ('phosphonatase') has been purified and shown to catalyse the cleavage of the C–P bond[65]. The degradation of phosphonolipids is thus better understood at present than is their biosynthesis.

5. FUNCTION

The precise function of phosphonolipids is unknown. It has been suggested[8, 49] that the presence of phosphonates may confer some advantage in phosphorus storage or in coping with the wide ranges of concentrations of cations in the marine environment. There is, in addition, a certain pattern in the distribution and biochemical properties of these compounds, which may suggest another possible function. This may lie in their ability to resist, in part, hydrolysis by enzymes, as viewed against the background of their distribution, both in nature and within the cell. It is clear that the C–P bond in phosphonolipids makes them immune to attack by phospholipase-D. Since the polar group of the phospholipid is likely to be oriented towards the aqueous medium at the membrane surface, membranes containing phosphonolipids may be more able to resist certain types of enzymatic attack. This point is made by Kennedy and Thompson[66] who showed that, in the outer membrane of Tetrahymena over 60% of the total lipids contain the C–P bond -- an enrichment of more than twice the mean cell value. Similarly Smith et al.[22] found that the lipid of the Tetrahymena cilia contained twice as much AEP as the total cell lipid, and that some lipid components of cilia were 90–100% pure phosphonolipid. Thompson[54] also observed that, during autolysis, phospholipids in Tetrahymena were degraded several times faster than phosphonolipids. Rosenthal and co-workers reported that certain phosphonate analogues of phospholipids, which they synthesized, inhibit some phospholipases (A and C) as well as phosphatidate phosphohydrolase[67–71]. Thus, in a hostile or competitive environment, outer membranes containing phosphonolipids may be of advantage in conferring resistance to enzymatic attack.

If this postulate is accepted, it may be asked whether this property fits in with the pattern of phosphonolipid distribution in nature and whether there is a common feature

References p. 342

amongst the representatives of the three phyla, that would make the possession of added resistance a special, and perhaps vital, advantage. One such common feature appears to have emerged: the members of these three phyla present to their environment a living, naked membrane. Unlike other forms of life, they lack a protective coating, such as cellulose, chitin, or squamous epithelium, and may derive some protection from the presence of phosphonolipids in their membranes.

Mastalerz[6] expressed the opinion that C–P compounds pre-date the first life-forms. This is credible in view of the reducing atmosphere which existed on earth at that stage. Gulick[72] maintains that the first forms of life would have utilised exclusively compounds of phosphorus at an oxidation state lower than that of orthophosphate. As the oxidation level of the atmosphere increased orthophosphate became prevalent and biochemical systems based on it, evolved. If the retention of phosphonolipids were advantageous to the survival of the 'naked' species, and the supplies of reduced forms of phosphorus were dwindling, selection may have favoured a mutation that would allow phosphonate synthesis from phosphate. Naturally, no such selection pressures would apply to species already protected by adequate external features such as cell walls, chitin etc. This hypothesis cannot account, however, for the presence of phosphonates in mycobacteria[18], in which the cell wall would provide adequate protection for the cell membrane.
Horiguchi[78] has recently presented evidence for a new biosynthetic pathway in Tetrahymena in which PEP is converted to AEP via phosphonoacetaldehyde.

ACKNOWLEDGEMENTS

Two workers have given invaluable help to many of the authors mentioned in the preceding pages: Erich Baer, who with his colleagues, synthesized a large variety of phosphonolipids, and A.F. (Tex) Isbell, who prepared a large number of various aminophosphonates[77]. Both have generously made samples available and thus helped greatly in the identification of naturally occurring compounds.

REFERENCES

1 E. Baer and N.Z Stanacev, *J. Biol. Chem.*, 239 (1964) 3209.
2 M. Horiguchi and M. Kandatsu, *Nature*, 184 (1959) 901.
3 M. Horiguchi, *J. Agr. Chem. Soc. Japan*, 40 (1966) R25.
4 L.D. Quin, in M. Grayson and E.J. Griffith (Eds.), *Topics in Phosphorus Chemistry*, Interscience, New York, 1967, p. 23.
5 J.S. Kittredge and E. Roberts, *Science*, 164 (1969) 37.
6 P. Mastalerz, *Postepy Biochem.*, 15 (1969) 151.
7 A.J. Slotboom and P.P.M. Bonsen, *Chem. Phys. Lipids*, 5 (1970) 301.
8 J.S. Kittredge, E. Roberts and D.G. Simonsen, *Biochemistry*, 1 (1962) 624.
9 G. Rouser, G. Kritchevsky, D. Heller and E. Lieber, *J. Am. Oil Chem. Soc.*, 40 (1963) 425.
10 C.R. Liang and H. Rosenberg, *Biochim. Biophys. Acta*, 125 (1966) 548.
11 G.A. Thompson Jr., *Biochemistry*, 6 (1967) 2015.
12 R.M.C. Dawson and P. Kemp, *Biochem. J.*, 105 (1967) 837.
13 J.S. Kittredge, A.F. Isbell and R.R. Hughes, *Biochemistry*, 6 (1967) 289.
14 J.S. Kittredge, (Abstract), *VII Intern. Congr. Biochem., Tokyo,* 1967, p. 453.

15 T. Hori and I. Arakawa, *Biochim. Biophys. Acta*, 176 (1969) 898.
16 A. Hayashi, F. Matsuura and T. Matsubara, *Biochim. Biophys. Acta*, 176 (1969) 208.
17 L.D. Quin and F.A. Shelburne, *J. Marine Res.*, 27 (1969) 73.
18 G.R. Sarma, V. Chandramouli and T.A. Venkitasubramanian, *Biochim. Biophys. Acta*, 218 (1970) 561.
19 M. Sugita and T. Hori, *J. Biochem. (Tokyo)*. 69 (1971) 1149.
20 G.K. Chacko and D.J. Hanahan, *Biochim. Biophys. Acta*, 176 (1969) 190.
21 M. Kandatsu and M. Horiguchi, *Proc. Meet. Jap. Soc. Agr. Chem., Sapporo, 1964.*
22 J.D. Smith, W.R. Snyder and J.H. Law, *Biochem. Biophys. Res. Commun.*, 39 (1970) 1163.
23 H.E. Carter and R.C. Gaver, *Biochem. Biophys. Res. Commun.*, 29 (1967) 886.
24 M.W. Baldwin and J. Braven, *J. Marine Biol. Ass. U.K.*, 48 (1968) 603.
25 G. Simon and G. Rouser, *Lipids*, 2 (1967) 55.
26 T. Hori, I. Arakawa and M. Sugita, *J. Biochem. (Tokyo)*, 62 (1967) 67; (Abstract) *VII Intern. Congr. Biochem., Tokyo, 1967,* p. 834.
27 M. Sugita, I. Arakawa, T. Hori and Y. Sawada, *Seikagaku*, 40 (1968) 158.
28 T. Hori, M. Sugita and O. Itasaka, *J. Biochem. (Tokyo)*, 65 (1969) 451.
29 A. Hayashi, T. Matsubara and Y. Mishima, *J. Fac. Sci. Tech. Kinki Univ.*, 2 (1967) 39
30 C.R. Liang and K.P. Strickland, *Can. J. Biochem.*, 47 (1969) 85.
31 C.R. Liang, M. Segura and K.P. Strickland, *Can. J. Biochem.*, 48 (1970) 580.
32 A.J. de Koning, *Nature*, 210 (1966) 113.
33 G. Simon and G. Rouser, *Lipids*, 4 (1969) 607.
34 M. Kandatsu and M. Horiguchi, *Agr. Biol. Chem.*, 29 (1965) 781.
35 H. Shimizu, Y. Kakimoto, T. Nakajima, A. Kanazawa and I. Sano, *Nature*, 207 (1965) 1197.
36 M. Kandatsu, M. Horiguchi and M. Tamari, *Agr. Biol. Chem.*, 29 (1965) 779.
37 L.D. Quin, *Biochemistry*, 4 (1965) 324.
38 A.U. Alam and S.H. Bishop, (Abstract) *Div. Biol. Chem. 156th Meet. Am. Chem. Soc., Atlantic City, 1968.*
39 J.A. Alhadeff and G.D. Daves Jr., *Biochemistry*, 9 (1970) 4866.
40 R.G. Bridges and J. Ricketts, *Nature*, 211 (1966) 199.
41 L.L. Bieber, *Biochim. Biophys. Acta*, 152 (1968) 778.
42 C.R. Liang and H. Rosenberg, *Comp. Biochem. Physiol.*, 25 (1968) 673.
43 O. Itasaka, T. Hori and M. Sugita, *Biochim. Biophys. Acta*, 176 (1969) 783.
44 A.J. de Koning, *Biochim. Biophys. Acta*, 202 (1970) 187.
45 H. Rosenberg, *Nature*, 203 (1964) 299.
46 C.R. Liang and H. Rosenberg, *Biochim. Biophys. Acta*, 156 (1968) 437.
47 A. Trebst and F. Geike, *Z. Naturforsch.*, 22(b) (1967) 989.
48 W.A. Warren, *Biochim. Biophys. Acta*, 156 (1968) 340.
49 J.S. Kittredge and R.R. Hughes, *Biochemistry*, 3 (1964) 991.
50 M. Horiguchi, J.S. Kittredge and E. Roberts, *Biochim. Biophys. Acta*, 165 (1968) 164.
51 J.D. Smith and J.H. Law, *Biochemistry*, 9 (1970) 2152.
52 C.R. Liang, *Ph.D. Thesis*, Austral. Natl. Univ., Canberra, 1967.
53 E.A. Dennis and E.P. Kennedy, *J. Lipid Res.*, 11 (1970) 394.
54 G.A. Thompson Jr., *Biochim. Biophys. Acta*, 176 (1969) 330.
55 L.F. Borkenhagen, E.P. Kennedy and L. Fielding, *J. Biol. Chem.*, 236 (1961) PC 28.
56 K. Saito and K. Mukoyama, *Biochim. Biophys. Acta*, 164 (1968) 596.
57 T. Hori, I. Arakawa, M. Sugita and O. Itasaka, *J. Biochem. (Tokyo)*, 64 (1968) 533.
58 T. Hori, M. Sugita and O. Itasaka, *Mem. Shiga Univ.*, 19 (1969) 39.
59 L.D. Zeleznick, T.C. Myers and E.B. Titchener, *Biochim. Biophys. Acta*, 78 (1963) 546.
60 P. Mastalerz, Z. Wieczorek and M. Kochman, *Acta Biochim. Polon.*, 12 (1965) 151.
61 D.R. Harkness, *J. Bacteriol.*, 92 (1966) 623.
62 H. Rosenberg and J.M. La Naúze, *Biochim. Biophys. Acta*, 141 (1967) 79.
63 J.M. La Nauze and H. Rosenberg, *Biochim. Biophys. Acta*, 165 (1968) 438.
64 A.M. Lacoste and E. Neuzil, *Compt. Rend.*, 269 (1969) 254.
65 J.M. La Nauze, H. Rosenberg and D.C. Shaw, *Biochim. Biophys. Acta*, 212 (1970) 332.
66 K.E. Kennedy and G.A. Thompson Jr., *Science*, 168 (1970) 989.
67 A.F. Rosenthal and M. Pousada, *Biochim. Biophys. Acta*, 125 (1966) 265.

68 A.F. Rosenthal and S.C.H. Han, *Biochim. Biophys. Acta*, 152 (1968) 96.
69 A.F. Rosenthal and M. Pousada, *Biochim. Biophys. Acta*, 164 (1968) 226.
70 A.F. Rosenthal, S.V. Chodsky and S.C.H. Han, *Biochim. Biophys. Acta*, 187 (1969) 385.
71 A.F. Rosenthal and S.C.H. Han, *Biochim. Biophys. Acta*, 218 (1970) 213.
72 Å. Gulick, *Am. Scientist*, 43 (1955) 479.
73 V.M. Kapoulas, *Biochim. Biophys. Acta*, 176 (1969) 324.
74 K.A. Karlsson, *Biochem. Biophys. Res. Commun.*, 39 (1970) 847.
75 T. Glonek, T.O. Henderson, R.L. Hildebrand and T.C. Myers, *Science*, 169 (1970) 192.
76 C. Benezra, S.K. Pavanaram and E. Baer, *Can. J. Biochem.*, 48 (1970) 991.
77 A.F. Isbell, (Abstract) *VII Intern. Congr. Biochem., Tokyo, 1967*, p. 447.
78 M. Horiguchi, *Biochim. Biophys. Acta*, 261 (1972) 102.
79 C.V. Viswanathan and H. Rosenberg, *J. Lipid Res.*, 14 (1973) 327.

Chapter 13

PHOSPHOLIPIDS AND THE HEPATO-PORTAL SYSTEM

R. COLEMAN

Department of Biochemistry, University of Birmingham, P.O. Box 363, Birmingham B15 2TT, (Great Britain)

The liver plays a supportive role in the absorption of dietary lipids, by supplying bile, a fluid not only rich in bile salts but also in phospholipids. The cells of the intestinal mucosa take up the partially hydrolysed lipids from a stabilized dispersion, reassemble them and secrete them as lipoproteins, the chylomicrons.

The liver also removes carbohydrates and other metabolites from the bloodstream, converts them to lipids and secretes them into the circulation as lipoproteins. These lipoproteins possess a significant phospholipid content.

Associations of neutral lipids in water are unstable and form increasingly large droplets, which above a certain size would tend to clog the smaller blood vessels. Consequently the form in which neutral lipids are transported is as smaller stabilized associations with a more hydrophilic surface. Phospholipids play an important part in this stabilization process, probably due to their amphipathic character.

The phospholipid metabolism of all cells reflects the processes of membrane synthesis, degradation, modification and repair. The phospholipid metabolism of particular types of cells also reflects processes peculiar to these cells. The phospholipid metabolism of both intestinal mucosa and liver cells must, therefore, reflect the supply of phospholipids for export as part of the vehicle for triacylglycerol transport. In addition, the phospholipid metabolism of liver must reflect the export of phospholipid in the bile. Although the phospholipids of the liver cells are a complex mixture of phospholipid classes, phosphatidylcholine is the principal phospholipid exported in both the lipoproteins and in bile. Its export by the liver is considerable, amounting to the equivalent (in man) of up to 15–20% of the phosphatidylcholine in the liver per day.

1. FAT ABSORPTION

For an account of the development of concepts in this field, the reader should consult the previous edition of this book. There are also several recent reviews giving extensive coverage of various aspects of this subject[1–7].

Intraluminal triacylglycerol hydrolysis

The mixture of dietary acylglycerols, bile and pancreatic juice, aided by the churning action of the jejunum, rapidly form an emulsion of lipid droplets 500–1000 mμ in diameter. This represents an ideal medium for the action of pancreatic lipase, a true lipolytic enzyme whose activity is enhanced in the presence of bile salts[8]. The enzyme hydrolyses the ester bonds on the 1- and 3-positions of the triacylglycerol molecule yielding predominantly monoacylglycerol and free fatty acids[9]. In the intestinal lumen, the principal end product of triacylglycerol hydrolysis appears to be 2-monoacylglycerol in both man[10] and rat[11]; though about one-third of the dietary acylglycerol may be fully hydrolysed[6], presumably due to the action of pancreatic cholesterol esterase upon the 2-monoacylglycerol[12].

Intraluminal phospholipid hydrolysis

Since they are important constituents of biological membranes and of soluble lipoproteins, phospholipids are present in the diet in appreciable amounts, and they are also added to the luminal contents in bile and as components of the membranes of desquamated intestinal cells. The contribution from bile, principally as phosphatidylcholine, is considerable (up to 2 g per day in man[13]).

Pancreatic juice contains a phospholipase A_2, which is secreted as an inactive zymogen. Activation occurs in the intestinal lumen, probably through tryptic digestion, to yield the active enzyme: a single polypeptide chain of molecular weight 13 800 (refs. 14–16). Its activity is greatest towards acidic phospholipids and phosphatidylethanolamine. In the presence of bile salts, when conditions approximate more closely to those prevailing in the intestinal lumen, the hydrolysis of phosphatidylcholine is stimulated[14, 17].

Pancreatic juice also possesses phospholipase A_1 activity, but this is probably due to the action of pancreatic lipase upon phospholipids rather than to the presence of a second true phospholipase[17]. Analysis of luminal contents, however, showed that 1-acyl-*sn*-glycero-3-phosphorylcholine was the predominant phospholipid formed[18] and therefore the action of pancreatic lipase upon lecithin must be of relatively minor importance *in vivo*.

The brush border of intestinal cells contains several enzymes capable of hydrolysing phospholipids: a phospholipase A_2 which is activated by bile salts[19], a lysolecithinase[19], a sphingomyelinase[20] (EC.3.1.4.3) and a ceramidase[20]. As both pancreatic juice and pancreatic homogenates show little ability to degrade sphingomyelin, and since the sphingomyelinase and ceramidase are isolated with the brush border fraction, it is likely that the brush border is the main site of sphingomyelin hydrolysis during digestion[19]. Both ceramide and free sphingosine have been detected in significant amounts in luminal contents however[20], suggesting that hydrolysis, though a property of the brush border, actually occurs within the lumen. Apart from the sphingomyelinase, there appears to be no direct evidence for phospholipase C or D activities.

Physical state of the intraluminal phase

When bile is delivered to the duodenum it contains mixed micelles of bile salts, phospholipids and cholesterol. The relative proportions of these components in bile vary within the limits of a phase diagram[21], but are in molecular proportions approximately in the region of 13:3:1. In man, analyses of intraluminal contents during fat absorption have shown the presence of two phases in equilibrium[22]: an oil phase consisting of droplets of the size range 5 000–10 000 Å and an isotropic phase containing micelles of dimensions in the region of 30–100 Å. The oil phase contains predominantly non-polar lipids: triacylglycerols, diacylglycerols, cholesterol, and unionized fatty acids; the micellar phase contains bile salts, 2-monoacylglycerols and partially ionized fatty acids, but only small amounts of the more non-polar lipids[21–23]. It is thought that as the triacylglycerol in the emulsion is hydrolysed to progressively more hydrophilic products (2-monoacylglycerol and fatty acid), that these then leave the oil phase and enter the primary bile salt micelle forming a swelling system[21, 23, 24]. This swelling micelle is capable of incorporating small quantities of less polar lipids: cholesterol, fat-soluble vitamins, even triacylglycerols and hydrocarbons. The absorption of lipids is depressed in the absence of bile (e.g. in bile fistula) and the absorption of the more non-polar lipids is increased in the presence of dietary hydrolysable fat.

Whilst phospholipid is a prominent component of the original primary bile micelle, the presence of phospholipid (lecithin or lysolecithin) in the swollen micelle referred to above is far less certain[23] since analysis of the micellar phase[22], did not include phospholipid estimations.

Uptake of lipids from the intraluminal phase

The demonstration, by electron microscopy, of small membrane-bounded droplets in the cell, together with droplets occasionally seen between the microvilli, led several investigators to suggest pinocytosis of droplet fat as a method by which lipid might be taken into the cell and subsequently appear as membrane-bounded droplets[25–28]. This idea is now less favoured as a major contributor to intracellular droplet fat: from quantitative considerations[6], from the behaviour and fate of inert microparticulate substances during fat absorption (e.g. silver and ferritin)[29], and from electron microscopic studies of the absorption of micellar lipids[30, 31].

Micellar lipid is readily absorbed and metabolised by the intestinal mucosa[32]. Studies of this process at 0° showed that though fat was taken up readily, droplet fat did not appear in the cytoplasm. When the tissue was subsequently incubated at 37° in saline, droplet fat could then be demonstrated in the cytoplasm. This suggested that at least part of the intracellular droplet fat was derived from subsequent metabolism of the absorbed fat rather than by a direct uptake into the cell[30]. In support of this are the observations of the virtual absence of fat droplets in the terminal web during these experiments[30], and the localization of the droplet fat apparently within the endoplasmic reticulum[29].

Whilst it appears that lipid can be taken from the micellar solution present in the intestinal lumen, it does not necessarily follow that the whole contents of the micelle must enter the cell together during lipid uptake. The absorption of lipids occurs predominantly in the jejunum and proximal ileum, rather than in the distal ileum; whereas uptake of bile salts occurs principally in the ileum[1, 33–36]. Moreover, bile salt absorption is selective for particular bile salts[37]. A few experiments might be thought to point to absorption of mixed micelles, but it would then require excretion of bile salts to explain the differential topography of bile salt absorption. (See ref. 6 for discussion.)

A close proximity of the micellar lipid with the microvillar membrane has been observed[31], but the actual mode of passage of lipid from the micelles into the cell is a matter of much speculation. Several workers have considered that lipids, as monomolecular species, might diffuse from the micelle, dissolve in the membrane and from there diffuse into the cell by a passive process[4, 5, 21, 38–42]. Experiments with heat-inactivated[32] or chilled[30] mucosa have suggested that the initial lipid binding and penetration are enzyme-independent processes. The concentration gradient for inward diffusion would then be maintained by the re-esterification of lipid and its removal from the cell, processes which require energy. The overall rate-limiting step in this system is thought to be the micellar solubilisation of lipids prior to their uptake into the cell[43, 44], rather than the intracellular esterification and output processes[30].

Intracellular metabolism of acylglycerols

Whereas the lipids resulting from intraluminal hydrolysis are predominantly 2-monoacylglycerol and fatty acids, much of the lipid leaving the intestine in the form of chylomicrons, is triacylglycerol. It is clear that much resynthesis has taken place.

There are two main pathways by which this has been achieved: the monoacylglycerol pathway and the glycerophosphate-phosphatidic acid pathway[5]. The monoacylglycerol pathway of Clark and Hübscher[46] is quantitatively the more important pathway in the rat and probably in man, but is of less importance in the sheep[6]. This pathway involves reacylation of the 2-monoacylglycerol directly and its synthetase system has recently been studied in a purified form[46–47].

The second pathway utilizes a phospholipid intermediate, phosphatidic acid, the turnover of the phosphate moiety of which is normally high and is stimulated further by the ingestion of fatty acids[48]. This pathway appears to utilize a specific phosphatidate phosphohydrolase, EC. 3.1.3.4, which acts on membrane-bound, biosynthetically produced phosphatidate rather than on aqueous phosphatidate suspension[49–51]. Glycerophosphate is required for the operation of this pathway, and in the intestinal mucosa, this is supplied principally by glycolysis rather than by phosphorylation of glycerol. Two other routes have been described by which the intestinal mucosa may synthesize phosphatidic acid[7]. The first is via the acylation of dihydroxyacetone phosphate with subsequent reduction to lysophosphatidic acid and further acylation; the second via the phosphorylation of monoacylglycerol to give lysophosphatidic acid. The relative importance of these routes is not yet established.

The partitioning of fatty acids, by chain length, between portal blood and chyle was outlined in the previous edition of this book. 2-Monoacylglycerols, of short- and medium-chain fatty acids, are more readily hydrolysed than the 2-monoacylglycerols of long-chain fatty acids. This is true for both intracellular monoacylglycerol[52] and triacylglycerol[53] lipases of the mucosal cells and even for pancreatic lipase[6]. Further differentiation is probably due to the specificity of the two major esterifying pathways[6, 7, 54] which results in little esterification of short- and medium-chain fatty acids and their passage in unesterified form into the portal blood, where they are taken up by albumin[55].

Long-chain fatty acids appear principally as triacylglycerols in the chyle, the fatty acid positioning on the molecule being partly explained by the specificities of the esterifying systems[6, 7, 54], and the fatty acid composition being a balance of the exogenous and endogenous contributions[6, 7]. The extent of hydrolysis of long-chain 2-monoacylglycerols by mucosal intracellular lipases is unknown, but seems to be sufficient to support the fatty acid requirements of patients with a-*beta*lipoproteinaemia[55].

Intracellular metabolism of phospholipids

Early experiments with ^{32}P-labelled phospholipids[56], [^{14}C]palmitoyl-phospholipids[57] and doubly labelled phospholipids[58] suggested that some of the dietary phospholipid could be recovered as phospholipid in the chyle, since 40% of [^{32}P] lecithin fed to rats was recovered in the chylomicrons[59].

Feeding experiments with 1-[^{3}H] acyl-[^{32}P] lecithin, and 2-[^{14}C] acyl-[^{32}P] lecithin, showed that a much greater proportion of the 1-acyl label than the 2-acyl label was recovered in the chyle lecithin. The ^{3}H:^{32}P ratio was the same in the chyle as in the fed material whereas the ^{14}C:^{32}P ratio was greatly depressed. When [^{3}H] choline-[^{32}P] lecithin was fed, the ^{3}H:^{32}P ratio in the chyle lecithin was the same as that in the diet. In addition, when 1-[^{3}H] acyl-*sn*-glycero-3-[^{32}P] phosphorylcholine was fed, some 40% was recovered in the chyle as 1-[^{3}H] acyl-[^{32}P] lecithin[59]. These elegant experiments establish that a large proportion of the lecithin is hydrolysed to lysolecithin (1-acyl) and that this is re-acylated before being incorporated into chylomicrons[53]. The finding of enzymatic re-acylation mechanisms in the intestinal mucosa[60–63] supports this concept.

Since only 13–40% of the ^{32}P label was recovered in the chyle lecithin in the above experiments[59], more extensive metabolism of some of the dietary lecithin might be suspected. Other phospholipid metabolizing enzymes have been detected in mucosal preparations and these may play some role in these processes, though their individual quantitative significance remains to be established. These include phospholipase A and lysolecithinase activities[19, 62, 64–67] and an enzyme which converts two molecules of lysolecithin to lecithin and *sn*-glycero-3-phosphorylcholine[19]. Most of the fatty acids lost from dietary lecithin during digestion and absorption, appeared in the chyle as triacylglycerols[59, 68]. Fatty acids in biliary lecithins may be handled somewhat differently however; since [^{3}H]oleate derived from biliary lecithin primarily entered the portal blood, mainly as phospholipid[69]. The *sn*-glycero-3-phosphorylcholine resulting from phospholipid breakdown is probably broken down further via a phosphodiesterase which releases choline and glycero-3-phosphate. Feeding experiments with *sn*-glycero-3-

References p. 370

phosphoryl-[*Me*-^{3}H] choline established that this compound was predominantly hydrolysed, and only 8% reappeared in lecithin[68].

In the experiments quoted above[59–68], recoveries of dietary lecithin in the chyle never amounted to more than 40% of the dose fed. Even with increasing levels of phospholipid in the diet, the specific activity of chyle lecithin was always considerably lower than that of dietary lecithin. Calculations suggested that at least 60% of the chyle phospholipid was provided by endogenous sources, i.e. from bile, from blood phospholipids or from intestinal biosynthesis[59]. Mechanisms for lecithin biosynthesis '*de novo*' have been described in intestinal mucosa[70, 71], and during triacylglycerol absorption a four-fold stimulation of the turnover of the phosphate moiety of lecithin was observed[72]. In addition to the stimulation in turnover, a net increase in phospholipid appears in the lymph during triacylglycerol absorption[59, 73].

Much of what has been said above refers primarily to the metabolism of phosphatidylcholine. Feeding studies have shown that during sphingomyelin absorption ceramide and sphingosine are broken down within the intestinal cell, and that little of the original sphingomyelin molecule was found as sphingomyelin in the lymph. Most of the fatty acids of the sphingomyelin appeared in lymph triacylglycerols or in lymph lecithin. The sphingosine was partially converted to fatty acid, which also appeared in lymph triacylglycerol. A proportion of the choline originally in the sphingomyelin appeared in chyle lecithin. From these studies it would appear that dietary sphingomyelin is almost fully metabolised, and that much of the lymph sphingomyelin is obtained by endogenous biosynthesis[74].

A phospholipase specific for phosphatidylinositol has been described in the soluble fraction of guinea pig intestinal mucosa homogenates[75]. This enzyme yields diacylglycerol and is, therefore, of the phospholipase C type.

2. CHYLOMICRON METABOLISM

Thoracic duct lymph contains small amounts of phospholipid in the pre-fed state; but most of this is not associated with chylomicrons[59]. Ultracentrifugation of this lymph reveals the presence of lipoproteins of sedimentation characteristics (S_f 20–400, peak S_f 102) and overall lipid composition (triacylglycerol 67%, phospholipid 24%, cholesterol 6%) similar to plasma VLDL (very low density lipoprotein). These lipoproteins contribute to the plasma VLDL; but their lipids are derived from intestinal metabolism (bile lipids and intestinal biosynthesis), rather than from plasma or liver lipids[76]. The function of these VLDL leaving the intestine is probably to transport endogenous lipid material, mostly cholesterol[76]. In contrast to the chylomicrons, VLDL concentration in the lymph changes little during fat absorption; though VLDL compositions changes in response to changes in dietary lipid composition[77]. A further difference from chylomicrons is their slower rate of removal from the circulation[77].

The phospholipid content of the chyle rises dramatically after a meal[59, 73], and this rise is almost solely due to the increase in chylomicron phospholipid. This increase in lipid in the chyle is due to the production of increased numbers of chylomicrons, which may cause a rise in plasma chylomicron counts at the peak of fat absorption, to as much as five times the fasting level[78]. Chylomicrons, together with the shorter-chain fatty acids, account for most of the exogenous lipid leaving the intestine.

Chylomicron composition and structure

The composition and properties of mammalian chylomicrons are summarized in Table 1. Their density is largely dictated by their predominant component, triacylglycerol, which causes them to float during the ultracentrifugation of serum or plasma. Phospholipid analysis of dog chylomicrons showed that the principal phospholipid was phosphatidylcholine[86] (75%); and in this context one may recall the high turnover rate of phosphatidylcholine during fat absorption[72].

The values given in Table 1 are merely a guide as to the average composition and properties of the chylomicron population. It is clear that chylomicron particles must exist with a range of sizes and compositions. Fractionation of a chylomicron population showed that not only was the population heterogenous with regard to particle size, but also that the smaller chylomicrons contained a higher concentration of protein, phospholipid and cholesterol than the larger ones[81]. Calculations based upon the relative amounts of these components indicated that there was sufficient to form a surface monolayer[87]. As the particle size decreases, the ratio of the components needed to make such a surface layer to triacylglycerol might be expected to increase[81].

Electron microscopy of intact chylomicrons showed them to consist of a lipid core surrounded by a dense outer layer[81, 87]. Freezing and thawing or repeated rotary evaporation of chylomicron suspensions, produced two fractions: a triacylglycerol-rich fraction thought to represent the oily core, and a fraction thought to represent the surface coat[88, 89] The latter fraction contained phospholipid, protein, and cholesterol, but little cholesterol ester or triacylglycerol. The structure of the chylomicron, therefore, is probably that of a large non-polar core around which there is a stabilizing layer (probably a monolayer) of phospholipid and cholesterol; which makes interactions with a loose network of proteins[81, 83].

TABLE 1

PROPERTIES OF HUMAN CHYLOMICRONS[a]

Physical properties		*Composition*	*Range of values*
Floatation Coefficient	$S_f > 400$	Protein (% by weight)	2 (0.5– 2.5)
Density (g/ml)	<0.95	Lipid (% by weight)	98 (97.5–99.5)
Electrophoretic mobility	origin		
Molecular weight mean	10^{10}	Triglyceride (% of lipid)	90 (81–97)
(range 10^9–10^{12})		Phospholipid (% of lipid)	4 (2– 9)
Diameter Å (principal range)	1500–2000	Cholesterol (% of lipid)	2 (1– 3)
(overall range 1200–11000)		Cholesterol ester (% of lipid)	4 (1– 5)

[a]Derived from Refs. 79–85, 149.

Chylomicron formation

The formation of chylomicrons in the intestinal cell is under active investigation at the present time. Protein biosynthesis by the mucosal cell itself, has been demonstrated in several studies, and may be important in chylomicron formation[90–94]. Administration of puromycin, acetoxycyclohexamide or carbon tetrachloride, caused a depression in long-chain fatty acid uptake and in chylomicron release from the intestine[55, 95–97]. Since these inhibitors also affect protein biosynthesis in the liver, it was possible that depressed chylomicron formation was due to defective production of a precursor protein from liver. Other experiments with puromycin suggested that its effect may not be due primarily to depressed intestinal protein synthesis, but to depressed gastric and intestinal motility and lymph flow[98]. In the presence of orotic acid, which interferes with lipoprotein production in the liver, there is no impairment of chylomicron production by the intestinal mucosa, and no interference with long-chain fatty acid absorption[99–100]. This strongly suggests that chylomicron protein is formed primarily in the intestinal mucosa itself; and supports immunological findings that a new protein species (which is not found in fasting lymph) appears, and follows the increase in chyle triacylglycerol[83, 101].

The lipid components of the chylomicron are formed in the endoplasmic reticulum[7]. The origin of the protein, whether from membrane-bound ribosomes directly, or from soluble proteins[94], is not clear; but it would appear that the surface coat of the chylomicron may be made from intracellular components rather than from pinched-off plasma membrane[87, 89]. Evidence from electron microscopy suggested the formation of droplets within the cisternae and their accumulation within Golgi vacuoles[29–32, 102, 103]. There still remains, however, the problem of externalization; for which reversed pinocytosis (which has been implicated also in lipoprotein release by the liver) is a likely candidate[54, 103].

Chylomicrons in the circulation

Chylomicrons do not enter the circulatory system directly, probably because they are too large to penetrate the capillary endothelial membrane[104]. The peak of chylomicron concentration in man occurs approximately 2–3 h after a meal[78]; during this time the concentration in the plasma may rise to 30 mg/ml[84]. The level then drops rapidly, indicating clearance, and there is some evidence that large chylomicrons are cleared most rapidly[105]. Using chylomicrons containing labelled acylglycerols, the half life of clearance was found to be about 15 min[83, 85, 105]. Clearance of chylomicron phospholipid is less rapid, and is complicated by phospholipid exchange between the various types of circulating lipoproteins[106].

After leaving the lymph, chylomicrons are modified in the circulation picking up high density lipoprotein (HDL) which changes their electrophoretic mobility[83, 85]. There is also a decrease in phospholipid content and an increase in cholesterol ester[81, 83], which may be due to the activation of the enzyme lecithin:cholesterol acyltransferase (LCAT) by HDL[83, 107]. In addition, some component of HDL is also an activator of lipoprotein lipase (LL), which can hydrolyse and therefore facilitate the removal of triacylglycerols from the chylomicron[108].

As triacylglycerol is lost, the chylomicron surface would be reduced, and therefore the activated LCAT may be a device for removing surface lecithin and cholesterol[83]. Cholesterol ester and lysolecithin would be produced; the cholesterol ester may then move into the lipoprotein core or be taken up by another lipoprotein, and the lysolecithin, which is readily diffusible, might then leave the lipoprotein to be taken up by albumin, other lipoproteins, or cells[83]. In line with these ideas, it is observed that congenital absence of HDL results in delayed chylomicron clearance and elevated plasma triacylglycerol levels[83, 85].

Chylomicron clearance

Administration of heparin was observed to cause accelerated clearing of turbid plasma due to alimentary lipaemia. Under these conditions there was an accelerated hydrolysis of triacylglycerol and a decrease in the number of chylomicrons[109]. This effect of heparin has been shown to be due to its action upon LL, and is two-fold; both releasing the enzyme from binding sites, and activating it[110]. LL is involved in the normal removal of chylomicron acylglycerols, and is different from other lipases in that its preferred substrate is triacylglycerol contained in a lipoprotein rather than free triacylglycerol[104]. The products of this hydrolysis are ultimately glycerol and fatty acids. A component of HDL incorporated into the chylomicron surface appears to enhance LL activity[83].

LL is found on the surface of capillary endothelial cells of several tissues, and also on the surface of adipose tissue cells. It is probably responsible for the hydrolysis of chylomicron triacylglycerols in many tissues especially adipose tissue, heart and skeletal muscle[54, 83], chylomicron triacylglycerol has been shown to be an energy source in all three tissues. The amount of this enzyme in some tissues is under hormonal control; the activity in mammary gland tissue increases at parturition, whilst the activity in adipose tissue declines at this time. Its activity in adipose tissue and skeletal muscle alters with the nutritional state of the animal[112].

The phospholipid content of the chylomicrons is lost during clearance, probably resulting from the action of LCAT. Details of its fate and also that of the protein components are far from clear.

Role of the liver in chylomicron metabolism

The role of the liver in the direct processing of chylomicrons, is far less clear than that of the extrahepatic tissues. Using chylomicrons containing labelled triacylglycerol, early experiments showed that up to 40% of the label was present in the liver 10–15 min after infusion[113–115]. Morphological studies carried out at this time showed fat droplets in the parenchymal rather than the reticuloendothelial cells[116, 117]. In these short time periods, the use of doubly labelled acylglycerols (fatty acid/glycerol), indicated that little hydrolysis had occurred. These experiments did not establish that the lipid had in fact entered the cell, and it was, therefore, possible that the results might have represented lipid trapped in the extracellular spaces of the liver. The endothelial lining of the sinusoids is sufficiently open to allow this and electron microscopy demon-

References p. 370

strated that, in addition to the fat droplets appearing inside the cells, droplets were also sequestered in the space of Disse. Reverse perfusion in these short-term experiments washed triacylglycerols out of the liver still in the form of chylomicrons[119]. In other experiments with isolated perfused livers, little uptake or metabolism (oxidation) of [^{14}C] triacylglycerol was observed in the absence of heparin, even under conditions in which chylomicrons were sequestered[115].

In experiments with isolated livers, perfused with medium containing heparin, a much greater uptake and oxidation of chylomicron [^{14}C] triacylglycerol was observed[120]. The change in the labelling ratio in doubly labelled triacylglycerol suggested that hydrolysis had occurred. Even though some chylomicron triacylglycerol could be washed out of the liver, at the same time, evidence was obtained for incorporation of fatty acids into endogenous liver lipids[119]. It was therefore suggested that after trapping, the chylomicrons were significantly hydrolysed in the extracellular space.

Liver contains no LL under normal conditions of measurement[122], although some activity can be observed at high heparin concentrations[123]. It is, therefore, unlikely that this enzyme plays much part directly in chylomicron hydrolysis by the liver[124], although the LL activity in extrahepatic tissues may be a generator of those chylomicron fatty acids taken up by the liver *in vivo*[115, 119, 121].

Both liver slices and isolated liver cells have been shown to take up chylomicrons and to bring about the hydrolysis and further metabolism of chylomicron lipids[124–127]. Since extrahepatic LL cannot be present in these circumstances and since protamine sulphate, an inhibitor of LL had no effect upon this hydrolysis, the presence of a liver lipase was suspected[128]. The properties of such an enzyme have been described and it appears to be a constituent of the plasma membrane[128]. The use of an alternative technique of cell isolation, however, gave cells which bound chylomicrons poorly and had low lipolytic activity[129]. Thus the role of the liver in the direct binding and hydrolysis of chylomicron triacylglycerol remains an open question.

3. UPTAKE OF LIPIDS BY THE LIVER

Fatty acids

Liver cells readily take up non-esterified fatty acids (NEFA). These fatty acids may be derived from lipolysis at the liver cell surface, from intestinal absorption (short and medium chain length), from lipolysis of circulating triacylglycerols in the extrahepatic tissues, or by mobilization from adipose tissue. In the last three cases, plasma albumin serves as the vehicle for fatty acid transport.

Studies into the mechanism of this uptake have differentiated the process into two parts: the first an energy-independent binding to the surface membrane[126, 130]; the second an esterification, on which continued uptake then depends[130, 131]. The concentration of NEFA in liver cells is normally low[126, 132a].

On a long-term basis, it is triacylglycerol which accounts for most of the fatty acid uptake into liver cells, but studies using labelled fatty acids have implicated phospholipids in the process[131, 132b, 133]. Fatty acids were incorporated more rapidly into phospholipids than into triacylglycerols during fatty acid uptake[132b, 133], and the subcellular fraction showing the highest incorporation was the plasma membrane[131].

Using isolated cells, it was observed that the peak of [^{14}C] palmitate incorporation precedes both that of phosphatidic acid and of triacylglycerols. After about one minute the specific activity of the plasma membrane phosphatidylcholine was about 30 times higher than that of the total cellular phosphatidylcholine. The labelling in the plasma membrane then rapidly declined and could be recovered from other regions of the cell with an increasing proportion in the form of triacylglycerol. At all times, however, the labelling in the plasma membrane was principally in the form of phospholipid. Calculations based on the half-life of surface membrane phospholipid, showed that it might make a significant contribution to the fatty acid uptake of the cell[134].

Administration of [^{14}C] palmitate *in vivo* to rats showed, that of the liver phospholipids, phosphatidylcholine initially bore the highest labelling[134]. In another series of experiments conducted over extremely short time periods, the labelling of phosphatidylcholine, though rapid, was small, and was accompanied by a much more pronounced incorporation into phosphatidic acid[135]. The reasons for the difference in labelling patterns in these experiments are not clear, but may be related to fatty acid concentration or the availability of appropriate receptor molecules.

Esterification with palmitic acid at the cell surface requires the presence not only of appropriate receptor molecules, but also the enzymatic esterifying apparatus. Enzymes activating fatty acids have been reported in surface preparations from liver[136], though this is disputed by other workers[137, 138]. Likewise, demonstrations of transacylase activities in the plasma membrane[131, 139, 140], have been questioned[140]. Some of these differences may be due to kinetic considerations[140], but more likely, due to the use of plasma membrane preparations containing different amounts of sinusoidal and canalicular regions.

The acyl acceptor used in some experiments *in vitro* has been endogenous lipid[131, 134, 140], and in these experiments phosphatidylcholine (and phosphatidylethanolamine) were the principal early products suggesting the acylation of an endogenous lysophospholipid. Endogenous lyso-compounds could arise by the action of phospholipases located in the surface membrane itself. Recently phospholipase A_1 activity has been observed in rat liver plasma membrane preparations[142, 143], and in this connection it is interesting to note that all the phospholipids synthesized in one recent short term study with palmitate contained the palmitic acid in position 1 (ref. 135); whereas those synthesized in a parallel short term study with linoleic acid contained the label in position 2 (ref. 144).

After initial incorporation into phospholipids, labelled fatty acids then appear in triacylglycerols. This represents further metabolism of the intact phospholipid and phosphatidic acid, and might involve deacylation, or removal of the polar head groups of the phospholipids and phosphatidic acid. The sequence, relative importance, and subcellular location of these changes are subjects for future study.

References p. 370

Phospholipids

The addition of exogenous acceptor molecules in the form of 1- or 2-acyl lysophospholipids or *sn*-glycerol 3-phosphate, stimulated fatty acid uptake in studies *in vitro*[134, 139, 141]. In experiments using isolated liver preparations 1-acyl-*sn*-glycero-3-phosphorylcholine substantially stimulated the incorporation of fatty acid. This substrate might be expected to be representative of the lysophospholipids of blood produced by the action of LCAT upon lecithin. Injection of 1-acyl *sn*-glycero-3-phosphorylcholine showed that its half life was only 10 min which is very short when compared with other plasma phospholipids[60]. The principal method of removal of this compound was by an acylation reaction, and the liver was shown to have a greater rate of uptake than other tissues for both this compound and for 1-acyl-*sn*-glycero-3-phosphorylethanolamine, indicating that the liver plays a prominent role in preventing the build up of lytic concentrations of lysophospholipids in the circulation.

4. THE LIVER AND PLASMA LIPOPROTEINS

Plasma contains about 0.5 g of lipid per 100 ml, of which phospholipid accounts for approximately one third[84, 104]: the lipids are almost entirely associated with protein. Current evidence has strongly suggested that the intestinal mucosa is the site of origin of the chylomicrons and small amounts of VLDL and LDL[85, 93, 145]. For the remaining lipoproteins, isotopic evidence clearly implicates the liver as the principal site of component biosynthesis and final assembly. Several recent reviews may be consulted for those aspects of lipoprotein characterization and metabolism which cannot be presented here for reasons of space[83–85, 114, 146, 149].

Four main classes of plasma lipoproteins are recognized and classified according to their density range:- chylomicrons, very low density lipoproteins (VLDL), low density lipoproteins (LDL) and high density lipoproteins (HDL). The NEFA–albumin complex is not normally classed as a lipoprotein.

Physical and chemical characteristics

Analysis of lipoproteins (Table 2) indicates that each class represents a range of entities rather than a single species. By employing other separation procedures to compounds of a given class, fractions of slightly different composition, molecular weight, and properties have been obtained. The size and composition of the higher molecular weight classes (chylomicrons, VLDL, LDL) are continuously variable; whereas several distinct species of HDL appear to exist: HDL, HDL_2, VHDL[85]. VHDL (density >1.21), in contrast to all the other lipoprotein species, contains lysolecithin rather than lecithin as the principal phospholipid[85, 150].

Certain trends in the lipid composition can be noted: The triacylglycerol content as a proportion of the total lipid increases with decreasing density, whereas the phospholipid and total cholesterol content as a proportion of the total lipid decrease with decreasing

TABLE 2

PROPERTIES OF LIPOPROTEINS DERIVED FROM THE HUMAN LIVER[a]

	HDL *(High density)*	*LDL* *(Low density)*	*VLDL* *(Very low density)*
Electrophoresis (paper)	α	β	pre-β
S_f range (in 1.063 g/ml)	–	0–20	20–400
Density range (g/ml)	1.21–1.063	1.063–1.006	1.006–0.95
M wt ($\cdot 10^6$)	0.2–0.4	2–4	5–100
Diameter Å[b]	~84 (350 length)	~220	300–800
Shape[d]	–		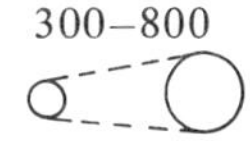
Protein % wt	50	22–20[c]	12 ... 5[c]
Lipid % wt	50	78–80	88 ... 95
Triacylglycerol % wt	5	10–12	45 ... 65
Phospholipid % wt	22	22–20	23 ... 13
Cholesterol % wt	3	10– 8	7 ... 5
Cholesterol ester % wt	17	36–37	13 ... 10
NEFA % wt	3	1	2
Approximate concentration mg/100 ml			
Males (~ 45 yrs)	280	400	140
Females (~ 45 yrs)	400	350	70

[a]Derived from refs. 82–85, 149. [b]HDL is rod-shaped. The size range of LDL is 170–260Å; the size range of VLDL is much wider. [c]The variation in analytical figures partly represents the variation of composition with size. [d]Strictly, the smallest circle under VLDL should be larger than the largerst LDL circle.

density. Whilst the gross fatty acid composition of the lipoproteins are different, this represents not so much a fundamental property of the lipoprotein; but rather the relative contributions of the constituent lipid classes (triacylglycerols, phospholipids, cholesterol esters).

Structure

Two basic types of structure for plasma lipoprotein have been suggested. The first, containing less than 30% of protein and therefore representing chylomicrons, VLDL and LDL, is thought to be an essentially micellar arrangement of a hydrophobic core of the more non-polar lipids (triacylglycerols, cholesterol esters) surrounded by a hydrophilic layer of phospholipid, cholesterol and protein. The evidence for this type of structure has been discussed briefly in relation to the chylomicron, and is documented more fully elsewhere[85, 147]. The second type of structure has been termed pseudomolecular; and in this the lipid is thought to be located in the interstices of a folded polypeptide chain[85]. HDL and VHDL are thought to have structures of this type.

The amount of covalent bonding between the proteins and the lipids is small[151], and although the overall composition appears characteristic in terms of lipid class, the individual lipid molecules are in dynamic exchange between lipoprotein classes and with cells. Triacylglycerols, phospholipids and cholesterol exchange more readily than cholesterol esters. This exchange, coupled to the activity of LCAT, explains the longer half-life of cholesterol in the circulation as compared to the other lipids.

The proteins of each type of lipoprotein have been studied in terms of amino acid composition, end group analysis and immunological behaviour both intact or as apo-proteins. From these studies, valuable clues have become apparent as to the relationship of each type of lipoprotein with the other lipoproteins and to their origins.

Metabolism of VLDL

Chylomicrons contain lipid (triacylglycerol), derived mainly from exogenous sources, and are produced in response to the demand for lipid transport consequent on fat absorption. The lipids of VLDL are primarily of endogenous origin, and are generated by the metabolic response of the liver: either from conversion of carbohydrates, from the reprocessing of chylomicron triacylglycerol, or from NEFA resulting from excess adipose tissue lipolysis. VLDL formation and levels, therefore, bear a close relationship to nutritional and metabolic state.

During fasting VLDL accounts for most of the plasma triacylglycerol and is synthesized mainly utilizing the fatty acids mobilized from adipose tissue. Diets high in fat but low in carbohydrate lower the VLDL and plasma triacylglycerol levels, but produce large oscillation of triacylglycerol (as chylomicrons) following a meal. On the other hand, diets high in carbohydrate but low in fat cause a maintained high VLDL level[148]. VLDL production is also stimulated by intake of medium chain length fatty acids[152].

The removal of triacylglycerol from VLDL is probably via a hydrolytic process involving lipoprotein lipase. The fatty acids are then taken up for storage, oxidation, or other metabolism: by adipose tissue, heart, skeletal muscle and possibly other tissues[83, 85, 110]. Material with floatation properties similar to LDL and HDL, appears to be produced during intensive hydrolysis of VLDL *in vivo*, following administration of heparin. These LDL are similar in lipid composition to other LDL, but the HDL formed differ in lipid composition from circulating HDL. Normalization may occur in the circulation due to the action of LCAT, which is particularly active with HDL lipids[85, 107]

The precursor–product relationship between VLDL and LDL is further supported by tracer studies *in vivo* using VLDL labelled with respect to protein[153, 154], and also from studies of rapid shifts of lipoprotein pattern which occur in certain metabolic situations[85]. The relationship between LDL and VLDL can be seen in other ways: (a) a-*beta*lipoproteinaemia, a genetic disease is characterised by the absence of LDL and VLDL (and chylomicrons) in the plasma[83, 85, 148]; (b) inhibition of protein biosynthesis by actinomycin or puromycin causes depressed synthesis of both VLDL and LDL; (c) orotic acid inhibits release of VLDL and LDL from normal and perfused livers. Since VLDL has immunological properties in addition to those of LDL, a more complex structure is indicated for the former[83, 155].

Whilst it is now clear that catabolism of VLDL can give rise (amongst other products) to LDL, the possible role of LDL in the formation of VLDL has been a matter of some speculation, i.e. can circulating LDL serve as a vehicle used directly by the liver as the basis for assembly of VLDL[83; 85, 146, 154]? This uncertainty has been resolved recently in a very clear fashion by experiments which demonstrated that whilst VLDL labelled in the protein components could act as the precursor of circulating LDL, protein-labelled LDL did not act as the precursor for VLDL[154].

VLDL in the circulation contains protein which is immunologically identical with HDL apo-proteins. Rapid metabolic shifts in lipoprotein patterns, due to breakdown of VLDL, are sometimes accompanied by increased levels of HDL, suggesting that HDL in the circulation may be derived from VLDL catabolism. Experiments with puromycin and orotic acid suggest, however, that the biosynthesis and release of HDL apo-proteins may be independent of VLDL production, and in a congenital deficiency of HDL apo-proteins (Tangier disease), VLDL is found in the plasma but with a different electrophoretic mobility from the normal[83, 85]. Circulating VLDL rapidly picks up and exchanges both protein and lipid components of circulating HDL and so a true precursor–product relationship between VLDL and HDL is much more difficult to establish than in the case of VLDL and LDL[154].

A speculative interpretation of the catabolism of VLDL is presented in Fig. 1.

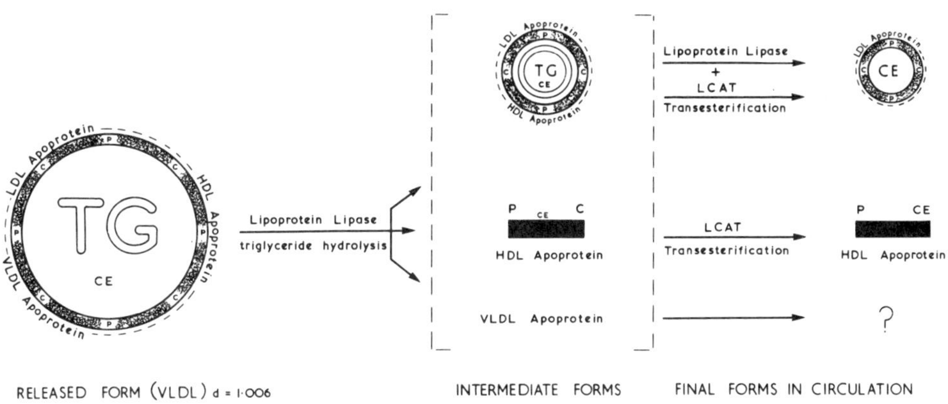

Fig. 1. Possible interrelationships of plasma lipoproteins derived from liver. Sizes approximate to relative sizes of lipoproteins (man). Abbreviations: VLDL, LDL, HDL, LCAT see text; P, phospholipid; C, cholesterol; CE, cholesterol ester; TG, triacylglycerol. Taken partly from ref. 154.

Site of biosynthesis of VLDL

The biosynthesis of lipoproteins occurs in the liver, all the individual components: protein, phospholipid, cholesterol and triacylglycerol can be made by the rough endoplasmic reticulum. Electron microscope studies of liver have demonstrated the formation of small osmiophilic droplets, firstly within the endoplasmic reticulum and Golgi region and later in the space of Disse[156–162]. The production of these bodies was inhibited by puromycin, and radioautography using tritiated lipid precursors indicated that they contained lipids[163, 164]. Thus a packaging and secretion process[165], similar to that of many other substances may be involved in lipoprotein production by the liver.

Turnover of lipoprotein components

The half-lives of other plasma lipoproteins are considerably longer than that of the chylomicron. Apo-VLDL has an apparent half life of 2–3 h[154], whilst the protein components of HDL and LDL appear to have a half life in the circulation of about 3–4 days[83, 165, 168]. Uptake of HDL apoprotein has been demonstrated by the liver[168], but little is known of the fate of the other apoproteins[83].

In addition to exchange between the various lipoprotein classes, there is also a rapid interchange of lipids between the lipoproteins and cells. Plasma cholesterol exchanges readily with cholesterol in erythrocytes and other cells[169, 170]. Phospholipids, particularly those of HDL and LDL, exchange with erythrocyte phospholipids; phosphatidylcholine and sphingomyelin exchanging more readily than phosphatidylserine or phosphatidylethanolamine[171, 172]. Lysolecithin, which is mainly carried by VHDL, exchanges readily[171] but is acylated to lecithin by LCAT after exchange, and may be the mechanism for the renewal of oxidized acyl groups in erythrocyte lipids. Thus the lipoproteins may keep erythrocyte lipids in a functional state[83]. Supporting this idea, is the observation that changes in erythrocyte composition and properties are seen in genetic diseases of the plasma lipoproteins[83].

Functions of plasma lipoproteins

The plasma lipoproteins play important and obvious roles in triacylglycerol transport; VLDL production being in response to overall nutritional state and amounting normally to 25–50 g lipid/day in adult man; chylomicron production being in response to dietary lipid intake and normally amounting to 50–150 g lipid/day[148]. Both types of lipoprotein serve primarily to replenish the fat stores, though some may be used in the formation of other types of lipid and to provide energy via oxidation. The lipid destined to support energy metabolism is, however, mostly carried as NEFA bound to plasma albumin (normally 50–150 g lipid/day).

LDL and HDL may be important vehicles for the transport of cholesterol (both esterified and free). Cholesterol is made in many tissues and exchanges readily into lipoproteins in the unesterified form. Through the action of LCAT, cholesterol is converted to cholesterol ester and is ultimately taken up by the cells of the liver. In genetic deficiency of HDL (Tangier disease), cholesterol transport to the liver is abnormal and cholesterol esters accumulate in many reticuloendothelial tissues[83, 85]. The liver normally plays an important part in regulating gain or loss of cholesterol from the body: by balancing dietary intake against metabolic conversion to bile salts and through the loss of a proportion of these in the faeces[174]. Cholesterol flux in human plasma amounts to approx. 1g/day.

A further role of HDL may be in connection with its stimulatory effect upon lipoprotein lipase and LCAT activities: these enzymes are stimulated by the presence of apo-

HDL, the preferred substrate for LCAT activity is HDL lipid, low levels of HDL in some disease states are paralleled by low levels of LCAT[83, 85, 148]. The joint effect of HDL and LCAT may play an important role in the maintenance of surface–volume relationships in the micellar lipoproteins, and may also regulate the cholesterol content of erythrocytes[83, 85, 148].

5. FATTY LIVER

In spite of the large triacylglycerol flux resulting from liver metabolism, the overall amount of triacylglycerol present in normal liver stays relatively constant. Many disturbances of liver metabolism result in triacylglycerol accumulation and give rise to the condition of fatty liver. In its early stages fatty liver is reversible; but in severe cases may be accompanied by cirrhosis and necrosis.

Fatty livers are larger and paler than normal livers. The weight gain is due principally to an increase in triacylglycerol and water content. The triacylglycerol is often 10–30 times the normal value. Total phospholipid levels show little change, although the amount of phospholipid per g of tissue is lower due to the increase in triacylglycerol and water content[174]. For a more comprehensive discussion of fatty liver than can be presented here, the reader is referred to several recent reviews[54, 104, 174, 175].

Occurrence of fatty liver

Fatty liver is often observed in several clinical conditions in man: in chronic alcoholism, diabetes mellitus, kwashiorkor, pernicious anaemia, certain types of infections and sometimes in pregnancy[174]. It is also seen as a consequence of certain experimental dietary deficiencies or imbalances: (a) deficiency of choline or its precursors: betaine, methionine, thetine; (b) protein deficiency; (c) amino acid imbalance; (d) excessive fat or cholesterol in the diet; (e) excessive caloric intake; (e) early stages of starvation. The administration of several toxic substances also may give rise to fatty liver: phosphorus, arsenic, antimony, carbon tetrachloride, ethanol, chloroform, other chlorinated hydrocarbons, orotic acid, ethionine and several others, including high doses of cortisone[54, 174].

Causes of fatty liver

The dynamic balance of triacylglycerol concentration in the liver is the result of the interplay between the input of triacylglycerol forming materials and their utilization. The principal contributors to provision of liver triacylglycerol are: (a) endogenous biosynthesis of fatty acids from carbohydrate; (b) uptake of fatty acids resulting from mobilization from adipose tissue; (c) uptake of dietary fat from chylomicrons. Liver triacylglycerols are then utilized for: (a) conversion to other lipid molecules; (b) hydrolysis followed by oxidation, energy production and ketone body formation; (c) incorporation into VLDL and secretion into the plasma.

If the formation of triacylglycerol exceeds its utilization, a fatty liver is produced.

References p. 370

Therefore, any metabolic situation or administered compound which enhances the formation of liver triacylglycerol or inhibits its utilization, will give rise to the condition. The development of a fatty liver can be a relatively rapid phenomenon, significant increases in liver triacylglycerol levels have been observed within a few hours under appropriate circumstances[174].

In the case of fatty livers produced by ethanol, cortisone or noradrenaline administration, enhanced triacylglycerol synthesis is probably the determining factor; since plasma triacylglycerol levels are normal or elevated[54, 104]. Other factors, such as decreased fatty acid oxidation and decreased lipoprotein synthesis, may also make a contribution to the final picture.

The final liberation of triacylglycerol into plasma as VLDL requires the coordination of several metabolic processes: (a) biosynthesis or provision of phospholipids and cholesterol; (b) biosynthesis or provision of the appropriate apoproteins; (c) assembly; (d) secretion. Interference with any of these processes may lead to accumulation of triacylglycerol[104]. Plasma triacylglycerol levels are lower than normal when fatty livers are due to interference with lipoprotein production[54].

Interference with lipoprotein production

The biosynthesis of apoprotein is clearly affected by protein deficiency, amino acid imbalance and administration of puromycin. Ethionine probably also operates by interfering with protein biosynthesis via its effects on ATP metabolism and the depression of RNA synthesis[104]. Carbon tetrachloride (and probably other chlorinated hydrocarbons), are also thought to act by inhibiting protein biosynthesis; but the primary damage in this case may be due to peroxidation of phospholipids in the endoplasmic reticulum, causing membrane damage which results in decreased protein biosynthesis from membrane-bound ribosomes[176]. Carbon tetrachloride also has a rapid inhibitory effect upon phospholipid biosynthesis, and this may further inhibit lipoprotein formation[177].

Orotic acid-induced fatty liver appears not to result from an inhibition of general protein biosynthesis, since the biosynthesis of other plasma proteins (HDL and albumin) is normal. The metabolic lesion, therefore, must lie in the synthesis of specific apoprotein, the assembly process, or the secretion of the intact lipoprotein. The effect of orotic acid is highly specific for liver lipoproteins, since this substance does not affect the synthesis or secretion of lipoproteins from the intestinal mucosa[178].

The development of fatty livers resulting from diets deficient in lipotropic agents (choline and choline precursors) has been extensively studied[174]. After several days on a choline-deficient diet, plasma phospholipid and triacylglycerol levels fall suggesting an impairment in lipoprotein production, which can be confirmed by direct lipoprotein analysis. At the same time, triacylglycerol accumulation is noted in the liver[174], but there is no alteration in protein synthesis[179]. In shorter term experiments, the fall in plasma phospholipids was noted earlier than that for neutral fat, and was evident during the first day[180]. Little change in the overall phospholipid content of the liver was apparent in these early stages, but changes were noted in the relative activities of the various pathways of lecithin biosynthesis. The early effects of choline deficiency may be to reduce the synthesis of phospholipid specifically required for plasma lipoprotein assembly; or, to alter the phospholipid pattern of hepatic membranes, in such a way as to inhibit lipoprotein production[181]. Many secondary metabolic disturbances are present in long-term choline deficiency. During recovery, the plasma lipoprotein levels are rapidly restored though the other changes are reversed much more slowly[174].

6. BILE FORMATION

Bile is a complex secretion, whose function has been discussed in relation to the intraluminal stage of fat absorption. It is originally formed in the parenchymal cells and secreted into the bile canaliculi. The bile canaliculi are specialized spaces, formed between two or sometimes three neighbouring hepatocytes, and are sealed with a junctional complex to prevent the biliary secretion from bathing the cells. The surface membranes in the bile canaliculi are thrown up into a series of microvilli. Bile canaliculi unite with bile ductules which in turn lead to the bile ducts. The bile ducts are connected, via the common bile duct, with both the gall bladder (for storage) and the duodenum[13, 182].

Bile is thought to be secreted initially by the hepatocytes as a liquid containing bile salts, bile lipids (phospholipids and cholesterol), bilirubin and electrolytes. The bile ductule cells are thought to secrete a fluid of different composition and may reabsorb some of the components of the original bile. The reabsorption of water and electrolytes is continued in the gall bladder and mucoprotein is added[13, 182]. Perfusion of the liver with bile salts leads to increased secretion of bile salts, cholesterol and phospholipids: perfusion with materials such as fluorescein produces a watery bile low in bile salts, cholesterol and phospholipids[183, 184].

Composition of bile

Bile contains those bile salts which are synthesized in the liver (cholic acid and chenodeoxycholic acid), and also deoxycholic acid and lithocholic acid which have been produced in the intestine by bacterial modification and then returned to the liver (the enterohepatic circulation)[13]. The bile acids are mostly present as their conjugates with glycine and taurine, and are anions at the pH of the bile.

Bile salts are biosynthesized by the mitochondrial and microsomal fractions of liver homogenates, and the conjugation mechanisms occur in the microsomal fraction[185, 186]. The mechanism of secretion of the bile salts is a subject of much discussion and uncertainty[186–188], and although on morphological grounds, a packaging of material in the Golgi region has been suspected[187], the energy requirements and several other characteristics suggest the participation of active transport processes[13, 187].

Other prominent components of the bile are cholesterol (almost entirely in the non-esterified form), and phospholipid. The concentrations of these components in human hepatic and gall-bladder bile are given in Table 3. The phospholipid in the bile of several species is mainly lecithin (90%) with small quantities of other lipids[189–191].

References p. 370

TABLE 3

APPROXIMATE COMPOSITION OF HUMAN BILE[a]

	Hepatic bile	*Gall bladder bile*
Bile acids mg/100 ml	700	7000
Phospholipids mg/100 ml	300	3000
Cholesterol mg/100 ml	60	600
Volume ml/day	500–1000	–

[a]These simplified values are meant to convey a general impression only. Actual values vary greatly according to stage of secretion and dietary considerations. For values see Refs. 14, 104, 213, 248. Liver weight (man) approx. 1.5 kg.

Origin of biliary phospholipid

Biliary phospholipid is derived from the liver[192, 193]. The presence of phospholipid and cholesterol in bile has led to some interesting speculation with regard to the processes of bile salt externalization. In bile, the bile salts are present as mixed micelles with phospholipid and cholesterol[21, 183]. Infusion of bile salts into animals with biliary fistula or perfusion of isolated liver increases the output of both phospholipid and cholesterol, which rise in the same proportion, suggesting that both these lipids came from the same source[183]. Since biliary lecithin is not secreted by bile duct epithelial cells[193, 194], it is likely that cholesterol, phospholipid and bile salts are all derived from the same cells.

Since the rate of incorporation of radioactive precursor into biliary lecithin, differed from that into liver mitochondrial and microsomal lecithin[193, 194], it has been suggested that the source of the biliary phospholipid is the canalicular membrane[183]. Three possibilities were examined whereby this might have been achieved; the breaking of the canalicular membrane followed by its solubilization; the solubilization of the membrane during the passing of the bile salt; or the dissection out of the lecithin and cholesterol from the membrane leaving other phospholipids and the protein intact[183]. Of these, the first two suggestions are unlikely, since not only is the protein content of the bile low; but also the other lipid components of the cellular membranes are in reduced amount[183]. However, in cholestasis from a variety of causes, the microvilli of the bile canaliculi are smaller and less evident[182, 195], which may indicate some solubilization under these more abnormal conditions. The third possibility is considered to be the most likely under normal circumstances, i.e. the selective removal of some phospholipids from the lipid bilayer in the canalicular membrane[183].

In the last few years it has become possible to obtain preparations of surface membrane from liver tissue[196–207], many of which contain bile canaliculi recognisable by electron microscopy (Fig. 2). These preparations have a high lipid content and contain both lecithin and relatively large amounts of cholesterol[206, 208, 209]. Thus it is possible that

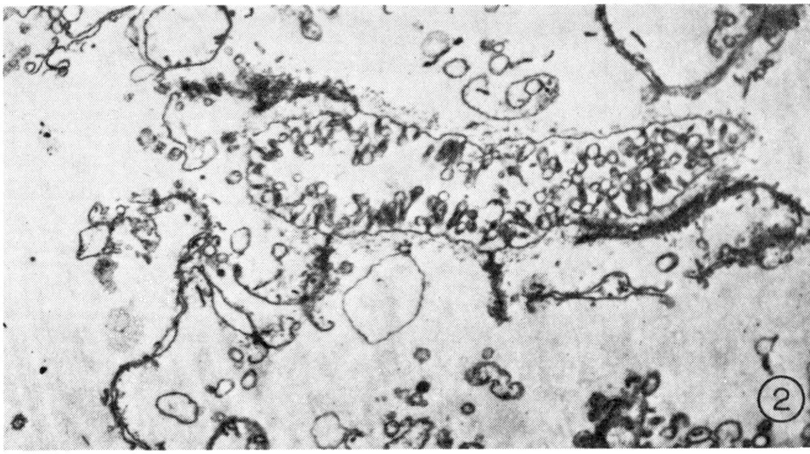

Fig. 2. Bile canaliculus in an isolated surface membrane preparation from guinea pig liver.

the surface membrane of the liver cell and more especially the canalicular region, could provide the lipid components of the bile.

If this membrane is to remain as a continuing structure in the cell, replacement of the lecithin and cholesterol would be required, thereby explaining some of the high turnover rates of these compounds in liver. It has been calculated that the canalicular membrane might turn over once every 10–100 min as a result of bile production[183].

Some instability of the bile canalicular membrane is indicated by the presence in bile of the enzymes alkaline phosphatase[13] and 5′-nucleotidase[210, 249], both of which have been demonstrated in the liver surface membrane preparations[197, 209, 211, 212]. Moreover, 5′-nucleotidase has been shown to be readily solubilized by bile salts without inactivation[212]. Both enzymes appear in enhanced levels in plasma during biliary obstruction[13, 211, 213, 214].

7. PHOSPHOLIPID IN RELATION TO OTHER ACTIVITIES OF THE LIVER

The liver has a high rate of turnover of phospholipids, as shown by incorporation studies with many lipid precursors. Some of the phospholipid synthesized is exported into the plasma as lipoprotein; and some into the bile as a constituent of the mixed micelle. In addition, a considerable proportion of the turnover is involved in the metabolism of the liver cell itself. By analogy with other cells, it might be expected that this would involve membrane and organelle biosynthesis, and modification, and recovery from metabolic or peroxidative damage. In many of these respects, the liver resembles other cells in the performance of these functions, which are covered in other chapters.

Some of this phospholipid turnover, however, is more intimately involved in the special functions of the liver and is therefore worthy of special mention: (a) in the response of the endoplasmic reticulum to drug administration; (b) in the role of the liver in supporting general metabolism during conditions of starvation and the response during re-feeding; and (c) in the regenerative power of the liver.

References p. 370

Drug metabolism

One of the more important functions of the liver is the conversion of many drugs and foreign compounds to more polar compounds as a prelude to excretion in the urine. Due to the existence of this metabolism, the effects of many compounds are relatively short-lived, and sometimes increased input is required to maintain therapeutic levels in the tissues. It has also been observed that pretreatment with one drug may lead to tolerance of a second, closely following drug. Drugs of this type induce an increase in the level of many drug-metabolizing enzymes (hydroxylases, demethylases and conjugases). In addition to the rise in the level of these enzymes, a proliferation of the endoplasmic reticulum of the cells is observed; indicating that the drug has set in motion a rather complex induction process, involving not only the increased production of drug metabolizing enzymes, but also the lipoprotein platform which supports them[215]. Its implications with respect to phospholipid metabolism are obvious.

Whilst drugs such as phenobarbitone cause an increase in hepatic membrane phospholipids, there is no associated increase in plasma phospholipid levels[216] indicating that this is probably a phenomenon of endogenously directed phospholipid metabolism. In the rat, the morphological changes are apparent on the second day and increase up to the fifth day, at which time smooth endoplasmic reticulum packs the cell. The phospholipid content per g of liver is increased by more than 50% (ref. 215). All classes of phospholipids are increased, but the greatest proportionate increase is observed in phosphatidylcholine[217], which also shows a high proportional rise in the rate of precursor incorporation following drug administration[217, 218]. An increase in ^{32}P incorporation into phospholipids is noted considerably earlier than increased drug metabolizing activity or protein content[215], and suggests a rapid stimulation of phospholipid biosynthesis. The overall phospholipid to protein ratio of the microsomal membranes, however, does not change during drug induction[219]. The stimulation of turnover and increase in phospholipid content, is greatest in rough and smooth I microsomes[219], and is virtually absent from mitochondria[215].

The stimulation in precursor incorporation[215–219], together with the effects of protein biosynthesis inhibitors on the system suggested that the induction process involved stimulation of biosynthesis, though inhibition of degradation may also contribute[217, 218]. In fasted animals, the stimulation of precursor incorporation was much less marked than in controls; thus in fasted rats, decreased degradation may play a more prominent part in the inductive processes[220].

Starvation

During starvation liver weight is decreased and the total amount of phospholipid appears to be decreased in approximate proportion[221], there is, however, little change in the total DNA content of the liver, and therefore the phospholipid/DNA ratio falls[222]. A loss in phospholipids resulting from de-differentiation of existing cells (i.e. loss of organelles) would lead to this result, and this is borne out by the morphological appearance of the liver during starvation.

The effects of starvation observed on lipid metabolism in liver have recently been summarized in detail[223]. Phospholipid loss may be due either to reduced synthesis or to enhanced degradation or a balance of the two.

When liver homogenates were prepared from starved animals, they showed a dramatic decrease in palmitate incorporation into phospholipids as compared to controls[223]. This study showed, during starvation, a decrease in the acylation of glycerophosphate, an increase in the further acylation of acylglycerols and a substantial increase in the level of phosphohydrolase[223]. These results suggest that either the increased acylation of acylglycerols might compete with phospholipid biosynthesis for a common intermediate or that the remaining stages of the formation of phospholipid from diacylglycerol are depressed. More work is obviously required to clarify further the events leading to the rather striking loss of phospholipid during starvation.

Regeneration

After partial hepatectomy, the lipid content of the remaining liver increases and this precedes the increase in mitotic activity[224]. The lipid which increases is principally triacylglycerol. Early work had suggested that phospholipid turnover was increased in parallel with mitotic activity[225, 226]; but in later studies with ^{32}P (ref. 227) or ^{14}C-acetate[224], no increased incorporation into phospholipids was observed. It was argued that the numbers of regenerating cells to mature cells was too small for any significant effect on phospholipid turnover to be observed.

8. PATHWAYS OF PHOSPHOLIPID METABOLISM IN LIVER

The liver possesses most of the pathways for the synthesis, modification or degradation of phospholipids. These include the acylation of glycerophosphate and dihydroxyacetone phosphate, the acylation of 1-and 2-lysolecithins, the formation of these lysolecithins by the actions of phospholipases A_1 and A_2; the exchange of serine, choline, ethanolamine and inositol, the CDPcholine and CDPethanolamine pathways, and the methylation pathway for lecithin formation[16, 17, 54, 219, 228]. This last pathway is far more important in liver than in many extrahepatic tissues, and *in vivo* contributes a significant amount (approx. 10–20%), to the total lecithin formation in this organ[229].

The variety of pathways for the metabolism of the phospholipids partially explains the different half lives observed for the various parts of the molecule[219, 227–234].

Phospholipid exchange

Discrepancies between the labelling pattern of phospholipids in *in vivo* and *in vitro* experiments, may be due to the exchange of phospholipids between cellular organelles and between soluble lipoproteins[235–244]. This exchange appears in many cases to be mediated by protein factors[237, 239, 240, 244], and supports the rapid equilibration of phospholipids between cellular organelles and, possibly, the cellular environment. This exchange may also transport newly synthesized phospholipids from the biosynthetic site, to the site where the phospholipid is to be utilized[244, 245]. (See also Chapter 9.)

The differences in phospholipid composition between both the subcellular organelles, the soluble lipoproteins and the bile phospholipids and the existence of phospholipid pools of differing rates of turnover cannot be readily explained by this phospholipid equilibration. They may be due in part to physicochemical considerations (e.g. specificity of the proteins or other neighbouring molecules) and also to secondary, local, enzymatic modifications[244, 245].

Phospholipid pools

Two pools of cholesterol with different rates of turnover are thought to exist in the liver; a structural pool of low turnover rate and a metabolic pool of higher turnover rate, suggestive of material more likely to be exported from the cell[219]. In early experiments studying ^{32}P incorporation, the rise, in specific radioactivity of biliary phospholipids slightly preceded that of total liver and plasma phospholipids[231]. Since this is very unlikely to mean that biliary lecithin could be the precursor of hepatic lecithin, it points to the existence of a rapidly turning over pool of lecithin supplying the biliary outflow. The final specific activity of the biliary lecithin (in radioprecursor experiments) exceeded that of the liver microsomal fraction[194]. The most rapidly labelled phospholipid in the bile was phosphatidylcholine; little of the ^{32}P label occurred in other (minor) biliary phospholipids. This is in sharp contrast to the more rapid labelling of the other phospholipids, in cellular organelles and in the plasma lipoproteins[194].

A different pool for biliary phosphatidylcholine is also suspected from fatty acid incorporation studies[193]. Using injected radioisotopically labelled materials it was shown that linoleoyl lecithin reached higher specific activities, with respect to ^{32}P and $[^{14}C]$-choline, than did arachidonoyl lecithin. Biliary linoleoyl lecithin reached higher specific activities than hepatic or plasma linoleoyl lecithin. In these studies, the specific activity of phosphatidylethanolamine was lower than that of the corresponding phosphatidylcholine, suggesting that methylation contributed little to the formation of biliary lecithin[193]. In addition, even the overall fatty acid composition of biliary, liver and plasma lecithins was different; biliary lecithins being predominantly palmitoyl–linoleoyl. Those of the liver and the plasma contained much greater amounts of stearoyl and arachidonoyl residues[191–193].

Since it has been suggested that the lecithins and cholesterol occurring in the bile are derived from the bile canalicular membrane, it is interesting in this context to note the lower degree of unsaturation in hepatic plasma membranes, than in the remaining cellular organelles[208, 246], and the relatively high content of palmitoyl linoleoyl lecithin in the overall plasma membrane preparation[247]. Specific purification of the bile canalicular membrane may demonstrate even higher levels of this precursor of biliary lecithin. The difference between the fatty acids of hepatic lecithins and plasma lecithin is not as marked, and is in keeping with the ready exchange noted between hepatic and plasma lecithin[244].

The proliferation of endoplasmic reticulum, observed following phenobarbitone administration, is accompanied by an increase in the amount of lecithin; but only minimal increases in the amount of other phospholipids. At the same time a three-fold stimulation of the methylation pathway for lecithin formation occurred, but there was little stimulation of the CDPcholine pathway[218]. This suggests that the methylation pathway may be used preferentially for the supply of structural lecithins.

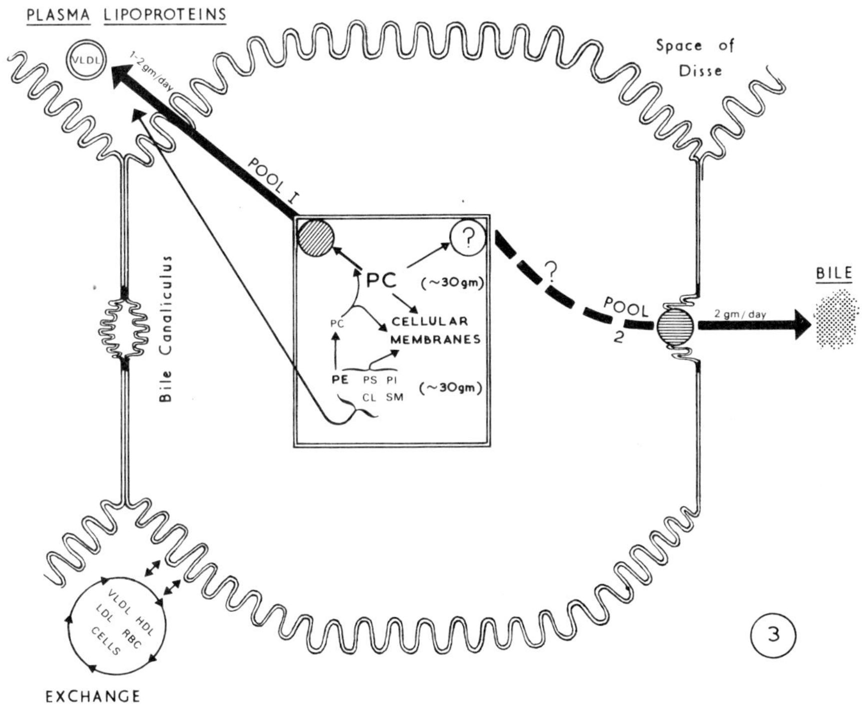

Fig. 3. Overall phospholipid metabolism of liver. A stylised hepatocyte is shown with the important morphological regions of the surface membrane indicated. Probable origins of appropriate pools of exported phospholipids are indicated as circles. Figures given are meant to be approximate only. Estimates are based on: appropriate phospholipid contents and compositions of liver, bile and plasma lipoproteins; liver weight, bile volume per day, plasma volume and lipoprotein half lives. Values may be found in earlier sections of text, in other chapters, and in the previous edition of this book.

9. OVERALL PHOSPHOLIPID METABOLISM IN LIVER

The liver is a large organ of complex function; it contains phospholipids mainly as components of the cellular membranes. Liver phospholipid metabolism reflects the turnover of these membranes; it also reflects the export of phospholipids in two secretions: plasma lipoproteins (as VLDL), and bile. In both secretions, the predominant phospholipid is lecithin. The quantity of lecithin exported, per day, is considerable: in man, it amounts to approx. 10–20% of the total weight of liver lecithin, divided roughly equally between the two routes.

Not only are these two secretions removed from the cell through different regions of the surface membrane: bile through the bile canaliculus, and plasma lipoprotein into the space of Disse; but also it appears that the lecithins are of distinct types and are drawn from separate pools. The selective mechanisms involved in this intriguing situation are, so far, imperfectly understood.

ACKNOWLEDGEMENTS

The author wishes to acknowledge with gratitude both the encouragement and valued criticism given to him in the past, by the late Professor G. Hübscher. His thanks are also due to Drs. D.N. Brindley, and C. Green for allowing him access to material prior to publication, and to Drs. H. Magnani, R.H. Michell and J.B. Finean for valued comment on some of the material in the manuscript.

REFERENCES

1 J.M. Johnston, *Advan. Lipid Res.*, 1 (1963) 105.
2 J.R. Senior, *J. Lipid Res.*, 5 (1964) 495.
3 K.J. Isselbacher, *Federation Proc.*, 26 (1967) 1420.
4 J.M. Johnston, in C.F. Code (Ed.), *Handbook of Physiology*, Sect. 6, Vol. III, Williams and Wilkins, Baltimore, 1968, p. 1353.
5 G. Hübscher, *Biochem. J.*, 114 (1969) 46 p.
6 G. Hübscher, in S.J. Wakil (Ed.), *Lipid Metabolism*, Vol. 1, Academic Press, New York, 1970, p. 279.
7 D.N. Brindley, in D.H. Smyth (Ed.), *Intestinal Absorption*, 1973 (in the press).
8 P. Desnuelle, *Advan. Enzymol.*, 23 (1961) 129.
9 B. Entressangles, H. Sari and P. Desnuelle, *Biochim. Biophys. Acta*, 125 (1966) 597.
10 A.F. Hofmann, *J. Lipid Res.*, 3 (1962) 391.
11 R. Reiser and H.C. Fu, *Biochim. Biophys. Acta*, 116 (1966) 563.
12 F.H. Mattson and R.A. Volpenhein, *J. Lipid Res.*, 7 (1966) 536.
13 T. Hargreaves, *The Liver and Bile Metabolism*, North Holland, Amsterdam, 1968.
14 G.H. de Haas, N.M. Postema, W. Nieuwenhuizen and L.L.M. van Deenen, *Biochim. Biophys. Acta*, 159 (1968) 103.
15 G.H. de Haas, N.M. Postema, W. Nieuwenhuizen and L.L.M. van Deenen, *Biochim. Biophys. Acta*, 159 (1968) 118.
16 W.J. Lennarz, *Ann. Rev. Biochem.*, 39 (1970) 359.
17 L.L.M. van Deenen and G.H. de Haas, *Ann. Rev. Biochem.*, 35 (1966) 157.
18 B. Borgström, *Acta Chem. Scand.*, 2 (1957) 749.
19 P.B. Subbiah and J. Ganguly, *Biochem. J.*, 118 (1970) 233.
20 A. Nilsson, *Biochim. Biophys. Acta*, 176 (1969) 339.
21 A.F. Hofman and D.M. Small, *Ann. Rev. Med.*, 18 (1967) 333.
22 A.F. Hofman and B. Borgström, *J. Clin. Invest.*, 43 (1964) 247.
23 A.F. Hofman, in C.F. Code (Ed.), *Handbook of Physiology*, Sect. 6, Vol. V, Williams and Wilkins, Baltimore, 1968, p. 2507.
24 D.M. Small, *Federation Proc.*, 29 (1969) 1320.
25 S.L. Palay and L.J. Karlin, *J.Biophys. Biochem. Cytol.*, 5 (1959) 363.
26 E.W. Strauss, *J. Cell Biol.*, 17 (1963) 597.
27 P. Millington and J.B. Finean, in A.C. Frazer (Ed.), *Biochemical Problems of Lipids*, Elsevier, Amsterdam, 1963, p. 116.
28 C.T. Ashworth and J.M. Johnston, *J. Lipid Res.*, 4 (1963) 454.
29 R.R. Cardell, S. Badenhausen and K.R. Porter, *J. Cell Biol.*, 34 (1967) 123.

30 E.W. Strauss, *J. Lipid Res.*, 7 (1966) 307.
31 C.T. Ashworth and J.F. Lawrence, *J. Lipid Res.*, 7 (1966) 465.
32 J.M. Johnston and B. Borgström, *Biochim. Biophys. Acta*, 84 (1964) 412.
33 L. Lack and I.M. Weiner, *Am. J. Physiol.*, 200 (1961) 313.
34 B. Borgström, G. Lundh and A. Hoffman, *Gastroenterology*, 45 (1963) 229.
35 M.C. Playoust and K.J. Isselbacher, *J. Clin. Invest.*, 43 (1964) 467.
36 W.J. Simmonds, A.F. Hofman and E. Theodor, *J. Clin. Invest.*, 46 (1967) 874.
37 J.M. Dietschy, *J. Lipid Res.*, 9 (1968) 297.
38 C.A. Hogben, *Ann. Rev. Physiol.*, 22 (1960) 381.
39 T.H. Wilson, *Intestinal Absorption*, Saunders, Philadelphia, 1962.
40 J.F. Danielli, in H.C. Meng (Ed.), *Lipid Transport*, Thomas, Springfield, Ill., 1963, p. 104.
41 E.L. Coe, *Nature*, 211 (1966) 80.
42 C.A. Hogben, *Gastroenterology*, 50 (1966) 51.
43 S.S. Raghavan and J. Ganguly, *Ind. J. Biochem.*, 4 (1967) 68.
44 J.D. Hamilton, J.P.W. Webb and A.M. Dawson, *Biochim. Biophys. Acta*, 176 (1969) 27.
45 B. Clark and G. Hübscher, *Nature*, 185 (1960) 35.
46 G.A. Rao and J.M. Johnston, *Biochim. Biophys. Acta*, 125 (1966) 465.
47 G.A. Rao and J.M. Johnston, *Biochim. Biophys. Acta*, 144 (1967) 25.
48 J.M. Johnston and J.H. Bearden, *Arch. Biochem. Biophys.*, 90 (1960) 57.
49 M.E. Smith, B. Sedgwick, D.N. Brindley and G. Hübscher, *European J. Biochem.*, 3 (1967) 70.
50 J.M. Johnston, G.A. Rao, P.A. Lowe and B.E. Schwarz, *Lipids*, 2 (1967) 14.
51 G. Hübscher, D.N. Brindley, M.E. Smith and B. Sedgwick, *Nature*, 216 (1967) 449.
52 R.R. Dinella, H.C. Meng and C.R. Park, *J. Biol. Chem.*, 235 (1960) 3076.
53 J.R. Senior and K.J. Isselbacher, *J. Clin. Invest.*, 42 (1963) 187.
54 B. Shapiro, *Ann. Rev. Biochem.*, 36 (1967) 247.
55 H.J. Kayden and M. Medick, *Biochim. Biophys. Acta*, 176 (1969) 37.
56 C. Artom and M.A. Swanson, *J. Biol. Chem.*, 175 (1948) 871.
57 B. Bloom, J.Y. Kiyasu, W.O. Reinhardt and I.L. Chaikoff, *Am. J. Physiol.*, 177 (1954) 84.
58 R. Blomstrand, *Acta Physiol. Scand.*, 34 (1955) 147.
59 R.O. Scow, Y. Stein and O. Stein, *J. Biol. Chem.*, 242 (1967) 4919.
60 Y. Stein and O. Stein, *Biochim. Biophys. Acta*, 116 (1966) 95.
61 A. Nilsson and B. Borgström, *Biochim. Biophys. Acta*, 137 (1967) 240.
62 P.V. Subbiah, P.S. Sastry and J. Ganguly, *Biochem. J.*, 113 (1969) 441.
63 P.V. Subbiah, P.S. Sastry and J. Ganguly, *Biochem. J.*, 118 (1970) 241.
64 B. Epstein and B. Shapiro, *Biochem. J.*, 71 (1959) 615.
65 A. Ottolenghi, *J. Lipid Res.*, 5 (1964) 532.
66 A. Ottolenghi, *Federation Proc.*, 28 (1969) 885.
67 J. Gallai-Hatchard and R.H.S. Thompson, *Biochim. Biophys. Acta*, 98 (1965) 128.
68 A. Nilsson, *Biochim. Biophys. Acta*, 152 (1968) 379.
69 P. Boucrot and J. Clement, *Biochim. Biophys. Acta*, 187 (1969) 59.
70 M.I. Gurr, D.N. Brindley and G. Hübscher, *Biochim. Biophys. Acta*, 98 (1965) 486.
71 D.N. Brindley and G. Hübscher, *Biochim. Biophys. Acta*, 106 (1965) 495.
72 M.I. Gurr, W.F.R. Pover, J.N. Hawthorne and A.C. Frazer, *Nature*, 197 (1963) 79.
73 J.L. Bollman, E.V. Flock, J.C. Cain and A. Grindlay, *Am. J. Physiol.*, 163 (1950 41.
74 A. Nilsson, *Biochim. Biophys. Acta*, 164 (1968) 575.
75 R.S. Atherton and J.N. Hawthorne, *European J. Biochem.*, 4 (1968) 68.
76 R.K. Ockner, F.B. Hughes and K.J. Isselbacher, *J. Clin. Invest.*, 48 (1969) 2079.
77 R.K. Ockner, F.B. Hughes and K.J. Isselbacher, *J..Clin. Invest.*, 48 (1969) 2367.
78 A.C. Frazer, *Ann. Rev. Biochem.*, 21 (1952) 245.
79 D.S. Frederickson, R.I. Levy and R.S. Lees, *New Engl. J. Med.*, 276 (1967) 34.
80 D.S. Frederickson, R.I. Levy and R.S. Lees, *New Engl. J. Med.*, 276 (1967) 94, 148, 215, 273.
81 D.B. Zilversmit, *Federation Proc.*, 26 (1967) 1599.
82 F.K. Hatch and R.S. Lees, *Advan. Lipid Res.*, 6 (1968) 1.
83 V.N. Schumaker and G.H. Adams, *Ann. Rev. Biochem.*, 38 (1969) 113.
84 J.L. Oncley and N.R. Harvie, *Proc. Natl. Acad. Sci. (U.S.)*, 64 (1969) 1107.
85 A.V. Nicholls, *Proc. Natl. Acad. Sci. (U.S.)*, 64 (1969) 1128.

86 O. Minari and D.B. Zilversmit, *J. Lipid Res.*, 4 (1963) 424.
87 M.M. Salpeter and D.B. Zilversmit, *J. Lipid Res.*, 9 (1968) 187.
88 D.B. Zilversmit, *J. Clin. Invest.*, 44 (1965) 1610.
89 D.B. Zilversmit, *J. Lipid Res.*, 9 (1968) 180.
90 K.J. Isselbacher and D.M. Budz, *Nature*, 200 (1963) 364.
91 J.H. Bragdon, *J. Lab. Clin. Med.*, 52 (1958) 564.
92 M. Rodbell, D.S. Frederickson and K. Ono, *J. Biol. Chem.*, 234 (1959) 567.
93 P.S. Roheim, L.I. Gidez and H.A. Eder, *J. Clin. Invest.*, 45 (1966) 297.
94 F.T. Hatch, Y. Aso, L.M. Hagopian and J.J. Rubenstein, *J. Biol. Chem.*, 241 (1966) 1655.
95 S.M. Sabesin and K.J. Isselbacher, *Science*, 147 (1965) 1149.
96 M.J. Greenberger, J.B. Rodgers and K.J. Isselbacher, *J. Clin. Invest.*, 45 (1966) 217.
97 D.H. Alpers and K.J. Isselbacher, *Biochim. Biophys. Acta*, 158 (1968) 414.
98 T.G. Redgrave and D.B. Zilversmit, *Am. J. Physiol.*, 217 (1969) 336.
99 H.G. Windmueller and R.I. Levy, *J. Biol. Chem.*, 243 (1968) 4878.
100 H.J. Kayden, M.I. Glassman and M. Medick, *Biochim. Biophys. Acta*, 176 (1969) 44.
101 P. Alaupovic, R.H. Furman, S.L. Walraven, W.H. Falor and H. Newman, *Circulation*, 36 (1967) II-2.
102 R.A. Jersild, *J. Cell Biol.*, 31 (1966) 413.
103 W.O. Dobbins, *Am. J. Clin. Nutr.*, 22 (1969) 257.
104 E.J. Masoro, *Physiological Chemistry of Lipids in Mammals*, Saunders, London, 1968.
105 H.S. Quarfordt and D.S. Goodman, *Biochim. Biophys. Acta*, 116 (1966) 382.
106 P.J. Nestel, R.J. Havel and A. Bezman, *J. Clin. Invest.*, 42 (1963) 1313.
107 J. Glomset, *J. Lipid Res.*, 9 (1968) 155.
108 A. Scanu, *J. Biol. Chem.*, 242 (1967) 711.
109 P.F. Hahn, *Science*, 98 (1943) 19.
110 D.S. Robinson, *Advan. Lipid Res.*, 1 (1963) 133.
111 D.S. Robinson and S. Otway, *Biochem. J.*, 100 (1966) 50P.
112 J.H. Bragdon and R.S. Gordon, *J. Clin. Invest.*, 37 (1958) 574.
113 D.S. Robinson, in H.C. Meng (Ed.), *Lipid Transport*, Thomas, Springfield, Ill., 1964, p. 194.
114 A. Scanu, *Advan. Lipid Res.*, 3 (1965) 63.
115 P. Mayes, *Biochem. J.*, 114 (1969) 47P.
116 N.R. DiLuzio, *Ann. N.Y. Acad. Sci.*, 88 (1960) 244.
117 R.G. Murray and S. Freeman, *J. Lab. Clin. Med.*, 38 (1951) 56.
118 O. Stein and Y. Stein, *Lab. Invest.*, 17 (1967) 436.
119 P. Belfrage, *Biochim. Biophys. Acta*, 125 (1966) 474.
120 B. Morris, *J. Physiol.*, 168 (1963) 564.
121 M.C. Schotz, B. Arnesjo and T. Olivecrona, *Biochim. Biophys. Acta*, 125 (1966) 485.
122 J.M. Felts and P.A. Mayes, *Nature*, 206 (1965) 195.
123 P.A. Mayes and J.M. Felts, *Biochem. J.*, 108 (1968) 483.
124 J.A. Ontko and D.B. Zilversmit, *J. Lipid Res.*, 8 (1967) 90.
125 C. Green and J.A. Webb, *Biochim. Biophys. Acta*, 84 (1964) 404.
126 J.A. Higgins and C. Green, *Biochem. J.*, 99 (1966) 631.
127 J.A. Ontko, *Biochim. Biophys. Acta*, 137 (1967) 13.
128 J.A. Higgins and C. Green, *Biochim. Biophys. Acta*, 144 (1967) 211.
129 J.M. Felts and M.N. Berry, *Biochim. Biophys. Acta*, 231 (1971) 1.
130 M. Vavrecka, R. Poledne and R. Petrasek, *Biochim. Biophys. Acta*, 125 (1966) 176.
131 J.A. Higgins and C. Green, *Biochem. J.*, 104 (1967) 26P.
132a G. Göransson and T. Olivecrona, *Acta Physiol. Scand.*, 62 (1964) 224.
132b N. Baker and M.C. Schotz, *J. Lipid Res.*, 8 (1967) 646.
133 M.H. Laudat, J. Koenig and P. Laudat, *Bull. Soc. Chim. Biol.*, 51 (1969) 105.
134 J.D. Wright and C. Green, *Biochem. J.*, 123 (1971) 837.
135 B. Akesson, J. Elovson and G. Arvidson, *Biochim. Biophys. Acta*, 218 (1970) 44.
136 S.V. Pande and J.F. Meade, *J. Biol. Chem.*, 243 (1968) 352.
137 M. Farstad, J. Bremer and K.R. Norum, *Biochim. Biophys. Acta*, 132 (1967) 492.
138 K. Lippel, J. Robinson and E.G. Trams, *Biochim. Biophys. Acta*, 206 (1970) 173.
139 Y. Stein, C. Widnell and O. Stein, *J. Cell Biol.*, 39 (1968) 185.
140 W.L. Stahl and E.G. Trams, *Biochim. Biophys. Acta*, 163 (1968) 459.

141 H. Eibl, E.E. Hill and W.E.M. Lands, *European J. Biochem.*, 9 (1969) 250.
142 J.D. Newkirk and M. Waite, *Biochim. Biophys. Acta*, 225 (1971) 224.
143 R.H. Michell and R. Coleman, unpublished observation.
144 B. Akesson, *Biochim. Biophys. Acta*, 218 (1971) 57.
145 H.G. Windmueller and R.I. Levy, *Circulation*, 36 (1967) II-43.
146 D.S. Frederickson, R.I. Levy and R.S. Lees, *New Engl. J. Med.*, 276 (1967) 94.
147 A. Gotto, *Proc. Natl. Acad. Sci. (U.S.)*, 64 (1969) 1119.
148 D.S. Frederickson, *Proc. Natl. Acad. Sci. (U.S.)*, 64 (1969) 1138.
149 R.J. Havel, *Advan. Int. Med.*, 15 (1969) 117.
150 V.P. Skipski, M. Barclay, R.K. Barclay, V.A. Fetzer, J.J. Good and F.M. Archibald, *Biochem. J.*, 104 (1967) 340.
151 W.R. Fisher and S. Gurin, *Science*, 143 (1964) 362.
152 R.H. Furman, R.P. Howard, O. Brusco and P. Alanpovic, *J. Lab. Clin. Med.*, 66 (1965) 912.
153 D. Gitlin, D.G. Cornwell, D. Nakasato, J.L. Oncley, W.L. Hughes and C.A. Janeway, *J. Clin. Invest.*, 37 (1958) 172.
154 R.I. Levy, personal communication.
155 R. Bütler, *Vox Sanguinis*, 12 (1967) 2.
156 O. Stein and Y. Stein, *Israel J. Med. Sci.*, 1 (1965) 378.
157 C.M. Baglio and E. Farber, *Federation Proc.*, 24 (1965) 556.
158 C.T. Ashworth, F. Wrightsman, B. Cooper and N.R. DiLuzio, *J. Lipid Res.*, 6 (1965) 258.
159 B. Lombardi, *Lab. Invest.*, 15 (1966) 1.
160 A.B. Novikoff, P.S. Ronheim and N. Quintana, *Lab. Invest.*, 15 (1966) 27.
161 A.L. Jones, N.B. Ruderman and M.G. Herrera, *J. Lipid Res.*, 8 (1967) 429.
162 R.L. Hamilton, D.M. Regen, M.E. Gray and V.S. Le Quire, *Lab. Invest.*, 16 (1967) 305.
163 O. Stein and Y. Stein, *Israel J. Med. Sci.*, 2 (1966) 239.
164 O. Stein and Y. Stein, *Exptl. Cell Res.*, 42 (1966) 198.
165 M. Schramm, *Ann. Rev. Biochem.*, 36 (1967) 307.
166 R.H. Furman, S.S. Sanbar, P. Alaupovic, R.H. Bradford and R.P. Howard, *J. Lab. Clin. Med.*, 63 (1964) 193.
167 K.W. Walton, P.J. Scott, P.W. Dykes and J.W.L. Davies, *Clin. Sci.*, 29 (1965) 217.
168 M. Hoffman and M. Fried, *Federation Proc.*, 27 (1968) 666.
169 J.M. Graham and C. Green, *Biochem. J.*, 103 (1967) 16C.
170 K.R. Bruckdorfer and C. Green, *Biochem. J.*, 104 (1967) 270.
171 T. Sakagami, O. Minari and T. Orii, *Biochim. Biophys. Acta*, 98 (1965) 111.
172 C.F. Reed, *J. Clin. Invest.*, 47 (1968) 749.
173 N.B. Myant, *Biochem. J.*, (1971) in the press.
174 C.C. Lucas and J.H. Ridout, *Progr. Chem. Fats*, 10 (1967) 1.
175 B. Lombardi, *Lab. Invest.*, 15 (1966) 1.
176 R.O. Recknagel and A.K. Ghoshal, *Lab. Invest.*, 15 (1966) 132.
177 Y. Shimizu, *J. Lipid Res.*, 10 (1969) 479.
178 H.J. Kayden, M.I. Glassman and M. Medic, *Biochim. Biophys. Acta*, 176 (1969) 44.
179 B. Lombardi and A. Oler, *Lab. Invest.*, 17 (1967) 308.
180 B. Rosenfeld and J.M. Lang, *J. Lipid Res.*, 7 (1966) 10.
181 B. Lombardi, P. Pani and F.F. Schlunk, *J. Lipid Res.*, 9 (1968) 437.
182 H. Popper, *Ann. Rev. Med.*, 19 (1968) 39.
183 D.M. Small, *Advan. Int. Med.*, 16 (1970) 243.
184 W.G.M. Hardison and T.I. Francis, *Gastroenterology*, 56 (1969) 1164.
185 D. Mendelsohn, L. Mendelsohn and E. Staple, *Biochemistry*, 5 (1966) 3194.
186 S. Bergström and H. Danielsson, in C.F. Code (Ed.), *Handbook of Physiology*, Section 6, Vol. V, Williams and Wilkins, Baltimore, 1968, p. 2391.
187 H.O. Wheeler, in C.F. Code (Ed.), *Handbook of Physiology*, Sect. 6, Vol. V, Williams and Wilkins, Baltimore, 1968 p. 2409.
188 L.S. Schanker, in C.F. Code (Ed.), *Handbook of Physiology*, Sect. 6, Vol. V, Williams and Wilkins, Baltimore, 1968, p. 2433.
189 P. Barton and J. Glover, in W. Taylor (Ed.), *The Biliary System*, Blackwell, London, 1965 p. 189.
190 H.L. Spitzer, E.L. Kyriakides and J.L. Balint, *Nature*, 204 (1964) 288.

191 D.R. Illingworth and J. Glover, *Biochem. J.*, 108 (1968) 48P.
192 J.A. Balint, E.C. Kyriakides, H.L. Spitzer and E.S. Morrison, *J. Lipid Res.*, 6 (1965) 96.
193 J.A. Balint, D.A. Beeler, D.H. Treble and H.L. Spitzer, *J. Lipid Res.*, 8 (1967) 486.
194 L. Swell, C. Entenman, G.F. Leong and R.J. Holloway, *Am. J. Physiol.*, 215 (1968) 1390.
195 F. Schaffner, H. Popper and V. Perez, *J. Lab. Clin. Med.*, 56 (1960) 623.
196 D.H. Neville, *J. Biophys. Biochem. Cytol.*, 8 (1960) 413.
197 P. Emmelot, C.J. Bos, E.L. Benedetti and P.L. Rümke, *Biochim. Biophys. Acta*, 90 (1964) 126.
198 M. Takeuchi and H. Terayama, *Exptl. Cell Res.*, 40 (1965) 32.
199 R. Coleman and J.B. Finean, *Biochim. Biophys. Acta*, 125 (1966) 197.
200 A.A. El-Aser, J.T.R. Fitzsimons, R.H. Hinton, E. Reid, E. Klucis and P. Alexander, *Biochim. Biophys. Acta*, 127 (1966) 553.
201 R. Coleman, R.H. Michell, J.B. Finean and J.N. Hawthorne, *Biochim. Biophys. Acta*, 135 (1967) 573.
202 J.M. Graham, J.A. Higgins and C. Green, *Biochim. Biophys. Acta*, 150 (1968) 303.
203 C.S. Song, A. Kappas and O. Bodansky, *Ann. N.Y. Acad. Sci.*, 166 (1969) 565.
204 H.M. Berman, W. Gram and M.A. Spirtes, *Biochim. Biophys. Acta*, 183 (1969) 10.
205 T.K. Ray, *Biochim. Biophys. Acta*, 196 (1970) 1.
206 W.H. Evans, *Biochem. J.*, 116 (1970) 833.
207 R.H. Hinton, M. Dobrota, J.T.R. Fitzsimons and E. Reid, *European J. Biochem.*, 12 (1970) 349.
208 R. Coleman, *Chem. Phys. Lipids*, 2 (1968) 144.
209 R. Coleman and J.B. Finean, in M. Florkin and E.H. Stotz (Eds.), *Comprehensive Biochemistry*, Vol. 23, Elsevier, Amsterdam, 1968 p. 99.
210 R. Coleman, unpublished observations.
211 V.H. Börning, J. Stepan, A. Horn, R. Giertler, G. Thiele, B. Vecerek, *Z. Physiol. Chem.*, 348 (1967) 1311.
212 P. Emmelot and C.J. Bos, *Biochim. Biophys. Acta*, 120 (1966) 369.
213 S.P.N. Sherlock, *Diseases of the Liver and Biliary System*, 3rd ed., Blackwell, Oxford, 1963, p. 40.
214 P.G. Hill and H.G. Sammons, *Quart. J. Med.*, 36 (1967) 457.
215 S. Orrhenius, J.L.E. Ericsson and L. Ernster, *J. Cell Biol.*, 25 (1965) 627.
216 A.L. Myasnikov, *Circulation*, 17 (1958) 99.
217 S. Orrhenius, *J. Cell Biol.*, 26 (1965) 725.
218 D.L. Young, G. Powell and W.O. McMillan, *J. Lipid Res.*, 12 (1971) 1.
219 H. Glaumann and G. Dallner, *J. Lipid Res.*, 9 (1968) 720.
220 J.L. Holtzman and J.R. Gillette, *Biochem. Biophys. Res. Commun.*, 24 (1966) 639.
221 R. Whitehead, *J. Biol. Chem.*, 139 (1941) 897.
222 R.H. Campbell and H.W. Kosterlitz, *J. Biol. Chem.*, 175 (1948) 989.
223 M. Vavrecka, M.P. Mitchell and G. Hübscher, *Biochem. J.*, 115 (1969) 139.
224 G. Bartsch and G.B. Gerber, *J. Lipid Res.*, 7 (1966) 204.
225 E. Levin, S. Albert and R.M. Johnson, *Arch. Biochem. Biophys.*, 64 (1956) 272.
226 E. Levin, R.M. Johnson and S. Albert, *Arch. Biochem. Biophys.*, 73 (1958) 247.
227 M.I. Gurr, C. Prottey and J.N. Hawthorne, *Biochim. Biophys. Acta*, 106 (1965) 357.
228 H. Goldfine, *Ann. Rev. Biochem.*, 37 (1968) 303.
229 P. Bjornstad and J. Bremer, *J. Lipid Res.*, 7 (1966) 38.
230 R.M.C. Dawson, *Biol. Rev. Camb. Phil. Soc.*, 32 (1957) 188.
231 D.B. Zilversmit and E.V. Hendel, *Arch.Biochem. Biophys.*, 73 (1958) 224.
232 M.J. Fletcher and D.R. Sanadi, *Biochim. Biophys. Acta*, 51 (1961) 356.
233 T. Omura, P. Siekevitz and G.E. Palade, *J. Biol. Chem.*, 242 (1967) 2389.
234 M. Pascaud, *Biochim. Biophys. Acta*, 84 (1964) 528.
235 K.W.A. Wirtz and D.B. Zilversmit, *J. Biol. Chem.*, 243 (1968) 3596.
236 K.W.A. Wirtz and D.B. Zilversmit, *Biochim. Biophys. Acta*, 187 (1969) 468.
237 K.W.A. Wirtz and D.B. Zilversmit, *Biochim. Biophys. Acta*, 193 (1969) 105.
238 K.W.A. Wirtz, L.M.G. van Golde and L.L.M. van Deenen, *Biochim. Biophys. Acta*, 218 (1970) 176.
239 M.C. Blok, K.W.A. Wirtz and G.L. Scherphof, *Biochim. Biophys. Acta*, 233 (1971) 61.
240 K.W.A. Wirtz and D.B. Zilversmit, *FEBS Letters*, 7 (1970) 44.
241 W.C. McMurray and R.M.C. Dawson, *Biochem. J.*, 112 (1969) 91.

242 F.B. Jungalwala and R.M.C. Dawson, *European J. Biochem.*, 12 (1970) 399.
243 M. Akiyama and T. Sakagami, *Biochim. Biophys. Acta*, 187 (1969) 105.
244 D.B. Zilversmit, *J. Lipid Res.*, 12 (1971) 36.
245 K.W.A. Wirtz, *Thesis*, University of Utrecht, 1971.
246 V.P. Skipski, M. Barclay, E.M. Archibald, O. Terebus-Kekish, E.S. Rechman and J.J. Good, *Life Sci.*, 4 (1965) 1673.
247 R.C. Pfleger, N.G. Anderson and F. Snyder, *Biochemistry*, 7 (1968) 2826.
248 C. Long (Ed.), *Biochemists' Handbook*, Spon, London, 1961.
249 P. Hill, *M. Sc. Thesis*, University of Birmingham, 1965.

Chapter 14

PHOSPHOLIPIDS AND THE NERVOUS SYSTEM

G.B. ANSELL

Department of Pharmacology (Preclinical), Medical School, Birmingham B15 2TJ (Great Britain)

In the previous edition of this book[1] the longest chapter was devoted to the nervous system. The fact that the current account is shorter reflects neither a decrease in enthusiasm for this topic nor, indeed, a lack of interest on the part of the editors. It is hoped that the chapter will give an up-to-date statement of present knowledge and indicate areas of ignorance, partial or total, since, with brain phospholipids as with many other aspects of neurochemistry, the descriptive stage is still with us. Recent useful reviews and accounts can be found in refs. 2–5.

1. THE PHOSPHOLIPID COMPOSITION OF THE NERVOUS SYSTEM

About 25% of the dry weight of brain tissue consists of phospholipids, the percentage composition of which is given in Chapter 16 where some information about the morphological distribution in grey and white matter can also be found. Determinations of the fatty acid and aldehyde composition of the phospholipids have been numerous over the past decade and information about the molecular species present is now accumulating[6–8]. No phospholipid is exclusive to the nervous system although nervous tissue is characterised by a large amount of ethanolamine plasmalogen; this lipid and its choline analogue are also found in relatively high concentrations in the other excitable tissues, cardiac and skeletal muscle.

Some eighteen years ago it looked as if considerable information about phospholipid distribution in the brain would be forthcoming from the application of the quantitative histochemical methods of Lowry[9] adapted for the phospholipid analysis of circumscribed brain areas[10]. This work has not been taken further and few phospholipid studies have stemmed from the elegant work of Hydén[11] on microdissected neurones and glia. Roots and Johnston[12, 13] manually fished out neurones from a dilute suspension of cells from the lateral vestibular nucleus of the ox brain and obtained sufficient material for a qualitative lipid analysis. Phosphatidylcholine, phosphatidylethanolamine, ethanolamine plasmalogen, and phosphatidylserine were the major phospholipids detected[13]. In the main, however, the neurochemist interested in phospholipids has turned his attention to the techniques of subcellular fractionation developed by Whittaker[14] and De Robertis[15] and more recently to neuronal perikarya and glial cells separated by bulk procedures.

Neurones and glia

By mild maceration of brain cortex, sieving it through pieces of mesh of decreasing aperture size and centrifuging the resultant suspension on Ficoll–sucrose gradients, it is possible to obtain fractions which are highly enriched either in nerve cell perikarya or in glial cells. This innovation is due to Rose[16]. 'Neurones' so obtained therefore lack their axons and are generally devoid of dendritic processes; the outer membranes from which these processes originated are believed to re-seal so that little cytoplasm is normally lost, though there is a loss of potassium[17]. It must be remembered, however, that for some neurones, the axon can account for 1000 times the volume of the cell body so that the perikaryon represents a tiny fraction of the unit. Glial cells usually retain their processes in this procedure.

The original method for the bulk separation[16] has been criticised by Cremer et al.[18] who stated that the 'neuronal' fraction was contaminated with fragmented capillaries and the glial fraction with nerve endings. It has since been modified by a number of workers, e.g. Blomstrand and Hamberger[19]. Fewster et al.[20] used a series of Ficoll gradients to separate glial cells from bovine white matter while Norton and Poduslo[21] successfully disaggregated brain tissue with trypsin prior to fractionation on sucrose gradients. Sinha and Rose[22] have compared the method of Norton and Poduslo with their own and concluded that although the preparations obtained by the latter method were of good appearance and purity they were lacking in viability as measured by certain metabolic parameters. Nevertheless, it would appear from a more recent paper by Norton and Poduslo[23] that their yields are somewhat better than those of Rose, i.e. 15–20% for neurones and 3.5% for glia (largely astrocytes) and the method is probably more satisfactory if lipid analysis is desired. The reader is referred to this paper[23] for a discussion of the rather complicated calculation by which their yields were computed.

The lipid analyses of neuronal perikarya, astrocytes and oligodendroglia by Norton and Poduslo[23, 24] (see Chapter 16) are in general agreement with those of Hamberger and Svennerholm[25] and Tamai et al.[26]. Lipid constitutes 24% of the dry weight of the neuronal cell body (43 pg/cell) of the rat whereas it constitutes 39% of that of the astroglial cell (230 pg/cell)[23]. The oligodendroglial cell contains only 7 pg lipid[24]. Neurones contain more lipid per unit weight than whole brain tissue; glia contain more cholesterol and phosphatidylserine than neuronal cell bodies which have a higher amount of phosphatidylinositol.

Fewster and Mead[27, 28] have determined the fatty acid and fatty aldehyde composition of the lipids in glial cells from white matter and Hamberger and Svennerholm[25] have also analysed the fatty acid components of certain lipids in glia and neuronal perikarya.

Subcellular fractions

Over a decade ago, Whittaker and De Robertis independently showed that the rate centrifugation procedure developed for liver tissue was insufficient to give an adequate separation of the subcellular components of brain tissue because pinched-off nerve endings ('synaptosomes') and myelin behave similarly to mitochondria. An excellent review of how

TABLE I

THE PHOSPHOLIPID COMPOSITION OF SUBCELLULAR FRACTIONS FROM WHOLE RAT BRAIN (μmole/g FRESH WEIGHT OF WHOLE BRAIN)

	Whole brain	*Myelin*	*Synaptosomes*	*Mitochondria*	*Microsomes*	*Soluble fraction*	*Nuclei and undisrupted cells*
Total phospholipid	60.0	19.8	3.6	5.0	8.5	3.6	2.9
Phosphatidylcholine	15.1	6.4	1.4	1.9	3.0	1.0	0.9
Choline plasmalogen	0.6	0.1	–	<0.1	<0.1	<0.1	<0.1
Phosphatidylethanolamine	7.4	3.2	0.5	1.3	1.6	0.8	1.0
Ethanolamine plasmalogen	15.8	6.4	0.6	0.5	1.2	0.8	0.45
Sphingomyelin	3.7	1.4	0.2	0.1	0.6	0.2	0.1
Phosphatidic acid	1.0	0.3	<0.1	–	0.3	–	–
Cardiolipin	0.6	0	–	0.4	0	–	–
Phosphatidylinositol	1.4	0.6	0.2	0.2	0.3	–	0.2
Alkylether phospholipid	1.6	0.1	–	0.1	0.3	0.1	0.1
Phosphatidylserine	4.7	1.7	0.5	0.3	0.9	–	0.1

These values are compiled from values given by Lapetina et al.[31], Dickerson[32] and Spanner (unpublished).

it proved possible to separate these fractions was published by Whittaker[14] in 1965 and, for an up-to-date account of more sophisticated advances, the reader is referred to ref. 29.

The fact that the crude mitochondrial fraction of brain tissue contains both myelin and synaptosomes renders valueless the earlier determinations of brain mitochondrial phospholipids which were detailed in Chapter 11 of the previous edition of this book. Eichberg et al.[30] were the first to give a detailed phospholipid analysis of the subcellular fractions of brain tissue prepared by more reliable techniques (see Chapter 16). Table 1 gives the phospholipid composition of rat brain subcellular fractions calculated in terms of the original fresh weight. Since recoveries are often low, because of losses during preparation, it has been more usual to express results as a percentage of the total phospholipid within a given fraction (Chapter 9, Chapter 16).

Myelin

Since $K^{\oplus}$ and $Na^{\oplus}$ ions do not move across the ensheathed part of the axonal membrane, the myelin sheath allows increased conduction velocity along the axon by making the local bioelectric current generated at a node of Ranvier act ahead at the next node and, in the words of Bunge[33] "the myelin sheath thus provides 'limited access' for current flow, making the myelinated axon the superhighway of the nervous system". This physiological property is that of the sheath as a whole and must depend on the characteristics of the investing membranes as well as the many tight windings that constitute the myelin proper. In peripheral nerve, electron microscopy has established that there is an inner and outer collar of Schwann cell cytoplasm that encloses the myelin and that the myelin itself is formed from the spiral disposition of the Schwann cell plasma membrane. There is considerable evidence[33] that the oligodendroglial cell in the central nervous system (CNS) plays a role similar to that of the Schwann cell peripherally (Fig. 1), though the oligodendroglial cell can contribute to more than one internode of myelin in different axons. Each expands at the node to enclose glial or Schwann cytoplasm. Some morphological differences between central and peripheral myelin are shown diagrammatically (Fig. 2). Myelin in the peripheral nerve can be regenerated and in many instances regeneration can occur in the CNS[33].

When brain tissue is lightly homogenised in isotonic media the myelin sheaths are broken into fragments and the compacted lipid-rich membranes of the myelin proper can be separated by suitable techniques. Myelin can be readily prepared from central white matter in large amounts and a great deal of work has been done on its lipid composition. An excellent account of the best way to prepare pure myelin has been given by Norton[35]; it is quite difficult to obtain material free from axoplasmic fragments but osmotic shock with hypotonic solutions is helpful. Furthermore, as Norton points out, the original method

Fig. 1. This diagram illustrates both known and hypothetical aspects of the mature myelin sheath and its relation to glial cells. The unit or plasma membrane (pm) is here designated as two lines separated by a space except in the mitochondrion, where it is represented by a single line. The inner mesaxon (im), formed when a glial process completes the initial turn round an axon (a) and starts a second, is retained after myelin formation is completed. Some cytoplasm of the glial process is present here. Cytoplasm is trapped occasionally at cy. On the fully formed sheath exterior, a bit of glial cytoplasm is also retained.

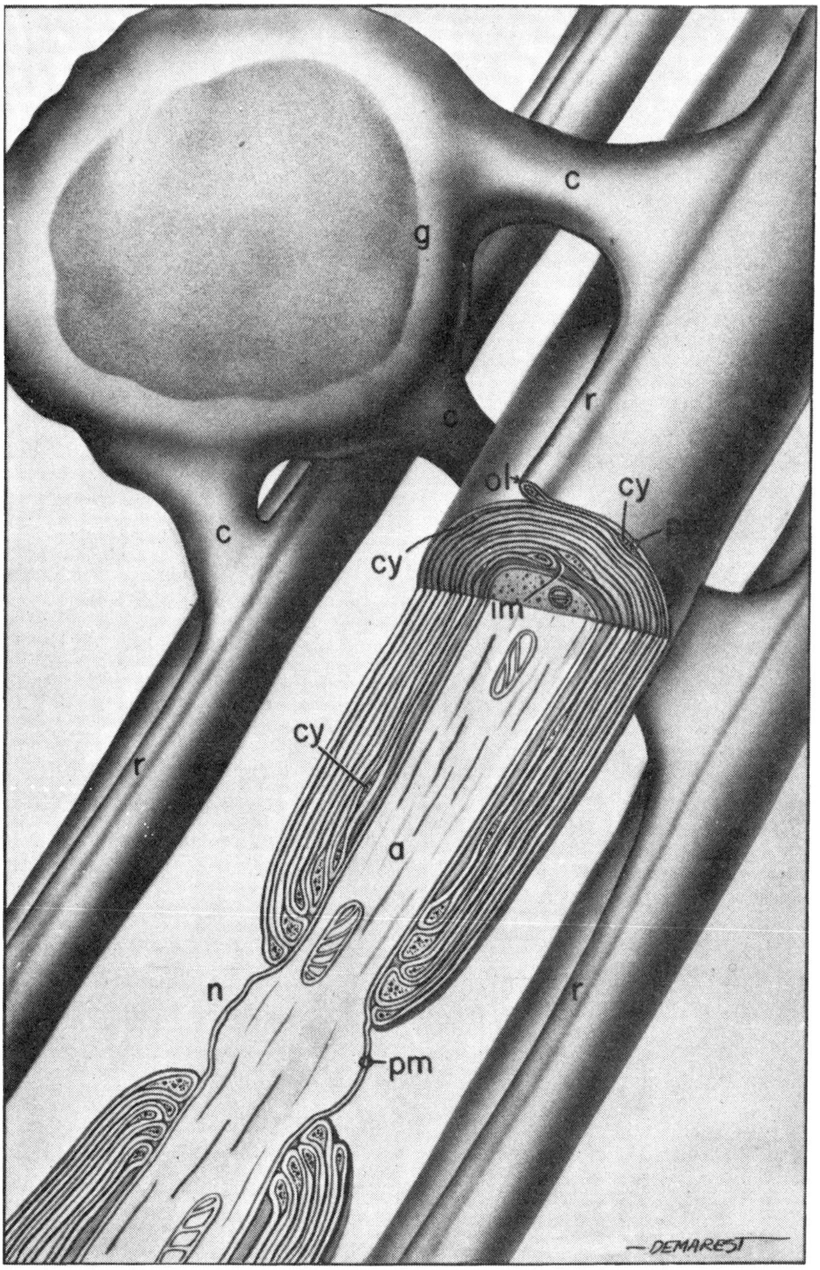

In transverse sections, this cytoplasm is confined to a loop (ol) of plasma membrane (pm) but, along the internode length, it forms a ridge (r) which is continuous with a glial cell body (g) at c. When viewed transversely, the sheath components are oriented in a spiral, only the innermost and outermost layers ending in loops; in the longitudinal plane, every myelin unit terminates in a separate loop near a node (n). Within these loops glial cytoplasm is also retained. (From ref. 34 by permission of the authors and the Rockefeller Press.)

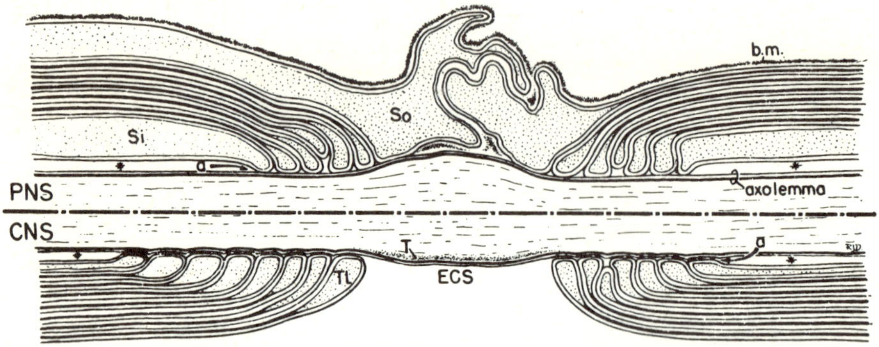

Fig. 2. A nodal region from PNS (above) and CNS (below). In the PNS the Schwann cell provides both an inner collar (Si) and an outer collar (So) of cytoplasm in relation to the compact myelin. The outer collar (So) is extended into the nodal region as a series of loosely interdigitating processes. Terminating loops of the compact myelin come into close apposition to the axolemma in the region near the node apparently providing some barrier (arrow at a) for movement of materials into or out of the periaxonal space (marked by *). The Schwann cell is covered externally by a basement membrane. In the CNS the myelin ends similarly in terminal loops (tl) near the node and there are periodic thickenings of the axolemma where the glial membrane is applied in the paranodal region. These may serve as diffusion barriers and thus confine the material in the periaxonal space (marked *) so that movement in the direction of the arrow at a would be restrained. At many CNS nodes there is considerable extracellular space (ECS). (From ref. 33 by permission of the author and the American Physiological Society.) CNS, central nervous system; PNS, peripheral nervous system.

devised by his group[36] is suitable for white matter from large animals but not for whole brains of small animals on which most metabolic investigations have been made.

Preparation of myelin from peripheral nerve e.g. peripheral myelinated axons which make up the sciatic nerve, is difficult because of the presence of connective and adipose tissue. O'Brien et al.[37] avoided these problems by using intradural spinal roots and a modified version of the method of Autilio et al.[36]. (For other variations see refs. 38 and 39.)

The criteria for myelin purity are complex. Obviously, the appearance of a large typical field under the electron microscope is very useful but there are difficulties in identifying small vesicles. Purified myelin should be 90% soluble in chloroform : methanol (2:1 v/v). It should be free from the plasma membrane marker enzymes acetylcholinesterase (EC 3.1.1.7) and 5′-nucleotidase (EC 3.1.3.5) which have been commonly reported in myelin preparations. There should be minimal levels of ATPase (EC 3.6.1.3), succinic dehydrogenase (EC 1.3.99.1) from mitochondria, and nucleic acids from endoplasmic reticulum and cell nuclei. The only enzyme apparently exclusively associated with myelin or the axolemma, which is impossible to separate from myelin, is 2,3′-cyclic nucleotide-3-phosphohydrolase[40, 41]. The association of this enzyme with myelin was confirmed by its virtual absence from brain tissue in which myelination had not taken place[42]. It must be pointed out, however, that the enzyme has now been detected in glial cells from tumours which are similar in some respects to oligodendroglia maintained in tissue culture[43].

Myelin accounts for about 50% of the dry weight of ox white matter and 65% of the total lipid therein[44]. Lipid accounts for 70% of the dry weight of both central myelin and that of peripheral nerve[37, 45]. The myelin of peripheral nerve (dorsal spinal roots) differs in composition from that of the CNS (see Chapter 16) in its lower concentration of cerebroside and higher concentration of sphingomyelin. The most abundant phospholipid in myelin from all sources so far examined is ethanolamine plasmalogen. Peripheral nerve also contains significant amounts of serine plasmalogen according to O'Brien et al.[37]. Cardiolipin is absent whereas triphosphoinositide (phosphatidylinositol diphosphate; TPI) has been thought unique to myelin[44, 46, 47], though more recent work indicates a small non-myelin pool (Chapter 15, p. 430).

The almost complete solubility of myelin in chloroform : methanol (2:1 v/v) is to a large extent accounted for by the presence of a proteolipid complex i.e. a lipoprotein soluble in lipid solvents (for earlier literature see Ansell and Hawthorne[1]) which accounts for 60% of the protein in myelin[48, 49]. The proteolipid protein is poor in acidic and basic amino acids and rich in the less polar amino acids (leucine, isoleucine, valine, glycine, proline, phenylalanine, and alanine) which helps to explain its association with lipid and consequent solubility of the complex in chloroform:methanol. It is believed to have a monomeric size of around 12 500 but dimers and trimers have been reported[50]. It may well have some structural function since proteolipid proteins from peripheral nerve myelin and other membranes have a similar amino acid composition[51]. An acid-soluble protein also exists in proteolipid[52], which has a similar amino acid composition to 'classical' proteolipid protein.

Another protein is found in myelin, the basic encephalitogenic protein (experimental allergic encephalitogenic protein, EAE protein, A_1 protein) which is water-soluble following acid extraction and has been extensively studied[53, 54]. Ox brain contains about 10 mg EAE protein/g wet weight[55], and it accounts for 30% of the protein of human brain myelin. It is of great interest because it produces allergic encephalomyelitis when injected into experimental animals (see p. 409). It is highly basic and unfolded with an isoelectric point > 12 and its amino acid sequence has been determined[53, 56]. It is mentioned here because Palmer and Dawson[57] found that a preparation of EAE protein, probably in its native state, could form a complex with triphosphoinositide of 0.145 μmole TPI/mg protein; this was comparable with the ratio of TPI to EAE protein in brain tissue[58]. The role of such a complex in myelin is, as yet, unknown though Palmer and Dawson[58] suggest that it may act as a centre for binding zwitterionic phospholipids and cholesterol. Other complexes of inositol lipids with proteins have been reported from time to time but these 'phosphatido-peptides' are probably artefacts formed during extraction processes[59].

Nuclei

Few determinations have been made of the phospholipids of brain cell nuclei and none on nuclei from neurones and glia separately. The most reliable values are those of Siakotos et al.[60] (Chapter 16). Ethanolamine and choline lipids account for more than 70% of the total. It is not possible to calculate the amount per g tissue reliably.

Mitochondria

There have been a number of determinations of the phospholipid composition of mitochondria from brain (Chapter 16) and the results have been expressed in a variety of ways. In Table 1 values for rat brain have been recalculated to give the amount of each lipid in the mitochondria in terms of the weight of fresh tissue to supplement the information given in Chapter 16 and Chapter 9. As might be expected the mitochondria are enriched in cardiolipin. The mitochondrial fraction as conventionally prepared contains the lysosomes which can be separated from it[61] but the phospholipid composition of the lysosomes has not been determined.

Microsomal fraction

It can be seen from Table 1 that the microsomal fraction is second only to myelin in the amount of ethanolamine plasmalogen it contains. Microsomes have been prepared separately from neuronal perikarya and glial cells[62].

Synaptosomes

As discussed in Chapter 15 homogenisation of neurone-containing areas of the brain in isotonic media shears off the pre-synaptic nerve terminals (*boutons terminaux*), of which some 16 000 may be attached to the outer membrane of a single neurone[63]. The nerve endings re-seal themselves and, as normally isolated, retain the sub-synaptic web and post-synaptic membrane. Within these 'synaptosomes' can be seen vesicles and intraterminal mitochondria. Values for the synaptosomes as a whole will thus reflect the mean of the phospholipid composition of all the components. The original values of Eichberg et al.[30] for the phospholipid composition of the synaptosomes of the guinea pig cortex are given in Chapter 16.

By careful use of hypo-osmotic shock it is possible to rupture the synaptosomal outer membrane and, by centrifugation on a stepwise sucrose gradient, to obtain sub-synaptosomal components, notably synaptic plasma membranes and vesicles. The best characterised preparations of synaptosomal plasma membranes are those reported by Cotman et al.[64] and Breckenridge et al.[7]. They were found to have a high lipid:protein ratio of 1:1 and to have higher molar ratios of cholesterol:phospholipid, much larger amounts of gangliosides and larger amounts of sphingomyelin than plasma membranes from other sources. The phospholipid composition of synaptosomal plasma membranes is given in Chapter 16. Breckenridge et al.[7] have found some unusual features in the fatty acid composition of the phospholipids from this material. The ethanolamine (diacyl and alkenyl acyl) and serine phospholipids contain 30% of docosahexaenoic acid ($C_{22:6}$ (n–3)) (cf. another excitable membrane from retinal rod outer segments[65]) while the fatty acid of the sphingomyelin consists of stearic acid almost exclusively. Palmitic acid accounts for 57% of the fatty acids of the phosphatidylcholine which might be partly dipalmitoyl lecithin since this has been found in the synaptosomal fraction by Hamberger and Svennerholm[25].

2. BIOSYNTHESIS OF PHOSPHOLIPIDS IN THE NERVOUS SYSTEM

The capacity of isolated brain tissue to synthesize complete phospholipids from small precursor molecules has been known for many years (for early studies, see ref. 1) and many of the pathways discussed in Chapter 4 are found in the brain and peripheral nerve. Knowledge of the pathways has been derived from numerous studies *in vivo* and studies *in vitro* with slices, homogenates and sub-cellular fractions. No specific studies on the different capacities of neurones and glial cells have been made apart from those of Freysz et al.[66] who isolated neuronal perikarya and glial cells from the cortex of rats injected intraperitoneally with [^{32}P]orthophosphate and showed that, in general, neuronal phospholipids turned over faster than those of glia. Some of the more prominent features of phospholipid synthesis, largely in the CNS, are set out below and discussed further in the section devoted to the developing nervous system.

Phosphatidic acid

The various pathways to this key intermediate phospholipid are discussed in Chapters 4 and 15. The two known pathways in the nervous system are the phosphorylation of 1,2-*sn*-diacyl-glycerol and the two-stage acylation of *sn*-glycerol 3-phosphate. The dihydroxyacetone pathway of Hajra and Agranoff[67] does not seem to have been investigated in brain. According to Lapetina and Hawthorne[68] diglyceride kinase is concentrated in the supernatant and microsomal fraction of brain, with the particulate enzyme possibly associated with surface membranes; it is more than 100 times as active as the system producing phosphatidic acid by the acylation of glycerophosphate. The special relevance of the diglyceride kinase reaction in nervous tissue is discussed in Chapter 15.

Presumably the formation of phosphatidic acid by the acylation of glycerophosphate[69] is the system by which the lipid is generated *de novo*. Mårtensson and Kanfer[69] solubilised the enzyme from a 'particulate' fraction of rat brain and found that it was stimulated 6-fold by a heat-stable factor in the 100 000 x *g* supernatant (cf. Smith and Hübscher[70] who found that a similar factor from liver supernatant was heat-*labile*). Complete acylation seems to occur since lysophosphatidic acid was not a major product of the reaction. Although these experiments strongly suggested that the acylation takes place in the mitochondrial fraction there is clear evidence that it also occurs in the microsomal fraction[71, 72] with the same K_m for glycerophosphate of $4 \cdot 10^{-4}$ *M*. In both fractions cytidine nucleotides, particularly CTP, inhibit the acylation[72]. This could form part of a control mechanism in phospholipid biosynthesis especially since CTP levels may be controlling RNA synthesis[73].

Phosphatidylcholine

The synthesis of the phospholipid in brain tissue by the cytidine pathway was first demonstrated in brain dispersions by McMurray et al.[74] and the presence of about 0.4 μmole of phosphorylcholine and 0.05 μmole CDPcholine per g of fresh tissue is now well established (Table 2). Choline kinase is found exclusively in the soluble fraction of the tissue as originally indicated by the work of McCaman and Cook[79], and confirmed by Ansell and Spanner[80]. At its optimum pH of 10 the enzyme has more than twenty times the activity of cholinephosphate cytidylyltransferase (EC 2.7.7.15) on a fresh weight basis[81]. Until recently the latter was also considered an enzyme of the soluble fraction which can be separated from the analogous enzyme which forms CDPethanolamine[82].

TABLE 2

LEVELS OF ETHANOLAMINE AND CHOLINE COMPOUNDS IN ADULT (>18 WEEKS) FEMALE WHOLE RAT BRAIN (μmole/g WET WEIGHT)

Choline	0.17[a]	Ethanolamine	1.35[b]
Phosphorylcholine	0.38	Phosphorylethanolamine	0.94
sn-Glycero-3-phosphorylcholine	0.04	*sn*-Glycero-3-phosphoryl-ethanolamine	0.25[c]
CDPcholine	0.05	CDPethanolamine	0.027
Phosphatidylcholine	14.7	Phosphatidylethanolamine	7.4
Choline plasmalogen	0.62	Ethanolamine plasmalogen	15.73
Sphingomyelin	3.65	1-alkyl-2-acyl-*sn*-glycero-3-phosphorylethanolamine	1.4–1.8

These values from the author's laboratory are for brains extracted within a few minutes of death. (The standard deviation was usually not greater than ± 10%.)

[a]Consistent values of half this have been obtained by Ewetz et al.[75] and values as low as 0.02 μmole/g wet weight for anaesthetised animals frozen in liquid nitrogen, by Eade et al.[76].

[b]According to Oja[77] there is a rapid post-mortem rise in free ethanolamine, and values for brain tissue removed from the open cranium of an anaesthetised animal, transferred to liquid air and extracted with perchloric acid, are very low.

[c]Values of <0.1 μmole/g were obtained for 4-week-old rats frozen whole in liquid nitrogen[78].

However, more recent studies have indicated its enrichment in the microsomal fraction[83]; it is high in cerebellar cortex and low in corpus callosum. The final step catalysed by the enzyme choline phosphotransferase (EC 2.7.8.2.) can transfer maximally 13 μmole of phosphorylcholine/g tissue/h according to McCaman and Cook[79]. Although it has been reported that diglycerides containing unsaturated fatty acids are more active as acceptors, this would seem unlikely (Chapter 5, p. 111). The overall rate of uptake of free choline into the total choline lipid in an incubated brain slice is relatively slow at about 0.05 μmole/g tissue/h[84] and the rate-limiting step could well be the formation of CDPcholine.

The phosphatidylcholine of brain mitochondria is rich in polyunsaturated fatty acids[85] which could be introduced by the acylation system described by Webster and Alpern[86], subsequent to the production of a 1-acyl-*sn*-glycero-3-phosphorylcholine by an endogenous phospholipase A_2. By analogy with liver[87, 88] this phospholipase should also be mitochondrial but its sub-cellular location in brain is as yet unknown.

There is no evidence that significant amounts of phosphatidylcholine can be formed in the brain by a step-wise methylation of phosphatidylethanolamine though this is a most important pathway in liver (Chapters 4 and 13). Very little labelling of the choline moiety of choline lipids was detected in the brain of rats injected intracerebrally with [1,2-^{14}C]-ethanolamine[89] or labelled dimethylethanolamine[90] though the formation of the phospholipids corresponding to these injected bases was rapid. Methylation has not been detected *in vitro*[91] and the fatty acid composition of the phosphatidylcholine of the brain microsomal fraction, where methylation would presumably occur[92], is completely different from that of the phosphatidylethanolamine[93].

A $Ca^{2\oplus}$-dependent, energy-independent exchange reaction by which ethanolamine or serine are incorporated into brain microsomal phospholipids (p. 388 and p. 391) was originally thought not to operate for choline. However, Miller and Dawson have now demonstrated a significant $Ca^{2\oplus}$-dependent exchange reaction for choline, largely in the microsomal fraction of guinea pig brain[94]. Kanfer[95] has confirmed this finding using a particulate fraction, and states that the system operates optimally at pH 9.0, and that –SH groups should be absent.

Ethanolamine phosphoglycerides

The cytidine pathway for the synthesis of phosphatidylethanolamine was first demonstrated for homogenates of brain tissue and a crude mitochondrial fraction by McMurray et al.[74] and the intermediate reactions demonstrated *in vivo* following the intracerebral injection of [1,2–^{14}C]ethanolamine by Ansell and Spanner[89]. Levels of free ethanolamine, phosphorylethanolamine and CDPethanolamine in rat brain are given in Table 2.

The ethanolaminephosphotransferase (EC 2.7.8.1) has been extensively studied recently[96, 97]. It is largely found in the microsomal fraction and the K_m for CDPethanolamine was about $2.5 \cdot 10^{-4} M$ and for the acceptor lipid about $2 \cdot 10^{-3} M$. The acceptor lipid could be either 1,2-diacyl-*sn*-glycerol or 1-alkenyl-2-acyl-*sn*-glycerol. Maximum rates of about 250 nmole of phosphatidylethanolamine or 80 nmole of ethanolamine plasmalogen/mg microsomal protein/h were found with both chicken and rat microsomal fractions. Some 1-alkyl-2-acyl-GPE was formed when cytidine phosphate [^{32}P] phosphorylethanolamine was incubated with brain homogenates[98]. The formation of a saturated ether lipid was not stimulated by the presence of 1,2,-dialkyl-*sn*-glycerol[96] though Radomińska-Pyrek and Horrocks[99] recently found a more natural acceptor (1-alkyl-2-acylglycerol) which stimulated the formation of the alkyl ether lipid. This is almost certainly crucial to the synthesis of brain ethanolamine plasmalogen (see p. 389). In any event, the experiments showed that the ethanolamine phosphotransferase was relatively unspecific towards the lipid acceptor. $Mn^{2\oplus}$ ions were more effective at low concentrations (5 m*M*) in stimulating the phosphotransferase than were $Mg^{2\oplus}$ ions[97], in contrast to their effect on cholinephosphotransferase activity[79].

The overall transfer of phosphorylethanolamine to ethanolamine lipids in homogenates of rat brain was found to be low[100]. Porcellati et al.[101] also found the conversion low in the presence of a chicken brain microsomal preparation though it was stimulated by the presence of the soluble fraction of the brain, a not surprising finding since the soluble fraction contains the ethanolamine phosphate cytidylyltransferase. As with phosphatidylcholine formation in brain, it is highly likely that the formation of CDPethanolamine is the rate-limiting step in the formation of phosphatidylethanolamine.

Although the phosphorylation of ethanolamine by brain homogenates was demonstrated 20 years ago[102] and shown to be a relatively rapid reaction *in vivo*[89]

(0.14 μmole/g brain/h), this reaction has not been studied in detail and it is not known whether the kinase responsible is the same as the choline kinase (EC 2.7.1.32). Certainly, the kinase of the soluble fraction of brain has a very low affinity for ethanolamine (K_m at pH 10 only about $10^{-2}M$; Spanner, unpublished). In fact, the origin of the phosphorylethanolamine moiety of ethanolamine lipids in the brain is perhaps more complicated than might be deduced from the studies mentioned above.

Firstly, there is no evidence that ethanolamine is formed in the brain by the decarboxylation of free or lipid-bound L-serine in spite of the fact that carbons C_2 and C_3 of L-serine provide the carbons of ethanolamine in the mammal[103]. The results of experiments with intraperitoneally[104] or intracerebrally[105] injected [^{14}C]serine were inconclusive. Formation of ethanolamine from L-serine can however, occur by the synthesis and subsequent phosphorylative catabolism of sphingosine since Stoffel and Henning[106] showed that when [1-^{3}H]*erythro*-DL-sphingosine ([1-^{3}H]4*t*-sphingenine) was injected intraperitoneally into myelinating rats, the radioactivity in the brain was almost all in lipid-bound ethanolamine and choline. Since no free radioactive sphingosine could be detected in the brain it was assumed that the sphingosine was being phosphorylated in the liver and then hydrolysed to hexadec-2*t*-en-1-al and phosphorylethanolamine[107]. Presumably this phosphorylethanolamine can itself be hydrolysed and free ethanolamine transferred to the brain, if, by analogy with phosphorylcholine[108], the phosphorylethanolamine does not survive as such in the blood stream. In any event, this could explain the origin of ethanolamine from C_2 and C_3 of L-serine, since these provide C_1 and C_2 of sphingosine. The recent experiments by Stoffel et al.[109] have shown that intracerebrally injected DL-dihydrosphingosine (sphinganine) can also yield phosphorylethanolamine which could presumably be fed directly to the cytidine pathway. A rationale for the observed rapid phosphorylation of free ethanolamine in brain[89] then becomes obscure.

Secondly it must be pointed out that the incorporation of ethanolamine into ethanolamine lipids can occur via a $Ca^{2\oplus}$-dependent exchange reaction first noted briefly by Ansell and Spanner[110] and studied at some length by Porcellati et al.[111]. This energy-independent reaction is most active in the microsomal fraction and might account for some of the incorporation of ethanolamine into the lipids of the brain after its intracerebral injection.

So far this account has been concerned with the method by which ethanolamine can be incorporated into ethanolamine-containing lipids in the brain and rests on the assumption that the cytidine pathway is extremely important. It would imply that the phosphatidylethanolamine, ethanolamine plasmalogen and the alkyl acyl ethanolamine phospholipid (which accounts for 10% of the ethanolamine-containing lipids of mammalian brain) are synthesized independently with a supply of appropriate neutral glyceride acceptors (cf. Joffe[112]). Free 1,2-diacylglycerol is present in brain and has a rapid turnover[113] but no free alkenyl acyl glycerol has been detected[114] though Bickerstaffe and Mead[115] suggested that the alkenyl ether linkage is synthesized at the glyceride stage in the brain. It now seems highly likely that the biosynthesis of the alkenyl ether lipid and the alkyl acyl lipid are closely related in brain tissue and that the saturated ether is the precursor of the unsaturated one. This was suggested by the more rapid labelling of the saturated ether lipid than the alkenyl ether lipid after the intracerebral injection of [1,2-^{14}C]ethanolamine[116, 117] and [^{14}C]glycerol[118]. The brilliant studies of Snyder and his colleagues have now established that the long-chain primary alcohols are the precursors of 1-alkyl-2-acyl-*sn*-glycerol 3-phosphates by a series of reactions initially involving dihydroxyacetone phosphate (Chapter 4). The phosphatidic acid analogue can be dephosphorylated to yield 1-alkyl-2-acyl-glycerol which can then react with CDPethanolamine or CDPcholine[119]. The enzymes necessary for the synthesis of *O*-alkyl glycerols have been demonstrated in brain by Snyder et al.[120]. Takahashi and Schmid showed that small amounts of long-chain alcohols are present in mammalian brain[121] and that, when [1-^{14}C]hexadecanol was injected intracerebrally, it was incorporated into *O*-alkyl and *O*-alkenyl phosphoglyceride at twice the rate of [1-^{14}C]palmitate[122]. Moreover, the latter could be reduced to the alcohol *in vivo* and no free long-chain aldehyde detected. The experiments also showed that the 1-acyl moiety could not be the precursor of the 1-alkenyl moiety to an *O*-alkenyl moiety this has been subsequently demonstrated by Bell et al.[123] using Schmid and Takahashi[122] did not absolutely demonstrate a direct conversion of an *O*-alkyl moiety to *O*-alkenyl moiety this has been subsequently demonstrated by Bell et al.[123] using [1-^{14}C]- and [^{18}O]hexadecanol. Radomińska-Pyrek and Horrocks[99] have now demonstrated

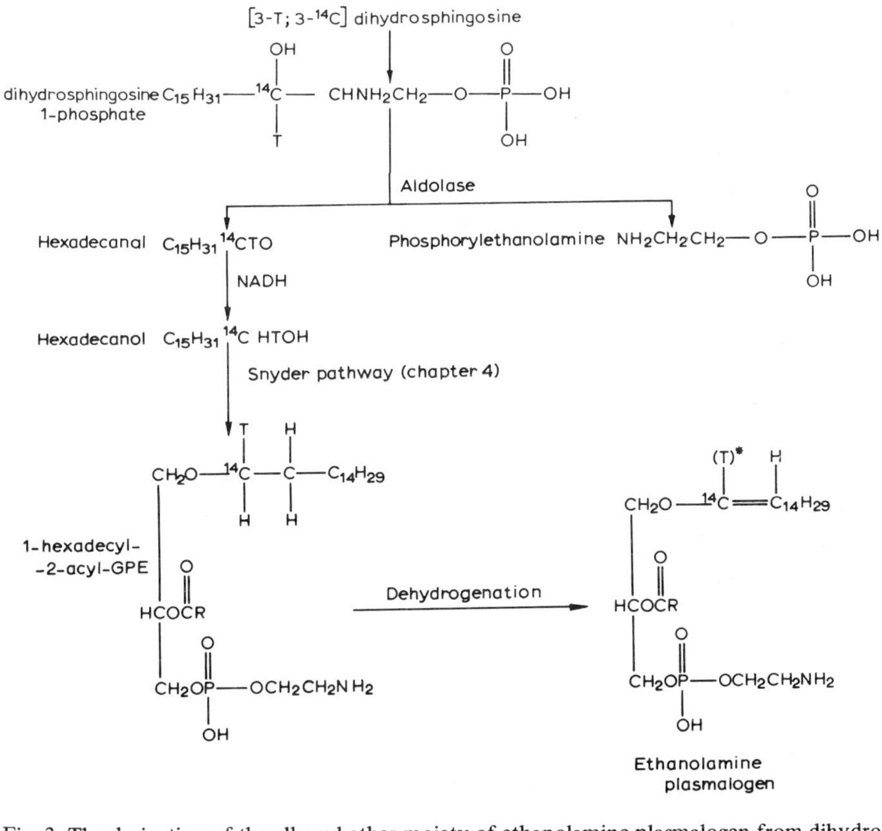

Fig. 3. The derivation of the alkenyl ether moiety of ethanolamine plasmalogen from dihydrosphingosine. The label at * is lowered by 50% on the assumption that the dehydrogenase does not distinguish between the two H atoms of C_1 of the alkyl ether chain. Modified from ref. 102: T, tritium.

that brain microsomal phosphotransferase can transfer [^{32}P]phosphorylethanolamine to added 1-alkyl-2-acyl-*sn*-glycerol to yield the alkyl acyl ethanolamine phospholipid and that a dehydrogenase is also present which catalyses the conversion of the complete alkyl ether lipid to the alkenyl ether one[124]. Curiously the conversion of the alkyl ether choline lipid does not take place[125].

The ramifications of ethanolamine plasmalogen biosynthesis have been further extended by the work of Stoffel and his collaborators[109]. It was pointed out on page 388 that sphingosine can be phosphorylated and the phosphate ester attacked by a lyase to yield phosphorylethanolamine and hexadec-2*t*-en-1-al; dihydrosphingosine yields palmitaldehyde. After the intraperitoneal injection of DL-[3-^{3}H]dihydrosphingosine into myelinating rats, 48% of the radioactivity in the liver lipid and 94% of that in the brain lipid was found in the alkenyl chain of the plasmalogen (palmitaldehyde)[109]. Intracerebrally injected DL-[3-^{3}H]-sphingosine, also heavily labelled the bound palmitaldehyde and, after the intracerebral injection of DL-[3-^{3}H; 3-^{14}C] dihydrosphingosine the doubly labelled aldehyde of the plasmalogen had a [^{3}H : ^{14}C] ratio, half that of the injected precursor. The loss of ^{3}H was interpreted as a formation of palmitaldehyde, its reduction to hexadecanol, subsequent formation of a hexadecyl glycerol ether and dehydrogenation of the alkyl ether to the alkenyl ether as shown in Fig. 3. If the aldehyde had been oxidised it would have lost its ^{3}H; if it had been incorporated intact its ^{14}C: ^{3}H ratio would have remained unchanged[109]. The final dehydrogenation stage has, of course, now been demonstrated *in vitro*[124].

These findings suggest a close relationship between sphingolipid metabolism, or more particularly sphingosine catabolism, and ethanolamine lipid biosynthesis in the brain, especially if the two systems have a close spatial relationship within the cell as suggested by Henning and Stoffel[126]. A useful review of sphingosine metabolism has appeared recently[127].

Sphingomyelin

The biosynthesis of this phospholipid in brain tissue has not been studied in any great detail since the original reaction between a *threo*-ceramide and CDPcholine was shown to occur in brain homogenates by Sribney and Kennedy[128], though it is now known that the natural *erythro*-sphingosine can serve as substrate (Chapter 4). An alternative pathway involving the acylation of sphingosyl-1-phosphorylcholine originally observed by Brady et al.[129] was shown to proceed more rapidly with acyl-CoA and the *erythro*-derivative than the *threo*[130] when the enzyme source was a crude mitochondrial fraction from brain, though the activity of the enzyme was low. The reaction of CDPcholine with *erythro*-sphingosine to yield sphingosyl-1-phosphorylcholine (i.e. via CDPcholine : sphingosine cholinephosphotransferase)[131] has not been confirmed[132].

When injected intracerebrally, *N*-[1-^{14}C]stearoyl sphingosine was partially hydrolysed but some intact ceramide could be found in sphingomyelin[133] suggesting the operation of the pathway of Sribney and Kennedy[128]. Free ceramides occur in brain tissue[134], and have been separated into three groups[135], one containing hydroxy fatty acids, the second a mixture of unsubstituted long-chain fatty acids, (e.g. lignoceric) and the third stearic acid. Morell and Radin[136] observed that the acyl-CoA : long-chain-base acyl transferase present in a microsomal preparation from mouse brain converted more effectively into ceramides, those fatty acids which are present in highest concentrations in brain sphingolipids. If the ceramide–CDPcholine pathway to sphingomyelin is important in brain it is likely therefore that its fatty acid composition is controlled at the stage of ceramide biosynthesis. The suggested formation of ceramides by the reversal of ceramidase action[137] seems unlikely[136].

No experiments *in vivo* to test the validity of alternative pathways to sphingomyelin have been reported; nor is there any information about the effect of varying the nature of the long-chain fatty acid in the ceramide. Kopaczyk and Radin[133] observed no incorporation of ceramides, tree from hydroxy acids, into cerebrosides, and Ställberg-Stenhagen and Svennerholm, in a careful study of sphingomyelin of human brain[138], found no hydroxy fatty acids in that phospholipid. These two observations suggest that the fatty acid composition of a particular ceramide will determine whether it is incorporated into cerebroside or sphingomyelin. It also appears from a recent study by Lastennet et al.[139] that sphingomyelins containing fatty acids of carbon number >20 have a turnover as measured by ^{32}P differing from those with carbon number <20 but the subcellular localisation of the sphingomyelin is also important (see p. 393).

A recent review of sphingomyelin metabolism has appeared[140].

Phosphatidylserine

As discussed in Chapter 4 the only known pathway for the formation of phosphatidylserine in animal tissues occurs by a $Ca^{2\oplus}$-dependent exchange reaction. This has been studied in detail by Porcellati et al.[111] and a rate of 120 nmole/g fresh brain tissue/h was found in the microsomal fraction where the activity resides. The K_m value was $3 \cdot 10^{-4} M$ and similar to that for ethanolamine; the two bases probably compete for the same enzyme. The mechanism of the reaction and the exact nature of the phospholipid which exchanges its base with serine or ethanolamine are, however, unknown.

Experiments on the incorporation of L-serine *in vivo* have not yielded much information because the amino acid is rapidly metabolised and, for example, readily provides the glycerol moiety of phospholipids as was briefly noted by Ansell and Spanner[104] (see also ref. 105). The ramifications of serine metabolism in brain have been discussed recently[126] and deserve further study, especially in relation to sphingosine metabolism.

3. CATABOLISM

The pathways by which phospholipids are degraded in the central and peripheral nervous systems are very incompletely understood and studies have been largely confined to choline phospholipids. Following earlier studies by Gallai-Hatchard et al.[141], Gatt and his colleagues[142] unequivocally demonstrated the presence of phospholipase A_1 (phosphatide acyl hydrolase, EC 3.1.1.4) in a crude particulate fraction from rat brain and prepared a partially purified enzyme from calf brain[143] which was found to be equally active towards phosphatidylcholine and phosphatidylethanolamine in the presence of Triton X-100 + Na taurocholate. Activity towards the acyl group in the 2-position was negligible and lysolecithins (1-palmitoyl-*sn*-glycero-3-phosphorylcholine and 2-stearoyl-*sn*-glycero-3-phosphorylcholine) were only slightly attacked. The phospholipase had a pH optimum of 4.0, a K_m of $8 \cdot 10^{-4} M$ and did not require $Ca^{2\oplus}$ ions for activity. Using 1-stearoyl-2-acyl-*sn*-glycero-3-phosphorylcholine, Cooper and Webster[144] showed that the phospholipase A_1 activity of a lipid-free preparation was highest in human cerebellar cortex (1120 nmole/g fresh weight/h) and lowest in sciatic nerve (75 nmole/g fresh weight/h); cerebral grey matter was slightly more active than white matter. They also showed that when saline extracts of acetone-dried powders of brain tissue were heated at 90° for 5 min most of this activity was lost but that a phospholipase A_2 was present which survived this treatment; it had an optimal activity of pH 5.0–5.3, did not require $Ca^{2\oplus}$ and was inhibited by pre-incubation with di-isopropylfluorophosphonate (DFP), unlike the A_2 enzyme from pancreas.

A lysolecithinase, enriched in the microsomal fraction of rat brain and equally active towards 1-acyl-*sn*-glycero-3-phosphorylcholine and the ethanolamine analogue has been described[145]. It was optimally active at pH 7.2–8.6 and required no cations for activity.

A phospholipase C active towards sphingomyelin was found in the 20 000 x g particles of rat brain tissue by Barnholz et al.[146]. It had a pH optimum of 5.0, a K_m of $1.3 \cdot 10^{-4} M$ and was more active towards sphingomyelin with the *erythro*- rather than the *threo*-configuration . Unlike other C enzymes (Chapter 4) it required no $Ca^{2\oplus}$ ions for actitivity and was inactive towards lecithin; it presumably resembles the comparable enzyme of liver[147]. Very recently a largely microsomal phospholipase C, active towards phosphatidylethanolamine has been detected in rat brain tissue; it is optimally active at pH 6.5 and is not dependent on bivalent cations for actitivity[148].

Many years ago Webster et al.[149] demonstrated a glycerophosphorylcholine diesterase (*sn*-glycero-3-phosphorylcholine glycerophosphohydrolase ,EC 3.1.4.2) in brain tissue which was optimally active at pH 9.5 and activated maximally by $Mg^{2\oplus}$ ions. Its subcellular distribution in brain is unknown but by analogy with kidney[150] is probably microsomal. Compared with kidney and spleen its activity is low but, like liver, whole brain tissue from the rat can hydrolyse 2 nmole/mg protein/min.

A $Mg^{2\oplus}$-dependent plasmalogenase in rat brain which hydrolyses the alkenyl ether linkage of ethanolamine plasmalogen[151] is enriched in white matter and the crude mitochondrial fraction[152]. A non-enzymatic, oxygen-dependent system for degrading this linkage has been found by Yavin and Gatt[153, 154] in rat brain supernatant; a ferrous-ascorbate complex reduces and forms an 'active' oxygen which attacks the alkenyl linkage. Its significance is obscure at the present time.

Undoubtedly, the enzymes hydrolysing phosphoinositides, particularly triphosphoinositide, are very much more active (up to 20 times) than other phospholipases in the brain. A TPI phosphomonoesterase which removes successively the monoester phosphate moieties of TPI to yield first diphosphoinositide (phosphatidylinositol monophosphate; DPI) and then phosphatidylinositol was described by Thompson and Dawson[155]. It requires $Mg^{2\oplus}$ ions[156] and appears to be mainly in the cell sap[157]. Very high rates of hydrolysis of TPI of 300–500 μmole/g fresh tissue/h have been reported for the monoesterase[158]

Brain tissue is also rich in a triphosphoinositide phosphodiesterase described by Thompson and Dawson[159] and subsequently studied at some length by Thompson and his colleagues. Under optimum conditions high rates of hydrolysis of 400–500 μmole/g tissue/h were obtained[160]. Thus the activities of both the mono- and diesterase *in vitro* are comparable with enzymes such as acetylcholinesterase and $Na^{\oplus}/K^{\oplus}$ ATPase.

The phosphodiesterase present in both the soluble fraction and 'particles' is active towards all three phosphoinositides but with different pH optima[161], and is interesting in that it is the most active phospholipase C-like activity in animal tissues. The major product of phosphoinositide hydrolysis is 1-stearoyl-2-arachidonyl-*sn*-glycerol, a diglyceride constituting some 60% of the total glyceride pool of rat brain (0.1 – 0.15 μmole/g fresh weight)[162], and there may be a special relationship between this diglyceride and phosphoinositide metabolism. (The other obvious source of a diglyceride pool is phosphatidic acid phosphatase activity). It is believed that acetylcholine specifically stimulates the phosphodiesterase cleavage of phosphoinositides and the enzyme may therefore be concerned with the action of acetylcholine at the receptor site (Chapter 15).

4. PHOSPHOLIPID METABOLISM *in vivo*

The early experiments by Dawson and Richter[163] and Ansell and Dohmen[164], who followed the metabolism of the phospholipids in whole mammalian brain over a period of 0–7 h after the intraperitoneal injection of [^{32}P]orthophosphate, showed that they were turning over at varying rates. Inositol lipids and phosphatidic acid showed a relatively high rate of incorporation whilst phosphatidylserine showed considerably less. In these experiments the metabolic activity was measured as a ratio of the radioactivity in the phospholipid to that in the acid-soluble fraction of which inorganic phosphate, ATP and phosphocreatine make up the greater part. At that time, little was known about the biosynthetic pathways, but it seemed reasonable to assume that the energy-rich phosphates were supplying the phosphate moiety of the phospholipids and these came rapidly into isotopic equilibrium as the labelled phosphate entered the brain[165]. What the experiments seemed to show was that, although the entry of phosphate into the brain was slow and hence the proportion of radioactivity in all brain components tiny compared with those of, say, liver, there was nevertheless a relatively rapid transfer of label from the acid-soluble phosphate in the brain to phospholipids (for a more detailed discussion of this point see ref. 1). If it is accepted that phospholipids are synthesized in the brain *in situ* then the rate of synthesis was highly significant and a *notional* average replacement time for the phosphate moiety of 70 h was indicated by an extrapolation of the much shorter exchange times of Dawson and Richter[163].

Subsequent investigations by Davison and his colleagues threw some doubt on the idea of a short half-life for phospholipids and other lipids in the brain. They showed that, when a variety of labelled precursors were injected intraperitoneally into immature rats and rabbits, radioactivity persisted in the brain, particularly in the white matter, for many months[166–171]. The experiments suggested that, once laid down in early life, the lipids were not subject to significant turnover in adult life. The slow turnover of adult rat brain phospholipids was also demonstrated by injecting adult animals with [^{32}P]orthophosphate. (For a more detailed discussion see refs. 1 and 3.) When it became possible to isolate rela-

References p. 415

tively pure myelin, experiments with intraperitoneally injected [1-^{14}C]glycerol indicated that phospholipid turnover in this fraction was particularly low[172]; it was also difficult to label the phospholipids of myelin in adult guinea pigs with intraperitoneally injected [^{32}P]orthophosphate, though phosphatidic acid and phosphatidylinositol were labelled[46]. All these experiments appeared to support the concept of a metabolically stable pool of brain phospholipids as suggested by the earlier experiments of Thompson and Ballou, who fed or injected rats with tritium oxide and obtained an average half-life for brain phospholipids of 90 days[173, 174]. The idea that the 'stable' contribution was myelin was attractive because the compact multilamellar mass round the axon might be expected to be metabolically inert (Figs. 1 and 2).

It is pertinent at this point to mention briefly the blood–brain barrier. The experiments of Dawson and Richter[163] and Ansell and Dohmen[164] were designed to take into account the observation that penetration of labelled phosphate from the blood into the brain was apparently slow. However, Davison and Dobbing[169] concluded from their experiments with 16-day-old rats that the delay in transfer from blood to brain was actually less than that from brain acid-soluble phosphate to brain phospholipid phosphate. In a series of penetrating reviews, Dobbing[175–177] has criticised the whole concept of the blood–brain barrier and argues cogently that the restricted entry of any component of the blood into the brain is not so much determined by the *physical* properties of the capillaries or attendant glia as by the *metabolic* properties of the tissue. Included in these are the many known types of active transport mechanisms operating in both directions, together with the varying metabolic activities within the cell which play such an important part in determining the net movement and exchange of substances between blood and brain. Thus a slow flux of phosphate through the phospholipids of the adult brain could be considered as a reflection of the relative lack of synthesis *de novo* or turnover of phospholipids in inert structures. Young developing brain tissue might be expected to remove from the blood large quantities of labelled phosphate. These ideas were apparently supported by the experiments of August et al.[178] and Smith and Eng[179]. On the other hand the results of Smith and Eng[179] also showed that there was significant labelling of phosphatidylcholine after the intraperitoneal injection of [1-^{14}C]acetate into adult rats which was confirmed when [U-^{14}C]-glucose was used as a precursor[180]. The half-life of the lecithin of myelin was 41 days compared with 18 days for phosphatidylinositol and >8 months for sphingomyelin. It was of interest that measurement of turnover by loss of radioactivity from myelin phospholipids was similar whether actively myelinating or adult rats were used. Extensive labelling of subcellular fractions of adult rat brain after the intracerebral injection of [1,2-^{14}C]ethanolamine, was demonstrated by Ansell and Spanner[89], the phosphatidylethanolamine and the ethanolamine plasmalogen of the 'small myelin' being particularly well labelled. Subsequent studies[181] showed a half-life of 21 days for the ethanolamine phospholipids of myelin as a whole. An even shorter half-life for the ethanolamine phospholipids of mouse brain myelin was found by Horrocks and co-workers using [^{14}C]ethanolamine[182] and [U-^{14}C] palmitic acid[183]. Lapetina et al.[184] demonstrated a rapid equilibration between the phospholipids of the membranous structures after the subarachnoid injection of [^{32}P]orthophosphate. In an extensive study with a number of labelled precursors Jungalwala and

Dawson[185] also demonstrated considerable labelling of myelin phospholipids and estimated half-lives of less than 20 days for phosphatidylethanolamine, ethanolamine plasmalogen and phosphatidylcholine. Although labelling in the microsomal fraction was initially very much higher than in the myelin, after 20 days the specific radioactivities of phospholipids of both fractions were falling at comparable rates. There was little evidence of persistence in the myelin and such persistence of labelled glycerol and ethanolamine that did occur could be due to the small continuous increase in the amount of myelin which occurs even in the adult rat[186]. Persistence of [^{32}P]orthophosphate could also be explained by recycling of the radioactive acid-soluble pool remaining in the brain. This was particularly noticeable in the myelinating rats when, presumably, the injected radioactive phosphate was in equilibrium with a very much smaller body pool.

From these experiments one can draw only the general conclusion that the inert pool of myelin phospholipid is probably much less than was originally thought. Furthermore the results of the various workers reflect not only the difficulty of interpreting turnover in the brain *in vivo* but also that of comparing intracerebral 'pulse labelling' with a continuous supply of label from an intraperitoneal injection.

The origin and transfer of brain phospholipids

It has been clear for some time that the microsomal fraction of brain contains the enzymatic equipment for, at the very least, the final stage of the synthesis of the major brain phospholipids. It has also been shown that, after the intraperitoneal or intracerebral injection of labelled precursors, the phospholipids of all the subcellular components become labelled[89, 184, 187] though sphingomyelin is poorly labelled by ^{32}P. These experiments also demonstrated that after short times the fractions became labelled in the sequence microsomes > mitochondria > myelin and it seemed likely that the phospholipids could be synthesized in the endoplasmic reticulum and in some way transferred to other components[182]. As discussed in detail in Chapter 9 this is known to occur in liver organelles and, in a very careful study, it has been recently demonstrated[94] not only that the capacity of brain mitochondria is limited to the synthesis of phosphatidic acid and diphosphatidylglycerol but that, *in vitro*, labelled phospholipid in microsomes could be transferred to mitochondria and *vice versa*[187a]. Microsomes could also donate to synaptosomes but interestingly not to myelin. The synaptosomal outer membrane and intraterminal mitochondria contained no cholinephosphotransferase activity whereas the synaptic vesicles and probably the intraterminal endoplasmic reticulum could transfer phosphorylcholine from CDPcholine to phosphatidylcholine. The inability to demonstrate a transfer from microsomes to myelin *in vitro*, whereas it would appear to occur *in vivo*[185], might be due to technical difficulties[187a].

If one accepts that the major site of phospholipid synthesis is the endoplasmic reticulum of mixed neurones and glia, though it has only been shown for phosphatidylcholine, then the phospholipid exchange has to be extremely rapid to explain the labelling of other brain organelles *in vivo*. To what extent the transport of phospholipids down the axon[188] is important in, for example, formation of phospholipids at the nerve

References p. 415

terminals, is unknown. Neither is there any evidence that phospholipids can be transferred from glial cells to neurones or *vice versa*. After an intraperitoneal injection of [^{32}P]orthophosphate into rats most neuronal phospholipids were found to be labelled faster than their counterparts in glia though this might be due to faster penetration of precursor[66]. Half-lives, measured by the rate of fall in specific radioactivity, were similar for both neurones and glia, e.g. 11 and 14 days respectively for phosphatidylcholine.

Although brain tissue can synthesize phospholipids *de novo* an extracerebral source of intact phospholipids cannot be discounted. It is known that labelled cholesterol can be carried intact via the bloodstream and deposited in membranes of rabbit brain[168]. A number of 1-^{14}C-labelled fatty acids, palmitic, oleic, linoleic and linolenic can also be taken directly from the bloodstream after feeding or injection and incorporated into complex lipids without prior degradation, though a conversion of linolenic to docosahexaenoic acid and linoleic to arachidonic acid occurs[189–192]. In a brief report Hoelzl and Franck[193] gave evidence for the incorporation of doubly labelled lecithin into brain after an intraperitoneal injection but Mead (personal communication) failed to demonstrate the transfer of dipalmitoyl lecithin.

Indirect evidence for the transfer of choline in a lipid-bound form to the brain has been obtained[90] and this could well be as lysophosphatidylcholine since Stein and Stein[194] showed that this lipid could be carried intact to various organs by the blood (they did not study brain). Very recently Illingworth et al.[195] have shown that labelled lysolecithin does apparently enter the brain from the blood unchanged and thereafter donates choline to a number of water-soluble choline compounds. In this connection it is of interest that 5–10% of plasma phospholipid is lysolecithin. The fact that injected free labelled choline can enter the brain from the blood[196] by a process which exhibits saturation kinetics[197] is not necessarily evidence that choline normally enters the brain in this form and may merely reflect the finding that the base can pass out of the brain[198]. All that is fairly certain is that the mammalian brain cannot synthesize choline and must receive it from the blood.

Phospholipid metabolism in cell culture

The use of cell culture offers the prospect of studying lipid metabolism in an isolated system for much longer periods than is possible with a slice or homogenate. Menkes[199] has found that brain explants remain metabolically active for several days in a chemically defined serum-free medium, and that they incorporate radioactive lipid precursors into a number of lipid fractions. Nevertheless the outgrowths were from multipolar glial cells rather than neurones and the presence of serum was necessary for maintenance longer than 3 days. Outgrowth was better also with explants from 3–6-day-old rats rather than old animals. Good incorporation of 1-^{14}C-labelled stearic, palmitic and lignoceric acids into ethanolamine and choline phosphoglycerides was obtained and these fatty acids also underwent desaturation. Such studies have been extended to biopsy samples from patients with demyelinating diseases and offer some scope for future studies (see also ref. 200).

5. PHOSPHOLIPIDS AND THE DEVELOPING NERVOUS SYSTEM

The growth of the nervous system and the brain in particular is a topic which, though once the province of anatomists and embryologists, has increasingly interested the biochemist and the amount of biochemical information published since the appearance of the proceedings of a symposium held in 1954 (ref. 201) is enormous. To obtain a fairly comprehensive picture the reader is referred to refs. 45, 202 and 203. The development of the brain is a sequential process anatomically and there is evidence that it is also sequential from a biochemical point of view. Furthermore, the rate of development varies considerably from species to species and the chronological extrapolation of events from one species to another is extremely difficult. Parturition itself has no significance for the development of the brain and the stages can be divided up as follows[202].

Period	*Process*	*Completion*
I	Neuronal multiplication	Guinea pig – 45th day of gestation Rat – 3rd post-natal day Man – 25th week of gestation
IIa	A 'growth spurt' i.e. a maturation period of axonal and dendritic growth, glial multiplication and myelination. Development of certain enzyme systems	Guinea pig – at birth Pig – 5 weeks after birth Rat – 3 weeks after birth Man – 2 years after birth
IIb	Later but overlapping period of growth	–
III	Mature state	–
IV	Senile regression	–

The events in IIa are obviously extremely complex and the deposition of myelin, if the most obvious feature, is not the only one since, for example, the proliferation of glial cells just before myelination is equally dramatic. Nevertheless the biochemistry of myelination has been extensively studied with reference to the lipid content of the end product, particularly in the rat, which is a convenient post-natal developer.

Changes in the levels and composition of phospholipids in the developing brain

The levels of phospholipids in the developing brain have been studied many times and details were given by Ansell and Hawthorne[1]. The magnitude of the changes in the lipids is shown in Fig. 4. The dramatic increases in certain phospholipids in the developing rat brain are shown in Fig. 5, which is taken from one of the most complete studies made on this species[204]. (Values for other species, including man, have been reviewed by Rouser and Yamamoto[205].) Phosphatidylethanolamine and phosphatidylcholine are more general components of cell membranes and thus their percentage increase is much less than for ethanolamine plasmalogen, triphosphoinositide and sphingomyelin which are characteristic

References p. 415

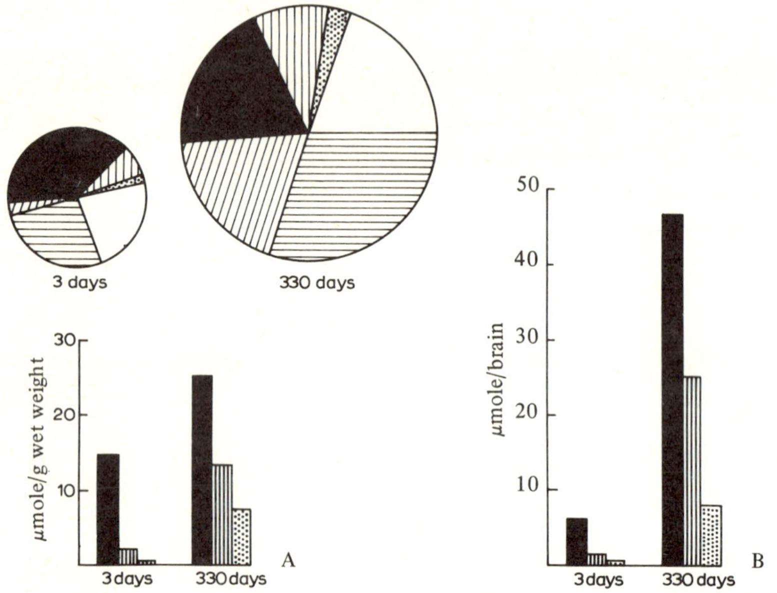

Fig. 4. A diagram to demonstrate the difference in levels of lipids in the brain of the male rat at 3 days and 330 days of age. The circles represent the difference in weight at the two ages and the molecular proportions of the lipids. Histograms A and B show, respectively, concentrations and absolute amounts of three phospholipids. ■, phosphatidylcholine; ▥, ethanolamine plasmalogen; ▨sphingomyelin; ▨, cerebroside; ▤, cholesterol; □, other lipids, largely phospholipids.

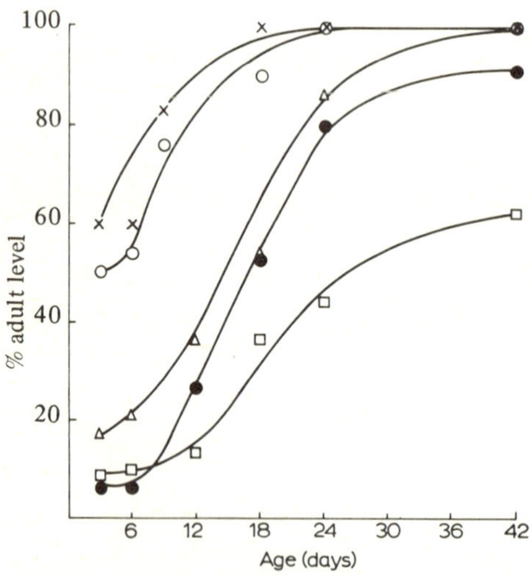

Fig. 5. The values of some phospholipids (µmole/g fresh weight) in the developing rat brain expressed as a percentage of the adult level. The values are those of Wells and Dittmer[204] and the figure is reproduced from ref. 210 by permission of Plenum Press.
x, phosphatidylcholine; ○, phosphatidylethanolamine; △, ethanolamine plasmalogen; □, phosphatidylinositol diphosphate; ●, sphingomyelin.

of isolated myelin. It might be concluded that no further increase in concentration of say ethanolamine plasmalogen occurs after 3–4 weeks but Fig. 6 shows that, if the concentration

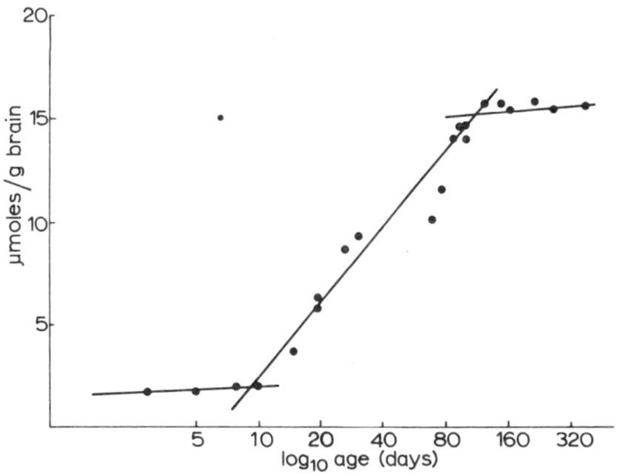

Fig. 6. The change in the concentration of ethanolamine plasmalogen in the brain of the rat with age. (Reproduced from ref. 210 by permission of Plenum Press.)

is plotted against $\log_{10}$ of the age as suggested by Rouser and Yamamoto[206] for human brain an increase in concentration occurs up to 100 days. Not only does the rat brain increase in size well into maturity but the actual concentration of myelin phospholipid rises, probably reflecting the finding that in this species myelination extends well into adult life[186]. This graph shows a sharp 'switching on' of ethanolamine plasmalogen deposition and an almost equally sharp 'switching off'.

In addition to changes in phospholipid levels, an alteration in the fatty acid composition occurs. Thus, in the developing rat brain the proportion of $C_{18:0}$ and $C_{18:1}$ fatty acids in phosphatidylcholine increases at the expense of $C_{16:0}$ and $C_{16:1}$ (ref. 207) and this is probably true for most glycerophospholipids[208]. The increase in $C_{18:0}$ and $C_{18:1}$ in phosphatidylcholine is due to the high level of these two fatty acids in this phospholipid in myelin which makes an increase in contribution to the total phosphatidylcholine of brain with increasing age[85]. The percentage of $C_{18-22\,(n-3)}$ acids in the ethanolamine lipids of human grey matter rises from 20% at birth to 35% in the adult; in white matter the level of these acids falls from about 17% to 8%[209]. Before myelination, $C_{18:0}$ predominates in the sphingomyelin of both grey and white matter and after myelination C_{22-26} and especially C_{24} fatty acids account for 70% of the total[138]. Complex changes also occur in the fatty acid composition of galactolipids. The general tendency during development is for the fatty acids to increase in chain length and degree of unsaturation, but the phospholipids of adult white matter contain a higher proportion of monoenoic acids and a lower proportion of polyenoic acids than those of grey matter[209]. The precise reason for this is unknown, though it could be related to the relatively compact structure of mature myelin[211] (see Fig. 1).

Myelin composition during development

The first report that the lipid composition of myelin changed during development[212] was treated with scepticism but has since been amply confirmed for several species, more

References p. 415

particularly for rats and mice[35, 213–215]. The essential point is that from about the age of 15 days the relative proportions of galactolipid, cholesterol and ethanolamine plasmalogen increase as the relative proportion of choline phosphoglycerides decreases (Fig. 7). The increases in the relative proportions of galactolipid and cholesterol to phospholipid are

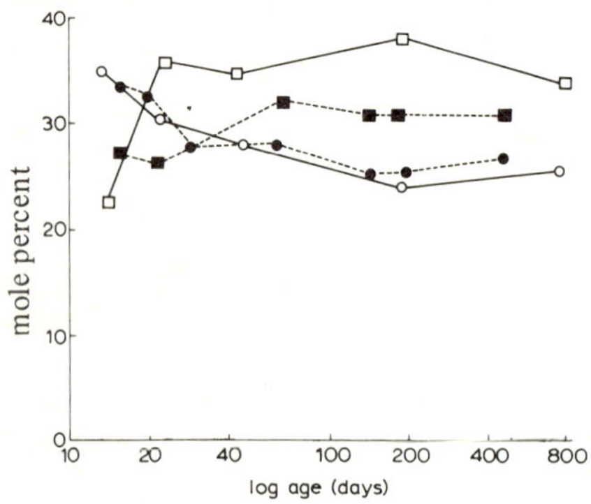

Fig. 7. The proportions of ethanolamine plasmalogen and choline phosphoglycerides in myelin phospholipids as a function of age. The circles represent choline phosphoglycerides and the squares ethanolamine plasmalogen. Open symbols are values for the mouse[212] and closed symbols are values for the rat[35]. (Reproduced from ref. 216 by permission of the author.)

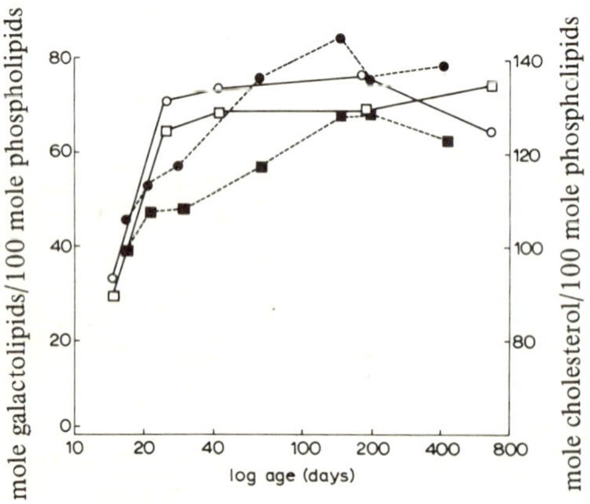

Fig. 8. The amounts of galactolipid and cholesterol relative to phospholipids in myelin as a function of age. The circles represent galactolipid and the squares cholesterol. The open symbols are values for the mouse[213] and the closed symbols are values for the rat[35]. (Reproduced from ref. 216 by permission of the author.)

nearly equal on a molar basis[216] (Fig. 8) and these increase even when myelination is apparently complete in the rat (cf. ref. 186).

Myelin prepared from the developing brains of rodents ('early myelin') and containing less cerebroside and more phospholipid than mature myelin, resembles cell membranes rather than mature myelin. It was proposed by Davison and his colleagues[217, 218], that this 'early myelin' was a mixture of mature myelin and 'pro-myelin' fractions deriving from the oligodendroglial plasma membrane. This 'pro-myelin' or 'myelin-like' component was then prepared[219] and shown to be poor in cerebrosides and to contain a higher proportion of shorter chain fatty acids in phosphoglycerides than mature myelin. Further studies[220] showed that the material resembled myelin in enzyme activity but differed in that the basic protein was absent. 'Early myelin' therefore could consist of the 'myelin-like' and compact myelin fraction from developing brain, the 'myelin-like' fraction being a transition state between the oligodendroglial plasma membrane and compact myelin. The 'myelin-like' fraction may also contain some fragments of oligodendroglial plasma membranes themselves[43].

Phospholipid metabolism in the developing brain

The changes in the levels and composition of phospholipids and other lipids in the developing mammalian brain must be the result of complex control mechanisms of which nothing is known. It is possible that control is partially mediated via precursor and enzyme levels but there have been very few studies of the levels of precursors in the brain during development. It is of some interest that the concentration of phosphorylethanolamine in the brain of the new-born rat is five times that of the 21-day-old animal[221]. The only other report on a change in precursor level is that of Wells and Dittmer[204] who found that the level of phosphatidylglycerophosphate, a precursor of phosphatidylglycerol (significant amounts of which occur in brain), rises during the pre-myelination period in the rat, stays constant during myelination and then falls rapidly on maturation; it is the only phospholipid to do so. This "is consistent with it having an intermediary role in a pathway with a rate-limiting reaction subsequent to its formation".[204]

Some investigations have been made of changes in levels of enzymes associated with phospholipid metabolism in development. Laatsch[222] found that the activity of glycerophosphate dehydrogenase (*sn*-glycerol 3-phosphate: NAD oxidoreductase, EC 1.1.1.8) in rat fore-brain rises rapidly from birth to reach a peak at 25 days of age, whereafter it remains constant. Since brain tissue contains no glycerol kinase[223] (ATP:glycerolphosphotransferase, EC 2.7.1.30) and since, at physiological pH, the dehydrogenase reaction is in favour of glycerophosphate production, the increase in the dehydrogenase could be interpreted as a response to an increased glycerophosphate requirement. Enzymes involved in phosphoinositide synthesis have also been studied. The optimal activity of phosphatidylinositol kinase precedes myelination whereas there is a definite parallel in the increase in diphosphoinositide kinase, the deposition of triphosphoinositide, which is associated with myelin, and the deposition of myelin itself[158]. The monoesterase which hydrolyses triphosphoinositide also increases with the rising concentration of its substrate and both enzymes are found, not in the myelin itself, but in the cell sap[224, 157].

Triphosphoinositide monoesterase is not the only catabolic enzyme which increases in amount as myelin lipids are deposited, since the plasmalogenase[225] and a cerebrosidase (cerebroside galactosidase)[226] also show the most rapid rate of increase during myelination. The significance of these observations remains to be established.

References p. 415

Dysmyelination

Although disorders of myelin structure are mostly found in demyelinating diseases there are certain faults in myelin which occur during the myelination period and there are certain human inborn errors of metabolism in which myelin has different properties from normal or may be completely lacking. In 1964 two myelin deficient mutants of the laboratory mouse were described, the Jimpy (Jp) and the Quaking (Qk)[227] and later another one, 'msd'[228] all of which have a common defect of myelination in the CNS but not the peripheral nervous system. These are murine leucodystrophies (the Jimpy mouse also exhibits sudanophilia in the white matter of the central nervous system). Both the Jimpy and the Quaking mouse have the same phenotype, characterised by abnormal gait, tremor and seizures which appear at the onset of myelination (12 days old), though they have different genotypes. The Quaking mutation is an autosomal recessive and the Jimpy mutation is an X-linked recessive. Since the Quaking mouse lives for about a year it has been studied more than the Jimpy strain which does not reach the adult state.

Analysis of the whole brain of the Quaking mouse showed a lack of 'characteristic' myelin lipids, cerebrosides, sulphatides and cholesterol[229]. A decrease in phospholipids was also found[229, 230] and Dawson and Clarke[231] demonstrated a significant lack of ethanolamine plasmalogen, TPI and phosphatidic acid, while proportions of other phospholipids increased (Table 3). More recently myelin isolated from the mutant brain has been compared

TABLE 3

THE PHOSPHOLIPID COMPOSITION OF WHOLE BRAIN AND MYELIN FROM NORMAL AND QUAKING MICE.

	Whole brain (adults)[a] *µmole/g brain*		*Myelin (42–60 days old)*[b] *µmole/g brain*	
	Normal	*Quaking*	*Normal*	*Quaking*
Lipid P	67.8	46.5	18.9	6.1
Phosphatidylcholine	22.0	17.8	7.0	3.4
Phosphatidylethanolamine	9.9	8.7	3.6	1.0
Ethanolamine plasmalogen	10.8	5.3	4.9	0.9
Phosphatidylserine	7.9	5.2	2.2	0.6
Phosphatidylinositol	1.5	1.5		
Triphosphoinositide	1.5	0.7		
Sphingomyelin	3.1	2.0	1.2	0.2

[a]Dawson and Clarke[231].
[b]Singh et al.[232].

with that of normal litter-mate controls[232–234] and found to be deficient in cerebroside, ethanolamine plasmalogen and sphingomyelin (Table 3). In addition the myelin from the deficient animals had a lower proportion of monounsaturated alkenyl and acyl groups in the plasmalogen and 24:1 hydroxyl and unsubstituted acyl groups in the cerebrosides. In many ways Quaking myelin resembles 'early myelin'[232]. It is likely that the biosynthesis of long-chain fatty acids, particularly non-hydroxylated ones, is impaired in some way[235] but the problem is more likely to be one of control than a specific enzymatic defect[236]. Furthermore, recent studies[234] indicate that there are compositional changes in the brain other than in the myelin.

The dysmyelination in the Jimpy mutant is more severe than in the Quaking mutant, with a greater reduction in myelin deposition. With the possible exception of ethanolamine plasmalogen, however, there are no significant changes in the phospholipids of whole brain[237, 238]. Nevertheless, it has been observed that the sphingomyelin in myelin from 16–24-day-old Jimpy mice was deficient in $C_{22:0}$, $C_{24:0}$ and particularly $C_{24:1}$ with a marked elevation of $C_{18:0}$ (ref. 239). The major deficiency was in cerebroside, emphasising the importance of this galactolipid in the formation of compact myelin in the brain. Such cerebrosides that were present were also deficient in long-chain fatty acids (greater than C_{19}).

Other inborn errors characterised by dysmyelination

There are a number of diseases, fortunately rare, affecting man and other species, in which myelin fails to develop normally, but this process is complicated by abnormal deposits of certain lipids (lipidoses). Other organs are often affected and the subject has been discussed at length[4, 240, 241, 242]. In very general terms it may be stated that, as far as the lipids are concerned, the most profound effects are seen in levels of sphingolipids, particularly galactolipids.

In *metachromatic leucodystrophy* (late infantile type) there is an abnormality of myelin which, though of apparently normal ultrastructure, exists in smaller amounts than usual. Metachromatic material accumulating in white matter and in the viscera was shown to be sulphatide[243–245]. Total phospholipid in white matter is reduced by 50% (ref. 246), and there is also loss of sphingomyelin. Analysis of isolated myelin showed a marked deficiency of phospholipids and increase in sulphatides[247, 248]; thus the sphingomyelin was found to be 50% of normal while the molar ratio of sulphatide to cerebroside was 20 : 1 compared to 1 : 4 in control myelin. The known enzyme deficient in this disease is a sulphatase which can remove sulphate from the sulphatide[249, 250]. Curiously enough, the fatty acid composition of cerebrosides and sulphatides in this condition was found to be almost normal whereas the sphingomyelin of the white matter is poor in C_{22-26} acids and rich in $C_{18:0}$, resembling sphingomyelin prior to myelination[138] (cf. Quaking and Jimpy mice).

In the autosomal recessive disease *globoid cell leucodystrophy* (Krabbe type) there is an enhanced cerebroside:sulphatide ratio in white matter[251]. Large globoid cells are present which are engorged with cerebrosides apparently in a crystalline state[252–255]. Sphingomyelin is reduced in amount and in white matter is much richer in C_{18} and poorer in C_{22-26} than that from normal tissue. A reduction in cerebroside sulphotransferase activity is found in tissues, which might partly account for the high cerebroside:sulphatide ratio[256], but the deficiency in galactocerebroside β-galactosidase in brain may be the more important primary one[257, 258]. Nevertheless these enzymes cannot be responsible for the abnormal fatty acid composition of the sphingomyelin.

References p. 415

Niemann-Pick disease is sometimes considered the prototype of the lipidoses, occurring usually in the first year of life and characterised by the large lipid-storing foam cells throughout the body, especially in the reticuloendothelial system. It is usually accompanied by neuronal destruction but cases have occurred without involving the nervous system[259]. There are probably more papers on this disease than individuals who have suffered from it, but it is of great historical interest since, as long ago as the 1930s, Klenk demonstrated that the lipid deposited in the viscera and the brain was sphingomyelin[260–262]. The amount in brain ranges from a little above normal to as much as 10 times normal. The fatty acid composition resembles that of cortical, not of white matter, the sphingomyelin being rich in C_{18} and poor in $C_{>20}$ acids (refs. 138, 263, 264). Other lipids are diminished and the whole picture resembles that of an immature brain. The accumulation of sphingomyelin might be accounted for by the severe deficiency of sphingomyelinase[265] though, as with globoid cell leucodystrophy, this cannot account for the abnormal fatty acid pattern.

Whether *Pelizaeus Merzbacher disease* is one of dysmyelination is a matter of opinion[240]. There is certainly a loss of white matter lipids[252, 266] and unsubstituted long-chain monounsaturated fatty acids are reduced.

One of the most interesting diseases in which myelin formation is disordered and which has been most studied from the biochemical viewpoint is *Refsum's disease* (hereditary ataxic polyneuropathy)[267]. This inborn error is characterised by severe peripheral neuropathy, cerebellar ataxia, and deafness. Peripheral nerves exhibit progressive degeneration of myelin often leading to complete demyelination of the sheaths. Klenk and Kahlke[268] demonstrated an accumulation of phytanic acid (3,7,11,15-tetramethyl hexadecanoic acid) in serum and tissue lipids. The acid is deposited in kidney, liver and heart but originally it was thought that the brain was relatively unaffected[269]. Phytanic acid is a normal constituent of the diet and the primary defect in Refsum subjects is probably the inability to alpha-oxidise and decarboxylate[270]. However, the relationship between the accumulation of phytanic acid and the demyelination of the nervous system is not clear.

In an extensive study on a single subject who had survived to the age of 57, MacBrinn and O'Brien[271] found deficiencies of ethanolamine lipids in grey matter and myelin and the accumulation of an ethanolamine lysophospholipid. Phytanic acid was present in the phosphoglycerides from grey and white matter of brain, spinal cord myelin and sciatic nerve, particularly in the 1-position of choline phosphoglyceride. Twenty percent of the fatty acid of the sciatic nerve was phytanate which is consistent with the fact that peripheral neuropathy is much more obvious than other symptoms. In many of the membranes of isolated central myelin abnormal lamellae were observed.

In one sense Refsum's disease is not dysmyelination. Presumably if phytanic acid was absent from the diet from birth the symptoms would not occur and myelin would develop normally; in any case symptoms often do not appear until adulthood. Perhaps there is a critical concentration of phytanic acid beyond which myelin cannot exist normally and 'falls apart'. The varying time of onset of the disease could be explained by the variable dietary intake and improvement occurs in subjects deprived of phytanic acid.

O'Brien proposed[272] that, if phytanic acid accumulated in myelin, it would tend to disrupt the packing of the hydrophobic tails in the bimolecular lipid leaflet because of the steric hindrance of the branched methyl groups. Surface films of phytanic acid are indeed 80% more highly expanded than those of palmitic acid. Even a small expansion of the myelin bilayer should result in a large diminution of interaction between adjacent lipid molecules[273]. It may be significant that myelin is rich in long chain C_{19-26} fatty acids, particularly in sphingolipids, which might normally enhance this packing and that any change in the amounts of these fatty acids may cause the myelin to become unstable. Lower levels of these fatty acids in the conditions mentioned earlier may also be partly responsible for dysmyelination. Since, however, a deficiency in long-chain fatty acids occurs in a variety of aetiologically unrelated disorders it is likely to be a secondary effect, not a primary one. Moreover, when *white matter* is analysed, too much emphasis must not be placed on lower levels of long-chain fatty acids since this may merely reflect the changed ratio between axoplasmic tissue and myelin in the abnormal white matter[138].

Thus dysmyelination, or the inability for normal myelin to develop, is found in a variety of inborn errors but, in the present state of knowledge it is impossible to classify them adequately. In some the lack of myelin is accompanied by increased deposits of certain lipids in the CNS and other organs and changes other than those in myelin occur. Where abnormal deposits of lipid occur this can be explained to some extent by the lack of certain catabolic enzymes. Furthermore, there are adult versions of some of the diseases where myelination may have been originally perfectly normal. Two other important lipidoses have not, it might be added, been mentioned, namely *Tay-Sachs* and *Gaucher's disease*, because phospholipid involvement in these conditions is probably unimportant.

There are two other inborn errors in which there is some defect of myelination but which are different from the leucodystrophies mentioned above in that the effects can be alleviated to some extent. These two conditions are phenylketonuria and leucinosis (maple-syrup urine disease).

In phenylketonuria, patients are normal at birth but deteriorate rapidly during infancy and, within 2 years of birth, the I.Q. is reduced to 30 or less. Thereafter, deterioration is slow but may continue into adult life. The damage to white matter can range from the negligible to the extensive[274–276]. Views vary as to whether it is a dysmyelination or a demyelination. In the adult patients there is considerable evidence of the latter because of sudanophilia[277]. Directly or indirectly, the lack of myelin, which is the only obvious morphological defect in this disease, must be an effect of the high concentration of phenylalanine since if this amino acid is not present in the diet mental retardation is absent. However, although lipid is reduced, the deficiency is largely in cerebroside and phospholipids are relatively unaffected[278]. The effects of the phenylalanine are not likely to be on lipid synthesis[279], but on the protein of the myelin proteolipid[280], though contrary views were expressed earlier[281].

The effect of some exogenous and endogenous factors on phospholipids in development

Protein-calorie deficiency

It has been known for a long time from experiments on animals and observations on man that in starvation the adult mammalian brain does not decrease significantly in size or composition and is therefore spared in comparison with say, skeletal muscle. The observations of Smith[282], that fasting 8-week-old rats for 3 days reduced the phospholipid levels in spinal cord, and of Joel et al.[283] that fasting hens for 15–16 days altered the levels of certain unsaturated fatty acids, do not affect this general conclusion. It is, however, reasonably certain that infantile undernutrition can be correlated with small adult stature, a reduced brain size and impaired intellectual capacity. These broader implications are outside the scope of this chapter and the reader is referred to two articles by Dobbing[284, 285]. Such are their socio-medical importance that numerous experiments have been carried out over the past decade on the effect of undernutrition on brain development, though careful experimental design is essential[286].

The growth spurt mentioned earlier (p. 397), which occurs at different times in different species, is a transient period during which an increase in brain weight and a number of other important developmental processes proceed at a rapid pace. It starts when the adult complement of neurones is present and is characterised by glial cell multiplication, myelination and the development of certain enzyme systems. Concurrently with these easily measured parameters, there is a substantial increase in dendritic complexity and it is probably during the growth spurt that most synaptic connections are established. Retardation of growth during this period results in quantitative deficits in the adult brain and the organ is therefore said to be vulnerable at this time. The concept of vulnerability has been thoroughly discussed by Dobbing[284–286] and this particular period should be distinguished from earlier periods in gestation when the brain is sensitive to drugs (e.g. thalidomide) and viral agents (e.g. Rubella) the morphological effects of which are rather more gross.

The observations of a number of workers[287–292] demonstrated that nutritional deprivation of the rat during suckling permanently reduced brain size and the levels of some brain components. Under-nutrition of hitherto 'privileged' rats subsequent to weaning had no permanent effect on brain size. The deprivation need be only slight e.g. by increasing the size of the suckling litter, and it has been found that levels of, say cholesterol, are reduced in excess of that accounted for by a decrease in brain size alone, so that the concentration is reduced[293]. In a very careful study of the somatosensory cortex of the rat[294] Bass et al. showed that a 21-day period of malnutrition from birth resulted in considerable histological abnormalities. During the first 10 post-natal days dendritic proliferation decreased, cortical growth was impaired and subsequent organisation affected so that glial cell migration and the formation of myelin was reduced.

Myelination is particularly affected and during rehabilitation following a 21-day period of malnutrition from birth a striking reduction in myelin components was observed both histologically and chemically[295]. This could be due to damage to glial cell precursors during the period of deprivation. When rats were underfed between 5 and 50 days of age a significant reduction in brain weight and total content of cholesterol and cerebrosides was found but the reduction in phospholipids was slight[289]. Plasmalogens, typical myelin phospholipids, were affected more than other phospholipids[296]. Not all investigators, however, have shown a permanent loss of myelin or myelin lipids[292, 297] and Guthrie and Brown[297] reported no change in phospholipids even with severe deprivation, though cholesterol was permanently reduced. In a study of the brains of malnourished infants, Fishman et al.[298] found a reduction in total lipids, particularly myelin lipids, resembling that produced in experimental animals. This could be attributed to a failure to produce oligodendrocytes.

Essential fatty acid deficiency

While protein-calorie deficiency produces interesting effects, a lack of essential fatty acids (EFA) during development is of greater interest to the lipid biochemist. Feeding weanling rats on an EFA-deficient diet produced only a slight reduction in the EFA levels in brain lipids and an increase of $C_{20:3\ (n-9)}$ fatty acids[299, 300]. More profound effects have been reported by Paoletti and co-workers who fed pregnant rats on a fat-free diet from 5 days before delivery so that the offspring were suckled on milk deficient in EFA[301]. Subsequently the young rats were fed on a diet deficient in EFA for one year. In addition to a reduction in body and brain weight there was a 10% decrease in the brain phospholipids in male rats; the reduction was less in females and no particular phospholipid was reduced so that the effect was unspecific. Although the total level of unsaturation in the fatty acid moieties remained constant the C_{20}, C_{22} and docosapentaenoic acids increased and arachidonic, docosatetraenoic and docosahexaenoic acids decreased in ethanolamine, choline and serine phosphoglycerides. In general terms this effect could be reversed by feeding the animals a control diet after 30 or 90 days on the deficient diet. Further studies[302] showed a particularly pronounced effect on the ethanolamine phosphoglycerides, namely a rise in n–9 trienes ($C_{20:3}$ and $C_{22:3}$) and a fall in n–6 tetraenes ($C_{20:4}$ and $C_{22:4}$) giving a rise in the triene/tetraene ratio, generally considered a biochemical index of EFA deficiency. The ratio in myelin ethanolamine phosphoglycerides was much higher than in microsomes and mitochondria[303] indicating a greater modification of its structure and reinforcing views on its relative lability (pp. 394–395). (See also ref. 304).

Svennerholm and his colleagues[305, 306] have argued that to study the effect of EFA deficiency alone, it must have commenced at least two generations previously and must be the only deficiency i.e. it should not be produced by a fat-free diet. Although, as a result of such a regimen, many offspring died, the brains of those that survived for 21 days were notable only for a lowered cerebroside level. Brain weights were not different from controls, and there were only minor increases in $C_{20:3\ (n-9)}$ and $C_{22:5\ (n-6)}$ fatty acids in the lipids, including phospholipids, of the deficient group. Ganglioside levels were normal indicating normal growth of neurones and astroglia. Thus the results published by Paoletti's group may have reflected calorie deficiency rather than an EFA deficiency.

Thyroid deficiency

Neonatal thyroidectomy leads to structural changes in the brain and impairment of behaviour[307]. The slower growth of the brain after 14 days of age is connected with a decrease in the average size of cells rather than a reduction in cell numbers[308, 309]. The degree of dendritic arborisation is reduced[310] and there may be fewer synaptic terminals. Balázs et al.[311] reported a loss of myelin which could be related to the decrease in available axonal surfaces. Phospholipid levels were less affected than cerebrosides; the deficiency in

References p. 415

the latter was permanent but phospholipid levels recovered. Thyroid deficiency, induced in pregnant rabbits with methylthiouracil, did not affect the deposition of brain phospholipid in foetal life but there was a post-natal decrease[312]. Because the effect on brain weight was greater, however, there appeared to be an increase in phospholipid *concentration*. Thyroxin treatment caused a slight increase in brain phospholipid levels *in vivo* and stimulated synthesis *in vitro*[313].

X-Radiation

The developing brain is sensitive to X-rays, a sensitivity which decreases with age. For the first few days after birth the rat brain is particularly radiosensitive in that cellular growth and myelination are severely inhibited[314]. After 750 R of X-radiation at 2 days of age the total lipid of the brain stem started to fall at 12 days of age compared with controls and did not subsequently recover[315]. (See also ref. 316.) Significantly, the specific activity of glycerol phosphate dehydrogenase (EC 1.1.1.8), an enzyme increasing in amount during myelination[222] and capable of producing glycerol 3-phosphate for phospholipid synthesis, was reduced by the irradiation.

6. DEMYELINATION

Primary demyelination

Primary demyelination, or loss of the myelin sheath, and at least an initial sparing of the axon and neurone, is a relatively rare occurrence. It does occur in certain human and experimental diseases which are distinguished from the dysmyelinating diseases in that, as far as is known, they are not genetically determined and the degeneration of the myelin occurs after being normally constituted. Of these disorders, multiple sclerosis is the most common. "It ranks high among organic nervous diseases affecting the white races. In the northerly parts of Great Britain at least one adult in every 1300 is affected"[317]. For the literature on rarer disorders the reader is referred to refs. 318 and 319. Multiple sclerosis is characterised by demyelination of the central tracts; grey translucent scars (plaques) appear which contain glial elements, axons and a few myelin fragments. It may, like experimental allergic encephalomyelitis (EAE) be an auto-immune disease or, according to more recent thinking, may be caused by a slow-acting virus. The suggestion that it is caused by a deficiency of EFA has not gained much support. (For references on aetiology see ref. 240.)

The loss of lipid from the plaques is to be expected since they are myelin-free[1]; among the phospholipids there is a disproportionate loss of ethanolamine plasmalogen. In 1953 Cumings[320] reported a loss of lipid in apparently normal white matter from multiple sclerotic brains, suggesting a very early biochemical lesion. This was confirmed by a number of workers[240] but in a recent report[321] it has been shown that the levels of most phospholipids in the apparently normal white matter were, with the exception of phosphatidylserine, very similar to controls. This careful study produced no evidence for a primary lipid defect in myelin. Such abnormalities as have been reported might well be explained by the presence of small plaques.

There have been no reports of enhanced levels of phospholipases or other lipid-catabolising enzymes in plaques apart from an enhanced level of a plasmalogenase[152]. Increased levels of lysolecithin have, however, been found[322]. It is interesting that oligodendroglia enlarge and multiply in multiple sclerosis, as if in preparation for a re-myelination which never in fact takes place (cf. Wallerian degeneration).

Experimental allergic encephalomyelitis is another condition which is probably a primary demyelination and has been extensively studied because, as its name suggests, it can be induced in experimental animals[240, 318, 319]. It has indeed been considered a possible model for other demyelinating diseases. It is induced by a specific protein (p. 383) and demyelination usually occurs 10 days after sensitisation. Many changes occur, some of which are not related to myelin. At the early acute stage there is a decreased capacity to synthesize myelin lipids of the spinal cord[323]; thereafter the capacity appears to be normal. The overall increase in capacity to synthesize the phospholipids of brain and spinal cord in the acute stage[324] may therefore be due to increased synthesis in non-myelin fractions[323]. According to Maggio et al.[325] there is a 6% increase in total brain phospholipid of the rat and a 30% decrease in sulphatides in the acute stage.

Certain toxic agents, cyanides, carbon monoxide and lead also cause a primary demyelination[240, 318] but no specific investigations on phospholipid changes appear to have been carried out.

Secondary demyelination

When a peripheral nerve is severed (neurotmesis) the portion distal to the section loses its ability to transmit a nerve impulse and degenerates; eventually sheath and axon disappear. The histological changes associated with this degeneration (Wallerian degeneration) have been extensively studied[318] and numerous investigations have also been made of lipid changes. The early work of Rossiter's group, based on older methods of analysis, clearly established the extensive phospholipid loss in the combined myelin sheaths and axons of the sciatic nerve trunk[1, 326]. It is generally agreed that few lipid changes can be detected in the distal section for the first week after severance but 50% of the phospholipid had disappeared by the 14th day from rabbit sciatic nerve and the loss of ethanolamine plasmalogen was particularly extensive[327, 328]. There was also an increase in the level of lysophosphatidylethanolamine[327, 328]. In the central nervous system degeneration proceeds at a slower rate. Thus, degeneration of the optic nerve[329] does not reach a maximum rate until 100 days after the enucleation of the eye.

In the early stages of Wallerian degeneration there is an increase of acid phosphatase and proteinase, probably of lysosomal origin[330]. There have been almost no studies of phospholipases in demyelinating tissue but, in the sub-acute degeneration of the monkey spinal cord caused by Vitamin B_{12} deficiency (a *primary* demyelination), an enhanced level of plasmalogenase was observed in the plaques compared with normal white matter[152].

The proximal stump of a sectioned nerve will show hypercellularity, the axon tip grows out and if conditions allow, links with the degenerating distal portion. The Schwann cells of the degenerated distal portion can then re-form myelin. A similar regeneration will occur after crushing a nerve (axonotmesis). There is some evidence that the distal Schwann

References p. 415

cells in degenerating nerves develop a greater capacity to synthesize phospholipids in anticipation of re-myelination[1]. As might be expected the incorporation of precursors into the phospholipids of regenerating nerves is greater than into controls[331].

Certain organophosphorus compounds with an anti-cholinesterase action, such as tri-*o*-cresylphosphate (TOCP) and di-isopropyl flurophosphonate (DFP) cause degeneration of the Wallerian type in peripheral and central nervous systems[332]. (Ingestion of an alcoholic beverage ('Ginger Jake') contaminated with TOCP caused some 20 000 cases of paralysis in the U.S.A. in 1930 and there was a similar occurrence as recently as 1959 in Morocco, this time from contaminated cooking-oil.) The degeneration is a type of neuropathy and the neurone dies[333]. The demyelination is not connected with the well-known anti-cholinesterase action of TOCP or DFP because other organophosphorus compounds with anti-cholinesterase activity do not cause demyelination[334]. Ataxia appears in the hen about two weeks after treatment with TOCP. During the degeneration of the sciatic nerve, the levels of lysophosphatidylethanolamine and lysophosphatidylcholine rise and there is a consistent increase in the $C_{18:1}$ acid in the glycerophospholipids.[335] However, no change in major classes of lipids or their fatty acid composition was found by Joel et al.[283] when degeneration was induced with bis-(monoisopropylamino) fluorophosphine oxide (Mipafox) (see also ref. 336).

A few studies have been made of the effect of the organophosphorus compounds on enzymes concerned with phospholipid metabolism. No change in the rate of incorporation of [^{32}P]orthophosphate into the total phospholipids of the CNS in TOCP-poisoned hens was found[337]. Shcltawy and Dawson[336] confirmed this for sciatic nerve but observed that the incorporation of ^{32}P into triphosphoinositide was stimulated within a day of TOCP poisoning. This could be due to Schwann cell proliferation rather than to the primary biochemical lesion produced by the poison. Although low doses of DFP have been reported to reduce the transfer of phosphorylcholine to phosphatidylcholine in the brain[338] no significant effects on the enzymes cholinephosphate cytidylyltransferase or choline phosphotransferase were found[339]. In view of the advances in our knowledge of cholinesterases obtained by the use of organophosphorus compounds it would seem that further studies on these compounds in relation to demyelination could lead to useful findings. There are, however, complications inherent in studies of the effects of toxic organophosphorus compounds on the metabolism of sciatic nerve[336].

7. SOME ASPECTS OF THE ROLE OF PHOSPHOLIPIDS IN THE FUNCTION OF THE NERVOUS SYSTEM

From the account so far, it will be apparent that although phospholipids are present in all membranes of the nervous system, particularly the myelin sheath, little can be said about their specific functional roles.

The possible roles of phosphoinositides and phosphatidic acid in synaptic transmission are treated fully in Chapter 15, and it has been reported that the 'responsive' phosphatidylinositol is tightly bound to the cholinergic receptor[340] (but see Chapter 15) which, according to the work of De Robertis and his colleagues[341–343], is itself conceivably a phospholipid-rich proteolipid. This proteolipid was isolated from mammalian nerve ending membranes and showed a high affinity for *d*-tubocurarine (curare) and atropine. Using a source of pure cholinergic nerve endings from *Torpedo* and *Electrophorus electricus* it was shown that the proteolipid obtained also had a high affinity binding for acetylcholine, hexamethonium and other cholinergic drugs, but not for atropine. Non-binding of atropine would be expected since the electroplax and the myoneuronal junctions contain only nicotinic and no muscarinic receptors. Proteolipid from mixed muscarinic and nicotinic central synapses could bind atropine quite well. In further experiments it has been shown that ultrathin artificial lipid membranes, to which had been added a small amount of this acetylcholine-binding proteolipid from *Electrophorus*[344], reacted to the addition of acetylcholine by a rapid and transient increase in conductance. The response was not elicited by choline and was reduced by *d*-tubocurarine. It must be stated at this juncture, however, that the receptor–ligand affinities or interactions deduced from all these experiments may be artifactual, arising from the use of Sephadex columns[345].

If the proteolipid is an important component of the receptor it remains to be seen what role, if any, its phospholipid component plays. Recently, Simpson[346] has investigated the use of phospholipase C to determine the possible involvement of phospholipids in the acetylcholine receptor of the neuromuscular junction. He could find no evidence that the removal of phospholipid polar groups by phospholipase C affected the action of a range of cholinergic blocking agents, hemicholinium, botulinum toxin, curare and DFP. Although an unsatisfactory parameter for the loss of phospholipid was adopted and phosphatidylinositol is not readily attacked by the particular phospholipase C used (though this may be irrelevant), it would appear from these experiments that phospholipid is not involved in the receptor site at the neuromuscular junction. On the other hand Bartels and Rosenberg[347] showed that venom phospholipase A irreversibly depolarised and blocked conduction in the electroplax of *Electrophorus* but only when there was extensive hydrolysis of phosphatidylcholine. This suggests either a specific role for this phospholipid in this particular organ or a dispersive role for the lysophosphatidylcholine produced.

Watkins[348] has remarked on the similarity between the polar head groups of phosphatidylcholine, phosphatidylethanolamine and phosphatidylserine and the charge distribution in three putative transmitters, acetylcholine, γ-aminobutyric acid and glutamic acid respectively. These transmitters might facilitate the dissociation of the phospholipid from its binding to protein, though this is not easily envisaged, and so promote depolarisation. There are other possible relationships or interactions between acetylcholine and phosphatidylcholine. The choline of both components is present in biological systems in the *gauche* form, not the anticipated *trans* form i.e. the preferred conformation is synclinal[349]. Furthermore, a substituted phosphorylcholine e.g. methyl-ethoxy-phosphorylcholine is an inhibitor, albeit a weak one, of acetylcholinesterase[350] presumably because its quaternary nitrogen can compete for the anionic site of the enzyme and the phosphate moiety for the esteratic site. Both acetylcholine and phosphorylcholine contain the O–C–C–N grouping and Sundaralingam[349] has suggested that the phosphorylcholine moiety of phosphatidylcholine, which is presumably near the surface of the membrane, could sequester acetylcholinesterase in an inert form.

References p. 415

It has been suggested that the polar head groups of phospholipids are also involved in axonal conduction. Of particular interest is the suggestion that dephosphorylation of polyphosphoinositides might allow the release of $Ca^{2\oplus}$ ions and lead to the depolarisation of the axonal membrane, while re-phosphorylation would allow for re-uptake of calcium and repolarisation[351, 352] (see also Chapter 15). Polyphosphoinositides certainly appear to control the binding of $Ca^{2\oplus}$ ions in the erythrocyte membrane[353] and the activity of the $Ca^{2\oplus}$-activated ATP-ase was higher in polyphosphoinositide-enriched membranes than in normal ones. If polyphosphoinositides are involved in ion transport, however, changes would be expected in the turnover of the phosphate groups on stimulation. Birnberger et al.[354] pre-labelled the TPI of the leg and claw nerve of the lobster with ^{32}P and showed that stimulation enhanced the release of radioactivity from the lipid. The loss was prevented by cyanide which did not, however, affect the conduction. Conversely, in experiments with vagus nerve, there was no effect on labelling of polyphosphoinositide when the action potential was blocked by tetrodotoxin[355, 356].

There have been a number of attempts to allocate a role in axonal conduction to other membrane phospholipids and the giant axon of the squid has been extensively used, even if it has not provided any answers. Rosenberg and his collaborators[357–361] showed that snake venom containing phospholipase A, applied externally to the squid axon, increased permeability, blocked conduction and rendered the axon sensitive to curare and acetylcholine. Although great losses of membrane phospholipid occurred, it was subsequently found that the effects were more likely to be due to the detergent action of the lysophospholipids produced by the phospholipase A[361]. In view of the possible importance of the polar head group, Rosenberg[362] tested the action of phospholipase C. With this enzyme, polar head groups are removed from phospholipids and the hydrolysis products are non-toxic. However, even when 50–60% of the phospholipid-phosphorus had disappeared from the membrane, axonal conduction and permeability were maintained. The phospholipid bases appeared in the external medium, not the axoplasm, and the diglycerides remained in the membrane. Abbott et al.[363] suggested that, in the experiments of Rosenberg, most of the phospholipid was lost from Schwann cells, not axolemma, and showed that application of a mixture of phospholipases A and B to the outside of the axon had no effect, but internal application suppressed the action potential. This suggested that phospholipids are essential for conduction but these results, like the earlier ones of Rosenberg, could be accounted for by the lytic action of lysophospholipids produced when phospholipase A is used.

There have been suggestions that an endogenous phospholipase may be responsible for the release of acetylcholine from vesicles. In model experiments, Heilbronn[364] has demonstrated that a phospholipase A from *Naja Naja siamensis* weakens and breaks down synaptic vesicles obtained from rat brain cortex and cholinergic vesicles from the electric organ of *Torpedo nobiliana*, thereby releasing acetylcholine. Electron micrographs of the treated vesicles suggested the formation of pieces of membrane, about one-third the circumference of the intact vesicles, indicating discrete areas sensitive either to the phospholipase or to the products of its action. The enzyme appeared to attack phosphatidylethanolamine first. Studies on the nerve ending membranes showed the presence of an endogenous phospholipase A which could be responsible for such a process *in vivo*. It had been shown earlier[365] that phospholipase C was much less effective than phospholipase A in releasing acetylcholine from cortex *slices*, suggesting that diglycerides can maintain the structure of the vesicles better than lysophospholipids (cf. effects of phospholipase A on axonal membranes, p. 412).

From the foregoing account it will be apparent that useful information about the function of phospholipids in the nervous system can be obtained from preparations which are, relatively speaking, structurally homogeneous. In recent years some interesting studies on monolayers, artificial membranes and erythrocyte membranes have yielded information about the action of certain drugs, particularly anaesthetics. The Overton-Meyer theory predicts that anaesthesia occurs when a lipid-soluble anaesthetic reaches such a concentration in an excitable membrane as to produce disorder. Specific chemical groupings in the anaesthetic are not required and this could explain the wide variety of chemically unrelated compounds which can induce anaesthesia and block action potentials.

In a pioneering study Skou[366] showed that members of the cocaine series (tertiary amines) in their undissociated form could penetrate into a monomolecular layer of lipids extracted from frog sciatic nerve. He found that there was a parallel between their ability to penetrate this layer and their blocking potencies as determined with intact nerves. Clements and Wilson[367] showed that the immediate effect of the anaesthetic was to expand the membrane. Benzyl alcohol has been used as a model anaesthetic and shown to penetrate membranes and monolayers. Metcalfe et al.[368] showed that the nuclear magnetic relaxation rate of benzyl alcohol was increased in the presence of erythrocyte membranes and concluded that this was caused by the restriction of the molecular motion of the alcohol induced by the structure of the membrane. Rotation increased as the anaesthetic concentration rose and it was suggested that alterations were induced in the membrane structure which might be correlated with physiological changes. Similar effects were obtained with membrane preparations from nervous tissue[369]. Alcohol anaesthetics and phenothiazine tranquilizers cause membrane expansion in intact erythrocytes and erythrocyte ghosts[370, 371]. The membrane appears to be stabilised and a close correlation between stabilisation and narcotic action was found. From studies on artificial phospholipid membranes (liposomes) Johnson and Bangham[372] concluded that anaesthetics increase the freedom of movement of groups in the lipid near the interface, sterically impeding the rearrangement which produces the sudden increase in permeability to $Na^{\oplus}$ ions on which the generation of the action potential depends.

The studies outlined above did not take particular note of interaction between the pharmacological agent and the charged polar head groups. These may be of some importance since the latter can bind $Ca^{2\oplus}$ ions which play a role in controlling the excitability of nervous tissue. Lowering the calcium concentration increases the permeability of membranes and it has been shown that local anaesthetics such as procaine, which are organic bases, can displace $Ca^{2\oplus}$ ions from the membrane i.e. antagonise calcium binding[373, 374]. Thus, the mechanism by which anaesthesia is produced need not necessarily be due merely to the physical presence of the anaesthetic in the membrane (cf. ref. 375). The displacement of $Ca^{2\oplus}$ ions from phospholipid monolayers by pharmacologically active bases has been extensively studied by Hauser and Dawson[376]. They found that the binding of $^{45}Ca^{2\oplus}$ to a monolayer of phosphatidylinositol at the air–water interface was maximal when the separation of the polar head groups was approximately the same as the diameter of a hydrated $Ca^{2\oplus}$ ion. Calcium could be displaced by a variety of organic bases at low concentrations (e.g. 0.4 μM chlorpromazine) provided that they were at least partially ionised and that they caused an increase in the pressure of the film. An increase in pressure indicated partial penetration of the film, not necessarily extending to the hydrophobic areas. However, as the authors point out, displacement of $Ca^{2\oplus}$ ions was caused by anaesthetics like cocaine, excitants like strychnine and bases which do not have pharmacological action, so that displacement of calcium may not necessarily be important in the development of anaesthesia. Scherphof et al.[377] have found that local anaesthetics of the procaine type inhibit the hydrolytic action of the calcium-dependent pancreatic phospholipase A towards membrane-bound mitochondrial lipids from liver and egg yolk suspensions. The inhibition was competitively antagonised by $Ca^{2\oplus}$ ions and paralleled the anaesthetic potency of a series of anaesthetics. Furthermore, 1 mM chlorpromazine inhibits the calcium-dependent enzyme degrading phosphatidylinositol in the thyroid gland of the sheep, presumably by displacing $Ca^{2\oplus}$ ions from the site of enzymatic attack[378]. It is, however, difficult to see how the actions of anaesthetics and other pharmacological agents are related to calcium-dependent phospholipases, since in nervous tissue these enzymes do not require calcium for activity (p. 392). Local anaesthetics may affect the metabolism of nerves in other ways. Thus 63 μM cinchocaine causes a 3–5-fold increase in the labelling of phosphatidylinositol from [^{32}P]orthophosphate in the isolated unstimulated vagus nerve of the rabbit[355].

Conclusion

From this account of phospholipids in the nervous system it can be seen that a considerable amount is known about the biosynthesis and catabolism of the molecules themselves and their distribution within the various structures of the nervous system. However, conclusions about the roles they play in the functioning of the tissue are largely impossible to draw. The possible roles of phosphatidic acid and the phosphoinositides are discussed in Chapter 15 but the functions of the other phospholipids are unknown.

Advances are most likely to come from studies on isolated neurones and glia, on organelles such as the synaptosome (particularly from a 'homogeneous' system) and myelin and on giant axons. These have only become broadly 'available' since the publication of the first edition of this book some eight years ago and the next decade should prove rewarding for those working on phospholipid function in neural tissues. Unfortunately we cannot always follow the advice of the King in *Alice in Wonderland* and 'Begin at the beginning and go on till you come to the end: then stop'; serendipity will undoubtedly continue to play its part.

ACKNOWLEDGMENTS

The author acknowledges with gratitude the help given to him by Dr. Sheila Spanner who tactfully drew his attention to references he would otherwise have missed. He also wishes to thank Professor P.B. Bradley and Dr. J. Dobbing for their criticisms and suggestions.

REFERENCES

1 G.B. Ansell and J.N. Hawthorne, *The Phospholipids: Chemistry, Metabolism and Function,* Elsevier, Amsterdam, 1964, p. 278.
2 A.N. Davison, in A.N. Davison and J. Dobbing (Eds.), *Applied Neurochemistry,* Blackwell, Oxford, 1968, p. 178.
3 A.N. Davison, in M. Florkin and E.H. Stotz (Eds.), *Comprehensive Biochemistry*, Vol. 18, Elsevier, Amsterdam, 1970, p. 293.
4 J. Eichberg, G. Hauser and M.L. Karnovsky, in G.H. Bourne (Ed.), *The Structure and Function of Nervous Tissue. III. Biochemistry and Disease*, Academic Press, New York, 1969, p. 185.
5 P.V. Johnston and B.I. Roots, *Nerve Membranes*, Pergamon, Oxford, 1972.
6 J.F. O'Brien and R.L. Geison, *J. Neurochem.*, 18 (1971) 1615.
7 W.C. Breckenridge, G. Gombos and I.G. Morgan, *Biochim. Biophys. Acta*, 266 (1972) 695.
8 B.J. Holub, A. Kuksis and W. Thompson, *J. Lipid. Res.*, 11 (1970) 558.
9 O.H. Lowry, N.R. Roberts, K.Y. Leiner, M-L. Wu and A.L. Farr, *J. Biol. Chem.*, 207 (1954) 1.
10 O.H. Lowry, N.R. Roberts, K.Y. Leiner, M-L. Wu, A.L. Farr and R.W. Albers, *J. Biol. Chem.*, 207 (1954) 39.
11 H. Hydén, in J. Brachet and A.E. Mirsky (Eds.), *The Cell*, Vol. IV, Academic Press, New York, 1960, p. 215.
12 B.I. Roots and P.V. Johnston, *Biochem. J.*, 94 (1965) 61.
13 P.V. Johnston and B.I. Roots, *Biochem. J.*, 98 (1966) 157.
14 V.P. Whittaker, *Progr. Biophys. Mol. Biol.*, 15 (1965) 39.
15 E. De Robertis, A. Pellegrino De Iraldi, G. Rodriguez De Lores-Arnaiz and L. Salganicoff, *J. Neurochem.*, 9 (1962) 23.
16 S.P.R. Rose, *Biochem. J.*, 102 (1967) 33.
17 H. Haljämae and A. Hamberger, *J. Neurochem.*, 18 (1971) 1903.
18 J.E. Cremer, P.V. Johnston, B.I. Roots and A.J. Trevor, *J. Neurochem.*, 15 (1968) 1361.
19 C. Blomstrand and A. Hamberger, *J. Neurochem.*, 16 (1969) 1401.
20 M.F. Fewster, A.B. Scheibel and J.F. Mead, *Brain Res.*, 6 (1967) 401.
21 W.T. Norton and S.E. Poduslo, *Science*, 167 (1970) 1144.
22 A.K. Sinha, and S.P.R. Rose, *Brain Res.*, 33 (1971) 205.
23 W.T. Norton and S.E. Poduslo, *J. Lipid Res.*, 12 (1971) 84.
24 S.E. Poduslo and W.T. Norton, *J. Neurochem.*, 19 (1972) 727.
25 A. Hamberger and L. Svennerholm, *J. Neurochem.*, 18 (1971) 1821.
26 Y. Tamai, S. Matsukawa and M. Satake, *Brain Res.*, 26 (1971) 149.
27 M.E. Fewster and J.F. Mead, *J. Neurochem.*, 15 (1968) 1041.
28 M.E. Fewster, and J.F. Mead, *J. Neurochem.*, 15 (1968) 1303.
29 V.P. Whittaker and L.A. Barker, in R. Fried (Ed.), *Methods of Neurochemistry, II,* Marcel Dekker, New York, 1972, p. 1.
30 J. Eichberg Jr., V.P. Whittaker and R.M.C. Dawson, *Biochem. J.*, 92 (1964) 91.
31 E.G. Lapetina, E.F. Soto and E. De Robertis, *J. Neurochem.*, 15 (1968) 437.
32 J.W.T. Dickerson, in A.N. Davison and J. Dobbing (Eds.), *Applied Neurochemistry*, Blackwell, Oxford, 1968, p. 48.
33 R.P. Bunge, *Physiol. Rev.*, 48 (1968) 197.
34 M.B. Bunge, R.P. Bunge and H. Ris, *J. Biochem. Biophys. Cytol.*, 10 (1961) 67.
35 W.T. Norton, in R. Paoletti and A.N. Davison (Eds.), *The Chemistry of Brain Development*, Plenum, New York, 1971, p. 327.
36 L.A. Autilio, W.T. Norton and R.D. Terry, *J. Neurochem.*, 11 (1964) 17.
37 J.S. O'Brien, E.L. Sampson and M.B. Stern, *J. Neurochem.*, 14 (1967) 357.
38 L.A. Horrocks, *J. Lipid Res.*, 8 (1967) 569.
39 C.W.M. Adams, Y.H. Abdulla, D.R. Turner and O.B. Bayliss, *Nature*, 220 (1968) 171.
40 T. Kurihara and Y. Tsukada, *J. Neurochem.*, 14 (1967) 1167.
41 R.W. Olafson, G.J. Drummond and J.F. Lee, *Can. J. Biochem.*, 47 (1969) 961.
42 T. Kurihara, J.L. Nussbaum and P. Mandel, *J. Neurochem.*, 17 (1970) 993.
43 J.P. Zanetta, P. Benda, G. Gombos and I.G. Morgan, *J. Neurochem.*, 19 (1972) 881.
44 W.T. Norton and L.A. Autilio, *J. Neurochem.*, 13 (1966) 213.

45 A.N. Davison and A. Peters, *Myelination*, Thomas, Springfield, 1970, p. 89.
46 J. Eichberg and R.M.C. Dawson, *Biochem. J.*, 96 (1965) 644.
47 A. Sheltawy and R.M.C. Dawson, *Biochem. J.*, 100 (1966) 12.
48 L. Autilio, *Federation Proc.*, 25 (1966) 764.
49 L.C. Mokrasch, *Life Sci.*, 6 (1967) 1905.
50 J. Folch-Pi and P.J. Stoffyn, *Ann. N.Y. Acad. Sci.*, 195 (1972) 86.
51 F. Wolfgram, *Biochim. Biophys. Acta*, 147 (1967) 383.
52 F. Wolfgram, *J. Neurochem.*, 13 (1966) 461.
53 P.R. Carnegie, *Biochem. J.*, 123 (1971) 57.
54 P.R. Carnegie, *Nature*, 229 (1971) 25.
55 R.E. Martenson and F.N. LeBaron, *J. Neurochem.*, 13 (1966) 1469.
56 E.H. Eylar and G.A. Hashim, *Proc. Natl. Acad. Sci. (U.S.)*, 61 (1968) 644.
57 F.B. Palmer and R.M.C. Dawson, *Biochem. J.*, 111 (1969) 629.
58 F.B. Palmer and R.M.C. Dawson, *Biochem. J.*, 111 (1969) 637.
59 R.M.C. Dawson, *Biochem. J.*, 97 (1965) 134.
60 A.N. Siakotos, G. Rouser and S. Fleischer, *Lipids*, 4 (1969) 234.
61 S. Spanner and G.B. Ansell, in E. Reid (Ed.), *Separations with Zonal Rotors*, Wolfson Bioanalytical Centre, Guildford, 1971, p. V3-1.
62 G. Goracci, C. Blomstrand, G. Arienti, A. Hamberger and G. Porcellati, *J. Neurochem.*, 1973, in the press.
63 J.T. Aitken and J.E. Bridger, *J. Anat.*, 95 (1961) 38.
64 C. Cotman, M.L. Blank, A. Moehl and F. Snyder, *Biochemistry*, 8 (1969) 4606.
65 R.E. Anderson and M.B. Maude, *Biochemistry*, 9 (1970) 3624.
66 L. Freysz, R. Bieth and P. Mandel, *J. Neurochem.*, 16 (1969) 1417.
67 A.K. Hajra and B.N. Agranoff, *J. Biol. Chem.*, 243 (1968) 3542.
68 E.G. Lapetina and J.N. Hawthorne, *Biochem. J.*, 122 (1971) 171.
69 E. Mårtensson and J. Kanfer, *J. Biol. Chem.*, 243 (1968) 497.
70 M.E. Smith and G. Hübscher, *Biochem. J.*, 101 (1966) 308.
71 E.S. De Jiménez and W.W. Cleland, *Biochim. Biophys. Acta*, 176 (1969) 685.
72 F. Possmayer and J.B. Mudd, *Biochim. Biophys. Acta*, 239 (1971) 217.
73 P. Mandel and S. Edel-Harth, *J. Neurochem.*, 13 (1966) 591.
74 W.C. McMurray, K.P. Strickland, J.F. Berry and R.J. Rossiter, *Biochem. J.*, 66 (1957) 634.
75 L. Ewetz, B. Sparf and B. Sörbo, in *In vitro Procedures with Radioisotopes in Medicine*, Intern. At. Energy Agency, Vienna, 1970, p. 175.
76 I. Eade, C. Hebb and S.P. Mann, *J. Neurochem.*, 20 (1973) 1499.
77 S.S. Oja, *Ann. Acad. Sci. Fenn. Ser. A. V. Medica*, 125 (1966) 7.
78 G.B. Ansell and J.M. Norman, *Biochem. J.*, 54 (1953) viii.
79 R.E. McCaman and K. Cook, *J. Biol. Chem.*, 241 (1966) 3390.
80 G.B. Ansell and S. Spanner, in J. Ganguly and R.M.S. Smellie (Eds.), *Current Trends in the Biochemistry of Lipids*, Academic Press, London, 1973, p. 151.
81 T. Chojnacki and G.B. Ansell, *J. Neurochem.*, 14 (1967) 413.
82 T. Chojnacki, A. Radomińska-Pyrek and T. Korzybski, *Acta Biochim. Polon.*, 14 (1967) 383.
83 M.V. Gomez, E.F. Domino, J.C. Santiago and O.Z. Sellinger, *Neurobiology*, 1 (1971) 103.
84 E.T. Pritchard, *Can. J. Biochem. Physiol.*, 40 (1962) 353.
85 T.R. Skrbic and J.N. Cumings, *J. Neurochem.*, 17 (1970) 85.
86 G.R. Webster and R.J. Alpern, *Biochem. J.*, 90 (1964) 35.
87 M. Waite and L.L.M. Van Deenen, *Biochim. Biophys. Acta*, 137 (1967) 498.
88 J. Nachbaur and P.M. Vignais, *Biochem. Biophys. Res. Commun.*, 33 (1968) 315.
89 G.B. Ansell and S. Spanner, *J. Neurochem.*, 14 (1967) 873.
90 G.B. Ansell and S. Spanner, *Biochem. J.*, 122 (1971) 741.
91 E.F. Marshall, T. Chojnacki and G.B. Ansell, *Biochem, J.*, 95 (1965) 30P.
92 J. Bremer and D.M. Greenberg, *Biochim. Biophys. Acta*, 46 (1961) 205.
93 G.Y. Sun and L.A. Horrocks, *Lipids*, 5 (1970) 1006.
94 E.K. Miller and R.M.C. Dawson, *Biochem. J.*, 126 (1972) 805.
95 J.N. Kanfer, *J. Lipid Res.*, 13 (1972) 468.
96 G. Porcellati, M.G. Biasion and M. Pirotta, *Lipids*, 5 (1970) 734.

97 G.B. Ansell and R.F. Metcalfe, *J. Neurochem.*, 18 (1971) 647.
98 G.B. Ansell, T. Chojnacki and R.F. Metcalfe, *J. Neurochem.*, 12 (1965) 649.
99 A. Radomińska-Pyrek and L.A. Horrocks, *J. Lipid Res.*, 13 (1972) 580.
100 G.B. Ansell and T. Chojnacki, *Biochem. J.*, 98 (1966) 303.
101 G. Porcellati, M.G. Biasion and G. Arienti, *Lipids*, 5 (1970) 725.
102 G.B. Ansell and R.M.C. Dawson, *Biochem. J.*, 50 (1951) 241.
103 H.R.V. Arnstein, *Biochem. J.*, 48 (1951) 27.
104 G.B. Ansell and S. Spanner, *Biochem. J.*, 84 (1962) 12P.
105 A.A. Abdel-Latif and L.G. Abood, *J. Neurochem.*, 13 (1966) 1189.
106 W. Stoffel and R. Henning, *Z. Physiol. Chem.*, 349 (1968) 1400.
107 W. Stoffel and G. Assmann, *Z. Physiol. Chem.*, 351 (1970) 1041.
108 R.F. Riley, *J. Biol. Chem.*, 153 (1944) 535.
109 W. Stoffel, D. LeKim and G. Heyn, *Z. Physiol. Chem.*, 351 (1970) 875.
110 G.B. Ansell and S. Spanner, *Biochem. J.*, 100 (1966) 50P.
111 G. Porcellati, G. Arienti, M. Pirotta and D. Giorgini, *J. Neurochem.*, 18 (1971) 1395.
112 S. Joffe, *J. Neurochem.*, 16 (1969) 715.
113 G.Y. Sun and L.A. Horrocks, *J. Neurochem.*, 16 (1969) 181.
114 L.A. Horrocks, in J. Domonkos, A. Fonyó, I. Husák and J. Szentágothai (Eds.), *Abstracts of the 3rd International Meeting of the International Society for Neurochemistry*, Akadémiai Kiadó, Budapest, 1971, p. 312.
115 R. Bickerstaffe and J.F. Mead, *Lipids*, 3 (1968) 317.
116 L.A. Horrocks and G.B. Ansell, *Lipids*, 2 (1967) 329.
117 L.A. Horrocks, in R. Paoletti, R. Fumagalli and C. Galli (Eds.), *Abstracts of the 2nd International Meeting of the International Society for Neurochemistry*, Tamburini, Milano, 1969, p. 221.
118 L.A. Horrocks, in F. Snyder (Ed.), *Ether Lipids: Chemistry and Biology*, Academic Press, New York, 1972, p. 171.
119 F. Snyder, M.L. Blank and B. Malone, *J. Biol. Chem.*, 245 (1970) 4016.
120 F. Snyder, M. Hibbs and B. Malone, *Biochim. Biophys. Acta*, 231 (1971) 409.
121 T. Takanashi and H.H.O. Schmid, *Chem. Phys. Lipids*, 4 (1970) 243.
122 H.H.O. Schmid and T. Takahashi, *J. Lipid Res.*, 11 (1970) 412.
123 O.E. Bell Jr., M.L. Blank and F. Snyder, *Biochim. Biophys. Acta*, 231 (1971) 579.
124 L.A. Horrocks and A. Radomińska-Pyrek, *FEBS Letters*, 22 (1972) 190.
125 H.H.O. Schmid, T. Muramatsu and K.L. Su, *Biochim. Biophys. Acta*, 270 (1972) 317.
126 R. Henning and W. Stoffel, *Z. Physiol. Chem.*, 350 (1969) 827.
127 W. Stoffel, *Chem. Phys. Lipids*, 5 (1970) 139.
128 M. Sribney and E.P. Kennedy, *J. Biol. Chem.*, 233 (1958) 1315.
129 R.O. Brady, R.M. Bradley, O.M. Young and H. Kaller, *J. Biol. Chem.*, 240 (1965) PC 3693.
130 Y. Fujino and T. Negishi, *Biochim. Biophys. Acta*, 152 (1968) 428.
131 Y. Fujino, T. Negishi and S. Ito, *Biochem. J.*, 109 (1968) 310.
132 W. Stoffel and G. Assmann, *Z. Physiol. Chem.*, 353 (1972) 65.
133 K.C. Kopaczyk and N.S. Radin, *J. Lipid Res.*, 6 (1965) 140.
134 J.S. O'Brien and G. Rouser, *J. Lipid Res.*, 5 (1964) 339.
135 E. Klenk and R.T.C. Huang, *Z. Physiol. Chem.*, 349 (1968) 451.
136 P. Morell and N.S. Radin, *J. Biol. Chem.*, 245 (1970) 342.
137 E. Yavin and S. Gatt, *Biochemistry*, 8 (1969) 1692.
138 S. Ställberg-Stenhagen and L. Svennerholm, *J. Lipid Res.*, 6 (1965) 146.
139 A. Lastennet, L. Freysz and P. Mandel, *J. Neurochem.*, 19 (1972) 831.
140 P. Morell and P. Braun, *J. Lipid Res.*, 13 (1972) 293.
141 J. Gallai-Hatchard, W.L. Magee, R.H.S. Thompson and G.R. Webster, *J. Neurochem.*, 9 (1962) 545.
142 S. Gatt, Y. Barenholz and A. Roitman, *Biochem. Biophys. Res. Commun.*, 24 (1966) 169.
143 S. Gatt, *Biochim. Biophys. Acta*, 159 (1968) 304.
144 M.F. Cooper and G.R. Webster, *J. Neurochem.*, 17 (1970) 1543.
145 Z. Leibovitz and S. Gatt, *Biochim. Biophys. Acta*, 164 (1968) 439.
146 Y. Barnholz, A. Roitman and S. Gatt, *J. Biol. Chem.*, 241 (1966) 3731.
147 J.N. Kanfer, O.M. Young, D. Shapiro and R.O. Brady, *J. Biol. Chem.*, 241 (1966) 1081.
148 D.J. Williams, S. Spanner and G.B. Ansell, *Trans. Biochem. Soc.*, 1 (1973) 466.
149 G.R. Webster, E.A. Marples and R.H.S. Thompson, *Biochem. J.*, 65 (1957) 374.

150 J.J. Baldwin and W.E. Cornatzer, *Biochim. Biophys. Acta*, 164 (1968) 195.
151 G.B. Ansell and S. Spanner, *Biochem. J.*, 94 (1965) 252.
152 G.B. Ansell and S. Spanner, *Biochem. J.*, 108 (1968) 207.
153 E. Yavin and S. Gatt, *European J. Biochem.*, 25 (1972) 431.
154 E. Yavin and S. Gatt, *European J. Biochem.*, 25 (1972) 437.
155 W. Thompson and R.M.C. Dawson, *Biochem. J.*, 91 (1964) 233.
156 R.M.C. Dawson and W. Thompson, *Biochem. J.*, 91 (1964) 244.
157 J.G. Salway, M. Kai and J.N. Hawthorne, *J. Neurochem.*, 14 (1967) 1013.
158 J.G. Salway, J.L. Harwood, M. Kai, G.L. White and J.N. Hawthorne, *J. Neurochem.*, 15 (1968) 221.
159 W. Thompson and R.M.C. Dawson, *Biochem. J.*, 91 (1964) 237.
160 K.M.W. Keough and W. Thompson, *J. Neurochem.*, 17 (1970) 1.
161 K.M.W. Keough and W. Thompson, *Biochim. Biophys. Acta*, 270 (1972) 324.
162 K.M.W. Keough, G. Macdonald and W. Thompson, *Biochim. Biophys. Acta*, 270 (1972) 337.
163 R.M.C. Dawson and D. Richter, *Proc. Roy. Soc. (London) B*, 137 (1950) 252.
164 G.B. Ansell and H. Dohmen, *J. Neurochem.*, 2 (1957) 1.
165 O. Lindberg and L. Ernster, *Biochem. J.*, 46 (1950) 43.
166 A.N. Davison, R.S. Morgan, M. Wajda and G. Payling Wright, *J. Neurochem.*, 4 (1959) 360.
167 A.N. Davison, J. Dobbing, R.S. Morgan and G. Payling Wright, *J. Neurochem.*, 3 (1958) 89.
168 A.N. Davison, J. Dobbing, R.S. Morgan and G. Payling Wright, *Lancet*, 1 (1959) 658.
169 A.N. Davison and J. Dobbing, *Biochem. J.*, 73 (1959) 701.
170 A.N. Davison and J. Dobbing, *Biochem. J.*, 75 (1960) 565.
171 A.N. Davison and J. Dobbing, *Biochem. J.*, 75 (1960) 571.
172 L.M. Cuzner, A.N. Davison and N.A. Gregson, *Ann. N.Y. Acad. Sci.*, 122 (1965) 86.
173 R.C. Thompson and J.E. Ballou, *J. Biol. Chem.*, 208 (1954) 883.
174 R.C. Thompson and J.E. Ballou, *J. Biol. Chem.*, 223 (1956) 795.
175 J. Dobbing, *Physiol. Rev.*, 41 (1961) 130.
176 J. Dobbing, *Guy's Hosp. Repts.*, 112 (1963) 267.
177 J. Dobbing, in A.N. Davison and J. Dobbing (Eds.), *Applied Neurochemistry*, Blackwell, Oxford, 1968, p. 317.
178 C. August, A.N. Davison and F. Maurice-Williams, *Biochem. J.*, 81 (1961) 8.
179 M.E. Smith and L.F. Eng, *J. Am. Oil Chem. Soc.*, 42 (1965) 1013.
180 M.E. Smith, *Biochim. Biophys. Acta*, 164 (1968) 285.
181 G.B. Ansell and S. Spanner, *J. Neurochem.*, 15 (1968) 1371.
182 L.A. Horrocks, *J. Neurochem.*, 16 (1969) 13.
183 G.Y. Sun and L.A. Horrocks, *J. Lipid Res.*, 14 (1973) 206.
184 E.G. Lapetina, G. Rodriguez De Lores-Arnaiz and E. De Robertis, *J. Neurochem.*, 16 (1969) 101.
185 F.B. Jungalwala and R.M.C. Dawson, *Biochem. J.*, 123 (1971) 683.
186 W.T. Norton, cited by A.N. Davison, *Scientific Basis of Medicine, Ann. Rev.* (1969) 220.
187 P. Mandel and J.L. Nussbaum, *J. Neurochem.*, 13 (1966) 629.
187a E.K. Miller and R.M.C. Dawson, *Biochem. J.*, 126 (1972) 823.
188 M. Miani, *Progr. Brain Res.*, 13 (1964) 115.
189 G.A. Dhopeshwarkar and J.F. Mead, *Biochim. Biophys. Acta*, 187 (1969) 461.
190 G.A. Dhopeshwarkar and J.F. Mead, *Biochim. Biophys. Acta*, 210 (1970) 250.
191 G.A. Dhopeshwarkar, C. Subramanian and J.F. Mead, *Biochim. Biophys. Acta*, 231 (1971) 8.
192 G.A. Dhopeshwarkar, C. Subramanian and J.F. Mead, *Biochim. Biophys. Acta*, 239 (1971) 162.
193 J. Hoelzl and H.P. Franck, in R. Paoletti, G. Fumagalli and C. Galli (Eds.), *Abstracts of the 2nd International Meeting of the International Society for Neurochemistry*, Tamburine, Milan, 1969, p. 219.
194 Y. Stein and O. Stein, *Biochim. Biophys. Acta*, 116 (1966) 95.
195 D.R. Illingworth and O.W. Portman, *Biochem. J.*, 130 (1972) 557.
196 J. Schuberth, B. Sparf and A. Sundwall, *J. Neurochem.*, 16 (1969) 695.
197 I. Diamond, *Arch. Neurol.*, 24 (1971) 333.
198 K. Dross and H. Kewitz, *Arch. Exptl. Pathol. Pharmakol.*, 274 (1972) 91.
199 J.H. Menkes, *J. Neurochem.*, 18 (1971) 1433.
200 J.H. Menkes, *Lipids*, 7 (1972) 135.
201 H. Waelsch (Ed.), *Biochemistry of the Developing Nervous System*, Academic Press, New York, 1955.

202 A.N. Davison and J. Dobbing, in A.N. Davison and J. Dobbing (Eds.), *Applied Neurochemistry*, Blackwell, Oxford, 1968, p. 253.
203 H. McIlwain and H. Bachelard, *Biochemistry and the Central Nervous System*, 3rd ed., Churchill, London, 1971.
204 M.A. Wells and J.C. Dittmer, *Biochemistry*, 6 (1967) 3169.
205 G. Rouser and A. Yamamoto, in A. Lajtha (Ed.), *Handbook of Neurochemistry*, Vol. I, Plenum, New York, 1969, p. 121.
206 G. Rouser and A. Yamamoto, *Lipids*, 3 (1968) 284.
207 E. Marshall, R. Fumagalli, R. Niemiro and R. Paoletti, *J. Neurochem.*, 13 (1966) 857.
208 Y. Kishimoto, W.E. Davies and N.S. Radin, *J. Lipid Res.*, 6 (1965) 532.
209 L. Svennerholm, *J. Lipid Res.*, 9 (1968) 570.
210 G.B. Ansell, in R. Paoletti and A.N. Davison (Eds.), *The Chemistry of Brain Development*, Plenum, New York, 1971, p. 63.
211 J.S. O'Brien, *Science*, 147 (1965) 1099.
212 L.A. Horrocks, R.J. Meckler and R.L. Collins, in G.B. Ansell (Ed.), *Variation in Chemical Composition of the Nervous System as Determined by Developmental and Genetic Factors*, Pergamon, Oxford, 1966, p. 46.
213 L.A. Horrocks, *J. Neurochem.*, 15 (1968) 483.
214 M.L. Cuzner and A.N. Davison, *Biochem. J.*, 106 (1968) 29.
215 L.F. Eng and E.P. Noble, *Lipids*, 3 (1968) 157.
216 L.A. Horrocks, *Progr. Brain Res.*, 1973, in the press.
217 A.N. Davison, M.L. Cuzner, N.L. Banik and J. Oxberry, *Nature*, 212 (1966) 1373.
218 N.L. Banik, M.J. Blunt and A.N. Davison, *J. Neurochem.*, 15 (1968) 471.
219 N.L. Banik and A.N. Davison, *Biochem. J.*, 115 (1969) 1051.
220 H.C. Agrawal, N.L. Banik, A.H. Bone, A.N. Davison, R.F. Mitchell and M. Spohn, *Biochem. J.*, 120 (1970) 635.
221 H.C. Agrawal, J.M. Davis and W.A. Himwich, *J. Neurochem.*, 13 (1966) 607.
222 R.H. Laatsch, *J. Neurochem.*, 9 (1962) 487.
223 O. Wieland and M. Suyter, *Biochem. Z.*, 329 (1957) 320.
224 M. Kai, J.G. Salway and J.N. Hawthorne, *Biochem. J.*, 106 (1968) 791.
225 G.B. Ansell and S. Spanner, in G.B. Ansell (Ed.), *Variation in Chemical Composition of the Nervous System as Determined by Developmental and Genetic Factors*, Pergamon, Oxford, 1966, p. 7.
226 D.M. Bowen and N.S. Radin, *J. Neurochem.*, 16 (1969) 501.
227 R.L. Sidman, M.M. Dickie and S.H. Appel, *Science*, 144 (1964) 309.
228 H. Meier and A.D. Macpike, *Exptl. Brain Res.*, 10 (1970) 512.
229 N.A. Baumann, C.M. Jacque, S.A. Pollet and M.L. Harpin, *European J. Biochem.*, 4 (1968) 340.
230 E.L. Hogan and K.C. Joseph, *J. Neurochem.*, 17 (1970) 1209.
231 R.M.C. Dawson and N. Clarke, *J. Neurochem.*, 18 (1971) 1313.
232 H. Singh, N. Spritz and B. Geyer, *J. Lipid Res.*, 12 (1971) 473.
233 S. Greenfield, W.T. Norton and P. Morell, *J. Neurochem.*, 18 (1971) 2119.
234 N.A. Gregson and J.M. Oxberry, *J. Neurochem.*, 19 (1972) 1065.
235 N.A. Baumann, M.L. Harpin and J.N. Bourré, *Nature*, 227 (1970) 960.
236 N.A. Baumann, J.M. Bourré, C. Jacque and S. Pollet, in *Lipids, Malnutrition and the Developing Brain* (CIBA Found. Symp.), Excerpta Medica, Amsterdam, 1972, p. 91.
237 J.L. Nussbaum, N. Neskovic and P. Mandel, *J. Neurochem.*, 16 (1969) 927.
238 E.L. Hogan, K.C. Joseph and G. Schmidt, *J. Neurochem.*, 17 (1970) 75.
239 J.L. Nussbaum, N. Neskovic and P. Mandel, *J. Neurochem.*, 18 (1971) 1529.
240 B. Gerstl, in J.N. Cumings (Ed.), *Biochemical Aspects of Nervous Diseases*, Plenum, London, 1972, p. 69.
241 J. Wilson, in A.N. Davison and J. Dobbing (Eds.), *Applied Neurochemistry*, Blackwell, Oxford, 1968, p. 401.
242 G-C. Guazzi and L. Van Bogaert, in G.H. Bourne (Ed.), *The Structure and Function of Nervous Tissue. III*, Academic Press, New York, 1969, p. 383.
243 J.H. Austin, *Proc. Soc. Exptl. Biol. Med.*, 100 (1959) 361.
244 H. Jatzkewitz, *Z. Physiol. Chem.*, 311 (1958) 279.
245 B. Hagberg, P. Sourander, L. Svennerholm and H. Voss, *Acta Paediat.*, 49 (1960) 135.

246 B. Hagberg, P. Sourander and L. Svennerholm, *Am. J. Dis. Child.*, 104 (1962) 644.
247 W.T. Norton and S.E. Poduslo, in G.B. Ansell (Ed.), *Variation in the Chemical Composition of the Nervous System as Determined by Developmental and Genetic Factors*, Pergamon, Oxford, 1966, p. 82.
248 J.N. Cumings, E.J. Thompson and H. Goodwin, *J. Neurochem.*, 15 (1968) 243.
249 J. Austin, D. Armstrong and L. Shearer, *Arch. Neurol. Chicago*, 13 (1965) 593.
250 E. Mehl and H. Jatzkewitz, *Biochem. Biophys. Res. Commun.*, 19 (1965) 407.
251 J.N. Cumings and B. Rozdilsky, *Neurology*, 15 (1965) 177.
252 W. Blackwood and J.N. Cumings, *J. Neurol. Neurosurg. Psychiat.*, 17 (1954) 33.
253 J.H. Austin, *J. Neurochem.*, 10 (1963) 921.
254 C.-M. Shaw and C.B. Carlson, *J. Neuropath. Exptl. Neurol.*, 29 (1970) 306.
255 H. Mei Liu, *J. Neuropath. Exptl. Neurol.*, 29 (1970) 441.
256 B.K. Bachhawat, J. Austin and D. Armstrong, *Biochem. J.*, 104 (1967) 15C.
257 J. Austin, K. Suzuki, D. Armstrong, R. Brady, B.K. Bachhawat, J. Schlenker and D. Stumpf, *Arch. Neurol. Chicago*, 23 (1970) 502.
258 K. Suzuki and Y. Suzuki, *Proc. Natl. Acad. Sci. (U.S.)*, 66 (1970) 302.
259 A.C. Crocker and S. Farber, *Medicine*, 37 (1958) 1.
260 E. Klenk, *Z. Physiol. Chem.*, 229 (1934) 151.
261 E. Klenk, *Z. Physiol. Chem.*, 235 (1935) 24.
262 E. Klenk, *Z. Physiol. Chem.*, 262 (1939–40) 128.
263 H. Pilz and H. Jatzkewitz, *J. Neurochem.*, 11 (1964) 603.
264 G. Rouser, G. Kritchevsky and C. Galli, *J. Am. Oil Chem. Soc.*, 42 (1965) 412.
265 R.O. Brady, J.N. Kanfer, M.B. Mock and D.S. Fredrickson, *Proc. Natl. Acad. Sci. (U.S.)*, 55 (1966) 366.
266 B. Gerstl, N. Malamud, R.B. Hayman and P.R. Bond, *J. Neurol. Neurosurg. Psychiat.*, 28 (1965) 540.
267 S. Refsum, *Acta Neurol. Scand. Suppl.*, 38 (1946)
268 E. Klenk and W. Kahlke, *Z. Physiol. Chem.*, 333 (1963) 133.
269 T.R. Skrbic and J.N. Cumings, *Clin. Chim. Acta*, 23 (1969) 17.
270 J. Wilson and R.H.S. Thompson, in J.N. Cumings (Ed.), *Biochemical Aspects of Nervous Diseases*, Plenum, London, 1972, p. 1.
271 M.C. MacBrinn and J.S. O'Brien, *J. Lipid Res.*, 9 (1968) 552.
272 J.S. O'Brien, *J. Theoret. Biol.*, 15 (1967) 307.
273 F.A. Vandenheuvel, *J. Am. Oil Chem. Soc.*, 40 (1963) 455.
274 L. Crome and C.M.B. Pare, *J. Ment. Sci.*, 106 (1960) 862.
275 G.A. Jervis, in F.L. Lyman (Ed.), *Phenylketonuria*, Thomas, Springfield, 1963, p. 96.
276 E.C. Alvord Jr., L.D. Stevenson, F.S. Vogel and R.L. Engle Jr., *J. Neuropath. Exptl. Neurol.*, 9 (1950) 298.
277 N. Malamud, *J. Neuropath. Exptl. Neurol.*, 25 (1966) 254.
278 J.N. Cumings, I.K. Grundt and T. Yanagihara, *J. Neurol. Neurosurg. Psychiat.*, 31 (1968) 334.
279 L. Crome, V. Tymms and L.I. Woolf, *J. Neurol. Neurosurg. Psychiat.*, 25 (1962) 143.
280 H.C. Agrawal, A.H. Bone and A.N. Davison, in H. Bickel, F.P. Hudson and L.I. Woolf (Eds.), *Phenylketonuria and some other Inborn Errors of Amino Acid Metabolism*, Thieme, Stuttgart, 1971, p. 121.
281 B. Gerstl, N. Malamud, L.F. Eng and R.B. Hayman, *Neurology*, 17 (1967) 51.
282 M.E. Smith, *J. Neurochem.*, 10 (1963) 531.
283 C.D. Joel, H.W. Moser, G. Majno and M.L. Karnovsky, *J. Neurochem.*, 14 (1967) 479.
284 J. Dobbing, in *Lipids, Malnutrition and the Developing Brain* (CIBA Found. Symp.), Excerpta Medica, Amsterdam, 1972, p. 9.
285 J. Dobbing and J.L. Smart, in S.A. Barnett (Ed.), *Ethology and Development* (Clinics in Developmental Medicine, No. 47), Heinemann, London, 1973, in the press.
286 J. Dobbing, in A.N. Davison and J. Dobbing (Eds.), *Applied Neurochemistry*, Blackwell, Oxford, 1968, p. 287.
287 A.R. Buchanan and J.E. Roberts, *Proc. Soc. Exptl. Biol. Med.*, 69 (1948) 101.
288 J. Dobbing, *Proc. Roy. Soc. B.*, 159 (1964) 503.
289 W.J. Culley and E.T. Mertz, *Proc. Soc. Exptl. Biol. Med.*, 118 (1965) 233.
290 J. Dobbing and E.M. Widdowson, *Brain*, 88 (1965) 357.

291 M. Winick and A. Noble, *J. Nutr.*, 89 (1966) 300.
292 J.W. Benton, H.W. Moser, P.R. Dodge and S. Carr, *Pediatrics*, 38 (1966) 801.
293 J. Dobbing and J. Sands, *Biol. Neonate*, 19 (1971) 363.
294 N.H. Bass, M.G. Netsky and E. Young, *Arch. Neurol.*, 23 (1970) 289.
295 N.H. Bass, M.G. Netsky and E. Young, *Arch. Neurol.*, 23 (1970) 303.
296 W. Culley, L. Yuan and E. Mertz, *Federation Proc.*, 25 (1966) 674.
297 H.A. Guthrie and M.L. Brown, *J. Nutr.*, 94 (1968) 419.
298 M.A. Fishman, A.L. Prensky and P.R. Dodge, *Nature*, 221 (1969) 552.
299 H. Mohrhauer and R.T. Holman, *J. Neurochem.*, 10 (1963) 523.
300 L. Rathborne, *Biochem. J.*, 97 (1965) 620.
301 C. Galli, H.B. White Jr., and R. Paoletti, *J. Neurochem.*, 17 (1970) 347.
302 H.B. White Jr., C. Galli and R. Paoletti, *J. Neurochem.*, 18 (1971) 869.
303 R. Paoletti and C. Galli, in *Lipids, Malnutrition and the Developing Brain* (CIBA Found. Symp.), Excerpta Medica, Amsterdam, 1972, p. 121.
304 G.Y. Sun, *J. Lipid Res.*, 13 (1972) 56.
305 C. Alling, Å. Bruce, I. Karlsson, O. Sapia and L. Svennerholm, *J. Nutr.*, 102 (1972) 773.
306 L. Svennerholm, C. Alling, Å. Bruce, I. Karlsson and O. Sapia, in *Lipids, Malnutrition and the Developing Brain* (CIBA Found. Symp.), Excerpta Medica, Amsterdam, 1972, p. 141.
307 J.T. Eayrs, in *Scientific Basis of Medicine, Annual Reviews*, Athlone, London, 1966, p. 317.
308 J.T. Eayrs and B. Goodhead, *J. Anat. (Lond.)*, 93 (1959) 385.
309 R. Balázs, S. Kovács, P. Teichgräber, W.A. Cocks and J.T. Eayrs, *J. Neurochem.*, 15 (1968) 1335.
310 J.T. Eayrs, *Acta Anat.*, 25 (1955) 160.
311 R. Balázs, B.W.L. Brooksbank, A.N. Davison, J.T. Eayrs and D.A. Wilson, *Brain Res.*, 15 (1969) 219.
312 A. Cuarón, J. Gamble, N.B. Myant and C. Osorio, *J. Physiol.*, 168 (1963) 613.
313 N.B. Myant and L.A. Cole, *J. Neurochem.*, 13 (1966) 1299.
314 O.A. Schjeide, J. Yamazaki, K. Haack, E. Ciminelli and C.D. Clemente, *Acta Radiol.*, 5 (1966) 185.
315 J. De Vellis, O.A. Schjeide and C.D. Clemente, *J. Neurochem.*, 14 (1967) 499.
316 O.A. Schjeide, R.I-San Lin and J. De Vellis, *Radiation Res.*, 33 (1968) 107.
317 D. McAlpine, N.D. Compston and C.E. Lumsden, *Multiple Sclerosis*, Livingstone, London, 1955.
318 C.W.M. Adams, M.Z.M. Ibrahim and S. Leibowitz, in C.W.M. Adams (Ed.), *Neurohistochemistry*, Elsevier, Amsterdam, 1965, p. 437.
319 Research in Demyelinating Diseases, *Ann. N.Y. Acad. Sci.*, 122, Art. 1. 1965.
320 J.N. Cumings, *Brain*, 76 (1953) 551.
321 C. Alling, M-T. Vanier and L. Svennerholm, *Brain Res.*, 35 (1971) 325.
322 G.R. Webster and R.H.S. Thompson, *Biochim. Biophys. Acta*, 63 (1962) 38.
323 M.E. Smith, *J. Neurochem.*, 16 (1969) 1099.
324 M.E. Smith, *Ann. N.Y. Acad. Sci.*, 122, Art. 1 (1965) 95.
325 B. Maggio, F.A. Cumar and H.J. Maccioni, *J. Neurochem.*, 19 (1972) 1031.
326 A.C. Johnson, A.R. McNabb and R.J. Rossiter, *Biochem. J.*, 45 (1949) 500.
327 J. Domonkos and L. Heiner, *J. Neurochem.*, 15 (1968) 87.
328 J. Domonkos and L. Heiner, *J. Neurochem.*, 15 (1968) 93.
329 R.E. McCaman and E. Robins, *J. Neurochem.*, 5 (1959) 18.
330 E. Holtzmann and A.B. Novikoff, *J. Cell.Biol.*, 27 (1965) 651.
331 D. Nicholls and R.J. Rossiter, *J. Neurochem.*, 11 (1964) 813.
332 J.B. Cavanagh, *J. Neurol. Neurosurg. Psychiat.*, 17 (1954) 163.
333 J.B. Cavanagh, *Guy's Hosp. Rept.*, 112 (1963) 303.
334 A.N. Davison, *Brit. J. Pharmacol.*, 8 (1953) 212.
335 J.F. Berry and W.H. Cevallos, *J. Neurochem.*, 13 (1966) 117.
336 A. Sheltawy and R.M.C. Dawson, *Biochem. J.*, 111 (1969) 157.
337 G.R. Webster, *Biochem. J.*, 57 (1954) 153.
338 W.L. Nelson and C.P. Barnum, *J. Neurochem.*, 6 (1960) 43.
339 G.B. Ansell and T. Chojnacki, *Progr. Biochem. Pharmacol.*, 3 (1967) 189.
340 G.G. Lunt, O.M. Canessa and E. De Robertis, *Nature, New Biology*, 230 (1971) 187.
341 E. De Robertis, *Science*, 171 (1971) 963.
342 C.J. Gómez, J.M. Pasquini, E.F. Soto and E. De Robertis, *J. Neurochem.*, 17 (1970) 1485.
343 E. De Robertis, and S. Fiszer de Plazas, *Biochim. Biophys. Acta*, 219 (1970) 388.

344 M. Parisi, E. Rivas and E. De Robertis, *Science*, 172 (1971) 56.
345 S.R. Levinson and R.D. Keynes, *Biochim. Biophys. Acta*, 288 (1972) 241.
346 L.L. Simpson, *J. Neurochem.*, 19 (1972) 499.
347 E. Bartels and P. Rosenberg, *J. Neurochem.*, 19 (1972) 1251.
348 J.C. Watkins, *J. Theoret. Biol.*, 9 (1965) 37.
349 M. Sundaralingam, *Nature*, 217 (1968) 35.
350 L.E. Tammelin, *Arkiv. Kemi*, 12 (1958) 287.
351 M. Kai and J.N. Hawthorne, *Ann. N.Y. Acad. Sci.*, 165 (1969) 761.
352 J.N. Hawthorne and M. Kai, in A. Lajtha (Ed.), *Handbook of Neurochemistry. III*, Plenum, New York, 1970, p. 491.
353 J.T. Buckley and J.N. Hawthorne, *J. Biol. Chem.*, 247 (1972) 7218.
354 A.C. Birnberger, K.L. Birnberger, S.G. Eliasson and P.C. Simpson, *J. Neurochem.*, 18 (1971) 1291.
355 J.G. Salway and I.E. Hughes, *J. Neurochem.*, 19 (1972) 1233.
356 J.N. Hawthorne and G.L. White, *Trans. Biochem. Soc.*, 1 (1973) 359.
357 P. Rosenberg and T.R. Podleski, *J. Pharmacol. Exptl. Ther.*, 137 (1962) 249.
358 P. Rosenberg and T.R. Podleski, *Biochim. Biophys. Acta*, 75 (1963) 104.
359 P. Rosenberg and F.C.G. Hoskin, *J. Gen. Physiol.*, 46 (1963) 1065.
360 P. Rosenberg and K.Y. Ng, *Biochim. Biophys. Acta*, 75 (1963) 116.
361 P. Rosenberg and E. Condrea, *Biochem. Pharmacol.*, 17 (1968) 2033.
362 P. Rosenberg, *Toxicon*, 8 (1970) 235.
363 N.J. Abbott, T. Deguchi, D.T. Frazier, K. Murayama, T. Narahashi, A. Ottolenghi and C.M. Wang, *J. Physiol.*, 220 (1972) 73.
364 E. Heilbronn, *Progr. Brain Res.*, 36 (1972) 29.
365 E. Heilbronn, *J. Neurochem.*, 16 (1969) 627.
366 J.C. Skou, *Acta Pharmacol Toxicol.*, 10 (1954) 325.
367 J.A. Clements and K.M. Wilson, *Proc. Natl. Acad. Sci. (U.S.)*, 48 (1962) 1008.
368 J.C. Metcalfe, P. Seeman and A.S.V. Burgen, *Mol. Pharmacol.*, 4 (1968) 87.
369 J.C. Metcalfe and A.S.V. Burgen, *Nature*, 220 (1968) 587.
370 P. Seeman, W.O. Kwant, T. Sauks and W. Argent, *Biochim. Biophys. Acta*, 183 (1969) 490.
371 P. Seeman, W.O. Kwant, and T. Sauks, *Biochim. Biophys. Acta*, 183 (1969) 499.
372 S.M. Johnson and A.D. Bangham, *Biochim. Biophys. Acta*, 193 (1969) 92.
373 M.S. Feinstein, *J. Gen. Physiol.*, 48 (1964) 357.
374 M.P. Blaustein, *Biochim. Biophys. Acta.*, 135 (1967) 653.
375 S. Ohki, *Biochim. Biophys. Acta*, 219 (1970) 18.
376 H. Hauser and R.M.C. Dawson, *Biochem. J.*, 109 (1968) 909.
377 G.L. Scherphof, A. Scarpa and A. Van Toorenenbergen, *Biochim. Biophys. Acta*, 270 (1972) 226.
378 F.B. Jungalwala, N. Freinkel and R.M.C. Dawson, *Biochem. J.*, 123 (1971) 19.

Chapter 15

PHOSPHOLIPID METABOLISM AND TRANSPORT OF MATERIALS ACROSS THE CELL MEMBRANE

J.N. HAWTHORNE*

Department of Chemistry, University of Calgary, Alberta (Canada)

In 1954, Hokin and Hokin[1] showed that secretion of amylase by pancreas slices was accompanied by an increased incorporation of $[^{32}P]P_i$ into phospholipids, principally phosphatidic acid and phosphatidylinositol. Since then, similar effects have been reported in a wide variety of tissues, including sympathetic ganglia, salivary gland, thyroid, adrenal medulla, salt glands of sea birds, activated lymphocytes and phagocytosing leucocytes. The one factor common to these very different systems seems to be the involvement of the cell plasma membrane in a transport process, for instance the secretion of hormones, or the uptake of material into the cell.

There is a considerable literature on this subject and since a complete coverage is not possible in one chapter, a selection has been made from the more recent work. For readers who want a more comprehensive account several reviews are available. Schramm[2] deals generally with the secretion of enzymes while Hokin[3] concentrates on the relation between phosphatidylinositol metabolism and zymogen secretion in pancreas. Other reviews[4, 5] deal particularly with phosphoinositide metabolism and function in nervous tissue.

1. TRANSMISSION PROCESSES IN NERVE TISSUE

Phospholipids seem to be involved in at least three ways in the physiological activity of nerve tissue. First, phosphatidylinositol and its precursor, phosphatidic acid, take part in the process of ganglionic transmission. Second, the polyphosphoinositides are concerned in some way with the functioning of the neuronal plasma membrane, including the axolemma. Third, phospholipid is needed for the activity of the 'sodium pump' ATPase in nervous tissue. In all three cases, membrane transport or permeability is involved.

(a) Phospholipid metabolism at the nerve ending

The present concept of chemical transmission at a synapase can be outlined as follows. The arrival of the nerve impulse at the nerve ending causes the release of transmitter (e.g. acetylcholine) from the vesicles in which it is stored into the synaptic cleft. The transmitter then reacts with a specific receptor in the postsynaptic membrane, causing

*Present address: Department of Biochemistry, Medical School, University Park, Nottingham (Great Britain).

a permeability change which leads to excitation, or in some cases inhibition. Biochemical studies of this process have been helped by the isolation of synaptosomes or nerve-ending particles from brain homogenates. The subject has been reviewed by Whittaker[6] and De Robertis[7]. Fig. 1 (p. 425) shows a diagram of the synapse and an electron micrograph of isolated synaptosomes.

Work on the role of phospholipids in synaptic transmission began with the observation of Hokin and Hokin[1, 8] that acetylcholine increased the incorporation of $[^{32}P]\,P_i$ into the phospholipids of brain slices. Phosphatidic acid and phosphatidylinositol were the lipids chiefly affected, though a slight increase in the labelling of phosphatidylcholine was also seen. Diphosphoinositide and triphosphoinositide were not affected by acetylcholine.

Work with subcellular fractions of brain[9, 10] showed a somewhat similar effect of acetylcholine, but in this case it was limited to phosphatidic acid. The incorporation of $[^{32}P]$orthophosphate into phosphatidylinositol was not affected. There are two major pathways for the biosynthesis of phosphatidic acid, the acylation of *sn*-glycero-3-phosphate (1) or the phosphorylation of diglyceride (2). Dihydroxyacetone phosphate may also be acylated and then reduced by NADPH to lysophosphatidic acid in liver and Ehrlich ascites tumour cells[11]. The importance of this route in forming phosphatidic acid in brain has yet to be determined.

sn-glycero-3-phosphoric acid + 2 acyl-CoA → phosphatidic acid + 2 CoASH (1)
1,2-diacyl-*sn*-glycerol + ATP → phosphatidic acid + ADP (2)

Using a crude microsomal fraction and $[^{32}P]$glycerophosphate as precursor, the formation of labelled phosphatidic acid was not stimulated by acetylcholine. However, the transmitter stimulated the production of phosphatidic acid from $[^{32}P]$ATP produced from labelled carbamyl phosphate in the system[10]. This suggested to the Hokins that reaction (2) but not reaction (1) was affected by acetylcholine. The stimulatory effect of acetylcholine on phosphatidic acid labelling by $[^{32}P]$ATP was abolished by freezing, deoxycholate addition and other procedures which damage membranes.

These observations were clarified by the work of Durell and Sodd[12, 13] who showed that the acetylcholine effect was due to synaptosomes present in the microsomal or cytoplasmic fractions used by the Hokins. Labelling of phosphatidic acid by external $[^{32}P]$ATP was not affected by acetylcholine. The sensitive pool was within the synaptosomes, where ATP was produced from $[^{32}P]$orthophosphate by the intra-terminal mitochondria. In the earlier experiments[10] added labelled ATP was presumably hydrolyzed to serve as a source of $[^{32}P]$orthophosphate for these synaptosomal mitochondria. The abolition of the acetylcholine effect by deoxycholate could now be explained by disruption of synaptosomes.

Durell, Garland and Friedel[14] have suggested that the primary effect of acetylcholine is to increase the breakdown of phosphatidylinositol to diglyceride by the phospholipase C route (3) (p. 426). The diglyceride would then increase formation of phosphatidic acid by reaction (2).

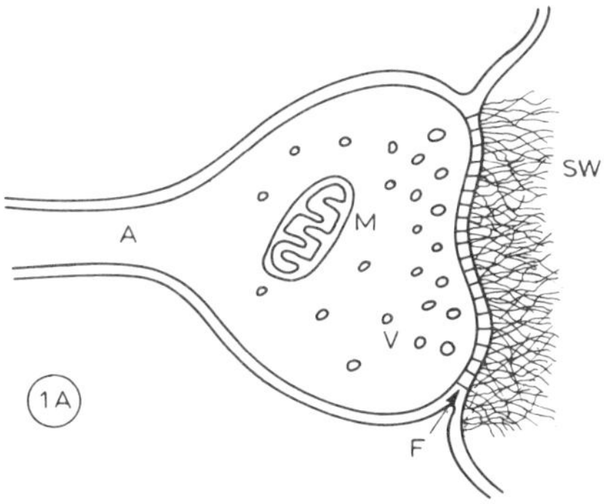

Fig. 1A. Diagram of a synaptic region. A, axon; M, intra-terminal mitochondrion; V, transmitter vesicles; F, synaptic filaments; SW, sub-synaptic web within the post-synaptic neurone.

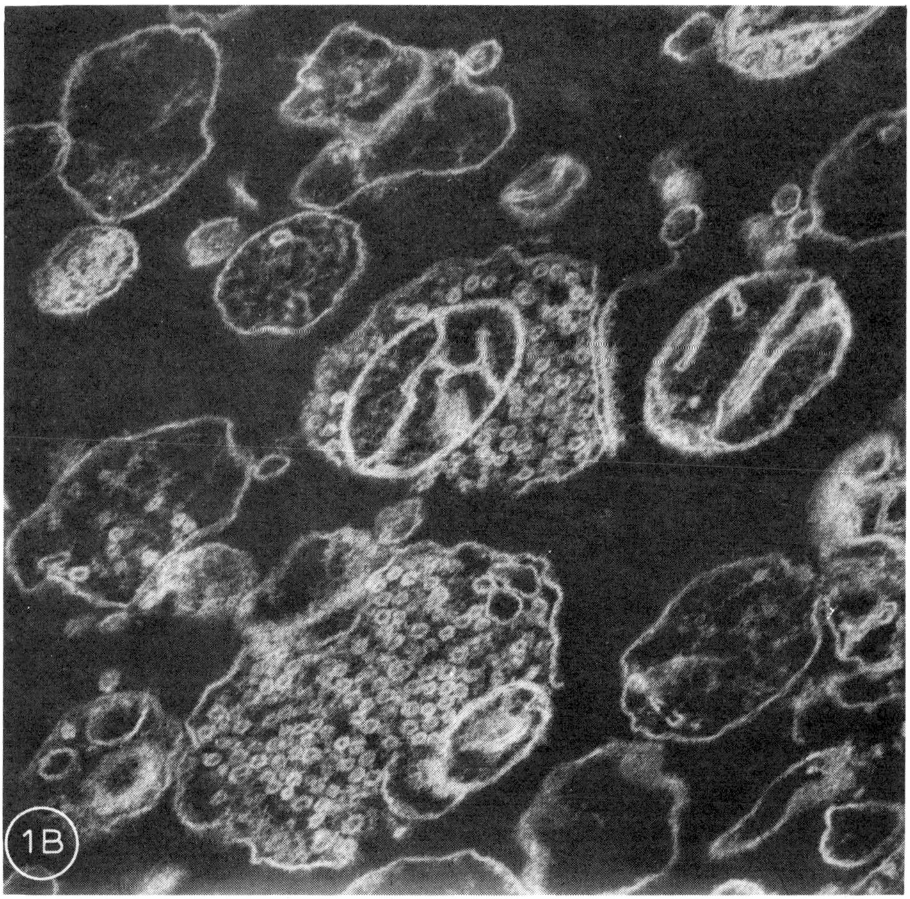

Fig. 1B. Electron micrograph of isolated synaptosomes. Note the post-synaptic membrane (length approx. 0.5 μ) attached to the right-hand side of the central synaptosome. (Kindly supplied by Dr. V.P. Whittaker of the Biochemistry Department, University of Cambridge.)

References p. 438

$$\text{phosphatidylinositol} + H_2O \rightarrow \text{1,2-diacyl-}sn\text{-glycerol} + \text{inositol-1-phosphate} \qquad (3)$$

The experimental evidence[15] leading to this suggestion is not very strong, however, depending on the radioactivity associated with an ill-defined mixture of insoluble barium salts.

Yagihara and Hawthorne[16] have recently made more detailed studies of the effect of acetylcholine on the metabolism of phospholipids in synaptosomes; again the major change was in phosphatidic acid*. The amount of this lipid decreased and its specific radioactivity rose after the incubation of synaptosomes with [^{32}P]P$_i$ in the presence of 10^{-4} *M* acetylcholine. Subsequent osmotic shock and preparation of sub-synaptosomal fractions showed that most of the labelled phosphatidic acid was associated with the synaptic vesicle fraction (unpublished work of the same authors). Another pool of radioactive phosphatidate was associated with the mitochondria, but only the vesicular lipid was affected by acetylcholine. Phosphatidylinositol was not well labelled by ^{32}P under the conditions used and the radioactitivy in the presence of acetylcholine was not significantly increased. It was suggested that the increased labelling of phosphatidate in experiments of this type might be due to reformation of vesicles containing the transmitter.

Though useful information can be obtained from such work with synaptosomes, it provides an inadequate model of synaptic transmission for several reasons. The post-synaptic neurone is not present, but only a fragment of its membrane attached to the nerve terminal. The full effect of acetylcholine on phosphatidylinositol is not seen. Finally, addition of acetylcholine to a nerve-ending preparation does not correctly reflect events *in vivo*, where acetylcholine is discharged from the nerve-ending to interact with the post-synaptic membrane. It seems quite clear though, that the action of acetylcholine is only seen where the plasma membrane of the synaptosome or neurone is intact[10,5]. The drug does not affect partially purified diglyceride kinase[17] and there are conflicting reports about its action on phosphatidylinositol inositol phosphohydrolase[16,108]. It may be that the effect of acetylcholine on one or other of these enzymes is mediated by cyclic AMP, but so far this has not been investigated. The phospholipid changes are not secondary to an increased incorporation of ^{32}P into the nucleotide fraction[16].

Many of the difficulties inherent in work with synaptosomes are avoided by using isolated sympathetic ganglia. Larrabee and his colleagues[18,19] have effectively exploited this technique. In their experiments the isolated ganglion is mounted in a special vessel containing electrodes for stimulation and for recording post-ganglionic action potentials. The ganglion is surrounded by oxygenated physiological saline containing suitable labelled precursors and is accompanied by a control ganglion taken from the same animal, this ganglion not being attached to electrodes. Rat superior cervical ganglia stimulated in this apparatus showed an increased incorporation of [^{32}P]orthophosphate into phosphatidylinositol, but not into phosphatidic acid, phosphatidylcholine or phosphatidylethanolamine. Stimulation also increased the incorporation of [^{14}C]glucose into phosphatidylinositol.

Further work[19] showed that the effect on phosphatidylinositol turnover was associated with synaptic transmission. It did not accompany the conduction of impulses

---

*** Schacht and Agranoff[107] made similar studies with a light synaptosome fraction and found increased labelling of both phosphatidic acid and phosphatidylinositol.**

along nerve trunks. The effect was abolished by tubocurarine, which blocks transmission without preventing acetylcholine release. This suggested that the lipid change occurred in a post-synaptic structure. A similar conclusion was reached by Hokin[20] who still found the acetylcholine effect on phosphatidylinositol turnover in slices from denervated ganglia in which the pre-synaptic elements had degenerated. There was still some acetylcholine-stimulated labelling of phosphatidic acid with ^{32}P. Radioautography designed to detect lipid-bound [^{3}H]inositol suggested that the increased labelling of phosphatidylinositol from [^{3}H]inositol in slices of cat superior cervical ganglia incubated with acetylcholine was associated with the cytoplasm of the post-synaptic neuronal cell body[21].

The phosphatidylinositol effect in cervical sympathetic ganglia has also been shown *in situ* by Larrabee[22]. In the first group of experiments both pre-ganglionic nerves were cut and one ganglion of the anaesthetized rat was stimulated electrically for 5.2 h. Both ganglia were then excised for lipid extraction and separation. To allow for differences in the labelling of phosphate precursors the activity was expressed as the ratio of ^{32}P in phosphatidylinositol to that in phosphatidylcholine, the latter compound being unaffected by stimulation. In the second group of experiments the natural activity of the sympathetic nervous system supplied the impulses to one ganglion, and the pre-ganglionic fibres of the other were cut. A small effect on phosphatidylinositol labelling was seen. This was increased in the third type of experiment by allowing the rats to breathe a mixture of 50% air and 50% nitrogen. The resulting anoxia increased the activity of the sympathetic nervous system. Combining the results from the second and third groups of experiments gave a mean increase of 12% in phosphatidylinositol labelling, which was significant statistically ($P = 0.002$).

The phosphatidylinositol effect has also been demonstrated by Pumphrey[23] in slices of guinea pig cerebral cortex stimulated electrically in the presence of [^{32}P]-orthophosphate. The relative specific radioactivity of this lipid rose from 13.6 to 25.3. Phosphatidic acid showed a smaller increase, of marginal statistical significance, while phosphatidylcholine, phosphatidylethanolamine (including plasmalogen) and cardiolipin showed decreases. Stimulation did not affect the concentration of phosphatidylinositol in the slices. The effects were not simply due to changes in the labelling of water-soluble precursor phosphates.

De Robertis[24] and his colleagues claim that they have isolated from nerve-ending membranes and electric organ a proteolipid which is the acetylcholine receptor. It is further claimed by Lunt, Canessa and De Robertis[25] that the pool of phosphatidylinositol affected by acetylcholine is tightly bound to the receptor proteolipid. These exciting claims call for further study since the data briefly presented so far[24,25] have some unusual features. While diphosphoinositide and triphospoinositide are tightly bound to brain proteins, phosphatidylinositol can be readily removed by neutral solvents[26]. Passage through a Sephadex LH20 column, the procedure used to purify the receptor proteolipid, should remove any phosphatidylinositol, though the polyphosphoinositides would remain. The Sephadex technique has been criticized by Levinson and Keynes[109].

References p. 438

Effects of other neurotransmitters

Hokin[27,28] has shown that several substances other than acetylcholine affect the incorporation of labelled phosphate into the phospholipids of guinea pig brain slices. Noradrenaline (0.1 and 1.0 m*M*) stimulated incorporation into both phosphatidylinositol and phosphatidic acid. A similar effect was found with 10 m*M* 5-hydroxytryptamine, though at a concentration of 1 m*M* this substance only affected phosphatidic acid. Inhibitory effects were found with 1 m*M* dopamine in slices from corpus striatum, but not in other areas of the brain. There was stimulation of phosphatidate labelling in the hypothalamus with 0.1 m*M* dopamine. Phosphate incorporation into phosphatidylinositol, diphosphoinositide and phosphatidic acid, but not into the other phosphatides, was inhibited by 1 m*M* γ-aminobutyric acid in cerebral cortex. It is not easy to interpret these varied effects, some of which required very high concentrations of the transmitter. In addition there is doubt about the role of some of these compounds as neurotransmitters[28]. Hokin suggested that excitatory transmission involves increased labelling of phosphatidate and phosphatidylinositol, while inhibitory transmission involves a decrease[28].

Anticholinergic psychomimetics, such as *N*-methyl-4-piperidyl-diphenyl glycolate increased the labelling by ^{32}P of several phospholipids in brain slices[29]. The effect was associated with the nerve-ending fraction, where both phosphatidic acid and phosphatidylinositol increased in specific radioactivity[30].

Physiological significance of the phospholipid changes

Since phospholipids are membrane components, the possibility that transmitters are altering the permeability of the post-synaptic membrane by changing its phospholipid composition springs to mind at once. It now seems clear that neither phosphatidic acid nor phosphatidylinositol are directly involved in the 'sodium pump' either as carriers or as modifiers of the conformation of a carrier protein[31]. It is still possible, however, that changes in these lipids could regulate ion transport indirectly by altering the properties of a membrane. Acetylcholine had no effect on the incorporation of labelled glycerol into phosphatidylinositol, while the incorporation of labelled phosphate and inositol was increased[32]. This is consistent with removal of the inositol phosphate moiety by an enzyme of the phospholipase C type, leaving diglyceride. Such a change would certainly alter the properties of the membrane containing the original phosphatidylinositol and the necessary enzyme is active in brain[33].

Before discussing the significance of the phospholipid changes, some apparent inconsistencies in the experimental results must be examined. Larrabee's work[18,19,22] comes nearest to the physiological situation and here ganglionic transmission is accompanied by increased turnover of phosphatidylinositol but not of phosphatidic acid. The same applies when cortex slices are stimulated electrically[23]. Application of acetylcholine, usually in concentrations of 0.1 m*M*, with an equivalent amount of eserine, stimulates labelling of both lipids in brain slices[1,8]. With the synaptosomal preparations, however, 0.1 m*M* acetylcholine stimulates only phosphatidic acid* with any certainty[13,16] though when cytoplasm and other organelles are present, phosphatidylinositol is also affected[34]. As already pointed out, the lack of a phosphatidylinositol effect with synaptosomes is

* But see ref. 107.

probably due to the absence of post-synaptic structures, but another important factor is the concentration of cholinergic drug present in the various experiments. Larrabee[35] has pointed out that the high concentrations above will depress rather than stimulate ganglia. One reason for the high concentrations of acetylcholine needed to produce effects on phospholipid metabolism in brain slices may be that eserine prevents its uptake[36]. The phosphonofluoridate Soman inhibits cholinesterase without preventing acetylcholine uptake[37]. It seems that at lower, physiological concentrations of acetylcholine only phosphatidylinositol shows the increased incorporation of phosphate. This is confirmed by experiments with carbamyl choline on brain cytoplasmic fractions[34].

A possible explanation[14] is that there is a small metabolically active pool of phosphatidic acid in equilibrium with ATP through diglyceride kinase. Low concentrations of acetylcholine would not increase the total radioactivity of this pool if as much phosphatidate as is newly synthesized is converted to phosphatidylinositol, or in other words if phosphatidate synthesis is rate-limiting. At higher concentrations of acetylcholine more phosphatidate is produced than can be converted to phosphatidylinositol, so the total radioactivity in both compounds increases.

Pre-ganglionic denervation, which leads to degeneration of the pre-synaptic structures, did not affect the phosphatidylinositol response to acetylcholine *in vitro* but considerably reduced the phosphatidic acid response[20]. This indicates that acetylcholine stimulates phosphatidate turnover in a pre-synaptic structure, presumably the nerve terminal, as well as in the post-synaptic cell body. Presumably this is the effect seen with isolated synaptosomes.

It has been suggested that this pre-synaptic 'phosphatidate effect' is associated with re-synthesis of transmitter-containing vesicles, since the labelled lipid was found in the vesicle membrane[16]. Hokin[5], on the other hand, considers that it may be associated with acetylcholine-induced release of transmitter from the vesicles, though recent studies[38,39] make this explanation unlikely.

The post-synaptic 'phosphatidylinositol effect' may be related to changes in the permeability of the post-synaptic membrane if we accept the results of De Robertis and his group[24,25]. If the effect is associated with the cytoplasm rather than the membrane of the post-synaptic cell body[21], then it may be part of some intracellular transport process[5], possibly related to axonal flow. At present, either view is possible.

Heilbronn[40] has suggested that acetylcholine release may result from the action of phospholipase A on the membrane of the synaptic vesicles. Formation of lysolecithin could cause fusion of vesicle membrane with external nerve ending membrane, leading to exocytosis, as has been suggested also by Howell and Lucy[41]. Phospholipase A has been detected in synaptosomal membranes[40]. If the enzyme requires calcium ions for activity, as do other enzymes of this type, an explanation would be provided for the calcium dependence of acetylcholine release[40].

(b) Polyphosphoinositide metabolism and function

The polyphosphoinositides, diphosphoinositide and triphosphoinositide, differ

References p. 438

markedly from phosphatidylinositol in metabolism and function. Because of their monoesterified phosphate groups the polyphosphoinositides are strongly bound to protein, at least in part by ionic bonds; they also have considerable affinity for calcium ions[4,42,43]. As judged by the incorporation of ^{32}P, phosphatidylinositol metabolism is more active in sympathetic ganglia than in nerve trunks, while the opposite is true of triphosphoinositide[44].

Traditionally, the polyphosphoinositides have been regarded as myelin constituents and there is good evidence for this. The myelin fraction contains most of the recovered di- and triphosphoinositide and their concentrations in purified myelin are 2–3 times greater than in whole brain[45]. Myelinated peripheral nerves contain higher concentrations than unmyelinated nerves[46] and the concentrations in brain rise during myelination[47]. Diphosphoinositide kinase and triphosphoinositide monoesterase, enzymes concerned with the metabolism of triphosphoinositide, increase in activity during myelination[48,49].

It seems clear, however, that there is a second, small, metabolically active pool of polyphosphoinositide which is not associated with myelin. Di- and triphosphoinositide are both present in small amounts before myelination takes place[47] and phosphatidylinositol kinase activity rises well before myelination[48]. Studies of *post mortem* triphosphoinositide breakdown and labelling with ^{32}P indicate that there are two metabolic pools[50]. Studies of sub-synaptosomal fractions show that phosphatidylinositol kinase is associated with the outer membrane of the nerve-ending, suggesting that diphosphoinositide synthesis may be important here and in the axolemma, which is a continuation of the same membrane[51]. Regional distribution figures show that some of the triphosphoinositide and much of the diphosphoinositide of brain occur in structures other than myelin[52].

There is an interesting parallel between the rates of synthesis of acetylcholine and triphosphoinositide, the same holding true for the hydrolysis of these two compounds[4]. Determined *in vitro* and expressed as nmoles per g fresh brain per min the synthetic rates are 50 and 26 respectively, while the hydrolytic enzymes acetylcholinesterase and triphosphoinositide phosphomonoesterase have rates of 6000 and 4000 respectively. The diesterase attacking triphosphoinositide is similarly active[53]. These figures suggest that the polyphosphoinositide is produced and rapidly destroyed at certain sites, as the ^{32}P incorporation studies also indicate.

A theory linking triphosphoinositide turnover with calcium binding and permeability changes in the axolemma during the passage of the action potential has been presented elsewhere[4]. Electrical stimulation of ganglia and nerve trunks appears to reduce the incorporation of ^{32}P into triphosphoinositide (M.G. Larrabee and G.L. White, unpublished work) though as might be expected, the effect is an elusive one. Hokin[5] points out that acetylcholine can inhibit the incorporation of ^{32}P into the diphosphoinositide of brain cytoplasmic fractions. Synaptosomes incubated *in vitro* with labelled inorganic phosphate actively synthesize diphosphoinositide and triphosphoinositide[16]. On the whole there seems to be no specific function of the polyphosphoinositides in synaptic transmission. Future work may show that their role is connected with permeability changes in the neuronal plasma membrane, including the axolemma.

(c) Phospholipids and the 'sodium pump' ATPase

The maintenance of high potassium ion and low sodium ion concentrations within the cell requires the expenditure of energy in the form of ATP. The enzyme system involved is localized in the plasma membrane and has been studied in erythrocytes, brain microsomal fractions, kidney and several other tissues. Studies with ouabain, the cardiac glycoside which inhibits the sodium pump, indicate that the essential enzyme system is an ATPase activated by $Na^{\oplus}$ and $K^{\oplus}$ ions. It is usually considered that the actual transfer of the monovalent cations across the cell membrane depends on a conformational change in a membrane protein which is part of the $Na^{\oplus}/K^{\oplus}$-ATPase, and that the change is caused by a phosphorylation. The reactions can be written as follows, their sum representing the hydrolysis of ATP by ATPase:

$$\text{ATP} + \text{carrier} \xrightarrow{Na^{\oplus}} \text{carrier-P} + \text{ADP} \tag{4}$$

$$\text{carrier-P} \xrightarrow{K^{\oplus}} \text{carrier} + P_i \tag{5}$$

The $Na^{\oplus}/K^{\oplus}$-ATPase appears to be a lipoprotein and an early suggestion[54] was that the phosphorylation involved diglyceride, which was converted to phosphatidic acid. It is now established that the phosphate group is attached to the γ-carboxyl of a glutamic acid residue in the protein. Nevertheless, phospholipid seems to be necessary for the activity of the sodium pump ATPase, though some of the studies in this area are more convincing than others. Several authors fail to realise that extraction of the enzyme system with organic solvents at room temperature could inactivate it because of the interaction of the solvents with the enzyme protein, whether or not activating lipid is removed.

Tanaka and Strickland[55] prepared ATPase from a brain membrane fraction by a method involving treatment with 0.3% sodium deoxycholate followed by ammonium sulphate precipitation and passage through Sephadex G-25 to remove the remaining deoxycholate. The $Na^{\oplus}/K^{\oplus}$-dependent ATPase activity was abolished by the deoxycholate treatment and could be restored by the addition of a commercial lecithin preparation. Later work[56] with brain and kidney $Na^{\oplus}/K^{\oplus}$-ATPases showed that contaminating phosphatidylserine was the effective activator, not lecithin. A similar conclusion was reached by Wheeler and Whittam[57], working with a crude microsomal enzyme from brain, which had been treated with deoxycholate. Phosphatidylserine restored activity, while phosphatidylethanolamine, phosphatidylcholine, phosphatidylinositol and phosphatidic acid were ineffective.

Other recent studies cast doubt on the specific requirement for phosphatidylserine. Lyso-phosphatidylcholine, monocetylphosphate or dicetylphosphate are also effective activators[58] and their roughly equal activity on a molar basis suggests that the precise structure of the lipid may not be important. The activating lipids are equally effective in molecular or micellar dispersions. They may be required to prevent aggregation of enzyme molecules rather than to produce a particular conformation of the enzyme protein. As in blood clotting, it may be that the surface properties, including the ζ-potential, of the

References p. 438

activating lipid are critical. The required surface properties could be obtained from a number of different lipids or lipid mixtures.

Noguchi and Freed[59] have provided further problems by extracting liquid nitrogen-treated brain with chloroform-methanol (1:1, by vol.) at −75°. The lipid-free residue had very little $Na^{\oplus}/K^{\oplus}$-ATPase, but activity was restored by re-addition of the lipid. Pure cholesterol was equally effective, but little activity was restored by the addition of phosphatidylethanolamine, phosphatidylserine or phosphatidylcholine. Sphingomyelin was quite inactive.

A highly purified $Na^{\oplus}/K^{\oplus}$-ATPase has been prepared from brain microsomal material in L.E. Hokin's laboratory[60], using sodium iodide treatment and solubilization with the non-ionic detergent Lubrol. The enzyme appeared to be about 50% pure and was almost completely inhibited by ouabain. It contained the expected sub-unit with a glutamyl phosphate residue (mol. wt. 94 000). There was no cholinesterase activity. The purified enzyme had the following composition (% dry wt.): protein, 44.9; phospholipid, 21.0; cholesterol, 6.6; hexose (from cerebroside?), 2.5; Lubrol, 16.5. During purification the phospholipid–protein ratio remained constant but the cholesterol content fell. Studies of the lipid requirement using this preparation could give more useful results than have been reported with whole brain or microsomal fractions.

2. THE SECRETION OF CATECHOLAMINES BY THE ADRENAL MEDULLA

The adrenal medulla contains a high concentration of the catecholamines adrenaline and noradrenaline stored in characteristic chromaffin granules. The biochemistry of these granules has been well reviewed by Smith[61]. In addition to the catecholamines (20% of dry wt.) the granules contain 15% ATP, 22% lipid (mostly phospholipids and cholesterol, 0.58 moles cholesterol per mole lipid phosphate), 0.16% calcium and soluble acidic proteins known as chromogranins. The lipid contains an unusually large amount of lysolecithin (17% of lipid P).

Hokin et al.[62] used acetylcholine and eserine to stimulate the secretion of catecholamines by slices of guinea pig adrenal medulla incubated in a medium containing [^{32}P]-orthophosphate. Increased incorporation of phosphate into phosphatidylinositol and phosphatidic acid was seen with 10^{-5} M acetylcholine, while the labelling of phosphatidylcholine and phosphatidylethanolamine was unaffected. Trifaró[63] obtained similar results with bovine adrenal medulla, though here there was an increase in the specific radioactivity of phosphatidylcholine as well as in those of phosphatidylinositol and phosphatidic acid. Incorporation of [^{14}C]glycerol was not affected by acetylcholine, nor were the tissue levels of nucleotide and orthophosphate. With ^{32}P, the specific radioactivities of nucleotide and orthophosphate pools were unaffected by acetylcholine. Stimulation did not change the total or individual phospholipid content of the tissue, including that of lysolecithin.

Trifaró[63] also observed that the acetylcholine-induced release of catecholamines reached a maximum after 10 min but that the labelling of the phospholipids only began after this time. When calcium salts were omitted from the medium acetylcholine produced no increase in catecholamine secretion, but the phospholipid changes still took place[64].

The results suggest that there is no direct correlation between the phospholipid effect and catecholamine release. The increased phospholipid synthesis might be part of a recovery process involving resynthesis of membranes. The same can be said of zymogen secretion in the pancreas, which will be discussed later.

Incubation of chromaffin granules with ATP causes release of their contents and hydrolysis of the ATP, some phosphorylation of the granule membrane taking place[65]. About half of the phosphorylated material is extracted by acidified chloroform-methanol and appears to be diphosphoinositide[66].

Secretion of catecholamines from the adrenal medulla involves fusion of the chromaffin granule membrane with the cell plasma membrane followed by release of the granule contents. The empty granule membranes are then detached from the plasma membrane and remain behind in the cell. Their subsequent fate is not clear at present. The unusually high content of lysolecithin in the granule membrane has prompted several studies on the phospholipases of the adrenal medulla. Winkler and Smith[67] have shown that this lysolecithin is formed by a phospholipase A_2 and that the lysosomes of the adrenal medulla contain phospholipases A_1 and A_2 (ref. 68). Phospholipases have not been detected in the chromaffin granule itself and these authors have suggested that the lysolecithin of the granule membrane may be formed by interaction with lysosomes in the Golgi apparatus, where both types of particle may originate. Stimulation of perfused adrenal glands with carbamylcholine causes release of phospholipases A_1 and A_2 together with the catecholamines and the evidence suggests that these are lysosomal phospholipases[69]. The high content of lysolecithin in the membrane of the chromaffin granule is likely to decrease its stability and may also help the process of fusion with the plasma membrane[41]. It is not unlikely therefore, that phospholipase action is involved at some point in the secretory process.

3. ZYMOGEN SECRETION BY THE PANCREAS

Stimulation of pancreatic acinar cells by pancreozymin or acetylcholine induces secretion of enzymes which is accompanied by phospholipid changes[3] similar to those already described in sympathetic ganglia or the adrenal medulla. The secretion process is not unlike that in adrenal medulla. The enzymes are stored, often as inactive zymogens, in special granules which are probably formed in the Golgi region of the cell. Stimulation causes the granule membrane to fuse with the acinar cell membrane and the contents of the granule are released. As in adrenal medulla, the secretion process requires calcium ions, though the phospholipid effect does not.

Pancreas is rich in lipid, mostly triglyceride[70], which is used as an energy source during secretion[71]. Prottey and Hawthorne[72] have studied the biosynthesis of phosphatidic acid and phosphatidylinositol in pancreas. Phosphatidic acid is formed by at least two routes, the acylation of glycerophosphate and the phosphorylation of diglyceride. Phosphatidylinositol biosynthesis follows the pathway via CDPdiglyceride. Neither acetylcholine nor pancreozymin had any effect on diglyceride kinase or any of the other enzymes involved in phosphatidylinositol biosynthesis in homogenates or subcellular

References p. 438

fractions of pancreas[73]. These secretagogues did not affect the labelling of ATP by [^{32}P]orthophosphate in pancreas slices which showed increased labelling of phosphatidylinositol.

Jamieson and Palade[74,75] have made comprehensive studies of the secretion process by electron microscopy. They conclude that the zymogen proteins synthesized on the rough endoplasmic recticulum are then transported to condensing vacuoles in the Golgi complex. These vacuoles become zymogen granules as their contents are concentrated. Inhibition of protein synthesis[76,77] did not affect the transport of zymogens to the granules and to the lumen by exocytosis. These transport processes required energy from oxidative phosphorylation, the ultimate substrate being triglyceride, as already mentioned. Since there seemed to be no synthesis of membrane protein during the secretory cycle, the authors suggested that the cell may re-utilize the granule membrane.

Rough and smooth microsomal fractions, zymogen granules and a plasmalemma fraction have been isolated from guinea pig pancreas[78] and their lipid and enzyme compositions studied[79,80]. The rough microsomal fraction corresponded to rough endoplasmic reticulum and had lower contents of cholesterol and sphingomyelin than the other membranes. It was concluded that there was no mixing of membranes when zymogens were transferred from the rough endoplasmic reticulum to the Golgi area. The smooth microsomal membranes, zymogen granule membrane and plasmalemma were related but different, to judge from lipid and enzyme determinations. The high lysolecithin and fatty acid content of the membranes was attributed to lipase action *post-mortem*. In an earlier study of the zymogen granule lipids, White and Hawthorne[73] also concluded that lysophospholipids were artefacts resulting from enzyme action. In addition to the lipase and phospholipase A_2 secreted by pancreas, the tissue contains a phospholipase A_1 which attacks both phosphatidylinositol and phosphatidylcholine[81]. This enzyme is not found in the zymogen granules and could be of lysosomal origin. The zymogen granule membrane and the chromaffin granule membrane of the adrenal medulla both resemble plasmalemma in their high content of cholesterol and sphingomyelin (see Table 1).

TABLE 1

LIPID COMPOSITION OF GRANULES AND PLASMALEMMA

	Sphingomyelin (% lipid P)	*Cholesterol/ phospholipid (molar ratio)*	*Reference*
Zymogen granule	8.9	0.56	73
Zymogen granule	23.5	0.55	79
Chromaffin granule	10.9	0.58	82
Plasmalemma (pancreas)	19.2	0.51	79

There is no reason, however, to think that the lysolecithin of the chromaffin granule is an artefact, since the adrenal medulla does not have high concentrations of phospholipase like the pancreas.

4. THYROTROPHIN AND THE THYROID GLAND

Pituitary thyrotrophin (TSH) causes thyroid slices to secrete the thyroid hormones and at the same time stimulates the incorporation of ^{32}P into phosphatidylinositol[83]. Scott, Mills and Freinkel[84] made some interesting observations with ^{14}C-labelled precursors, glucose, glycerol, palmitate and oleate. TSH increased the incorporation of ^{14}C from each of these substances into phosphatidylinositol specifically among the phospholipids. The same effect was found with their incorporation into *sn*-1,2-diglyceride but not 1,3-diglyceride or triglyceride, and provides some support for the theory that the phospholipid effect involves hydrolysis of phosphatidylinositol to diglyceride (see page 424). The phosphatidylinositol effect was also seen with isolated pig thyroid cells[85]. Cyclic AMP may act as a 'second messenger' when TSH stimulates thyroid tissue, but its dibutyryl derivative did not produce the same effect on phospholipids as TSH. A similar conclusion about dibutyryl cyclic AMP was reached by Schneider[86], who also showed that TSH did not affect the labelling of the terminal phosphate of ATP in thyroid slices. This is in keeping with the results from pancreas and brain, indicating that the increased incorporation of ^{32}P into phosphatidylinositol is not merely a reflection of increased ATP labelling.

Jungalwala, Freinkel and Dawson[87] studied the metabolism of phosphatidylinositol in pig thyroid, concluding that synthesis by the CDPdiglyceride route was microsomal and that the calcium-dependent phospholipase C type enzyme attacking phosphatidylinositol was in the supernatant fraction. Chlorpromazine (1 m*M*) completely inhibited both synthesis and breakdown of phosphatidylinositol. This drug also inhibits the TSH-induced secretion and phosphatidylinositol turnover in thyroid slices. Earlier studies[88] indicate that chlorpromazine effectively displaces calcium ions from a phosphatidylinositol monolayer. This may account for its inhibition of the catabolic enzyme and of secretion, since calcium ions are required for the secretion process too. Jungalwala et al.[87] also found a more rapid turnover of the inositol moiety than of the phosphate group in thyroidal microsomes. The effect required nucleotide coenzymes.

When thyroid slices were subjected to osmotic shock (transfer from a 290 mosM medium to a 96 mosM medium for 15 min) increased labelling of phosphatidylcholine and phosphatidylethanolamine from ^{32}P was seen[89]. There was a decrease in phosphatidylinositol labelling though the ATP pool became more radioactive, probably because of membrane damage which allowed ^{32}P to enter the slices more readily from the medium. It was considered that the increased synthesis of phosphatidylcholine and phosphatidylethanolamine might be associated with membrane repair. Sphingomyelinase action on the thyroid slices also caused a similar increase.

5. ACTIVATION OF LYMPHOCYTES BY PHYTOHAEMAGGLUTININ

Phytohaemagglutinin (PHA), a glycoprotein from kidney beans, sequentially activates RNA, protein and DNA synthesis and cell division in resting human lymphocytes cultured *in vitro*. This activation is accompanied by a 21-fold stimulation of ^{32}P incorporation into phosphatidylinositol over the first 30 min, while lecithin labelling increases only 0.08-fold[90]. The changes are not due to increased labelling of ATP. Since PHA promotes the secretion of γ-globulin by the lymphocytes, the phospholipid effect may once more be associated with a transport process.

6. PHAGOCYTOSIS AND PHOSPHOLIPID METABOLISM

Karnovsky and his colleagues[91,92] have shown that phagocytosis is accompanied by the increased incorporation of phosphate into phosphatidic acid, phosphatidylinositol and phosphatidylserine. Guinea pig polymorphonuclear leucocytes were made to engulf starch granules 1–3 μ in diameter for this study. Glycolysis is stimulated during phagocytosis by these cells, but the resulting increase in specific radioactivity of glycerophosphate was not sufficient to account for the phospholipid changes.

The process by which zymogen secretion takes place in pancreatic acinar cells is similar to phagocytosis, but in reverse. The same applies to the adrenal chromaffin cells. It seems quite likely that the phospholipid turnover associated with phagocytosis is involved with the changes taking place in the plasma membrane of the leucocytes.

7. PHOSPHOLIPID CHANGES IN DENERVATED MUSCLE

Graf, Hudson and Strickland[93,94] showed that denervation of rat gastrocnemius muscle caused an increase in the total phospholipid concentration but decreases in the concentration of cardiolipin and phosphatidylethanolamine. Experiments with ^{32}P indicated that ATP specific activity rose on denervation and that the rapidly labelled pools of phosphatidylinositol and the polyphosphoinositides were no longer detectable. The total amount of polyphosphoinositide phosphate was reduced to 29% of the original level 15 days after denervation.

In rat hemidiaphragm[95], denervation similarly depressed the incorporation of phosphate into phosphatidylinositol. The effect was seen in strips of muscle without motor end-plates but was slightly greater in strips containing moter end-plates.

8. GENERAL COMMENTS ON THE METABOLISM OF PHOSPHOLIPIDS DURING TRANSPORT PROCESSES

The work summarized in the previous sections serves to show that the increased turnover of phosphatidylinositol is associated with transport processes in a wide variety of tissues. It also accompanies secretion by salivary glands and by the salt-glands of seabirds. It is not seen, however, in the secretion of bicarbonate induced in pancreas by secretion, nor in the secretion of steroid hormones by the adrenal cortex. Absorption processes in the small intestine do not seem to involve it either, though in that tissue phosphatidylinositol incorporates phosphate more rapidly than any other phospholipid except phosphatidic acid[96]. Amino-acid transport in hamster small intestine is associated with increased incorporation of linoleate into phosphatidylcholine, but not with an increased *de novo* synthesis of this lipid[97].

In all the processes which have been discussed in this chapter, the increased synthesis of phosphatidylinositol and phosphatidic acid accompanies transport of material across the plasma membrane. The transport can be initiated by a chemical stimulus applied to this membrane and the process may involve opening and closing the membrane. Why should membrane breakdown and re-synthesis involve only two phospholipids and not all the membrane phospholipids? There is as yet no satisfactory answer to this question, but it is not unreasonable to suppose that the membrane is opened up at specific areas rich in phosphatidylinositol by an enzyme of the phospholipase C type specific for inositol lipids[98]. This would be a more economical process than the destruction of all the membrane phospholipids. In addition, mammalian tissues do not appear to possess an enzyme of the phospholipase C type attacking other phospholipids. Closure of the membrane presumably requires a supply of phosphatidylinositol transported from its site of synthesis on the endoplasmic reticulum. If this is the mechanism, studies with ^{32}P may only detect increased labelling of phosphatidylinositol in the membranes of the endoplasmic reticulum[21], as the unlabelled phosphatidylinositol which has been transferred to the plasmalemma is replaced. The fate of the inositol phosphate released by this type of phospholipase C has not been considered. An interesting recent study by Dawson et al.[99] indicates that it is not inositol 1-phosphate but the inositol 1:2-cyclic phosphate. The reaction is thus:

$$\text{phosphatidylinositol} \rightarrow \text{1, 2-diacyl-}sn\text{-glycerol} + \text{D-inositol 1:2-cyclic phosphate} \quad (6)$$

There is a specific phosphodiesterase in many tissues which converts D-inositol 1:2-cyclic phosphate to D-inositol-1-phosphate[100, 110].

Since the biosynthetic pathway incorporating ^{32}P into phosphatidylinositol differs radically from that labelling phosphatidylcholine and phosphatidylethanolamine it has sometimes been thought the ^{32}P incorporation studies might be misleading. The phosphate of phosphatidylinositol arises from glycerophosphate via phosphatidic acid and CDPdiglyceride or it can be derived more directly from ATP by the diglyceride kinase route to phosphatidic acid. The other phospholipids derive their phosphate from ATP via the phosphorylation of choline or ethanolamine and then the formation of CDPcholine or CDP-ethanolamine. In comparison with choline and ethanolamine phosphates, phosphatidic acid is likely to be highly labelled since it is used in glyceride synthesis as well. In addition, phosphatidic acid can be formed in mitochondrial membranes[101] where the diglyceride kinase route could use ATP formed in the mitochondria. Such considerations could explain differences in labelling patterns between phosphatidylinositol and the nitrogen-containing phospholipids but do not account for the stimulated incorporation of ^{32}P into phosphatidylinositol during secretion, since there is no parallel increase in ATP labelling in any case investigated so far.

9. DIPALMITOYL LECITHIN IN LUNG ALVEOLI

Gaseous exchanges across the alveolar membranes of the lung are very different transport processes from those considered so far in this chapter, but it seems appropriate

References p. 438

to mention a specific phospholipid requirement for the correct functioning of these membranes in certain species. Pattle and Thomas[102] first showed that alveolar washings contained material which was lipoprotein in nature and highly surface-active. This 'lung surfactant' derives its activity from a lecithin which is rich in palmitic acid. This surfactant lecithin lowered the surface tension of water to 0.5 dynes/cm while the total lecithin of lung only lowered it to 20 dynes/cm[103]. The surfactant lecithin contained 77% palmitic acid and a considerable amount of the dipalmitoyl species. Foetal lung is filled with fluid which is expelled at birth and the production of dipalmitoyl lecithin from acetate seems most active in the mature foetus and newborn lamb[104].

Alveoli in those species which produce lung surfactant have an inherent tendency to collapse and are stabilized by the action of the surfactant in reducing the tension at the membrane–air interface[105]. Infants with hyaline membrane disease have difficulty in breathing because of alveolar collapse due to lack of surfactant. Species with a lung structure not dependent on unstable alveoli (frog, turtle, chicken) lack the surfactant dipalmitoyl lecithin[106].

ACKNOWLEDGEMENTS

I am indebted to Dr. Angela White for a preliminary survey of the literature and to Miss Marion Martin for typing the manuscript.

REFERENCES

1 M.R. Hokin and L.E. Hokin, *J. Biol. Chem.*, 209 (1954) 549.
2 M. Schramm, *Ann. Rev. Biochem.*, 36 (1967) 307.
3 L.E. Hokin, *Intern. Rev. Cytol.*, 23 (1968) 187.
4 J.N. Hawthorne and M. Kai, in A. Lajtha (Ed.), *Handbook of Neurochemistry*, Vol. 3, Plenum, New York, 1970, p. 491.
5 L.E. Hokin, in G.H. Bourne (Ed.), *The Structure and Function of Nervous Tissue*, Vol. 3, Academic Press, New York, 1969, p. 161.
6 V.P. Whittaker, *Progr. Biophys. Mol. Biol.*, 15 (1965) 39.
7 E. de Robertis, *Science*, 156 (1967) 907.
8 L.E. Hokin and M.R. Hokin, *Biochim. Biophys. Acta*, 16 (1955) 229.
9 L.E. Hokin and M.R. Hokin, *J. Biol. Chem.*, 233 (1958) 822.
10 L.E. Hokin and M.R. Hokin, *J. Biol. Chem.*, 234 (1959) 1387.
11 B.W. Agranoff and A.K. Hajra, *Proc. Natl. Acad. Sci. (U.S.)*, 68 (1971) 411.
12 J. Durell and M.A. Sodd, *J. Biol. Chem.*, 239 (1964) 747.
13 J. Durell and M.A. Sodd, *J. Neurochem.*, 13 (1966) 487.
14 J. Durell, J.T. Garland and R.O. Friedel, *Science*, 165 (1969) 862.
15 J. Durell, M.A. Sodd and R.O. Friedel, *Life Sci.*, 7, II (1968) 363.
16 Y. Yagihara and J.N. Hawthorne, *J. Neurochem.*, 19 (1972) 355.
17 E.G. Lapetina and J.N. Hawthorne, *Biochem. J.*, 122 (1971) 171.
18 M.G. Larrabee, J.D. Klingman and W.S. Leicht, *J. Neurochem.*, 10 (1963) 549.
19 M.G. Larrabee and W.S. Leicht, *J. Neurochem.*, 12 (1965) 1.
20 L.E. Hokin, *J. Neurochem.*, 13 (1966) 179.
21 L.E. Hokin, *Proc. Natl. Acad. Sci. (U.S.)*, 53 (1965) 1369.
22 M.G. Larrabee, *J. Neurochem.*, 15 (1968) 803.
23 A.M. Pumphrey, *Biochem. J.*, 112 (1969) 61.

24 E. de Robertis, *Science*, 171 (1971) 963.
25 G.G. Lunt, O.M. Canessa and E. de Robertis, *Nature New Biol.*, 230 (1971) 187.
26 J.N. Hawthorne and P. Kemp, *Advan. Lipid Res.*, 2 (1964) 127.
27 M.R. Hokin, *J. Neurochem.*, 16 (1969) 127.
28 M.R. Hokin, *J. Neurochem.*, 17 (1970) 357.
29 L.G. Abood and J.H. Biel, *Intern. Rev. Neurobiol.*, 4 (1962) 217.
30 E. Heilbronn and L. Widlund, *J. Neurochem.*, 17 (1970) 1039.
31 H. McIlwain, *Chemical Exploration of the Brain*, Elsevier, Amsterdam, 1963, p. 147.
32 L.E. Hokin and M.R. Hokin, *J. Biol. Chem.*, 233 (1958) 818.
33 W. Thompson, *Can. J. Biochem.*, 45 (1967) 853.
34 C.M. Redman and L.E. Hokin, *J. Neurochem.*, 11 (1964) 155.
35 M.G. Larrabee, in S.S. Kety, E.V. Evarts and H.L. Williams (Eds.), *Sleep and Altered States of Consciousness*, Williams and Wilkins, Baltimore, 1967, p. 64.
36 J. Schuberth and A. Sundwall, *J. Neurochem.*, 14 (1967) 807.
37 R.L. Polak and M.M. Meeuws, *Biochem. Pharmacol.*, 15 (1966) 989.
38 B. Collier and H.S. Katz, *Brit. J. Pharmacol.*, 39 (1970) 428.
39 D.A. Brown, K.B. Jones, J.V. Halliwell and J.P. Quilliam, *Nature*, 226 (1970) 958.
40 E. Heilbronn, in P.B. Bradley and R.W. Brimblecombe (Eds.), *Biochemical and Pharmacological Mechanisms Underlying Behaviour*, Elsevier, Amsterdam, 1972, p. 29
41 J.I. Howell and J.A. Lucy, *FEBS Letters*, 4 (1969) 147.
42 H.S. Hendrickson and J.G. Fullington, *Biochemistry*, 4 (1965) 1599.
43 H.S. Hendrickson and J.L. Reinertsen, *Biochem. Biophys. Res. Commun.*, 44 (1971) 1258.
44 Y. Yagihara, J.G. Salway and J.N. Hawthorne, *J. Neurochem.*, 16 (1969) 1133.
45 J. Eichberg and R.M.C. Dawson, *Biochem. J.*, 96 (1965) 644.
46 A. Sheltawy and R.M.C. Dawson, *Biochem. J.*, 100 (1966) 12.
47 M.A. Wells and J.C. Dittmer, *Biochemistry*, 6 (1967) 3169.
48 J.G. Salway, J.L. Harwood, M. Kai, G.L. White and J.N. Hawthorne, *J. Neurochem.*, 15 (1968) 221.
49 J. Eichberg and G. Hauser, *Federation Proc.*, 26 (1967) 675.
50 A. Sheltawy and R.M.C. Dawson, *Biochem. J.*, 111 (1969) 147.
51 J.L. Harwood and J.N. Hawthorne, *J. Neurochem.*, 16 (1969) 1377.
52 J. Eichberg and G. Hauser, *Trans. Am. Soc. Neurochem.*, 1 (1970) 40.
53 K.M.W. Keough and W. Thompson, *J. Neurochem.*, 17 (1970) 1.
54 L.E. Hokin and M.R. Hokin, *Nature*, 184 (1959) 1068.
55 R. Tanaka and K.P. Strickland, *Arch. Biochem. Biophys.*, 111 (1965) 583.
56 L.J. Fenster and J.H. Copenhaver Jr., *Biochim. Biophys. Acta*, 137 (1967) 406.
57 K.P. Wheeler and R. Whittam, *Nature*, 225 (1970) 449.
58 R. Tanaka, T. Sakamoto and Y. Sakamoto, *J. Membrane Biol.*, 4 (1971) 42.
59 T. Noguchi and S. Freed, *Nature New Biol.*, 230 (1971) 148.
60 S. Uesugi, N.C. Dulak, J.F. Dixon, T.D. Hexum, J.L. Dahl, J.F. Perdue and L.E. Hokin, *J. Biol. Chem.*, 246 (1971) 531.
61 A.D. Smith, in P.N. Campbell (Ed.), *The Interaction of Drugs and Subcellular Components in Animal Cells*, Churchill, London, 1968, p. 239.
62 M.R. Hokin, B.G. Benfey and L.E. Hokin, *J. Biol. Chem.*, 233 (1958) 814.
63 J.M. Trifaró, *Mol. Pharmacol.*, 5 (1969) 382.
64 J.M. Trifaró, *Mol. Pharmacol.*, 5 (1969) 424.
65 J.M. Trifaró and J. Dworkind, *Mol. Pharmacol.*, 7 (1971) 52.
66 J.T. Buckley, Y.A. Lefebvre and J.N. Hawthorne, *Biochim. Biophys. Acta*, 239 (1971) 517.
67 H. Winkler and A.D. Smith, *Arch. Exptl. Pathol. Pharmakol.*, 261 (1968) 379.
68 A.D. Smith and H. Winkler, *Biochem. J.*, 108 (1968) 867.
69 A.D. Smith, *Biochem. J.*, 114 (1969) 72P.
70 C. Prottey and J.N. Hawthorne, *Biochem. J.*, 101 (1966) 191.
71 J.D. Jamieson and G.E. Palade, *J. Cell. Biol.*, 39 (1968) 589.

72 C. Prottey and J.N. Hawthorne, *Biochem. J.*, 105 (1967) 379.
73 D.A. White and J.N. Hawthorne, *Biochem. J.*, 120 (1970) 533.
74 J.D. Jamieson and G.E. Palade, *J. Cell. Biol.*, 34 (1967) 577.
75 J.D. Jamieson and G.E. Palade, *J. Cell. Biol.*, 34 (1967) 597.
76 J.D. Jamieson and G.E. Palade, *J. Cell. Biol.*, 39 (1968) 580.
77 J.D. Jamieson and G.E. Palade, *J. Cell. Biol.*, 48 (1971) 503.
78 J. Meldolesi, J.D. Jamieson and G.E. Palade, *J. Cell. Biol.*, 49 (1971) 109.
79 J. Meldolesi, J.D. Jamieson and G.E. Palade, *J. Cell. Biol.*, 49 (1971) 130.
80 J. Meldolesi, J.D. Jamieson and G.E. Palade, *J. Cell. Biol.*, 49 (1971) 150.
81 D.A. White, D.J. Pounder and J.N. Hawthorne, *Biochim. Biophys. Acta*, 242 (1971) 99.
82 H. Blaschko, H. Firemark, A.D. Smith and H. Winkler, *Biochem. J.*, 104 (1967) 545.
83 N. Freinkel, *Endocrinology*, 61 (1957) 448.
84 T.W. Scott, S.C. Mills and N. Freinkel, *Biochem. J.*, 109 (1968) 325.
85 T.W. Scott, N. Freinkel, J.H. Klein and M. Nitzan, *Endocrinology*, 87 (1970) 854.
86 P.B. Schneider, *J. Biol. Chem.*, 244 (1969) 4490.
87 F.B. Jungalwala, N. Freinkel and R.M.C. Dawson, *Biochem. J.*, 123 (1971) 19.
88 H. Hauser and R.M.C. Dawson, *Biochem. J.*, 109 (1968) 909.
89 P.B. Schneider, *J. Biol. Chem.*, 245 (1970) 6281.
90 D.B. Fischer and G.C. Mueller, *Proc. Natl. Acad. Sci. (U.S.)*, 60 (1968) 1396.
91 M.L. Karnovsky and D.F.H. Wallach, *J. Biol. Chem.*, 236 (1961) 1895.
92 M.L. Karnovsky, in R.M.C. Dawson and D.N. Rhodes (Eds.), *Metabolism and Physiological Significance of Lipids*, Wiley, New York, 1964, p. 501.
93 G.L.A. Graff, A.J. Hudson and K.P. Strickland, *Biochim. Biophys. Acta*, 104 (1965) 524.
94 G.L.A. Graff, A.J. Hudson and K.P. Strickland, *Biochim. Biophys. Acta*, 104 (1965) 543.
95 H.R. Steinberg and J. Durell, *J. Neurochem.*, 18 (1971) 277.
96 M.I. Gurr, W.F.R. Pover, J.N. Hawthorne and A.C. Frazer, *Nature*, 197 (1963) 79.
97 M.E. McLeod and R. Bressler, *Proc. Soc. Exptl. Biol. Med.*, 130 (1969) 268.
98 R.S. Atherton and J.N. Hawthorne, *European J. Biochem.*, 4 (1968) 68.
99 R.M.C. Dawson, N. Freinkel, F.B. Jungalwala and N. Clarke, *Biochem. J.*, 122 (1971) 605.
100 R.M.C. Dawson and N. Clarke, *Biochem. J.*, 127 (1972) 113.
101 E.H. Shephard and G. Hübscher, *Biochem. J.*, 113 (1969) 429.
102 R.E. Pattle and L.C. Thomas, *Nature*, 189 (1961) 844.
103 T.E. Morgan, T.N. Finley and H. Fialkow, *Biochim. Biophys. Acta*, 106 (1965) 403.
104 N. Chida and F.H. Adams, *J. Lipid Res.*, 8 (1967) 335.
105 R.E. Pattle, *Physiol. Rev.*, 45 (1965) 48.
106 W.R. Harlan Jr., J.H. Margraf and S.I. Said, *Am. J. Physiol.*, 211 (1966) 855.
107 J. Schacht and B.W. Agranoff, *J. Biol. Chem.*, 247 (1972) 771.
108 O. Canessa de Scarnati and G. Rodriguez de Lores Arnaiz, *Biochim. Biophys. Acta*, 270 (1972) 218.
109 S.R. Levinson and R.D. Keynes, *Biochim. Biophys. Acta*, 288 (1972) 241.
110 N. Clarke and R.M.C. Dawson, *Biochem. J.*, 130 (1972) 229.

Chapter 16 '

THE PHOSPHOLIPID COMPOSITION OF MAMMALIAN TISSUES

DAVID A. WHITE*

Department of Physiological Chemistry, The Johns Hopkins Medical School, Baltimore, Maryland 21205 (U.S.A.)

It is interesting to follow the progression from earlier studies on the phospholipid composition of mammalian whole tissues to the more detailed analyses of the present day. In an effort to correlate composition with membrane integrity and function (e.g. Getz[1]) emphasis has shifted from whole tissue analyses to analyses of the various membranous subcellular fractions with detailed studies not only of individual phospholipids and their overall fatty acid composition but also the molecular species of a particular phospholipid.

As discussed in Chapter 3, effective analysis of phospholipids from a tissue depends first on a quantitative extraction and removal of non-lipid contaminants followed by an accurate and reliable method of separation of intact phospholipids or of their derivatives. The data compiled in the following tables include representative values from the literature and references to other determinations of the same type. The variation in results probably reflects to some extent the methods used in extraction and separation, and for this reason the method of separation has been included. However, despite the differences in methods there is generally quite good agreement between analyses for a given tissue. The modern techniques of one- and two-dimensional thin-layer chromatography on silica impregnated with various ions[2] appear to be very reliable and reproducible with added advantages of speed and of requiring much less material for analysis than older methods. The work of Rouser and co-workers (for reviews see refs. 3, 4 and 5) has greatly contributed to knowledge and technique in this field.

The results of phospholipid analyses have been expressed here as percentage of total phospholipid to one place of decimals. Although this is perhaps unfair to some authors who have presented results to two places of decimals ± S.D. it allows easier comparison of values for the same tissue. Where standard deviations appear in the original reference this is indicated in the tables. The data compiled are essentially for tissues of mammalian species but where references have contained values from non-mammalian species these have also been included. However, no detailed compilation of phospholipids of tissues of non-mammalian species has been attempted.

A disadvantage of many thin-layer chromatographic procedures is that they do not allow measurement of individual plasmalogens and many analyses have not contained any

*Present address: Department of Biochemistry, The Medical School, University of Nottingham, Nottingham NG7 2RD (Great Britain).

reference to plasmalogen levels. In tissues where plasmalogen content is significant (e.g. brain, heart) and where it has been assayed, the values are included. Plasmalogen analyses of various rat tissues are given by Wood et al.[6] and a comprehensive review of alkyl- and alkenyl glycerol ethers has recently appeared (Snyder[7]).

Generally all mammalian tissues contain qualitatively the same phospholipids. Quantitative differences are self-evident from the tables and as whole tissue phospholipid composition reflects the contribution of the various subcellular fractions, discussion is left for Chapter 9.

Phospholipid is found almost exclusively in the membranous structures of the cell[8], very little appearing in the supernatant fraction[9]. An early review of the lipid composition of subcellular fractions was compiled by Fleischer and Rouser[10].

Fatty acids analyses are expressed as percentage by weight of total fatty acid derived from a given phospholipid. The use of specific phospholipases A_1 and A_2 has allowed the positional distribution of fatty acids of a particular phospholipid to be determined. Further, with the introduction of argentation chromatography (for a review of development and uses see Morris[11]) the separation of molecular species of modified and intact phospholipids according to the degree of unsaturation of their constituent fatty acids has been achieved. Thus comprehensive analyses of the fatty acids of phospholipids can now be carried out using a combination of specific phospholipases, argentation chromatography and gas-liquid chromatography of both acids and diglycerides. So far, these methods have been used with success only with phosphatidylcholine[12–15] and to some extent with phosphatidylethanolamine[16] and phosphatidylglycerol[17]. However, a complete analysis of mono-, di- and triphosphoinositides of bovine brain using the above methods and a specific brain phosphodiesterase has been reported by Holub, Kuksis and Thompson[18] and a report on the separation of molecular species of phosphatidylinositol has also recently appeared[19]. Unfortunately, space does not allow the inclusion of positional distribution and pairing of fatty acids as each reference would require several pages. Such complete analyses have been reported for phosphatidylcholine of heart[12], kidney[12], plasma[12], lung[14], brain[14], liver[14] and intestinal mucosa[15] of rat; lung, brain, liver and kidney of rabbit[14], pig[14] and cow[14] and human erythrocytes and plasma[13]. Thus Marai and Kuksis[13] have identified over sixty molecular species accounting for about 98% of total phosphatidylcholine of human plasma and erythrocytes, but only about thirty of these occurred in concentrations of over 1%. Generally the molecular species contained one saturated (C_1) and one unsaturated (C_2) fatty acid but there appeared to be no preferential pairing of any one saturated with any other unsaturated fatty acid. Since the two major unsaturated fatty acids of animal lecithins are linoleic (18:2) and arachidonic (20:4) acids the two major species are dienoic and tetraenoic. Although qualitatively similar, the molecular species of lecithin from various tissues of the same animal species differ significantly in composition, positional distribution and pairing of fatty acids. However, there were quite close similarities in the molecular composition of lecithins of lung and also brain irrespective of species, suggesting some tissue specificity[14]. Metabolic studies indicate that the dienoic and tetraenoic species turn over at differing rates and this has been attributed to the type of fatty acid involved in the turnover.

How much the different molecular species of phospholipids contribute to the essential structure and metabolic activities of membranes is the goal of future work. By varying the diet of rats Van Golde and Van Deenen[20] showed that the molecular species of rat liver phospholipids could be altered somewhat without any apparent changes in the physical properties of the tissue membranes. A similar effect on the fatty acids of *E. coli* was shown by varying the growth medium. It is to be hoped that detailed studies on the composition and metabolism of phospholipids in biological membranes and knowledge of lipid–lipid and lipid–protein interactions from both natural and model systems (Chapters 7 and 8) will relate the structure of a given phospholipid to its rôle in its natural membrane.

ACKNOWLEDGEMENTS

The author is indebted to the Wellcome Trust, London for a travel grant and to Miss Marion Martin and Miss JoAnn Lopresti for invaluable help in the preparation and typing of the manuscript.

EDITORIAL NOTE

Because of space limitations, only one set of data is presented for each tissue or fraction. A footnote gives references to other analyses. The editors regret that this selection was necessary. We certainly do not intend to imply that the figures quoted are necessarily better than those referred to in the footnotes.

NOTES ON THE TABLES

The methods of separation used in determining the phospholipid analyses quoted in Tables 1–19 are as follows:

A. Silicic acid column chromatography.
B. Deacylation of phospholipids followed by either: (*i*) two-dimensional paper chromatography, (*ii*) paper electrophoresis, or (*iii*) anion exchange column chromatography.
C. One-dimensional thin-layer chromatography.
D. Two-dimensional thin-layer chromatography.
E. One-dimensional chromatography on silicic acid-impregnated paper.
F. Two-dimensional chromatography on silicic acid-impregnated paper.
G. One-dimensional chromatography on formaldehyde-impregnated paper.

Because of the wide variation in solvent systems used by different workers no attempt has been made to quote any particular solvent system.

An asterisk denotes that mean values ± S.D. are given in the reference.

References p. 479

TABLE 1

THE PHOSPHOLIPID CONTENT OF TISSUES

Tissue	*Species*	*Reference*	*Lipid P mg/g wet weight*	*Other references*
Brain	Sheep	24	2.14	9
	Lamb	24	1.07	
	Rat	26	1.85	27, 28, 29
	Rabbit	30	2.35	
	Guinea pig	35	1.96	
Heart	Human	38	0.66	
	Bovine	38	0.45	42, 43
	Sheep	24	0.82	9
	Lamb	24	0.71–0.67	
	Rat	38	0.47	
	Mouse	38	0.56	
	Rabbit	30	0.83	
	Frog	38	0.35	
Liver	Human	45	1.32	44
	Bovine	45	0.80	
	Sheep	24	1.51	9
	Lamb	24	0.88	
	Rat	52	1.27	9, 45, 51
	Mouse	45	0.80	
	Frog	45	0.46	
Kidney	Human	45	0.58	
	Bovine	45	0.41	
	Sheep	24	1.05	9
	Lamb	24	0.55	
	Rat	45	1.31	51
	Mouse	45	0.83	
	Frog	45	0.37	
Spleen	Human	45	0.79	
	Bovine	54	0.63	45
	Sheep	9	1.11	
	Rat	45	0.55	
	Mouse	45	0.52	
Lung	Human	55	0.14	
	Bovine	55	0.48	
	Rat	55	0.56	
	Mouse	55	0.65	
	Dog	57	0.64	
Skeletal muscle	Human	38	0.53	
	Bovine (longissimus dorsi)	59	0.38	38
	Rat	38	0.35	
	Mouse	38	1.41	
	Guinea pig	58	0.42	
	Pigeon	59	0.24	
	Frog	38	0.13	

TABLE 1 (continued)

Tissue	*Species*	*Reference*	*Lipid P mg/g wet weight*
Testes	Rat	61	0.57
	Ram	62	0.48
Peripheral nerve	Sheep (sciatic)	71	1.67
	Bovine (splenic)	71	0.42
	Rabbit (sciatic)	71	1.95
	Monkey (sciatic)	71	1.67
	Hen (sciatic)	71	1.66
	Crab, claw	71	0.38
	leg	71	0.35
	Lobster, claw	71	0.41
	leg	71	0.52
Pancreas	Bovine	73	0.87
Thyroid	Bovine	75	0.24
Thymus	Rat	76	0.39–0.48
Adrenals	Sheep, medulla	9	2.26[a]
	cortex	9	1.39[a]
Pituitary	Human	78	0.61
	Bovine, anterior	78	0.95
	posterior	78	0.98
Skin	Guinea pig	86	0.32
	Mouse	87	0.26
Pineal	Human	82	0.66
	Bovine	82	1.04
	Sheep	82	0.96
Gastric mucosa	Frog	88	0.19
Corpus luteum	Ox	91	1.24
Adipose tissue	Bat	94	0.37
Bladder	Toad	93	0.31

[a]Converted from reference using water content of 75%.

References p. 479

TABLE 2

PHOSPHOLIPID COMPOSITION OF BRAIN

Source	*Human*			*Sheep*	*Lamb*	*Rat*		
	18-year-old male	*55-year-old male*					*Glia*	*Neurones*
		Grey matter	*White matter*					
Reference	*21*	*22*	*22*	*24*	*24*	*25*	*36*	*36*
Phosphatidylcholine	30.3	39.1	30.7	37.3	49.7	36.8	37.0	37.0
Choline plasmalogen				0.9	0.7		1.6	1.6
Lysophosphatidylcholine								
Phosphatidylethanolamine	36.2	40.0	34.1	7.7	13.3	36.4	19.0	17.1
Ethanolamine plasmalogen				16.5	11.8		14.2	13.9
Phosphatidylinositol	2.6			2.1	2.6	3.1	3.0	2.9
Diphosphoinositide						0.6[n]		
Triphosphoinositide						1.9[n]		
Phosphatidylserine	17.7	12.6	15.7	9.2	5.8	11.8	9.9	9.8
Sphingomyelin	13.2	8.3	19.5	12.8	4.2	5.7	5.2	5.6
Phosphatidylglycerol					0.2			
Diphosphatidylglycerol				2.0	0.3	2.2	5.2	5.7
Phosphatidic acid				2.6	0.9	1.2		
Other				3.0[a]	3.2[a]		3.3[g]	3.2[f]
Recovery	100.0	100.0	100.0	94.1	92.7	97.8	98.4	96.8
Method of separation	D*	E	E	B	B	F*	G	G

[a]alkyl ether lipid; [b]1.0% serine plasmalogen; [c]35.4% plasmalogen; [d]33.2% plasmalogen; [e]50.5% plasmalogen; [f]2.0% serine plasmalogen; [g]1.7% serine plasmalogen; [h]not determined; [j]56.0% plasmalogen and sphingomyelin; [k]calculated from ref. 181; 45 min *post mortem*, on ice; [m]calculated from ref. 181; thiopentone anaesthesia, brain fixed *in situ* with liquid N_2; [n]calculated from ref. 181; head frozen in liquid N_2 immediately after decapitation; [o]includes phosphatidylserine; [p]presumably cardiolipin, phosphatidic acid and triphosphoinositide; [q]9.7% plasmalogen; [r]10.7% plasmalogen; [s]16.2% plasmalogen.

Other brain analyses: human[23], sheep[9], rat[23,26–29], rabbit[23], bovine[32–34].

			Bovine				*Rabbit*	*Guinea pig*	*Pigeon*	*Dogfish*
Immature brain				*White matter frontal*		*Optic nerve*				
Neurones	*Astroglia*	*Oligo dendroglia*		*Cortex*	*Medulla*					
182	*182*	*183*	*31*	*23*	*23*	*37*	*30*	*35*	*23*	*23*
55.2	51.2	47.3	29.2	24.1	25.1	17.5	31.2	32.6	31.1	33.6
			0.0				1.0	tr		
							0.4			
25.2	28.6	22.5	12.1	42.7	35.9	8.4	12.7	12.8	37.9	28.1
			21.1				22.6	17.9		
6.8	4.9	6.6	3.2	12.5[o]	3.0	h	3.0	2.9	4.2	3.0
			0.2[k]					0.6[m]		
			3.2[k]					2.9[m]		
5.4	7.3	7.6	16.6		16.6	9.5	15.8[b]	12.3	10.7	8.1
4.4	5.2	11.4	12.5	16.9	16.6		12.4	8.3	10.4	8.2
			0.7		2.1	h	2.0	1.4		2.2
			0.5			h		0.9		
9.7[p]	10.7[p]	10.8[p]	2.1[a]					3.2[a]		
106.7[q]	107.8[r]	106.1[s]	98.0	96.2	99.3[c]	91.4[j]	101.1	92.5	94.3[d]	83.2[e]
			B	C	C	B	D	B	C	C

TABLE 3

PHOSPHOLIPID COMPOSITION OF HEART[a]

Source	*Rat*[b]	*Mouse*	*Bovine*	*Calf*	*Human*	*Frog*[c]	*Sheep*	*Lamb*	*Rabbit*
Reference	*38*	*38*	*31*	*25*	*38*	*38*	*24*	*24*	*30*
Phosphatidylcholine }	36.0	40.6	24.2	42.8	40.0	39.7	25.5	39.1	27.1
Choline plasmalogen }			17.5				20.4	7.6	15.1
Lysophosphatidylcholine	0.6	1.6		0.4	3.5	0.3			0.7
Phosphatidylethanolamine }	29.7	28.9	16.5	26.6	26.3	31.5	12.4	10.4	15.5
Ethanolamine plasmalogen }			11.0				11.0	14.7	15.4
Lysophosphatidylethanolamine					1.5				
Phosphatidylinositol	3.7	4.3	4.1	3.4	6.1	4.2	4.7	3.7	3.6
Phosphatidylserine	3.2	2.8	2.4	3.4	2.7	5.1	2.2	2.4	3.3[e]
Sphingomyelin	3.1	3.5	11.5	9.1	4.9	6.5	5.5	6.4	5.0
Phosphatidylglycerol	1.0	0.6		0.6	0.6	0.2			
Diphosphatidylglycerol	11.2	11.2	8.9	12.1	9.0	9.3	5.8	4.8	11.6
Phosphatidic acid	0.3	0.1	2.2		0.2	0.1	0.4	0.5	
Unidentified	11.3	6.3		0.5	5.0	3.1	3.2[d]	3.9[d]	1.0
Recovery	100.0	99.9	96.1	98.9	99.8	99.0	92.2	93.7	98.3
Method of separation	D*	D*	B	F*	D*	D*	B	B	D

[a]Values quoted from ref. are specifically for heart muscle; [b]Average of 5 values; [c]Average of 3 values; [d]Alkyl ether lipid; [e]0.7% phosphatidalserine.
Other heart analyses: rat[37], bovine[31,39–43], calf[25], sheep[9].

TABLE 4

PHOSPHOLIPID COMPOSITION OF LIVER

Source	*Human*	*Rat*	*Sheep*	*Foetal lamb*	*Bovine*	*Mouse*	*Frog*
Reference	*45*	*25*	*24*	*24*	*31*	*45*	*45*
Phosphatidylcholine	43.6	50.8	39.7	46.4	54.2	45.6	46.5
Choline plasmalogen			0.8	0.4	1.5		
Lysophosphatidylcholine	1.3	1.4				0.7	0.8
Phosphatidylethanolamine	27.9	25.2	27.9	21.0	9.4	25.1	25.6
Ethanolamine plasmalogen			3.6	2.0	3.6		
Phosphatidylinositol	8.6	7.2	8.2	5.6	7.9	7.7	7.3
Lysophosphatidylinositol		0.4					
Phosphatidylserine	3.1	3.2	2.3	3.7	4.2	3.9	3.9
Lysophosphatidylserine		0.2					
Sphingomyelin	4.6	4.2	5.0	8.7	5.8	4.7	5.1
Phosphatidylglycerol			5.3	0.8		0.1	0.2
Diphosphatidylglycerol	3.7	4.8	1.1	3.2[b]	4.1	4.7	2.0
Phosphatidic acid	0.7	0.4	0.8	1.5	2.2	0.1	0.2
Unidentified	5.5		2.1[a]	3.0[a]	0.5[a]	7.3	8.0
Recovery	99.0	97.8	96.8	96.2	93.4	99.9	99.6
Method of separation	D*	F*	B	B	B	D*	D*

[a]Alkyl ether lipid; [b]Includes cerebroside.
Other liver analyses: human[44,46], rat[9,37,45,47–52], sheep[9,53], bovine[32,45], mouse[50].

TABLE 5

PHOSPHOLIPID COMPOSITION OF KIDNEY

Source	*Rat*	*Human*	*Bovine*	*Mouse*	*Frog*	*Sheep*	*Lamb*
Reference	*45*	*45*	*45*	*45*	*45*	*24*	*24*
Phosphatidylcholine	34.3	33.1	32.4	35.6	34.3	35.8	36.9
Choline plasmalogen						3.2	1.1
Lysophosphatidylcholine	1.0	2.9	0.7	0.9	0.2		
Phosphatidylethanolamine	27.1	27.4	28.6	26.4	30.0	10.6	12.8
Ethanolamine plasmalogen						7.4	10.6
Lysophosphatidylethanolamine	0.3	3.0	0.3	0.0	0.0		
Phosphatidylinositol	5.9	5.5	7.2	5.7	5.6	4.6	3.7
Phosphatidylserine	7.3	6.4	7.5	7.0	6.0	3.2	5.1
Sphingomyelin	12.1	12.0	13.7	10.4	12.2	12.1	12.4
Phosphatidylglycerol	0.3	0.6	0.4	0.5	0.3		
Diphosphatidylglycerol	6.5	4.2	6.5	7.3	4.6	5.9	
Phosphatidic acid	0.3	0.4	0.2	0.2	0.3	2.3	1.1
Alkyl ether lipid							3.2
Unidentified	5.1	4.9	2.3	5.8	5.9	12.3	
Recovery	99.8	99.9	99.8	99.8	99.6	97.4	87.3
Method of separation	D*	D*	D*	D*	D*	B	B

Other kidney analyses: rat[51], sheep[9].

TABLE 6

PHOSPHOLIPID COMPOSITION OF SPLEEN

Source	*Rat*[a]	*Mouse*	*Human*	*Bovine*	*Sheep*
Reference	*45*	*45*	*45*	*45*	*9*
Phosphatidylcholine	41.9	41.5	41.2	36.0	40.5
Choline plasmalogen					3.0
Lysophosphatidylcholine	1.4	1.5	1.5	0.8	
Phosphatidylethanolamine	24.0	24.9	24.5	24.5	25.0
Ethanolamine plasmalogen					9.8
Lysophosphatidylethanolamine				0.3	
Phosphatidylinositol	5.5	5.8	4.4	4.3	1.6
Phosphatidylserine	8.0	7.9	8.3	12.0	2.1
Sphingomyelin	6.6	6.6	12.8	14.5	12.5
Phosphatidylglycerol	0.8	0.3	0.3		
Diphosphatidylglycerol	2.1	2.0	1.0	0.3	1.5
Phosphatidic acid	0.5	0.2	0.2	0.6	1.5
Unidentified	8.9	8.6	5.5	6.7	9.1
Recovery	99.6	99.3	99.7	100.0	94.6[b]
Method of separation	D*	D*	D*	D*	B

[a]Average of 4 values given in reference; [b]Includes 8.0% alkyl ether lipid.
Other spleen analysis: bovine[54].

TABLE 7

PHOSPHOLIPID COMPOSITION OF LUNG

Source	*Rat*[a]	*Rabbit*	*Mouse*	*Frog*	*Human*	*Bovine*	*Dog*	
							Whole lung	*Lung wash*
Reference	*55*	*56*	*55*	*55*	*55*	*55*	*57*	*57*
Phosphatidylcholine	44.9	50.1	43.7	42.6	47.5	39.5	58.3[b]	74.0[b]
Lysophosphatidylcholine	1.3		1.1	0.7	2.0	0.3		
Phosphatidylethanolamine	22.2	21.9	20.1	21.2	17.5	21.2	12.4[c]	16.0[d]
Lysophosphatidylethanolamine			0.0	0.2	0.6	0.2		
Phosphatidylinositol	3.9	4.0	4.1	3.4	3.2	3.3	} 13.7	} 2.8
Phosphatidylserine	9.0	4.0	8.1	7.8	7.0	9.4		
Sphingomyelin	10.0	17.9	9.3	10.8	11.1	16.1	10.6	2.0
Phosphatidylglycerol	2.2		2.4	1.8	2.5	2.0		
Diphosphatidylglycerol	1.1		1.1	0.9	1.0	1.0		
Phosphatidic acid	0.3		0.4	0.3	0.5	1.4		
Other	1.3	1.3	2.5	3.6	2.6	2.7	1.1	4.2
Recovery	96.2	99.2	92.8	93.3	95.5	97.1	96.1	99.0
Method of separation	D*	C	D*	D*	D*	D*	A	A

[a]Mean of 5 values; [b]Includes lysophosphatidylcholine; [c]6.2% phosphatidyldimethylethanolamine; [d]7.3% phosphatidyldimethylethanolamine.

TABLE 8

PHOSPHOLIPID COMPOSITION OF SKELETAL MUSCLE

Source	*Human*		*Guinea pig*	*Bovine*	*Rat*	*Mouse*	*Frog*	*Lobster*	*Scallop*	*Pink abalone*	*Sea urchin*	*Pigeon breast*	*Ox longissimus dorsi*
Reference	*58*		*58*	*38*	*38*	*38*	*38*	*38*	*38*	*38*	*38*	*59*	*59*
Phosphatidylcholine	55.7	48.0	50.0	46.5	51.1	52.4	55.2	54.6	35.4	40.4	47.5	24.0	35.2
Choline plasmalogen												11.6	13.6
Lysophosphatidylcholine	0.5	tr	0.3	0.7	2.7	0.6	0.4	1.1	0.5	0.7	1.4		
Phosphatidylethanolamine	21.9	26.4	27.7	26.6	22.2	25.8	29.0	23.6	25.6	26.9	23.1	2.3	5.4
Ethanolamine plasmalogen												9.7	13.3
Lysophosphatidylethanolamine	0.0		0.0	0.0	1.3	tr	tr	0.0	0.2	0.0	0.0		
Phosphatidylinositol	6.0	8.8	7.9	5.6	8.9	6.7	5.5	4.9	5.0	4.4	4.3	4.5	2.7
Phosphatidylserine	4.2	3.3	3.6	4.1	3.7	4.0	3.4	6.2	11.9	9.9	9.0	1.1	
Sphingomyelin	5.6	4.0	4.2	4.5	2.7	3.6	3.1	8.4	0.0	0.0	0.0	1.6	3.0
Phosphatidylglycerol		1.0		0.3	0.9	0.7	0.2	tr	0.1	tr	0.4		
Diphosphatidylglycerol	4.7	6.6	6.4	8.9	1.4	5.8	2.3	1.3	1.4	0.6	4.7	7.4	5.8
Phosphatidic acid	0.3	1.2	0.2	0.3	0.8	0.1	0.1	0.2	tr	0.3	0.3		
Other	0.8		0.3						16.8[a]	9.3[a]	3.6[b]	22.8	19.9
Recovery	99.7	99.3	100.6	97.5	95.7	99.7	99.2	100.3	96.9	92.5	94.3	85.0	98.9
Method of separation	D*	D*	D*	D*	D*	D*	D*	D*.	D*	D*	D*	B	B

[a]Ceramide aminoethylphosphonate; [b]Ceramide phosphorylethanolamine.
Other analyses: rat[59,60].

TABLE 9

PHOSPHOLIPID COMPOSITION OF ERYTHROCYTES AND HUMAN PLATELETS

Source	*Human*	*Rat*	*Rabbit*	*Pig*	*Dog*	*Horse*	*Sheep*	*Cow*	*Goat*	*Cat*	*Guinea pig*	*Hen*	*Platelets (human)*
Reference	*31*	*68*	*69*	*68*	*68*	*68*	*68*	*68*	*68*	*68*	*68*	*69*	*70*
Phosphatidylcholine	34.7[a]	47.5	33.9	23.3	46.9	42.4				30.5	41.1	42.3	38.4
Lysophosphatidylcholine		3.8	<0.3	0.9	1.8	1.7				tr	tr		0.7
Phosphatidylethanolamine	28.0[a]	21.5	31.9	29.7	22.4	24.3	26.2	29.1	27.9	22.2	24.6	29.1	27.0
Phosphatidylinositol	0	3.5	1.6	1.8	2.2	<0.3	2.9	3.7	4.6	7.4	2.4	5.5	5.0
Phosphatidylserine	14.3	10.8	12.2	17.8	15.4	18.0	14.1	19.3	20.8	13.2	16.8		10.1
Sphingomyelin	20.1	12.8	19.0	26.5	10.8	13.5	51.0	46.2	45.9	26.1	11.1	22.7	17.0
Phosphatidic acid	2.2	<0.3	1.6	<0.3	0.5	<0.3	<0.3	<0.3	<0.3	0.8	4.2		
Unidentified	1.5						4.8	1.7	0.8				1.9
Recovery	100.8	100.2	100.5	100.3	100.0	100.5	99.3	100.3	100.3	100.5	100.5	99.6	95.1
Method of separation	B	D	D*	D	D	D	D	D	D	D	D	D*	F*

[a] 1.2% choline plasmalogen and 10.4% ethanolamine plasmalogen.
Other analyses: human[64–67], rabbit[66], pig[67], cow[67]; human platelets[66].

TABLE 10

PHOSPHOLIPID COMPOSITION OF PERIPHERAL NERVE

Source	*Crab*		*Lobster*		*Cow Splenic*	*Rabbit Sciatic*	*Sheep Sciatic*	*Monkey Sciatic*	*Hen Sciatic*
	Claw	*Leg*	*Claw*	*Leg*					
Reference	*71*	*71*	*71*	*71*	*71*	*71*	*71*	*71*	*71*
Phosphatidylcholine	30.1	30.6	30.6	25.9	32.5	13.1	10.0	13.0	14.6
Choline plasmalogen		1.1	1.7						1.3
Phosphatidylethanolamine	8.3	7.1	6.1	5.1	5.0	5.4	3.8	3.9	4.1
Ethanolamine plasmalogen	14.7	14.4	8.1	14.8	11.9	24.8	22.1	26.9	30.3
Phosphatidylinositol	1.9	2.8	1.5	0.6	3.6	1.1	3.0	0.5	0.8
Diphosphoinositide			0.1	tr[a]		tr[c]	0.5	0.4	0.1
Triphosphoinositide	0.3	0.3	1.0	0.6	tr[b]	2.6	2.6	3.2	1.5
Phosphatidylserine	5.9	5.4	8.8	5.6	10.7	11.8	14.8	16.1	11.3
Sphingomyelin	13.1	11.0	18.3	12.8	14.2	19.7	25.5	30.2	22.1
Phosphatidylglycerol			0.7	1.6					
Diphosphatidylglycerol	0.1	0.3	0.2	0.4	0.5			0.2	0.2
Phosphatidic acid	4.3	3.1	2.0	3.3	0.5	1.3	2.2	0.2	1.3
Serine plasmalogen	2.1	0.9					2.6	1.5	
Alkyl ether lipid	14.7	15.3	12.2	11.9	1.4	2.6	1.9	1.6	4.4
Recovery	95.5	92.3	91.3	82.6	80.3	82.4	89.0	99.7	92.0
Method of separation	B	B	B	B	B	B	B	B	B

[a]0.02%; [b]0.05%; [c]0.01%.
Other analysis: chicken sciatic [72].

TABLE 11

PHOSPHOLIPID COMPOSITION OF PANCREAS, THYROID, THYMUS, PITUITARY AND ADRENALS

Tissue	*Pancreas*			*Thyroid*	*Thymus*		*Pituitary*				*Adrenals*				
Source	*Guinea pig*	*Bovine*	*Rat*	*Bovine*	*Rat*	*Bovine*	*Bovine*		*Human*	*Rat*	*Dog whole tissue*		*Sheep cortex*	*Sheep medulla*	*Bovine medulla*
							Anterior	*Posterior*							
Reference	*73*	*73*	*37*	*75*	*76*	*77*	*78*[b]	*78*[c]	*78*[d]	*79*	*80*	*80*	*9*	*9*	*81*[a]
Phosphatidylcholine	50.3	51.1	46.8	44.3	49.5	57.2	37.1	35.6	37.7	31.5	37.0	34.0	40.0	50.2	30.6
Choline plasmalogen													1.4	1.8	
Lysophosphatidylcholine					1.1					4.5					11.6
Phosphatidylethanolamine	17.8	17.0	9.9	25.0	19.1	34.2	32.5	25.3	25.0	28.0[g]	24.0	28.0	10.5	9.1	15.1
Ethanolamine plasmalogen													15.5	15.0	
Phosphatidylinositol	7.9	8.7	8.7	11.7	11.9	0.2	5.9	6.7	4.2	15.5	8.7	6.3	5.8	5.5	18.0
Phosphatidylserine	4.4	4.2	5.3	2.4	5.9	0.3	2.0	7.3	5.7		3.7	3.3	2.7	3.0	2.3
Sphingomyelin			25.4	15.0	10.1	5.8	8.8	11.2	11.4	10.5	4.1	3.6	2.6	3.4	10.2
Diphosphatidylglycerol					1.0	2.1					5.5	5.6	2.9		
Phosphatidic acid		2.0	tr		1.3						3.2	5.2	3.6	4.8	2.5
Other	19.6[e]	16.6[f]		1.7			3.1	2.1	7.2	10[h]			21.7	15.0	9.4
Recovery	100.0	100.0	100.0	100.1	99.9	99.8	89.4	88.2	91.2	100	86.2	86.0	106.7	107.4	99.7
Method of separation	B	B	B	D	D	C	D	D	D	C	A	A	B	B	C*

[a]Average of 7 determinations; [b]Average of 6 determinations; [c]Average of 8 determinations; [d]Average of 4 determinations; [e]Includes 8% plasmalogen and 9.3% acid- and alkali-stable lipid; [f]Includes 5.9% plasmalogen and 8.8% acid- and alkali-stable lipid; [g]Includes lyso PE (3%) and PS; [h]Acidic phospholipid.
Other data: bovine pancreas[74], rat thymus[76].

TABLE 12

PHOSPHOLIPID COMPOSITION OF INTESTINAL MUCOSA, CORPUS LUTEUM, MAMMARY GLAND, MILK, SKIN AND AORTA

Tissue	*Intestinal mucosa*	*Corpus luteum*[a]	*Mammary gland*	*Milk*	*Aorta*			*Skin*	
Species	*Rat*	*Bovine*	*Bovine*	*Bovine*	*Human*	*Bovine*	*Rabbit*	*Guinea pig*	*Mouse*
Reference	*37*	*91*	*92*	*92*	*84*	*84*	*85*	*86*	*87*
Phosphatidylcholine	53.5	54.7[b]	52.3	33.9	30.2	33.0	34.1	41.9[d]	42.3[e]
Lysophosphatidylcholine					1.2	0.4		19.4	
Phosphatidylethanolamine	25.4	22.5[b]	23.6	42.4	17.3	24.4	22.2	25.0	24.3[e]
Phosphatidylinositol	4.1	5.9	7.6	4.0	2.9	5.1	} 6.1		5.8
Phosphatidylserine	5.2	4.4	3.2	2.0	7.5	10.1			14.1[e]
Sphingomyelin	5.8	7.0	7.6	17.7	38.1	22.2	33.5	19.4	7.8
Diphosphatidylglycerol	2.8	3.7			0.7	1.8			
Phosphatidic acid	3.2	1.1			0.6	0.8	4.0	2.3	5.7
Unknown		1.2	3.1[c]		1.4	2.2			
Recovery	100.0	100.5	97.4	100.0	99.9	100.0	99.9	108.0	100.0
Method of separation	B	B	D*	D*	D*	D*	C*	C	B*

[a]Average of 2 values quoted; [b]Choline plasmalogen 1.6%, ethanolamine plasmalogen 8.8%; [c]Possibly diphosphatidylglycerol; [d]Includes PS and PI; [e]Plasmalogens included: choline 3.1%, ethanolamine 7.7%, serine 1.7%.

TABLE 13

PHOSPHOLIPID COMPOSITION OF OTHER TISSUES

Tissue	*Species (Reference)*
Testis	Ram (62) rat (61)
Head of epididymis	Rat (61)
Tail of epididymis	Rat (61)
Spermatozoa (testicular)	Ram (62, 63)
Spermatozoa (ejaculated)	Ram (63)
Pineal	Human, bovine, sheep (82)
Hypothalamus:	
chiasma opticum	Rat (83)
area anterior	Rat (83)
nucleus supraopticus	Rat (83)
nucleus paraventricularis	Rat (83)
Bone, cancellous	Bovine (95)
Bone, compact	Bovine (95, 96)
Epiphyseal cartilage: resting, proliferating, hypertrophic, calcified	Bovine (95)
Red bone marrow	Human (97), pig (98), rabbit (99), cow (99), sheep (99)
Eye fractions: retina, iris, choroid, sclera, cornea, vitreous body	Bovine (102)
Eye fractions: inner and outer cortex, nucleus	Bovine (101)
Retinal rod outer segments	Bovine (103, 104)
Lens	Man, rhesus monkey, sheep, ox, rabbit, chicken, pig, cod fish, octopus (100), rat (100, 101)
Gastric mucosa	Bullfrog (88)
Bile	Sheep, lamb (89), bovine (89, 90)
Brown adipose tissue	Bat (94)
Bladder	Toad (93)

TABLE 14

PHOSPHOLIPIDS OF SUBCELLULAR FRACTIONS: NUCLEI

Tissue	*Brain*			*Liver*			*Kidney*	*Uterus*	*Ovary*	*Erythrocyte (nuclear membrane)*
Source	*Guinea pig*	*Bovine*	*Human*	*Rat*	*Bovine*	*Pig*	*Rat*	*Rat*	*Rat*	*Hen*
Reference	*35*	*105*	*105*	*37*	*107*	*106*	*51*	*51*	*51*	*69*
Phosphatidylcholine	33.6	53.1	47.6	57.3	54.5	58.2	19.0	38.0	36.5	53.0
Lysophosphatidylcholine			2.5		1.6	1.0				
Phosphatidylethanolamine	31.4[a]	20.8	17.8	26.1	22.5	25.0	19.0	8.0	9.0	26.0
Phosphatidylinositol	2.6	7.9	5.0	3.9	7.6	8.8				7.1
Phosphatidylserine	11.1	5.0	9.3	5.5	6.0	4.8	15.0	20.0	16.5	
Sphingomyelin	7.1	3.0	10.7	6.3	5.8	2.0	11.0	4.0	0	13.0
Phosphatidylglycerol		tr	0.6							
Diphosphatidylglycerol	2.3		0.3		2.1					
Phosphatidic acid	0.7	0.6	1.2	tr						
Other	2.8	9.4	2.8	1.1			36.0	30.0	38.0	
Recovery	91.6	99.8	98.2	100.2	100.1	99.8	100.0	100.0	100.0	99.1
Method of separation	B	D*	D*	B	D*	D*	E	E	E	D

[a]Value includes 17.4% ethanolamine plasmalogen.

Other data: rat liver[47,50,51,106].

References p. 479

TABLE 15

PHOSPHOLIPIDS OF SUBCELLULAR FRACTIONS: MITOCHONDRIA

Tissue	*Heart*					*Liver*			
Source	*Human*	*Pig*	*Rat*	*Bovine*	*Sheep*	*Rat*			*Mouse*
						Whole mitochondria	*Inner[a] membrane*	*Outer[a] membrane*	
Reference	*108*	*109*	*110*	*43*	*9*	*189*	*189*	*189*	*50*
Phosphatidylcholine	36	36.2	55	15.4	16.6	40.5	37.9	45.4	43.9
Choline plasmalogen				27.8	26.3				
Lysophosphatidylcholine	2			1.9		1.4	0.6	1.8	
Phosphatidylethanolamine	28	25.3	32	14.8	6.4	34.7[b]	38.3	30.4	33.4
Ethanolamine plasmalogen				14.0	9.8				
Lysophosphatidylethanolamine				2.9					
Phosphatidylinositol	6	14.3	5		0.9	6.6	2.0	8.4	4.4
Phosphatidylserine	5	2.8	4	1.0	0.8				
Sphingomyelin	6	5.6		2.8	2.3	2.4	0.8	4.4	
Phosphatidylglycerol			2						
Diphosphatidylglycerol	} 17[a]	13.3	2	19.8	14.5	14.8[c]	20.4[c]	9.5[c]	18.3
Phosphatidic acid					1.5				
Other		2.7			11.4				
Recovery	100	100.1	100	100.5	90.6	100.4	100.0	99.9	100.0
Method of separation	C*	E	B	A	B	C, D*	C, D*	C, D*	D*

[a]Prepared by digitonin treatment; reference also quotes values for membranes prepared by "osmotic shock"; [b]Includes 0.9% phosphatidylserine; [c]Includes phosphatidic acid.
Other data: rat heart[111], bovine heart[110, 112–114], sheep heart[110], rat liver[9, 37, 50, 51, 119, 115–118], rat brain[118, 119], rat kidney[51].

		Brain			*Kidney*			*Spleen*	*Pancreas*	*Adrenal cortex*	*Adrenal medulla*	*Ovary*	*Uterus*	*Muscle (skeletal)*
Bovine	*Sheep*	*Rat*	*Bovine*	*Guinea pig*	*Rat*	*Sheep*	*Cow*	*Sheep*	*Bovine*	*Sheep*	*Bovine*	*Rat*		*Mouse*
113	*9*	*28*	*120*	*35*	*110*	*9*	*113*	*9*	*74*	*9*	*121*	*51*	*51*	*122*
43.4	48.0	50.0	28.5	40.0	41	37.1	39.6	30.5	51.2	35.0	44	52.5	57.0	41.4
	0.5		0.2			3.7		2.9		0.7				
									3.1		1.7			1.5
34.5	19.0	32.1	18.5	23.3	30	16.8	38.1	6.5	22.8	10.6	34.9	17.5	11.0	32.8
	1.5		8.7	9.4		5.4		9.8		11.2				
	4.3	13.4	2.9	5.5	14	3.9	(3)	1.1 }	8.4	5.1 }	7.6			5.7
	0.9		11.3	5.9	1	2.3	tr	1.3 }		1.6 }		11.0	6.5	3.2
	2.7		17.3	3.7		4.8		6.9	4.0	1.6	3.6	6.5	15.0	3.3
			0.5		3									
17.2	14.1	1.6	4.6	11.1	9	12.5	19.2	7.3 }	9.2[c]	6.0 }	7.6[c]			12.5
	1.5		0.7	0.6		1.8		1.5 }		4.6 }				
4.9	10.0	16.8	1.3	2.2		7.4	3.5	23.4		16.5		12.5	10.5	
100.0	105	103.9	93.4	101.7	98	97.2	100.4	92.2	97.7	92.9	99.7	100.0	100.0	100.4
D*	B	A	B	B	B	B	D*	B	C*	B	C*	E	E	D*

References p. 479

TABLE 16

PHOSPHOLIPIDS OF SUBCELLULAR FRACTIONS: MICROSOMAL FRACTIONS

Tissue	*Heart*				*Liver*						*Kidney*	
Source	*Human*	*Bovine*	*Pig*	*Rat*	*Sheep*	*Rat*			*Mouse*	*Pig*	*Sheep*	*Rat*
						Whole microsomes[e]	*Smooth e.r.*[e]	*Rough e.r.*				
Reference	*108*	*43*	*109*	*111*	*9*	*189*	*189*	*189*	*50*	*106*	*9*	*51*
Phosphatidylcholine	35	19.0	41.8	38.2	61.3	59.1	55.4	60.2	58.0	59.9	33.3	44.0
Choline plasmalogen		25.9			2.1						4.5	
Lysophosphatidylcholine	2	6.5	8.0	1.0		2.0	1.9	1.5	1.4	tr		
Phosphatidylethanolamine	26	5.8	21.2	30.3	13.8	24.1[f]	21.5[f]	23.3[f]	24.9	27.4	12.5	14.0
Ethanolamine plasmalogen		21.8			0.8						4.6	
Lysophosphatidylethanolamine		1.5										
Phosphatidylinositol	9	1.0	3.9	6.2	7.0	9.2	6.7	9.0	6.5	6.5	4.6	
Phosphatidylserine	6	1.0	3.6	4.7	1.6				2.9	3.2	5.6	11.0
Sphingomyelin	10	8.9	16.5	4.7	3.6	4.2	12.3	3.9	6.3	2.9	22.5	19.0
Diphosphatidylglycerol } Phosphatidic acid	12[g]	9.2	2.0	15.2[g]	1.7	1.0[g]	1.9[g]	1.2[g]			1.4	
Phosphatidic acid					1.4						2.1	
Other			2.7		9.4						5.9	11.5
Recovery	100	100.6	99.7	100.3	102.7	99.6	99.7	99.1	100.0	99.9	97.1	100.0
Method of separation	C*	A	E	C*	B	C,D*	C,D*	C,D*	D*	D*	B	E

[a]14% plasmalogen; [b]19.6% plasmalogen; [c]15.7% plasmalogen; [d]10.9% plasmalogen; [e]Microsomes and subfractions prepared in 5 m*M* $MgCl_2$; reference also gives data for fractions prepared in 1 m*M* $MgCl_2$; [f]includes phosphatidylserine; [g]includes phosphatidic acid.

Other data: rat liver[9, 37, 50, 51, 106, 110, 116, 123, 124], rat kidney[110], rat brain[28, 29, 119], human brain[125].

Brain								*Adrenal cortex*	*Adrenal medulla*	*Pancreas*			*Uterus*	*Ovary*	*Muscle (skeletal)*
Sheep	*Guinea pig*	*Rat*	*Human*	*Rabbit*	*Pigeon*	*Dogfish*	*Frog*	*Sheep*	*Bovine*	*Bovine*	*Guinea pig (rough e.r.)*	*Guinea pig (smooth e.r.)*	*Rat*	*Rat*	*Mouse*
9	*35*	*23*	*23*	*23*	*23*	*23*	*23*	*9*	*121*	*74*	*126*	*126*	*51*	*51*	*122*
39.8	41.4	34.7	43.3	27.9	36.5	40.0	39.4	57.2	42.2	49.1	34.7	26.8	29.5	53.0	61.1
0.8								2.6							
									2.0	5.2	13.1	13.1			2.6
18.5	13.0	33.6	20.2	33.1	36.1	22.7	31.5	5.4	33.0	29.9	35.8	36.5	23.5	11.0	6.2
6.4	14.8							9.8							15.7
3.7	4.8	3.3	7.3	} 19.5		8.3	8.9	3.0	} 12.1	} 3.1	} 5.6	4.8			6.8
11.6	11.9	10.7	12.8		19.9	13.2	11.0	1.1					20.5	8.0	2.5
11.8	7.5	7.3	14.1	13.1	9.8	12.2	11.0	2.5	9.3	6.4	3.5	14.2	17.5	12.0	3.3
1.0	0.4	3.3	2.3	6.4		2.0	5.8	1.5	} 1.2[g]	} 3.8[g]					2.0
1.7	0							4.9							
4.9	1.9							14.1			7.5	7.5	9.0	16.0	
100.3	95.5	92.9[a]	100.0[c]	100.0[b]	102.3	98.4[d]	107.7	102.4	99.8	97.5	100.2	102.9	100.0	100.0	100.4
B	B	C	C	C	C	C	C	B	C*	C*	C*	C*	E	E	D*

References p. 479

TABLE 17

PHOSPHOLIPID COMPOSITION OF SUBCELLULAR FRACTIONS: MISCELLANEOUS

Fraction	*Supernatant*							*Plasma membrane*			
Tissue	*Liver*		*Kidney*	*Brain*		*Uterus*	*Ovary*	*Liver*	*Pancreas*	*Mammary gland*	*Erythrocyte*
Source	*Sheep*	*Rat*	*Rat*	*Sheep*	*Rat*	*Rat*	*Rat*	*Rat*	*Guinea pig*	*Bovine*	*Hen*
Reference	*9*	*9*	*51*	*9*	*29*	*51*	*51*	*50*	*126*	*128*	*69*
Phosphatidylcholine	66.0	45	32.0	43.8	53.6	10.0	41.0	46.1	24.7	27.0	30.3
Choline plasmalogen	0.4										
Lysophosphatidylcholine								0.5	7.1		
Phosphatidylethanolamine	8.1	6.5	17.5	19.6	17.8	15.0	13.5	24.7	34.4	29.4	30.7
Ethanolamine plasmalogen	0.3	1.3									
Phosphatidylinositol	3.1	3.1		3.0	14.3			6.7	4.7 (PI + PS)	13.2	3.7
Phosphatidylserine			17.0	9.1	10.7	13.0	9.5	4.2		8.4	
Sphingomyelin	2.3	6.7	8.0	4.8	3.6	22.0	11.0	16.8	19.2	22.0	30.7
Phosphatidylglycerol											
Diphosphatidylglycerol	0.4	0.6		0.5				1.0			
Phosphatidic acid	5.8	18.2		18.2							
Other	10.8	9.7	25.5	5.2		40.0	25.0		8.8		
Recovery	97.2	91.1	100.0	104.2	100.0	100.0	100.0	100.0	98.9	100.0	100.4
Method of separation	B	B	E	B	D	E	E	D*	C*	D*	D

[a]Diphosphatidylglycerol was determined to be 0.3% by silica gel paper chromatography;
[b]Diphosphatidylglycerol was determined to be 1.3% by silica gel paper chromatography;
[c]15–19% 1-alkenyl-2-acyl PE; 15% diacyl.
Other data: rat liver supernatant[51], rat liver plasma membrane[117, 123, 127, 189, 190], rat liver lysosomes[189].

Synaptosome	*Synaptic vesicles*		*Lysosomes*	*Golgi apparatus*	*Membrane*	*Granule*	*Zymogen granule*		*Chromaffin granule*	*Myofibrils*	
Brain			*Liver*	*Liver*	*Blood platelets*	*Blood platelets*	*Pancreas*		*Adrenal medulla*	*Muscle*	
Rat	*Rat and Bovine*	*Guinea pig*	*Rat*	*Rat*	*Human*	*Human*	*Bovine*	*Guinea pig*	*Bovine*	*Bovine (heart)*	*Mouse (skeletal)*
184–186	*186–188*	*35*	*110*	*123*	*70*	*70*	*74*	*126*	*121*	*43*	*122*
42–44	35–45	40.7	45	45.3	38.3	31.8	41.9	21.5	26.0	17.4	52.8
		0.6								27.0	
1	**4**			5.9	0.4	1.9	6.6	9.1	16.8	2.0	1.9
} 33–36	} 30–34[c]	13.1	24	17.0	27.2	28.7	34.8	31.5	36.1	10.5	7.3
		12.0								15.1	21.3
				6.3						1.0	
4	3–5	5.9	8	8.7	6.1	3.9	} 4.4	} 5.0	} 9.2	5.7	6.6
8–13	12	10.2		4.2	9.9	11.7				3.7	2.0
5–7	10–11	12.3		12.3	16.9	20.2	8.9	23.5	10.9	9.0	2.7
			2								
		0.0	20				} 2.8		} 0.6	8.5	5.4
1–2	0–7.6	2.3									
		3.2			1.2[a]	1.8[b]		9.0			
94–107	94–118.6	100.3	98	100.7	100.0	100.0	99.4	99.6	99.9	99.9	100.0
		B	B	B*	F*	F*	C*	C*	A	A	D*

References p. 479

TABLE 18

PHOSPHOLIPID COMPOSITION OF MYELIN FROM CENTRAL NERVOUS SYSTEM

Tissue	*Brain white matter*								*Brain grey matter (heavy)*	*Whole brain (heavy)*	*(light)*	*(light)*	*(heavy)*	*(light)*	*Optic nerve*
Source	*Human*	*Bovine*	*Rat*	*Guinea pig*	*Rabbit*	*Pigeon*	*Dogfish*	*Frog*	*Human*	*Human*		*Bovine*		*Rat*	*Bovine*
Reference	*23*	*33*	*23*	*35*	*23*	*23*	*23*	*23*	*129*	*129*	*129*	*129*	*129*	*29*	*133*
Phosphatidylcholine	} 21.0	23.9	26.2	32.1	18.5	25.5	24.5	39.2	21.0	20.5	21.8	19.1	18.4	32.1	23.0
Choline plasmalogen		0.9													
Phosphatidyl-ethanolamine	} 51.9	10.1	44.9	14.1	35.8	37.9	44.2	39.9	30.6	33.8	35.4	40.8	36.1	39.3	42.4
Ethanolamine plasmalogen		33.5		24.4											
Lysophosphatidyl-ethanolamine									5.5	2.2	2.0	0.6	0.2		
Phosphatidylinositol	1.6	2.4	5.1	2.9	2.6	2.5	2.2	2.2	2.0	1.7	1.8	1.1	1.8	6.7	
Phosphatidylserine	18.8	14.5	15.6	14.1	15.1	9.3	8.3	7.6	20.5	18.6	18.8	15.5	18.0	14.6	15.3
Sphingomyelin	19.4	15.4	8.6	7.3	16.5	8.0	14.6	9.3	11.7	15.0	16.3	18.7	15.0	7.3	19.3
Diphosphatidyl-glycerol	0.4		2.7		7.6	8.9	6.5	1.8							
Phosphatidic acid				1.3					0.5	1.3	1.1	1.1	1.7		
Other		0.6[d]		2.3					0.3[a]	1.0[a]	1.0[a]	0.2[a]	0.4[a]		
Recovery	113.1[c]	101.3	103.1[e]	98.5	96.1[f]	99.1[g]	100.6	100.0[h]	96.9[b]	97.2[b]	99.6[b]	100.8[b]	100.0[b]	100.0	100.0
Method of separation	C*	C	C*	B	C	C	C	C	D*	D*	D*	D*	D*	D	A

[a]Lysobisphosphatidic acid; [b]Total recovery includes unidentified lipids; [c]30.7% plasmalogen; [d]Serine plasmalogen; [e]49.2% plasmalogen; [f]33.8% plasmalogen; [g]50.4% plasmalogen; [h]50.5% plasmalogen.

Other data: human white matter[22, 125, 129, 130], bovine white matter[23, 34, 129, 131–133], rat white matter[119, 132], guinea pig and rabbit white matter[132].

TABLE 19

PHOSPHOLIPID COMPOSITION OF PERIPHERAL NERVOUS SYSTEM MYELIN

Tissue	*Sciatic nerve*		*Spinal root*	*Intradural root*	*Spinal cord*		
Source	*Guinea pig*	*Rat*	*Bovine*	*Bovine*	*Human*	*Bovine*	*Frog*
Reference	*132*	*132*	*133*	*134*	*129*	*129*	*129*
Phosphatidylcholine			24.4	22.6	18.6	19.7	30.6
Choline plasmalogen	21	24					
Lysophosphatidylcholine							
Phosphatidylethanolamine			29.4	28.2	36.8	38.4	37.0
Ethanolamine plasmalogen	45	37					
Lysophosphatidylethanolamine					0.3		2.4
Phosphatidylinositol		19			1.6	1.2	1.6
Phosphatidylserine	34		15.6	14.0	15.9	15.3	7.1
Sphingomyelin		20	30.6	29.1	20.7	17.6	4.8
Phosphatidic acid					1.4	1.3	3.0
Other					0.3[a]	0.9[a]	
Recovery	100	100	100.0	93.9	100.6[b]	99.7[b]	97.1[b]
Method of separation	C	C	A	A	D*	D*	D*

[a]Lysobisphosphatidic acid; [b]Total includes unidentified phospholipids.

TABLE 20

FATTY ACIDS OF PHOSPHATIDYLCHOLINE OF WHOLE TISSUE AND SUBCELLULAR FRACTIONS

Tissue	*Ref.*	*Fatty acids (as % total by weight)*																		
		14:0	*15:0*	*16:0*	*16:1*	*17:0*	*18:0*	*18:1*	*18:2*	*18:3*	*20:0*	*20:1*	*20:2*	*20:3*	*20:4*	*20:5*	*22:0*	*22:4*	*22:5*	*22:6*
Brain, human (white matter)	136	1.3		34.3	1.0		13.4	45.2	0.4			1.1			1.3				0.3	0.1
Brain, human (grey matter)	136	2.9		45.0	3.1		9.3	31.4	0.4			0.7			4.1					3.1
Brain, rat	139	0.3		45.0	1.4		13.8	32.3				0.4			5.1					
Brain, chicken	140	0.7		51.0	tr		16.5	26.3	0.7						3.9			tr	0.4	0.6
Liver, bovine	142			16.1	2.4		35.3	16.6	9.2	1.3				4.4	5.9	1.5			5.8	0.9
Liver, rat	148	0.5		29.7	1.0		16.8	10.4	16.8					1.5	18.3					3.4
Liver, rabbit	142			29.3	1.8		17.3	16.6	35.5											
Liver, pig	142			24.8	1.3		28.8	14.7	13.3	tr				1.5	10.9	1.2			2.2	1.3
Liver, chicken	140	tr		32.9	1.0		19.5	26.5	10.3						5.4			0.4	0.5	2.9
Liver, mink (chicken fed)	144			24	3		15	15.5	16.5			0.5			11	0.5			2	5.5
Liver, mink (mackerel fed)	144			26	1.5		16	10.5	3			2			5.5	4.5			2.5	22.5
Heart, bovine	43			19.1	2.5		7.2	18.9	31.0	7.2				2.3	7.1	2.6			1.5	
Erythrocytes, human	64			31.2		0.7	11.8	18.9	22.8		0.2	0.5	1.1	1.9	6.7	0.3			0.8	2.1
Serum, human	153	0.7		43.2	1.0		13.3	11.1	16.4		0.8				9.2					
Plasma, human	154			26.1		0.5	13.6	12.9	24.7		0.1	0.5	0.7	3.5	10.9	0.5		0.5	1.1	4.1
Plasma, bovine	155			19	1		29	16	23	1				4	7					
Plasma, rat	147			27.8	0.9		25.6	8.3	19.8						18.0					
Platelets, human	66			23.8			16.0	20.6	11.5		0.8	2.2		2.5	14.7	2.2		0.3	3.9	
Bile, bovine	89			33.2	6.4		11.8	24.5	7.4	4.4										
Bile, sheep	89			29.3	6.0		13.6	34.5	5.5	1.7										
Bile, rat	147			43.6	1.2		8.3	6.3	30.8						9.7					
Lung, pig	156	1.3		47.1	2.8		8.5	26.2	6.3	0.5				0.9	4.9					
Lung wash, dog	57	4	tr	66	5	1	7	9	6						tr					
Spleen, pig	156	0.4		30.0	0.8		14.7	24.8	9.5	tr				3.9	← 12.3 →				0.7	
Kidney, pig	156	0.2		29.0	0.7		16.0	26.3	18.0	tr				1.6	6.9	tr			0.6	

TABLE 20 (continued)

Tissue	Ref.	Fatty acids (as % total by weight)																		
		14:0	15:0	16:0	16:1	17:0	18:0	18:1	18:2	18:3	20:0	20:1	20:2	20:3	20:4	20:5	22:0	22:4	22:5	22:6
Adipose tissue, rat	144			26	2		16	14.5	6.5			2.5			8	9			1.0	8.5
Mammary cells, bovine[a]	159	17.5	2.8	32.5	1.4	0.9	10.0	18.2	11.3	tr										
Muscle, mouse	160	0.4	0.3	45.3	0.5	1.2	12.8	12.8	17.8	1.0					6.5					
Muscle (breast), pigeon	156	tr		24.4	1.8		17.9	22.6	24.1					1.6	← 3.8 →				2.0	0.5
Muscle, trout	156	1.6		26.2	2.9		2.7	11.0	1.6					0.4	← 10.9 →				3.0	38.5
Muscle, tuna	143	0.7		19.0	2.1		5.0	7.6	0.7	0.8					2.7	7.1		tr	tr	55
Thyroid, rat[b]	76	0.4	0.4	32.1	1.1		17.5	24.8	8.1		1.9						0.6	0.5		17.2
Egg, hen	155			32.0	1.0		16.0	30	17.0						3.0					
Sciatic nerve, chicken[c]	72			38.6	1.7		10.8	27.2	2.2		1.6	0.7					5.6			
Mitochondria,																				
heart, human	108	1.4		39.3	2.8		8.6	21.2	10.2		tr				12.5					
bovine	43			22.6	2.2		5.5	13.5	36.7	2.5				2.5	9.7	9.7			1.1	
rat	111	0.3		19.8	0.5		26.6	9.7	14.0						21.5					3.8
liver, rat	189	0.4		27.0	3.9		21.6	13.0	12.4					1.3	17.7					2.9
outer membrane[d]	189	0.4		27.0	4.1		21.0	13.5	13.5					1.1	15.7					3.5
inner membrane[d]	189	0.3		27.1	3.6		18.0	16.2	15.8					1.0	18.5					3.8
brain, rat	139	0.1		47.1	1.0		12.2	30.5				0.5			5.6					1.6
Microsomes																				
heart, human	108	1.7		40.9	3.1		9.7	21.4	9.3		tr				9.5					
bovine	43			21.3	3.6		10.3	14.7	31.8	2.5				6.2	9.4					
rat	111	0.2		23.1	0.7		28.2	10.5	14.4		0.1				18.1					2.4
liver, rat	189	0.8		24.5	3.3		21.0	12.3	17.7					1.2	15.8					2.9
smooth e.r.	189	0.4		28.6	3.1		26.5	10.6	14.9					1.4	14.0					0.7
rough e.r.	189	0.5		22.7	3.6		22.0	11.1	16.1					1.8	19.7					2.9
brain, rat	139	0.4		56.6	0.8		9.8	26.7			0.6				3.7					0.6

References p. 479

TABLE 20 (continued)

Tissue	*Ref.*	*Fatty acids (as % total by weight)*																		
		14:0	*15:0*	*16:0*	*16:1*	*17:0*	*18:0*	*18:1*	*18:2*	*18:3*	*20:0*	*20:1*	*20:2*	*20:3*	*20:4*	*20:5*	*22:0*	*22:4*	*22:5*	*22:6*
Supernatant, brain, rat	139	0.1		38.7	1.5		14.7	35.5				1.2			7.3					
Plasma membrane,																				
erythrocyte, hen	69			25.8	1.5		10.0	19.8	29.8					4.0	1.6					
Plasma membrane,																				
liver, rat	123	0.9		36.9			31.2	6.4	12.9	tr	tr				11.1					
Platelet membrane,																				
blood, human	70			31.4	1.4		15.7	27.4	7.5		0.5				12.5					
Lysosomes	189			32.8	2.9		34.9	10.2	8.1					1.1	8.4					1.6
Nuclear membrane,																				
erythrocyte, hen	69			21.3	1.5		13.2	21.8	29.6					3.7	0.5				1.0	2.9
liver, hen	107			16.0			9.4	34.1	39.1						1.4					
Golgi apparatus,																				
liver, rat	123	0.9		34.7			22.5	8.7	18.1	tr	tr				14.5					
Granules,																				
blood platelets,																				
human	70			31.5	1.8		15.6	27.5	7.2		0.9				11.8					
Myelin,																				
CNS, human	136	0.4		40.1	0.8		6.1	51.6	0.6						0.4					
whole brain, rat	139	0.2		34.7	1.1		17.3	42.6				0.8			0.9					

[a]Other fatty acids include 10:0 3%, 12:0 5.3%; [b]Other values include 22:1 11.7%; 22:2 0.4%; [c]Other values include 24:0 3.0%; 24:1 7.0%; [d]Mitochondrial membranes were obtained by digitonin treatment. Reference also gives figures for membranes prepared by osmotic shock.

Other data: human brain[135,137,138], rat cerebrum[139], bovine liver[141,143], rat liver[52,142–147], human erythrocytes[13,65,66,149,150], human serum[152], human plasma[13,150], bovine plasma[150], human platelets[70], rabbit and bovine lens[157], bovine retinal rod[103], bovine iris[158], bovine pituitary[78], dog adrenal[80], bovine pineal[82], menhaden and salmon muscle[143], hen egg[142,143,161], bovine heart mitochondria[10], rat liver mitochondria[116,177,178], rat brain mitochondria[28], rat liver microsomes[123,177,178], rat brain microsomes[28], bovine mammary gland plasma membrane[127], rat liver plasma membrane[127,189], bovine corpus callosum myelin[131], bovine optic nerve and white matter myelin[133], human PNS myelin[136], bovine spinal root myelin[134], rat, guinea pig and rabbit cerebrum myelin[139].

TABLE 21

FATTY ACIDS OF DIPHOSPHATIDYLGLYCEROL OF WHOLE TISSUE AND SUBCELLULAR FRACTIONS

Tissue	Ref.	Fatty acid (as % total fatty acid by weight)														
		12:0	*14:0*	*16:0*	*16:1*	*18:0*	*18:1*	*18:2*	*18:3*	*20:0*	*20:1*	*20:2*	*20:3*	*20:4*	*22:5*	*22:6*
Brain, bovine[a]	172		0.8	6.5		3.2	54.0	17.8					1.8	5.9		
Brain, rat	173			13.5		46.3	27.8	7.8								
Heart, human	174	←	0.3 →	6.7	3.0		7.3	82.2	0.5	←		tr				→
Heart, bovine	175		0.2	0.6	1.1	6.6	8.1	77.2	1.3		0.8			4.1		
Heart, rat	173			5.5		7.3	9.5	74.8								
Liver, bovine	141			1.7		1.8	8.1	70	15.5							
Liver, rat	162			4.4	2.6	1.6	14.9	77.5								
Kidney, rat	173			8.4		14.0	16.0	60.5								
Kidney, pig	156			5.8	3.1	1.5	←	74.0	→							
Muscle, rat	173			13.7		40.2	20.4	23.3								
Mitochondria, heart, human[b]	108	0.5	1.6	5.9	5.9	3.7	14.7	67.6								
rat[b]	111	0.1	1.7	7.7	1.9	3.5	10.3	65.6				1.9		0.5		
Mitochondria, heart, bovine	43			1.3	2.5	tr	9.0	84.0	2.7				·	·		
liver rat	189		0.2	7.0	7.6	3.6	19.9	58.8					1.2	1.8		
inner membrane[d]	189		0.5	7.7	6.2	4.8	19.2	57.8					2.4	1.4		
inner membrane[e]	189		1.0	6.4	6.9	3.9	20.3	59.4					2.5	0.6		
Mitochondria, brain, rat[c]	28		0.8	12.5	5.0	21.3	18.6	10.3		1.8	1.3			7.5		7.6
Microsomes, heart, human[b]	108	0.4	3.6	14.1	6.7	8.6	21.2	45.4								
rat	111	3.3	2.0	11.2	1.7	7.8	9.5	48.0				4.1		0.4		
bovine	43			3.3	3.7	2.9	10.7	76.7	2.6							

[a]Also 15:0 0.3%, 17:0 1.3%; [b]Includes phosphatidic acid; [c]Diphosphatidylglycerol sample contained approx. 20% PE; [d]Prepared by osmotic shock; [e]Prepared by digitonin treatment.

Other data: bovine heart[42, 174, 176], rat liver[173], rat thymus[76], lung, spleen and testes[173], pineal[82], dog adrenal[80], mouse muscle[160], pigeon breast muscle[156], trout muscle[156], rat liver mitochondria[116, 117, 180].

TABLE 22

FATTY ACID COMPOSITION OF PHOSPHATIDIC ACID, PHOSPHATIDYLGLYCEROL AND LYSO-COMPOUNDS OF WHOLE TISSUE AND SUBCELLULAR FRACTIONS

Tissue	*Ref.*	*Fatty acid (as % total fatty acid by weight)*															
		14:0	*16:0*	*16:1*	*18:0*	*18:1*	*18:2*	*18:3*	*20:0*	*20:1*	*20:2*	*20:3*	*20:4*	*22:0*	*22:4*	*22:5*	*22:6*
Phosphatidic acid																	
Mouse muscle[b]	160	5.2	15.9	3.2	13.5	19.2	8.8	2.6	2.4				tr				
Rat liver	162		32.8		14.8	23.1	21.7						6.9				
Myelin, bovine corpus callosum	131		6.1	1.0	35.6	40.2	2.3						9.1				1.7
Phosphatidylglycerol																	
Microsomes, brain, rat	28	3.2	12.1	2.3	24.5	21.6	2.3		1.3	1.6			6.6				8.4
Mitochondria, liver, rat	179	0.7	12.0	1.8	14.1	20.9	19.7				13.4	6.5	10.0				
Lysophosphatidylcholine																	
Human erythrocyte	149	tr	47.0	4.0	26.0	12.0	← 1.0 →		5.0				1.0				
Human serum	153	1.8	62.2	0.8	13.1	4.4	2.6		4.4				8.0				
Human plasma	154		39.5		15.2	13.5	21.1	0.3	0.3		0.6		3.4		1.0	0.2	1.4
Microsomes, liver, rat	123[a]	3.4	24.4	tr	23.7	1.7	29.4						1.7				
Golgi, liver, rat	123[a]	3.7	31.7	5.3	17.4	17.2	15.2						9.6				
Plasma membrane, liver, rat	123	2.1	37.6	tr	27.1	21.2	9.3										
Lysophosphatidylethanolamine																	
Golgi, liver, rat	123[a]	5.2	22.3	5.4	12.4	32.8	2.8		tr				1.3	1.6[c]			
Plasma membrane, liver, rat	123[a]	1.1	18.2	tr	16.9	3.0	42.2		tr				0.9	1.6[d]			

[a]These values quoted for 18:2 are for combined 18:2 and 18:2(i) given in reference; [b]Other fatty acids include: 9:0 5.8%, 10:0 1.3%, 11:0 2.3%, 12:0 3.4%, 13:0 0.6%, 15:0 0.9%, 17:0 3.1%, 17:1 1.5%; [c]Also 24:0 16.3%; [d]Also 24:0 12.7%.

Other data: Lysophosphatidylcholine: dog adrenal[80], bovine bile[89], rat thymus[76], rat liver plasma membrane[127]. Phosphatidic acid: rat thymus[76].

TABLE 23

FATTY ACIDS OF PHOSPHATIDYLETHANOLAMINE OF WHOLE TISSUE AND SUBCELLULAR FRACTIONS

Tissue	*Ref.*	*Fatty acid (as % total fatty acid by weight)*																			
		12:0	*14:0*	*16:0*	*16:1*	*18:0*	*18:1*	*18:2*	*18:3*	*20:0*	*20:1*	*20:2*	*20:3*	*20:4*	*20:5*	*22:0*	*22:3*	*22:4*	*22:5*	*22:6*	*24:0*
Brain, human,																					
white matter[b]	135			5.9	0.4	30.4	8.7	0.5	0.1		0.5		1.4	13.2				8.3	2.0	28.6	
grey matter[b]	135			6.2	1.1	13.8	43.2	0.5	0.1		6.0		1.4	7.9				13.4	0.8	3.0	
Brain, bovine,																					
grey matter[b]	138		0.3	8.4	0.8	28.5	13.2	0.2			0.2			13.2					6.7	27.3	
Brain, chicken[b]	140		1.6	17.9[a]	1.1	28.5[a]	12.4	0.4						12.6				4.4	2.0	17.5	
Liver, rat	144			25	2	14	6	6			0.5			10	9				2	20.5	
Liver, chicken	140		tr	16.2[a]	0.8	29.6	13.6	13.6						17.2				1.2	0.8	6.4	
Liver, mink																					
(chicken fed)	144			18.0	2	18.5	8	8			0.5			15.5	3.5				3.5	14.5	
Liver, mink																					
(mackerel fed)	144			27	1	17.5	8	2.5			2			4	2.5				2	28	
Heart, bovine	43			3.6	1.0	21.4	6.9	22.0	4.2				2.3	30.5	6.3				1.8		
Erythrocyte, human	64			12.9		11.5	18.1	7.1		0.3	0.7	1.3	1.5	23.7	1.2			7.5	5.3	8.2	0.3
Erythrocyte, pig	151			23.4	tr	11.5	51.5	13.6													
Plasma, human	154			8.0		15.4	7.9	15.8		1.3	0.8	1.2	0.3	26.3	1.4			1.9	3.9	12.5	0.3
Platelets, human	66			8.5		17.2	6.7	1.9		2.3	2.0		1.1	37.2	3.6			5.3	10.4	3.4	
Muscle, mouse	160		1.1	5.5	0.3	20.6	12.1		7.3	7.0				12.3		19.5					
Lung wash, dog	57		1	31	3	17	31	3	2					12							
Adipose tissue, rat	144			11.5	2	17	15	3			2.5			10	8.5				2.5	19.5	
Mammary cells, bovine	159		4.9	23.0	1.5	37.4	22.0	3.3	0.4												
Egg, hen	161		0.4	19.8	1.2	39.5	←	20.1	→					← 9.0	→			←	10.0 →		
Sciatic nerve, chicken	72		1.9	24.6	3.5	10.0	38.4	2.9			1.6					2.0			1.0	1.0	
Mitochondria,																					
heart, human	108	tr	tr	6.1	2.7	26.6	7.1	3.4						37.5							
bovine	10			1.1	1.5	38.1	3.8	15.4	2.6					33.4		2.5					
rat	111		0.1	8.2	0.8	37.9	7.8	8.9				0.2		24.1		0.2			0.8	5.5	0.4
liver, rat	189		0.3	26.6	3.2	27.3	12.0	5.4						22.0						3.2	
inner membrane[c]	189		1.2	19.6	1.0	24.8	13.0	8.5					1.9	23.5						5.6	

References p. 479

TABLE 23 (Continued)

Tissue	Ref.	Fatty acid (as % total fatty acid by weight)																			
		12:0	14:0	16:0	16:1	18:0	18:1	18:2	18:3	20:0	20:1	20:2	20:3	20:4	20:5	22:0	22:3	22:4	22:5	22:6	24:0
outer membrane[c]	189		0.6	25.3	2.5	25.5	10.0	4.7					0.1	22.8						8.0	
brain, rat	28			8.7	1.0	32.4	30.3	1.0			3.2			10.9					3.1	9.0	
Microsomes,																					
heart, human	108	0.4	0.6	8.3	2.7	22.3	9.3	4.2						37.1							
bovine	43			4.5	tr	18.5	7.1	22.0	tr				7.7	35.6	4.6						
liver, rat	189		tr	22.6	2.3	23.4	9.8	10.3					tr	23.1						7.2	
smooth, e.r.	189		0.2	26.1	2.7	25.5	9.9	10.9					0.7	15.8						7.2	
rough e.r.	189		0.3	21.5	1.5	23.5	8.3	10.2						22.9						11.3	
brain, rat	28		0.4	9.3		41.8	25.8	tr			1.6			8.7					2.4	9.1	
Plasma membrane,																					
liver, rat	123	tr	1.9	25.8	0.2	33.2	10.4	8.7	tr					17.9							
erythrocyte, hen	69			12.2	4.8	14.0	13.4	33.7					8.3	1.2			1.5		1.4	5.8	
Lysosomes, liver, rat	189			28.5	4.6	27.2	12.3	5.4					1.3	13.4						7.0	
Platelet membrane,																					
blood, human	70			11.7	0.2	31.6	5.8	2.1	1.4	0.4				28.9				5.6	2.0	1.6	
Nuclear membrane,																					
erythrocyte, hen	69			11.4	4.3	14.0	15.2	34.1					7.5				1.0		1.8	7.1	
liver, bovine	107			6.6		38.0	14.0	11.3						19.5		2.6		5.3		2.8	
Golgi, liver, rat	123		0.7	33.5	0.4	31.8	5.1	10.0	tr					18.3							
Platelet granules,																					
blood, human	70			12.2	0.5	32.3	5.5	2.2	1.5	0.5				31.6				3.8	1.1	1.2	
Myelin[b],																					
CNS, human	136		0.4	6.5	0.4	7.7	72.5	tr			3.9			1.6					4.6		0.6
white matter, bovine	133		0.6	6.8	1.1	5.1	46.6	0.4	0.6	tr	6.2	2.3	3.6	6.2				2.7		12.2	
PNS, human	136		1.2	21.0	1.8	5.0	42.0	1.1			4.0			2.0							5.7

[a]16:0 and 18:0 values include 16:0A and 18:0A; [b]Brain has large amounts of ethanolamine plasmalogen (Table 2). These figures refer to the total fatty acids of ethanolamine phospholipids, i.e. they include the plasmalogen acyl groups; [c]Prepared by digitonin treatment.

Other data: human brain[136,137], bovine brain[163,164], rat liver[16,52,145,147] human erythrocyte[65,66,149], human serum[152,153], human platelets[70], rabbit and bovine lens[157], bovine retinal rod[103], bovine iris[158], dog adrenal[80], bovine pituitary[78], bovine pineal[82], rat thymus[76], bovine heart mitochondria[43], rat liver mitochondria[116,177,178], rat heart microsomes[111], rat liver microsomes[123,177,178], bovine mammary gland plasma membrane[128], rat liver plasma membrane[127], bovine myelin[131,133,134].

TABLE 24

FATTY ACIDS OF PHOSPHATIDYLINOSITOL OF WHOLE TISSUE AND SUBCELLULAR FRACTIONS

Tissue	*Ref.*	*As % total fatty acid by weight*																
		12:0	*14:0*	*16:0*	*16:1*	*18:0*	*18:1*	*18:2*	*18:3*	*20:0*	*20:1*	*20:3*	*20:4*	*22:0*	*22:3*	*22:4*	*22:5*	*22:6*
Brain, human (white matter)[a]	135			15.9	2.1	32.4	19.5	0.5	0.3		0.7	5.1	17.3			2.4	tr	1.1
Brain, human (grey matter)	70			14.0	1.8	34.9	12.6	0.6	0.1		0.3	1.8	28.2			1.5	tr	4.0
Brain, bovine[b]	18		tr	7.9	0.2	37.7	10.0	1.2			0.5	4.8	33.8			tr	tr	1.0
Brain, cat[c]	167		1.2	10.5	2.9	31.4	10.5	3.7	0.8	0.6			30.8	0.9				
Heart, bovine	168		tr	8	tr	53	10	12	1			1	14					
Liver, bovine[d]	142			1.9		46.3	10.2	8.6	2.5			6.0	7.1				7.1	
Liver, rat	168		tr	11.0	2	59	6	3				1	27					
Mitochondria, heart, human	108	0.3	0.5	5.5	2.9	41.2	8.1	4.2		tr			26.7					
rat[e]	111	0.1	0.3	8.2	0.8	37.9	7.8	8.9		0.2			24.1				0.3	5.5
liver, rat	189			26.3	5.8	38.4	14.0	4.2					7.6					3.2
outer membrane[g]	189			30.2	6.0	35.3	14.4	8.2					5.9					
outer membrane[h]	189		2.0	27.3	3.8	30.6	18.2	5.1					8.6					4.5
Microsomes, heart, human	108	0.2	0.3	6.3	2.9	41.2	12.1	4.3		tr			22.0	1.7	1.2			
rat[f]	111	0.1	0.6	8.1	0.8	41.6	8.1	6.9					25.0	0.1			0.3	2.9
liver, rat	189		2.5	19.3	1.8	45.0	7.2	3.2					21.4					
smooth e.r.	189			16.4	6.0	45.4	7.4	2.3				2.0	18.8					
rough e.r.	189			16.5	5.5	43.2	7.5	3.4				1.2	22.8					
Plasma membrane																		
liver, rat	123	tr	tr	14.0		49.0	2.3	1.6					10.2					
erythrocyte, hen	69			11.5		30.8	13.4	23.3				8.3	1.8		1.5		1.4	2.5
Lysosomes	189			36.3	7.5	27.1	13.5	2.4				2.1	11.1					
Golgi apparatus, liver, rat	123	tr	3.9	36.3		19.9	21.9	1.6					10.2					
Myelin, corpus callosum, bovine[j]	131			15.2	2.6	16.8	27.9	3.0		4.9	6.1		7.3			1.6		4.2

[a]18:4 1.2%; 22:4 1.2%; [b]22:3 2.2%; [c]12:0 9.8%; 17:0 1.0%; [d]20:5 4.3%; [e]20:5 0.2%; 24:0 0.4%; [f]20:2 0.3%; 24:0 0.2%; [g]membranes prepared by osmotic shock treatment; [h]membranes prepared by digitonin treatment; [j]23:0 5.3%.

Other data: bovine brain[165, 166], bovine heart[42] and liver[168], rat liver[149], human platelets[66, 70], bovine pineal[82], mouse muscle[160], pigeon pancreas[169], rat thymus[76], rat liver mitochondria[116, 177], rat liver microsomes[123, 177], bovine mammary gland plasma membrane[128], human platelet membrane[70] and granule[70], hen erythrocyte nuclear membrane[69], bovine liver nuclear membrane[107].

References p. 479

TABLE 25

FATTY ACIDS OF POLYPHOSPHOINOSITIDES OF BOVINE BRAIN

	Ref.	*Fatty acid (% by weight total fatty acid)*																
		14:0	*16:0*	*16:1*	*17:0*	*18:0*	*18:1*	*18:2*	*18:3*	*20:1*	*20:2*	*20:3*	*20:4*	>20:4	22:0	22:2	22:3	22:4
Diphosphoinositide	18	tr	5.2	0.5		37.3	14.5	1.2		1.5	1.7	9.9	22.5			0.5	5.5	tr
Diphosphoinositide	170		5.0			37.9	13.0	1.2	1.8	1.6	2.9	5.4	25.9	7.5				
Diphosphoinositide	166		6.7	0.5	1.2	35.9	14.7			0.8	1.0	3.4	26.4		5.0			4.2
Triphosphoinositide	18	tr	4.3	tr		36.8	14.0	0.6		1.2	2.0	10.4	26.4			1.1	3.1	0.1
Triphosphoinositide	170		4.8			42.1	13.7	0.4	0.7	0.7	3.7	4.0	24.3	5.6				
Triphosphoinositide	170		6.0			34.9	12.3	1.2	1.3	1.3	2.8	6.1	23.5	8.2				
Triphosphoinositide	166		8.0	0.8	1.3	39.8	18.1			1.0		3.2	21.9		3.6			2.3
Triphosphoinositide	166		6.7	0.5	1.5	35.2	15.6	1.2		1.4	0.9	3.8	26.4		4.3			2.4

TABLE 26

FATTY ACIDS OF PHOSPHATIDYLSERINE OF WHOLE TISSUE AND SUBCELLULAR FRACTIONS

Tissue	*Ref.*	*As % total fatty acid by weight*															
		14:0	*16:0*	*16:1*	*18:0*	*18:1*	*18:2*	*20:0*	*20:1*	*20:2*	*20:3*	*20:4*	*20:5*	*22:4*	*22:5*	*22:6*	*24:X*
Brain, human (white matter)	135		1.8	0.6	45.7	38.1	0.2		3.5		0.9	1.4		2.0	0.3	0.9	2.4
Brain, human (grey matter)	135		2.9	0.6	43.6	10.7	0.1		0.7		0.9	2.0		5.0	2.1	30.7	0.4
Liver, rat	148	0.9	22.8	2.2	15.3	15.0	30.9				0.8	7.6				2.0	
Erythrocyte, human	64		2.7		37.5	8.1	3.1	0.3	0.4	0.8	2.6	24.2	0.3	4.0	4.9	10.1	1.1
Plasma, human	154		7.8		32.6	11.6	8.6	0.9	0.8	1.6	4.2	25.1	0.2	1.2	1.2	2.3	0.5
Platelets, human	70		1.5	0.7	44.9	26.7		1.6				22.6			tr	tr	
Muscle, mouse	160	0.4	3.0	1.1	51.3	13.0	5.2	2.0				5.6					
Sciatic nerve, chicken[a]	72		20.8	1.2	5.9	43.4	6.7		5.5			2.8			7.5	1.5	1.8
Mitochondria, heart, rat	111	0.6	8.4	1.4	36.3	8.8	6.6	0.4		0.7		13.6	1.5		2.1	11.5	0.5
Microsomes, heart, rat	111	0.6	6.8	1.1	44.5	8.9	5.8	0.4		0.4		13.2			0.7	11.7	0.7
liver, rat,	123	9.0	11.1	1.0	4.7	21.8	52.3					tr					
Plasma membrane, liver, rat	123	1.5	38.7	tr	46.1	8.4	1.0					4.2					
Nuclear membrane, liver, bovine[b]	107		2.2		54.5	28.2	2.7					2.9		3.1			
Golgi apparatus, liver, rat	123	7.1	29.6	11.8	8.2	40.3	2.9					tr					
Myelin, CNS, human	136	0.3	2.6	0.6	40.0	43.3	tr		3.6			4.7			2.3	2.3	
white matter, bovine	133	0.7	3.6	1.4	30.6	43.4	1.8	0.6	5.6	1.8	0.7	1.3		2.6		3.0	
PNS, human	136	1.4	7.2	1.3	17.6	63.5	0.6		2.7			tr			1.2	0.7	

[a]22:1 1.3%; [b]22:0 6.5%.

Other data: human brain white matter[136, 137] and grey matter[136, 137], bovine brain[163], chick brain[140], human erythrocyte[65, 149], human platelets[66], rabbit and bovine lens[157], bovine retinal rod outer segments[103], bovine iris[158], bovine pineal[82], rat thymus[76]; human heart mitochondria[108], human heart microsomes[108], bovine mammary gland plasma membrane[128], human platelet membrane and granules[70], bovine corpus callosum myelin[131], bovine optic nerve myelin[133] and spinal root myelin[134].

References p. 479

TABLE 27

FATTY ACID COMPOSITION OF SPHINGOMYELIN OF WHOLE TISSUE AND SUBCELLULAR FRACTIONS

Tissue	*Ref.*	*Fatty acid (as % total fatty acid by weight)*																
		14:0	*16:0*	*16:1*	*18:0*	*18:1*	*18:2*	*18:3*	*20:0*	*20:1*	*20:2*	*20:4*	*22:0*	*22:5*	*23:0*	*24:0*	*24:1*	*26:1*
Brain, human (white matter)	22		10.2		20.1	6.5			1.1				1.7		2.8	6.9	30.2	5.5
Brain, human (grey matter)[a]	22		6.8		61.9	2.6			2.7				1.1		1.5	1.9	12.0	2.4
Liver, rat	52		21.6	0.6	9.9	4.1	2.4		1.8				13.9		8.5	23.6	13.4	
Plasma, human[b]	154		26.2		8.2	1.8	1.1		4.1	0.5	0.7	1.6	14.5	0.7	7.0	11.5	18.3	tr
Platelets, human[c]	70		24.6	1.2	5.8	1.3			9.4			tr	29.3		1.2	10.9	12.8	
Muscle, mouse[d]	160	1.5	12.5	1.8	18.5	7.7	19.5	3.1	2.0			9.9						
Sciatic nerve, chicken[e]	72	2.1	21.8	5.7	22.3	19.2	2.6		1.3	0.5			5.3			3.8	9.3	
Mitochondria, heart, human	108	3.1	41.2	3.4	16.6	13.8	6.2		2.2			6.9						
rat	111	0.6	26.6	1.0	23.9	9.5	11.2		2.8		0.2	9.9	3.2	4.1		0.1		
Microsomes, heart, human	108	2.9	40.8	5.5	16.0	12.4	4.6		1.8			4.5	0.5			3.2		
rat	111	1.2	27.0	1.2	22.1	8.9	7.4		7.0		0.5	5.2	7.1	3.9		1.0		
liver, rat	123	tr	26.0		7.2	12.5	13.9		tr			3.7				29.3		
Plasma membrane, liver, rat	123	tr	24.3		11.6	tr	tr		2.1			11.4	7.3			37.9		
Platelet membrane, blood, human	70		21.9	0.2	5.6	0.7	0.1		9.8			31.5	5.8		2.1	7.7	12.0	
Golgi apparatus, liver, rat	123	tr	24.8		10.0	1.1	tr		1.7			9.4	13.6			35.5		
Platelet granules, blood, human	70		22.5	0.3	5.7	0,8	0.2		9.6	tr		29.2	4.0		2.1	7.3	15.4	
Myelin, CNS, human	136		5.4		33.6	0.4			0.5				0.8		1.4	8.0	40.0	
PNS, human	136		5.0		12.0	0.2			3.3				15.0		6.5	22.0	31.0	
Nuclear membrane, liver, bovine	107		30.1		13.4	2.4			10.2				20.6			23.3		

[a]22:1 1.3%; [b]24:2 2.0%; [c]22:2 2.5%; [d]9.0 4.9%; 10:0 3.4%; 11:0 2.4%; 12:0 2.6%; 14:0 1.5%; 14:1 1.2%; 17:0 2.4%; [e]14:1 1.4%.

Other data: bovine brain[171], rat liver[148], human erythrocytes[64, 150], bovine erythrocytes[150], pig erythrocytes[151], human serum[152, 153], human and bovine plasma[150], rabbit and bovine lens[157], bovine iris[158], human, sheep and ox pineal[82], bovine mammary cells[159], rat thymus[76]; rat liver plasma membrane[127], bovine corpus callosum myelin[131], bovine optic nerve and white matter myelin[133], human white matter myelin[179], bovine spinal root myelin[134].

REFERENCES

1 G.S. Getz, *Advan. Lipid Res.*, 8 (1970) 175.
2 O. Renkonen and P. Varo, in G.V. Marinetti (Ed.), *Lipid Chromatographic Analysis*, Vol. 1, Chapter 3, Marcel Dekker, New York, 1967, p. 41.
3 G. Rouser, G. Kritchevsky and A. Yamamoto, in G.V. Marinetti (Ed.), *Lipid Chromatographic Analysis*, Vol. 1, Chapter 3, Marcel Dekker, New York, 1967, p. 99.
4 G. Rouser, G.J. Nelson, S. Fleischer and G. Simon, in D. Chapman (Ed.), *Biological Membranes*, Chapter 2, Academic Press, New York, 1968, p. 5.
5 G. Rouser, G. Kritchevsky, A. Yamamoto, G. Simon, C. Galli and A.J. Bauman, in S.P. Colowick and N.O. Kaplan (Eds.), *Methods in Enzymology*, Vol. XIV, Academic Press, New York, 1969, p. 272.
6 R. Wood, R.D. Harlow and F. Snyder, *Biochim. Biophys. Acta*, 176 (1969) 641.
7 F. Snyder, *Progr. Chem. Fats Lipids*, 10 (1969) 287.
8 R.O. Brady and E.G. Trams, *Ann. Rev. Biochem.*, 33 (1964) 75.
9 G.S. Getz, W. Bartley, D. Lurie and B.M. Notton, *Biochim. Biophys. Acta*, 152 (1968) 325.
10 S. Fleischer and G. Rouser, *J. Am. Oil Chem. Soc.*, 42 (1965) 588
11 L.J. Morris, *J. Lipid Res.*, 7 (1966) 717.
12 A. Kuksis, W.C. Breckenridge, L. Marai and O. Stachnyk, *J. Lipid Res.*, 10 (1969) 25.
13 L. Marai and A. Kuksis, *J. Lipid Res.*, 10 (1969) 141.
14 A. Montfoort, L.M.G. van Golde and L.L.M. van Deenen, *Biochim. Biophys. Acta*, 231 (1971) 335.
15 M. Yurkowski and B. Walker, *Biochim. Biophys. Acta*, 231 (1971) 145.
16 S.M. Hopkins, G. Sheehan and R.L. Lyman, *Biochim. Biophys. Acta*, 164 (1968) 272.
17 F. Haverkate and L.L.M. van Deenen, *Biochim. Biophys. Acta*, 106 (1965) 78.
18 B.J. Holub, A. Kuksis and W. Thompson, *J. Lipid Res.*, 11 (1970) 558.
19 M.G. Luthra and A. Sheltawy, *Biochem. J.*, 126 (1972) 1231.
20 L.M.G. van Golde and L.L.M. van Deenen, *Biochim. Biophys. Acta*, 125 (1966) 496.
21 G. Rouser and Y. Yamamoto, *Lipids*, 3 (1968) 284.
22 J.S. O'Brien and E.L. Sampson, *J. Lipid Res.*, 6 (1965) 537.
23 M.L. Cuzner, A.N. Davison and N.A. Gregson, *J. Neurochem.*, 12 (1965) 469.
24 T.W. Scott, B.P. Setchell and J.M. Bassett, *Biochem. J.*, 104 (1967) 1040.
25 R.E. Wuthier, *J. Lipid Res.*, 7 (1966) 544.
26 M.L. Cuzner and A.N. Davison, *Biochem. J.*, 106 (1968) 29.
27 M.A. Wells and J.C. Dittmer, *Biochemistry*, 6 (1967) 3169.
28 L.A. Biran and W. Bartley, *Biochem. J.*, 79 (1961) 159.
29 N. Herschkowitz, G.M. McKhann and E.M. Shooter, *J. Neurochem.*, 15 (1968) 161.
30 K. Owens, *Biochem. J.*, 100 (1966) 354.
31 R.M.C. Dawson, N. Hemington and J.B. Davenport, *Biochem. J.*, 84 (1962) 497.
32 R. Letters, *Biochem. J.*, 93 (1964) 313.
33 W.T. Norton and L.A. Autilio, *J. Neurochem.*, 13 (1966) 213.
34 W.T. Norton and L.A. Autilio, *Ann. N.Y. Acad. Sci.*, 122 (1965) 77.
35 J. Eichberg, V.P. Whittaker and R.M.C. Dawson, *Biochem. J.*, 92 (1964) 91.
36 L. Freysz, R. Bieth, C. Judes, M. Sensenbrenner, M. Jacob and P. Mandel, *J. Neurochem.*, 15 (1968) 307.
37 M.I. Gurr, C. Prottey and J.N. Hawthorne, *Biochim. Biophys. Acta*, 106 (1965) 357.
38 G. Simon and G. Rouser, *Lipids*, 4 (1969) 607.
39 M.R. Das and G. Rouser, *Lipids*, 2 (1967) 1.
40 G.V. Marinetti, J. Erbland and J. Kochen, *Federation Proc.*, 16 (1957) 837
41 J.N. Youngs and W.E. Cornatzer, *Comp. Biochem. Physiol.*, 9 (1963) 257.
42 G.M. Gray and M.G. Macfarlane, *Biochem. J.*, 70 (1958) 409.
43 L.W. Wheeldon, Z. Schumert and D.A. Turner, *J. Lipid Res.*, 6 (1965) 481.
44 P.O. Kwiterovich, H.R. Sloan and D.S. Frederickson, *J. Lipid Res.*, 11 (1970) 322.
45 G. Rouser, G. Simon and G. Kritchevsky, *Lipids*, 4 (1969) 599.
46 G. Rouser, G. Kritchevsky, A. Yamamoto, A.G. Knudson and G. Simon, *Lipids*, 3 (1968) 287.
47 M.I. Gurr, J.B. Finean and J.N. Hawthorne, *Biochim. Biophys. Acta*, 70 (1963) 406.

48 F. Parker and N.F. Peterson, *J. Lipid Res.*, 6 (1965) 455.
49 V.P. Skipski, R.T. Peterson and M. Barclay, *Biochem. J.*, 90 (1964) 374.
50 L.D. Bergelson, E.V. Dyatlovitskaya, T.I. Torkhovskaya, I.B. Sorokina and N.P. Gorkova, *Biochim. Biophys. Acta*, 210 (1970) 287.
51 J.J. Biezenski, T.H. Spaet and A.L. Gordon, *Biochim. Biophys. Acta*, 70 (1963) 75.
52 G. Fex, *Biochim. Biophys. Acta*, 231 (1971) 161.
53 J.A. Peters and L.M. Smith, *Biochem. J.*, 92 (1964) 379.
54 G.M. Gray, *Biochem. J.*, 77 (1960) 82.
55 C.F. Baxter, G. Rouser and G. Simon, *Lipids*, 4 (1969) 243.
56 L. Gluck, M.V. Kulovich and S.J. Brody, *J. Lipid Res.*, 7 (1966) 570.
57 T.E. Morgan, T.N. Finley and H. Fialkow, *Biochim. Biophys. Acta*, 106 (1965) 403.
58 H. Hof and R.G. Simon, *Lipids*, 5 (1970) 485.
59 J.B. Davenport, *Biochem. J.*, 90 (1964) 116.
60 J. Boichot and G. Di Constanzo, *Bull. Soc. Chim. Biol.*, 50 (1968) 1711.
61 T.W. Scott, R.M.C. Dawson and I.W. Rowlands, *Biochem. J.*, 87 (1963) 507.
62 T.W. Scott and B.P. Setchell, *Biochem. J.*, 107 (1968) 273.
63 T.W. Scott, J.W. Voglmayr and B.P. Setchell, *Biochem. J.*, 102 (1967) 456.
64 J.T. Dodge and G.B. Phillips, *J. Lipid Res.*, 8 (1967) 667.
65 P. Ways and D.J. Hanahan, *J. Lipid Res.*, 5 (1964) 318.
66 P. Cohen and A. Derksen, *Brit. J. Haematol.*, 17 (1969) 359.
67 T. Sato and T. Fujii, *Chem. Pharm. Bull.*, 19 (1971) 377.
68 G.J. Nelson, *Biochim. Biophys. Acta*, 144 (1967) 221.
69 H. Kleinig, H. Zentgraf, P. Comes and J. Städler, *J. Biol. Chem.*, 246 (1971) 2996.
70 A.J. Marcus, H.L. Ullman and L.B. Safier, *J. Lipid Res.*, 10 (1969) 108.
71 A. Sheltawy and R.M.C. Dawson, *Biochem. J.*, 100 (1966) 12.
72 J.F. Berry and W.H. Cevallos, *J. Neurochem.*, 13 (1966) 117.
73 C. Prottey and J.N. Hawthorne, *Biochem. J.*, 101 (1966) 191.
74 D.A. White and J.N. Hawthorne, *Biochem. J.*, 120 (1970) 533.
75 V. Macchia and I. Pastan, *Biochim. Biophys. Acta*, 152 (1968) 704.
76 D. Abramson and M. Blecher, *Biochim. Biophys. Acta*, 98 (1965) 117.
77 H.G. Rose and J.H. Frenster, *Biochim. Biophys. Acta*, 106 (1965) 577.
78 H. Singh and K.K. Carroll, *Lipids*, 5 (1970) 121.
79 J. Clément, G. Clément and M. Fontaine, *Compt. Rend. Soc. Biol.*, 157 (1963) 1716.
80 T.-C. Lo Chang and C.C. Sweeley, *Biochemistry*, 2 (1963) 592.
81 J. Trifaro, *Mol. Pharmacol.*, 5 (1969) 382.
82 J. Basinska, P.S. Sastry and H.C. Stancer, *J. Neurochem.*, 16 (1969) 707.
83 R.C. Roozemond, *J. Neurochem.*, 17 (1970) 179.
84 G. Rouser and R.D. Solomon, *Lipids*, 4 (1969) 232.
85 T. Fujii and T. Sato, *Chem. Pharm. Bull.*, 19 (1971) 385.
86 V.J.W. Long and H. Yardley, *J. Invest. Derm.*, 54 (1970) 174.
87 A. Tickner, *Biochem. J.*, 88 (1963) 80.
88 D.K. Kasbekar, G.M. Forte and J.G. Forte, *Biochim. Biophys. Acta*, 163 (1968) 1.
89 E.P. Adams and T.J. Heath, *Biochim. Biophys. Acta*, 70 (1963) 688.
90 A. Yamamoto and G. Rouser, *Biochim. Biol. Sper.*, 6 (1967) 135.
91 T.W. Scott, W. Hansel and L.E. Donaldson, *Biochem. J.*, 108 (1968) 317.
92 J.G. Parsons and S. Patton, *J. Lipid Res.*, 8 (1967) 696.
93 J. De Graeff, E.F. Dempsey, L.D.F. Lameyer and A. Leaf, *Biochim. Biophys. Acta*, 106 (1965) 155.
94 H.J. Wells, M. Makta, W.W. Wells and P.H. Krutzsch, *Biochim. Biophys. Acta*, 98 (1965) 269.
95 R.E. Wuthier, *J. Lipid Res.*, 9 (1968) 68.
96 I.M. Shapiro, *Calc. Tiss. Res.*, 5 (1970) 21.
97 M. Wajda, *Biochem. J.*, 95 (1965) 252.
98 R. Pietruszko, *Biochim. Biophys. Acta*, 64 (1962) 562.
99 E. Mulder, J. De Gier and L.L.M. van Deenen, *Biochim. Biophys. Acta*, 59 (1964) 502.
100 R.M. Broekhuyse, *Biochim. Biophys. Acta*, 218 (1971) 546.
101 R.M. Broekhuyse and J.H. Veerkamp, *Biochim. Biophys. Acta*, 152 (1968) 316.

102 R.M. Broekhuyse, *Biochim. Biophys. Acta*, 152 (1968) 307.
103 R.E. Anderson and M.B. Maude, *Biochemistry*, 9 (1970) 3624.
104 N.C. Nielsen, S. Fleischer and D.G. McConnell, *Biochim. Biophys. Acta*, 211 (1970) 10.
105 A.N. Siakotos, G. Rouser and S. Fleischer, *Lipids*, 4 (1969) 234.
106 H. Kleinig, *J. Cell Biol.*, 46 (1970) 396.
107 T.W. Keenan, R. Berezney, L.K. Funk and F.L. Crane, *Biochim. Biophys. Acta*, 203 (1970) 547.
108 J. Gloster and P. Harris, *Cardiovascular Res.*, 3 (1969) 45.
109 G.V. Marinetti, J. Erbland and E. Stotz, *J. Biol. Chem.*, 233 (1958) 562.
110 E.H. Strickland and A.A. Benson, *Arch. Biochem. Biophys.*, 88 (1960) 344.
111 J. Gloster and P. Harris, *Cardiovascular Res.*, 4 (1970) 1.
112 T. Gulick-Krzywicki, E. Rivas and V. Luzzati, *J. Mol. Biol.*, 27 (1967) 303.
113 S. Fleischer, G. Rouser, B. Fleischer, A. Casu and G. Kritchevsky, *J. Lipid Res.*, 8 (1967) 170.
114 S. Fleischer, H. Klouwen and G. Brierley, *J. Biol. Chem.*, 236 (1961) 2936.
115 A. Sheltawy, *Biochem. J.*, 95 (1965) 561.
116 G.S. Getz, W. Bartley, F. Stirpe, B.M. Notton and A. Renshaw, *Biochem. J.*, 83 (1962) 181.
117 B.J. Dod and G.M. Gray, *Biochim. Biophys. Acta*, 150 (1968) 397.
118 S. Løvtrup and L. Svennerholm, *Exptl. Cell. Res.*, 29 (1963) 298.
119 P. Mandel and J.L. Nussbaum, *J. Neurochem.*, 13 (1966) 629.
120 P. Parsons and R.E. Basford, *J. Neurochem.*, 14 (1967) 823.
121 H. Blaschko, H. Firemark, A.D. Smith and H. Winkler, *Biochem. J.*, 104 (1967) 545.
122 K. Owens and B.P. Hughes, *J. Lipid Res.*, 11 (1970) 486.
123 T.W. Keenan and D.J. Morré, *Biochemistry*, 9 (1970) 19.
124 G. Dallner, P. Siekevitz and G.E. Palade, *J. Cell. Biol.*, 30 (1966) 73.
125 J.N. Cumings, E.J. Thompson and H. Goodwin, *J. Neurochem.*, 15 (1968) 243.
126 J. Meldolesi, J.D. Jamieson and G.E. Palade, *J. Cell. Biol.*, 49 (1971) 130.
127 R.C. Pfleger, N.G. Anderson and F. Snyder, *Biochemistry*, 7 (1968) 2826.
128 T.W. Keenan, D.J. Morré, D.E. Olson, W.N. Younghans and S. Patton, *J. Cell. Biol.*, 44 (1970) 80.
129 A.N. Siakotos, G. Rouser and S. Fleischer, *Lipids*, 4 (1969) 239.
130 E.J. Thompson, H. Goodwin and J.N. Cumings, *Nature*, 215 (1967) 168.
131 F.A. Manzoli, S. Stefoni, L. Manzoli-Guidotti and M. Barbieri, *FEBS Letters*, 10 (1970) 317.
132 M.J. Evans and J.B. Finean, *J. Neurochem.*, 12 (1965) 729.
133 M.C. MacBrinn and J.S. O'Brien, *J. Neurochem.*, 16 (1969) 7.
134 J.S. O'Brien, E.L. Sampson and M.B. Stern, *J. Neurochem.*, 14 (1967) 357.
135 L. Svennerholm, *J. Lipid Res.*, 9 (1968) 570.
136 J.S. O'Brien and E.L. Sampson, *J. Lipid Res.*, 6 (1965) 545.
137 J.S. O'Brien, D.F. Fillerup and J.F. Mead, *J. Lipid Res.*, 5 (1964) 329.
138 H. Yabuuchi and J.S. O'Brien, *J. Lipid Res.*, 9 (1968) 65.
139 T.R. Skrbic and J.N. Cumings, *J. Neurochem.*, 17 (1970) 85.
140 K. Miyamoto, L.M. Stephanides and J. Bernsohn, *J. Lipid Res.*, 7 (1966) 664.
141 M.G. Macfarlane, *Biochem. J.*, 78 (1961) 44.
142 G.A.E. Arvidson, *J. Lipid Res.*, 6 (1965) 574.
143 D.B. Menzel and H.S. Olcott, *Biochim. Biophys. Acta*, 84 (1964) 133
144 H. Brockerhoff, R.J. Hoyle and P.C. Hwang, *Biochim. Biophys. Acta*, 144 (1967) 541.
145 H. Kanoh, *Biochim. Biophys. Acta*, 176 (1969) 756.
146 E.C. Kyriakides and J.A. Balint, *J. Lipid Res.*, 9 (1968) 142.
147 J.A. Balint, D.A. Beeler, D.H. Treble and H.L. Spitzer, *J. Lipid Res.*, 8 (1967) 486.
148 G.S. Getz, W. Bartley, F. Stirpe, B.M. Notton and A. Renshaw, *Biochem. J.*, 80 (1961) 176.
149 G. Soula, P. Valdiguié and L. Douste-Blazy, *Bull. Soc. Chim. Biol.*, 50 (1968) 1331.
150 D.J. Hanahan, R.M. Watts and D. Pappajohn, *J. Lipid Res.*, 1 (1960) 421.
151 D.J. Hanahan, J.E. Ekholm and B. Benson, *Biochim. Biophys. Acta*, 231 (1971) 343.
152 G.J. Nelson and N.K. Freeman, *J. Biol. Chem.*, 235 (1960) 578.
153 M.A. Antar, M.A. Ohlson and M.O. Osborn, *Biochem. J.*, 105 (1967) 117.
154 G.B. Phillips and J.T. Dodge, *J. Lipid Res.*, 8 (1967) 676.
155 D.J. Hanahan, H. Brockerhoff and E.J. Barron, *J. Biol. Chem.*, 235 (1960) 1917.
156 G.M. Gray and M.G. Macfarlane, *Biochem. J.*, 81 (1961) 480.

157 R.E. Anderson, M.B. Maude and G.L. Feldman, *Biochim. Biophys. Acta*, 187 (1969) 345.
158 T.W. Culp, P.W. Tucker, C.R. Ratliff and F.F. Hall, *Biochim. Biophys. Acta*, 218 (1970) 259.
159 J.E. Kinsella and R.D. McCarthy, *Biochim. Biophys. Acta*, 164 (1968) 530.
160 C.E. Eybel and R.G. Simon, *Lipids*, 5 (1969) 590.
161 J.C. Hawke, *Biochem. J.*, 71 (1959) 588.
162 F. Possmayer, G.L. Scherphof, T.M.A.R. Dubbelman, L.M.G. van Golde and L.L.M. van Deenen, *Biochim. Biophys. Acta*, 176 (1969) 95.
163 E. Klenk and P. Böhm, *Z. Physiol. Chem.*, 288 (1951) 98.
164 H. Debuch, *Z. Physiol. Chem.*, 304 (1956) 109.
165 R.S. Atherton and J.N. Hawthorne, *European J. Biochem.*, 4 (1968) 68.
166 H.S. Hendrickson and C.E. Ballou, *J. Biol. Chem.*, 239 (1964) 1369.
167 R.J. Rossiter and F.B. Palmer, *Biochem. Z.*, 342 (1965) 337.
168 H. Brockerhoff, *Arch. Biochem. Biophys.*, 93 (1961) 641.
169 R.W. Keenan and L.E. Hokin, *Biochim. Biophys. Acta*, 84 (1964) 458.
170 W. Thompson, *Biochim. Biophys. Acta*, 187 (1969) 150.
171 J.S. O'Brien and G. Rouser, *J. Lipid Res.*, 5 (1964) 339.
172 H. Yabuuchi and J.S. O'Brien, *J. Neurochem.*, 15 (1968) 1383.
173 S. Courtade, G.V. Marinetti and E. Stotz, *Biochim. Biophys. Acta*, 137 (1967) 121.
174 H. Okuyama and S. Nojima, *J. Biochem. (Tokyo)*, 57 (1965) 529.
175 H.G. Rose, *Biochim. Biophys. Acta*, 84 (1964) 109.
176 G.V. Marinetti, *Biochim. Biophys. Acta*, 84 (1964) 55.
177 A. Yamamoto, M. Isozaki, K. Hirayama and Y. Sakai, *J. Lipid Res.*, 6 (1965) 295.
178 M.G. Macfarlane, G.M. Gray and W. Wheeldon, *Biochem. J.*, 77 (1960) 626.
179 Y. Kishimoto, B.W. Agranoff, N.S. Radin and R.M. Burton, *J. Neurochem.*, 16 (1969) 397.
180 G.M. Gray, *Biochim. Biophys. Acta*, 84 (1964) 35.
181 R.M.C. Dawson and J. Eichberg, *Biochem. J.*, 96 (1965) 634.
182 W.T. Norton and S.E. Poduslo, *J. Lipid Res.*, 12 (1971) 84.
183 S.E. Poduslo and W.T. Norton, *J. Neurochem.*, 19 (1972) 727.
184 C. Cotman, M.L. Blank, A. Moehl and F. Snyder, *Biochemistry*, 8 (1969) 4606.
185 W.C. Breckenridge, G. Gombos and I.G. Morgan, *Biochim. Biophys. Acta*, 266 (1972) 695.
186 Y. Yagihara, J.E. Bleasdale and J.N. Hawthorne (unpublished).
187 V.P. Whittaker, *Ann. N.Y. Acad. Sci.*, 137 (1966) 982.
188 G. Rouser and A. Yamamoto, in A. Lajtha (Ed.), *Handbook of Neurochemistry*, Vol. 1, Plenum, New York, 1969, p. 121.
189 A. Colbeau, J. Nachbaur and P.M. Vignais, *Biochim. Biophys. Acta*, 249 (1971) 462.
190 R.P. van Hoeven and P. Emmelot, *J. Membrane Biol.*, 9 (1972) 105.

SUBJECT INDEX*

Acetal phosphatidate, original isolation, 9
Acetolysis, of phospholipids, 38
Acetylcholine, PA metabolism and, 424
–, phospholipid turnover and, 424
Acid hydrolysis, of phospholipids, 37
Acyl-carrier protein, lyso-PA synthesis and, 290
–, PA synthesis and, 68
Acyl-CoA:lysophospholipid transferase, physicochemical factors affecting activity, 109
–, specificity, 109
Acyldihydroxyacetone phosphate, isolation, 32
N-acyl ethanolamine phosphoglyceride, characterisation, 24
–, isolation, 24
–, structure, 24
Acylglycerols, intracellular metabolism, 348
–, intraluminal hydrolysis, 346
Acyltransferases, in erythrocyte membrane, 83
–, of *E. coli,* 290
–, metabolic turnover of phospholipids, 81
–, physicochemical factors affecting activity, 109
–, of plants, 268
–, specificity, 109
–, –, for acyl CoA, 82
–, –, for fatty acids, 82
Adipose tissue, phospholipid composition, 457
Adrenal gland, phospholipid composition, 456
Adrenal medulla, PA effect in, 432
–, PI effect in, 432
–, phospholipid turnover in, 432
Aerobic shift, phospholipid metabolism in *H. parainfluenzae* and, 321
Agrobacteriae, PC in, 295
Agrobacterium tumefaciens, PC biosynthesis in, 298
Alanylphosphatidylglycerol, biosynthesis in *M. laidlawii,* 303
Albumin, phospholipid membranes and, 164
Alkaline alcoholysis, of phospholipids, 37
Alkyldihydroxyacetone phosphate, isolation, 32
Alumina, chromatography, 46
Amino-acyl phosphatidylglycerols, biosynthesis, 302
–, function, 305
–, metabolism, 304
–, in micro-organisms, 302

* Abbreviations, see p. viii.

Amino-acyl-tRNA, in amino-acyl PG synthesis, 303
2-Aminoethylphosphonic acid (AEP), *see* Phosphonolipids
–, original isolation, 333
Amphipathic substances, phospholipid packing and, 102
Anaesthetics, interaction with membranes, 413
Animal cells, PA biosynthesis in, 68
–, PS biosynthesis in, 73
Animal tissues, phospholipid metabolism in, 67
Antibiotics, as ion carriers, *see* Ionophores
Aorta, phospholipid composition, 457
ATP:choline phosphotransferase, in plants, 269
ATPase, mitochondrial, 227
–, of sarcoplasmic reticulum, 230
–, sodium pump, 431

Bacteria, amino-acyl phosphatidylglycerols in, 302
–, methyl derivatives of PE in, 296
–, PA biosynthesis in, 68
–, PE biosynthesis in, 71
Bacteroides melaninogenicus, ceramide phosphorylethanolamine in, 314
–, ceramide phosphorylglycerol in, 314
–, phosphosphingolipids in, 314
Bile, composition, 347, 363
–, formation, 363
–, micelle, 347
–, phospholipid composition, 364, 457
–, salts, 346
–, –, absorption, 348
Black lipid films, *see* Phospholipid bilayers
Bladder, phospholipid composition, 457
Brain, development, phospholipid changes in, 397
–, phospholipid composition, 379, 447
–, phospholipid metabolism in, 393, 401
–, phospholipid transfer in, 395

Calcium, phospholipid interactions, 414
Cardiolipin, (*see also* Diphosphatidylglycerol)
–, biosynthesis in plants, 269
–, composition of subcellular fractions, 459–467
–, composition of tissues, 447–458
–, fatty acid composition, in animals, 471

Cardiolipin (*continued*)
–, –, in plants, 261–266
–, occurrence, 27
–, original isolation, 27
–, in plant tissues, 254
–, preparation, 27
–, structure, 27
CDPcholine, in PC biosynthesis, 70
CDPcholine:ceramide cholinephosphotransferase, physicochemical factors affecting activity, 111
–, specificity, 111
CDPcholine:1,2-diglyceride cholinephosphotransferase, *see* Choline phosphotransferase
CDPdiglyceride, biosynthesis in micro-organisms, 292
–, in CL biosynthesis, 76
–, chemical synthesis, 19
–, in PG biosynthesis, 76
–, in PI biosynthesis, 72
–, specificity, 112
–, structure, 19
CDPethanolamine, in PE biosynthesis, 71
CDPethanolamine:1,2-diglyceride ethanolaminephosphotransferase, *see* Ethanolamine phosphotransferase
Cell culture, phospholipid metabolism in, 396
Cephalin, (*see also* Phosphatidylethanolamine)
–, fractionation, 7
–, original isolation, 4
Ceramidase, action, 88
–, in intestinal cells, 346
Ceramide phosphoglycolipids, in yeasts, 315
Ceramide phosphorylcholine, *see* Sphingomyelin
Ceramide phosphorylethanolamine, in *B. melaninogenicus*, 314
–, characterisation, 35
–, occurrence, 35
Ceramide phosphorylglycerol, in *B. melaninogenicus*, 314
–, characterisation, 35
–, isolation, 35
Cerebroside, brain levels during development, 398
–, –, during dietary deficiencies, 407
Chenodeoxycholic acid, 363
Cholesterol, in bile, 363
–, brain levels during development, 398
–, in chylomicrons, 351
–, interactions with phospholipids, 136
–, phospholipid membranes and, 159
–, pools in liver, 368
Cholesterol, (*continued*)
–, transport by lipoproteins, 360
Cholesterol ester, in chylomicrons, 352
Cholic acid, 363
Choline, determination of, 59
–, discovery, 2
–, transfer to brain, 396
Choline phosphoglyceride, alk-1-enyl acyl derivative, 19
–, –, chemical synthesis, 21
–, –, isolation, 21
–, –, occurrence, 21
–, 1-alkyl 2-acyl derivative, 19
–, –, characterisation, 21
–, –, occurrence, 21
–, diacyl derivative (PC), 19
–, –, chemical synthesis, 20
–, –, occurrence, 20
–, –, preparation, 20
–, –, structure, 20
–, monoacyl derivative, 19
–, –, chemical synthesis, 20
–, –, occurrence, 20
–, –, preparation, 20
Choline phosphotransferase, physicochemical factors affecting activity, 111
–, in plants, 269, 271
–, specificity, 111
Chromaffin granule, 432
Chromatium strain D, chromatophore formation in, 323
Chromatography, alumina and, 46
–, DEAE cellulose and, 52
–, early work, 6
–, gas-liquid, 55
–, hydroxylapatite and, 54
–, paper, 54
–, silica impregnated paper, 51
–, silicic acid and, 46
–, TEAE cellulose and, 53
–, thin layer, 48
–, –, R_F values of phospholipids, 49, 50
Chromatophores, in Chromatium strain D, 323
–, in *R. spheroides*, 322
Chyle, and fat absorption, 349
Chylomicrons, 345
–, in circulation, 352
–, clearance, 353
–, composition, 351
–, formation, 352
–, and HDL, 352
–, LCAT of, 352
–, metabolism, 350
–, –, and liver, 353
–, phospholipids in, 351

Chylomicrons, (*continued*)
–, protein in, 351
–, structure, 351, 357
Classification, of phosphodiol lipids, 11
–, of phosphoglycerides, 10
–, of phosphoglycolipids, 11
–, of phospholipids, 10
–, of phosphonolipids, 333
–, of phosphosphingolipids, 11
Clostridia, methyl derivatives of PE in, 296, 297
Clostridium butyricum, lyso-PA synthesis in, 290
–, –, phosphatidyl-*N*-methyl ethanolamine synthesis in, 300
Corpus luteum, phospholipid composition of, 457
CTP:cholinephosphate cytidylyltransferase, in plants, 269, 271
–, specificity, 110
CTP:ethanolaminephosphate cytidylyl-transferase, specificity, 110
CTP:phosphatidic acid cytidylyltransferase, physicochemical factors affecting activity, 112
–, in plants, 267
–, specificity, 112
Cyclopropane phospholipid synthetase, physicochemical factors affecting activity, 113
–, specificity, 113
Cytochrome induction, 318
Cytochrome c, phospholipid interaction, 224
–, phospholipid membranes and, 162

Deacylase, physicochemical factors affecting activity, 108
–, specificity, 108
DEAE cellulose, chromatography and, 52
Deoxycholic acid, 363
Diacylglycerol phosphoglyceride, identification, 28
–, occurrence, 28
O-Diglucosylglycerol phosphoglyceride, characterisation, 32
–, purification, 32
Diglyceride kinase, physicochemical factors affecting activity, 112
–, PA synthesis and, 69, 92
–, in plants, 268
–, specificity, 112
Dihydrosphingosine, biosynthesis, 79
–, degradation, 93
Dihydroxyacetone phosphate, in glycerol ether phospholipid biosynthesis, 74
–, in PA synthesis, 69
Dimannophosphoinositides, biosynthesis, 309
N,N-dimethylethanolamine phosphoglyceride, chemical synthesis, 24
–, isolation, 24
–, occurrence, 24
–, structure, 24
Dipalmitoyl lecithin, in lung alveoli, 437
Diphosphatidylglycerol, (*see also* Cardiolipin)
–, biosynthesis, 76
–, – in *E. coli*, 77
–, breakdown in *H. parainfluenza*, 293
–, fatty acid composition, 471
Diphosphoinositide (DPI), (*see also* Polyphosphoinositides)
–, biosynthesis, 72
–, isolation, 30
–, occurrence, 30
–, structure, 27
"Diphosphoinositide fraction", characterisation, 30
–, original isolation, 9
Diphosphoinositide kinase, in TPI biosynthesis, 72
Drug metabolism, 366
–, endoplasmic reticulum and, 366
–, the liver and, 366
–, phospholipids and, 366
Duodenum, 347
Dysmyelination, 402
–, metachromatic leucodystrophy, 403
–, globoid cell leucodystrophy, 403
–, Niemann–Pick disease, 404
–, Pelizaeus–Merzbacher disease, 404
–, Refsum's disease, 404

Electron microscopy, of membranes, 186
Electron transport system, induction and phospholipid metabolism in micro-organisms, 318
–, reconstitution by phospholipids, 225
Endoplasmic reticulum, drug metabolism and, 366
–, phospholipid composition, 214
–, phospholipid: protein composition during maturation, 245
–, phospholipid synthesis and, 231
Enzyme action on phospholipids, 97
Erythrocyte, acyltransferases in, 83
–, phospholipid composition, 454
Escherichia coli, acyltransferases of, 290
–, CL synthesis in, 77
–, fatty acid composition of lipids of, 316, 324

Escherichia coli, (continued)
–, β-galactosidase in, 324
–, β-galactosidase acetyltransferase in, 324
–, glycerol-requiring mutant of, 324
–, linoleate in, 324
–, oleate in, 324
–, PA synthesis in, 289
–, PE turnover in, 294
–, PG biosynthesis in, 293
–, phosphatase activity, 291
–, phosphotransferase system in, 326
–, phosphatidylglycerophosphate synthesis in, 293
–, phospholipases in, 294
–, sugar transport in, 326
–, unsaturated fatty acid auxotroph, 324
Ethanolamine, determination of, 59
Ethanolamine phosphoglycerides, alkyl-acyl derivative, occurrence, 23
–, –, structure, 23
–, alkyl-1-enyl, acyl derivative (EP), occurrence, 23
–, –, preparation, 23
–, –, structure, 22
–, diacyl derivative (PE), chemical synthesis, 22
–, –, occurrence, 22
–, –, preparation, 22
–, monoacyl derivative (lyso-PE), occurrence, 21
Ethanolamine phosphotransferase, specificity, 111
Ethanolamine plasmalogen, biosynthesis in nervous tissue, 389
–, brain levels during development, 398
–, in normal and quaking mice, 402
–, sphingolipid metabolism and, 390
Etherase, action, 90
–, in rat liver, 90

Fat absorption, 345
–, partition theory of, 348
Fatty acid, acyltransferase specificity, 82
Fatty acid, composition of CL, 471
–, – of lipids during brain development, 399
–, – of lipids, electron transport induction and, 319
–, – of lipids of *E. coli,* 316, 324
–, – of lyso compounds, 472
–, – of PA, 472
–, – of PC, 468
–, – of PE, 473
–, – of PG, 472
–, – of PI, 475
–, – of PS, 477
–, –, phospholipid turnover and, 242
–, – of plant phospholipids, 261–266
–, – of polyphosphoinositides, 476
Fatty acid, (*continued*)
–, – of SM, 478
–, deficiency, brain lipid metabolism and, 407
–, metabolism, phospholipids and, 282
–, non-esterified, 354
–, separation, 57
–, uptake by liver, 354
Fatty aldehyde, separation, 57
Fatty liver, causes, 361
–, lipoprotein production and, 362
–, occurrence, 361

β-galactosidase, in *E. coli,* 324
Galactoside acetyltransferase, in *E. coli,* 324
β-galactoside, carrier, 325
–, transport in *E. coli,* 324
Gastric mucosa, phospholipid composition, 457
GDPmannose, and PI mannoside synthesis, 309
Glucosaminylglycerol phosphoglyceride, characterisation, 28
–, isolation, 28
–, in micro-organisms, 305
Glucosaminyl phosphatidylglycerol, *see* Glucosaminylglycerol phosphoglyceride
Glucose-6-phosphatase, 230
Glucose phosphoglyceride, occurrence, 31
–, preparation, 31
Glycerol ether phospholipids, biosynthesis, 73
Glycerol phosphoglycerides, *see* Phosphatidylglycerol
Glycerophosphate–phosphatidic acid pathway, 348
Glycerophosphate phosphoglycerides, chemical synthesis, 26
–, occurrence, 26
–, structure, 26
Glycerylphosphoryl diglucosyl diglyceride, derivatives of, in micro-organisms, 312
Glycosyl transglycosyl transferase, in Gram-negative bacteria, 326
–, PE and, 326
Golgi apparatus, phospholipid composition, 216
Gram-negative bacteria, glycosyl transglycosyl transferase in, 326
–, lipopolysaccharides in, 326

Haemophilus parainfluenza, CL in, 293
–, PE in, 294
–, phospholipid metabolism in, 321
Heart, phospholipid composition, 448
Heparin, 353
Hepato-portal system, phospholipids and, 345
High density lipoprotein (HDL), chemical properties, 356
–, cholesterol transport and, 360

High density lipoprotein, (*continued*)
–, chylomicrons and, 352
–, LCAT activity and, 360
–, physical properties, 356
–, structure, 357
–, VLDL and, 359
N-2-Hydroxyethyl alanine phosphoglyceride, isolation, 25
–, structure, 25
Hydroxylapatite, chromatography and, 54

Ileum, 348
Inositol, determination, 55
–, in PI biosynthesis, 72
Inositol diphosphate phosphoglyceride, *see* Triphosphoinositide
Inositol monophosphate phosphoglyceride, *see* Diphosphoinositide
Inositol phosphatidyl transferase, specificity, 112
Inositol phosphoglyceride, *see also* Phosphatidylinositol)
–, monomannosyl-hexamannosyl structure, 31
Intestinal cells, ceramidase of, 346
–, lysolecithinase of, 346
–, phospholipase A_2 of, 346
–, sphingomyelinase of, 346
Intestinal mucosa, phosphatidate phosphohydrolase in, 348
–, PC metabolism in, 349
–, phospholipase C in, 350
–, phospholipid composition, 457
–, phospholipid metabolism, 345, 349
–, SM metabolism, 350
Intraluminal phase, physical state, 347
–, uptake of lipids from, 347
Intraluminal phospholipid hydrolysis, 346
Intraluminal triacylglycerol hydrolysis, 346
Ionophores, 157
Isotropic phase, of intraluminal contents, 347

Jejunum, 346, 348

Kidney, phospholipid composition, 450

Lac-operon, 324
Langmuir trough, 145
LCAT, action, 83
(*see also* Phosphatidylcholine: cholesterol acyl transferase)
–, in chylomicrons, 352
–, HDL and, 360
Leucocyte, PI effect in, 436
Leucodystrophy, 403

Linoleate, in *E. coli*, 324
Lipid, classification, 1
–, compound, 1
–, derived, 1
–, extraction from plants, 267
–, protein interactions, 171
–, simple, 1
–, uptake by liver, 354
Lipopolysaccharide, biosynthesis in Gram-negative bacteria, 326
Lipoprotein, (*see also* HDL, LDL, VLDL)
–, fatty liver and, 362
–, phospholipid exchange and, 236
–, soluble, 190
β-Lipoproteinaemia, 358
Lipoprotein lipase, 352
Liposome, (*see also* Phospholipid vesicles)
–, multilamellar vesicles, 151
–, smectic mesophase, 151
–, unilamellar vesicles, 151
Lithocholic acid, 363
Liver, cholesterol pools in, 368
–, chylomicron metabolism and, 353
–, drug metabolism and, 366
–, etherase in, 90
–, PC pools in, 368
–, phospholipase A_1 in, 355
–, phospholipid composition, 449
–, – metabolism, 345, 367
–, – pools in, 368
–, – turnover, 365
–, plasma lipoproteins and, 356
–, regeneration, 367
–, starvation and, 366
–, uptake of fatty acids, 354
–, uptake of lipids, 354
Low density lipoprotein (LDL), chemical properties, 356
–, physical properties, 356
–, structure, 357
–, VLDL and, 358
Lung, phospholipid composition, 452
Lymphocyte, PI effect in, 435
Lysolecithinase, in intestinal cells, 346
Lysophospholipase, *see* Phospholipase B
Lysophospholipids, composition of subcellular fractions, 459–467
–, composition of tissues, 447–458
–, fatty acid composition, 472
Lysophosphatidic acid in *C. butyricum*, 290
Lysyl-phosphatidylglycerol, in *S. aureus*, 302

Mammary gland, phospholipid composition, 457
Membrane, biogenesis, 240
–, biological, 171

Membrane, (*continued*)
–, electron microscopy, 186
–, extraction, 174
–, optical studies, 181
–, phospholipid function in, 223, 229, 323
–, protein extraction, 161
–, – interaction with phospholipid membrane, 161
–, reconstituted, 194
–, solubilisation, 174
–, spectroscopic studies, 181
–, transport: role of phospholipids, 229
–, X-ray diffraction, 177
Mesomorphism, lyotropic, 123
–, phase diagram for PC-water system, 126
–, structure of lyotropic phases, 124
–, thermotropic, 120
N-methylethanolamine phosphoglycerides, chemical synthesis, 24
–, isolation, 24
–, occurrence, 24
–, structure, 24
Micro-organisms, phospholipids of, 289
–, phospholipid metabolism in, 316
Microbial membrane, phospholipid function in, 323
Microsomal membrane, phospholipid composition, 462
–, –, in rat brain, 379, 384
–, phospholipid function in, 229
–, –, turnover in, 243
–, protein turnover in, 243
Milk, phospholipid composition, 457
Mitochondria, ATPase of, 227
–, function of phospholipids of, 223
–, half-life of, 242
–, phospholipid composition, 209, 460
–, –, in rat brain, 379, 383
–, phospholipid synthesis, 232
–, phospholipid turnover, 243
–, protein turnover, 243
Monoacylglycerol pathway, 348
Monoacylglycerol phosphoglyceride, identification, 28
–, occurrence, 28
Monomolecular films, *see* Phospholipid monolayer
Mueller membranes, *see* Phospholipid bilayer
Multilamellar vesicles, 151
Multiple sclerosis, primary demyelination, 408
Muscle, denervated, phospholipid changes in, 436
Mycobacterium, PI mannoside biosynthesis in, 309
Mycoglycolipid, in yeasts, 315
Mycoplasma laidlawii, alanyl phosphatidylglycerol biosynthesis in, 303
Myelin, composition during brain development, 399
–, phospholipid composition, 379, 380, 466, 467
–, phospholipid labelling, 394
Myelination, inborn errors and, 403
–, nutrition and, 405

Nerve, phospholipid composition, 455
Nerve ending, phospholipid metabolism in, 423
Nervous system, phospholipids and, 372
–, development, phospholipids and, 397
Nervous tissue, phospholipid biosynthesis in, 384
Neurospora crassa, PC biosynthesis in, 300
Neurotransmitters, phospholipid turnover and, 428
Niemann–Pick disease, 404
Nomenclature, of lipids containing glycerol, 12
–, of phospholipids, 12
–, of sphingolipids, 15
Non-esterified fatty acid (NEFA), 354
–, albumin complex, 356
Nuclear membranes, phospholipid composition, 219
Nucleus, phospholipid composition, 459
–, –, in brain cells, 379, 383
Nutrition, brain development and, 405

Oil phase, of intraluminal contents, 347
Oleate, in *E. coli,* 324
Oxidation reactions, of phospholipids, 37

Pancreas, PA effect in, 433
–, PI effect in, 433
–, phospholipid composition, 456
–, – turnover, 433
–, secretion and zymogen granule, 433
Pancreatic cholesterol esterase, 346
Pancreatic juice, 346
Pancreatic lipase, 346
Pelizaeus–Merzbacher disease, 404
Phagocytosis, 436
Phenobarbitol, 244
Phenobarbitone, 366
Phosphatase, activity in *E. coli,* 291
Phosphatidate phosphohydrolase, action, 92
–, activity in intestinal mucosa, 348
–, physicochemical factors affecting activity, 108
–, specificity, 108

Phosphatidic acid (PA), analogues of, 17
–, biosynthesis, 68
–, –, in animal cells, 68
–, –, in bacterial systems, 68
–, –, in micro-organisms, 289
–, –, in nervous tissue, 385
–, –, in plants, 268
–, chemical synthesis, 19
–, composition of subcellular fractions, 459–467
–, composition of tissues, 447–458
–, effect in adrenal medulla, 432
–, effect in pancreas, 433
–, fatty acid composition, 472
–, –, in plants, 261–266
–, fatty acid positioning, 68
–, isolation, 17, 45
–, metabolism in synaptosomes, 424
–, original isolation, 9
–, in plant tissues, 254
–, triacylglycerol synthesis and, 348
Phosphatidic acid phosphatase, *see* Phosphatidate phosphohydrolase
Phosphatidic acid synthetase system, physico-chemical factors affecting activity, 109
–, specificity, 109
Phosphatidylcholine (PC), absorption, 349
–, in Agrobacteriae, 295
–, biosynthesis, 69
–, – in *A. tumefaciens,* 298
–, – in nervous system, 385
–, – in *N. crassa,* 300
–, – in plants, 270
–, – in *S. cerevisiae,* 300
–, – in *T. pyriformis,* 301
–, catabolism in micro-organisms, 302
–, choline confirmation, 129
–, composition of subcellular fractions, 459–467
–, composition of tissues, 447–458
–, in developing brain, 398
–, discovery, 2
–, export by liver, 345
–, fatty acid composition, 468
–, –, in plant tissues, 261–266
–, fatty acid uptake in liver and, 355
–, isolation, 45
–, metabolism in intestinal mucosa, 349
–, in micro-organisms, 295, 296
–, phase diagram of PC–water system, 126, 128
–, in plant tissues, 254
–, pools in liver, 368
–, in Protozoa, 298
–, purification, 45
–, in Rhodopseudomonae, 295
Phosphatidylcholine:cholesterol acyl transferase (LCAT), physicochemical factors affecting activity, 109
–, specificity, 109
Phosphatidylethanolamine (PE), biosynthesis, 71
–, – in bacteria, 294
–, – in nervous tissue, 387
–, – in plants, 270
–, composition of subcellular fractions, 459–467
–, – of tissues, 447–458
–, in developing brain, 398
–, fatty acid composition, 473
–, –, in plant tissues, 261–266
–, isolation, 45
–, lipopolysaccharide biosynthesis and, 326
–, methylation pathway, 77
–, methyl derivatives of, in bacteria, 296
–, in micro-organisms, 293
–, original isolation, 9
–, in plant tissues, 254
–, as precursor of PC, 77
–, purification, 45
–, turnover in *E. coli,* 294
–, turnover in *H. parainfluenza,* 294
Phosphatidylethanolamine: *N*-methyltransferase, specificity, 113
Phosphatidylglycerol (PG), *O*-amino acid esters of, characterisation, 28
–, –, isolation, 28
–, –, occurrence, 29
–, amino-acyl derivatives of, *see* Amino-acyl phosphatidylglycerols
–, biosynthesis, 75
–, – in *E. coli,* 293
–, – in plants, 269
–, chemical synthesis, 26
–, composition of subcellular fractions, 459–467
–, composition of tissues, 447–458
–, dialkyl analogue of, 26
–, diether derivatives of, in micro-organisms, 306
–, diglucosyl derivatives of, in micro-organisms, 306
–, fatty acid composition, 472
–, –, in plant tissues, 261–266
–, isolation, 26, 45
–, occurrence, 26
–, phosphotransferase system and, 326
–, in plant tissues, 254
–, as precursor of phosphatidylglycero-phosphate, 77

Phosphatidylglycerol (*continued*)
–, structure, 25
–, sugar transport and, 326
Phosphatidylglucose, *see* Glucose phosphoglyceride
Phosphatidylglycerol phosphoglyceride, *see* Cardiolipin and diphosphatidylglycerol
Phosphatidylglycerophosphatase, specificity, 109
Phosphatidylglycerophosphate, biosynthesis in *E. coli*, 293
–, chemical synthesis, 26
–, dialkyl analogue of, 26
–, in PG synthesis, 76
–, occurrence, 26
–, structure, 26
–, synthetase, specificity, 112
Phosphatidylinositol (PI), biosynthesis in plants, 270
–, composition of subcellular fractions, 459–467
–, composition of tissues, 447–458
–, effect in adrenal medulla, 432
–, – in leucocytes, 436
–, – in lymphocytes, 435
–, – in pancreas, 433
–, fatty acid composition, 475
–, –, in plant tissues, 261–266
–, occurrence, 30
–, in plant tissues, 254
–, preparation, 29, 45
–, separation from PS, 46, 48
–, structure, 29
–, turnover in sympathetic ganglia, 426
–, – in thyroid, 435
Phosphatidylinositol diphosphate, *see* Triphosphoinositide
Phosphatidylinositol kinase, 72
Phosphatidylinositol mannosides, biosynthesis in micro-organisms, 309
–, – in *P. shermanii*, 312
–, chemical characterisation, 307
–, metabolism in micro-organisms, 312
–, in Mycobacterium, 307
Phosphatidylinositol monophosphate, *see* Diphosphoinositide
Phosphatidyl-*N*-methylethanolamine, biosynthesis in *C. butyricum*, 300
Phosphatidyl-*N*-monomethylethanolamine:*N*-methyltransferase, specificity, 113
Phosphatidylserine (PS), biosynthesis in animal cells, 73
–, – in nervous tissue, 391
Phosphatidylserine, (*continued*)
–, – in plants, 270
–, chemical synthesis, 25
–, composition of subcellular fractions, 459–467
–, – of tissues, 447–458
–, fatty acid composition, 477
–, –, in plants, 261–266
–, isolation, 45
–, occurrence, 25
–, original isolation, 9
–, in plant tissues, 254
–, as precursor of PE, 77
–, purification, 45
–, sodium pump ATPase and, 431
–, structure, 25
Phosphodiol lipids, classification, 11
–, isolation, 32
Phosphoglycerides, chemical synthesis, 17
–, classification, 10
–, definition, 11
–, isolation, 45
–, structure, 17
Phosphoglycolipids, classification, 11
Phosphoinositides, (*see also* Diphosphoinositide, Phosphatidylinositol and Triphosphoinositide)
–, biosynthesis, 72
–, composition of subcellular fractions, 459–467
–, – of tissues, 447–458
–, fatty acid composition, 475, 476
–, original isolation, 9
–, resolution, 54
Phospholipase A_1, action, 87
–, in liver plasma membrane, 355
–, localisation, 87
–, in nervous tissue, 391, 392
–, in pancreatic juice, 346
–, physicochemical factors affecting activity, 106
–, plant lipid catabolism and, 276
–, specificity, 106
Phospholipase A_2, action, 86
–, characterisation of pancreatic enzyme, 87
–, in intestinal cells, 346
–, in pancreatic juice, 346
–, physicochemical factors affecting activity, 106
–, specificity, 87, 106
Phospholipase B, action, 88
–, localisation, 88
–, physicochemical factors affecting activity, 107
–, specificity, 107

Phospholipase C, action, 91
–, in intestinal mucosa, 350
–, localisation, 91
–, in nervous tissue, 392
–, physicochemical factors affecting activity, 107
–, plant lipid catabolism and, 274
–, specificity, 91, 107
–, SM specific, 392
Phospholipase D, physicochemical factors affecting activity, 107
–, plant lipid biosynthesis and, 272
–, plant lipid catabolism and, 274
–, reaction, 275
–, specificity, 107
Phospholipases, action, effect of lipid solvent, 101
–, action on monomolecular films, 101
–, in *E. coli*, 294
–, in nervous tissue, 391
–, neurotransmitter release and, 412
Phospholipid, acetolysis of, 38
–, acid hydrolysis of, 37
–, action of hydriodic acid on, 38
–, alkaline alcoholysis of, 37
–, amphipathic substances and, 102
–, analysis of, 43, 61
–, base analysis, 59
–, base exchange of, 78
– bilayer, action of ionophores, 157
––, action of uncouplers, 159
––, binding of metallic ions, 156
––, electrical capacitance, 150
––, electrical resistance, 150
––, excitability, 157
––, formation, 148
––, interfacial tension, 149
––, intrinsic properties, 148
––, ionic permeability, 155
––, –, effect of acyl chain length, 159
––, –, effect of cholesterol, 160
––, –, effect of unsaturation, 159
––, methodology, 148
––, surface energy, 149
––, thickness, 149
–, in bile, origin, 364
–, chemical synthesis, 17
–, chemistry, 9
–, chromatography, *see* Chromatography
–, classification, 10
–, composition of animal tissues, 447–458
–, – of plant tissues, 254
–, classification, 10
–, cytochrome c interaction, 224
–, degradation, 86
Phospholipid, (*continued*)
–, detection, 51
–, discovery, 2
–, drug metabolism and, 366
–, dysmyelination and, 402
–, exchange between membranes, 236
–, exchange processes in plants, 279
–, extraction of, 43
–, extraction from plants, 267
–, fatty acid composition, 469–478
–, fatty acid metabolism and, 282
–, fatty acid uptake by liver and, 355
–, hepato-portal system and, 345
–, hydrolytic reagents and, 37
–, interactions, with cholesterol, 136
–, –, with membranes, 171
–, –, with metal ions, 132
–, –, with protein, 171
–, intracellular metabolism, 231, 349
–, isolation, 17
– membrane, cholesterol and, 159
––, cytochrome c and, 162
––, excitability, 157
––, interaction, with ions, 155
––, –, with pharmacological compounds, 164
––, –, with protein, 161
––, ionophores, 157
––, permeability, 155
–, membrane transport and, 229
–, melting points, 119
–, metabolic errors and, 402
–, metabolic interconversion, 77
–, metabolic turnover, 81
–, in micro-organisms, 289
–, as model membranes, 143
– monolayer, enzyme action on, 101
––, intrinsic properties, 144
––, ionic binding, 156
––, ionic permeability, 160
––, methodology, 144
––, surface potential, definition, 144
––, –, measurement, 145
––, surface pressure, definition, 144
––, –, measurement, 144
––, surface viscosity, 145
–, nomenclature, 12
–, oxidation reactions, 37
–, packing, 102
–, phase transitions of, 129
–, physical chemistry, 117
–, purification, 45
–, qualitative analysis, 51
–, reduction by lithium aluminium hydride, 38
–, separation, 43, 58

Phospholipid, (*continued*)
–, solid state behaviour, 117
–, "solutions of", in water, 139
–, structures, 17
–, in subcellular organelles, 205
–, synthetic, surface properties of, 146
–, transfer between subcellular organelles, 395
–, transmembranal transport and, 423
– turnover, assessment of, 240
––, fatty acid composition and, 242
––, rate, 84
– vesicles, electrical resistance, 155
––, formation, 150
––, ionic binding, 156
––, ionic permeability, 155
––, –, effect of acyl chain length, 159
––, –, effect of cholesterol, 160
––, –, effect of unsaturation, 159
––, methodology, 150
––, properties, 154
––, spectrin and, 161
––, thickness of lipid layer, 153
Phosphonolipids, biosynthesis, 337
–, chemistry, 333
–, classification, 333
–, degradation, 340
–, discovery, 333
–, distribution, 335, 336
–, function, 341
Phosphoryl choline, biosynthesis, 70
–, – in nervous tissue, 385
–, nervous tissue function and, 412
–, PC biosynthesis and, 70
Phosphoryl ethanolamine, biosynthesis in nervous tissue, 388
–, PE biosynthesis and, 71
Phosphoryl glycerophosphate, characterisation, 35
–, isolation, 35
Phosphosphingolipids, in *B. melaninogenicus,* 314
–, classification, 11
–, definition, 11
–, long chain bases, isolation, 34
–, –, structure, 34
–, in micro-organisms, 314
–, structure, 32
Phosphotransferase system, in *E. coli,* 326
–, in *S. aureus,* 325
Photosynthesis, phospholipids in, 280
Photosynthetic plants, phospholipid metabolism in, 253
Phytanic acid, in Refsum's disease, 404
Phytoglycolipid, structure, 36
Phytohaemagglutinin, 435
Pinocytosis, 347
Pituitary gland, phospholipid composition, 456
Plants, phospholipid biosynthesis in, 268
–, – catabolism in, 274
–, – composition, 254
–, – fatty acid composition, 261–266
–, – function in, 280
–, – metabolism in, 253
–, ––, subcellular localization, 278
Plasma lipoproteins, chemical characteristics, 356
–, functions, 360
–, liver and, 356
–, physical characteristics, 356
–, structure, 357
–, turnover of components, 360
Plasmalogen, biosynthesis, 73
–, in brain development, 398
–, chemical synthesis, 21
–, composition of subcellular fractions, 459–467
–, composition of tissues, 447–458
–, degradation of alk-1-enyl bond, 89, 108
Plasmalogenase, in brain development, 401
–, in nervous tissue, 392
Plasma membrane, phospholipid composition of, 218
Platelets, phospholipid composition of, 454
Polyphosphoinositides, (*see also* Diphosphoinositide and Triphosphoinositide), fatty acid composition, 476
–, in nervous tissue, 429
Polyphosphoinositide phosphomonoesterase, 92
Propionibacterium shermannii, PI mannoside biosynthesis in, 312
Protagon, controversy, 4
–, original concept, 5
(M)-Protein, 325
Protein, calorie deficiency, brain development and, 405
–, in chylomicrons, 351
–, phospholipid interaction, 171
–, – membrane interaction, 162
–, turnover in microsomal membranes, 243
–, – in mitochondrial membranes, 243
Protozoa, PC in, 298

Quaking mice, 402

Receptor proteolipid, in nervous system, 411
Refsum's disease, 404
Rhodopseudomonae, PC in, 295
Rhodopseudomonas spheroides, chromatophore formation in, 322
–, phospholipid metabolism during anaerobic shift, 322

Saccharomyces cerevisiae, PC biosynthesis in, 300
–, ceramide phosphoglycolipids in, 315
Sarcoplasmic reticulum, ATPase of, 230
L-Serine, in PS biosynthesis, 73
Serine phosphoglycerides, chemical synthesis, 25
–, occurrence, 25
–, structure, 24
Silica-impregnated paper, in chromatography, 51
Silicic acid, in chromatography, 46
Skeletal muscle, phospholipid composition, 453
Skin, phospholipid composition, 457
Sonication, of multilamellar particles, 151
–, of ovalecithin, 100
Species variation, of phospholipid composition of subcellular fractions, 459–467
–, of phospholipid composition of tissues, 447–458
Spectrin, 161
Sphingolipid (*see also* Phosphosphingolipid), nomenclature, 15
Sphingomyelin (SM), biosynthesis, 78
–, – in nervous tissue, 390
–, in brain development, 398
–, chemical synthesis, 35
–, composition of subcellular fractions, 459–467
–, – of tissues, 447–458
–, fatty acid composition of, 478
–, metabolism in intestinal mucosa, 350
–, occurrence, 35
–, original isolation, 4
–, preparation, 34
–, structure, 35
Sphingomyelinase, in intestinal cells, 346
–, physicochemical factors affecting activity, 108
–, specificity, 108
Sphingosine, biosynthesis, 79
–, determination, 60
–, isolation, 34
–, structure, 34
Spleen, phospholipid composition of, 451
Staphylococcus aureus, fatty acid composition of lipids during electron transport induction, 319
–, glycerol auxotroph, 325
–, lysyl phosphatidylglycerol biosynthesis in, 302
–, phospholipid composition and electron transport induction, 318
–, phosphotransferase system in, 325
Streptococcus, glycerylphosphoryl diglucosyl diglyceride in, 312
Subcellular fractions, fatty acid composition of phospholipids, 469–478
–, phospholipid composition, 459–467
–, of rat brain, phospholipid composition, 379
Subcellular organelles, phospholipids in, 205
Sugar transport, in *E. coli,* 324
Sympathetic ganglia, PI turnover in, 426
Synaptosome, PA metabolism in, 424
–, phospholipid composition, 384

Tangier disease, 359, 360
TEAE cellulose, in chromatography, 53
Temperature, fatty acid composition of lipids of *E. coli* and, 316
Temperature shifts, phospholipid metabolism in micro-organisms and, 316
Tetrahymena, PC biosynthesis in, 301
–, phosphonolipid biosynthesis in, 337
–, temperature effect on membrane, 318
Tetrasaccharide phytoglycolipid, structure, 36
Thymus, phospholipid composition, 456
Thyroid gland, deficiency, brain lipid metabolism and, 407
–, phospholipid composition, 456
–, phospholipid turnover in, 435
Thyrotrophin (TSH), 435
Tissue variation, of fatty acid content of phospholipids, 469–478
Transacylation, plant lipid biosynthesis and, 273
Triacylglycerol, dynamic balance in liver, 361
Triphosphoinositide (TPI), biosynthesis, 72
–, isolation, 30
–, occurrence, 30
–, structure, 30
Triphosphoinositide phosphodiesterase, 392, 401
Triphosphoinositide phosphomonoesterase, 392, 401

Unilamellar vesicles, 151
Unsaturated fatty acids, in *E. coli,* 324

Valinomycin, conformation, 158
–, ion transport and, 158
–, phospholipid membrane and, 156
Very high density lipoprotein (VHDL), structure, 357
Very low density lipoprotein (VLDL), 350
–, chemical properties, 356
–, metabolism, 358
–, physical properties, 356
–, relationship with HDL, 359
–, relationship with LDL, 358

VLDL, (*continued*)
–, site of biosynthesis, 359
–, structure, 357

Wallerian degeneration, secondary demyelination, 409

X-ray diffraction, of intact membranes, 177
–, of phospholipids, 117
–, of protein-phospholipid complexes, 192

X-ray irradiation, brain development and, 408

Yeasts, (*see also S. cerevisiae*)
–, ceramide phosphoglycolipids in, 315
–, mycoglycolipids in, 315
–, PA synthesis in, 290

Zymogen granule, secretion, 433